Wave Scattering in Complex Media: From Theory to Applications

NATO Science Series

A Series presenting the results of scientific meetings supported under the NATO Science Programme.

The Series is published by IOS Press, Amsterdam, and Kluwer Academic Publishers in conjunction with the NATO Scientific Affairs Division

Sub-Series

I. Life and Behavioural Sciences	IOS Press
II. Mathematics, Physics and Chemistry	Kluwer Academic Publishers
III. Computer and Systems Science	IOS Press
IV. Earth and Environmental Sciences	Kluwer Academic Publishers
V. Science and Technology Policy	IOS Press

The NATO Science Series continues the series of books published formerly as the NATO ASI Series.

The NATO Science Programme offers support for collaboration in civil science between scientists of countries of the Euro-Atlantic Partnership Council. The types of scientific meeting generally supported are "Advanced Study Institutes" and "Advanced Research Workshops", although other types of meeting are supported from time to time. The NATO Science Series collects together the results of these meetings. The meetings are co-organized bij scientists from NATO countries and scientists from NATO's Partner countries – countries of the CIS and Central and Eastern Europe.

Advanced Study Institutes are high-level tutorial courses offering in-depth study of latest advances in a field.
Advanced Research Workshops are expert meetings aimed at critical assessment of a field, and identification of directions for future action.

As a consequence of the restructuring of the NATO Science Programme in 1999, the NATO Science Series has been re-organised and there are currently Five Sub-series as noted above. Please consult the following web sites for information on previous volumes published in the Series, as well as details of earlier Sub-series.

http://www.nato.int/science
http://www.wkap.nl
http://www.iospress.nl
http://www.wtv-books.de/nato-pco.htm

Series II: Mathematics, Physics and Chemistry – Vol. 107

Wave Scattering in Complex Media: From Theory to Applications

edited by

Bart A. van Tiggelen

Laboratoire de Physique et Modélisation des Milieux Condensés, CNRS,
Université Joseph Fourier, Grenoble, France

and

Sergey E. Skipetrov

Moscow State University, Russia and
Laboratoire de Physique et Modélisation des Milieux Condensés, CNRS,
Université Joseph Fourier, Grenoble, France

Kluwer Academic Publishers

Dordrecht / Boston / London

Published in cooperation with NATO Scientific Affairs Division

Proceedings of the NATO Advanced Study Institute on
Wave Scattering in Complex Media: From Theory to Applications
Cargèse, Corsica, France
10–22 June 2002

A C.I.P. Catalogue record for this book is available from the Library of Congress.

ISBN 1-4020-1393-0 (HB)
ISBN 1-4020-1394-9 (PB)

Published by Kluwer Academic Publishers,
P.O. Box 17, 3300 AA Dordrecht, The Netherlands.

Sold and distributed in North, Central and South America
by Kluwer Academic Publishers,
101 Philip Drive, Norwell, MA 02061, U.S.A.

In all other countries, sold and distributed
by Kluwer Academic Publishers,
P.O. Box 322, 3300 AH Dordrecht, The Netherlands.

Printed on acid-free paper

Printed in the Netherlands.

This volume is dedicated to the 65th anniversary of Roger Maynard without whom the field "Waves in Complex Media" would have been different.

Table of Contents

Chapter VI. OPTICS OF SOFT CONDENSED MATTER

Chapter VII. WAVE SCATTERING IN NATURAL MEDIA

Chapter VIII. COMMUNICATION IN A DISORDERED WORLD

Preface

The NATO Advanced Study Institute *"Wave Scattering in Complex Media: From Theory to Applications"* was organized in June 2002, at the *Institut des Etudes Scientifiques* in Cargèse, on the French Mediterranean island Corsica. This book is a collection of invited lectures and contributed talks that were presented at this Summer School. Also a few poster contributions have been included.

The organization of this big project was an initiative of the French Research Group PRIMA (*PRopagation et Imagerie en Milieu Aléatoire*), supported by the French CNRS. After having organised a similar project in March 1998 at the *Ecole de Physique* in Les Houches (French Alpes) a lot of enthusiasm existed for a follow-up. This time we wanted to put a stronger emphasis on applications, such as remote sensing, imaging, optics of liquid crystals, and seismology, and the many new hot topics of wave diffusion in complex media, like the random laser, left-handed materials, phononic and photonic materials, "communication in a disordered world", optics with cold atoms, and "imaging without a source". The School also offered lectures on well-established subjects such as mesoscopic physics, time-reversal, and Anderson localization. Perhaps less accent was put on photonic bandgap materials and their many applications, that got full attention at a recent excellent NATO ASI organized by Prof. Costas Soukoulis in 2000.

We finally selected, among the roughly 200 inscriptions, 109 participants (among which 8 observers) from 22 countries, with most participants coming from France, USA and Russia. Limited by lodging capacity and funding we had to turn down many young students, especially from France and Russia. The maximum participation also put a severe pressure on the organization of the Institute to find adequate lodging for all these people, to feed them at noon, and to cope continuously with the many problems and questions they bring along. They have not disappointed us: Brigitte, Chantal, Pierre-Eric, Natalie and Joseph-Antoine, you were the best!

The new NATO rule of 40% participation from NATO partner countries or Mediterranean Dialogue Countries complicated the organization considerably, but finally accomplished the aimed result that many young researchers from these countries could participate in this high-level international School for the first time, and exchange ideas and future projects with the lecturers. Our main problem was that no good infrastructure existed in these countries to lobby and select high level young students, except perhaps in Poland and in Israel. Thanks to joint efforts of several invited lecturers and a flexible attitude at the Scientific and Environmental Affairs Division of NATO in Brussels (Dr. F. Pedrazzini), we finally reached

Outdoor poster session. Photo by Gabriel Cwilich

a 32% participation from NATO partner countries, a number that as far as we know has never been reached for any international School on this topic held in the EC or in the USA.

Although our project was mainly financed by the NATO grant, it also benefitted considerably from national funding. The Research Group PRIMA, directed by Prof. Claude Boccara and Prof. Roger Maynard, participated for about 8%, and what is perhaps equally important, used their national and international connections to achieve high scientific quality. We thank the *Formation Permanente* (Prof. Jean Laforest) of the French CNRS (its department SPM for Physics and Mathematics as well as its department SDU for Earth Science and Sciences of the Universe) for granting this School the label *"Ecole Thématique du CNRS"* which facilitated full support of 15 CNRS employees, plus one visiting professor (Prof. Gabriel Cwilich). Financial support from the French Ministry for Research (Prof. Jean-Michel Dion) enabled us to support almost all other French participants, including 1 director and 3 invited speakers from France (Prof. Mathias Fink, Prof. Michel Campillo, Prof. Roger Maynard). The French embassee in Moscow (Prof. B. Fleutiaux and Prof. Michel Zigone) financed the trip of 3 brilliant Russian students to France and was as such a big support in our struggle to meet the NATO criteria. Finally, the US National Science Foundation kindly supported two american Ph.D. students. And let us not forget to mention that extra NATO funds were made available to support one Turkish participant, two Greek and two Portuguese participants. A demand for support at the European Community was rejected because "the content of the School is too dispersed which could create confusion for the training purposes". Unfortunately, though advertised at loud in their guide for proposers, interdisciplinarity is not recognized by the European Community.

We want to conclude by thanking the Scientific Committee — Profs. Eric Akkermans, Azriel Z. Genack, Yurii A. Kravtsov, Ad Lagendijk, Roger Maynard, and Arkadiusz Orłowski — for their crucial role in achieving the high scientific level of the School, and the 19 lecturers for accepting our invitation, most of them within several days. We think that the overall ambiance during the two weeks in Cargèse has been excellent, with no major incidents, sunny weather, good physics, and even the world championships soccer coming in directly by satellite. We thank the director of the "Institut d'Etudes Scientifiques", Dr. Elisabeth Dubois-Violette, for supporting our project already at a very early stage, and of course NATO for their generous support. At this very day, a new Research Group IMCODE ("Imagerie, Communication et Désordre") is created in France so who knows, we will be back...

Grenoble, February 1, 2003
Bart van Tiggelen and Sergey Skipetrov

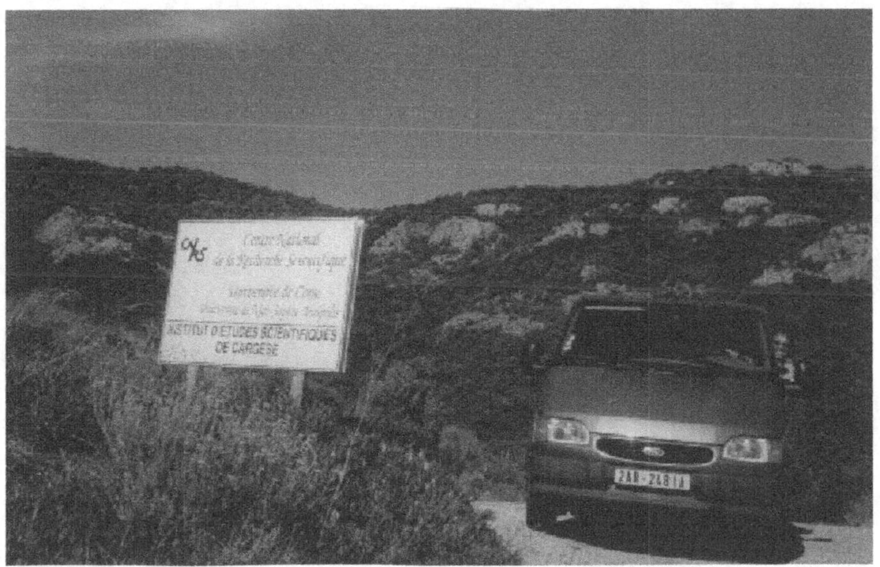

Photo by Lucienne Zwaan

The scientific directors

Bart A. **VAN TIGGELEN**, Grenoble, France
Sergey E. **SKIPETROV**, Moscow, Russia

The lecturers

Eric **AKKERMANS**, Haifa, Israel
Yurii **BARABANENKOV**, Moscow, Russia
Michel **CAMPILLO**, Grenoble, France
Martin **ČOPIČ**, Ljubljana, Slovenia
Mathias **FINK**, Paris, France
Valentin **FREILIKHER**, Ramat-Gan, Israel
Azriel **GENACK**, New York, USA
Yurii **KRAVTSOV**, Moscow, Russia
Vladimir **KUZMIN**, St. Petersburg, Russia
Ad **LAGENDIJK**, Enschede, the Netherlands
Georg **MARET**, Konstanz, Germany
Roger **MAYNARD**, Grenoble, France
Michael **MISHCHENKO**, New York, USA
Arkadiusz **ORŁOWSKI**, Warsaw, Poland
John **PAGE**, Manitoba, Canada
Vadim **ROMANOV**, St. Petersburg, Russia
Costas **SOUKOULIS**, Iowa, USA/Heraklion, Greece
Richard **WEAVER**, Urbana-Champaign, USA
Diederik **WIERSMA**, Florence, Italy

The participants

Berengère **ABOU**, Paris, France
Elena **AKSENOVA**, St. Petersburg, Russia
Matija **AVSEK**, Ljubljana, Slovenia
Sandrina **BARBOSA**, Lisboa, Portugal
Gerrit **BAUER**, Delft, the Netherlands
Viatcheslav **BELYI**, Moscow, Russia
Remy **BERTHET**, Paris, France
Temel **BILICI**, Istanbul, Turkey
Konstantin **BLIOKH**, Kharkov, Ukraine
Claude **BOCCARA**, Paris, France
Jorge **BRAVO-ABAD**, Madrid, Spain
Boris **BRET**, Amsterdam, the Netherlands
John **BURKHARDT**, Annapolis, USA
Andrey **CHABANOV**, New York, USA
Sébastien **CLERC**, Cannes La Bocca, France
Gabriel **CWILICH**, New York, USA
Julien **DE LA GORGUE DE ROSNY**, Paris, France
Sylvain **DOUTE**, Grenoble, France
Guillaume **DUPUIS**, Paris, France
Roman **EGORCHENKOV**, Moscow, Russia
Christoph **EISENMANN**, Konstanz, Germany
Nicolas **FERLAY**, Aubiere, France
Mauro **FERREIRA**, Delft, the Netherlands
Piotr **FITURNY**, Warszaw, Poland
Antoine **FOLACCI**, Corte, France
Luis **FROUFE-PÉREZ**, Madrid, Spain
Josselin **GARNIER**, Toulouse, France
Alexandre **GRET**, Golden, USA
Matt **HANEY**, Golden, USA
Darko **HANZEL**, Ljubljana, Slovenia
Ursula **ITURRARÀN-VIVEROS**, Mexico, Mexico
Maria **KAFESAKI**, Heraklion, Greece
Olga **KRAVTSENYUK**, St. Petersburg, Russia
Maria **KUSHNIR**, Haifa, Israel
Guillaume **LABEYRIE**, Valbonne, France
David **LACOSTE**, Paris, France
Eric **LAROSE**, Grenoble, France
Pavel **LITVINOV**, Kharkov, Ukraine
Peter **LODAHL**, Enschede, the Netherlands
Valery **LOIKO**, Minsk, Belarus
Fernando **LOPEZ-TEJEIRO**, Madrid, Spain
Evgenii **MAKEEV**, Moscow, Russia
Giuliano **MALLOCI**, Cagliari, Italy
Ludovic **MARGERIN**, Grenoble, France
Pia **MASSATCH-DAHLGREN**, Lausanne, Switzerland
Igor **MEGLKINSKI**, Saratov, Russia
Alenka **MERTELJ**, Ljubljana, Slovenia
Christian **MINATURA**, Valbonne, France
Faez **MIRI**, Konstanz, Germany
Alexander **MORKOTUN**, Moscow, Russia
Karri **MUINONEN**, Pino Torinese, Italy
Sushil **MUJUMDAR**, Florence, Italy
Evgenii **MUKHAY**, Moscow, Russia
Cord **MÜLLER**, Dresden, Germany
Tobias **MÜLLER**, Berlin, Germany
Francisco **NEVES**, Portalegre, Portugal
Xiaohui **NI**, New York, USA
Manuel **NIETO-VESPERINAS**, Madrid, Spain
Michel **PACILLI**, Nice, France
Raluca Sorina **PENCIU**, Bucharest, Romania
Felipe **PINHEIRO**, Grenoble, France
Daniel **REINKE**, Konstanz, Germany
Jorge **RIPOLL**, Heraklion, Greece
Luis **ROJAS**, Fribourg, Switzerland
Philippe **ROUX**, San Diego, USA
Marian **RUSEK**, Warszaw, Poland
Juanjo **SAENZ**, Madrid, Spain
Gregory **SAMELSOHN**, Ramat-Gan, Israel
Kirill **SANDORMIRSKI**, Minsk, Belarus
Riccardo **SAPIENZA**, Florence, Italy
Frank **SCHEFFOLD**, Fribourg, Switzerland
Patrick **SEBBAH**, Nice, France
Juliette **SELB**, Paris, France
Vladimir **SHALAEV**, West-Lafayette, USA
Nir **SHILOAH**, Ness-Ziona, Israel
Olivier **SIGWARTH**, Paris, France
Rudolf **SPRIK**, Amsterdam, the Netherlands
Shakhzod **TAKHIROV**, Tashkent, Uzbekistan
Arnaud **TOURIN**, Paris, France
Laurent **VABRE**, Paris, France
Kasper **VAN WIJK**, Golden, USA
Ben **VEIHELMAN**, Amsterdam, the Netherlands
Gorden **VIDEEN**, Adelphi, USA
Jia **XIAOPING**, Paris, France
Yaacov **YUNGER**, New York, USA
Zhao-Qing **ZHANG**, Hong-Kong, China

Photo by Eric Grossi and Valentin Freilikher

CHAPTER I
WAVES
IN COMPLEX MEDIA

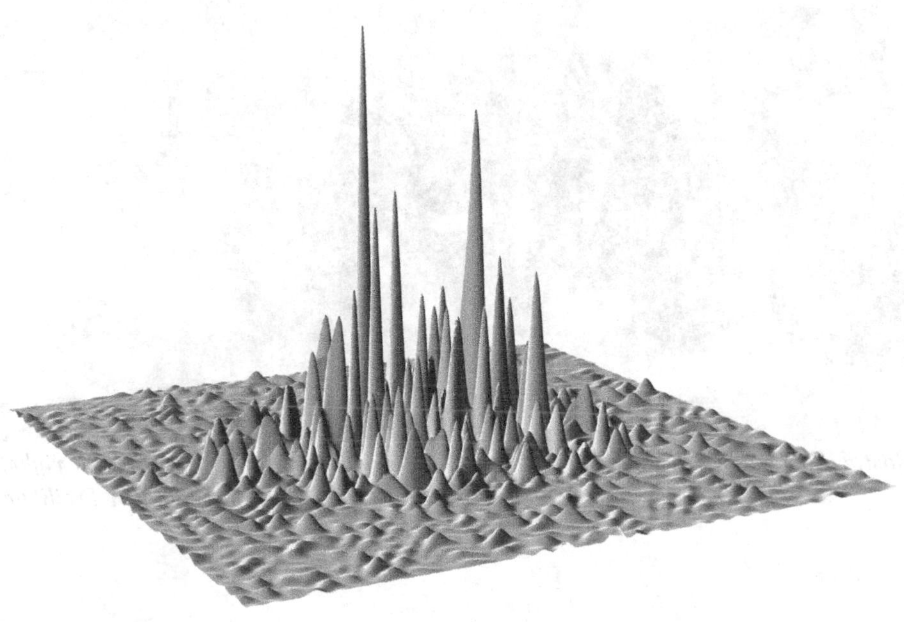

Calculated intensity distribution on the output surface of a monodomain nematic liquid crystal that is illuminated from the back side by a laser beam. The overall profile is anisotropic due to the anisotropy in the diffusion constant. The highly irregular fine structure is a bulk speckle pattern. Image provided by Diederik Wiersma, European Laboratory for Non-linear Spectroscopy, Sesto-Fiorentino (Florence), Italy. www.complexphotonics.com.

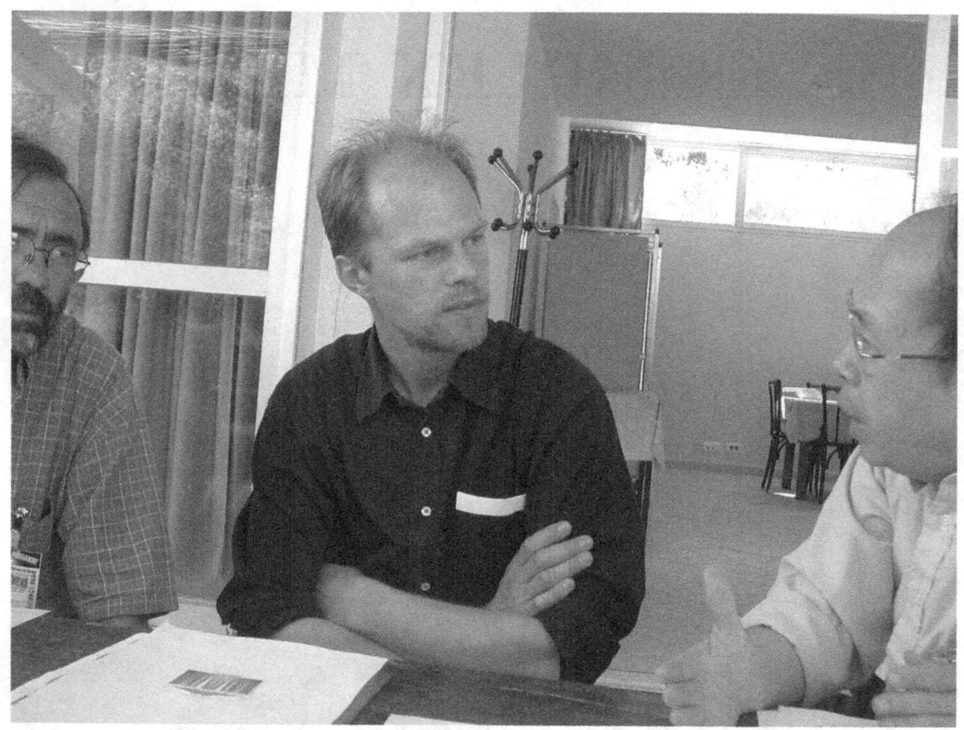

Costas Soukoulis, Diederik Wiersma, and Zhao-Qing Zhang (from left to right).
Photo by Valentin Freilikher

LIGHT TRANSPORT IN COMPLEX PHOTONIC SYSTEMS

From Random Lasers to Photonic Crystals

D.S. WIERSMA, S. GOTTARDO, R. SAPIENZA, S. MUJUMDAR, S. CAVALIERI, M. COLOCCI AND R. RIGHINI
European Laboratory for Non-linear Spectroscopy and INFM
Via Nello Carrara 1, 50019 Sesto-Fiorentino (Florence), Italy
www.complexphotonics.org

L. DAL NEGRO, C. OTON, M. GHULINYAN, Z. GABURRO AND L. PAVESI
INFM and Dept. of Physics, Univ. of Trento, via Sommarive 14, Povo (TN), Italy

F. ALIEV
Univ. of Puerto Rico, Dept of Physics, PO Box 23343 Rio Pedros, San Juan, Puerto Rico

AND

P.M. JOHNSON, A. LAGENDIJK AND W.L. VOS
Univ. of Amsterdam, Valckenierstraat 65, 1018 XE Amsterdam, The Netherlands and Dept. of Appl. Physics & MESA+ Research Inst., Univ. of Twente, P.O. Box 217, Enschede, The Netherlands

1. Introduction

The transport of light in complex dielectric materials is a rich and fascinating topic of research. With complex dielectrics we intend dielectric structures with an index of refraction that has variations on a length scales that is very roughly comparable to the wavelength. Such structures strongly scatter light. A possible building block for constructing a complex dielectric is a micro sphere of diameter comparable to the wavelength and of a certain refractive index that is different from its surrounding medium. The single scattering from such a sphere has a rich structure due to internal resonances in the sphere, but its behaviour is well-understood and can be calculated using the formalism of Mie-scattering [1]. A complex dielectric material can then be realized by micro-assembly of several micro spheres.

B.A. van Tiggelen and S.E. Skipetrov (eds.),
Wave Scattering in Complex Media: From Theory to Applications, 3–20.

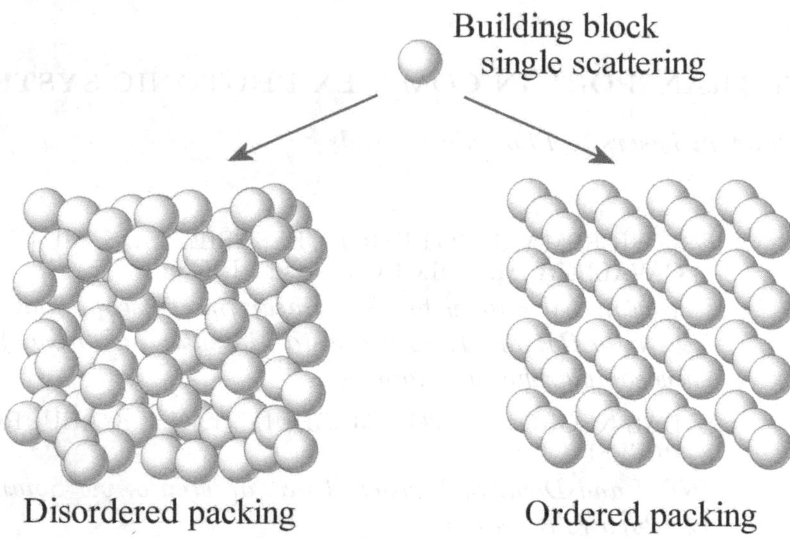

Figure 1. Micro assembly of a complex photonic system. The two extremes are fully disordered assembly (left) leading to random multiple light scattering and ordered assembly (right) resulting in a photonic crystal or possibly a photonic band gap material.

The spheres can be assembled in various ways, with as two opposite possibilities a completely disordered packing and a fully ordered assembly. (See Fig. 1.) Even though the same spheres with the same single scattering properties are used, their cumulative behaviour after assembly will depend heavily on the way the spheres are packed together. This is due to the interference between the scattered waves and the way the waves are multiply scattered from one sphere to another. If the spheres are packed according to a crystal-like structure then the interference will be constructive only in certain well defined directions, giving rise to Bragg refraction and reflection. In the disordered case the light waves will perform a random walk from one sphere to the other. The occurrence of interference effects is now less obvious to understand, however also in random systems interference effects turn out to be very important.

Interference of light in random dielectric systems influences the transport of light in a way that is similar to the interference that occurs for electrons when they propagate in disordered conducting materials. As a result, several interference phenomena that are known to occur for electrons appear to have their counterpart in optics as well [2]. Interesting examples are correlations and memory effects in laser speckle [3], universal conductance fluctuations of light [4], weak localization [5], and Anderson localization [6]. In the case of Anderson localization the interference effects are so strong that the transport comes to a halt and the light becomes localized in randomly distributed modes inside the system. Interference effects in multiple scattering can furthermore be used to study the dynamics of optically dense colloidal systems [7].

Also in ordered systems interference can give rise to dramatic effects. If the scattering of the spheres that constitute a photonic crystal is strong enough (that is the refractive index contrast between the spheres and their surrounding medium is large and their diameter is resonant with the wavelength) the interference can become destructive in all directions, for a certain range of frequencies. In analogy with the behaviour of electrons in semiconductors this range of optical frequencies is referred to as a photonic band gap [8, 9]. Inside a photonic band gap the density of light modes becomes zero, which means that even vacuum fluctuations are suppressed. A small impurity inside such a photonic band gap material will give rise to a localized mode around this impurity.

The micro assembly of complex photonic materials as depicted in Fig. 1 is concerned with three dimensional structures. The same principle can be applied to lower dimensional systems. In the case of 1D structures one uses multi-layers of different refractive index and thickness that are stacked either periodically or randomly, or via any other desired packing rule. (See Fig. 2.) The behaviour of light in three dimensional systems is often difficult to describe theoretically. The advantage of lower dimensional structures is that an analytical theoretical description is often available, facilitating the interpretation of experimental results. Results on lower dimensional structures can then be used to learn more about the complex behaviour of three dimensional systems.

Whereas the knowledge on the propagation of light waves in completely ordered and disordered structures is now rapidly improving, little is known about the behaviour of optical waves in the huge intermediate regime between total order and disorder. An example of a partially ordered system that we will discuss in this paper is a liquid crystal in the nematic phase. Here the single scattering element is anisotropic due to directional ordering of the liquid crystal molecules along a common axis. Another example that we will discuss is that of light propagation in quasi-crystals in which the scattering elements are assembled in a non-periodic but deterministic way. To construct a quasi-crystal we resort to 1D systems. On the two extremes of order and disorder we will discuss random laser action in amplifying random materials and photonic crystals. In particular we will discuss, in both cases, the role of liquid crystal infiltration to control the scattering strength and thereby the optical properties of these complex systems.

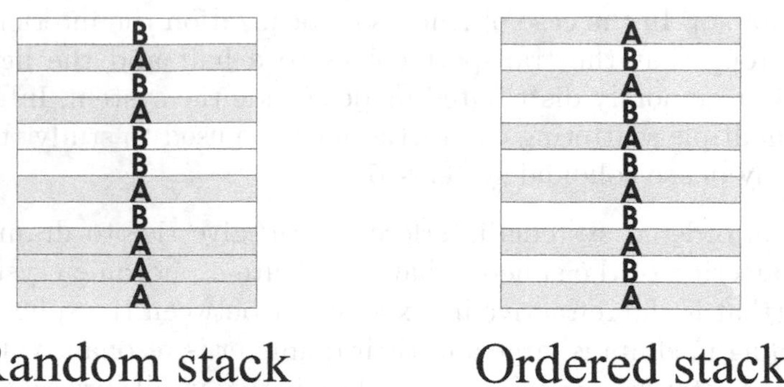

Figure 2. One-dimensional complex photonic systems. By stacking two types of layers (A and B), one can obtain random or ordered one-dimensional structures. In principle any desired stacking rule can be used which allows to explore the regime in between complete order and disorder.

2. Disorder

Light waves in disordered materials perform a random walk which leads to a multiple scattering process. The optical properties of random systems are full of surprises and apart from the before mentioned interference effects several other interesting phenomena occur in these systems that do not necessarily depend on interference. Nice examples are optical magneto resistance [10], the photonic Hall effect [11], and optical NTC resistance [12]. The knowledge on light diffusion in random systems is furthermore being applied successfully in medical imaging [13].

Of particular interest for photonic applications are disordered materials that provide optical amplification via stimulated emission. If the gain in such amplifying random media becomes larger than the loss through the boundaries, the system exhibits random laser action [14]. Such materials can be realized, for instance, by powdering of a laser crystal or by introduction of laser dye in various random media. It was shown that the emission of a random laser is narrow banded [15] and can exhibit laser spiking [16]. Theoretical studies furthermore show that a random laser source has interesting photon statistics that are neither those of a regular laser nor those of a common light bulb [17]. Recent experiments on Zinc-oxide powders aimed at combining random laser action with Anderson localization effects [18].

A crucial parameter in all multiple light scattering experiments is the scattering strength of the material expressed as the diffusion coefficient or transport mean free path. In many experimental studies one would like to be able to vary the scattering strength of a sample without modifying its

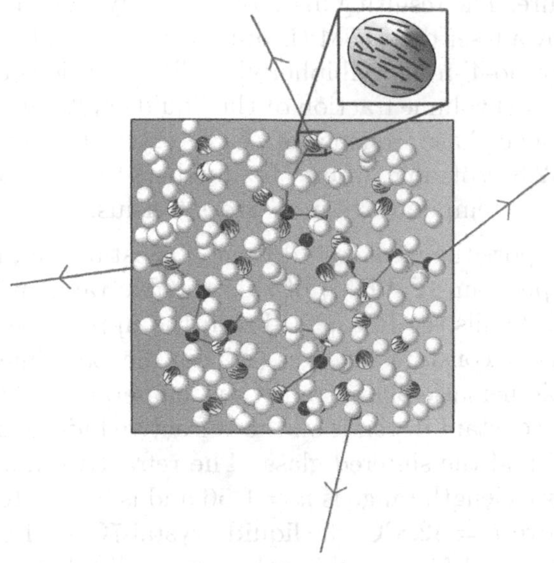

Figure 3. An amplifying disordered material with tunable diffusion coefficient. The laser dye is dissolved in liquid crystal and is excited by an external laser beam to provide optical gain. The refractive index contrast between glass powder and liquid crystal depends on temperature and hence the scattering strength can be temperature controlled.

other properties. We have found a simple way to obtain external control over the diffusion constant of a random sample. Liquid crystals have the beautiful property that they go through various partially ordered phases when heated. The index of refraction is different in every liquid crystal phase. Of special interest is the nematic phase due to its birefringence in the index of refraction (the index of refraction depends on the propagation direction and polarization of a light wave), which disappears when the liquid crystal is heated into the isotropic phase [19]. By infiltrating a random sample with a liquid crystal, one can obtain a system of which the diffusion constant strongly depends on temperature [20]. This is similar to, for instance, the concept of smart screens based on polymer dispersed liquid crystals that change their opacity with temperature [21]. For a random laser, having control over the diffusion constant has important consequences. The lasing threshold depends on the diffusion constant of the random material. This means that if we have a temperature-dependent diffusion constant, we are able to bring the random laser above and below threshold by changing its temperature.

We realized such a temperature-tunable random laser in the following way. Various types of glasses were ground into a fine powder and sintered under high pressure. The resulting discs of randomly assembled glass grains were infiltrated by a laser dye DCM (Lambdachrome 6500) dissolved in the liquid crystal 4-cyano-4'-n-heptylbiphenyl (7CB) in various concentrations. (See Fig. 3.) The final volume fraction of the liquid crystal in the sample was about 0.26 vol%. The phase sequence of 7CB is crystalline' (15.0) crystalline (30.0) nematic (42.8) isotropic, where the numbers between brackets denote the phase transition temperatures in degrees Celsius.

In Fig. 4 we report the measured diffusion constant and the bandwidth of the emission spectrum (FWHM) of this tunable random laser material. For experimental details see Ref. [22]. From the upper graph in Fig. 4 we see that the diffusion constant diverges (within the experimental accuracy) above the isotropic-nematic phase transition temperature of the liquid crystal. The diffusion constant depends on the refractive index contrast between the liquid crystal and the sintered glass. The refractive indices of SK11 in the 600-800 nm wavelength range is $n = 1.56$ and is hardly temperature dependent. Just above $T = 42.8°$C, the liquid crystal 7CB is isotropic (I) with a refractive index $n = 1.56$, and this value gradually decreases with rising temperature. Between $T = 30.0°$C and 42.8°C, this liquid crystal is nematic (N) and locally birefringent with refractive index $n_{\mathrm{o}} = 1.52$ and $n_{\mathrm{e}} = 1.68$ at $T = 36.0°$C, and with increasing birefringence at decreasing temperature [23]. The observed strong decrease of D when lowering the temperature into the nematic region is therefore due to the enhanced refractive index contrast between liquid crystal and sintered glass. The divergence of the diffusion constant above 42.5 degrees is due to partial refractive index matching between SK11 and 7CB in the isotropic phase. By choosing different liquid crystal / glass combinations, one can obtain different tuning curves for the diffusion constant. Note that the phase behaviour of a liquid crystal inside sintered glass is different from the typical phase behaviour of bulk liquid crystal and that the nematic-isotropic phase transition is smeared out. This behaviour is common for liquid crystals in confined geometries like porous glasses and is due to interaction between the liquid crystal molecules and the surface of the porous host [24].

The random laser emission from the liquid crystal / dye infiltrated sintered glass was characterized by exciting the samples from their front interface with a frequency doubled Q-switched Nd:YAG laser, operating at 10 Hz repetition rate. The emission spectrum was recorded by collimating the diffuse emission from either the front or the rear sample surface onto the input slit of a single grating spectrometer equipped with a cooled and gated optical multi channel analyzer to provide single shot spectra. At high powers the Q-switch of the laser was operated in single shot mode to prevent cumulative heating (and consequent damage) of the sample. From the lower graph in Fig. 4 we see that indeed there is a strong effect of temperature on

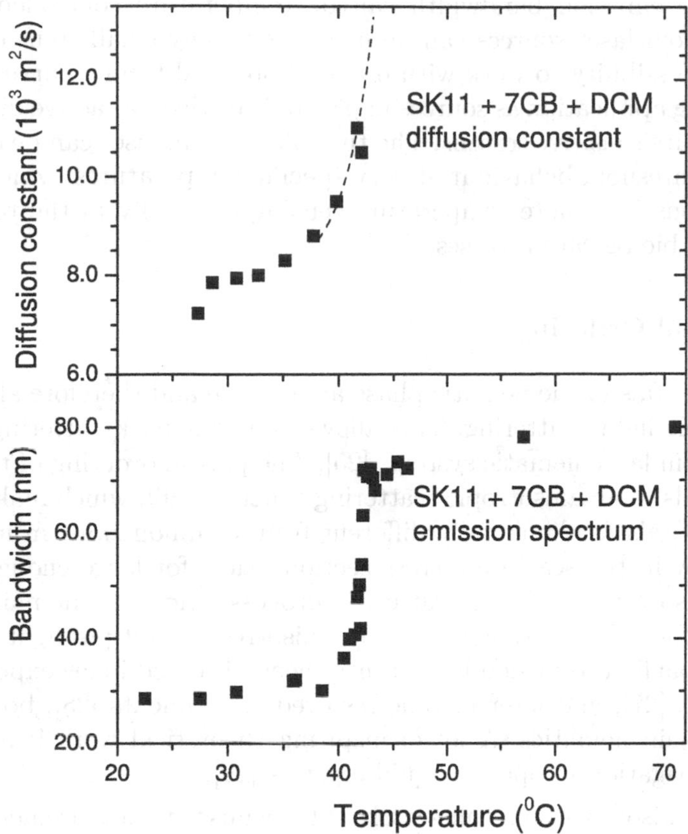

Figure 4. Diffusion constant (upper) and bandwidth of emission spectrum (lower) of sintered SK11 glass powder infiltrated with liquid crystal 7CB and laser dye DCM. Volume fraction liquid crystal in sintered glass 0.26, overall dye concentration in the sample 1.1 mmol/l. Sample thickness 1.3 mm, excitation beam diameter on sample 1.9 mm. Excitation pulse energy 4.5 mJ, pulse duration 14 ns. There is a strong decrease of bandwidth below 42.5°C. This corresponds to an increase in scattering (lowering of the diffusion constant). Below 42.5°C the scattering strength of the material becomes large enough to bring the random laser above threshold.

emission spectrum. The main feature is a strong decrease of the bandwidth of emission below 42.5°C. The transition temperature corresponds within the experimental error with the nematic-isotropic phase transition temperature of 7CB. Below this temperature the scattering is strong enough to bring the random laser above threshold. By heating the liquid crystal into the isotropic phase the diffusion constant of the sample increases (the opacity decreases) and and the random laser action disappears.

The temperature-tunable random laser provides a new light source of which the emission bandwidth can be temperature controlled. The fact that random laser sources can be made extremely small (tens of microns) and the possibility to work with different spectral tuning curves allows for interesting application as sources in photonic devices, as active displays, and temperature sensitive screens. The tunable random laser can be designed to have its threshold behaviour at very specific temperatures, which opens up applications in remote temperature sensing especially in the temperature regime of biological processes.

3. Partial Ordering

Liquid crystals in the nematic phase are opaque and therefore also give rise to multiple light scattering. This allows coherent backscattering to be observed from large nematic systems [25]. The partial ordering of the nematic phase leads to an anisotropic scattering function [26], which makes nematic liquid crystals fundamentally different from common random media. This anisotropy in the scattering cross section leads, for large enough samples, to an anisotropic multiple scattering process, and monodomain nematics are therefore ideal systems to study anisotropic multiple light scattering. Anisotropic light diffusion has recently been observed in cw experiments by Kao *et al.* [27], and later in time-resolved experiments [28], both on large monodomain nematics. A lot of inspiring theoretical work is available on light propagation in opaque liquid crystals [29].

Comparison between time-resolved transmission measurements and static transmission measurements on monodomain nematic LC's reveals a difference in the observed anisotropy whereas similar measurements on strongly anisotropic disordered GaP networks provide the same value [30]. The diffusion constant in an isotropic system can be written as the square of the average step size of the underlying random walk process (after correction for forward scattering) divided by the average time Δt it takes the random walker to cover this step: $D = \ell^2/(3\Delta t)$. For an anisotropic system like a liquid crystal the step length and time, in principle, both become anisotropic and their value will depend on the propagation direction of the random walker. However, for sufficiently large systems, one should assume that Δt takes some average value, which allows to write the perpendicular and parallel values of the diffusion constant in terms of one average Δt. (This is actually a necessary requirement to make the diffusion approximation in the first place.) It is clear that under these assumptions the *anisotropy* in the diffusion constant $D_\parallel/D_\perp$ can be calculated from the *anisotropy* in the transport mean free path, since then we can write: $D_\parallel/D_\perp = \ell_\parallel^2/\ell_\perp^2$. In static experiments one measures the transport mean free path of the system whereas the diffusion constant is a dynamical property and can be measured only in time-resolved experiments. However, under the assumptions above, one can derive the *anisotropy* in the diffusion constant from the measured *anisotropy* in the mean free path, as was done in the static experiments of Refs. [27, 28, 30]. The above assumptions might not be valid for multiple scattering processes with non-Gaussian statistics (like Levy-flights) in which the diffusion approximation breaks down or for systems with long range correlations. It is not clear if the diffusion approximation is correct at all for nematic liquid crystals. An unambiguous way to measure the transport mean free path is via coherent backscattering [5] which is, however, technically very challenging for liquid crystals due to their long mean free paths. The angular resolution in the measurements reported in Ref. [25] was not sufficient to resolve the width of the backscattering cone and therefore did not allow to determine the mean free paths.

4. Quasi-Crystals

Quasi-crystals form another class of fascinating systems in between fully ordered and completely disordered. Quasi-crystals are non-periodic structures that are constructed following a deterministic generation rule [31]. If made from dielectric material, the resulting structure has fascinating optical properties. Quasi-crystals of the Fibonacci type, for instance, exhibit an energy spectrum that consists of a self-similar Cantor set with zero Lebesgue measure [32]. The transmission spectrum of a Fibonacci system also contains forbidden frequency regions called 'pseudo band gaps' similar to the band gaps of a photonic crystal [33]. In the frequency regime outside these Fibonacci band gaps, the light waves are critically localized. In contrast with the fully disordered (Anderson) localized case, these critically localized states decay weaker than exponentially, most likely by a power law, and have a rich self-similar structure [34]. This makes these systems very interesting for light localization studies, as proposed by Kohmoto *et al.* [35]. The first Fibonacci sequence for electron transport studies was realized by Merlin *et al.* [36], which was followed by several experiments and theoretical studies on electron propagation in these systems [37]. The experimental work on light transport in this fascinating class of structures is limited so far. Important pioneering experiments were performed by Gellerman *et al.* [38] who observed self-similarity in the transmission spectrum of Fibonacci dielectric multi-layers and by Hattori *et al.* [39] who measured the Fibonacci dispersion curves.

A Fibonacci quasi-crystal is a deterministic aperiodic structure that is formed by stacking two different compounds A and B according to the Fibonacci generation scheme: $S_{j+1} = \{S_{j-1}S_j\}$ for $j \geq 1$; with $S_0 = \{B\}$ and $S_1 = \{A\}$. The lower order sequences are $S_2 = \{BA\}$, $S_3 = \{ABA\}$,

$S_4 = \{BAABA\}$, etc. We have realized one-dimensional Fibonacci structures from porous Silicon by stacking 233 layers using two types of layers A and B according to the Fibonacci packing scheme, leading to a 12-th order Fibonacci system [40]. To study the transport properties of the Fibonacci band edge modes we have performed time-resolved transmission experiments using a fixed Mach-Zehnder interferometer coupled with a Michelson interferometer to measure the interferometric cross correlation of the transmitted pulse with a reference pulse. This technique provides both the amplitude and phase information of the transmission through the sample. For further details we refer to Ref. [41]. As laser source we used a tunable parametric oscillator, pumped by a fast (200 fs pulse duration) Ti:Sapphire laser.

The measured pulse envelope of the time-resolved transmittance is plotted in Fig. 5. The transmission spectrum of the sample is given in the top graph, together with the laser spectra corresponding to three time-resolved measurements. When the incoming pulse is resonant with one transmission peak, the pulse is significantly delayed and stretched (I). This stretching becomes surprisingly strong close to the band edge (III). In addition to the delay and stretching, when the spectrum of the laser pulse overlaps with two adjacent narrow transmission modes a strongly oscillatory behaviour is observed (II). These oscillations can be interpreted as due to beating between individual band edge modes. Indeed the frequency of the oscillations corresponds to the frequency difference between the peaks in the transmission spectrum. The strong pulse delay leads to a group velocity suppression that turns out to be three times larger than that observed in three dimensional photonic crystals made of colloidal polystyrene spheres [42]. The response in the time domain can be calculated from the inverse Fourier transformation of the product of the complex transmission coefficient and the incident pulse envelope [43]. Taking exactly the experimental incidence pulse envelope (FWHM: 63 cm^{-1}), coherent beating and pulse stretching are very well reproduced, as can be seen from the theoretical curves in the right column of Fig. 5.

The large group velocity reduction and pulse stretching that we find experimentally is only observed close to the band edge. The band edge is also the region where the periodic like features (band gap) of the Fibonacci system go over into its disorder properties (critically localized states). In Fig. 6 we have plotted the calculated intensity distribution inside the sample as a function of frequency. The Fibonacci pseudo band gap is clearly visible in the lower part of the plot. Just above the pseudo band gap the band edge modes are visible. The insets show the normalized field intensity distributions for several frequencies. The incoming field is normalized to 1. The first inset (a) shows the exponentially decaying intensity of light that

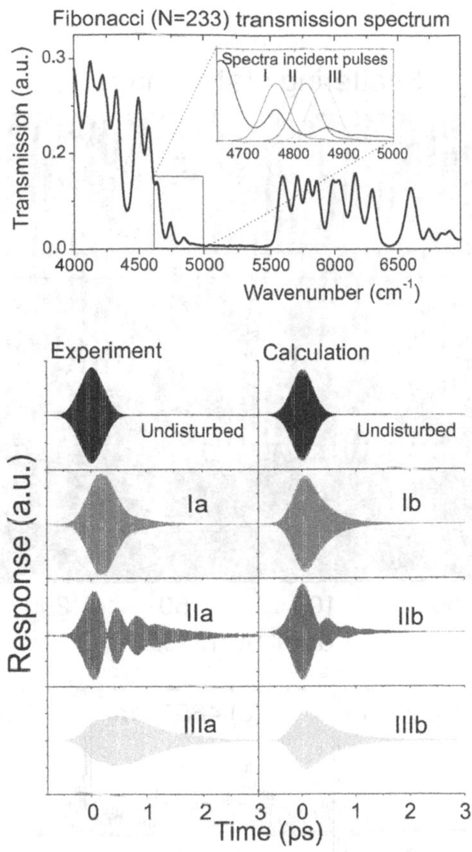

Figure 5. Experimental data and calculation of the transmission through Fibonacci samples at four different frequencies. Also the undisturbed pulse which has only passed through the Si substrate and not the Fibonacci sample is plotted for comparison. When the laser pulse is resonant with one band edge state the transmitted intensity is strongly delayed and stretched. When two band edge states are excited, mode beating is observed. The top graph shows the transmission spectrum of the sample together with the three laser spectra corresponding to the measurements I-III.

is incident in the band gap region, whereas the other insets (b–f) show the intensity distribution in the band edge region when moving away from the band gap. (Insets (b) and (d) correspond to the first two transmission maxima whereas (c) is taken in between these maxima.) The first and second order band edge resonances (with only one and two maxima respectively) are suppressed due to the minor drift in layer thickness and porosity.

The resonances of the band edge states are sharp enough to allow for mode beating when adjacent modes are excited simultaneously, as observed

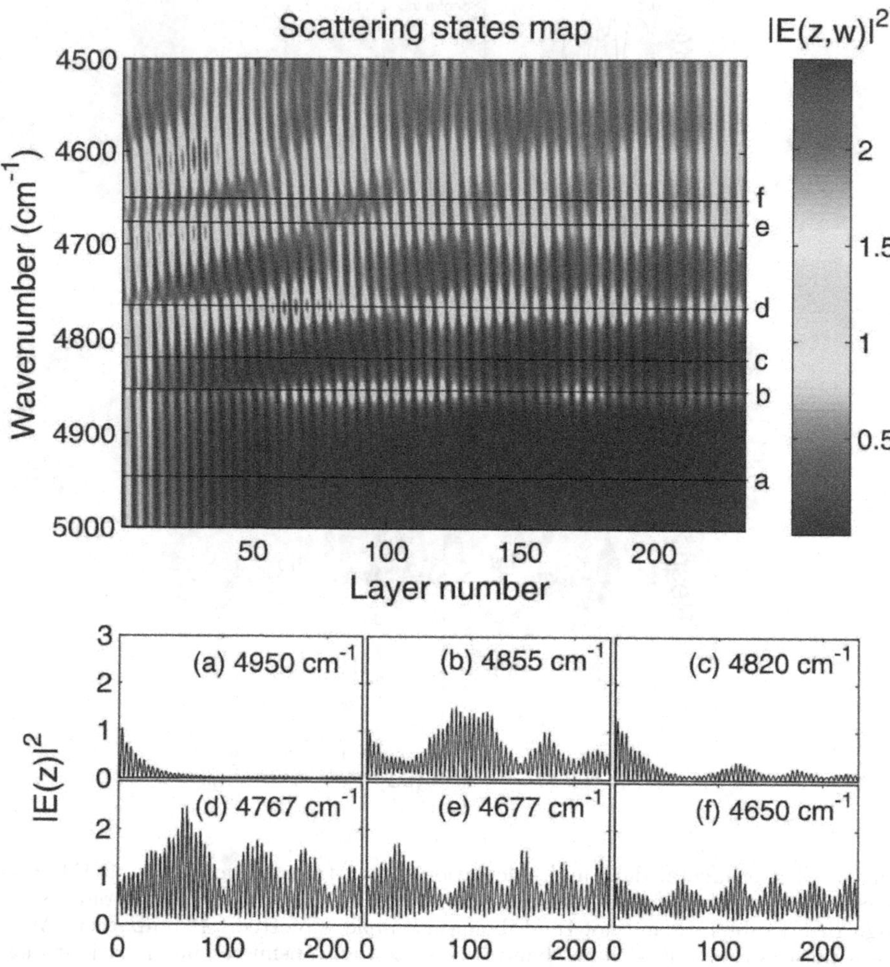

Figure 6. Calculated intensity distribution inside the Fibonacci quasi-crystal sample as used in the experiments. X-axis: layer number, y-axis: wavenumber of the incident light. This means that any horizontal cut through the graph represents an intensity distribution at one specific frequency. The insets show the intensity distributions at several frequencies. (Horizontal cuts are indicated by black lines.) The input intensity of the electric field has been normalized to unity. Note that the intensity inside the sample can become larger than one, due to internal resonances.

in the experiments. The distributions that we find in Fig. 6 have a notable similarity to the band edge resonances occurring in photonic crystals [44] but are less regular. Band edge resonances in (finite-size) photonic crystals are due to a transient standing wave that is formed inside the sample and can temporarily store a substantial amount of energy. This is consistent with a large group velocity reduction and strong pulse stretching as observed in our experiments. Since this transient standing wave is formed from reflection by the sample boundaries, it has the characteristic intensity distribution of the various harmonics of a standing wave. Band edge resonances in photonic crystals are not localized states since their extension scales linearly with the system size and they do not decay to zero [44]. In contrast, the Fibonacci band edge resonances will decay via a power law due to their critically localized nature. Fibonacci systems can provide an interesting alternative to regular photonic crystals for the realization of photonic devices like e.g. optical filters with a self-similar spectrum and a high wavelength selectivity in the band edge region. Another interesting future application of these materials could be realized in the field of random lasers, where the Fibonacci band edge resonances could serve as a new type of complex cavity that provides the feedback for laser action.

5. Photonic Crystals with Liquid Crystal Infiltration

If we assemble a complex dielectric material in a periodic way we obtain a crystal like structure that under appropriate conditions can exhibit a photonic band gap. For applications of photonic band gap materials as photonic devices it is very useful to have external control over their band structure. In section 2 we have seen how liquid crystal infiltration can be used to tune the diffusion constant of a disordered complex system. If liquid crystal infiltration is used in (ordered) photonic crystal structures it could allow to control its photonic band gap, as was proposed by Busch and John [45]. Temperature control of the band gap of two dimensional photonic crystals has been demonstrated experimentally, and shifts as large as 70 nm in the central wavelength of the band gap were observed [46]. Electric field tuning of the photonic stop band in opals was explored experimentally but the effects observed so far were limited by surface anchoring of the liquid crystal [47, 48].

One might expect that inverse opal structures, having spherical voids, are favourable for external field switching since a spherical void does not impose a preferred alignment direction. In addition, at fixed lattice constant, the inverse opal voids are bigger than those of a direct opal. To further exploit the possibilities of electric field tuning of photonic crystals we characterized the stop band of infiltrated titania inverse opals.

The titania inverse opals were based on self-organized direct opals of monodisperse polystyrene spheres. These direct opals were infiltrated with TiO_2, via a precursor solution (tetra-propoxy-titane). Calcination of the

TiO_2 by heating to 450°C subsequently also removes the polystyrene and leaves a clean titania inverse opal (air spheres in a TiO_2 backbone). See for more details Ref. [49]. The lattice constant a of the resulting titania inverse opals was 451 nm and their pore radius was 160 nm.

The titania inverse opal was infiltrated overnight with the liquid crystals 5CB or E7. The infiltrated samples were mounted between two glass slides coated with a thin transparent layer of Indium doped Tin Oxide (ITO) on the glass surface facing the sample. Before mounting the glass slides, part of the ITO coating was carefully removed in order to avoid regions of facing ITO layers without sample in between. The sample thickness equals the distance between glass plates and ranges from 0.2 to 0.5 mm in our case.

The electric conductance of the tin oxide allows to apply an electric field over the sample perpendicular to the sample plane. Without the electric field the optical properties of the sample are expected to be isotropic, whereas at high enough field the nematic director should obtain some global alignment and hence the stop band of the infiltrated photonic crystal is expected to shift towards higher wavelengths. Since the sample is a very good insulator (much better than air), and facing ITO layers always have sample material in between, we can apply an electric field as high as 24 V/μm to the sample (ac at 500 Hz).

A good way to characterize the stop band of a photonic crystal is via angular and wavelength resolved reflection measurements. A stop band appears as a wavelength regime where the reflectivity is high. In Fig. 7 we report angular resolved reflection spectra of a liquid crystal infiltrated titania inverse opal. If the system were to exhibit a complete photonic band gap, there would exist a wavelength region in which all reflection spectra overlap. This is not the case for our liquid crystal infiltrated samples as can be seen in Fig. 7. Although the density of states in un-infiltrated titania inverse opals is strongly reduced [50], after liquid crystal infiltration the refractive index contrast is diminished. This is the reason that the stop bands in Fig. 7 are not overlapping.

When we apply the electric field, the position of the stop bands is not changed considerably within the accuracy of our experiment (resolution about 1 nm). When we apply the maximum electric field of 24 V/μm we observe only a minor increase of the reflected intensity. (See inset of Fig. 7.) These data might indicate also a tiny red shift of the reflection peak which however can not be considered significant. This increase and red shift could be mistakenly interpreted as a switching effect of the stop band of the sample by the electric field. We observe that it sets in slowly after applying the electric field, which suggests an interpretation in terms of a heating effect of the sample and not a field effect. Heating effects on the reflection spectra from infiltrated photonic crystals were studied by e.g. Mertens *et*

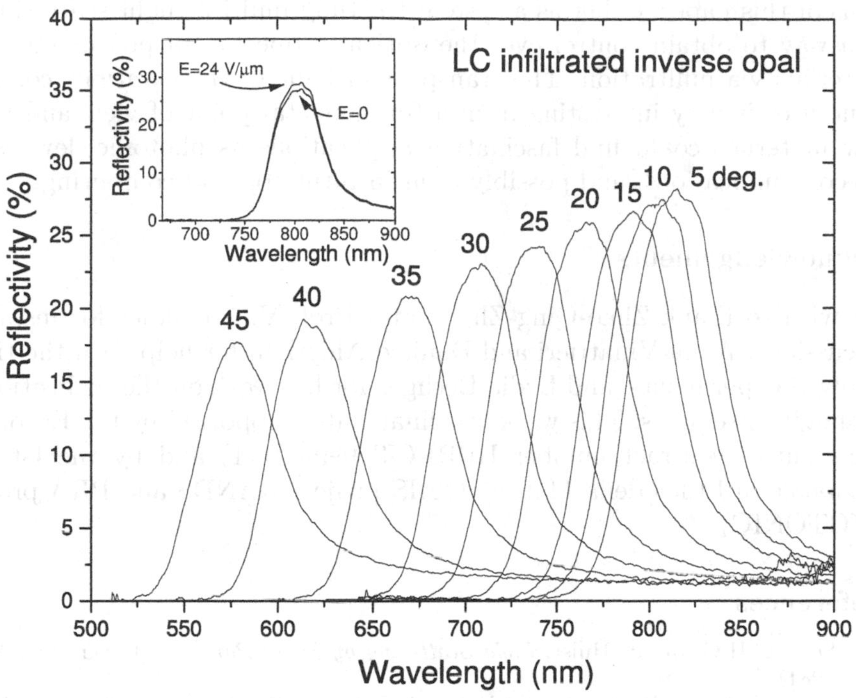

Figure 7. Reflection spectra of titania inverse opals infiltrated with the liquid crystal E7. The emission of a broad-band lamp was collimated via an optical fiber on the sample (spot size on the sample surface 100 μm), and the reflected light was detected via an optical fiber by a spectrometer (spectral resolution 1 nm). The incoming and detection angle are equal and given in the graph with respect to the sample normal. The inset shows the tiny increase of the reflection peak at 10 degrees upon switching on the electric field. (Field strength 24 V/μm.)

al. [51]. Apparently the alignment of the liquid crystal due to the electric field is too small to induce any appreciable global anisotropy in our system. A possible improvement of the coupling with the external electric (or even magnetic) field could be obtained by a low-concentration addition of highly dielectric (or magnetic) nano rods suspended in the liquid crystal.

6. Conclusions

We have discussed several examples of light transport in complex photonic structures, going from disordered systems to partially ordered materials, quasi-crystals, and fully ordered systems. The vast regime between completely ordered and disordered structures is very rich and has yet to be fully explored. Liquid crystals have been a common theme in several sections of this paper, either as a system to study multiple light scattering or as a way to obtain control over the optical properties of porous photonic materials via infiltration. The transport of light in these various complex structures is very interesting from a fundamental point of view and these new materials could find fascinating applications as photonic devices, in telecommunications, and possibly even in future optical computing.

Acknowledgements

We wish to thank Zhao-Qing Zhang and Prof. V. Freilikher for inspiring discussions, Anna Vinattieri and Daniele Alderighi for help with the time-resolved experiments, and Lydia Bechger for her work on the realization of titania inverse opals. This work was financially supported by the European community (contract number HPRI-CT1999-00111) and by the Istituto Nazionale di Fisica della Materia (PAIS project RANDS and PRA project PHOTONIC).

References

1. See e.g. H.C. van de Hulst, *Light Scattering by Small Particles* (Dover, New York, 1981).
2. See for instance: P. Sheng, *Introduction to Wave Scattering, Localization, and Mesoscopic Phenomena* (Academic Press, San Diego, 1995).
3. I. Freund, M. Rosenbluh, and Shechao Feng, Phys. Rev. Lett. **61**, 2328 (1988); Shechao Feng, Ch. Kane, P.A. Lee, and A. D. Stone, Phys. Rev. Lett. **61**, 834 (1988); N. Garcia and A.Z. Genack, Phys. Rev. Lett. **63**, 1678 (1989); M.P. van Albada, J.F. de Boer, and A. Lagendijk, Phys. Rev. Lett. **64**, 2787 (1990). P. Sebbah, B. Hu, A.Z. Genack, R. Pnini, and B. Shapiro, Phys. Rev. Lett. **88**, 123901 (2002).
4. F. Scheffold and G. Maret, Phys. Rev. Lett. **81**, 5800 (1998).
5. Y. Kuga and A. Ishimaru, J. Opt. Soc. Am. A **8**, 831 (1984); M.P. van Albada and A. Lagendijk, Phys. Rev. Lett. **55**, 2692 (1985); P.E. Wolf and G. Maret, Phys. Rev. Lett. **55**, 2696 (1985).
6. S. John, Phys. Rev. Lett. **53**, 2169 (1984); P.W. Anderson, Philos. Mag. B **52**, 505 (1985); R. Dalichaouch, J.P. Armstrong, S. Schultz, P.M. Platzman, and S.L. McCall, Nature **354**, 53 (1991); A.Z. Genack and N. Garcia, Phys. Rev. Lett. **66**, 2064 (1991); D.S. Wiersma, P. Bartolini, A. Lagendijk, and R. Righini, Nature **390**, 671 (1997); A.A. Chabanov and A.Z. Genack, Phys. Rev. Lett. **87**, 233903 (2001).
7. G. Maret and P.E. Wolf, Z. Phys. B **65**, 409 (1987); D.J. Pine, D.A. Weitz, P.M. Chaikin, and E. Herbolzheimer, Phys. Rev. Lett. **60**, 1134 (1988); S. Fraden and G. Maret, Phys. Rev. Lett. **65**, 512 (1990).
8. E. Yablonovitch, Phys. Rev. Lett. **58**, 2059 (1987); S. John, Phys. Rev. Lett. **58**, 2486 (1987).
9. *Photonic Bandgap Materials*, edited by C.M. Soukoulis (Kluwer, Dordrecht, 1996); J.D. Joannopoulos, R.D. Meade, and J.N. Winn, *Photonic Crystals* (Princeton University Press, Princeton, NJ, 1995).
10. A. Sparenberg, G.L.J.A. Rikken, and B.A. van Tiggelen, Phys. Rev. Lett. **79**, 757 (1997).
11. B.A. van Tiggelen, Phys. Rev. Lett. **75**, 422 (1995); G.L.J.A. Rikken and B.A. van Tiggelen, Nature **381**, 54 (1996).
12. D.S. Wiersma, Mol. Cryst. and Liq. Cryst. **375**, 15 (2002).

13. A. Yodh and B. Chance, Physics Today **48**(3), 34 (1995).
14. V.S. Letokhov, Zh. Eksp. Teor. Fiz. **53**, 1442 (1967) [Sov. Phys. JETP **26**, 835 (1968)]; A.Y. Zyuzin, Phys. Rev. E **51**, 5274 (1995); S. John and G. Pang, Phys. Rev. A **54**, 3642 (1996); D.S. Wiersma and A. Lagendijk, Phys. Rev. E **54**, 4256 (1996); G.A. Berger, M. Kempe, and A.Z. Genack, Phys. Rev. E **56**, 6118 (1997); Xunya Jiang and C.M. Soukoulis, Phys. Rev. Lett. **85**, 70 (2000). G. van Soest, F.J. Poelwijk, R. Sprik, and A. Lagendijk, Phys. Rev. Lett. **86**, 1522 (2001).
15. N.M. Lawandy, R.M. Balachandran, A.S.L. Gomes, and E. Sauvin, Nature **368**, 436 (1994); but see also the discussion on weak scattering in: D.S. Wiersma, M. van Albada, A. Lagendijk, Nature **373**, 203 (1995).
16. C. Gouedard, D. Husson, C. Sauteret, F. Auzel, and A.J. Migus, J. Opt. Soc. Am. B **10**, 2358 (1993).
17. C.W.J. Beenakker, Phys. Rev. Lett. **81**, 1829 (1998).
18. H. Cao, Y.G. Zhao, S.T. Ho, E.W. Seelig, Q.H. Wang, and R.P.H. Chang, Phys. Rev. Lett. **82**, 2278 (1999); H. Cao, Y. Ling, J.Y. Xu, C.Q. Cao, and Prem Kumar, Phys. Rev. Lett. **86**, 4524 (2001); but see also Y. Sun, J.B. Ketterson, and G.K.L. Wong, Appl. Phys. Lett. **77**, 2322 (2000).
19. P.G. de Gennes and J. Prost, *The Physics of Liquid Crystals,* 2nd edition (Oxford, New York, 1993); S. Chandrasekhar, *Liquid Crystals* (Cambridge Univ. Press, Cambridge, 1977).
20. D.S. Wiersma, M. Colocci, R. Righini, and F. Aliev, Phys. Rev. B **64**, 144208 (2001).
21. See for instance *Liquid Crystal Dispersions*, P. S. Drzaic, (World Scientific, Singapore, 1995); A. Mertelj, L. Spindler, and M. Copic, Phys. Rev. E **56**, 549 (1997); J.H.M. Neijzen, H.M.J. Boots, F.A.M.A. Paulissen, M.B. van der Mark, and H.J. Cornelissen, Liq. Cryst. **22**, 255 (1997); L. Leclercq, U. Maschke, B. Ewen, X. Coqueret, L. Mechernene, and M. Benmouna, Liq. Cryst. **26**, 415 (1999); T. Bellini, N.A. Clark, V. Degiorgio, F. Mantegazza, and G. Natale, Phys. Rev. E **57**, 2996 (1998).
22. D.S. Wiersma and S. Cavalieri, Nature **414**, 708 (2001); D.S. Wiersma and S. Cavalieri, Phys. Rev. E **66**, 056612 (2002).
23. Refractive index values at $\lambda = 632.8$ nm. See D.A. Dunmur, M.R. Manterfield, W.H. Miller, and J.K. Dunleavy, Mol. Cryst. Liq. Cryst. **45**, 127 (1978).
24. *Liquid Crystals in Complex Geometries Formed by Polymer and Porous Networks*, edited by G.P. Crawford and S. Zumer (Taylor & Francis, London, 1996); T. Bellini, N.A. Clark, C.D. Muzny, L. Wu, C.W. Garland, D.W. Schaefer, and B.J. Olivier, Phys. Rev. Lett. **69**, 788 (1992).
25. D.V. Vlasov, L.A. Zubkov, N.V. Orekhova, and V.P. Romanov, Pis'ma Zh. Eksp. Teor. Fiz. **48**, 86 (1988) [JETP Lett. **48**, 91 (1988)]; H.K.M. Vithana, L. Asfaw, and D.L. Johnson, Phys. Rev. Lett. **70**, 3561 (1993).
26. D. Langevin, Solid State Comm. **14**, 435 (1974); D. Langevin and M.-A. Bouchiat, J. Phys. (Paris) C **1**, 197 (1975); A.Y. Val'kov and V.P. Romanov, Zh. Eksp. Teor. Viz. **82**, 1777 (1982) [Sov. Phys. JETP **56**, 1028 (1983)].
27. M.H. Kao, K.A. Jester, A.G. Yodh, and P.J. Collings, Phys. Rev. Lett. **77**, 2233 (1996).
28. D.S. Wiersma, A. Muzzi, M. Colocci, and R. Righini, Phys. Rev. Lett. **83**, 4321 (1999).
29. V.P. Romanov and A.N. Shalaginov, Opt. Spectrosc. (USSR) **64**, 774 (1988); B.A. van Tiggelen, R. Maynard, and A. Heiderich, Phys. Rev. Lett. **77**, 639 (1996); H. Stark and T.C. Lubensky, Phys. Rev. Lett. **77**, 2229 (1996); B.A. van Tiggelen and H. Stark, Rev. Mod. Phys. **72**, 1017 (2000).
30. P.M. Johnson, B.P.J. Bret, J.G. Rivas, J.J. Kelly, and A. Lagendijk, Phys. Rev. Lett. **89**, 243901 (2002).
31. T. Fujiwara and T. Ogawa, *Quasicrystals* (Springer Verlag, Berlin, 1990).
32. G. Gumbs and M.K. Ali, Phys. Rev. Lett. **60**, 1081 (1988).
33. See e.g. Franco Nori and J. P. Rodriguez, Phys. Rev. B **34**, 2207 (1986); R.B. Capaz,

B. Koiller, and S.L.A. de Queiroz, Phys. Rev. B **42**, 6402 (1990).
34. T. Fujiwara, M. Kohmoto, and T. Tokihiro, Phys. Rev. B **40**, 7413 (1989); C.M. Soukoulis and E.N. Economou, Phys. Rev. Lett. **48**, 1043 (1982).
35. M. Kohmoto, B. Sutherland, and K. Iguchi, Phys. Rev. Lett. **58**, 2436 (1987); B. Sutherland and M. Kohmoto, Phys. Rev. B **36**, 5877 (1987).
36. R. Merlin, K. Bajema, R. Clarke, F.Y. Juang, and P.K. Bhattacharya, Phys. Rev. Lett. **55**, 1768 (1985).
37. J.B. Sokoloff Phys. Rev. Lett. **58**, 2267 (1987); Ch.Wang and R.A. Barrio, Phys. Rev. Lett. **61**, 191 (1988); E. Maciá and F. Domínguez-Adame, Phys. Rev. Lett. **76**, 2957 (1996); F. Piéchon, Phys. Rev. Lett. **76**, 4372 (1996); X. Huang and Ch. Gong, Phys. Rev. B **58**, 739 (1998); F. Steinbach, A. Ossipov, Tsampikos Kottos, and T. Geisel, Phys. Rev. Lett. **85**, 4426 (2000).
38. W. Gellermann, M. Kohmoto, B. Sutherland, and P.C. Taylor, Phys. Rev. Lett. **72**, 633 (1994).
39. T. Hattori, N. Tsurumachi, S. Kawato, and H. Nakatsuka, Phys. Rev. B **50**, 4220, (1994).
40. L. dal Negro, C.J. Oton, Z. Gaburro, L. Pavesi, P. Johnson, A. Lagendijk, R. Righini, M. Colocci, D.S. Wiersma, Phys. Rev. Lett. **90**, 055501 (2003).
41. R.H.J. Kop and R. Sprik, Rev. Sci. Instrum. **66**, 5459 (1995).
42. A. Imhof, W.L. Vos, R. Sprik, and A. Lagendijk, Phys. Rev. Lett. **83**, 2942 (1999).
43. F.L. Pedrotti and L.S. Pedrotti, *Introduction to Optics* (Prentice-Hall, 1987); A. Kavokin, G. Malpuech, A. Di Carlo, P. Lugli, and F. Rossi, Phys. Rev. B **61**, 4413 (2000).
44. M. Scalora *et al.*, Phys. Rev. E **54**, R1078 (1996).
45. K. Busch and S. John, Phys. Rev. Lett. **83**, 967 (1999).
46. K. Yoshino, Y. Shimoda, Y. Kawagishi, K. Nakayama, and M. Ozaki, Appl. Phys. Lett. **75**, 932 (1999); S.W. Leonard, J.P. Mondia, H.M. van Driel, O. Toader, S. John, K. Busch, A. Birner, U. Gosele, and V. Lehmann, Phys. Rev. B **61**, R2389 (2000).
47. D. Kang, J.E. Maclennan, N.A. Clark, A.A. Zakhidov, and R.H. Baughman, Phys. Rev. Lett. **86**, 4052 (2001).
48. Y. Shimoda, M. Ozaki, and K. Yoshino, Appl. Phys. Lett. **79**, 3627 (2001); Q.B. Meng, C.H. Fu, S. Hayami, Z.Z. Gu, O. Sato, and A. Fujishima, J. Appl. Phys. **89**, 5794 (2001); P. Mach, P. Wiltzius, M. Megens, D.A. Weitz, Keng-hui Lin, T.C. Lubensky, and A.G. Yodh, Phys. Rev. E **65**, 031720 (2002).
49. J.E.G.J. Wijnhoven and W.L. Vos, Science **281**, 802 (1998); J.E.G.J. Wijnhoven, L. Bechger, and W.L. Vos, Chem. Mater. **13**, 4486 (2001).
50. A.F. Koenderink, L. Bechger, H.P. Schriemer, A. Lagendijk, and W.L. Vos, Phys. Rev. Lett. **88**, 143903 (2002).
51. G. Mertens, T. Röder, R. Schweins, K. Huber, and Heinz-S. Kitzerow, Appl. Phys. Lett. **80**, 1885 (2002).

PROPAGATION OF LIGHT IN STRONGLY DISORDERED PHOTONIC MATERIALS AND RANDOM LASERS

P. LODAHL[1], G. VAN SOEST[2], J. GÓMEZ RIVAS[2],
R. SPRIK[2] AND A. LAGENDIJK[1]

[1] *Department of Applied Physics and MESA+ Research Institute, University of Twente, P.O. Box 217, 7500 AE Enschede, The Netherlands*

[2] *Van der Waals-Zeeman Institute, University of Amsterdam, Valckenierstraat 65, 1018 XE Amsterdam, The Netherlands*

1. Introduction

The quest for developing ever stronger interacting photonic materials has caused the birth of a widely expanding research field. Successful achievement of very strong elastic interaction between light and matter would allow demonstration of conceptually new physical phenomena, among which Anderson localization [1] and a full photonic bandgap [2, 3] are prime examples. Experimental realization of these phenomena would open new avenues in control and manipulation of light and matter [4], and could lead to fascinating new developments in quantum optics.

Several parameters determine the quality of a photonic material depending on the application. The interaction of light and the photonic material is mediated by scattering, and strong scattering is ensured by having a high refractive index contrast of the scattering particle with respect to the background material. Furthermore, the particle size of the scatterer relative to the wavelength of light is critical. Most efficient interaction is found in the Mie scattering regime where the size of the scatterer is comparable to the wavelength. Finally, the amount of optical absorption present in the photonic material often turns out to be a sensitive parameter that may hamper the observation of the delicate interference phenomena.

An alternative way of tuning the optical properties of a multiple scattering media is to add optical gain. This can be done for instance by doping

B.A. van Tiggelen and S.E. Skipetrov (eds.),
Wave Scattering in Complex Media: From Theory to Applications, 21–44.

the material with active molecules. The inclusion of gain induces a variety of new phenomena [5, 6]. In the case of a random multiple scattering media a threshold for lasing action and frequency narrowing can be observed. For that reason the systems have been called random lasers. This point of view is further supported by recent measurements proving that the photon statistics of a random laser is poissonian above threshold [7].

In the current contribution we will present recent experimental work on multiple light scattering in strongly photonic materials. The first part concerns porous gallium phosphide which is the strongest scattering medium for visible light reported to date. It is demonstrated how the photonic strength can be tuned sensitively during sample preparation by controlling the etching potential and the density of impurities in the sample. In the second part of the paper light propagation in random lasers is discussed. Experimental investigations of speckle statistics and enhanced backscattering in random lasers are supplemented by a new rate-equation model that explains the observations convincingly.

2. Fabrication and Tuning of Strongly Scattering Porous Gallium Phosphide

Recently a new strongly interacting photonic material was developed: porous gallium phosphide (GaP) [8]. It was realized that electrochemical etching of bulk GaP could provide a highly isotropic random network of holes making this porous structure an excellent candidate for experiments on light localization [9]. The high refractive index of GaP ($n = 3.3$) and the remarkable observation that optical absorption is undetectable in the samples give great promises. Furthermore, the physical size of the scatterers can be varied in the fabrication process which allows sensitive tuning of the photonic strength of GaP. In the current paper a detailed description of the electrochemical etching process of GaP is given, and measurements of the photonic strength are presented. A recent publication concerned the dependence of the photonic strength on the dopant density of the n-type GaP crystals [10]; here it is demonstrated that the etching potential V provides another important handle for tuning.

2.1. ELECTROCHEMICAL ETCHING OF GALLIUM PHOSPHIDE

Bulk n-type GaP can be etched electrochemically generating random porous structures. Etching occurs through the following chemical reaction of GaP in an acid electrolyte:

$$(\mathrm{GaP})_{\mathrm{n}} + 6\mathrm{h}^{+} \rightarrow (\mathrm{GaP})_{\mathrm{n}-1} + \mathrm{Ga}^{3+} + \mathrm{P}^{3+} \,. \tag{1}$$

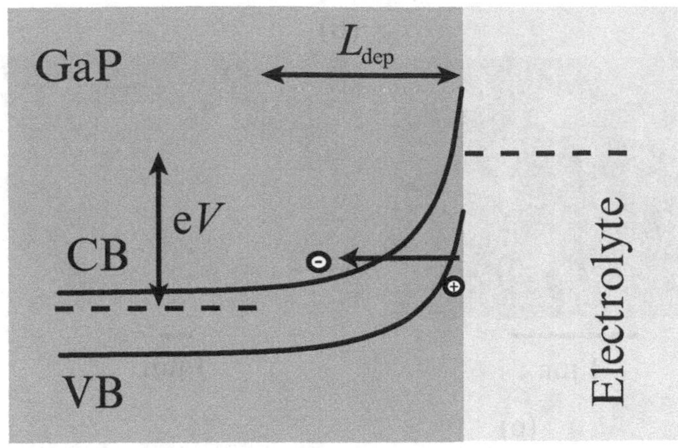

Figure 1. Interface between GaP and the electrolyte at a potential V. Electrons in the depletion layer $L_{\rm dep}$ may tunnel from the valence band (VB) to the conduction band (CB). The generated holes are used for the dissolution of the semiconductor at the interface.

The process requires the presence of valence-band holes that can be generated by applying a strong positive electric potential V on the semiconductor. This induces bending of the valence and conduction bands close to the semiconductor-electrolyte interface, c.f. Fig. 1. Hence electrons can tunnel from the valence to the conduction band generating valence band holes to drive the etching reaction (1). The region of bending of the valence and conduction bands is referred to as the depletion layer and has a spatial extent given by [11]

$$L_{\rm dep} = \left(\frac{2\epsilon\epsilon_0}{{\rm e}N}(V - V_{\rm fb}) \right)^{1/2} , \qquad (2)$$

where $\epsilon = 11$ is the dielectric constant of GaP, ϵ_0 is the vacuum permittivity, e is the electron charge, N is the donor concentration, and $V_{\rm fb}$ is the flat-band potential. For a given semiconductor electrode, the flat-band potential depends mainly on the nature of the electrolyte. In our experiments (aqueous 0.5M H_2SO_4 solution) we have $V_{\rm fb} \simeq -1.2$ V [12].

It is important to investigate in detail how the porous structure is formed during the etching since this provides valuable knowledge about how to control the process. We used 300 μm thick commercially available GaP wafers doped with different concentrations of sulfur impurities. The etched porous structure was studied with high resolution SEM images, where typical examples are displayed in Fig. 2. The etching is seen to start at specific pits on the surface where defects are present. These surface defects may lead to a local enhancement of the electric field helping the hole generation

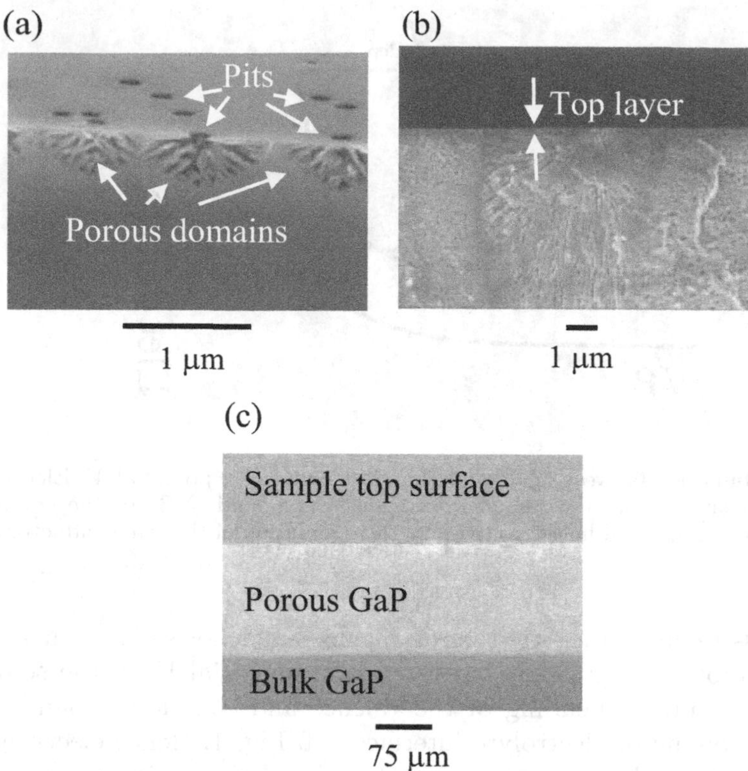

Figure 2. Electron-microscope (SEM) photographs of porous GaP: (a) cleaved cross section of a sample in which etching was stopped at its initial stage, (b) side view of a porous structure, (c) cleaved cross section of a layer of porous GaP with a thickness of 203 μm.

[12]. The pits can be clearly seen in the SEM photograph (a) of Fig. 2. This image shows a cleaved cross section of a porous GaP sample in which etching was stopped at the initial stage. From an initial pit a pore is formed. This pore branches leading to a porous domain that expands hemispherically. The porous domains, originating from different surface defects, can be seen in Fig. 2(a). Once the adjacent porous domains meet, a porous layer grows at a constant rate. The SEM photograph of Fig. 2(b) corresponds to a side view of a fully etched porous structure, while Fig. 2(c) is a cleaved cross section, on a larger spatial scale, of a porous GaP layer. We observe that there is a thin layer of bulk GaP on top of the porous structure, with only a few pits remaining where the pore formation was initiated [see Figs. 2(a) and (b)]. The presence of this top layer complicates the analysis of optical experiments on porous GaP substantially and can advantageously be removed by photochemical etching [13].

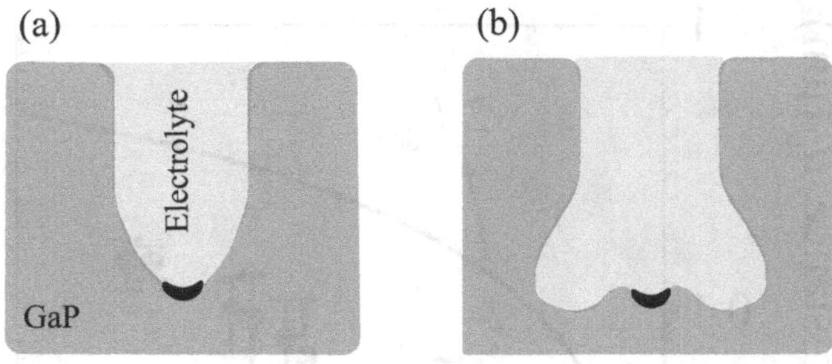

Figure 3. Schematic representation of the pore branching. Figure (a) represents a pore. The strongest electric field is at the pore tip due to the large curvature. The etching rate is higher in this region and it can be passivated, as represented by the black area. Figure (b) shows how the etching can proceed in the non-passive region leading to the branching of the pore.

2.2. PHOTONIC STRENGTH

The strength of the photonic coupling between light and material is quantified by the scattering mean free path ℓ_s, which is the average distance between two consecutive scattering events. To the lowest order approximation, i.e. in the independent scattering approximation, the scattering mean free path is given by

$$\ell_s = \frac{1}{\rho \sigma_s}, \tag{3}$$

where ρ is the density of scatterers and σ_s is the scattering cross section. Strong scattering is achieved by optimizing both ρ and σ_s. In our samples the density of scatterers is determined by the inter-pore distance d_p and the radius of the pores r, while the scattering cross section depends solely on the pore radius. These parameters can be tuned by controlling the etching potential and the doping concentration. The latter dependence was investigated in depth in [10], while we in this section present measurements on the photonic strength versus etching potential.

The inter-pore distance d_p in electrochemically etched n-type semiconductors is determined by the extent of the depletion layer L_{dep} [12, 14, 15]; and this distance is always smaller than $2L_{\text{dep}}$. If the depletion layers of adjacent pores overlap, the electric field is reduced and the etching stops. The dissolution process only occurs in regions where the field is enhanced, i.e. predominantly at the pore tip where the field enhancement is highest. Above a critical etching rate, the reaction (1) may be locally passivated, as limited by the formation of an oxide layer on the semiconductor-electrolyte interface. However, the etching can proceed close to the pore tip causing a branching of the pore and eventually the formation of the random network

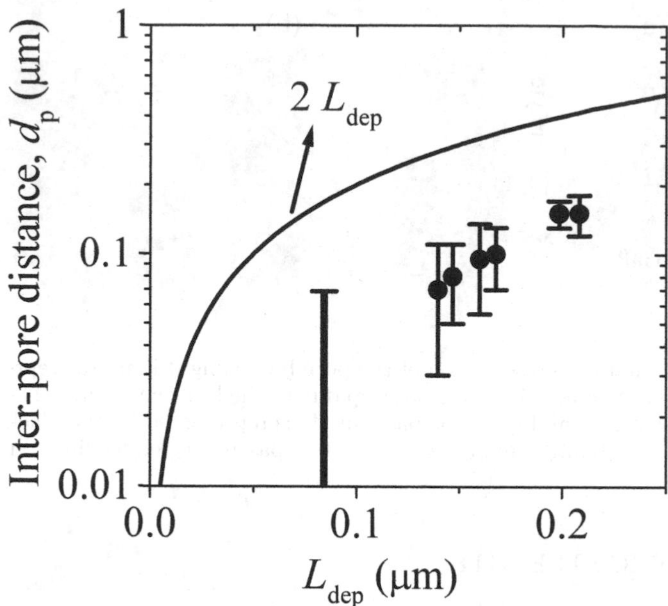

Figure 4. Inter-pore distance in porous GaP versus the extent of the depletion layer L_{dep}. The solid line represents $2L_{dep}$.

structure. This branching process is summarized in Fig. 3. Figure 4 shows the inter-pore distance of porous GaP plotted as a function of L_{dep}. The inter-pore distance is defined as the average distance between adjacent pore walls, as was estimated with a marked gauge from high magnification SEM images. As expected we observe that the pore distance increases with the extent of the depletion layer and that $2L_{dep} \geq d_p$, where $2L_{dep}$ is marked with a solid curve in Fig. 4.

As discussed in Ref. [15], the extent of the depletion layer also determines the average pore radius in n-type semiconductors. Larger pores are formed when L_{dep} is increased. Note that L_{dep} depends on the applied potential during etching [see Eq. (2)]. Therefore both the inter-pore distance and the pore radius will be changed when varying V, i.e. both the density of scatterers and the scattering cross section are changed. It will be shown below that increasing the potential V will lead to stronger scattering. This implies that the dominating contribution is the increased scattering cross section that overwhelms the reduction from the decrease in scatterer density.

A direct measurement of ℓ_s in strongly scattering samples is very intriguing due to the exponential decay of coherent transmission with the sample thickness [16]. Therefore, we extract the scattering strength from enhanced backscattering (EBS) measurements [17]. The width of the EBS cone is inversely proportional to the transport mean free path ℓ, defined as the average distance the light travels in the medium before the propagation direction is completely randomized [18]. In general the transport mean free path ℓ is larger than ℓ_s, and the two quantities only coincide in a medium formed by isotropic scatterers.

Figure 5 shows results of EBS measurements at a wavelength of 633 nm on two porous GaP samples with $N = 5 \times 10^{17}$ cm^{-3}, etched at 15 V (squares) and at 16.6 V (circles). The latter is the highest potential at which etching is possible. Above this potential the GaP electrode is passivated and etching stops. The EBS cone of the sample etched at the high potential is significantly wider, indicating stronger scattering. The extracted transport mean free paths are $\ell = 0.22 \pm 0.02$ μm (for the sample etched at 16.6 V) and $\ell = 0.43 \pm 0.03$ μm (for the sample etched at 15 V), as obtained from the fits of the EBS cones using diffusion theory [19] (the solid lines in Fig. 5). Internal reflection at the surface of the material was included in the fits with values of the effective refractive index obtained by measuring the angular-resolved transmission [20]. These measurements promotes porous GaP to the strongest scattering elastic media for visible light found to date.

We note that the measurements of the transport mean free path are consistent with the average pore radius of the samples extracted from SEM images. The sample etched at 15 V has an average pore radius $r = 0.05 \pm 0.02$ μm while the sample etched at 16.6 V has $r = 0.095 \pm 0.02$ μm. Thus, increasing the etching potential implies a structure with larger pores that scatters light more effectively since the pore radius approaches the wavelength of light. The inter-pore distance is found to be the same for the two different etching potentials within the error bars of our measurements (0.15 ± 0.03 μm).

2.3. CONCLUSIONS

Electrochemically etched porous GaP is the strongest scattering medium reported for visible light. Understanding the etching process allows tailoring the scattering strength of the materials. Important tuning parameters are the etching potential and the doping concentration in the GaP wafer. In this section a detailed account for the etching process was given and the tuning characteristics as a function of the etching potential were investigated. Based on enhanced backscattering measurements the transport mean free path was extracted. The detailed control of the etching process should allow future preparation of even stronger scattering samples potentially reaching the boundary for Anderson localization. The lack of detectable optical absorption in GaP would hopefully allow clear studies of the onset of localization.

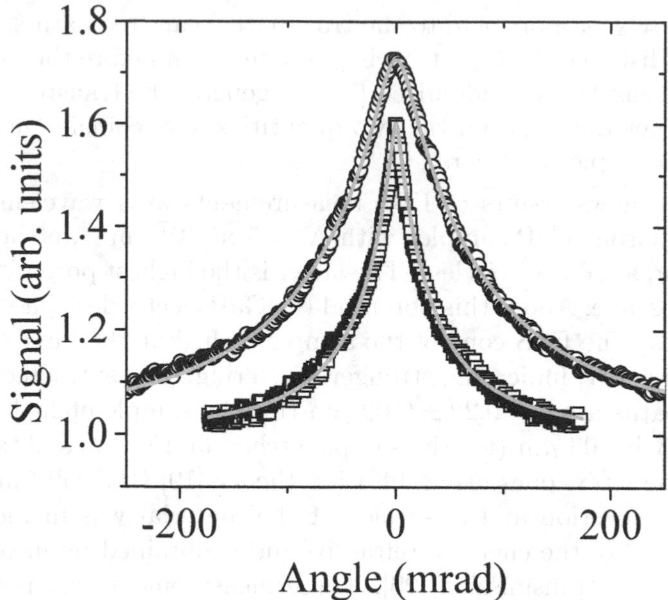

Figure 5. EBS measurements of porous GaP ($N = 5 \times 10^{17}$ cm^{-3}) anodically etched at 15 V (squares) and at 16.6 V (circles). The solid lines through the symbols are fits using diffusion theory.

3. Propagation of Light in Random Lasers

Random lasers are strongly scattering media with optical gain. These systems have many features in common with more conventional lasers based on an optical gain medium enclosed in a cavity to enhance stimulated emission. For example, a threshold for lasing action and associated frequency narrow emission have been observed in random lasers based on multiply scattering colloidal dispersions in dyes [6, 21]. Evidently the optical properties of random lasers are quite different from conventional lasers: the propagation of pump and fluorescence light is diffusive, and there is no "preferred direction" in feedback and loss processes. The ambiguities as to what exactly constitutes the loss in a random laser, how optical feedback works if it is non-directional, and the theoretical prediction of an intensity divergence will be discussed in the following. Experimental clarification is obtained by measuring speckle statistics and enhanced backscattering from a random medium with gain.

Gain in a multiple scattering medium can be included directly in the diffusion equation

$$\frac{\partial W}{\partial t} = D\nabla^2 W + \kappa_{\mathrm{g}} W + S\,, \tag{4}$$

where W is the energy density of the light, D the diffusion constant for light in the medium, κ_{g} is the gain coefficient, and S the source term. Solutions of Eq. (4) exist only for a finite sized system [5]. The maximum volume for which a solution can be found is called the critical volume V_{cr}. The physical origin of this behavior is the existence of infinitely long diffusive paths which will be amplified to infinite intensities in a gain medium. This results in a diverging intensity whenever the probability for a path to be scattered out of the system is compensated by the gain. Chances of leaving the medium are smaller for larger systems, so the smaller the gain the larger V_{cr}. The critical thickness is $L_{\mathrm{cr}} = \pi\sqrt{\ell\ell_{\mathrm{g}}/3}$, where ℓ the mean free path for scattering and ℓ_{g} the gain length.

Equation (4) is not sufficient to describe the lasing and threshold behavior of a random laser in general, hence the temporal and spatial dependence of the light fields must also be considered. A simple set of equations including these effects is given by [22, 23]

$$\begin{aligned}
\frac{\partial W_\ell}{\partial t} &= D\nabla_z^2 W_\ell + (\sigma_{\mathrm{e}} c n_1 - \sigma_{\mathrm{r}} c n_0)\, W_\ell + \frac{\beta}{\tau} n_1, \\
\frac{\partial W_{\mathrm{p}}}{\partial t} &= D\nabla_z^2 W_{\mathrm{p}} - \sigma_{\mathrm{a}} c n_0 W_{\mathrm{p}} + \frac{1}{\ell} I_{\mathrm{in}}, \\
\frac{\partial n_1}{\partial t} &= \sigma_{\mathrm{a}} c n_0 W_{\mathrm{p}} - (\sigma_{\mathrm{e}} c n_1 - \sigma_{\mathrm{r}} c n_0)\, W_\ell - \frac{1}{\tau} n_1.
\end{aligned} \tag{5}$$

Here W_ℓ and W_{p} are the space and time dependent "laser" and pump light densities where the spatial dependence has been restricted to 1D as expressed by the z-coordinate. The intensity W_ℓ represents only the light taking part in the amplification process; not all the emitted luminescence. The parameter β in Eqs. (5) is the fraction of spontaneous emitted light contributing to the laser light [24]. The ground state and excited state populations are n_0 and n_1, satisfying $n_0 + n_1 = n$, the total dye concentration. The cross-sections $\sigma_{\mathrm{e}}, \sigma_{\mathrm{a}}$, and σ_{r} describe emission, absorption, and reabsorption of the laser and pump light. Finally, τ is the relaxation time for spontaneous emission, and c is the speed of light in vacuum.

Equations (5) can be solved numerically (see [22, 25] for details) to obtain insight in the coupling between the diffusive part and the level dynamics. A characteristic example is given in Fig. 6. The dependence of the dynamic quantities on the position in the medium shows that only a relatively thin ($< 10\ell$) layer develops with significant population inversion at the arrival of a pump pulse. This layer provides the net gain and also serves as a source for spontaneous emission. The z-dependence of the inversion is exponential and non vanishing only where the pump light can reach it through the absorbing and scattering medium.

The spatial and temporal dependence of the population inversion has an immediate influence on the gain that the probe light will experience when diffusing through the medium. We will show in section 3.3 how this influences enhanced backscattering in a medium with gain. A close investigation of the temporal behavior reveals [22] that the dynamics of a random laser is very similar to that of a conventional laser system. For instance the relaxation oscillations [26] observed after arrival of a pump pulse (see Fig. 7) are well known in conventional lasers. The relaxation oscillation dynamics is found to force the system towards a well defined equilibrium, thus resolving the divergence problem found in the diffusion equation [Eq. (4)]. The random laser threshold we find is obtained from a transport formalism taking into account the local gain in a population rate equation. This shows that the "laser view" [27, 28, 29, 30] and the "multiple scattering view" [31, 32, 33, 34] on random lasers can be united into one. We note that the conclusions presented here are in agreement with other theoretical work (see, e.g., [5, 27, 29, 35, 36, 37]) based on diffusive transport and gain dynamics obtained by, e.g., Monte Carlo calculations including spectral dependence. Recent observations of sharp spikes in the emission spectrum of random lasers [38] could be due to interference effects in the scattering medium [39]. Interference effects are not included in Eqs. (5), but are discussed by Pacilli *et al.* [40].

3.2. SPECKLE IN RANDOM LASERS

Speckle is the strongly fluctuating, grainy intensity pattern resulting from the interference of a randomly scattered coherent wave. It can be observed in space, time, and frequency. Some statistical characteristics of the speckle pattern contain information about the transport process [41]. We discuss here only spatial speckles, recorded as the fluctuating intensity versus emission angle.

If a coherent plane wave falls on a rough surface, a speckle pattern can be seen on a screen positioned at some distance from the scattering object. The observed intensity $I = |E|^2$ follows a Rayleigh distribution

$$P(I) = \frac{1}{\langle I \rangle} e^{-\frac{I}{\langle I \rangle}} . \qquad (6)$$

The typical speckle spot size depends on the characteristic distance along the screen on which the fields that contribute to the speckle dephase.

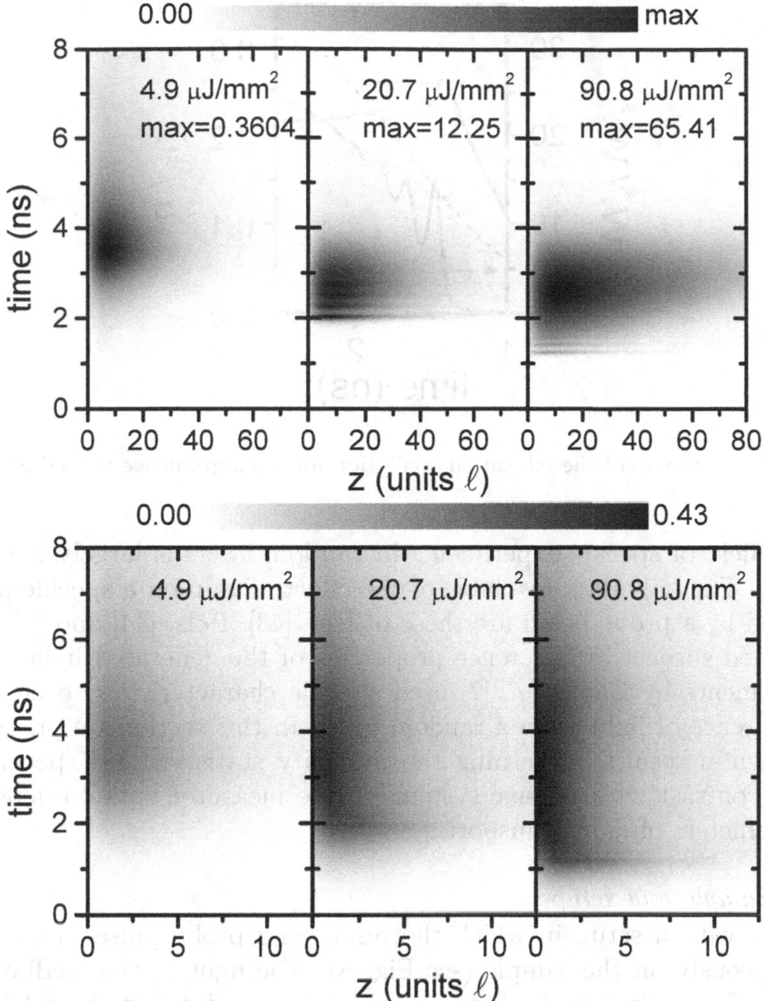

Figure 6. Simulation results of Eqs. (5) for a parameter choice characteristic for Sulforhodamine B in titania dispersions [25]. Top: Density of laser light $\tilde{w}_\ell = W_\ell/(\sigma_e c\tau)$ in a random laser as a function of time t (vertical axis) and position z (horizontal axis) for the three different pump fluences indicated in the figures. The grayscale runs from zero to the maximum density (indicated in the plots). The threshold pump fluence is 10 μJ/mm^2. Bottom: Inversion n_1/n in a random laser as a function of time t (vertical axis) and position z (horizontal axis) for the three different pump fluences indicated in the figures. The maximum inversion obtained for above threshold pump fluences (> 10 μJ/mm^2) is about 0.43.

The largest path length difference at a spot on the screen is caused by the partial waves arriving from opposite ends of the illuminated region of the scattering surface. Consequently, the typical angular size of a speckle spot is λ/d [42], if d is the diameter of the illuminated region.

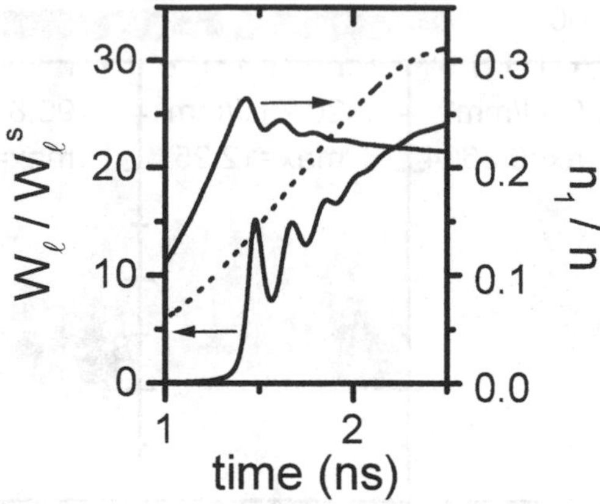

Figure 7. Close-up of the relaxation oscillation for a pump fluence of 53.6 μJ/mm^2.

The field of speckle experiments in random lasers is largely uncharted territory. The only data available of the effect of gain on a speckle pattern produced by a probe beam are those of Ref. [43]. Refs. [44] and [45] study the related subject of coherence properties of the generated light. Recent measurements by Cao *et al.* [7] used speckle characteristics to determine the coherence of light from a random laser. In this section we present our experimental results concerning the intensity statistics and speckle spot size. In contrast with passive systems, these measurements do depend on the parameters of light transport.

3.2.1. *Sample and setup*

We work with a setup in which the pump and probe pulses are incident simultaneously on the sample (see Fig. 8). The light in the medium then consists of amplified probe light, in which we are interested, and fluorescence. Since it is favorable to do the experiments at large amplification, and the maxima of the gain curve $\sigma_e(\lambda)$ and the fluorescence spectrum $L(\lambda)$ of dye are usually close in wavelength, the probe and fluorescence can not be separated spectrally or temporally. However, the coherence and the associated speckle pattern are different and will be used to discriminate between fluorescence and amplified light.

The pump and probe pulses are produced by one laser system consisting of an optical parametric oscillator (OPO) pumped by a Q-switched Nd:YAG laser (see section 3.3.1). For the speckle experiment we probe with the output of the OPO at 480 nm, and the residual doubled Nd:YAG light at 532 nm is used as pump. The samples consist of titania suspensions doped with 2 mM Coumarin 6 dye. The suspension is contained in a round plastic container with dimensions 6 mm (depth) $\times$ 10 mm (diameter), covered

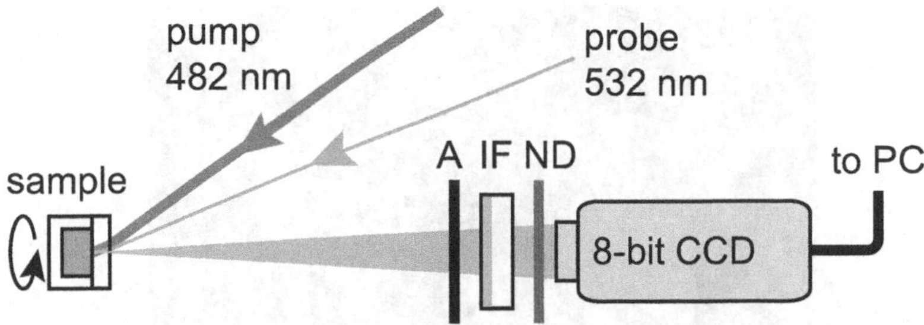

Figure 8. Schematic of the setup for speckle experiments. The pump (diameter 2 mm) and probe (diameter 0.8 mm) reach the sample simultaneously. The sample is mounted on a motor, spinning it slowly to prevent sedimentation. The scattered and amplified probe light is collected on an 8-bit 752 × 582 pixel CCD camera (Kappa CF 8/1 FMC), through an aperture (A), a 532 nm interference filter (IF), and one or more neutral density filters. The distance between sample and camera is 10 cm.

with a 4 mm thick glass window. The sample is rotating slowly to prevent sedimentation and dye degradation. The transport mean free path of the sample is found to be 10 μm from enhanced backscattering measurements. The threshold pump intensity is $I_{\mathrm{p}} = 0.22$ mJ/mm^2.

The speckle is recorded on a Kappa CF 8/1 FMC 8-bit CCD camera. The light from the sample passes through an aperture, blocking stray light, a 532 nm interference filter with a transmission FWHM of 1.0 nm to remove most of the fluorescence, and a neutral density filter. The image is a single shot exposure, because the liquid suspension leads to continuous speckle changes. The probe beam diameter is 0.8 mm. This produces a speckle that can be resolved well on the camera, with a large number of speckle spots. The pump spot is chosen to be larger, 2 mm, to provide a large amplifying region for the probe light to propagate in. Figure 8 is a schematic of the experimental arrangement of sample and detection.

3.2.2. *Intensity statistics*

The Rayleigh distribution of speckles is a very robust phenomenon. As such, measurements of the speckle intensity statistics is not an effective method to obtain new information about random lasers. However, it does provide a measurement of the degree of amplification of the incident probe, a property that is otherwise difficult to quantify.

We make use of the coherence of the Nd:YAG frequency doubled output as a probe pulse, producing a speckle with good visibility. Figure 9(a) shows an image of the speckle pattern in scattered and amplified probe light from a Coumarin6/TiO_2 random laser. The drawback of the large pump spot is a larger fluorescence component in the image.

The intensity histogram of the speckle in Fig. 9(a) is shown in Fig. 9(b).

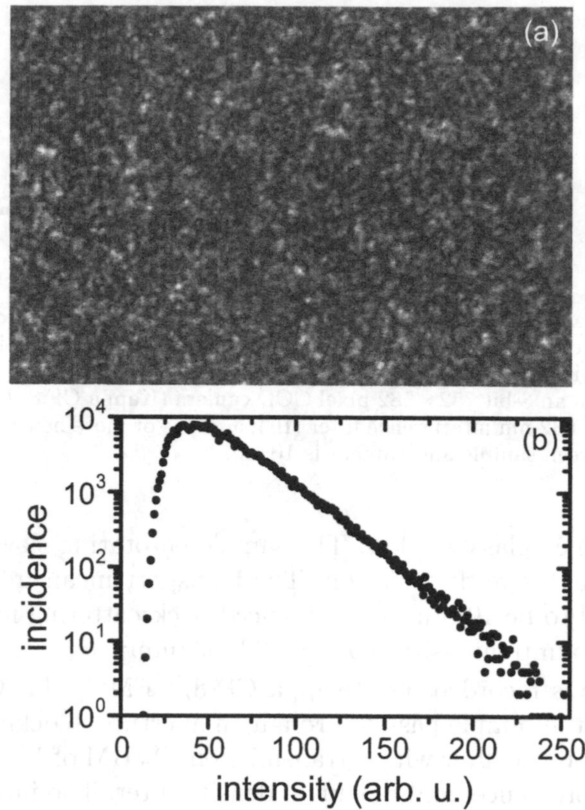

Figure 9. (a) Speckle pattern on an 8-bit CCD camera of frequency-doubled Nd:YAG scattered and amplified by a 2 mM Coumarin 6 solution in hexylene glycol with TiO_2 scatterers, $\ell = 10\ \mu$m. The pump pulse from the OPO of wavelength 482 nm has an energy of 0.32 mJ/mm^2. (b) Intensity histogram of the image in (a). The Rayleigh distribution is offset by a background of incoherent fluorescence. The slope of the linear decrease is $1/\langle I \rangle$.

The histogram has an unusual feature: it only starts to show Rayleigh statistics above intensity 50. The lower intensities are incoherent fluorescence, giving each pixel an offset. The average intensity can be determined from the slope of the exponential decrease for higher intensities.

We extract the average intensity of the amplified probe from the slope of the intensity statistics plotted in the manner of Fig. 9. The dependence of the average intensity of the speckle, or the intensity of the amplified probe, on pump fluence is plotted in Fig. 10.

3.2.3. *Speckle spot size*

A measurement of the speckle spot size yields the transverse dimension of the coherent source on the scattering surface. In a random laser this may

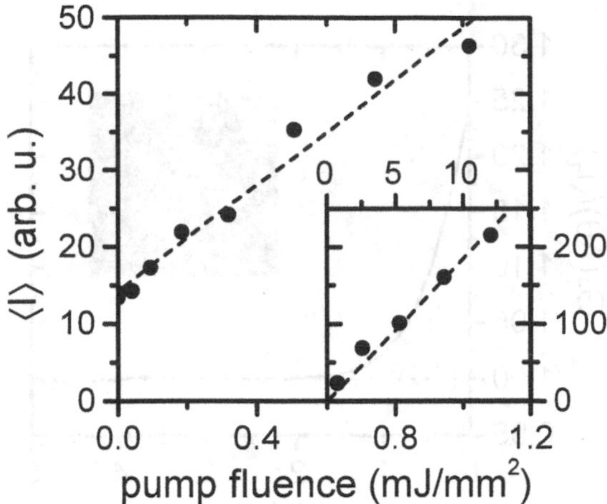

Figure 10. Average intensity derived from Rayleigh statistics as a function of pump fluence for a fixed probe intensity of 59 μJ/mm^2. The amplified intensity grows linearly with pump fluence. For comparison, we plot the mean intensity as a function of the *probe* fluence (in units of 0.01 mJ/mm^2) in the inset. The pump intensity is high: 2.9 mJ/mm^2. The lines are linear fits to the data.

very well depend on the pump energy, since a larger gain allows the light to spread further, thus enhancing long paths.

The spot size is measured by calculating the two-dimensional intensity autocorrelate $G_{\mathrm{I}}(\Delta\theta) = \langle I(\theta)I(\theta+\Delta\theta)\rangle$ of a speckle pattern as in Fig. 9(a). The contrast between the maximum at zero and the value at large $\Delta\theta$ is a factor 2: $G_{\mathrm{I}}(\Delta\theta \gg \lambda/d) = \langle I\rangle^2$, and $G_{\mathrm{I}}(0) = 2\langle I\rangle^2$.

The diffuse light source is partly incoherent due to the contribution of fluorescence. The consequences for the intensity autocorrelate are shown in Fig. 11, showing a cross section through the autocorrelate of the data in Fig. 9(a). The speckle contrast is reduced [46] to $1 + (\langle I\rangle/\langle I_{\mathrm{T}}\rangle)^2$, where $\langle I\rangle$ is the average intensity of the amplified probe and $\langle I_{\mathrm{T}}\rangle = \langle I\rangle + I_{\mathrm{F}}$ is the average total intensity, including the fluorescence intensity I_{F}. For the data in Fig. 9, $\langle I\rangle \approx I_{\mathrm{F}}$ and so the contrast in the autocorrelate is reduced to a factor 1.25.

The resulting $G_{\mathrm{T}}(\Delta\theta)$ [$= G_{I_{\mathrm{T}}}(\Delta\theta)$] is normalized to $\langle I_{\mathrm{T}}\rangle$ and analyzed quantitatively by modelling it with a function

$$G_{\mathrm{m}}(\Delta\theta) = 1 + \left[\frac{\langle I\rangle}{\langle I_{\mathrm{T}}\rangle}\right]^2 e^{-\Delta\theta/\theta_{\mathrm{c}}} . \qquad (7)$$

$\langle I\rangle$ and $\langle I_{\mathrm{T}}\rangle$ are determined directly from the data, so θ_{c} is the only free parameter in a fit to G_{m}, providing a way to determine the speckle spot size.

In Fig. 12(a) θ_{c} is plotted as a function of pump fluence. The speckle spots are found to shrink as the pump fluence is increased from 0 to

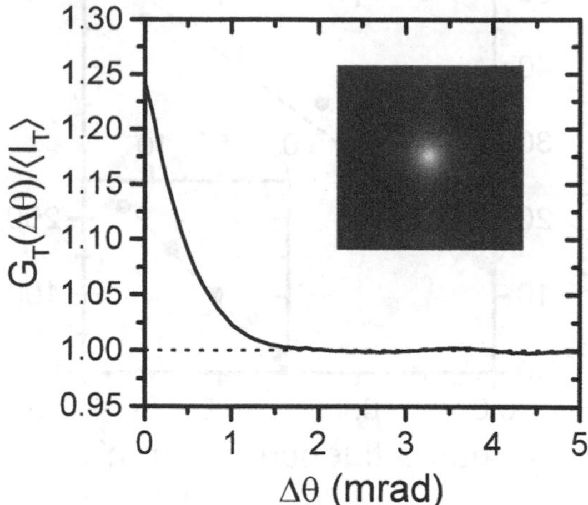

Figure 11. A cross section through the autocorrelate $G_{\mathrm{T}}(\Delta\theta)$ of the data in Fig. 9, normalized to the average total intensity $\langle I_{\mathrm{T}}\rangle$. The contrast is reduced from 2 to 1.25 due to the incoherent background. The two dimensional autocorrelate itself is shown as the inset.

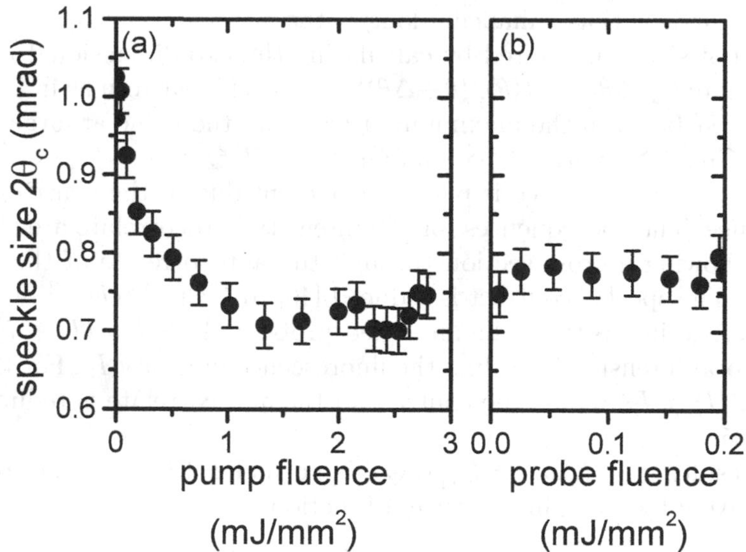

Figure 12. Characteristic decay angle of the autocorrelate. (a) For large pump fluence the speckle spots get significantly smaller compared to the case without gain, indicating that the amplification assists the spatial spreading of the probe light. The probe fluence is 59 μJ/mm^2. (b) The speckle size as a function of probe energy is found to be constant for a fixed pump fluence of 2.9 mJ/mm^2, hence demonstrating the role of the amplification in (a).

1 mJ/mm^2, after which their size is approximately constant. The optical resolution of the imaging system is ≈ 0.1 mrad. Apparently the source of diffuse light producing the speckle becomes larger if the pump fluence is larger. This is consistent with the notion that mainly the long paths are amplified in a random laser. If only the intensity is increased, without actually changing the amount of amplification, the speckle size is constant, c.f. Fig. 12(b).

Without the pump, the speckle size is set by the probe beam diameter of 0.8 mm (80ℓ), and for the highest pump energies the equivalent source size increases to approximately 1.5 times this value. For high pump fluence the speckle does not get smaller. This is consistent with the result obtained in Section 3.1 with Eqs. (5).

3.3. ENHANCED BACKSCATTERING IN RANDOM LASERS

In this section we describe enhanced backscattering measurements that probe the gain dynamics in a random laser. We focus specifically on the laser threshold and experimentally investigate its role for the light propagation. Enhanced backscattering (EBS) has evolved from being a subject of study in itself [47] into a tool that can be utilized for studies of wave transport in random media in a very precise and quantitative manner [9]. The shape of the backscatter cone is determined by the characteristic transport distance of the light in the medium. The exponential amplification of the intensity with path length in a gain medium results in a larger contribution of long light paths compared to light in passive materials. The long light paths constitute the top of the EBS cone: a relatively larger contribution of long paths yields a sharper and narrower EBS line shape [32, 48]. This sensitivity to long paths makes EBS particularly well-suited for testing the alleged divergence behavior.

3.3.1. *Experimental details*

The setup is similar to the one used in the speckle experiment (see section 3.2). Our measurements of EBS in high gain amplifying random media are performed with samples consisting of 220 nm diameter TiO_2 colloidal particles suspended in 1.0 mM Sulforhodamine B laser dye in methanol. When measuring EBS it is important to average out the speckle, which has a much larger intensity variation than the cone, and so will obscure it. A source with short coherence length produces a speckle with less contrast, so it is easier to average. In this case the short coherence length of the OPO is actually an advantage, and the use of the OPO allows us to take Sulforhodamine B as a gain medium, which is easier to work with than Coumarin 6. The scattered light is collected through an interference filter and a fo-

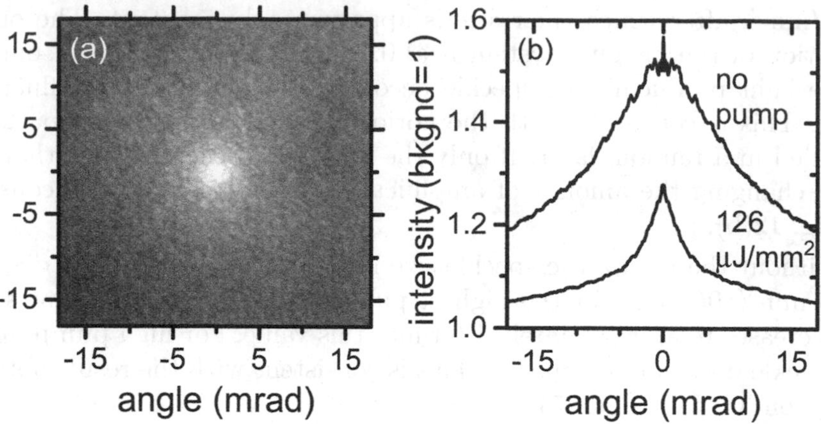

Figure 13. (a) CCD image of scattered probe light for a pump fluence of 126 μJ/mm^2. The exposure is the sum of 72 shots, averaging out most of the speckle. (b) The EBS cone derived from image (a) (bottom curve) by averaging over the azimuthal angle around the top. The resulting curve is mirrored around $\theta = 0$ and the background is normalized. For low intensities residual speckle may be a problem, especially around the top of the cone where the amount of pixels contributing to the average is small. As an example the cone without pump is shown (top curve). $E < 2$ due to single scattering, stray light, and fluorescence [c.f. Fig. 9(b)].

cusing lens on the CCD camera to record the EBS cone. We accumulate 51 to 204 different speckles (realizations) in each exposure, depending on the collected intensity. The angular resolution is 1 mrad, as limited by the probe beam divergence.

3.3.2. *Experimental results*

In Fig. 13 a typical example of a measurement is shown. The image clearly shows the larger intensity near the backscattering direction. For the analysis we manually find the center of the peak and integrate on concentric circles around it. From the backscattering cone measured without pump, taking into account the reabsorption, we infer that the transport mean free path $\ell = 3$ μm. Earlier EBS experiments [32, 49] have been performed with materials in which the laser threshold could not be reached. In our sample the laser threshold is found to be at a pump fluence of 10 μJ/mm^2. The salient features of the influence of gain can be seen in Fig. 13. Both the width of the cone and the enhancement factor E become smaller with increasing pump energy. The enhancement factor has a (gain-independent) value of maximally 2 that is diminished by angle-independent contributions to the intensity. The cone width is related to the transport length: for a cone without gain it scales $\propto \ell^{-1}$, and this allows determination of the transport mean free path. The larger fraction of long light paths at high gain reduces the width of the EBS cone, but there appears to be no sign of a divergence or a sudden change in behavior at the threshold crossing. After the initial cone narrowing, the width saturates far above threshold at a value that is a third of the width of the cone without gain [25]. The decrease of E from 1.65 to 1.25 is due to the incoherent fluorescence component in the collected light, which becomes stronger for higher pump energies.

3.3.3. *Comparison with theory*

For a comparison with theory we need to extract EBS cones from the calculated time- and position-dependent gain profile data. The z-dependence of $\kappa_g(z,t)$ in EBS can be treated with the method due to Deng *et al.* [50]. The basis is an expansion of a solution of the diffusion equation in eigenmodes leading to a static form factor of the EBS intensity $\gamma_E(\theta)$.

In the analysis of $\kappa_g(z)$ the time variation is a subtle issue. The similarity of time scales of gain dynamics and light transport makes it very difficult to solve the time-dependent EBS cone in a varying gain-profile. Extending the method outlined above to $\kappa_g(z,t)$ would mean that the ϕ_n and ε_n become time-dependent. We chose instead to use the averaging property by the time-integrated detection method in the experiment to simplify the analysis.

Since the EBS process itself samples the medium on the time scale needed to build up a cone, it only senses slow variations in $n_1(z)$. The longest paths that contribute in an experiment with an angular resolution of 1 mrad have a separation between entrance and exit points of $d = 10^3\lambda \approx 600$ μm. The diffusive transport time over this distance is $d^2/D \approx 1.7$ ns. We mimic this property by low-pass filtering the data, and subsequently averaging $n_1(z,t)$ in time windows of length d^2/D, and use this mean inversion profile to calculate a "partial" EBS cone. The partial cones are summed, each weighted with the mean probe intensity in its window. This procedure largely overcomes the dominance of the nearly critical $\kappa_g(z)$ occurring in the relaxation oscillation: long paths, needed for the divergence to happen, do not have the time to build up in the ≈ 50 ps that the "supercritical" inversion lasts. This demonstrates once again that the dynamic picture, although allowing for high inversion densities, prevents the explosion.

3.3.4. *Discussion*

The comparison between theory and experiment for EBS cones is displayed in Fig. 14. The agreement between experimental data and the theoretical description is excellent for low and high pump energies. Initially, the width of the EBS cone drops quickly with increasing pump pulse energy. Far above threshold, the FWHM saturates (at ≈ 10 mrad, depending on system

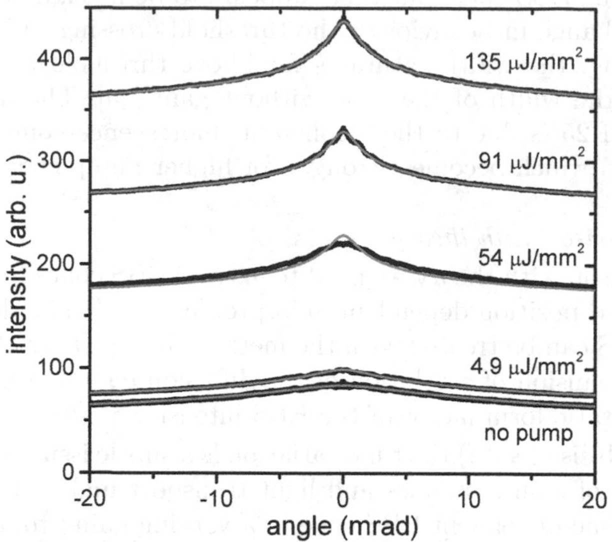

Figure 14. Black points: enhanced backscattering cones for pump fluences ranging from 0 to 135 μJ/mm^2. Gray lines: cones calculated from the dynamic theory. The experimental results are accurately reproduced by the theory, except for intermediate pump energies (approximately 20–70 μJ/mm^2; one example shown) where the relaxation oscillations dominate the temporal inversion profile.

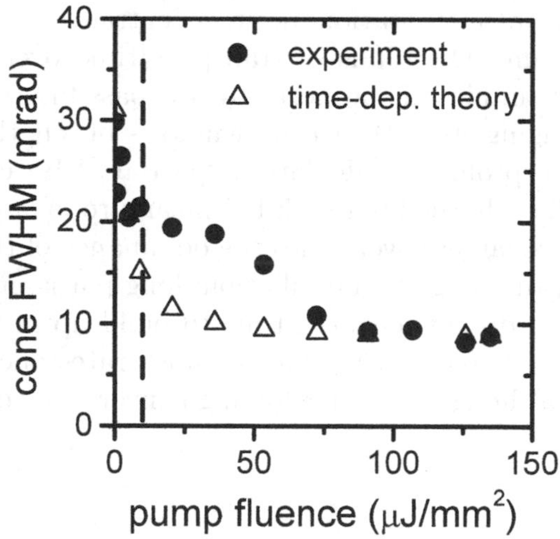

Figure 15. The full widths at half maximum of the backscatter cones as a function of pump fluence. The dashed line indicates the threshold obtained from an independent measurement. The circles are experimental data and the triangles are the corresponding results from simulations.

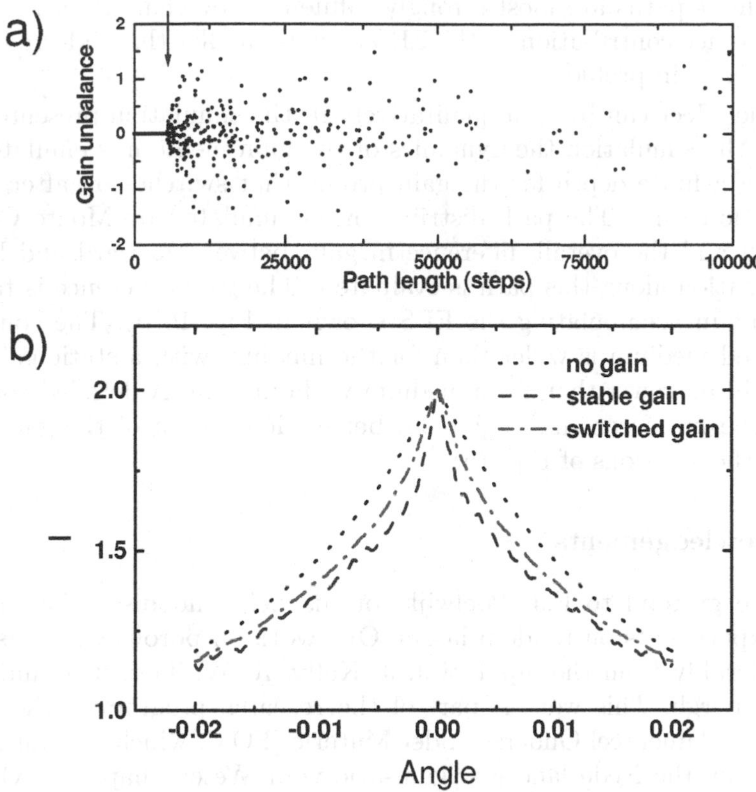

Figure 16. Simulation of the effect on EBS of a difference in gain profile for forward and reverse paths. (a) Unbalance between the forward and backward gain for some paths generated by Monte Carlo simulation. The gain was switched off after a time indicated by the arrow. (b) The EBS for the medium without gain (dotted), a static gain (dash), and a switched gain (dash-dot).

parameters) due to the pump-independence of the excited state population $n_1(z)$ above threshold.

For pump fluences between 20 and 70 μJ/mm^2 the theory deviates from the experimental results, see Fig. 15. This discrepancy is due to the entanglement of the time scales of transport and variation of $n_1(z)$. Since $n_1(z)$ changes faster than the time needed for the formation of a backscatter cone, the reversibility of transport in the medium is affected. A wave traversing the medium along a certain path experiences a spatiotemporal gain profile that is in principle different from the profile seen by the wave in the reversed path. This reduces the interference contrast in the scattered light, as the two waves no longer have equal amplitudes when exiting the medium. This unbalance is especially prominent just above threshold where the long-lived oscillations make up an important part of the temporal gain profile.

Long light paths are most strongly influenced by changes in $n_1(z)$. Their interference contribution to the EBS cone is smaller than inferred from the averaged gain profile.

The effect can be seen qualitatively in the simulation presented in Fig. 16. In the simulation the dynamics of the gain medium is emulated by assuming a finite depth for the gain profile that switches off after a certain relaxation time. The path distribution is simulated by Monte Carlo generation and the overall difference in gain between forward and backward propagation along this path is evaluated. The gain difference is taken into account in accumulating the EBS shown in Fig. 16(b). The cone for the switched medium is wider than for the medium with a static gain profile, and still narrower than for a medium without gain. A detailed comparison with the measurements requires a better description of the gain medium using the solutions of Eqs. (5).

Acknowledgements

We are grateful to F.J. Poelwijk for his help and discussions regarding our experiments on random lasers. Our work on porous GaP has profited considerably from the input of J. J. Kelly, R. W. Tjerkstra, and D. Vanmaekelbergh. This work is part of the research program of the Stichting voor Fundamenteel Onderzoek der Materie FOM, which is financially supported by the Nederlandse Organisatie voor Wetenschappelijk Onderzoek NWO.

References

1. P.W. Anderson, Phys. Rev. **109**, 1492 (1958).
2. E. Yablonovitch, Phys. Rev. Lett. **58**, 2059 (1987).
3. S. John, Phys. Rev. Lett. **58**, 2486 (1987).
4. *Photonic Crystals and Light Localization in the* 21st *Century*, edited by C.M. Soukoulis (Kluwer Academic Publishers, Dordrecht, 2001).
5. V.S. Letokhov, Sov. Phys. JETP **26**, 835 (1968).
6. N.M. Lawandy, R.M. Balachandran, A.S.L. Gomes, and E. Sauvain, Nature (London) **368**, 436 (1994).
7. H. Cao, Y. Ling, J.Y. Xu, C.Q. Cao, and P. Kumar, Phys. Rev. Lett., **86**, 4524 (2001).
8. F.J.P. Schuurmans, D. Vanmaekelbergh, J. van de Lagemaat, and A. Lagendijk, Science **284**, 141 (1999).
9. F.J.P. Schuurmans, M. Megens, D. Vanmaekelbergh, and A. Lagendijk, Phys. Rev. Lett. **83**, 2183 (1999).
10. J. Gómez Rivas, A. Lagendijk, R.W. Tjerkstra, D. Vanmaekelbergh, and J.J. Kelly, Appl. Phys. Lett. **80**, 4498 (2002).
11. S.M. Sze, *Physics of Semiconductor Devices* (Wiley-intersceince, New York, 1981).
12. B.H. Erné, D. Vanmaekelbergh, and J.J. Kelly, J. Electrochem. Soc. **143**, 305 (1996).
13. J. Gómez Rivas, *Light in Strongly Scattering Semiconductors*, Ph.D. thesis, Uni-

versity of Amsterdam (2002). Available at: http://www.tn.utwente.nl/cops/.
14. P.C. Searson, J.M. Macaulay, and F.M. Ross, J. Appl. Phys. **72**, 253 (1992).
15. X.G. Zhang, J. Electrochem. Soc. **138**, 3750 (1991).
16. J. Gómez Rivas, R. Sprik, A. Lagendijk, L.D. Noordam, and C.W. Rella, Phys. Rev. E **62**, R4540 (2000); Phys. Rev. E **63**, 046613 (2001).
17. D.S. Wiersma, M.P. van Albada, and A. Lagendijk, Rev. Sci. Instrum. **66**, 5473 (1995).
18. E. Akkermans, P.E. Wolf, and R. Maynard, Phys. Rev. Lett. **56**, 1471 (1986).
19. M.B. van der Mark, M.P. van Albada, and A. Lagendijk, Phys. Rev. B. **37**, 3575 (1988).
20. J. Gómez Rivas, D.H. Dau, A. Imhof, R. Sprik, B.P.J. Bret, P.M. Johnson, T.W. Hijmans, and A. Lagendijk, submitted (2002).
21. G. van Soest, M. Tomita, and A. Lagendijk, Opt. Lett. **24**, 306 (1999).
22. G. van Soest, F.J. Poelwijk, R. Sprik, and A. Lagendijk, Phys. Rev. Lett. **86**, 1522 (2001).
23. G. van Soest, F.J. Poelwijk, and A. Lagendijk, Phys. Rev. E **65**, 046603 (2002).
24. G. van Soest and A. Lagendijk, Phys. Rev. E 65, 047601 (2002).
25. G. van Soest, *Experiments on Random Lasers*, Ph.D. thesis, University of Amsterdam (2002). Available at: http://www.tn.utwente.nl/cops/.
26. See e.g. A.E. Siegman, *Lasers*, (University Science Books, Mill Valley, 1986).
27. M. Siddique, R.R. Alfano, G.A. Berger, M. Kempe, and A.Z. Genack, Opt. Lett. **21**, 450 (1996).
28. R.M. Balachandran, N.M. Lawandy, and J.A. Moon, Opt. Lett. **22**, 319 (1997).
29. W.L. Sha, C.-H. Liu, F. Liu, and R.R. Alfano, Opt. Lett. **21**, 1277 (1996).
30. R.M. Balachandran, A.E. Perkins, and N.M. Lawandy, Opt. Lett. **21**, 650 (1996).
31. A.Yu. Zyuzin, JETP Lett. **61**, 990 (1995).
32. D.S. Wiersma, M.P. van Albada, and A. Lagendijk, Phys. Rev. Lett. **75**, 1739 (1995).
33. A.A. Burkov and A.Yu. Zyuzin, Phys. Rev. B, **55**, 5736 (1997).
34. X. Jiang and C.M. Soukoulis, Phys. Rev. B **59**, 6159 (1999).
35. G.A. Berger, M. Kempe, and A.Z. Genack, Phys. Rev. E **56**, 6118 (1997).
36. R.M. Balachandran and N.M. Lawandy, Opt. Lett. **21**, 1603 (1996).
37. S. John and G. Pang, Phys. Rev. A **54**, 3642 (1996).
38. H. Cao, J.Y. Xu, S.-H. Chang, S.T. Ho, E.W. Seelig, X. Liu, and R.P.H. Chang, Phys. Rev. Lett. **84**, 5584 (2000).
39. X. Jiang and C.M. Soukoulis, Phys. Rev. Lett. **85**, **70** (2000).
40. M. Pacilli, P. Sebbah, and C. Vanneste, Laser action in the regime of strong localization, in this volume
41. A.Z. Genack, Fluctuations, correlation and average transport of electromagnetic radiation in random media, in: P. Sheng (Ed.), *Scattering and Localization of Classical Waves in Random Media*, World Scientific Series on Directions in Condensed Matter Physics, vol. 8, (World Scientific, Singapore, 1990).
42. J.W. Goodman, Statistical properties of laser speckle patterns, sect. 2.1, 2.2 (on intensity statistics) and 2.5 (on speckle spot size) in: J.C. Dainty (Ed.), *Laser Speckle and Related Phenomena*, 2nd ed., Topics in Applied Physics, vol. 9, (Springer, Berlin, 1984).
43. D.S. Wiersma and A. Lagendijk, Physica A **241**, 82 (1997).
44. C. Gouedard, D. Husson, C. Sauteret, F. Auzel, and A. Migus, J. Opt. Soc. Am. B **10**, 2358 (1993).
45. M.A. Noginov, S.U. Egarievwe, N. Noginova, H.J. Caulfield, and J.C. Wang, Opt. Mat. **12**, 127 (1999).
46. G. Parry, Speckle patterns in partially coherent light, sect. 3.1 and 3.2 in: J.C. Dainty (Ed.), *op. cit.* [42].
47. Y. Kuga and A. Ishimaru, J. Opt. Soc. Am. A **8**, 831 (1984); M.P. van Albada and A. Lagendijk, Phys. Rev. Lett. **55**, 2692 (1985); P.E. Wolf and G. Maret, *ibid.* **55**,

2696 (1985).
48. A.Yu. Zyuzin, Europhys. Lett. **26**, 517 (1994).
49. P.C. de Oliveira, A.E. Perkins, and N.M. Lawandy, Opt. Lett. **21**, 1685 (1996).
50. W. Deng, D.S. Wiersma, and Z.Q. Zhang, Phys. Rev. B **56**, 178 (1997).

WEAK LOCALISATION OF LIGHT BY ATOMS WITH QUANTUM INTERNAL STRUCTURE

CORD A. MÜLLER
Max-Planck-Institut für Physik komplexer Systeme
Nöthnitzer Str. 38, D – 01187 Dresden
AND
CHRISTIAN MINIATURA
Laboratoire Ondes et Désordre (FRE 2302 du CNRS)
1361, route des Lucioles, F–06560 Valbonne

1. Introduction

1.1. WHY DO WE STUDY QUANTUM TRANSPORT?

As a commonplace, we could say that all simple problems in physics have been solved a long time ago, and that we are tempted to turn to the challenging field called, rather pompously, "wave transport in complex systems". Here, "wave" is meant to emphasize the influence of *interference* (which is truly quantum mechanical for massive particles like electrons or atoms). "Transport" implies that we are interested in situations *out of thermodynamic equilibrium* (but not too far, so that linear response theory applies). Finally, a system will be called "complex" whenever it is *disordered*, or *strongly interacting*, or *chaotic*. Our choice of this field is motivated by two aspects: on the side of applied physics, all remote sensing techniques need to incorporate the multiple scattering of waves in turbid media, and the miniaturisation process in semi-conductor industry arrives at length scales where the control of quantum interference becomes crucial. On the academic side of more fundamental physics, we would like to understand and enjoy the predictive power of the best physical theory available today: quantum theory.

The classical paradigm of transport in a disordered environment is *diffusion*. In general, a quantity $n(x,t)$ in space and time (think of a particle concentration) at equilibrium is distributed with a constant density n_0. If the quantity is locally conserved, a small variation $\delta n(x,t)$ then obeys the

B.A. van Tiggelen and S.E. Skipetrov (eds.),
Wave Scattering in Complex Media: From Theory to Applications, 45–58.

diffusion equation, $(\partial_t - D_0\nabla^2)\delta n(x,t) = 0$; D_0 is the diffusion constant. Its solution in Fourier space

$$\delta\hat{n}(q,t) = \delta\hat{n}(q,0)\exp(-D_0 q^2 t) \tag{1}$$

shows that very smooth fluctuations in real space (with small wave numbers $q \to 0$) are removed on very long time scales $1/D_0 q^2 \to \infty$. Indeed, because of the local conservation law, a small surplus $\delta n > 0$ cannot just simply disappear, but has to reach a place far away with a depleted density $\delta n < 0$ in order to restore the equilibrium situation $\delta n = 0$.

Classical Boltzmann diffusion theory has been used successfully to describe electric conduction in metals (first by Drude), or the diffusion of light intensity through stellar atmospheres or interstellar clouds (notably by Schuster and Chandrasekhar) [1]. However, at the microscopic level, one has to deal with waves. As an example, the quantum picture of electrons moving in a perfectly periodic crystal pinpoints the key role of interference: an electron, initially confined in a well, can resonantly tunnel through the lattice, yielding Bloch energy bands. In trying to understand the interplay between disorder and interference, P. Anderson showed that sufficiently strong disorder can suppress the quantum diffusion (leading to a vanishing diffusion constant $D = 0$), a phenomenon baptised *localisation* [2]. Later, it was realized that even far from the regime of (strong) localisation, diffusive transport is affected by interference: this so-called *weak localisation* reduces the Boltzmann diffusion constant, $D = D_0 - \delta D$, by the constructive pairwise interference of amplitudes associated with time-reversed scattering paths. Mesoscopic physics, namely the study of interference effects in wave transport through random media, was born [3].

1.2. WHY DO WE USE PHOTONS AND ATOMS?

The theory of localisation has been first developed in the condensed matter physics community. But as electrons are charged particles with a very strong and long-range Coulomb interaction, the pure one-particle picture of Anderson localisation has never been observed experimentally as far as we know. On the other hand, the radar physics community first had realized that interference of counter-propagating amplitudes can play an important role in the multiple scattering of electromagnetic waves [4]. Taking the better of the two worlds, S. John and P. Anderson suggested to study localisation using light or other non-interacting classical waves. Light scattering allows the use of modern lasers with excellent coherence properties as well as an accurate analysis of direction and polarisation. A large number of turbid media has thus been studied, from Saturn's rings to semi-conductor powders [5].

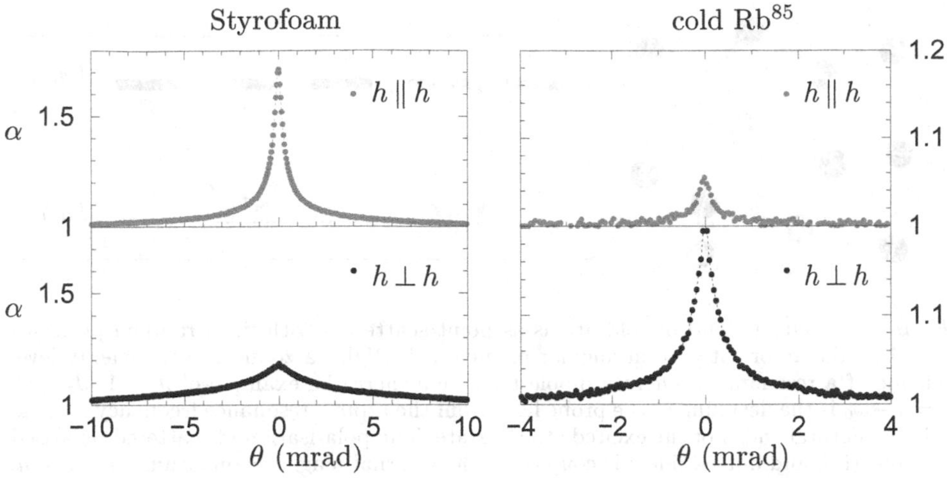

Figure 1. Experimental CBS enhancement α vs. the angle θ with respect to the backscattering direction for a classical medium (Styrofoam, left) and cold rubidium atoms (right), in the channel of preserved helicity ($h \parallel h$, top) and flipped helicity ($h \perp h$, bottom). Data generously provided by G. Labeyrie.

Today, laser-cooled atoms as light scatterers come close to a theoretician's dream team: they are perfectly identical (monodisperse) point particles. But more than that (theoretical convenience hardly ever justifies expensive experiments): their specific properties permit to study new regimes which are characterised, for example, by their quantum internal structure (as discussed in the present contribution), the finite width of the atomic fluorescence spectrum, non-linearities such as the saturation of an atomic transition, and the mechanical acceleration of the atom due to light scattering. Even more, at sufficiently low temperature, the atoms themselves become matter waves and can in turn be used to probe quantum transport, all the way down to the degenerate quantum regime of Bose-Einstein condensation (see, for example, the recent beautiful experimental realisation of the Mott-Hubbard quantum phase transition by the Munich BEC group [6]).

1.3. WHY DO WE NEED TO DESCRIBE AN INTERNAL STRUCTURE?

The first experiment of coherent backscattering of light from a cloud of cold atoms in 1999 [7, 8] yielded a surprising result (see Fig. 1): the observed interference peak in the channel of preserved helicity ($h \parallel h$) shows only a maximum enhancement of about 1.05, far below the classically expected factor of 2.0 due to reciprocity [9]. Also, this $h \parallel h$ enhancement is much

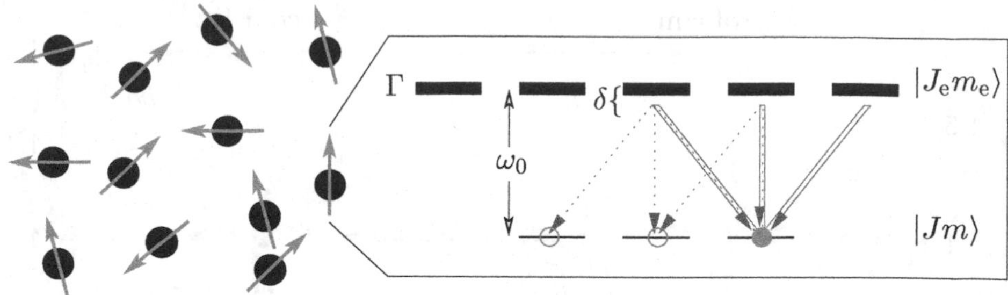

Figure 2. Left: a cloud of cold atoms as point scatterers with fixed random positions $\boldsymbol{r}$ and randomly oriented total angular momenta J. Right: a zoom into the energy level scheme of a resonant degenerate dipole transition, here the example of $J = 1$, $J_e = 2$. $\delta = \omega - \omega_0$ is the detuning of the probe light from the atomic resonance frequency ω_0, and Γ is the natural width of the excited atomic state. The polarisation of scattered polarised photons (full and dotted lines) is coupled to the internal magnetic quantum numbers m.

smaller than the measured value of 1.2 in the channel of flipped helicity ($h \perp h$). It was soon realized that the *degeneracy* of the probed atomic dipole transition $J = 3 \to J_e = 4$ is responsible for an *imbalance of CBS amplitudes* and therefore reduces the measurable enhancement factor [10]. We thus have to generalise the theory of the multiple scattering of polarised light by point dipoles to the case of an arbitrary atomic transition $J \to J_e$ [11]. This theory indeed explains the observed enhancement factors and shall be described in the following.

2. Multiple Scattering of a Photon by Atoms with Internal Degeneracy

2.1. THE ONE-PHOTON TRANSITION MATRIX

Consider a cloud of laser-cooled atoms confined in a standard magneto-optical trap. The cooling is such that their velocity spread v is much smaller than the Doppler velocity Γ/k (Γ is the natural width of the excited atomic state). Therefore, we can neglect the Doppler effect and may assume that the atoms' positions $\boldsymbol{r}_\alpha, \alpha = 1, \ldots, N$, remain fixed on the light-scattering time scale. On the other hand, the velocity spread should be much larger than the recoil velocity $v_{\text{rec}} = \hbar k/M$ (where M is the atomic mass) for the scattering of a photon of wave-vector k. This allows us to treat the positions as classical random variables and to follow the standard diagrammatic approach to describe multiple scattering (see [12] and references therein). The CBS probe beam with incident wave-vector $\boldsymbol{k}$, polarisation $\boldsymbol{\varepsilon}$ and frequency ω excites a closed atomic dipole transition defined by a ground state with total angular momentum J and an excited state J_e with frequency ω_0. In the absence of a magnetic field, these two levels with internal quantum numbers m and m_e are respectively $(2J+1)$- and $(2J_e+1)$-fold degenerate (see Fig. 2).

In order to minimise saturation effects and optical pumping, the experiment is performed at low laser intensity, so that we can consider the light scattering in the limit of one-photon Fock states $|\boldsymbol{k}\boldsymbol{\varepsilon}\rangle$ (in this notation, the transversality $\boldsymbol{\varepsilon}\cdot\boldsymbol{k}=0$ is understood). The transition amplitude for the scattering of an incident photon $|\boldsymbol{k}\boldsymbol{\varepsilon}\rangle$ into an emitted photon $|\boldsymbol{k}'\boldsymbol{\varepsilon}'\rangle$ by a single atom is

$\boldsymbol{k}\,\boldsymbol{\varepsilon}$ Jm' Jm $\boldsymbol{k}'\boldsymbol{\varepsilon}'$

$$= \langle Jm'|\bar{\boldsymbol{\varepsilon}}'\cdot \boldsymbol{t}(\omega)\cdot\boldsymbol{\varepsilon}|Jm\rangle\, e^{i(\boldsymbol{k}-\boldsymbol{k}')\cdot\boldsymbol{r}}. \quad (2)$$

Note that the exponential with the classical external degrees of freedom $\boldsymbol{r}$ is factorised from the matrix element with the internal quantum numbers. Therefore, the usual multiple scattering formalism applies [12]. But we have to analyse carefully the role of the internal degrees of freedom. We see in (2) that the incident and emitted polarisation vectors $\boldsymbol{\varepsilon}$ and $\bar{\boldsymbol{\varepsilon}}'$ (the bar denotes complex conjugation) are coupled by the transition matrix $t_{ij}(m,m';\omega)=\langle Jm|t_{ij}(\omega)|Jm\rangle$. As any 3×3 matrix, it can be decomposed into its irreducible components with respect to rotations,

$$t_{ij} = \underbrace{\frac{1}{3}\delta_{ij}t_{kk}}_{t_{ij}^{(0)}} + \underbrace{\frac{1}{2}(t_{ij}-t_{ji})}_{t_{ij}^{(1)}} + \underbrace{\frac{1}{2}(t_{ij}+t_{ji}) - \frac{1}{3}\delta_{ij}t_{kk}}_{t_{ij}^{(2)}} \quad (3)$$

its scalar part or trace $t^{(0)}$, its antisymmetric part $t^{(1)}$ and its traceless symmetric part $t^{(2)}$ (summation over repeated indices is understood). It is easy to verify that in the non-degenerate limit $J=0$, corresponding to the classical case of an isotropic dipole, only the scalar component $t_{ij}^{(0)}$ exists. So our task in the following will be to determine how the non-scalar components $t_{ij}^{(1,2)}$ influence the light propagation.

2.2. AVERAGE PROPAGATION INSIDE THE EFFECTIVE MEDIUM

In a disordered environment, the actual propagation depends on the precise realisation of disorder and yields, for example, an interference "fingerprint" or speckle-pattern. Universal properties can only be obtained on average. The average propagation of a light mode $|\boldsymbol{k}\boldsymbol{\varepsilon}\rangle$ inside the medium is described by the average propagator $\langle G_{ij}(\boldsymbol{k},\omega)\rangle$ where the average $\langle\ldots\rangle$ traces out the matter degrees of freedom. Following the standard procedure, we determine the average propagator in terms of the vacuum propagator

g_0 and the self-energy Σ with the aid of the Dyson equation (tensor indices are omitted for brevity)

$$\langle G \rangle = g_0 + g_0 \Sigma \langle G \rangle = \frac{1}{g_0^{-1} - \Sigma}. \tag{4}$$

For a sufficiently dilute scattering medium such that $n\lambda^3 \ll 1$ (n is the atomic number density with typical values of 6×10^{10} cm^{-3}, and λ is the optical wavelength, typically 780 nm in the experiments done so far), the independent scattering approximation applies, and the self-energy is simply the average single-scattering t-matrix: $\Sigma_{ij}(\omega) = N \langle t_{ij}(\omega) \rangle$. We suppose that the atomic scatterers are *distributed uniformly* over their internal states such that rotational invariance is restored on average: $\Sigma_{ij}(\omega) = \delta_{ij}\Sigma(\omega)$. This is intuitively clear since under a scalar average only the scalar component can survive. Now, it can be checked that the scalar component is precisely the one that mimics a classical isotropic dipole:

$$\Sigma(\omega) = n M_J \frac{3\pi}{k^2} \frac{\Gamma/2}{\delta + i\Gamma/2}. \tag{5}$$

Here, the quantum internal structure only enters through the scalar factor $M_J = (2J_e + 1)/3(2J + 1)$ with the non-degenerate limit $M_0 = 1$. The elastic mean free time is defined as $\tau = -(2\mathrm{Im}\Sigma(\omega))^{-1}$ and yields the mean free path $\ell = \tau$ in our units, where $c \equiv 1$. By virtue of the optical theorem, the mean free path is related to the total cross-section for elastic scattering by the usual Boltzmann expression $\ell = 1/n\sigma$.

Obviously, the quantum internal structure (or contribution of non-scalar components of the t-matrix) has disappeared under the scalar average over internal states, and we recover a standard scalar theory. Have we been too optimistic in hoping that the internal structure can explain the dramatic reduction of interference in the CBS signal? Well, of course not: the experimental signal in Fig. 1 is the *average intensity* which must be distinguished from the square of the average amplitude. More technically speaking, after calculating the average amplitude $\langle G \rangle$ with the Dyson equation (4), we have to turn to the average intensity $\langle \overline{G} G \rangle$.

2.3. THE AVERAGE INTENSITY VERTEX

In a dilute medium $n\lambda^3 \ll 1$, the building block for the multiple scattering series is the average single scattering intensity vertex $\langle t_{ij} \bar{t}_{kl} \rangle$ [13]. It connects two amplitudes with their respective polarisation vectors and can thus be written as a rank-four tensor,

$$\langle t_{ij}(\omega) \bar{t}_{kl}(\omega) \rangle = u(\omega) \begin{array}{c} i \overline{\quad} j \\ l \underline{\quad} k \end{array}. \tag{6}$$

Here, the four point polarisation vertex can be calculated analytically using standard methods of irreducible tensor operators [14]. It is simply given as the sum of the three pairwise contractions,

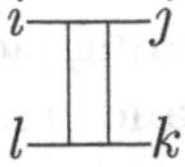

$$\begin{matrix} i & j \\ l & k \end{matrix} = w_1\delta_{ij}\delta_{kl} + w_2\delta_{ik}\delta_{jl} + w_3\delta_{il}\delta_{jk} \tag{7}$$

where the coefficients $w_1 = \frac{1}{3}(s_0 - s_2)$, $w_2 = \frac{1}{2}(s_2 - s_1)$ and $w_3 = \frac{1}{2}(s_2 + s_1)$ are given in terms of the squared $6J$-symbols

$$s_K = 3(2J_\mathrm{e} + 1)\begin{Bmatrix} 1 & 1 & K \\ J & J & J_\mathrm{e} \end{Bmatrix}^2 \tag{8}$$

associated with the irreducible t-matrix components $t^{(K)}$ of order $K = 0, 1, 2$. For the classical isotropic dipole $J = 0$, the coefficients become $(w_1, w_2, w_3) = (1, 0, 0)$ and yield the purely horizontal contraction $\delta_{ij}\delta_{kl}$. We see therefore that including the quantum internal structure is equivalent to replacing the simple dotted line in the usual diagrams [12] by the generalised vertex (7) with the richer topology of a *ribbon*:

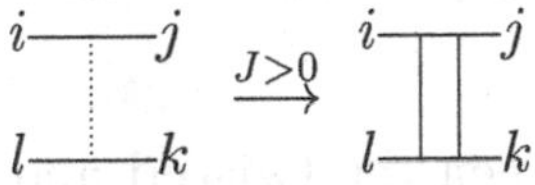

$$\begin{matrix} i & j \\ l & k \end{matrix} \quad \xrightarrow{J>0} \quad \begin{matrix} i & j \\ l & k \end{matrix} \tag{9}$$

2.4. SUMMATION OF LADDER AND CROSSED SERIES

Following the standard diagrammatic approach, we have to sum the so-called series of ladder diagrams

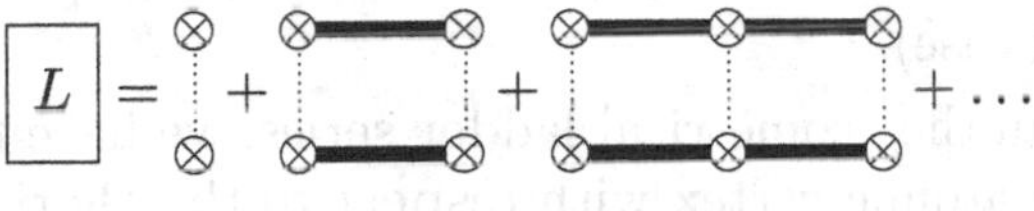

$$\boxed{L} = \ + \ + \ + \dots \tag{10}$$

that describe the average intensity neglecting all interference terms: the direct amplitude (upper line) is scattered by the same scatterers as the conjugate amplitude (lower line), as indicated by the dotted lines. At least formally, one recognises a geometrical series that may be summed up analytically once the single scattering vertex (first term on the r.h.s.) and the square of the average propagators (thick lines) are known. The interference correction associated with CBS and weak localisation is contained in the so-called maximally crossed diagrams

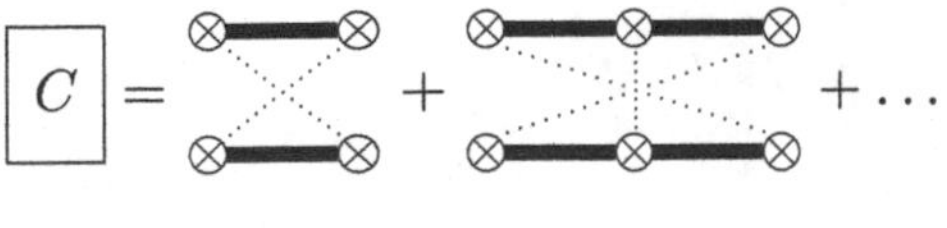

$$\boxed{C} = \ + \ + \dots \tag{11}$$

that describe the propagation of the direct and conjugate amplitude along the same scattering path, but in opposite directions. For classical point scatterers, the crossed and ladder diagrams are closely related by reciprocity: by turning around the lower line of a maximally crossed diagram, the connecting lines straighten out and yield the corresponding ladder diagram. By carefully counting incident and emitted momenta and polarisation indices, one shows that the diagrams are rigorously equal for scattering in the backward direction ($\boldsymbol{k}' = -\boldsymbol{k}$) and in the parallel polarisation channels $\bar{\boldsymbol{\varepsilon}}' = \boldsymbol{\varepsilon}$. This identity justifies the CBS enhancement factor of 2.0 in the helicity-preserving polarisation channel where, for spherically-symmetric scatterers, the single scattering contribution is absent.

In our case of a quantum internal structure, see (9), the classical vertex has to be substituted by the ribbon vertex. But now the correspondence between ladder and crossed diagrams is spoiled: when the bottom line of a crossed diagram is turned around, the connecting ribbons are twisted:

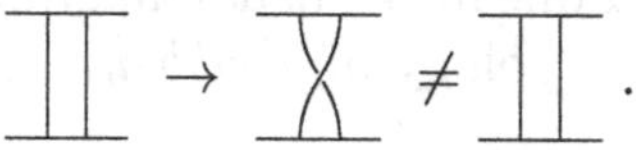

$$\mathbb{I} \to \mathbb{X} \neq \mathbb{I} \, . \tag{12}$$

The evident difference between the twisted and the straight ribbon shows that the quantum internal structure of the atomic scatterer indeed affects the interference corrections to the average intensity. More quantitatively, twisting the vertex is equivalent to the exchange of the coefficients w_2 and w_3 associated with the diagonal and vertical contractions. Once we have calculated the ladder series, we can then obtain the crossed series by simply replacing $w_2 \leftrightarrow w_3$ (and rearranging the momenta and polarisation vectors as in the classical case).

In order to sum the geometrical ladder series, we have to determine the eigenvalues of the atomic vertex with respect to the "horizontal" direction of summation. Using a basis of projectors $\mathsf{T}^{(K)}$ onto irreducible eigenmodes $[\varepsilon_i \bar{\varepsilon}_l]^{(K)}$ of the field polarisation matrix, we obtain a decomposition of the form

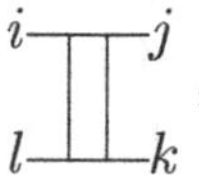

$$\begin{matrix} i & & j \\ & \mathbb{I} & \\ l & & k \end{matrix} = \sum_K \lambda_K(J, J_\mathrm{e}) \mathsf{T}^{(K)}_{il,jk} \tag{13}$$

where the eigenvalues λ_K are simple functions of the w...

Taking advantage of our simple substitution rule $w_2 \leftrightarrow w_3$, the twisted vertex of the crossed series then gives crossed eigenvalues χ_K such that

$$\begin{matrix} i & \text{—} & j \\ & \times & \\ l & \text{—} & k \end{matrix} = \sum_K \chi_K(J, J_\mathrm{e}) \mathsf{T}^{(K)}_{il,jk}. \tag{15}$$

Having diagonalised the scattering vertex, we have to treat also the transverse propagation between atoms. The actual calculation is rather cumbersome because it involves momentum-dependent eigenvalues and projectors; details can be found in [15]. But our approach permits us to sum analytically the ladder and crossed series, for the full transverse vector field with arbitrary polarisation, and for arbitrary atomic transitions. It is a first step towards a generalisation of the existing multiple scattering theories, either of scalar waves by anisotropic point-scatterers [16, 17], or of vector waves by isotropic Rayleigh scatterers [18, 19].

3. Bulk Transport Properties

3.1. DIFFUSION AND DEPOLARISATION

The summed ladder propagator describes the average intensity distribution inside the bulk medium (*i.e.* in the absence of boundaries). To gain qualitative insight, we can simplify the exact expressions by the diffusion approximation (retaining terms of order q^2). The crucial ingredients are the atomic eigenvalues λ_K and χ_K as well as the eigenvalues $b_0 = 1$, $b_1 = \frac{1}{2}$ and $b_2 = \frac{7}{10}$ associated with the transverse propagation. The ladder propagator in the long-time limit then takes the form

$$L(q,t) = \sum_K l_K \exp\left[-D_K q^2 t - t/\tau_\mathrm{p}(K)\right]. \tag{16}$$

Apart from the factor l_K with no importance here, we first notice an exponential decay that was anticipated above in the general diffusion picture (1). We obtain the diffusion constant $D_K \approx D_0 = \ell v_\mathrm{tr}/3$ in terms of the well-known transport velocity $v_\mathrm{tr} = \ell/\tau_\mathrm{tr}$ of resonant point scatterers [12]. But there is a second exponential decay, with a characteristic polarisation relaxation time

$$\tau_\mathrm{p}(K) = \frac{\tau_\mathrm{tr}}{1/b_K\lambda_K - 1}. \tag{17}$$

The scalar field mode ($K = 0$) describes the total intensity. Its relaxation time is infinite ($\tau_\mathrm{p}(0) = \infty$) for all atomic transitions (since $b_0 = \lambda_0 = 1$), and we recover a purely diffusive behaviour as required by energy conservation. The non-scalar field modes are exponentially damped on finite, and

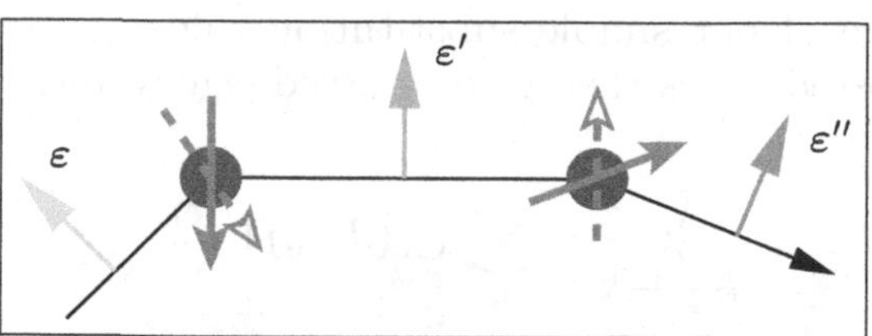

Figure 3. The transversality of propagation depolarises the multiply scattered light. This depolarisation is enhanced by an atomic quantum internal structure $J > 0$ (as discussed in sec. 3.1). Furthermore, the internal degeneracy leads to dephasing of the interference correction (as discussed in sec. 3.2).

in fact rather short time scales $\tau_{\mathrm{p}}(1,2) \leq \tau_{\mathrm{tr}}$: this simply says that a well-defined field polarisation is scrambled in the course of multiple scattering. With increasing degeneracy $J > 0$ of the atomic transition, these times get even shorter ($\lambda_K < 1$), which nicely confirms the intuitive picture that random transitions between different atomic Zeeman substates enhance the depolarisation.

3.2. WEAK LOCALISATION AND STRONG DEPHASING

Under the same approximations, the sum of crossed diagrams yields a contribution

$$C(q_c, t) = \sum_K c_K \exp\left[-D_K q_c^2 t - t/\tau_{\mathrm{p}}(K) - t/\tau_\phi(K)\right] \tag{18}$$

as a function of the total momentum $q_c = ||\boldsymbol{k} + \boldsymbol{k}'||$. One first recognises the same exponential decay as in the ladder contribution (17), indicating that depolarisation also affects the coherent contribution. But there is an additional source of exponential damping described by dephasing times [20]

$$\tau_\phi(K) = \frac{b_K \tau_{\mathrm{tr}}}{1/\chi_K - 1/\lambda_K} \tag{19}$$

which we define precisely as the damping times *with respect to* the incoherent depolarisation times. Of particular interest is the dephasing time $\tau_\phi(0)$ of the scalar mode or intensity. For atomic transitions of the type $J_{\mathrm{e}} = J + 1$, we find explicitly

$$\tau_\phi(0) = \frac{\tau_{\mathrm{tr}}}{J(2J+3)}. \tag{20}$$

The interference is only preserved ($\tau_\phi = \infty$) for classical dipoles $J = 0$. For the least possible degeneracy $J = \frac{1}{2}$, the dephasing time is already as short as $\tau_{\rm tr}/2$ and decreases as $1/J^2$. Even if the exact expression becomes meaningless for larger J (since the diffusion approximation is certainly not justified on time scales shorter than the transport time), it is evident that the internal structure destroys the interference very effectively. Therefore, in the presence of a quantum internal structure, we only expect a (very, very) weak localisation correction to the Boltzmann diffusion constant of light propagation.

4. Coherent Backscattering

We wish to calculate the CBS peak [21] analytically for arbitrary atomic transitions and therefore choose the simplest possible geometry of the scattering medium, a semi-infinite half-space. Having calculated the bulk propagator $F_0(\boldsymbol{r}_1 - \boldsymbol{r}_2)$ (either the ladder or the crossed contribution), we define the corresponding propagator for the semi-infinite half space by using the method of images, $F(\boldsymbol{r}_1, \boldsymbol{r}_2) = F_0(\boldsymbol{r}_1 - \boldsymbol{r}_2) - F_0(\boldsymbol{r}_1 - \boldsymbol{r}_{2'})$. The image point $2'$ is defined with respect to a mirror plane lying at a distance z_0 outside the boundary of the medium. Following the habits of the field, we use the so-called "skin layer depth" $z_0 \approx 0.7121\,\ell$ pertaining to the exact solution of the homogeneous Milne equation for vector waves and point dipole scatterers [18, 19].

By this approach, we can calculate the CBS enhancement factors and peak shapes *beyond the diffusion approximation* (which is crucial whenever only short paths contribute). The CBS enhancement factor α is plotted in Fig. 4 as a function of J for transitions of the type $J \to J+1$. In the classical limit $J = 0$, we recover values of $\alpha = 2.00$, 1.76, 1.24 and 1.12 for the different polarisation channels which are in excellent agreement with the exact values obtained by a solution of the vector Milne equation by the Wiener-Hopf method [18, 19]. This indicates that the method of images can be used with success even for signals involving short scattering paths (like in the perpendicular channels) provided that the *exact* propagator (beyond the diffusion approximation) be used.

Figure 4 shows that the least internal degeneracy reduces the CBS interference dramatically: the perfect factor of 2.0 in the $h \parallel h$ channel plunges down to 1.04, well below the other three polarisation channels. At this point, our theory indeed explains the astonishing experimental result shown in Fig. 1 that has motivated this work. Furthermore, an experiment using cold Strontium atoms without internal degeneracy ($J = 0$) has recently confirmed that excellent enhancement factors in agreement with the theoretical predictions are obtained [22].

Figure 5 shows a comparison between calculated CBS peak shapes for isotropic point scatterers (left side) and atoms with $J = 3 \to J_{\rm e} = 4$

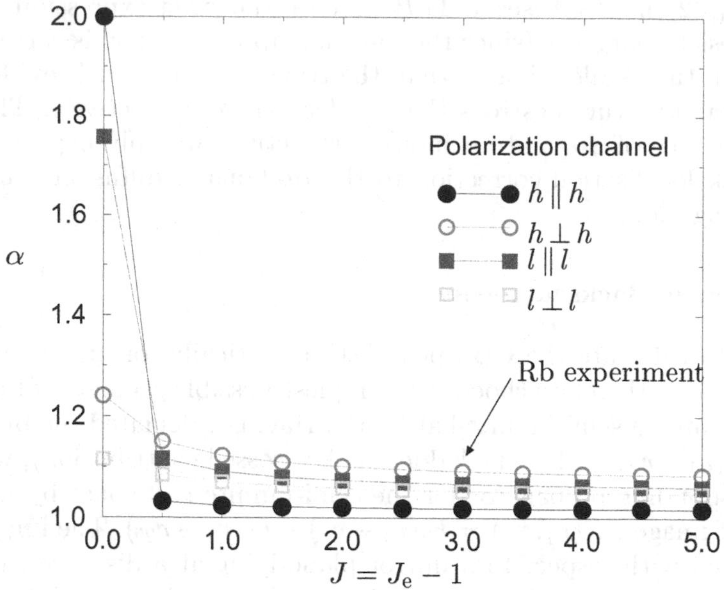

Figure 4. Calculated CBS enhancement factors at exact backscattering from a homogeneous half-space of atoms with a degenerate dipole transition $J \to J+1$ for the four usual channels of preserved ($\parallel$) and flipped ($\perp$) linear polarisation (l) or helicity (h). Contrary to the classical case, for atoms with $J > 0$ the best enhancement is expected in the $h \perp h$ channel.

(right side). The full lines are the sum of all scattering orders, and the dotted lines indicate the double scattering contribution (which is known in closed form [14]). Whereas the CBS peak in the classical case and parallel polarisation channels contains very high orders of scattering (corresponding to long scattering paths), the atomic internal degeneracy cuts off these contributions (as indicated already by the dephasing times (19)) and yields only very small peaks.

A quantitative comparison of the theory to the experimental results needs to take into account the finite geometry of the actual atomic cloud (roughly spherical, with a Gaussian density distribution). This means that analytical results are out of reach, and a numerical approach has to be taken. A Monte Carlo simulation of photon trajectories in various geometries has been realized by D. Delande and yields results which are in good agreement with the experimental data [23].

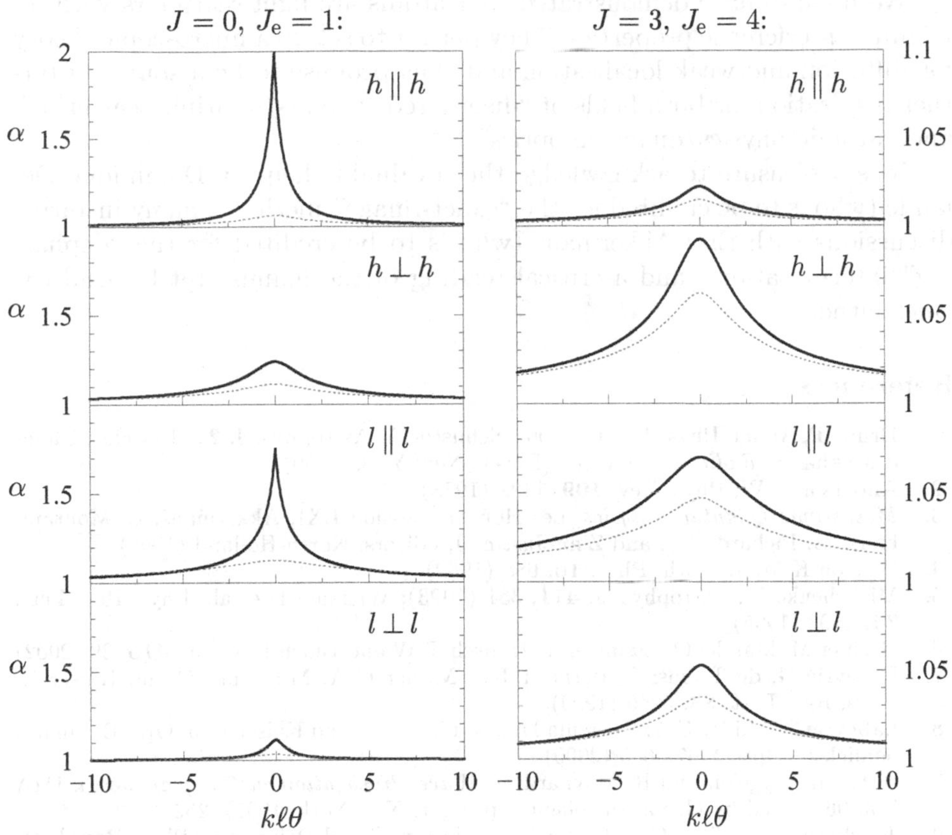

Figure 5. Calculated CBS enhancement factor α as a function of the reduced backscattering angle $k\ell\theta$ for isotropic dipole scatterer (left) and atoms with a degenerate $J = 3 \rightarrow J_e = 4$ transition (right). Full line: sum of all scattering orders. Dotted line: double scattering contribution.

5. Concluding Remarks and Acknowledgements

In summary, we have presented an analytical theory of the multiple scattering of polarised photons by resonant atomic dipole transitions with arbitrary degeneracy. We have shown how the usual diagrammatic approach can be generalised by using an intensity vertex with a "ribbon" topology that breaks the equivalence of ladder and crossed diagrams. The theoretical CBS peak heights reproduce the experimental results: the quantum internal structure indeed reduces the CBS interference drastically. Inasmuch as weak localisation acts as a precursor for Anderson localisation, our results indicate that in order to reach the strong localisation regime with cold atoms, the use of a non-degenerate transition is highly recommendable.

We hope to have demonstrated that atoms are light scatterers with intriguing interference properties. They permit to set up a microscopic theory for diffusion and weak localisation and thus promise to be a source of further inspiration for both fields of "disordered systems/multiple scattering" and "atomic physics/quantum optics".

It is a pleasure to acknowledge the invaluable help by Dominique Delande (who is to be credited for the "exact-image" method), many inspiring discussions with Eric Akkermans (who is to be credited for the "dephasing" interpretation), and a critical reading of the manuscript by Andreas Buchleitner.

References

1. Drude P, Annal. Phys. **1**, 566 (1900); Schuster A, Astrophys. J. **21**, 1 (1905); Chandrasekhar S, *Radiative Transfer* (Dover, New York, 1960)
2. Anderson P W, Phys. Rev. **109**, 1492 (1958)
3. *Mesoscopic Quantum Physics*, Les Houches session LXI, Akkermans, E, Montambaux, G. Pichard, J L, and Zinn-Justin, J, editors, North-Holland (1995)
4. Watson K M, J. Math. Phys. **10**, 688 (1969)
5. Mishchenko M, Astrophys. J. **411**, 351 (1993); Wiersma D et al., Phys. Rev. Lett. **74**, 4193 (1995)
6. Greiner M, Mandel O, Esslinger T, Hänsch T W and Bloch I, Nature **415**, 39 (2002)
7. Labeyrie G, de Tomasi F, Bernard J-C, Müller C A, Miniatura C and Kaiser R, Phys. Rev. Lett. **83**, 5266 (1999)
8. Labeyrie G, Müller C A, Wiersma D S, Miniatura C and Kaiser R, J. Opt. B: Quant. Semiclass. Opt. **2**, 672-685 (2000)
9. B.A. van Tiggelen and R. Maynard, in *Wave Propagation in Complex media*, IMA Vol. 96, edited by G. Papanicolaou (Springer, New York, 1997), 252
10. Jonckheere T, Müller C A, Kaiser R, Miniatura C and Delande D, Phys. Rev. Lett. **85**, 4269 (2000)
11. Müller C A, PhD thesis (Universities of Munich/Nice – Sophia Antipolis, 2001), `http://www.ub.uni-muenchen.de/elektronische_dissertationen/physik/Mueller_Cord.pdf`
12. van Rossum M C W and Nieuwenhuizen T M, Rev. Mod. Phys. **70**, 313 (1999)
13. Lagendijk A and van Tiggelen B A, Phys. Rep. **270**, 143 (1996)
14. Müller C A, Jonckheere T, Miniatura C and Delande D, Phys. Rev. A **64**, 053804 (2001)
15. Müller C A and Miniatura C, submitted, `physics/0205029`
16. Ozrin V D, Phys. Lett. A **162** 341 (1992)
17. Amic E, Luck J M and Nieuwenhuizen T, J. Phys. A: Math. Gen. **29**, 4915 (1996)
18. Ozrin V D, Waves Rand. Media **2** 141 (1992)
19. Amic E, Luck J and Nieuwenhuizen T, J. Phys. I France **7** 445 (1997)
20. Akkermans E, Miniatura C and Müller C A, submitted, `cond-mat/0206298`
21. Akkermans E, Wolf P and Maynard R, J. Phys. Fr. **49**, 77 (1988); van der Mark M, van Albada M and Lagendijk A, Phys. Rev. B **37**, 3575 (1988)
22. Bidel Y, Klappauf B, Bernard J-C, Delande D, Labeyrie G, Miniatura C, Wilkowski D and Kaiser R, Phys. Rev. Lett. **88** 203902 (2002)
23. Labeyrie G, Delande D, Müller C A, Miniatura C and Kaiser R, in preparation (2002)

LIGHT PROPAGATION IN CHIRAL AND MAGNETOCHIRAL RANDOM MEDIA

The Impact of Broken Symmetries

F.A. PINHEIRO AND B.A. VAN TIGGELEN
CNRS/Laboratoire de Physique et Modélisation des Milieux Condensés, Université Joseph Fourier
Maison des Magistères, B.P. 166
38042 Grenoble Cedex 9, France.

1. Introduction

The complex nature of wave transport, which constituted the central theme of the Cargèse meeting, can arise from several distinct sources. These sources can be generally identified as a complex media that support wave propagation and/or some complex constraint to wave transport. As examples of complex media that have recently attracted a lot of attention, one can mention cold atomic gases [1], strongly disordered media [2], photonic band gap materials [3] and natural environments, such as the Earth's crust [4] and biological tissues [5], intrinsically complex media that are now being studied using multiple scattering techniques. Much of this progress is supported by several interdisciplinary collaborations, motivated mainly by the numerous and potential applications [6].

The third main source of complexity in the study of wave propagation is broken symmetry. It imposes an important constraint to wave propagation in random media since it can dramatically influence observables of wave transport, giving rise to a new class of ondulatory phenomena. In the case of electromagnetic waves in particular, the presence of an external magnetic field induces the breaking of time-reversal symmetry and influences in a fundamental way interference effects of light multiply scattered by means of the Faraday effect [7]. The Faraday effect was also shown both theoretically [8] and experimentally [9] to be responsible for a magneto-transverse light diffusion in random media subject to an external magnetic field, a behavior that bears a strong phenomenological resemblance to the electronic Hall effect. The impact of time-reversal breaking in the propaga-

B.A. van Tiggelen and S.E. Skipetrov (eds.),
Wave Scattering in Complex Media: From Theory to Applications, 59–74.

tion of acoustic waves can be appreciated in the experiments carried out by Roux and Fink [10], where liquid vortices play a similar role as the magnetic field in the optical case. Recent studies of random matrix theory [11] also confirm the great sensibility of wave propagation in inhomogeneous media to broken time-reversal.

Mirror symmetry is another fundamental symmetry of nature playing a decisive role in wave propagation. Media that lack mirror symmetry are called *chiral* media, a term which was first coined by Lord Kelvin [12]:

"I call any geometrical figure, or group of points, *chiral*, and say it has chirality if its image in a plane mirror, ideally realized, cannot be brought to coincide with itself." In homogeneous optical media, the breaking of mirror symmetry manifests itself as the well-known natural optical activity [13]. The effects of chirality in multiple light scattering have recently been investigated, both experimentally [14] and numerically [15]. It was shown that multiply scattered light in chiral media achieves a non-zero degree of circular polarization in transmission [15]. This fact may have potential applications in imaging, mainly medical imaging, since sugar is chiral. When parity and time-reversal symmetries are simultaneously broken, a cross effect between the magnetic and natural optical activities, the so called *magnetochiral* (MC) effect, can occur [16, 17]. The MC effect has only recently been observed in homogeneous media, both in absorption [18] and in refraction [19, 20].

The present contribution is devoted to some novel manifestations that show up in the macroscopic transport properties of light in *inhomogeneous* media after the breaking of mirror and/or time-reversal symmetries. First, we shall focus on media with broken parity only, for which we adopt a microscopic treatment to describe electromagnetic wave propagation. In order to generate chirality in inhomogeneous media, we develop a model consisting of radiative pointlike scatterers distributed in chiral geometries. In Sec. 2 we derive a microscopic expression for the chiral transport mean free path ℓ_C^*, which is a *pseudolength* (i.e., a pseudoscalar with the dimension of a length), that we have calculated for some simple chiral systems. This microscopic treatment is also used to investigate the polarization properties of a diffuse beam emerging from a slab made of chiral media, with the confirmation that a residual circular polarization is obtained, in contrast to the achiral case where the emerging radiation must be completely unpolarized.

Secondly, in order to investigate MC effects in multiple scattering, we add an external magnetic field. To this end, we assume that the pointlike scatterers are Faraday active. Within this model, magnetochirality is a *collective* effect built up by multiple scattering: an *assembly* of Faraday active particles distributed in a *chiral* configuration should generate MC effects since in such a system both mirror and time-reversal symmetries are broken.

In Sec. 3 we show that multiple light scattering in MC media constitutes a new optical manifestation of chirality, in addition to the well-known natural optical activity. An optical parameter is introduced to quantify the chirality associated with spatial configuration of the scatterers, which we have calculated for some simple chiral systems.

Despite the important role played by chirality in several areas of research and its ubiquity in nature, only recently real progress has been made in the attempt to quantify or measure it. One of the first of them is due to Gilat [21]. Osipov *et al.* [22] have developed a molecular measure of chirality based on the behavior of the response functions that characterize molecular optical activity. Harris *et al.* [23, 24] have proposed a new elegant way, based on group theory, to measure the degree of chirality exhibited by a geometrical object. They have constructed a variety of rotationally invariant pseudoscalars and have applied them to quantify the degree of chirality of molecules of arbitrary shape, showing how these parameters govern a particular observable, such as the pitch of a cholesteric liquid crystal [23, 24] and the optical rotatory power [25]. The choice for a chiral "order" parameter is certainly not unique [24], but must always be a pseudoscalar *invariant under rotations*. This last fact guarantees that no rotation of the object exists that maps the mirror image of a chiral object onto itself. In Sec. 3 we will explore the MC effects in random media in order to construct an alternative (optical) way to measure the "amount" of chirality and compare it to the one proposed by Harris *et al.* [23, 24].

2. Light Transport in Chiral Random Media: a Microscopic Approach

2.1. CHIRAL MEAN FREE PATH: A MANIFESTATION OF CHIRALITY IN LIGHT DIFFUSION

In the theory of vector wave diffusion [26], the specific intensity tensor, whose components I_{ij} describe the polarization state of the diffuse radiation in the direction $\mathbf{p}$, can be expressed as [27]:

$$\mathbf{I}_{\mathbf{p}}(\mathbf{q},\chi) = i\left[\mathbf{G}_{\mathbf{p}} - \mathbf{G}^*_{\mathbf{p}}\right] - i\mathbf{G}_{\mathbf{p}} \cdot \mathbf{\Gamma}_{\mathbf{p}}(\mathbf{q},\chi) \cdot \mathbf{G}^*_{\mathbf{p}}, \tag{1}$$

where $\mathbf{q}$ is the Fourier conjugate of the position vector and $\mathbf{G}_{\mathbf{p}}(\chi) = [\omega^2 - p^2 + \mathbf{pp} - \mathbf{\Sigma}_{\mathbf{p}}(\chi)]^{-1}$ is the Dyson Green's tensor ($\mathbf{\Sigma}_{\mathbf{p}}(\chi)$ being the second rank mass operator). Notice that we have introduced a pseudoscalar χ associated with the chiral geometry of the scatterer [23, 24]. The second-rank tensor $\mathbf{\Gamma}_{\mathbf{p}}$ describes the weak angular dependence of the diffuse energy flow and obeys the integral equation [27]:

$$\mathbf{\Gamma}_{\mathbf{p}}(\mathbf{q},\chi) = \mathbf{L}_{\mathbf{p}}(\mathbf{q}) + \sum_{\mathbf{p}'} \mathbf{\Gamma}_{\mathbf{p}'}(\mathbf{q},\chi) \cdot \mathbf{G}_{\mathbf{p}'} \otimes \mathbf{G}^*_{\mathbf{p}'} \cdot \mathbf{U}_{\mathbf{pp}'}\,, \tag{2}$$

with the (four-rank) irreducible vertex $\mathrm{U}_{\mathbf{pp}'}(\mathbf{q},\chi)$ and the current tensor $\mathbf{L_p}(\mathbf{q}) \equiv 2(\mathbf{p}\cdot\mathbf{q})\,\mathbf{U}-\mathbf{pq}-\mathbf{qp}$ (where $\mathbf{U}$ is the unit tensor). In an isotropic and achiral media subject to weak disorder, the well-known Boltzmann results applies, $\mathbf{\Gamma}_{ij\mathbf{p}}(\mathbf{q}) = (\mathbf{p}\cdot\mathbf{q})\gamma_0\delta_{ij}$ with $\gamma_0 = 2/(1-\langle\cos\theta\rangle)$. For a low density ρ of the scatterers both the mass-operator $\mathbf{\Sigma_p}(\chi)$, and the irreducible vertex $\mathrm{U}_{\mathbf{pp}'}(\mathbf{q},\chi)$ are related to the T-matrix of one independent scatterer [26],

$$\begin{aligned} \mathbf{\Sigma_p}(\chi) &= n\mathbf{T_{pp}}(\chi) & \text{(3a)} \\ \mathrm{U}_{\mathbf{pp}'}(\mathbf{q},\chi) &= n\mathbf{T}_{\mathbf{p}+\mathbf{p}'+}(\chi)\otimes\mathbf{T}^*_{\mathbf{p}-\mathbf{p}'-}(\chi)\,. & \text{(3b)} \end{aligned}$$

We introduced $\mathbf{p}\pm \equiv \mathbf{p}\pm\frac{1}{2}\mathbf{q}$ and have implicitly applied rotational averaging of both expressions. For a chiral object is $\mathbf{\Sigma_p} \neq \mathbf{\Sigma_{-p}}$.

Our aim is to explicitly solve Eq. (2) in order to derive an exact microscopic equation for the chiral transport mean free path ℓ^*_C. In an isotropic chiral media, the tensor $\mathbf{\Gamma_p}$ (neglecting longitudinal terms) takes the form:

$$\mathbf{\Gamma_p}(\mathbf{q},\chi) = \gamma_0(\mathbf{p}\cdot\mathbf{q})\mathbf{\Delta_p} + \gamma_C(\chi)(\mathbf{p}\cdot\mathbf{q})\mathbf{\Phi}_{\widehat{\mathbf{p}}}, \qquad (4)$$

where $(\mathbf{\Delta_p})_{ij} \equiv \delta_{ij} - p_ip_j/p^2$ is the projector upon the space of transverse polarization (normal to $\mathbf{p}$) and $\mathbf{\Phi}$ is the antisymmetric Hermitian tensor $\Phi_{ij} = i\epsilon_{ijk}\widehat{p}_k$ (ϵ_{ijk} being the Levi-Civita tensor). It is easy to check from Eq. (2) that $\mathbf{\Gamma_p}(\mathbf{q},\chi)$ must obey the following symmetry relations:

$$\mathbf{\Gamma_p}(\mathbf{q},\chi) = -\mathbf{\Gamma_{-p}}(\mathbf{q},-\chi) = \mathbf{\Gamma_{-p}}(-\mathbf{q},-\chi), \qquad (5)$$

showing that $\gamma_C(\chi)$ is a pseudoscalar, contrary to γ_0. In order to obtain the coefficients γ_0 and γ_C in Eq. (4), let us separate the mass operator $\mathbf{\Sigma_p}$ in a symmetric (S) and an antisymmetric (A) part:

$$\Delta\mathbf{\Sigma_p} \equiv -\left(\mathrm{Im}\mathbf{\Sigma}^{\mathrm{S}}_{\mathbf{p}} + \mathrm{Im}\mathbf{\Sigma}^{\mathrm{A}}_{\mathbf{p}}\right) = -\mathrm{Im}\Sigma^0_{\mathbf{p}}\left(\mathbf{U} + \xi\mathbf{\Phi}_{\widehat{\mathbf{p}}}\right), \qquad (6)$$

where $\mathbf{\Sigma}^0_{\mathbf{p}}$ is given by Eq. (3a). The pseudoscalar ξ determines the rotatory power and the spatial dichroism of the effective medium. After inserting Eq. (4) into Eq. (2) and performing the operations $\int\frac{d\widehat{\mathbf{p}}}{4\pi}\mathrm{Tr}\mathbf{\Delta_p}\cdot$ and $\int\frac{d\widehat{\mathbf{p}}}{4\pi}\mathrm{Tr}\mathbf{\Phi_p}\cdot$, respectively, we get the two following coupled equations for γ_0 and γ_C:

$$\begin{aligned} \gamma_0 &= 2 + \chi\left(\frac{\gamma_0 + \xi\gamma_C}{1+\xi^2}\right) + \chi_C\left(\frac{\gamma_C - \xi\gamma_0}{1+\xi^2}\right), & \text{(7a)} \\ \gamma_C &= -\chi_C\left(\frac{\gamma_0 + \xi\gamma_C}{1+\xi^2}\right) - \chi_{CC}\left(\frac{\gamma_C - \xi\gamma_0}{1+\xi^2}\right), & \text{(7b)} \end{aligned}$$

where we have put $\mathbf{q} = \hat{\mathbf{p}}$ and used that $\mathbf{G_p}(\mathbf{q}) \cdot \mathbf{\Delta_p} \cdot \mathbf{G}^*_\mathbf{p}(\mathbf{q}) = \Delta\mathbf{G_p} \cdot (\Delta\mathbf{\Sigma_p})^{-1}$. The real-valued parameters χ, χ_C and χ_{CC} are given by:

$$\chi \equiv \frac{1}{8\pi\sigma_0}\int\frac{d\widehat{\mathbf{p}}}{4\pi}\int\frac{d\widehat{\mathbf{p}}'}{4\pi}\mathrm{Tr}\left[\mathbf{\Delta_p}\cdot\mathbf{T_{pp'}}\cdot\mathbf{\Delta_{p'}}\cdot\mathbf{T}^*_{\mathbf{pp'}}\right](\widehat{\mathbf{p}}\cdot\widehat{\mathbf{p}}'), \tag{8a}$$

$$\chi_C \equiv \frac{1}{8\pi\sigma_0}\int\frac{d\widehat{\mathbf{p}}}{4\pi}\int\frac{d\widehat{\mathbf{p}}'}{4\pi}\mathrm{Tr}\left[\mathbf{\Phi}_{\widehat{\mathbf{p}}}\cdot\mathbf{T_{pp'}}\cdot\mathbf{\Delta_{p'}}\cdot\mathbf{T}^*_{\mathbf{pp'}}\right](\widehat{\mathbf{p}}\cdot\widehat{\mathbf{p}}'), \tag{8b}$$

$$\chi_{CC} \equiv \frac{1}{8\pi\sigma_0}\int\frac{d\widehat{\mathbf{p}}}{4\pi}\int\frac{d\widehat{\mathbf{p}}'}{4\pi}\mathrm{Tr}\left[\mathbf{\Phi}_{\widehat{\mathbf{p}}}\cdot\mathbf{T_{pp'}}\cdot\mathbf{\Phi}_{\widehat{\mathbf{p}}'}\cdot\mathbf{T}^*_{\mathbf{pp'}}\right](\widehat{\mathbf{p}}\cdot\widehat{\mathbf{p}}'), \tag{8c}$$

where we have applied the optical theorem $\rho\sigma_r = -\mathrm{Im}\Sigma^0_\mathbf{p}$ (σ_r being the extinction cross section of one scatterer) [28]. It is important realize that the parameters χ and χ_{CC} are scalars whereas χ_C is a pseudoscalar. The knowledge of the scattering T matrix for a chiral configuration of particles enables us to calculate ξ, χ, χ_C, χ_{CC} and, consequently, γ_0 and γ_C.

Given γ_0 and γ_C, the diffuse transmission in chiral media, described by the specific intensity (1), is expressed as:

$$\mathbf{I_p} \sim \mathbf{\Delta_p} + \frac{i}{2}\ell_s(\widehat{\mathbf{p}}\cdot\mathbf{q})\left[\left(\frac{\gamma_0+\xi\gamma_C}{1+\xi^2}\right)\mathbf{\Delta_p} + \left(\frac{\gamma_C-\xi\gamma_0}{1+\xi^2}\right)\mathbf{\Phi}_{\widehat{\mathbf{p}}}\right], \tag{9}$$

where Eqs. (2,4,7a,7b) and the definitions (8a,8b,8c) have been used. In Eq. (9) $\ell_s = 1/\rho\sigma_r$ is the conventional scattering mean free path. Equation (9) suggests to define a transport mean free path ℓ^*:

$$\ell^* \equiv \frac{\ell_s}{2}\left(\frac{\gamma_0+\xi\gamma_C}{1+\xi^2}\right), \tag{10}$$

and, analogously, a *chiral transport mean free path* ℓ^*_C according to:

$$\ell^*_C \equiv \frac{\ell_s}{2}\left(\frac{\gamma_C-\xi\gamma_0}{1+\xi^2}\right). \tag{11}$$

Notice that in the absence of chirality (i.e., for $\xi = 0$, $\chi_C = 0$ and thus $\gamma_C = 0$), ℓ^* reduces to the usual transport mean free path $\ell_s/(1-\langle\cos\theta\rangle)$.

The chiral transport mean free path ℓ^*_C must vanish when the light is scattered from objects containing only two or three point particles, since those systems are necessarily achiral. A chiral system must have at least four point particles and the simplest of them is the so-called "twisted H", depicted in Fig. 1, which has the convenient property that its chirality can be controlled by varying the angle γ between its arms [23, 24].

In Fig. 2 we have numerically calculated the ratio ℓ^*_C/ℓ^* for light scattering by four pointlike scatterers placed at the vertices of this "twisted H". Notice that ℓ^*_C vanishes when the system is achiral, namely at the angles

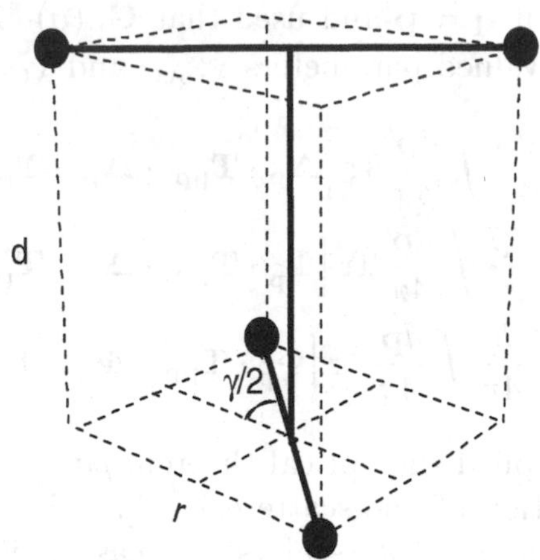

Figure 1. Four point scatterers located at the vertices of the simplest chiral geometry: the so-called "twisted H". The coordinates of the scatterers are: $[\pm\frac{r}{2}\cos(\frac{\gamma}{2}), \pm\frac{r}{2}\sin(\frac{\gamma}{2}), \pm\frac{d}{2}]$.

$\gamma = n\pi/2$, with n integer. In addition we observe that, for the set of parameters used, ℓ_C^* follows the sinusoidal behavior of the "chiral order parameter" $\psi \sim \sin(2\gamma)$ proposed by Harris *et al.* [23, 24]. This fact demonstrates that transport properties of light in random media are indeed quantified by the degree of chirality ψ exhibited by the system.

2.2. CIRCULAR POLARIZATION OF THE DIFFUSE BEAM

We will now apply the previous results to investigate the physical importance of the chiral mean free path ℓ_C^* in multiple light scattering inside a slab with length L, focusing on the polarization properties of the emerging diffuse radiation. The emerging intensity I from the slab is obtained by evaluating the specific intensity (i.e., the Fourier transform of Eq. (9)) at the output $z = L$. After applying the usual radiative boundary conditions for the current $\mathbf{J}$ ($\mathbf{J}(z = 0) = 1$ and $\mathbf{J}(z = L) = 0$) at both sides of the slab [29], the diffuse emerging radiation with circular polarization σ is expressed as:

$$I_\sigma(\theta, L) = \frac{4}{3}\frac{\ell^*}{L + \frac{4}{3}\ell^*}\left[1 + \frac{3}{2}\cos\theta\left(1 + \frac{\ell_C^*}{\ell^*}\sigma\right)\right], \tag{12}$$

where θ denotes the angle between the wavevector associated with the emerging diffuse radiation and the slab normal. The values $\sigma = \pm 1$ are associated with the helicity of the light wave (corresponding to right and left circularly polarized light). Equation (12) implies that the emerging

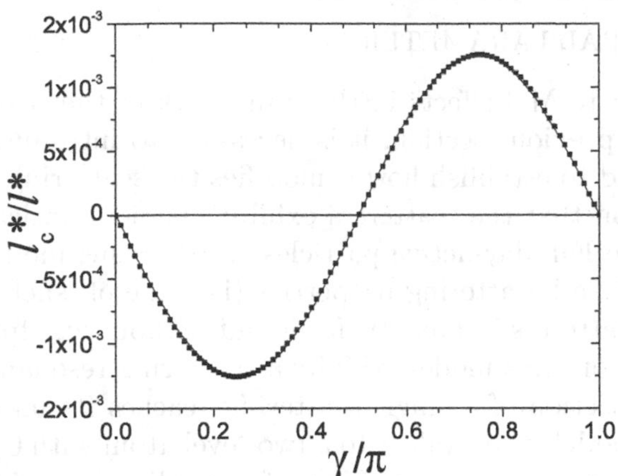

Figure 2. The ratio between the chiral transport mean free path and the transport mean free path ℓ_C^*/ℓ^* calculated for light scattered by four resonant pointlike scatterers at the vertices of the so called twisted H (main axis d and secondary axis r) as a function of the angle between its arms γ (marked by squares). The parameters used are $kr = 3.0$ and $kd = 9.0$ with k the light wavenumber. The solid curve corresponds to the chiral parameter $\psi \sim \sin(2\gamma)$ introduced in Refs. [23,24].

diffuse radiation has a *residual degree of circular polarization*. This effect is quantified by the Stokes parameter $V(\theta) = I_+(\theta) - I_-(\theta)$ for circular polarization:

$$\frac{V(\theta)}{I(\theta)} = \frac{\frac{3}{2}\cos\theta}{1+\frac{3}{2}\cos\theta}\left(\frac{\ell_C^*}{\ell^*}\right), \tag{13}$$

where $I(\theta) = I_+(\theta) + I_-(\theta)$. Our microscopic result (13) qualitatively agrees with previous Monte Carlo simulations [15], in which it was shown that multiple scattered light in chiral media achieves a degree of circular polarization, a fact that may have potential applications for imaging [15]. This result is not expected to happen in achiral media, where any incident polarization state of light rapidly depolarizes due to multiple scattering. It should be mentioned that the order of magnitude of the circular intensity difference predicted by Eq. (13) is strongly dependent on the amount of scattering. If the scattering is not too weak, the maximal values of the ratio ℓ_C^*/ℓ^* for four scatterers distributed at the vertices of the "twisted H", for instance, can be of the order of 10^{-2}.

3. Transport in Magnetochiral Media

3.1. A NEW CHIRAL PARAMETER

In order to generate MC effects in the framework of the scattering model presented in the previous section, it is necessary to introduce an external magnetic field and to establish how it modifies the scattering. To do so, we assume for now on that the scatterers exhibit magneto-optical effects, i.e., they are pointlike Faraday-active particles. Within this model, it is possible to find exactly all scattering properties (i.e., the off shell T matrix) of any number of scatterers in analytic form and without any further approximation. In addition, this model exhibits a scattering resonance, where the scattering cross section of a single scattering reaches its maximum value. Physically, this model is equivalent to a two-level atom with $0-1$ transition subject to the Zeeman effect [30]. Since, for small values of the magnetic field $\mathbf{B}$, the MC effect is small and linear, we will only consider the linear term in $\mathbf{B}$ and write for the T matrix [31]:

$$\mathbf{t}(\mathbf{B},\omega) = t_0\mathbf{U} + t_1\mathbf{\Phi}, \tag{14}$$

where $\Phi_{ij} = i\epsilon_{ijk}\widehat{B_k}$. The parameter t_0 is the ordinary Rayleigh T matrix, $t_0 = -4\pi\Gamma\omega^2/(\omega_0^2 - \omega^2 - 2i\Gamma\omega^3/3c_0)$, with ω_0 the resonance frequency and Γ the resonance linewidth. The quantity t_1 is defined as $t_1 = -\mu(B)kt_0^2/6\pi$, with $\mu(B)$ a magneto-optical parameter which, for an atom, is related to the Zeeman splitting $zB = \Delta E_z/\hbar$ (typically, $\mu = zB/\Gamma$). In addition, t_1 is related to the Faraday rotation inside the particle [31]. The knowledge of the single scattering T matrix in Eq. (14) allows us to obtain the total $3N \times 3N$ T matrix of an assembly of N magneto-optical scatterers situated at the positions $\mathbf{r}_1, \mathbf{r}_2, ..., \mathbf{r}_N$:

$$\mathbf{T}_{\mathbf{k},\mathbf{k}'} = \begin{pmatrix} e^{i\mathbf{k}\cdot\mathbf{r}_1} \\ \vdots \\ e^{i\mathbf{k}\cdot\mathbf{r}_N} \end{pmatrix}^* \mathbf{t}\cdot(\mathbf{U} - \mathbf{G}\cdot\mathbf{t})^{-1}\cdot \begin{pmatrix} e^{i\mathbf{k}'\cdot\mathbf{r}_1} \\ \vdots \\ e^{i\mathbf{k}'\cdot\mathbf{r}_N} \end{pmatrix}, \tag{15}$$

where $\mathbf{k}$ and $\mathbf{k}'$ are, respectively, the incident and the scattered wave vectors, and $|\mathbf{k}'| = |\mathbf{k}| = k$. The elements of the $3N \times 3N$ $\mathbf{G}$-matrix are equal to the Green functions calculated from the relative positions of the scatterers. The set of Eqs. 15 form a system of linear equations that determines the electric field scattered by the magneto-optical particles, and its solution requires the diagonalization of the $3N \times 3N$ scattering matrix [32]:

$$\mathbf{M}(k) \equiv \mathbf{t}(k)\cdot[\mathbf{U} - \mathbf{G}(k)\cdot\mathbf{t}(k)]^{-1}. \tag{16}$$

The numerical diagonalization of the scattering matrix $\mathbf{M}(k)$ in Eq. (16) [32] provides the scattered field amplitude for particles distributed in an arbitrary spatial configuration:

$$f(\sigma\mathbf{k} \to \sigma'\mathbf{k}') = \mathbf{g}_\sigma^* \cdot \mathbf{f}(\mathbf{k} \to \mathbf{k}') \cdot \mathbf{g}_{\sigma'} \quad \text{with}$$
$$\mathbf{f}(\mathbf{k} \to \mathbf{k}') = \sum_{NM} \mathbf{M}^{NM} \exp(i\mathbf{k} \cdot \mathbf{r}_N) \exp(-i\mathbf{k}' \cdot \mathbf{r}_M), \tag{17}$$

where $\mathbf{g}_\sigma$ and $\mathbf{g}_{\sigma'}$ are, respectively, the incident and scattered polarization vectors (expressed in the circular polarization basis) and the sum is performed over all scatterers. Since the Faraday effect is a small perturbation to the scattering, we will expand $\mathbf{f}$ *linearly* in the magnetic field, $\mathbf{f} = \mathbf{f}^0 + \mathbf{f}^1$, with $\mathbf{f}^1$ given by:

$$\begin{aligned} \mathbf{f}^1(\mathbf{k} \to \mathbf{k}', \mathbf{B}) &= -\frac{k}{4\pi}\mu(B) \sum_{NMN'} (\mathbf{M}_0^{NM} \cdot \mathbf{\Phi} \cdot \mathbf{M}_0^{MN'}) \exp(i\mathbf{k} \cdot \mathbf{r}_N) \\ &\times \exp(-i\mathbf{k}' \cdot \mathbf{r}_{N'}), \end{aligned} \tag{18}$$

where $\mathbf{M}_0$ is the 3×3 scattering matrix for $\mathbf{B} = 0$. The corresponding *magneto extinction cross section* $\sigma^1_{ext}(\mathbf{B}, \mathbf{k})$, linear in the magnetic field and integrated over all scattered angles, can be obtained from the optical theorem [28]:

$$\begin{aligned} \sigma^1_{ext}(\mathbf{B}, \mathbf{k}) &= -\frac{1}{k}\mathrm{Im} \sum_\sigma f^1(\sigma\mathbf{k} \to \sigma\mathbf{k}) \\ &= \frac{\mu(B)}{4\pi}\mathrm{Im} \sum_{NMN'} \mathrm{Tr}\left(\mathbf{M}_0^{NM} \cdot \mathbf{\Phi} \cdot \mathbf{M}_0^{MN'} \cdot \mathbf{\Delta}_\mathbf{k}\right) \\ &\times \exp\left[i\mathbf{k} \cdot (\mathbf{r}_N - \mathbf{r}_{N'})\right], \end{aligned} \tag{19}$$

where $(\mathbf{\Delta}_\mathbf{k})_{ij} \equiv \delta_{ij} - k_i k_j / k^2$. Notice that we have traced over polarization, which means that the Faraday effect itself cancels and that only MC effects remain. Eq. (19) can be regarded as the magneto dichroism of the multiple scattering system, associated with the energy removed from the incident beam due the application of an external magnetic field.

Until now, nothing was said about how the light scattering properties of magneto-optically active particles are affected by the chiral configuration and, more important, how these properties can be related to a measure of the degree of chirality. We recall that any chiral measure must be rotationally invariant. This implies that we must perform an *average* over *all solid rotations* of the positions $\{\mathbf{r}_N\}$ in (19) in order to estimate magneto-optical chirality. The angularly averaged magneto extinction cross section in MC media must necessarily have the form:

$$\langle \sigma^1_{ext}(\mathbf{B}, \mathbf{k}) \rangle_{4\pi} = g\mu(B)(\widehat{\mathbf{B}} \cdot \widehat{\mathbf{k}}), \tag{20}$$

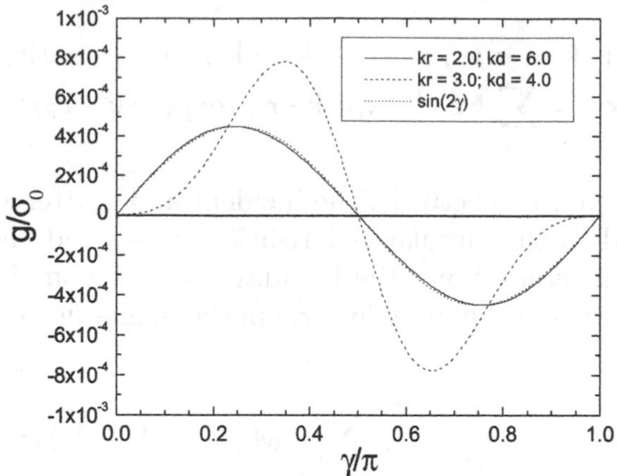

Figure 3. The on-resonance optical chiral parameter g plotted as a function of the angle γ between the "twisted H" arms. The solid curve corresponds to the values $kr = 2.0$ and $kd = 6.0$ (with k the light wavenumber). The dashed curve corresponds to the values $kr = 3.0$ and $kd = 4.0$. The dotted curve presents the chiral parameter $\psi \sim \sin(2\gamma)$ introduced in Refs. [23,24]. g was normalized by $\sigma_0 = N\sigma_r$, where $N = 4$ and $\sigma_r = 3\lambda_0^2/2\pi$.

where g is some pseudoscalar and the brackets $\langle ... \rangle_{4\pi}$ denote the average over all solid rotations. Equation (20) is manifestly rotationally invariant and obeys the two fundamental symmetry relations (parity and reciprocity) in MC media, as can be easily verified. Using Eqs. (19) and (20) and performing the angular average in (20) analytically, we obtain the following expression for g [33]:

$$\begin{aligned} g &= \frac{1}{2\pi}\mathrm{Re} \sum_{N'<N,M} \mathrm{Tr}\left(\mathbf{M}_0^{NM} \cdot \mathbf{\Phi}_{\widehat{\mathbf{r}}_{NN'}} \cdot \mathbf{M}_0^{MN'}\right) \left[j_1(kr_{NN'}) - \frac{j_2(kr_{NN'})}{kr_{NN'}}\right] \\ &+ \left(\widehat{\mathbf{r}}_{NN'} \cdot \mathbf{M}_0^{NM} \cdot \mathbf{\Phi}_{\widehat{\mathbf{r}}_{NN'}} \cdot \mathbf{M}_0^{MN'} \cdot \widehat{\mathbf{r}}_{NN'}\right) \\ &\times \; j_3(kr_{NN'}) \,, \end{aligned} \tag{21}$$

where $j_n(x)$ is the spherical Bessel function of the first kind of order n. Expression (21) represents a pseudoscalar, since it changes sign upon a mirror operation $\{\mathbf{r}_N\} \to \{-\mathbf{r}_N\}$.

In the following, we will numerically calculate g for some simple chiral systems.

3.2. PROBING THE CHIRALITY OF SIMPLE CHIRAL SYSTEMS

In Fig. 3 we have numerically calculated the parameter g defined in Eq. (21) for four magneto-optically active pointlike scatterers placed at the vertices of the twisted H (see Fig. 1) as a function of the angle γ for two different values of the wavelength λ of the incident light. Each scatterer was set on resonance with the incident radiation, i.e., we set $\lambda = \lambda_0$. In addition, g was normalized by the normal extinction cross section of the system, which can, for the set parameters used here, be adequately approximated by $\sigma_0 = N\sigma_r$ (where $N = 4$ and $\sigma_r = 3\lambda_0^2/2\pi$ is the on resonance extinction cross section for one scatterer) since we are in the independent scattering regime. In Fig. 3 we see that g exhibits an oscillatory behavior as a function of the angle γ and *vanishes exactly at the configurations for which the H is achiral.* These configurations correspond to the angles $\gamma = n\pi/2$, with n integer. If the value of the incident wavelength is modified, the dependence of g on γ changes, as one can see in Fig. 3, but g still vanishes at the same angles. For different values of the wavelength λ, g may also vanish at other values of γ, but the zeros imposed by symmetry (at the achiral configurations) remain unchanged. For comparison, we show by a dotted line the chiral parameter $\psi \sim \sin(2\gamma)$ for the twisted H proposed by Harris *et al.* [23, 24] which, for the set of values $kr = 2.0$ and $kd = 6.0$, nicely follows our optical parameter g.

A system composed of a large number of randomly distributed particles will in general be chiral. Our purpose will be to quantify the degree of chirality of this kind of system using the scattering parameter g defined in Eq. (21). We have numerically calculated g for 1000 different random configurations of N scatterers distributed in a sphere of radius R and volume V, and studied the behavior of g as a function of N for two distinct situations: increasing N upon keeping V constant and increasing N upon keeping the density $\rho = N/V$ of the scatterers constant.

In Fig. 4, we show a typical histogram for the values of g for 1000 realizations of $N = 10$ scatterers randomly distributed in a sphere. The histogram shows a normal distribution, as expected on the basis of the law of large numbers. Since we have a large number of realizations of the disorder, the mirror image of any configuration is equally probable, so that $\langle g\rangle = 0$. We will consider the variance $\Delta g \equiv \sqrt{\langle g^2\rangle - \langle g\rangle^2}$ of g as a candidate to probe the typical degree of chirality of an arbitrary random configuration.

In Fig. 5, we plot the values of Δg as a function of the number of particles distributed in a sphere with ρ constant. The variances Δg were determined by taking the full width at half maximum of the Gaussian fit to the histograms of g for 1000 different realizations of the disorder (see Fig. 4). One relevant dimensionless parameter in the problem is the quantity $\zeta \equiv \frac{4\pi\rho}{k^3}$, which is essentially the number of particles per cubic wavelength. For the value of ρ chosen in Fig. 5 we have $\zeta \approx 0.01$, which means that we are in the so called independent scattering regime and that only the first orders of scattering are relevant. This allows us to normalize

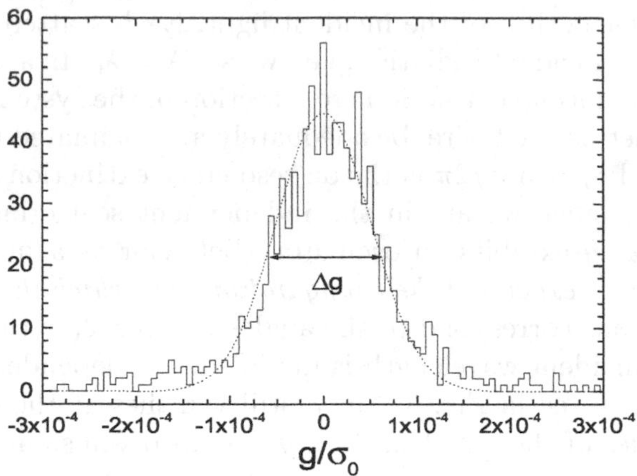

Figure 4. A typical histogram for the on-resonance values of g associated with 1000 different realizations of the positions of $N = 10$ scatterers randomly distributed in a sphere of density ρ constant. The values of g were normalized by σ_0 and the parameter ζ, defined in the text, equals $\zeta \approx 0.01$. The dotted curve corresponds to a Gaussian fit to the data.

Δg in Fig. 5 by $\sigma_0 = N\sigma_r$. As expected, Δg vanishes for $N = 2, 3$, since it is impossible to generate a chiral configuration with only two or three particles. The normalized Δg increases until it reaches, for roughly $N = 30$ particles in the sphere, an asymptotic value of approximately $\frac{\Delta g}{\sigma_0} \approx 1.35$ x 10^{-4}. For comparison, we also show in Fig. 5 the behavior of the variance $\Delta\psi$ of the chiral parameter ψ proposed by Harris *et al.* [23, 24] as a function of N. The comparison between the two curves in Fig. 5 clearly reveals that the two chiral measures g and ψ are, statistically speaking, *proportional.* The observed relation of proportionality between Δg and $\Delta\psi$ is given by:

$$\frac{\Delta g}{\sigma_0} \propto \frac{\Delta\psi}{R^8}\lambda^3, \tag{22}$$

showing that we can establish a relation between the optical magneto-chiral extinction g and the purely geometrical factor ψ/R^8 by multiplying the latter by the optical quantity λ^3. This demonstrates that the correspondence between g and ψ is valid not only for the "twisted H" case discussed before, but applies more generally to random systems. This also confirms that the parameter g, associated with MC effects in multiple scattering, is indeed an appropriate quantity to probe the chirality of optical systems.

In Fig. 6, Δg and $\Delta\psi$ are shown as a function of N if V is kept constant, i.e., where ρ was varied. In this case, $\zeta \approx 0.00025N$. We observe that both Δg and $\Delta\psi$ increase linearly with the number of scatterers. The comparison

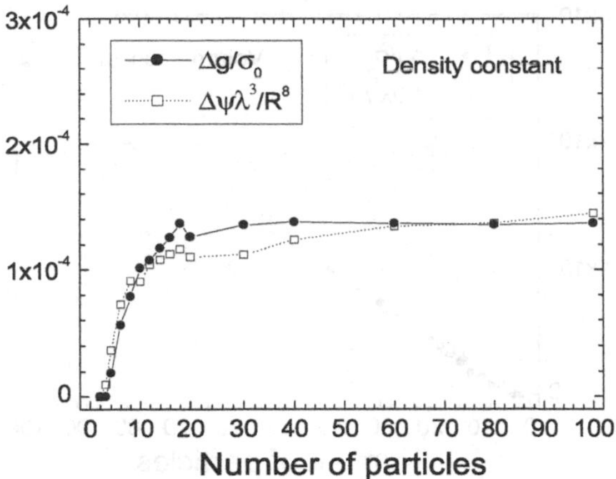

Figure 5. The variances Δg and $\Delta\psi$ of the chiral parameters g (full circles and solid line) and ψ (empty squares and dotted line), obtained from 1000 different realizations of the disorder, as a function of the number N of scatterers randomly distributed in a sphere of density constant and with $\zeta \approx 0.01$. The values g were calculated on-resonance and normalized by $\sigma_0 = N\sigma_r$. In order to show dimensionless quantities, the values of ψ were multiplied by λ^3/R^8 (where λ is the light wavelength and R is the radius of the sphere). This reveals the proportionality relation $\frac{\Delta g}{\sigma_0} \propto \frac{\Delta\psi}{R^8}\lambda^3$ between Δg and $\Delta\psi$. To allow a better comparison between Δg and $\Delta\psi$, we have multiplied the values of $\Delta\psi$ by an appropriate constant numerical factor. The lines are just a guide for the eyes.

between Δg and $\Delta\psi$ in Fig. 6 confirms, for V constant, the linear relation between these two chiral parameters. In order to understand the linear dependence of Δg on N we recall that we need at least four particles to constitute a chiral system. This suggests to consider the chiral parameter g as the sum of many random contributions of groups of four particles, with random sign but with fixed absolute value g_4. Since $\langle g\rangle = 0$ and since $\begin{pmatrix} N \\ 4 \end{pmatrix}$ distinct forms exist to group four particles together, we can estimate the normalized Δg as:

$$\frac{\Delta g}{\sigma_0} = \frac{\sqrt{\langle g^2\rangle}}{N\sigma_r} \sim \frac{1}{N}\sqrt{\begin{pmatrix} N \\ 4 \end{pmatrix}(g_4)^2}. \tag{23}$$

Taking the limit $N \to \infty$ and noticing that by Eq. (21) g_4 scales as $g_4/\sigma_0 \sim 1/(kR)^3$, we can rewrite Eq. (23) as:

$$\frac{\Delta g}{\sigma_0} \sim \frac{\sqrt{N(N-1)(N-2)(N-3)}}{N(kR)^3} \sim \frac{N}{(kR)^3} + ... \sim \rho\lambda^3 + ... \quad . \tag{24}$$

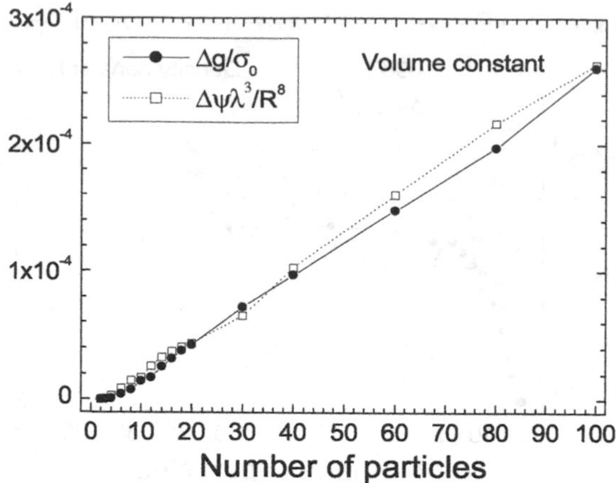

Figure 6. As in Fig. 5, but now with N scatterers distributed in a sphere at constant volume and with $\zeta \approx 0.00025N$.

This heuristic argument confirms the linear dependence of Δg on N observed in Fig. 6 for large N. However, for larger densities (typically for $\zeta \approx 1$) this relation is lost.

3.3. A PHOTONIC SUPERCURRENT IN MAGNETOCHIRAL RANDOM MEDIA?

In media where both mirror and time-reversal symmetries are broken, the existence of a "super" current of the type $\mathbf{J}_C(\mathbf{r},t) = D_C\mathbf{B}\,\nu(\mathbf{r},t)$ becomes symmetry allowed. Here $\nu(\mathbf{r},t)$ is the electromagnetic energy density. D_C must be a pseudoscalar to take over the role of the spatial gradient in Fick's law $\mathbf{J}(\mathbf{r},t) = -\mathbf{D}(\mathbf{B})\cdot\nabla\nu(\mathbf{r},t)$. The "super" current $\mathbf{J}_C$ would thus be a macroscopic manifestation of microscopically broken parity, just like the pitch in a cholesteric liquid crystal [23, 24], the optical rotatory power [25] and the scattering in MC random media as discussed previously. It has formally the same structure as the London current $\mathbf{J}_L = e\mathbf{A}\,\nu$ in a superconductor, with both $\mathbf{J}_L$ and $\mathbf{A}$ parity-odd vectors. $\mathbf{J}_L$ is imposed by gauge invariance, contrary to the normal diffuse electronic (Ohmic) current, that is described by the gauge invariant Fick's law. For the MC current $\mathbf{J}_C$ no such a strong gauge argument is available as it involves the observable $\mathbf{B}$ itself. We have derived a microscopic expression for $\mathbf{J}_C$ and have calculated it using our model for the MC effect in multiple scattering [34]. Unfortunately, we have arrived at the conclusion that, within such a model, the "super" current vanishes. We strongly suspect that a strong physical argument or a group theoretical element exists that forbids the existence of a photonic supercurrent.

Acknowledgments

We gratefully acknowledge G. Rikken, G. Wagnière, G. Düchs, M. Rusek and A. Orlowski for fruitful discussions. One of us (F.A.P.) also wishes to thank CNPq/Brazil for financial support.

References

1. R. Kaiser, in *Diffuse Waves in Complex Media*, edited by J.-P. Fouque, NATO Science Series (Kluwer, Dordrecht, 1999); G. Labeyrie *et al.*, Phys. Rev. Lett. **83**, 5266 (1999); Y. Bidel *et al.*, Phys. Rev. Lett. **88**, 203902 (2002).
2. A. Lagendijk *et al.*, in *Photonic Crystals and Light Localization in the 21st Century*, edited by C. M. Soukoulis , NATO Science Series (Kluwer, Dordrecht, 2001).
3. *Photonic Crystals and Light Localization in the 21st Century*, edited by C. M. Soukoulis , NATO Science Series (Kluwer, Dordrecht, 2001).
4. R. Hennino *et al.*, Phys. Rev. Lett. **86**, 3447 (2001).
5. J. Virmont and G. Ledanois, in *New Aspects of Electromagnetic and Acoustic Wave Diffusion*, edited by POAN Research Group (Springer, Heidelberg, 1998).
6. *Diffuse Waves in Complex Media*, edited by J.-P. Fouque, NATO Science Series (Kluwer, Dordrecht, 1999); *New Aspects of Electromagnetic and Acoustic Wave Diffusion*, edited by POAN Research Group (Springer, Heidelberg, 1998); *Waves and Imaging Through Complex Media*, edited by P. Sebbah (Kluwer, Dordrecht, 2001).
7. R. Lenke and G. Maret, Eur. Phys. J. B **17**, 171 (2000).
8. B. A. van Tiggelen, Phys. Rev. Lett. **75**, 422 (1995).
9. G. L. J. A. Rikken and B. A. van Tiggelen, Nature **381**, 54 (1996).
10. P. Roux and M. Fink, Europhys. Lett. **32**, 25 (1995).
11. C. W. J. Beenakker, Rev. Mod. Phys. **69**, 731 (1997).
12. W. Thomson, *Baltimore Lectures on Molecular Dynamics and the Wave Theory of Light*, (C. J. Clay, London, 1904), p. 619.
13. L. D. Barron, *Molecular Light Scattering and Optical Activity*, (Cambridge University Press, Cambridge, 1982).
14. M. P. Silverman, W. Strange, J. Badoz and I. A. Vitkin, Opt. Commun. **132**, 410 (1996).
15. B. P. Ablitt *et al.*, Waves in Random Media **9**, 561 (1999).
16. M. P. Groenewege, Mol. Phys. **5**, 541 (1962).
17. G. Wagnière, Chem. Phys. **245**, 165 (1999).
18. G. L. J. A. Rikken and E. Raupach, Nature (London) **390**, 493 (1997).
19. P. Kleindienst and G. H. Wagnière, Chem. Phys. Lett. **288**, 89 (1998).
20. M. Vallet *et al.*, Phys. Rev. Lett. **87**, 183003 (2001).
21. G. Gilat, J. Phys. A: Math. Gen. **22**, L545 (1989).
22. M. A. Osipov, B. T. Pickup and D. A. Dunmur, Mol. Phys. **84**, 1193 (1995).
23. A. B. Harris, R. D. Kamien and T. C. Lubensky, Phys. Rev. Lett. **78**, 1476 (1997); 2867(E).
24. A. B. Harris, R. D. Kamien and T. C. Lubensky, Rev. Mod. Phys. **71**, 1745 (1999).
25. M. S. Spector *et al.*, Phys. Rev. E **61**, 3977 (2000).
26. P. Sheng, *Wave Scattering, Localization and Mesoscopic Phenomena* (Academic, San Diego, 1995).
27. G. D. Mahan, *Many-Particle Physics* (Plenum, New York, 1981).
28. R. G. Newton, *Scattering Theory of Waves and Particles* (Springer Verlag, New York, 1982).

29. A. Ishimaru, *Wave Propagation and Scattering in Random Media*, Vol. 1 (Academic Press, San Diego, 1978).
30. G. Labeyrie, C. Miniatura and R. Kaiser, Phys. Rev. A **64**, 033402 (2001).
31. B. A. van Tiggelen, R. Maynard and T. M. Nieuwenhuizen, Phys. Rev. E **53**, 2881 (1996).
32. M. Rusek and A. Orlowski, Phys. Rev. E **51**, R2763 (1995); M. Rusek and A. Orlowski, Phys. Rev. E **56**, 6090 (1997).
33. F. A. Pinheiro and B. A. van Tiggelen, Phys. Rev. E **66**, 016607 (2002).
34. F. A. Pinheiro and B. A. van Tiggelen, J. Opt. Soc. Am. A **20**, 99 (2003).

DIFFUSE WAVES IN NONLINEAR DISORDERED MEDIA

S.E. SKIPETROV AND R. MAYNARD
Laboratoire de Physique et Modélisation des Milieux Condensés
CNRS, 25 Avenue des Martyrs, 38042 Grenoble, France

1. Introduction

The field of multiple scattering of classical waves (electromagnetic, acoustic and elastic waves, etc.) in disordered media has revived in the eighties [1]–[6], when the far-reaching analogies between the diffuse transport of waves and electrons in mesoscopic systems have been realized (see Ref. [7] for a comprehensive review of the latter issue). Using classical waves to study such phenomena as weak and strong localization, mesoscopic correlations and universal conductance fluctuations (see Refs. [8]–[10] for reviews) appears to be advantageous in many aspects: no need for low temperatures and small samples, better control of the experimental apparatus, possibility of more sensitive measurements, etc. In addition, experiments with classical waves can be readily performed in the *linear* regime, excluding interaction between scattered waves and hence simplifying the interpretation of experimental data, whereas the electron-electron interaction is always present in the realm of mesoscopic electronics and cannot simply be 'turned off', introducing significant difficulties in the theoretical model [11].

The 'interaction' of classical waves, analogous in some sense to the electron-electron interaction, can come about if the waves propagate in a *nonlinear* medium. Such an interaction is a matter of scientific enquiry in the fields of nonlinear optics [12], nonlinear acoustics [13], etc. Nowadays, the latter are rapidly developing scientific disciplines on their own right, having considerable fundamental importance and numerous practical applications. Various nonlinear phenomena (self-action: self-phase modulation and self-focusing of pulses and beams, harmonic generation, shock wave formation, etc.) occur in *homogeneous* nonlinear media depending on the type and strength of the nonlinearity. The effect of *weak* disorder on the propagation of nonlinear waves has also been studied both theoretically (treating it as a weak perturbation) and experimentally (e.g., for laser beam

B.A. van Tiggelen and S.E. Skipetrov (eds.),
Wave Scattering in Complex Media: From Theory to Applications, 75–97.

propagation through atmospheric turbulence) [14]. Unfortunately, only few experiments [15] have been performed up to now on multiple, *diffuse* scattering of classical waves in nonlinear media. Theoretical analysis of the problem is more advanced (see the bibliography of Ref. [16]), but is still insufficient to stimulate further experimental effort. Meanwhile, diffuse waves seem to be a good candidate for a detailed study of the combined effect of disorder and nonlinearity (or interaction) on wave (or quantum particle) propagation, as compared to interacting electrons in disordered mesoscopic samples: the strength and the type of nonlinear wave 'interaction' can be readily controlled and the nonlinear term in the wave equation is often of simple algebraic form. Despite these important simplifications, diffusion of classical waves in nonlinear media remains an involved problem and is far from being solved. The purpose of the present paper is to review our recent theoretical results [16]–[19] which are particularly susceptible to stimulate the reader's interest, further theoretical analysis, and, hopefully, new experiments.

For the sake of concreteness, we restrict our consideration to the problem of self-action of a scalar monochromatic wave (frequency ω) in a nonlinear disordered medium. This is described by the following wave equation:

$$\left\{ \nabla^2 - \frac{1}{c^2}\frac{\partial^2}{\partial t^2}\left[1 + \delta\varepsilon(\mathbf{r}) + \Delta\varepsilon_{\mathrm{NL}}(\mathbf{r}, t)\right] \right\} E(\mathbf{r}, t) = J(\mathbf{r}, t), \tag{1}$$

where $E(\mathbf{r}, t)$ is the wave amplitude, c is the speed of the wave in the average medium, $J(\mathbf{r}, t) = J_0(\mathbf{r}) \exp(-i\omega t)$ is a monochromatic source term, $\delta\varepsilon(\mathbf{r})$ is the fractional fluctuation of the linear dielectric constant, and the nonlinear part of the dielectric constant $\Delta\varepsilon_{\mathrm{NL}}(\mathbf{r}, t)$ depends on the intensity $I(\mathbf{r}, t) = |E(\mathbf{r}, t)|^2$ of the wave. Eq. (1) describes, e.g., propagation of light in a medium with intensity-dependent refractive index and (as the reader might already have noted) we adopt the 'optical' terminology from here on. Furthermore, we consider E in Eq. (1) to be a complex quantity, thus neglecting the (possible) generation of the third harmonics, and restrict ourselves to a weak nonlinearity (see below for the definition of 'weakness').

There are two fundamental questions to consider concerning the propagation of waves in a disordered medium with a weak nonlinearity. First, one can ask about the effect of nonlinearity on the phenomena known for waves in linear disordered media. We partially answer this question in Sec. 4, where we summarize the results of calculation of the angular correlation functions of scattered waves and the analysis of the coherent backscattering cone in a nonlinear disordered medium. The second question is more challenging: can the weak nonlinearity give rise to new physical phenomena which are not present in the linear medium? The answer to the latter question is 'yes' as we demonstrate in Sec. 5 where the instability of diffuse, multiple-scattered waves in a disordered medium with a weak nonlinearity is considered. In order to facilitate the presentation of our main results, we start by a very brief review of principal results known for waves in linear disordered (Sec. 2) and homogeneous nonlinear (Sec. 3) media.

2. Waves in Linear Disordered Media

Waves propagating in a *linear* disordered medium are described by Eq. (1) with $\Delta\varepsilon_{NL}(\mathbf{r},t) = 0$. Despite its linearity, this problem is a rather involved one and it attracts a lot of attention (see, e.g., the articles in the present book and Refs. [9] and [10]). In what follows, we are interested uniquely in the diffusion regime of wave propagation that is realized for $k_0\ell \gg 1$, where $k_0 = \omega/c$ and ℓ is the mean free path due to disorder. To simplify still further the consideration, we assume that the correlation length of $\delta\varepsilon(\mathbf{r})$ is much shorter than the wavelength λ and adopt the model of the Gaussian white-noise disorder: $\langle\delta\varepsilon(\mathbf{r})\delta\varepsilon(\mathbf{r}_1)\rangle = 4\pi/(k_0^4\ell)\delta(\mathbf{r}-\mathbf{r}_1)$, so that the scattering and the transport mean free paths coincide. Besides, we consider a disordered medium without absorption (i.e. both $\delta\varepsilon$ and $\Delta\varepsilon_{NL}$ are real), except if the opposite is specified explicitly. In the regions of space located at distances greater than ℓ from the sources of waves and the boundaries of the sample, the average intensity $\langle I(\mathbf{r})\rangle$ then obeys a diffusion equation [20]:

$$\frac{\partial}{\partial t}\langle I(\mathbf{r},t)\rangle = D\nabla^2\langle I(\mathbf{r},t)\rangle + S(\mathbf{r},t), \tag{2}$$

where $D = c\ell/3$ is the diffusion constant and $S(\mathbf{r},t)$ is the source term. In the following, we assume a time-independent source term in Eq. (2): $S(\mathbf{r},t) = S(\mathbf{r})$, yielding $I(\mathbf{r},t) = I(\mathbf{r})$.

The average intensity $\langle I(\mathbf{r})\rangle$ given by Eq. (2) is not sufficient to describe the intensity $I(\mathbf{r})$ of the wave, since the latter exhibits large fluctuations: $\delta I(\mathbf{r}) = I(\mathbf{r}) - \langle I(\mathbf{r})\rangle \sim \langle I(\mathbf{r})\rangle$ [20, 2]. The spatial correlation function of $\delta I(\mathbf{r},t)$ contains a short-range ($\Delta r < \ell$) but strong (~ 1) contribution [2]:

$$\langle\delta I(\mathbf{r})\delta I(\mathbf{r}_1)\rangle = \langle I(\mathbf{r})\rangle\langle I(\mathbf{r}_1)\rangle\left[\frac{\sin(k_0\Delta r)}{k_0\Delta r}\right]^2\exp\left(-\frac{\Delta r}{\ell}\right) \tag{3}$$

and a long-range ($\Delta r > \ell$) but weak [$\sim (k_0\ell)^{-2} \ll 1$] contribution [3, 4]:

$$\langle\delta I(\mathbf{r})\delta I(\mathbf{r}_1)\rangle \sim \frac{c}{k_0^2}I_0^2 G(\mathbf{r},\mathbf{r}_1) \sim \frac{I_0^2}{k_0^2\ell\Delta r}, \tag{4}$$

where $\Delta\mathbf{r} = \mathbf{r} - \mathbf{r}_1$, I_0 is a typical value of $\langle I(\mathbf{r})\rangle$, $G(\mathbf{r},\mathbf{r}_1)$ is the Green's function of Eq. (2) with $(\partial/\partial t) = 0$, and the last expression in Eq. (4) is obtained assuming that $G(\mathbf{r}, \mathbf{r}_1)$ is equal to its value in the infinite medium: $G(\mathbf{r}, \mathbf{r}_1) = G_0(\mathbf{r}, \mathbf{r}_1) = (4\pi D \Delta r)^{-1}$. The short-range contribution to the intensity correlation function (3) describes the speckle structure of the spatial intensity distribution in disordered media, while the long-range contribution (4) testifies that different speckle spots are correlated (weakly, since $k_0\ell \gg 1$). The correlation functions given by Eqs. (3) and (4) are commonly referred to as $C^{(1)}$ and $C^{(2)}$, respectively. Other contributions to the intensity correlation function can exist: $C^{(3)}$ is a factor $(k_0\ell)^{-2} \ll 1$ smaller than $C^{(2)}$ [9] and $C^{(0)} \sim (k_0\ell)^{-1}$ is non-vanishing only if the source of waves has the size smaller or of the order of the wavelength λ [21]. We do not consider the two latter contributions here.

Short- and long-range correlation functions of the field and intensity fluctuations can be also defined in the angular domain. For our purposes, it will be sufficient to consider only the short-range correlations $C^{(1)}$. If a plane wave is incident on a disordered slab of thickness $L \gg \ell$ (slab surfaces being perpendicular to the z-axis) with a wave vector $\mathbf{k}_a = \{\mathbf{q}_a, k_{az}\}$, where $\mathbf{q}_a = \{k_{ax}, k_{ay}\}$, the field (intensity) of the wave transmitted through the slab with a wave vector $\mathbf{k}_b$ is $E(\mathbf{k}_a, \mathbf{k}_b)$ $[I(\mathbf{k}_a, \mathbf{k}_b) = |E(\mathbf{k}_a, \mathbf{k}_b)|^2]$. If the above experiment is repeated for different incoming and outgoing wave vectors $\mathbf{k}'_a$ and $\mathbf{k}'_b$, respectively, the normalized correlation function of scattered fields

$$C(\mathbf{k}_a, \mathbf{k}_b; \mathbf{k}'_a, \mathbf{k}'_b) \equiv C_{aba'b'} = \frac{\langle E(\mathbf{k}_a, \mathbf{k}_b) E^*(\mathbf{k}'_a, \mathbf{k}'_b) \rangle}{\langle I(\mathbf{k}_a, \mathbf{k}_b) \rangle^{1/2} \langle I(\mathbf{k}'_a, \mathbf{k}'_b) \rangle^{1/2}} \tag{5}$$

appears to have a rather simple form [8]:

$$C_{aba'b'} = \delta_{\Delta \mathbf{q}_a, \Delta \mathbf{q}_b} F_{\mathrm{T}}(\Delta q_a L), \tag{6}$$

where $F_{\mathrm{T}}(x) = x/\sinh x$, $\Delta \mathbf{q}_a = \mathbf{q}_a - \mathbf{q}'_a$ (and similarly for $\Delta \mathbf{q}_b$), and $\Delta q_a \ell \ll 1$ is assumed.

The field correlation in reflection, when both incident and scattered waves are on the same side of the slab, is [8, 17]

$$C_{aba'b'} \simeq \delta_{\Delta \mathbf{q}_a, \Delta \mathbf{q}_b} \left[F_{\mathrm{R}}(\Delta \mathbf{q}_a) + F_{\mathrm{R}}(\mathbf{q}_a + \mathbf{q}_b + \Delta \mathbf{q}_a) \right], \tag{7}$$

where

$$F_{\mathrm{R}}(\mathbf{q}) \simeq (1 + 2\zeta) - 2(1 + \zeta)^3 q\ell + \ldots, \tag{8}$$

$\zeta \sim 1$ is the extrapolation factor ($\zeta = 2/3$ in the diffusion approximation), $q\ell \ll 1$, and the semi-infinite disordered medium is considered. The short-range correlation function of intensity fluctuations, $C^{(1)}$, in transmission (reflection) geometry is obtained by taking the square of the absolute value of Eq. (6) [Eq. (7)].

Finally, a well-studied phenomenon in the linear disordered medium is the coherent backscattering [1], consisting in a two-fold (with respect to the incoherent, diffuse background) enhancement of the average scattered intensity in the direction of exact backscattering ($\mathbf{k}_b = -\mathbf{k}_a$). For normal incidence ($\mathbf{q}_a = 0$), the angular line shape of the coherent backscattering cone can be approximately expressed using the function F_{R} defined in Eq. (8):

$$\frac{\langle I(\mathbf{q})\rangle}{\langle I_{\mathrm{inc}}\rangle} \simeq \frac{F_{\mathrm{R}}(1/L_{\mathrm{a}}) + F_{\mathrm{R}}[(q^2 + 1/L_{\mathrm{a}}^2)^{1/2}]}{F_{\mathrm{R}}(1/L_{\mathrm{a}})}, \tag{9}$$

where $\mathbf{q} \equiv \mathbf{q}_b$ and we have taken into account the absorption of light in the medium by introducing the macroscopic absorption length $L_{\mathrm{a}} \gg \ell$.

3. Waves in Homogeneous Nonlinear Media

If $\delta\varepsilon(\mathbf{r}, t) = 0$, Eq. (1) describes the propagation of waves in a *homogeneous* nonlinear medium (no scattering) with an intensity-dependent dielectric constant [12]. Below we consider the case of Kerr nonlinearity. If the nonlinear response of the medium is instantaneous and local, the dependence of $\Delta\varepsilon_{\mathrm{NL}}(\mathbf{r}, t)$ on $I(\mathbf{r}, t)$ is rather simple: $\Delta\varepsilon_{\mathrm{NL}}(\mathbf{r}, t) = \varepsilon_2 I(\mathbf{r}, t)$, where ε_2 is a nonlinear coefficient. In general, however, the nonlinear response is not instantaneous and can be modeled by a phenomenological equation of Debye type for $\Delta\varepsilon_{\mathrm{NL}}$:

$$\tau_{\mathrm{NL}} \frac{\partial}{\partial t} \Delta\varepsilon_{\mathrm{NL}}(\mathbf{r}, t) = -\Delta\varepsilon_{\mathrm{NL}}(\mathbf{r}, t) + \varepsilon_2 I(\mathbf{r}, t), \tag{10}$$

where τ_{NL} is the response time of the nonlinearity. Moreover, the nonlocality of the nonlinear response of the medium can be taken into account by considering $\Delta\varepsilon_{\mathrm{NL}}(\mathbf{r}, t)$ obeying an equation of diffusion type:

$$\tau_{\mathrm{NL}} \frac{\partial}{\partial t} \Delta\varepsilon_{\mathrm{NL}}(\mathbf{r}, t) = (a_{\mathrm{NL}})^2 \nabla^2 \left[\Delta\varepsilon_{\mathrm{NL}}(\mathbf{r}, t)\right] - \Delta\varepsilon_{\mathrm{NL}}(\mathbf{r}, t) + \varepsilon_2 I(\mathbf{r}, t), \tag{11}$$

where a_{NL} is some characteristic length, describing the degree of nonlocality of the nonlinear response.

The physics of nonlinear optical phenomena in Kerr media is rather rich [12] and it is not our purpose to review it here. In the framework of our study, we would like, however, to call the attention of the reader to the following two simple but illustrative examples. First, if a nonlinear Kerr medium, illuminated by a monochromatic plane wave of intensity I_0, is put in an optical resonator, the feedback provided by the resonator leads to the instability of the steady-state solution $E(\mathbf{r}, t) = E_0(\mathbf{r}) \exp(-i\omega t)$ and the intensity $I = |E|^2$ of the wave develops a time dependence [22]. This phenomenon occurs only at I_0 exceeding some threshold and the dynamics of $I(\mathbf{r},t)$ becomes progressively more complex as I_0 increases. The second example concerns two counter-propagating light waves in a Kerr medium. Again, the steady-state solution develops an instability for the intensities of the waves exceeding a threshold [23]. The above simple examples suggest that the unstable regimes are fundamental in nonlinear optics and this appears to be indeed true: instabilities, self-oscillations, pattern formation, and spatio-temporal chaos are observable in many nonlinear optical systems (see, e.g., Refs. [24]–[26] for reviews). Even incoherent light beams have been recently shown to exhibit pattern formation and 'optical turbulence' due to the so-called modulation instability [27], indicating that unstable behavior does not require the perfect coherence of the underlying wave field.

4. Angular Correlations and Coherent Backscattering for Waves in Nonlinear Disordered Media

After having briefly reviewed the key features of wave propagation in linear disordered and homogeneous nonlinear media, we are now in a position to attack the central problem of the present paper: diffuse wave propagation in nonlinear disordered media. A particularly pictorial description of the latter can be developed if the nonlinearity is assumed to be weak enough to have no significant effect on the general diffuse character of wave propagation in the medium and the values of the diffusion constant D and the mean free path ℓ. This requires the 'nonlinear scattering length' ℓ_{NL} due to the nonlinear term in Eq. (1) to be much larger than ℓ. ℓ_{NL} can be estimated by assuming the short-range intensity correlation function in a nonlinear disordered medium to be approximately the same is in the linear one [see Eq. (3)]. Calculating the total scattering crossection of statistically isotropic fluctuations of the nonlinear dielectric constant $\varepsilon_2 \delta I(\mathbf{r})$ in the Born approximation, we obtain

$$\ell_{\mathrm{NL}}^{-1} \propto k_0^2 \int_0^{2k_0} \Phi(\kappa)\kappa d\kappa, \tag{12}$$

where

$$\Phi(\mathbf{K}) \propto \varepsilon_2^2 \int \langle \delta I(\mathbf{r}) \delta I(\mathbf{r}+\Delta\mathbf{r}) \rangle \exp(-i\mathbf{K}\Delta\mathbf{r}) d^3\Delta\mathbf{r}, \tag{13}$$

and hence $\ell_{\mathrm{NL}} \sim (\Delta n^2 k_0)^{-1}$, where $\Delta n = n_2 I_0$ is the typical value of the nonlinear correction to the refractive index, $n_2 = \varepsilon_2/2$, and I_0 is the typical value of $\langle I(\mathbf{r},t)\rangle$. The condition of weak nonlinearity, $\ell \ll \ell_{\mathrm{NL}}$, then becomes

$$\Delta n^2 k_0 \ell \ll 1, \tag{14}$$

where, we recall, $k_0\ell \gg 1$. We adopt this condition throughout the rest of the paper.

We start the analysis of the effect of nonlinearity on multiple scattering of waves in disordered media by considering the angular correlation of scattered wave fields, $C_{aba'b'}$, as defined in Eq. (5) (Sec. 2), assuming the two incident waves to have different amplitudes (A for the wave with the wave vector $\mathbf{k}_a$ and A' for the wave with the wave vector $\mathbf{k}'_a$). Due to the nonlinearity of the medium, $C_{aba'b'}$ will now depend not only on $\mathbf{k}_a$, $\mathbf{k}'_a$, $\mathbf{k}_b$, and $\mathbf{k}'_b$, but also on A and A'. For simplicity, we put $A' \to 0$ (so that the effect of the nonlinearity on the propagation of the second wave is negligible) and limit ourselves to the case of instantaneous ($\tau_{\text{NL}} = 0$) local ($a_{\text{NL}} = 0$) nonlinearity.

The physical origin of the nonlinearity-induced modification of $C_{aba'b'}$ can be easily understood in the framework of the path-integral picture of wave propagation. The wave of amplitude A, propagating along some wave path of length s through a nonlinear disordered medium, acquires an additional, 'nonlinear' phase shift $\Delta\phi_{\text{NL}}$ with respect to the wave of amplitude $A' \to 0$, propagating in an essentially linear medium:

$$\Delta\phi_{\text{NL}} = k_0 n_2 \int_0^s ds_1 I(\mathbf{r}_1), \tag{15}$$

the average value of the latter being

$$\langle \Delta\phi_{\text{NL}} \rangle \approx k_0 n_2 A^2 s, \tag{16}$$

where we assume $\langle I(\mathbf{r}_1) \rangle \approx A^2$. In a linear medium, a similar phase difference arises between two waves at different frequencies $\omega \neq \omega'$:

$$\Delta\phi_{\Delta\omega} = (\Delta\omega/c)s, \tag{17}$$

where $\Delta\omega = \omega - \omega' \ll \omega$. It is well-known that the latter dephasing modifies the angular correlation function of scattered fields, which becomes (in transmission) [8]

$$C_{aba'b'} = \delta_{\Delta\mathbf{q}_a, \Delta\mathbf{q}_b} F_{\text{T}} \left[(\Delta q_a^2 - 2i\gamma^2)^{1/2} L \right], \tag{18}$$

where $\gamma^2 = 3\Delta\omega/(2\ell c)$.

The formal analogy between Eqs. (16) and (17) suggests that as long as the nonlinearity affects only the phase of the wave and not its amplitude (i.e. for a weak nonlinearity), the correlation function $C_{aba'b'}$ in a nonlinear disordered medium should be approximately given by Eq. (18), where $\Delta\omega$ is replaced by $\omega n_2 A^2$. This is indeed confirmed by a perturbative diagrammatic calculation [17] [see the diagram of Fig. 1(a)], that allows one

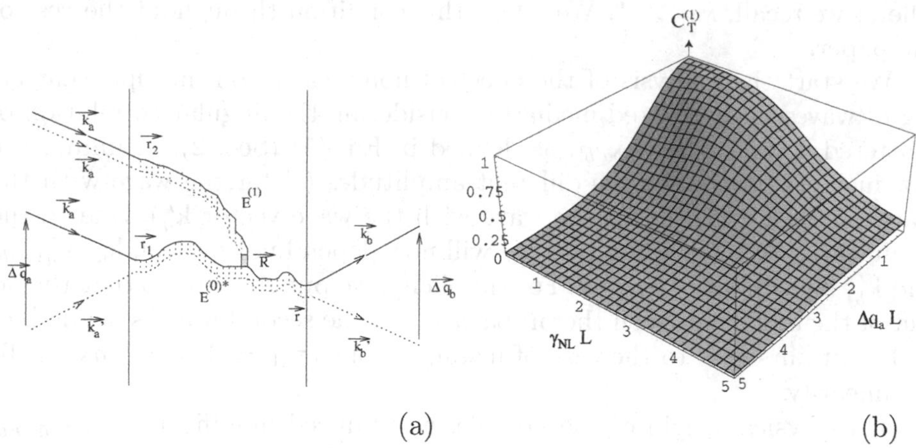

(a) (b)

Figure 1. (a) The diagram contributing to the angular correlation function of scattered wave fields in a nonlinear disordered medium. (b) Short-range intensity correlation function for a diffuse wave transmitted through a slab. The figures are from Ref. [17].

to obtain the lowest term in the series expansion of $C_{aba'b'}$ in $n_2A^2 \ll 1$. Applying then a 'natural' ansatz $f(x) - [i\epsilon^2/(2x)]f'(x) \simeq f[(x^2 - i\epsilon^2)^{1/2}]$ yields [17]:

$$C_{aba'b'} = \delta_{\Delta\mathbf{q}_a,\Delta\mathbf{q}_b} F_{\mathrm{T}} \left\{ \left[\Delta q_a^2 L^2 - 2i(L/\xi)^2 \right]^{1/2} \right\}, \tag{19}$$

where ξ is a new characteristic length, characterizing the damping of correlation due to the nonlinear effects:

$$\xi \approx \sqrt{\frac{\ell}{k_0 \Delta n}}. \tag{20}$$

Here $\Delta n \approx n_2 A^2$ is the typical value of the nonlinear part of the refractive index. The physical meaning of ξ is clear: it is the characteristic length for the loss of phase coherence between two waves following the same diffusion path, one with a vanishing amplitude $A' \to 0$ and the other one with a finite amplitude A. By analogy with the case of frequency correlation [Eq. (18)], γ_{NL} can be defined as $\gamma_{\mathrm{NL}} = 1/\xi$. The short-range correlation function of intensity fluctuations, $C_{\mathrm{T}}^{(1)}$, can be obtained by taking a square of the absolute value of Eq. (19). We show this correlation function in Fig. 1(b).

Similarly to the case of transmission geometry, correlation between diffusely reflected waves can also be calculated taking into account the additional dephasing due to the nonlinearity of the medium. Again, assuming that the two incident waves have different amplitudes A and $A' \to 0$, one

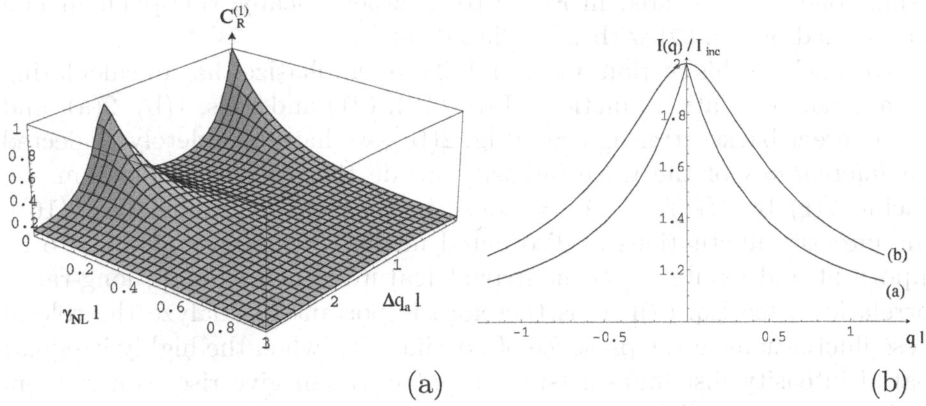

Figure 2. (a) Short-range angular correlation function of the fluctuations of diffusely reflected intensity in a semi-infinite nonlinear disordered medium. (b) Line shape of the coherent backscattering cone in a semi-infinite nonlinear disordered medium for $L_{\rm a}^{\rm NL} \to \infty$ [vanishing linear and nonlinear absorption, curve (a)] and $L_{\rm a}^{\rm NL} = 10\ell$ [curve (b)]. The figures are from Ref. [17].

finds for a semi-infinite disordered medium [17]

$$\begin{aligned} C_{aba'b'} &\simeq \delta_{\Delta\mathbf{q}_a,\Delta\mathbf{q}_b} \left\{ F_{\rm R} \left[\left(\Delta\mathbf{q}_a^2 - 2i(1+\zeta)/\xi^2 \right)^{1/2} \right] \right. \\ &+ \left. F_{\rm R} \left[\left(|\mathbf{q}_a + \mathbf{q}_b + \Delta\mathbf{q}_a|^2 - 2i(1+\zeta)/\xi^2 \right)^{1/2} \right] \right\}. \end{aligned} \quad (21)$$

This result is to be compared to the 'linear' result (7). In Fig. 2(a) we show the experimentally accessible short-range correlation function of intensity fluctuations, $C_{\rm R}^{(1)}$, obtained by taking a square of the absolute value of Eq. (21). Just as in a linear medium, two peaks occur at $\Delta\mathbf{q}_a = 0$ and $\mathbf{q}_a + \mathbf{q}_b + \Delta\mathbf{q}_a = 0$, respectively.

Finally, the coherent backscattering cone can be calculated at the same level of approximation [17]. If the nonlinear coefficient ε_2 in Eq. (1) is purely real, no deviation from the linear result (9) is found within the present theoretical framework. The contribution of the crossed diagrams to the intensity of backscattered wave is, however, modified if ε_2 has a small imaginary part (i.e. in the presence of nonlinear absorption). This phenomenon can be described by introducing a generalized absorption length $L_{\rm a}^{\rm NL}$, accounting for both linear and nonlinear absorption:

$$\frac{1}{(L_{\rm a}^{\rm NL})^2} \approx \frac{1}{L_{\rm a}^2} + \frac{k_0}{\ell}(1+\zeta)A^2 {\rm Im}\varepsilon_2. \quad (22)$$

Additional absorption, introduced by the nonlinear term in Eq. (1), will lead to the rounding of the triangular peak shape of the coherent backscattering cone, as illustrated in Fig. 2(b). Mathematically, this phenomenon is described by Eq. (9) with L_a replaced by L_a^{NL}.

To conclude this section, we would like to emphasize that in calculating the angular correlation functions [Eqs. (19), (21) and Figs. 1(b), 2(a)] and the coherent backscattering cone [Fig. 2(b)], we have completely neglected the *fluctuations* of the wave intensity inside the disordered medium, replacing $I(\mathbf{r})$ by $\langle I(\mathbf{r})\rangle \approx A^2 = \text{const}$ in the very beginning [Eq. (16)]. The intensity fluctuations in disordered media are, however, known to be important and exhibit some nontrivial features [such as, e.g., long-range correlations, see Eq. (4)]. It is therefore important to analyze the role of these fluctuations in the presence of nonlinearity, when the highly irregular spatial intensity distribution (speckle pattern) can give rise to a random modification of the dielectric constant (and, consequently, refractive index). Such an analysis is presented in the following section.

5. Instability of Diffuse Waves in Nonlinear Disordered Media

As we already mentioned in Sec. 3, unstable regimes are encountered in many nonlinear optical systems. It is therefore natural to expect that they may appear in a disordered nonlinear medium as well. The origin of the instability in this case is relatively easy to understand in the framework of the path-integral picture of wave propagation [16].

5.1. HEURISTIC DESCRIPTION

In the path-integral picture, the diffusely scattered wave field at some position $\mathbf{r}$ inside the disordered medium is represented as a sum of partial waves traveling along various diffuse paths. Due to the nonlinearity of the medium, the phase $\phi(\mathbf{r}, t)$ of a given partial wave arriving at $\mathbf{r}$ at time t depends on the spatio-temporal distribution of intensity $I(\mathbf{r}, t)$ (speckle pattern) inside the medium. At the same time, $I(\mathbf{r}, t)$ is a result of interference of many partial waves, and, therefore, $I(\mathbf{r}, t)$ is sensitive to their phases $\phi(\mathbf{r}, t)$. This leads to a sort of feedback mechanism: The phases of partial waves depend on the speckle pattern $I(\mathbf{r}, t)$, while the latter depends on the phases. It appears that such a feedback can destabilize the time-independent speckle pattern $I(\mathbf{r})$, leading to $I(\mathbf{r}, t)$ that fluctuates spontaneously with time, if the nonlinearity is strong enough [16]. As we show below, this phenomenon can be understood in reasonably simple terms in the case of local ($a_{NL} = 0$) and instantaneous ($\tau_{NL} = 0$) Kerr nonlinearity.

Consider a partial wave traveling along a typical diffuse path of length $s_0(t) \sim L^2/\ell$ from the source of waves to some point $\mathbf{r}$ located at a distance L from the source. We assume that the path length $s_0(t)$ can vary slowly with time due to, e.g., the motion of scattering centers [corresponding to a time-dependent disorder $\delta\varepsilon(\mathbf{r},t)$ in Eq. (1)], while the initial and final points of the path are fixed. The variations of s_0 with time are assumed to be slow (i.e. s_0 does not change significantly during the time $T_D = L^2/D$ required for the wave to cover the distance of order s_0). The phase $\phi(\mathbf{r},t)$ of the wave contains a 'linear' contribution $\phi_L(\mathbf{r},t) = k_0 s_0(t)$ and the 'nonlinear' one

$$\phi_{NL}(\mathbf{r},t) = k_0 \int_0^{s_0(t)} ds_1 [\Delta\varepsilon_{NL}(\mathbf{r}_1,t_1)/2] = k_0 n_2 \int_0^{s_0(t)} ds_1 I(\mathbf{r}_1,t_1), \quad (23)$$

where $\Delta\varepsilon_{NL}(\mathbf{r}_1,t_1)/2 = n_2 I(\mathbf{r}_1,t_1)$ is the nonlinear part of the refractive index, s_1 is a curvilinear coordinate of the point $\mathbf{r}_1$ along the path, $t_1 = t - [s_0(t) - s_1]/c$ is the time of wave passage through $\mathbf{r}_1$, and the integrals are along the path.

Let the fluctuating part of the linear dielectric constant $\delta\varepsilon(\mathbf{r},t)$ to change slightly during some time interval Δt: $\delta\varepsilon(\mathbf{r},t) \to \delta\varepsilon(\mathbf{r},t+\Delta t) = \delta\varepsilon(\mathbf{r},t) + \Delta[\delta\varepsilon(\mathbf{r},t)]$, leading to corresponding changes $\Delta I(\mathbf{r},t,\Delta t)$ and $\Delta\phi(\mathbf{r},t,\Delta t)$ of intensity and phase, respectively. The latter can be, in principle, found from Eq. (1). $\Delta\phi$ contains the linear and nonlinear parts, $\Delta\phi_L$ and $\Delta\phi_{NL}$, respectively. Under very general assumptions, one can show that $\langle\Delta I\rangle = 0$ and $\langle\Delta\phi\rangle = 0$, where the averaging is over $\delta\varepsilon(\mathbf{r},t)$, $\Delta[\delta\varepsilon(\mathbf{r},t)]$, and over all paths of the same length s_0. The second moment $\langle\Delta\phi^2\rangle = \langle\Delta\phi_L^2\rangle + \langle\Delta\phi_{NL}^2\rangle > 0$, where $\langle\Delta\phi_L^2\rangle$ is calculated in the framework of the so-called diffusing-wave spectroscopy (DWS) [5] and depends on the way in which $\delta\varepsilon(\mathbf{r},t)$ is modified. If, for example, the disordered medium is a suspension of Brownian particles (particle diffusion coefficient D_B), $\langle\Delta\phi_L^2\rangle = (\Delta t/\tau_0)(s_0/\ell)$, where $\tau_0 = (4k_0^2 D_B)^{-1}$ [5]. For the variance of the nonlinear phase difference, Eq. (23) yields

$$\langle\Delta\phi_{NL}^2\rangle = k_0^2 n_2^2 \int_0^{s_0} ds_1 \int_0^{s_0} ds_2 \langle\Delta I(\mathbf{r}_1,t_1)\Delta I(\mathbf{r}_2,t_2)\rangle, \quad (24)$$

where $t_2 = t_1 + \Delta t$ and both integrals are along the same diffusion path. Obviously,

$$\begin{aligned} \langle\Delta I(\mathbf{r}_1,t_1)\Delta I(\mathbf{r}_2,t_2)\rangle &= 2\left[\langle\delta I(\mathbf{r}_1,t)\delta I(\mathbf{r}_2,t)\rangle - \langle\delta I(\mathbf{r}_1,t)\delta I(\mathbf{r}_2,t+\Delta t)\rangle\right] \\ &\simeq \langle\Delta\phi_L^2\rangle \langle\delta I(\mathbf{r}_1,t)\delta I(\mathbf{r}_2,t)\rangle, \end{aligned} \quad (25)$$

where the second line is obtained by assuming that $\langle\delta I(\mathbf{r}_1,t)\delta I(\mathbf{r}_2,t+\Delta t)\rangle$ in the nonlinear medium is close to its value in the linear one (a sort of perturbation theory), replacing $\langle\delta I(\mathbf{r}_1,t)\delta I(\mathbf{r}_2,t+\Delta t)\rangle$ by the 'linear' result:

$$\langle\delta I(\mathbf{r}_1,t)\delta I(\mathbf{r}_2,t+\Delta t)\rangle \simeq \langle\delta I(\mathbf{r}_1,t)\delta I(\mathbf{r}_2,t)\rangle \exp\left[-(1/2)\langle\Delta\phi_L^2\rangle\right], \quad (26)$$

and assuming $\langle\Delta\phi_L^2\rangle \ll 1$.

Substituting Eq. (25) into Eq. (24) and changing the variables of integration to $s = (s_1 + s_2)/2$ and $\Delta s = s_1 - s_2$, we obtain

$$\langle \Delta\phi_{\rm NL}^2 \rangle = \langle \Delta\phi_{\rm L}^2 \rangle \, k_0^2 n_2^2 \int_0^{s_0} ds \int_{-s}^{s} d\Delta s \, \langle \delta I(\mathbf{r}_1) \delta I(\mathbf{r}_2) \rangle \, . \tag{27}$$

As we discussed in Sec. 2, the correlation function of intensity fluctuations $\langle \delta I(\mathbf{r}_1) \delta I(\mathbf{r}_2) \rangle$ in Eq. (27) contains two contributions: a short-range one [Eq. (3)] and a long-range one [Eq. (4)]. To perform the integration in Eq. (27), we replace $\Delta r = |\mathbf{r}_1 - \mathbf{r}_2|$ in the expression (3) for the short-range correlation function by Δs, assuming the wave path to be ballistic at distances shorter than ℓ. In contrast, for $\Delta r > \ell$ the wave path is diffusive, and hence Δr in the expression (4) for the long-range correlation function can be substituted by $(\Delta s \ell)^{1/2}$. Performing then integrations in Eq. (27) yields

$$\langle \Delta\phi_{\rm NL}^2 \rangle \simeq p \langle \Delta\phi_{\rm L}^2 \rangle \, , \tag{28}$$

where we introduce the *bifurcation parameter* [16]

$$p = \Delta n^2 \left(\frac{L}{\ell} \right)^2 \left(k_0 \ell + \frac{L}{\ell} \right) , \tag{29}$$

and the numerical factors of order unity are omitted.

Note that Eq. (29) is a sum of two contributions. The first one, $\Delta n^2 (L/\ell)^2 k_0 \ell$, originates from the short-range correlation of intensity fluctuations [Eq. (3)], while the second contribution, $\Delta n^2 (L/\ell)^3$, is due to the long-range correlation [Eq. (4)]. If $p \ll 1$, the nonlinear term $\langle \Delta\phi_{\rm NL}^2 \rangle$ represents just a small correction to the linear one $\langle \Delta\phi_{\rm L}^2 \rangle$ and our perturbation approach is likely to be valid. In contrast, the above perturbation theory diverges for $p > 1$, since $\langle \Delta\phi_{\rm NL}^2 \rangle$ becomes larger than $\langle \Delta\phi_{\rm L}^2 \rangle$. This suggests that $p \simeq 1$ is a critical point beyond which (i.e. for $p > 1$) the physics of the nonlinear problem is no longer similar to the physics of the linear one and hence the perturbation theory cannot be used. Although the mere breakdown of the perturbation theory does not permit to draw any far-reaching conclusions about the speckle pattern beyond the critical point $p \simeq 1$, a more rigorous analysis (see Secs. 5.2 and 5.3) shows that $p \simeq 1$ defines the instability threshold of the multiple-scattering speckle pattern, and that at $p > 1$ the latter should exhibit spontaneous temporal fluctuations even for a time-independent disorder $\delta\varepsilon(\mathbf{r})$ (i.e. for vanishing mobility of scattering centers: $D_{\rm B} \to 0$ and $\tau_0 \to \infty$ in the case of Brownian motion).

It is pertinent to note the extensive nature of the bifurcation parameter p. According to Eq. (29), weak nonlinearity can be efficiently compensated by a sufficiently large sample size L and $p \sim 1$ can be reached even at vanishing Δn, provided that L is large enough. In the limit of large sample size $L \gg k_0 \ell^2$, one gets $p \simeq \Delta n^2 (L/\ell)^3$ (as also found in Ref. [28]). This result is due to the long-range correlation (4) of intensity fluctuations which dominates in the integral of Eq. (27) because of its slow decrease ($\propto |\mathbf{r}_1 - \mathbf{r}_2|^{-1}$) and consequent extension over the whole disordered sample. Another remarkable feature of Eq. (29) is the independence of p of the sign of Δn. The phenomenon of the speckle pattern instability is therefore expected to develop in a similar way for both positive and negative nonlinear coefficients n_2. This is not common for the instabilities in nonlinear systems without disorder [22]–[27] since the latter are often related to the self-focusing effect, arising at $n_2 > 0$ only. The instability discussed in the present paper is of different type and has nothing to do with the self-focusing [which is negligible if the condition (14) is satisfied]. This does not mean, however, that the phenomena similar to those discussed in the present paper do not exist in homogeneous nonlinear media. In fact, it is easy to see that Eq. (23) describes nothing but the *self-phase modulation* [12] of the multiple-scattering speckle pattern. It is worth noting that the development of the self-phase modulation in a disordered medium appears to be rather different from that in the homogeneous case. Indeed, in a homogeneous medium of size L the nonlinear phase shift is *deterministic* and $\phi_{\text{NL}} = k_0 n_2 I_0 L$ for a wave of intensity I_0, while in the case of diffuse multiple scattering ϕ_{NL} is *random* with the average value $\langle \phi_{\text{NL}} \rangle \simeq k_0 n_2 I_0 (L^2/\ell)$ and variance $\langle \phi_{\text{NL}}^2 \rangle - \langle \phi_{\text{NL}} \rangle^2 \simeq p$. The threshold of the speckle pattern instability $p \simeq 1$ is then simply the point where the sample-to-sample fluctuations of the nonlinear phase shift become of order unity.

To conclude this subsection, we would like to comment on the recent results on modulation instability (MI) of incoherent light beams in homogeneous nonlinear media [27]. In this case, a 2D speckle pattern (speckled beam) with a controlled degree of spatial and temporal coherence is 'prepared' by sending a coherent laser beam on a rotating diffuser. The beam is then incident on a homogeneous nonlinear medium (inorganic photorefractive crystal). Pattern formation and 'optical turbulence' is observed if the intensity of the beam exceeds a specific threshold, determined by the degree of coherence of the beam (coherent MI has no threshold). Although the study presented in this paper is also concerned with the instability of speckle patterns in nonlinear media, it differs essentially from that of Ref. [27] at least in the following important aspects: (a) the initially coherent wave loses its spatial coherence *inside* the medium, due to the multiple scattering on heterogeneities of the refractive index, (b) the temporal coherence of the incident light is perfect (in the case of incoherent MI [27], the coherence time of the incident beam is much shorter than the response time of the medium τ_{NL}), and (c) the incident light beam is completely destroyed by the scattering after a distance $\sim \ell$ and light propagation is diffusive in the bulk of the sample. In addition, diffuse multiple scattering of light results in a distributed feedback mechanism, absent in the case of MI.

5.2. PATH-INTEGRAL PICTURE

A mathematically rigorous formulation of the heuristic analysis presented in the previous subsection has been performed in Ref. [16]. In addition to the condition of weak nonlinearity (14) we anticipate that the expected spontaneous fluctuations of the speckle pattern beyond the instability threshold will be slow enough or, more precisely, that the typical time of the speckle pattern dynamics will be much longer than the typical time between two successive scattering events ℓ/c.

For the sake of the mathematical simplicity, we consider a plane monochromatic wave incident on the stationary [$\delta\varepsilon(\mathbf{r})$ is time-independent] semi-infinite disordered medium ($L \to \infty$) with a finite (macroscopic) absorption length $L_a \gg \ell$ and a local ($a_{NL} = 0$) instantaneous ($\tau_{NL} = 0$) Kerr nonlinearity. It appears that instead of replacing the intensity correlation function $\langle \delta I(\mathbf{r}_1, t) \delta I(\mathbf{r}_2, t + \Delta t) \rangle$ in Eq. (25) by its 'linear' value (and thus constructing a sort of perturbation theory), one can obtain a self-consistent description of the speckle instability. The analysis [16] leads to a self-consistent equation for $\beta = -\ln g_1$, where $g_1(\tau) = \langle E(t) E^*(t+\tau) \rangle / \left\langle |E(t)|^2 \right\rangle$ is the time autocorrelation function of diffusely reflected wave, and g_1 denotes $\lim_{\tau\to\infty} g_1(\tau)$:

$$\exp(-\beta) = F(\beta), \tag{30}$$

where $F(\beta)$ is a known monotonically decaying function and $F(0) = 1$.

Fig. 3(a) illustrates the graphical solution of Eq. (30) for some realistic values of parameters. At strong enough absorption [two upper curves in Fig. 3(a)], a unique solution $\beta = \beta_1 = 0$ exists, corresponding to $g_1 = \exp(-\beta) = 1$. This means that the time autocorrelation function of diffusely reflected wave $g_1(\tau)$ does not decay with time, remaining equal to 1 for all τ (even for $\tau \to \infty$). The speckle pattern is therefore stationary (i.e. does not change with time). In contrast, for weak absorption [three lower curves in Fig. 3(a)] a second solution $\beta = \beta_2 > 0$ arises, corresponding to $g_1 = \exp(-\beta) < 1$. Eq. (30) now has two fixed points: $\beta = \beta_1 = 0$ and $\beta = \beta_2 > 0$. It is seen from Fig. 3(a) (see thick solid lines with arrows) that an iterative solution of Eq. (30) starting at some $\beta > 0$ converges to $\beta = \beta_1 = 0$ for strong enough absorption and to $\beta = \beta_2 > 0$ for weak absorption. $\beta = \beta_2 > 0$ is therefore a stationary point of Eq. (30) in the

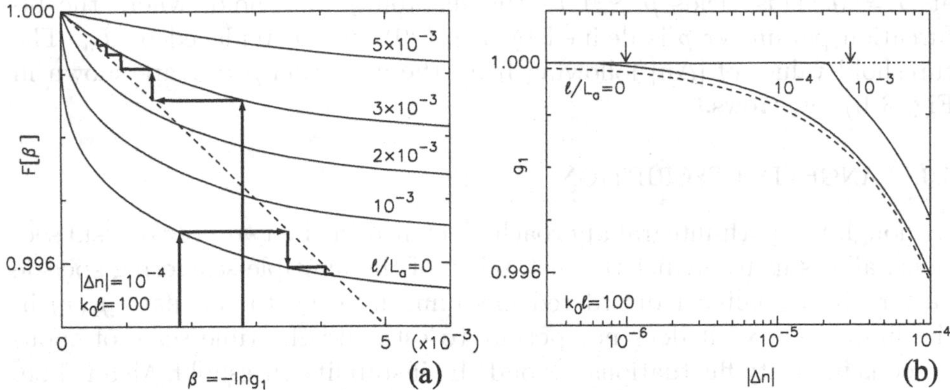

Figure 3. (a) Graphical solution of Eq. (30) at $k_0\ell = 100$, $|\Delta n| = 10^{-4}$ and the values of $\ell/L_{\rm a}$ indicated near each curve. The solid lines show $F(\beta)$, the dashed line is $\exp(-\beta)$. Thick solid lines with arrows illustrate the iterative solution of Eq. (30). (b) Bifurcation diagram of the speckle pattern in a semi-infinite disordered medium for $k_0\ell = 100$ and the values of $\ell/L_{\rm a}$ indicated near the each curve. The threshold values of $|\Delta n|$, following from the condition $p = 1$ at $\ell/L_{\rm a} > 0$, are indicated by arrows. In the absence of absorption, $\ell/L_{\rm a} = 0$ (dashed line), there is no threshold and the speckle pattern is unstable at any (even infinitely small!) $|\Delta n|$. The figures are from Ref. [16].

latter case, corresponding to the physically realizable solution beyond the instability threshold. $\beta > 0$ and $g_1 < 1$ signifies that the time autocorrelation function of diffusely reflected wave $g_1(\tau)$ decays with τ despite our assumption of stationary disorder [time-independent $\delta\varepsilon(\mathbf{r})$], although the present analysis does not allow us to estimate the characteristic time scale of this decrease. Decaying $g_1(\tau)$ corresponds to an *unstable* (i.e. time-varying) speckle pattern. Note that in contrast to the fluctuations of the speckle pattern due to the motion of scattering centers, employed in DWS to study the dynamics of the medium [5], the speckle dynamics in a stationary nonlinear random medium is *spontaneous*, i.e. it does not originate from the motion of scattering centers (since the latter are immobile) but is intrinsic for the underlying nonlinear wave equation (1). Spontaneous speckle dynamics is irreversible and hence the instability threshold $p \simeq 1$ is the point where the time-reversal symmetry is spontaneously broken for the multiple-scattered waves.

Fig. 3(b) shows the time autocorrelation function $g_1 = \exp(-\beta)$ obtained by solving Eq. (30) at $k_0\ell = 100$ and several values of $\ell/L_{\rm a}$. The transition from the stable ($g_1 = 1$) to unstable ($g_1 < 1$) regime is clearly seen when $\ell/L_{\rm a} > 0$. At $\ell/L_{\rm a} = 0$ (no absorption), we find that $g_1 < 1$ at any, even infinitely small $|\Delta n|$. The border between stable ($g_1 = 1$) and unstable ($g_1 < 1$) regimes can be found analytically by requiring $(\partial/\partial\beta)F(\beta) = -1$

at $\beta = 0$. This yields $p \simeq 1$ as the instability threshold, where the bifurcation parameter p is defined by Eq. (29) with L replaced by L_a. The threshold values of $|\Delta n|$ following from the condition $p = 1$ are shown in Fig. 3(b) by arrows.

5.3. LANGEVIN DESCRIPTION

Although the path-integral approach, sketched in the two previous subsections, allows us to predict the instability of the multiple-scattering speckle pattern in a nonlinear disordered medium and even to calculate $g_1(\tau)$ in the limit $\tau \to \infty$, it does not permit to estimate the time scale of spontaneous intensity fluctuations beyond the instability threshold. Meanwhile, this issue is of primary importance in view of the possible experimental observation of the instability phenomenon. In this subsection, we show that the spontaneous speckle dynamics can be studied by using the Langevin approach [18, 19].

As in the previous subsections, we assume the nonlinearity to be local ($a_{\mathrm{NL}} = 0$), but consider arbitrary nonlinearity response time τ_{NL}. The nonlinear part of the dielectric constant is assumed to be governed by Eq. (10). In addition, we neglect the absorption (assuming $L_a \gg L$) and restrict our consideration to the case when the development of the speckle pattern instability is dominated by the long-range intensity correlations. This requires the sample size to be large enough [$L \gg k_0\ell^2$, see Eq. (29) and its accompanying discussion] and the speckle dynamics to be slow (time scale of spontaneous intensity fluctuations $\tau \gg T_{\mathrm{D}}[(k_0\ell^2)/L]^2$). The latter condition follows from Eq. (27), where the limits of integration over Δs should be set to $\mp c\tau$, if $c\tau \ll s_0$ (or, equivalently, $\tau \ll T_{\mathrm{D}}$), since the correlation has simply no time to establish beyond these limits due to the finite speed of wave propagation. Integration in Eq. (27) then yields

$$\langle \Delta\phi_{\mathrm{NL}}^2 \rangle \simeq \Delta n^2 \left(\frac{L}{\ell}\right)^2 \left[k_0\ell + \frac{L}{\ell}\left(\frac{\tau}{T_{\mathrm{D}}}\right)^{1/2} \right] \langle \Delta\phi_{\mathrm{L}}^2 \rangle, \tag{31}$$

where the first term in square brackets [the same as in Eq. (29)] results from the short-range correlation (3), while the second term is due to the long-range one (4). Requiring that the second term dominates the first, we obtain $\tau \gg T_{\mathrm{D}}[(k_0\ell^2)/L]^2$ as stated above.

The Langevin equation for the intensity fluctuation $\delta I(\mathbf{r}, t)$ reads [4]:

$$\frac{\partial}{\partial t}\delta I(\mathbf{r}, t) - D\nabla^2 \delta I(\mathbf{r}, t) = -\nabla \cdot \mathbf{j}_{\mathrm{ext}}(\mathbf{r}, t), \tag{32}$$

where $\mathbf{j}_{\mathrm{ext}}(\mathbf{r}, t)$ are random external Langevin currents that have zero mean and a correlation function $\left\langle j_{\mathrm{ext}}^{(i)}(\mathbf{r}, t) j_{\mathrm{ext}}^{(j)}(\mathbf{r}_1, t_1) \right\rangle$ given by the diagram (i)

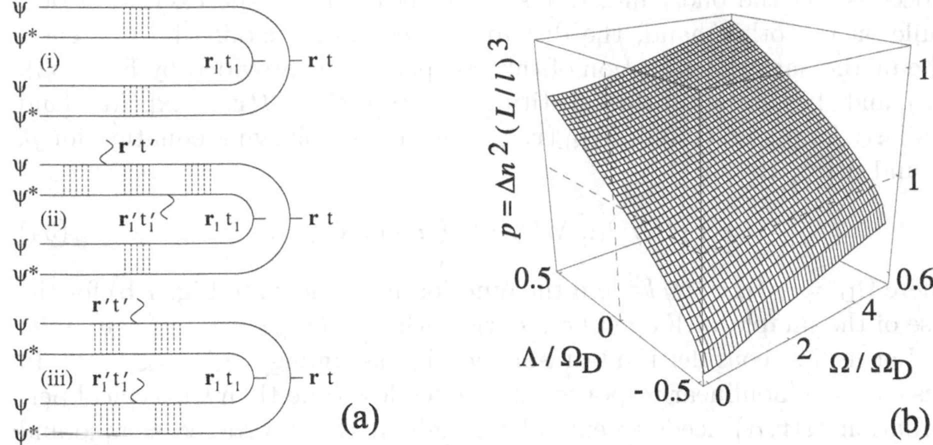

Figure 4. (a) Diagrams contributing to the correlation functions of Langevin currents $\mathbf{j}_{\text{ext}}(\mathbf{r}, t)$ [diagram (i)] and random response functions $\mathbf{q}(\mathbf{r}, \mathbf{r}', \Delta t)$ [diagrams (ii) and (iii)]. The diagrams (ii) and (iii) are obtained by the functional differentiation of the diagram (i) with respect to the dielectric constant of the disordered medium. Wavy lines in the diagrams (ii) and (iii) denote k_0^2 factors. (b) Surface describing the stability of the multiple-scattering speckle pattern in a disordered medium with fast nonlinearity ($\Omega_D \tau_{NL} \ll 1$). For given frequency Ω and bifurcation parameter p, the surface shown in the figure allows one to determine the value of the Lyapunov exponent Λ. If $\Lambda > 0$, the speckle pattern is unstable with respect to excitations at frequency Ω. The figures are from Ref. [18].

of Fig. 4(a). Being a 'fingerprint' of disorder, the Langevin currents will be modified if we modify the dielectric constant of the medium. In our case, the linear part of the dielectric constant is fixed [$\delta\varepsilon(\mathbf{r})$ in Eq. (1) is time-independent], but its nonlinear part $\Delta\varepsilon_{\text{NL}}(\mathbf{r}, t)$ can vary with time, and it is relatively easy to obtain the following dynamic equation for $\mathbf{j}_{\text{ext}}(\mathbf{r}, t)$ [18, 19]:

$$\frac{\partial}{\partial t}\mathbf{j}_{\text{ext}}(\mathbf{r}, t) = \int_V d^3\mathbf{r}' \int_0^\infty d\Delta t\, \mathbf{q}(\mathbf{r}, \mathbf{r}', \Delta t)\, \frac{\partial}{\partial t}\Delta\varepsilon_{\text{NL}}(\mathbf{r}', t - \Delta t), \qquad (33)$$

where $\mathbf{q}(\mathbf{r}, \mathbf{r}', \Delta t)$ is a random response function with zero mean and the correlation function $\langle q^{(i)}(\mathbf{r}, \mathbf{r}', \Delta t) q^{(j)*}(\mathbf{r}_1, \mathbf{r}_1', \Delta t_1)\rangle$ given by a sum of the diagrams (ii) and (iii) of Fig. 4(a). Eqs. (32) and (33) together with Eq. (10) for $\Delta\varepsilon_{\text{NL}}$ form a self-consistent system of equations for the stability analysis of the speckle pattern.

Consider now an infinitesimal periodic excitation of the stationary speckle pattern: $\delta I(\mathbf{r}, t) = \delta I(\mathbf{r}, \alpha) \exp(\alpha t)$, where $\alpha = i\Omega + \Lambda \neq 0$ and $\Omega > 0$. Such an excitation can be either damped or amplified, depending on the sign of the Lyapunov exponent Λ. The value of Λ is determined by two competing processes: on the one hand, diffusion tends to smear the excitation out, while on the other hand, the distributed feedback sustains its existence. The mathematical description of this competition is provided by Eqs. (32), (33), and (10) that after the substitution of $\delta I(\mathbf{r},t) = \delta I(\mathbf{r},\alpha)\exp(i\alpha t)$ [and similarly for $\mathbf{j}_{\text{ext}}(\mathbf{r},t)$ and $\Delta\varepsilon_{\text{NL}}(\mathbf{r},t)$] lead to the following equation for p, Ω, and Λ [18, 19]:

$$p \simeq F_1\left(\Omega/\Omega_{\text{D}}, \Lambda/\Omega_{\text{D}}\right) F_2\left(\Omega\tau_{\text{NL}}, \Lambda\tau_{\text{NL}}\right), \tag{34}$$

where $\Omega_{\text{D}} = 1/T_{\text{D}} = D/L^2$ and the function F_1 is shown in Fig. 4(b) for the case of the sample with open boundaries, while $F_2(x,y) = 1 + x^2 + y^2 + 2y$.

Let us first consider the fast nonlinearity, assuming $\tau_{\text{NL}}\Omega_{\text{D}} \ll 1$ [18]. In this case, the nonlinear response takes much less time than the typical perturbation $\delta I(\mathbf{r},\alpha)$ needs to extend throughout the disordered sample and we can set $F_2\left(\Omega\tau_{\text{NL}}, \Lambda\tau_{\text{NL}}\right) \simeq 1$ in Eq. (34). The stability of the speckle pattern with respect to periodic excitations is then described by Fig. 4(b). For a given frequency Ω, the sign of the Lyapunov exponent Λ depends on the value of p. Excitations at frequencies Ω corresponding to $\Lambda < 0$ are damped exponentially and thus soon disappear. In contrast, excitations at frequencies Ω corresponding to $\Lambda > 0$ are exponentially amplified, which signifies the instability of the speckle pattern with respect to excitations at such frequencies. Noting that Λ is always negative for $p < 1$, we conclude that all excitation are damped in this case and the speckle pattern is absolutely stable. In an experiment, any spontaneous excitation of the static speckle pattern will be suppressed and the speckle pattern will be independent of time: $\delta I(\mathbf{r},t) = \delta I(\mathbf{r})$, as in the linear case. When $p > 1$, an interval of frequencies $0 < \Omega < \Omega_{\text{max}}$ starts to open up with $\Lambda > 0$. The speckle pattern thus becomes unstable with respect to excitations at low frequencies. In an experiment, any spontaneous excitation of the static speckle pattern at frequency $\Omega \in (0, \Omega_{\text{max}})$ will be amplified and one will observe a time-varying speckle pattern $\delta I(\mathbf{r},t)$. Note that the absolute instability threshold $p \simeq 1$ agrees with the results of Secs. 5.1 and 5.2, while completely different calculation techniques have been applied in the three cases.

Fig. 5(a) shows the frequency-dependent instability threshold for the disordered sample with open boundaries and fast nonlinear response. Detailed analysis [18] of Eq. (34) shows that the frequency-dependent threshold value of p scales as $1 + C_1(\Omega/\Omega_{\text{D}})^2$ (where $C_1 \sim 1$ is a numerical constant) at $\Omega \ll \Omega_{\text{D}}$ and as $(\Omega/\Omega_{\text{D}})^{1/2}$ at $\Omega \gg \Omega_{\text{D}}$. The latter result can also be obtained from Eq. (31) by setting $\tau \sim 1/\Omega$ and applying $\langle\Delta\phi_{\text{NL}}^2\rangle > \langle\Delta\phi_{\text{L}}^2\rangle$ as the instability condition. For a given value of $p > 1$, the maximum excited frequency is $\Omega_{\text{max}} \sim \Omega_{\text{D}}(p-1)^{1/2}$ for $p - 1 \ll 1$ and $\Omega_{\text{max}} \sim \Omega_{\text{D}}p^2$ for $p - 1 > 1$.

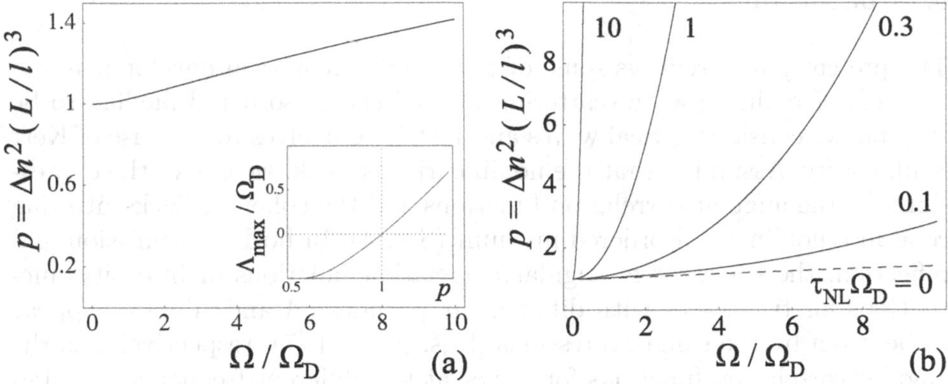

Figure 5. (a) Frequency-dependent instability threshold for a disordered sample with instantaneous nonlinear response and open boundaries. The inset shows the maximal Lyapunov exponent as a function of the bifurcation parameter p. (b) The same as (a) but for a sample with a nonzero nonlinearity response time τ_{NL}. The dashed line shows the result corresponding to $\tau_{\mathrm{NL}}\Omega_{\mathrm{D}} = 0$ [the same as the solid line in the panel (a)]. The figures are from Ref. [18] (a) and [19] (b).

Fig. 5(b) illustrates the effect of noninstantaneous nonlinearity on the frequency-dependent instability threshold for several values of $\tau_{\mathrm{NL}}\Omega_{\mathrm{D}}$ [19]. Obviously, slow nonlinear response of the medium rises the instability threshold for high-frequency excitations, while the absolute instability threshold $p \simeq 1$ remains the same, independent of the nonlinearity response time τ_{NL}. The reason for this is that just above $p = 1$ the speckle pattern becomes unstable with respect to excitations at very low frequencies, for which $\Omega \ll 1/\tau_{\mathrm{NL}}$ is always fulfilled and which, therefore, are not sensitive to the value of τ_{NL} [mathematically, $F_2(\Omega\tau_{\mathrm{NL}}, \Lambda\tau_{\mathrm{NL}}) \simeq 1$ in Eq. (34)]. In the limit of slow nonlinearity ($\tau_{\mathrm{NL}}\Omega_{\mathrm{D}} \gg 1$) and for $1 < p - 1 \ll 1$ we can set $F_1(\Omega/\Omega_{\mathrm{D}}, \Lambda/\Omega_{\mathrm{D}}) \simeq 1$ in Eq. (34) and find the threshold value of p to scale as $1 + (\Omega\tau_{\mathrm{NL}})^2$ and $\Omega_{\max} \sim \tau_{\mathrm{NL}}^{-1}(p-1)^{1/2}$.

As follows from the above analysis, a continuous low-frequency spectrum of frequencies $(0, \Omega_{\max})$ is excited at $p > 1$. In addition, the Lyapunov exponent Λ appears to decrease monotonically with Ω, favoring no specific frequency Ω just above the threshold [19]. This allows us to hypothesize that at $p = 1$ the speckle pattern undergoes a transition from a stationary to chaotic state. Such a behavior should be contrasted from the 'route to chaos' through a sequence of bifurcations, characteristic of many nonlinear and, in particular, optical systems [22]–[26].

6. Conclusion

The present paper reviews some of the recent theoretical developments in the field of multiple wave scattering in nonlinear disordered media. To be specific, we consider optical waves and restrict ourselves to the case of Kerr nonlinearity. Assuming that the nonlinearity is weak, we derive the expressions for the angular correlation functions and the coherent backscattering cone in a nonlinear disordered medium (Sec. 4). In both transmission and reflection, the short-range angular correlation functions of intensity fluctuations for two waves with different amplitudes (A and $A' \to 0$) appear to be given by the same expressions [Eqs. (6) and (7), respectively] as the angular correlation functions for waves at two different frequencies (ω and $\omega' = \omega - \Delta\omega$) in a linear medium, with $\Delta\omega$ replaced by $2\ell c/(3\xi^2)$, where ξ is a new *nonlinear* characteristic length defined by Eq. (20). The coherent backscattering cone is not affected by the nonlinearity, as long as the nonlinear coefficient ε_2 in Eq. (1) is purely real. If ε_2 has an imaginary part (which corresponds to the nonlinear absorption), the line shape of the cone is given by the same expression (9) as in an absorbing linear medium, where the linear macroscopic absorption length L_{a} should be replaced by the *generalized* absorption length $L_{\mathrm{a}}^{\mathrm{NL}}$ defined by Eq. (22).

For the nonlinearity strength exceeding a threshold $p \simeq 1$ [with the bifurcation parameter p given by Eq. (29)], we predict a new phenomenon — temporal instability of the multiple-scattering speckle pattern — to take place (Sec. 5). The instability is due to a combined effect of the nonlinear self-phase modulation and the distributed feedback mechanism provided by multiple scattering and should manifest itself in spontaneous fluctuations of the speckle pattern with time. Since the spontaneous dynamics of the speckle pattern is irreversible, the time-reversal symmetry is spontaneously broken when p surpasses 1. The important feature of our result is the extensive nature of the instability threshold, leading to an interesting possibility of obtaining unstable regimes even at very weak nonlinearities, provided that the disordered sample is large enough. To study the dynamics of multiple-scattering speckle patterns beyond the instability threshold, we generalize the Langevin description of wave diffusion in disordered media (Sec. 5.3). Explicit expressions for the characteristic time scale of spontaneous intensity fluctuations are derived with account for the noninstantaneous nature of the nonlinearity. The results of this study allow us to hypothesize that the dynamics of the speckle pattern may become chaotic immediately beyond the instability threshold, and that the cascade of bifurcations, typical for chaotic transitions in many known nonlinear systems [22]–[26], might not be present in the considered case of diffuse waves.

Finally, we discuss the experimental implications of our results. With common nonlinear materials, such as, e.g., carbon disulfide, $n_2 \simeq 3 \times 10^{-14}$ cm^2/W can be realized [12]. At $I \simeq 1$ MW/cm^2 this yields $\Delta n \simeq 3 \times 10^{-8}$ and the characteristic length ξ defined in Eq. (20) is $\xi \simeq 2$ cm for $\ell \simeq 100$ μm and $\lambda \simeq 0.5$ μm. As follows from Eq. (19), $\xi \lesssim L$ is required to observe a sizeable effect of the nonlinearity on the angular correlation function of transmitted light, and hence using a 2 cm-thick disordered sample should suffice to make the effect of nonlinearity measurable in, e.g., a dense suspension of carbon disulfide particles. In contrast, $\xi \lesssim \ell$ is necessary to observe the effect of the nonlinearity on the angular correlation function in the reflection geometry. This requires a stronger nonlinearity. In nematic liquid crystals, for example, Δn up to 0.1 is achievable (see, e.g., Ref. [29]). Taking $\Delta n = 10^{-3}$ and $\ell \sim 1$ mm [30], we get $\xi \sim 1$ mm $\sim \ell$. In the case of the coherent backscattering cone, very accurate experimental techniques developed recently to measure the angular dependence of backscattered light [31] give a hope that the effect of the nonlinearity can be observable without any particular problem. Finally, observation of the temporal instability of multiple-scattering speckle patterns, predicted in Sec. 5, will require, first of all, large enough sample size L and as low absorption as possible. In the absence of absorption (or for the macroscopic absorption length $L_a \gtrsim L$), $\Delta n_{\text{NL}} \sim 10^{-2}$ (realistic in nematic liquid crystals [29]) and $L/\ell \sim 20$ will suffice to get $p \simeq 1$ and reach the instability threshold. We note that in liquid crystals, the nonlinearity is due to the reorientation of molecules under the influence of the electric field of the electromagnetic wave and hence is essentially noninstantaneous. This emphasizes the importance of including the noninstantaneous nature of nonlinear response in our analysis (Sec. 5.3).

References

1. Van Albada, M.P. and Lagendijk, A. (1985) Observation of weak localization of light in a random medium, *Phys. Rev. Lett.* **55,** 2692–2695; Wolf, P.-E. and Maret, G. (1985) Weak localization and coherent backscattering of photons in disordered media, *Phys. Rev. Lett.* **55,** 2696–2699.
2. Shapiro, B. (1986) Large intensity fluctuations for wave propagation in random media, *Phys. Rev. Lett.* **57,** 2168–2171.
3. Stephen, M.J. and Cwilich, G. (1987) Intensity correlation functions and fluctuations in light scattered from a random medium, *Phys. Rev. Lett.* **59,** 285–287.
4. Zyuzin, A.Yu. and Spivak, B.Z. (1987) Langevin description of mesoscopic fluctuations in disordered media, *Sov. Phys. JETP* **66,** 560–566; Pnini, R. and Shapiro B. (1989) Fluctuations in transmission of waves through disordered slabs, *Phys. Rev. B* **39,** 6986–6994.
5. Maret, G. and Wolf, P.-E. (1987) Multiple light scattering from disordered media. The effect of Brownian motion of scatterers, *Z. Phys. B* **65,** 409–413; Pine, D.J., Weitz, D.A., Chaikin, P.M., and Herbolzheimer, E. (1988) Diffusing wave spectroscopy, *Phys. Rev. Lett.* **60,** 1134–1137.
6. Sheng, P. (ed.) (1990) *Scattering and Localization of Classical Waves in Random*

Media, World Scientific, Singapore.
7. Lagendijk, A. and Van Tiggelen, B.A. (1996) Resonant multiple scattering of light, *Phys. Rep.* **270,** 143–215.
8. Berkovits, R. and Feng S. (1994) Correlations in coherent multiple scattering, *Phys. Rep.* **238,** 135–172.
9. Van Rossum, M.C.W. and Nieuwenhuizen, Th.M. (1999) Multiple scattering of classical waves: microscopy, mesoscopy, and diffusion, *Rev. Mod. Phys.* **71,** 313–371.
10. Sebbah, P. (ed.) (2001) *Waves and Imaging through Complex Media,* Kluwer Academic Publishers, Dordrecht.
11. Altshuler, B.L., Lee, P.A., and Webb, R.A. (eds.) (1991) *Mesoscopic Phenomena in Solids,* Elsevier, Amsterdam; S. Datta (1995) *Electronic transport in mesoscopic systems,* Cambridge University Press, Cambridge.
12. Boyd, R.W. (2002) *Nonlinear Optics,* Academic Press, New York; Bloembergen, N. (1996) *Nonlinear Optics,* World Scientific, Singapore.
13. Hamilton, M.F. and Blackstock, D.T. (1998) *Nonlinear Acoustics,* Academic Press, New York; Morse, P.M. and Ingard, K.U. (1986) *Theoretical Acoustics,* Ch. 14, Princeton University Press, Princeton.
14. Kandidov, V.P. (1996) Monte Carlo method in nonlinear statistical optics, *Physics-Uspekhi* **39,** 1243–1272; Yahel, R.Z. (1990) Turbulence effects on high energy laser beam propagation in the atmosphere, *Appl. Opt.* **29,** 3088–3095; Strohbehn, J.W. (ed.) (1978) *Laser Beam Propagation in the Atmosphere,* Springer-Verlag, Berlin.
15. De Boer, J.F., Lagendijk, A., Sprik, R., and Feng, S. (1993) Transmission and reflection correlations of second harmonic waves in nonlinear random media, *Phys. Rev. Lett.* **71,** 3947–3950; Yoo, K.M., Lee, S., Takiguchi, Y., and Alfano, R.R. (1989) Search for the effect of weak photon localization in second-harmonic waves generated in a disordered anisotropic nonlinear medium, *Opt. Lett.* **14,** 800–801.
16. Skipetrov, S.E. and Maynard, R. (2000) Instabilities of waves in nonlinear disordered media, *Phys. Rev. Lett.* **85,** 736–739; Skipetrov, S.E. (2001) Temporal fluctuations of waves in weakly nonlinear disordered media, *Phys. Rev. E* **63,** 056614.
17. Bressoux, R. and Maynard, R. (2000) On the speckle correlation in nonlinear random media, *Europhys. Lett.* **50,** 460–465; Bressoux, R. and Maynard, R. (2001) Speckle correlations and coherent backscattering in nonlinear random media, in Ref. [10], 445–453.
18. Skipetrov, S.E. (2003) Langevin description of speckle dynamics in nonlinear disordered media, *Phys. Rev. E* **67,** 016601.
19. Skipetrov, S.E. (2003) Instability of speckle patterns in random media with noninstantaneous Kerr nonlinearity, *Opt. Lett.,* to appear.
20. Ishimaru, A. (1978) *Wave Propagation and Scattering in Random Media,* Academic Press, New York.
21. Shapiro, B. (1999) New type of intensity correlation in random media, *Phys. Rev. Lett.* **83,** 4733–4735; Skipetrov, S.E. and Maynard, R. (2000) Nonuniversal correlations in multiple scattering, *Phys. Rev. B* **62,** 886–891.
22. Ikeda, K., Daido, H., and Akimoto, O. (1980) Optical turbulence: Chaotic behavior of transmitted light from a ring cavity, *Phys. Rev. Lett.* **45,** 709–712; Nakatsuka, H., Asaka, S., Itoh, H., Ikeda, K., and Matsuoka M. (1983) Observation of bifurcation to chaos in an all-optical bistable system, *Phys. Rev. Lett.* **50,** 109–112.
23. Silberberg, Y. and Bar Joseph, I. (1982) Instabilities, self-oscillation, and chaos in a simple nonlinear optical interaction, *Phys. Rev. Lett.* **48,** 1541–1543; Silberberg, Y. and Bar Joseph, I. (1984) Optical instabilities in a nonlinear Kerr medium, *J. Opt. Soc. Am. B* **1,** 662–670.
24. Gibbs, H.M. (1985) *Optical Bistability: Controlling Light With Light,* Academic Press, New York.
25. Arecchi, F.T., Boccaletti, S., and Ramazza, P. (1999) Pattern formation and competition in nonlinear optics, *Phys. Rep.* **318,** 1–83.
26. Vorontsov, M.A. and Miller, W.B. (eds.) (1999) *Self-Organization in Optical Systems*

and Applications in Information Technology, Springer Verlag, Berlin.

27. Soljacic, M., Segev, M., Coskun, T., Christodoulides, D.N., and Vishwanath, A. (2000) Modulation instability of incoherent beams in noninstantaneous nonlinear media, *Phys. Rev. Lett.* **84,** 467–470; Kip, D., Soljacic, M., Segev, M., Eugenieva, E., and Christodoulides, D.N. (2000) Modulation instability and pattern formation in spatially incoherent light beams, *Science* **290,** 495–498; Klinger, J., Martin, H., and Chen, Z. (2001) Experiments on induced modulational instability of an incoherent optical beam, *Opt. Lett.* **26,** 271–273; Chen, Z., Sears, S.M., Martin, H., Christodoulides, D.N., and Segev, M. (2002) Clustering of solitons in weakly correlated wavefronts, *Proc. Nat. Acad. Sci.* **99,** 5223–5227.
28. Spivak, B. and Zyuzin, A. (2000) Mesoscopic sensitivity of speckles in disordered nonlinear media to changes of the scattering potential, *Phys. Rev. Lett.* **84,** 1970–1973.
29. Muenster, R., Jarasch, M., Zhuang, X., and Shen, Y.R. (1997) Dye-induced enhancement of optical nonlinearity in liquids and liquid crystals, *Phys. Rev. Lett.* **78,** 42–45.
30. Kao, M.H., Jester, K.A., Yodh, A.G., and Collings, P.J. (1996) Observation of light diffusion and correlation transport in nematic liquid crystals, *Phys. Rev. Lett.* **77,** 2233–2236.
31. Wiersma, D.S., Van Albada, M.P., and Lagendijk, A. (1995) An accurate technique to record the angular distribution of backscattered light, *Rev. Sci. Inst.* **66,** 5473–5476.

Olivier Sigwarth, Ad Lagendijk, Sergey Skipetrov, Valery Loiko, and Bart van Tiggelen (from left to right). Photo by Valery Loiko

CHAPTER II

MESOSCOPIC PHYSICS AND LOCALIZATION

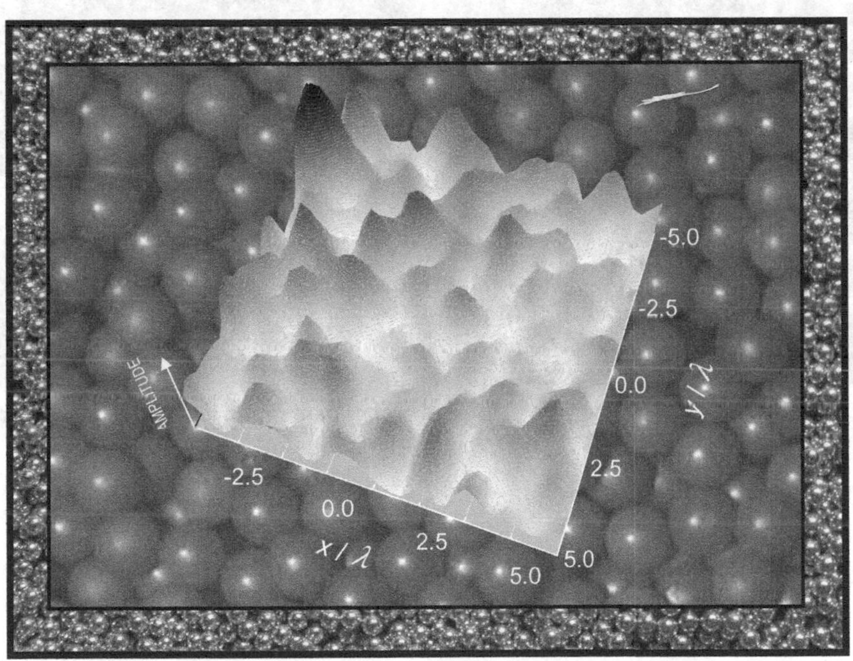

Near-field speckle pattern a few millimeters above the surface of a random suspension of beads in water (center), surrounded by two images of the beads. Image provided by John Page, University of Manitoba, Winnipeg, Canada.

Eric Akkermans. Photo by Valentin Freilikher

COHERENT EFFECTS IN THE MULTIPLE SCATTERING OF LIGHT IN RANDOM MEDIA

E. AKKERMANS
Department of Physics, Technion, 32000 Haifa, Israel

AND

G. MONTAMBAUX
Laboratoire de physique des solides, Université Paris-Sud F-91405 Orsay Cedex, France

Abstract. We review some of the characteristic features of the coherent multiple scattering of scalar electromagnetic waves in random media. The probability of quantum diffusion is defined and calculated up to the contribution of the cooperon. We show that there are additional corrections at the order of the cooperon which restore the normalization of the probability. We study also the angular and temporal (diffusive wave spectroscopy) correlation functions of speckle patterns. More particularly, we obtain a closed expression of the contribution to the time correlation function which is equivalent to the universal conductance fluctuations. Finally, the notion of dephasing is discussed and implemented for the case of the dephasing induced by the internal Zeeman degrees of freedom of cold atomic gases.

1. Introduction

This contribution aims to give a general survey of the field of coherent multiple scattering of light by random media. This field crosses through many topics in physics. Nevertheless, apart from details which are specific to each particular physical problem, there is a large amount of common features which characterize coherent multiple scattering of waves ranging from mesoscopic metals to astrophysics. A unified description of the basic features of coherent multiple scattering have been presented elsewhere [1, 2]. In this short review, we shall mostly focus on the behavior of elec-

B.A. van Tiggelen and S.E. Skipetrov (eds.),
Wave Scattering in Complex Media: From Theory to Applications, 101–124.

tromagnetic waves scattering either in suspensions of classical scatterers or in quantum systems such as cold atomic gases.

In the next section we shall describe two models for randomness which are most frequently used and discuss their relation with the cross section of an individual scatterer. Then, we shall review the main results about the disorder average amplitude in multiple scattering in the weak scattering limit. A central quantity which is ubiquitous in all the physical results is *the probability of quantum diffusion.* We shall discuss it quite in details and subsequently apply it to the study of correlation functions in speckle patterns, the time correlation function in the multiple scattering limit (diffusive wave spectroscopy) and to a discussion of some processes leading to dephasing *i.e.* to the existence of a finite phase coherence time in the multiple scattering of photons by cold atomic gases. We shall then discuss this kind of dephasing and compare it to other mechanisms in mesoscopic metallic systems.

2. Scalar Waves in Random Media

We consider an electromagnetic wave which propagates in a non dissipative and non magnetic heterogeneous dielectric medium characterized by the real and positive dielectric constant $\epsilon(\mathbf{r}) = \overline{\epsilon}+\delta\epsilon(\mathbf{r})$. It fluctuates around the average value $\overline{\epsilon}$. By writing the corresponding Maxwell equations, we obtain a wave equation for the electric field. In a first step, we shall disregard the polarization of the field and consider the case of a scalar electric field described by the complex function $\psi(\mathbf{r})$ and solution of the Helmholtz equation

$$-\Delta\psi(\mathbf{r}) - k_0^2\mu(\mathbf{r})\psi(\mathbf{r}) = k_0^2\psi(\mathbf{r}) \tag{1}$$

The quantity $\mu(\mathbf{r}) = \delta\epsilon/\overline{\epsilon}$ is the relative fluctuation of the dielectric constant and $k_0 = \overline{n}\omega/c$ where $\overline{n} = \sqrt{\frac{\overline{\epsilon}}{\epsilon_0}}$ is the average refraction index. This equation is similar to the Schrödinger equation for a free particle in a potential $V(\mathbf{r})$. Here, the disorder potential is $V(\mathbf{r}) = -k_0^2\mu(\mathbf{r})$. It is proportional to the square of the frequency so that, unlike electronic systems, a lowering of the frequency leads to a weaker effect of the disorder. The disorder potential is a random and continuous function of the position. We can choose the origin of the energies so that $\langle V(\mathbf{r})\rangle = 0$ where $\langle\cdots\rangle$ accounts for the average over the configurations of the disorder. A simple approximation consists in assuming that $V(\mathbf{r})$ is a Gaussian random variable. Therefore, only the second cumulant does not vanish. Moreover, we shall assume that the wavelength is much larger than the correlation length of the potential so that the two-point correlation function may be written as

$$\langle V(\mathbf{r})V(\mathbf{r}')\rangle = B\delta(\mathbf{r}-\mathbf{r}') \tag{2}$$

where B is a constant. This model corresponds to the so-called white noise limit.

The white noise model does not contain information about the microscopic nature of the disorder. Another model first introduced by Edwards [3] describes the disorder potential as a collection of N_i identical localized scatterers each of them characterized by the scattering potential $v(\mathbf{r})$, namely

$$V(\mathbf{r}) = \sum_{j=1}^{N_i} v(\mathbf{r} - \mathbf{r}_j) \quad . \tag{3}$$

The density $n_i = \frac{N_i}{\Omega}$ is taken to be constant in the limit $\Omega \to \infty$. We shall moreover consider the scattering potential $v(\mathbf{r})$ to be central and short range, *i.e.*, $v(\mathbf{r} - \mathbf{r}_j) = v_0 \delta(\mathbf{r} - \mathbf{r}_j)$. In the limit of large densities $n_i \to \infty$ of weak scatterers $v_0 \to 0$ but with a constant value for $n_i v_0^2$, the Edwards model is equivalent to the white noise model provided we have the equality $B = n_i v_0^2$.

The total cross section is given, in the Born approximation, by $\sigma = v_0^2/4\pi$, so that the parameter B of the white noise model is simply related to the cross section of an individual scatterer:

$$B = 4\pi n_i \sigma \tag{4}$$

However, it is important to keep in mind the following point. In order to obtain the Edwards limit of a δ potential, we must consider the limits of both an infinite strength U_0 and a vanishing range b so that the combination $v_0 = 4\pi b^3 U_0/3$ is constant. Strictly speaking, the Born approximation breaks down in this limit. However, in the limit of a white noise model where $v_0 \to 0$, the Born approximation remains valid.

3. The Average Amplitude of a Multiply Scattered Wave

In the presence of sources $j(\mathbf{r})$ of the electric field, the solutions of the Helmholtz equation (1) may be written,

$$\psi(\mathbf{r}) = \int d\mathbf{r}_i j(\mathbf{r}_i) G(\mathbf{r}_i, \mathbf{r}, k_0) \tag{5}$$

where the Green function $G(\mathbf{r}_i, \mathbf{r}, k_0)$ is solution of [4]

$$\left(\Delta_{\mathbf{r}} + k_0^2(1 + \mu(\mathbf{r}))\right) G(\mathbf{r}_i, \mathbf{r}, k_0) = \delta(\mathbf{r} - \mathbf{r}_i) \tag{6}$$

This Green function can be expressed in terms of the free Green function G_0 solution of the previous equation but in the absence of the scattering

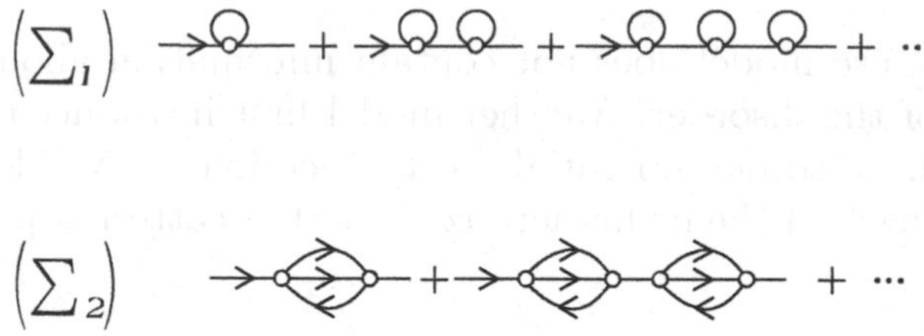

Figure 1. Series of terms respectively generated by Σ_1 and Σ_2 in the self-energy.

potential, *i.e.*, for $\mu(\mathbf{r}) = 0$,

$$G(\mathbf{r}_i, \mathbf{r}, k_0) = G_0(\mathbf{r}_i, \mathbf{r}, k_0) - k_0^2 \int d\mathbf{r}' G(\mathbf{r}_i, \mathbf{r}', k_0)\mu(\mathbf{r}')G_0(\mathbf{r}', \mathbf{r}, k_0) \qquad (7)$$

It is possible to write a formal multiple scattering expansion by iterating the previous relation. By taking the average over a white noise potential, we restore the translational invariance, so that the Fourier transform of the average Green function can be written

$$\overline{G}(\mathbf{k}) = G_0(\mathbf{k}) + \overline{G}(\mathbf{k})\Sigma(\mathbf{k}, \epsilon)G_0(\mathbf{k}) \qquad (8)$$

The function $\Sigma(\mathbf{k}, \epsilon)$ is called the self-energy and it represents the sum of all the irreducible scattering events.

The perturbative expansion of the self-energy is in terms of the parameter $n_i v_0^2$. For weak scattering the main contribution is obtained by keeping only the first contribution Σ_1 in Fig. 1. The imaginary part of the corresponding expression of the self-energy defines the elastic mean free path l_e, namely

$$\frac{1}{l_e} = -\frac{1}{k_0}\mathrm{Im}\Sigma_1(\mathbf{k}, k_0) = n_i \sigma \qquad (9)$$

where σ is the cross section defined in (4). The first neglected term Σ_2 in Fig. 1 gives a correction

$$\mathrm{Im}\Sigma_2(k) = \frac{\pi}{2kl_e}\mathrm{Im}\Sigma_1(k) \qquad (10)$$

Hence, it allows to identify the small dimensionless parameter $1/k_0 l_e$ which characterizes the weak scattering limit that we shall consider all along this paper. By keeping only the contribution of Σ_1, we neglect all the interference effects between scatterers (see Fig. 1).

Inserting (9) into (8) and performing a Fourier transform, one finds that the average Green function is given by

$$\overline{G}(\mathbf{r}_i, \mathbf{r}, \epsilon) = G_0(\mathbf{r}_i, \mathbf{r}, \epsilon)\ e^{-|\mathbf{r}-\mathbf{r}_i|/2l_e} = -\frac{e^{ik_0 R}}{4\pi R}\ e^{-R/2l_e} \tag{11}$$

where $R = |\mathbf{r} - \mathbf{r}_i|$. This expression corresponds to the retarded average Green function. The advanced one is obtained by changing the sign of k_0 in the previous expression.

4. The Probability of Quantum Diffusion

The quantities of physical interest are usually related not to the average Green function but instead to the so-called *probability of quantum diffusion* $P(\mathbf{r}, \mathbf{r}', t)$ for a wave packet to propagate between the points $\mathbf{r}$ and $\mathbf{r}'$ in a time t. The average over the disorder of the Fourier transform of this probability is given by [1, 2]

$$P(\mathbf{r}, \mathbf{r}', \omega) = \frac{4\pi}{c}\overline{G^R(\mathbf{r}, \mathbf{r}', \omega_0)G^A(\mathbf{r}', \mathbf{r}, \omega_0 - \omega)} \tag{12}$$

This probability is normalized to unity, namely

$$\int P(\mathbf{r}, \mathbf{r}', t)d\mathbf{r}' = 1 \tag{13}$$

or equivalently

$$\int P(\mathbf{r}, \mathbf{r}', \omega)d\mathbf{r}' = \frac{i}{\omega} \tag{14}$$

Three main contributions do appear in the probability $P(\mathbf{r}, \mathbf{r}', t)$:

- the probability to propagate between $\mathbf{r}$ and $\mathbf{r}'$ without scattering.
- the probability to propagate between $\mathbf{r}$ and $\mathbf{r}'$ by an incoherent sequence of multiple scattering. We shall call *diffuson* this contribution.
- the probability to go from $\mathbf{r}$ to $\mathbf{r}'$ by a coherent multiple scattering sequence. We shall call *cooperon* this coherent contribution to the probability.

The first contribution is obtained by replacing in (12) the average over the product of the two Green functions by the product of the averaged Green functions. This contribution known as the Drude-Boltzmann term rewrites as

$$P_0(\mathbf{r}, \mathbf{r}', \omega) = \frac{e^{i\omega R/c - R/l_e}}{4\pi R^2 c} \tag{15}$$

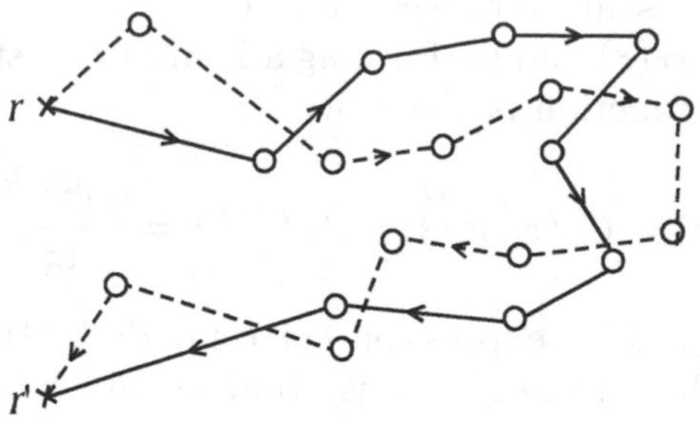

Figure 2. Typical trajectories associated respectively to the retarded (G^R) (solid line) and advanced (G^A) (dashed line) Green functions.

where c is the group velocity of the wave. It appears clearly that, at this approximation, the average probability is not normalized.

4.1. THE DIFFUSON

The second contribution to the probability accounts for multiple scattering in the weak disorder limit $kl_e \gg 1$. Following the previous description of the average Green function, we associate to each possible sequence $\mathcal{C}$ of independent effective collisions (Fig. 2) a complex amplitude $A(\mathbf{r}, \mathbf{r}', \mathcal{C})$. Hence, the Green function is given by the sum of such complex amplitudes [5].

In order to evaluate the probability which appears as a product of two Green functions, we notice first the following two points:

- due to the short range of the scattering potential, the set of scatterers entering in the multiple scattering sequences for both G^R and G^A must be identical.
- the average distance between two elastic collisions is set by the elastic mean free path $l_e \gg \lambda$. Therefore, if any two scattering sequences differ by even one collision event, the phase difference between the two complex amplitudes, which measures the difference of path lengths in units of λ will be very large and then the corresponding probability will vanish on average.

Therefore, we shall retain only the contributions of the type represented in

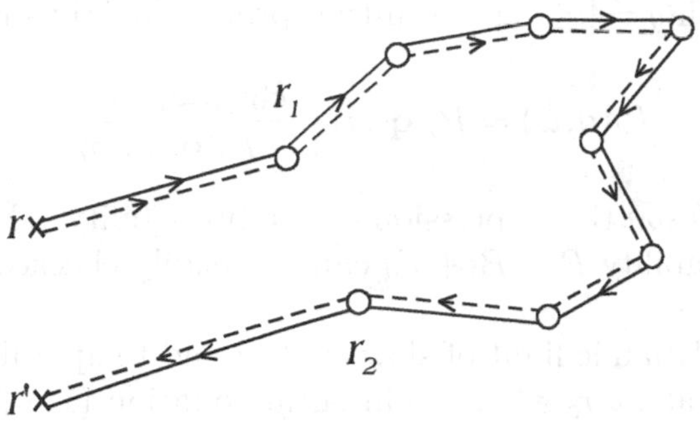

Figure 3. Multiple scattering trajectories which contribute to the average $\overline{G^R G^A}$.

Fig. 3 for which the corresponding probability $P_d(\mathbf{r}, \mathbf{r}', \omega)$ is

$$P_d(\mathbf{r}, \mathbf{r}', \omega) = \frac{4\pi}{c} \int \overline{G}^R_\epsilon(\mathbf{r}, \mathbf{r}_1) \overline{G}^A_{\epsilon-\omega}(\mathbf{r}_1, \mathbf{r}) \overline{G}^R_\epsilon(\mathbf{r}_2, \mathbf{r}') \overline{G}^A_{\epsilon-\omega}(\mathbf{r}', \mathbf{r}_2) \times \Gamma_\omega(\mathbf{r}_1, \mathbf{r}_2) d\mathbf{r}_1 d\mathbf{r}_2 \quad (16)$$

This expression involves two contributions. The first one

$$\overline{G}^R_\epsilon(\mathbf{r}, \mathbf{r}_1) \overline{G}^R_\epsilon(\mathbf{r}_2, \mathbf{r}') \overline{G}^A_{\epsilon-\omega}(\mathbf{r}_1, \mathbf{r}) \overline{G}^A_{\epsilon-\omega}(\mathbf{r}', \mathbf{r}_2)$$

describes the average propagation between any two points $\mathbf{r}$ and $\mathbf{r}'$ of the medium and the first $(\mathbf{r}_1)$ (respectively the last $(\mathbf{r}_2)$) collision event of the multiple scattering sequence. The second contribution defines $\Gamma_\omega(\mathbf{r}_1, \mathbf{r}_2)$ which we shall call the average *structure factor* of the scattering medium. In a sense, it generalizes to the multiple scattering situation the usual two-point correlation function in the single scattering case. Relying again on the assumption of independent collisions allows to write for $\Gamma_\omega(\mathbf{r}_1, \mathbf{r}_2)$ the integral equation

$$\Gamma_\omega(\mathbf{r}_1, \mathbf{r}_2) = \frac{4\pi}{l_e} \delta(\mathbf{r}_1 - \mathbf{r}_2) + \frac{4\pi}{l_e} \int \overline{G}^R_\epsilon(\mathbf{r}_1, \mathbf{r}) \overline{G}^A_{\epsilon-\omega}(\mathbf{r}, \mathbf{r}_1) \Gamma_\omega(\mathbf{r}, \mathbf{r}_2) d\mathbf{r} \quad (17)$$

This equation can be solved exactly for some geometries. For the infinite three dimensional space, we obtain for the structure factor the expression

$$\Gamma_\omega(\mathbf{q}) = \frac{4\pi}{l_e} \frac{1}{1 - P_0(\mathbf{q}, \omega)/\tau_e} \quad (18)$$

where $P_0(\mathbf{q},\omega)$ is the Fourier transform of (15) and it is given by $\frac{1}{qv}$ arctan $\frac{ql_e}{1-i\omega\tau_e}$ with $q=|\mathbf{q}|$. The resulting probability thus rewrites

$$P_d(\mathbf{q},\omega) = P_0(\mathbf{q},\omega)\frac{P_0(\mathbf{q},\omega)/\tau_e}{1-P_0(\mathbf{q},\omega)/\tau_e} \tag{19}$$

We shall call *diffuson* this expression of the probability. The normalization of the total probability $P = P_0 + P_d$ can be readily checked namely $P(\mathbf{q} = 0,\omega) = \frac{i}{\omega}$.

In the hydrodynamic limit of slow spatial and temporal variations, *i.e.*, for $|\mathbf{r}-\mathbf{r}_2| \gg l_e$ and $\omega\tau_e \ll 1$, the integral equation (17) for Γ_ω simplifies and is solution of the diffusion equation

$$[-i\omega - D\Delta_{\mathbf{r}_1}]\,\Gamma_\omega(\mathbf{r}_1,\mathbf{r}_2) = \frac{4\pi c}{l_e^2}\delta(\mathbf{r}_1-\mathbf{r}_2) \tag{20}$$

where $D = \frac{1}{d}\frac{l_e^2}{\tau_e} = \frac{1}{d}v^2\tau_e$ is the diffusion coefficient. At this approximation, P_d and Γ_ω are related by

$$P_d(\mathbf{r},\mathbf{r}',\omega) \simeq \frac{l_e^2}{4\pi c}\Gamma_\omega(\mathbf{r},\mathbf{r}') \tag{21}$$

so that P_d, as well, obeys a diffusion equation.

4.2. THE COOPERON

The two previous contributions, namely the Drude-Boltzmann term and the diffuson which takes into account multiple scattering, provide a normalized expression of the probability. Therefore, we may think having exhausted all the contributions. But, consider now the multiple scattering sequences like those represented in Fig. 4. It corresponds to the product of Green functions of the kind we have considered before. But now the two identical trajectories are time reversed. It is clear that if these trajectories are closed on themselves, there is no phase difference left between them. This requires time-reversal invariance namely $G^{R,A}(\mathbf{r},\mathbf{r}',t) = G^{R,A}(\mathbf{r}',\mathbf{r},t)$. We shall call X_c the contribution of this process to the total probability. It can be evaluated as before and is given by

$$\begin{aligned} X_c(\mathbf{r},\mathbf{r}',\omega) &= \frac{4\pi}{c}\int \overline{G}_\epsilon^R(\mathbf{r},\mathbf{r}_1)\overline{G}_\epsilon^R(\mathbf{r}_2,\mathbf{r}')\overline{G}_{\epsilon-\omega}^A(\mathbf{r}',\mathbf{r}_1)\overline{G}_{\epsilon-\omega}^A(\mathbf{r}_2,\mathbf{r}) \\ &\quad \times \Gamma'_\omega(\mathbf{r}_1,\mathbf{r}_2)d\mathbf{r}_1 d\mathbf{r}_2 \end{aligned} \tag{22}$$

In the case of time reversal invariance we have,

$$\overline{G}_\epsilon^R(\mathbf{r}_1,\mathbf{r})\overline{G}_\epsilon^A(\mathbf{r}_1,\mathbf{r}) = \overline{G}_\epsilon^R(\mathbf{r}_1,\mathbf{r})\overline{G}_\epsilon^A(\mathbf{r},\mathbf{r}_1) \tag{23}$$

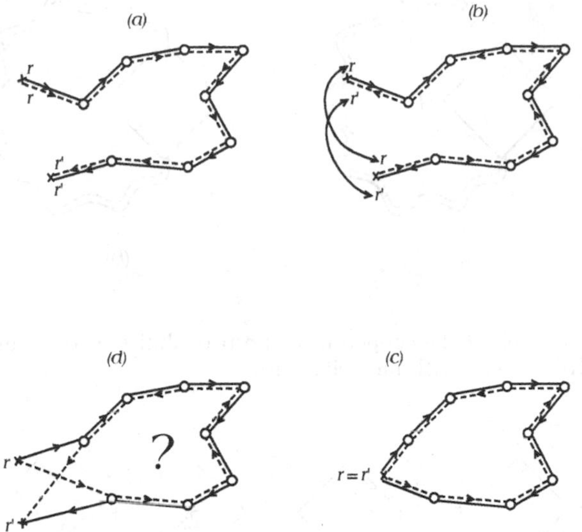

Figure 4. a) Classical contribution to the probability. b) By changing the direction of propagation of one of the trajectories there is still no overall dephasing provided that the two points **r** and **r**′ coincide (c). d) If $\mathbf{r} \neq \mathbf{r}'$, there is a finite dephasing between the two trajectories.

so that the new structure factor $\Gamma'_\omega(\mathbf{r}_1, \mathbf{r}_2) = \Gamma_\omega(\mathbf{r}_1, \mathbf{r}_2)$. In the diffusion approximation, by calculating the integrals (16) and (22), we obtain

$$X_c(\mathbf{r}, \mathbf{r}', \omega) = P_d(\mathbf{r}, \mathbf{r}, \omega) \left(\frac{\sin kR}{kR}\right)^2 e^{-R/l_e} \tag{24}$$

with $R = |\mathbf{r} - \mathbf{r}'|$. For $R = 0$, *i.e.*, for $\mathbf{r} = \mathbf{r}'$, we have

$$X_c(\mathbf{r}, \mathbf{r}, \omega) = P_d(\mathbf{r}, \mathbf{r}, \omega) \tag{25}$$

namely, the probability to come back to the initial point is *twice* the value given by the diffuson. Nevertheless, it should be noticed that unlike the diffuson P_d and the structure factor Γ_ω, the cooperon X_c does not obey a diffusion equation.

4.3. NORMALIZATION OF THE QUANTUM PROBABILITY

We have seen that the contributions of both the Drude-Boltzmann term and the diffuson end up with a normalized probability of quantum diffusion. Therefore, it is expected that the additional contribution of the

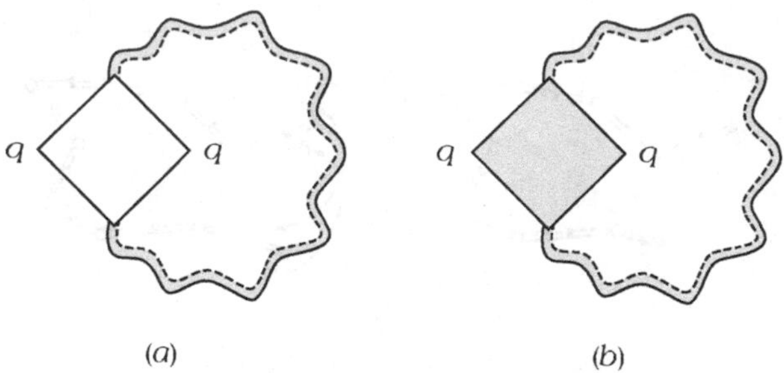

Figure 5. Contribution of the cooperon to the probability a) diagram for $X_c(\mathbf{q}, \omega)$. b) Dressing of the Hikami box with impurity lines.

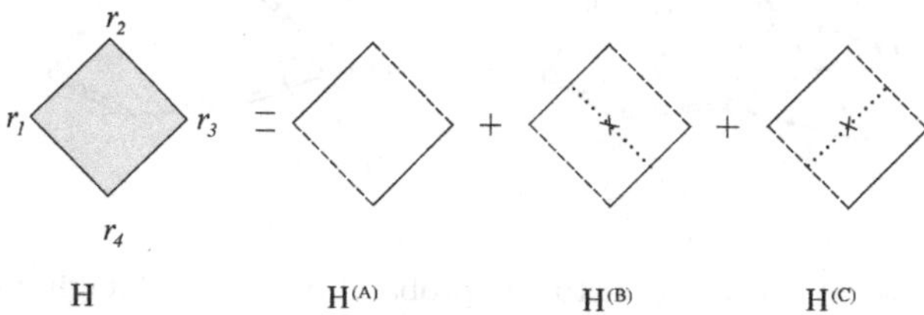

Figure 6. Hikami box.

cooperon will break this normalization. Indeed it appears, as a result of the previous calculation of the cooperon, that the probability is doubled at the origin in a small volume $\lambda^{d-1} l_e$. The relative volume of this enhancement is $\lambda^{d-1} l_e/(Dt)^{d/2}$. It has a maximum for $t \simeq \tau_e$ which corresponds to $1/(kl_e)^{d-1}$. But there exist other terms which contribute as well to the cooperon and which have not been taken into account so far. To identify them, we need to keep in mind the contribution to the cooperon represented by the diagram of Fig. 5. The box describes the interference between four amplitudes. This box can be dressed by one impurity line in two possible ways (Fig. 6). These two additional contributions are of the same order, therefore we should consider the three of them for the full calculation of the cooperon and eventually the normalization of the probability. The first diagram, *i.e.*, the cooperon gives the contribution

$$X_c(\mathbf{r}, \mathbf{r}', \omega) = \frac{4\pi}{c} H^{(A)}(\mathbf{r} - \mathbf{r}') \Gamma'(\mathbf{r}, \mathbf{r}', \omega) \tag{26}$$

where

$$H^{(A)}(\mathbf{R}) = \left(\frac{l_e}{4\pi}\right)^2 \left(\frac{\sin kR}{kR}\right)^2 e^{-R/l_e} \tag{27}$$

The two other diagrams have a negative contribution which at short distances $r \ll l_e$ is given by

$$H^{(B)}(\mathbf{R}) = H^{(C)}(\mathbf{R}) \propto -\left(\frac{l_e}{4\pi}\right)^2 \frac{1}{kl_e} \frac{\mathrm{Si}(2kR)}{kR} e^{-R/l_e} \tag{28}$$

where Si is the Sine integral function. These two diagrams give indeed at short distances a negligible contribution in comparison to (26). Hence, the return probability to the origin is doubled as obtained before. From the Fourier transforms

$$\begin{aligned} H^{(A)}(\mathbf{q}) &= \left(\frac{l_e}{4\pi}\right)^2 a(q) \\ H^{(B)}(\mathbf{q}) = H^{(C)}(\mathbf{q}) &= -\frac{c^2}{16\pi l_e} a(q)^2 \end{aligned} \tag{29}$$

where

$$a(q) = \frac{\pi}{k^2 q} \left[\arctan(2k-q)l_e + 2\arctan q l_e - \arctan(2k+q)l_e\right] \tag{30}$$

we obtain that the sum of these three contributions is

$$\int H(\mathbf{R}) d\mathbf{R} = 0 \tag{31}$$

Therefore the contribution of the cooperon which is to enhance the return probability to the origin does not change the normalization of the total probability which is achieved by a small reduction in the wings, *i.e.*, far enough from the origin although of the order of the elastic mean free path l_e.

5. Correlations in Speckle Patterns

Thus far, we have studied the disorder average value of the probability. The averaging procedure is obtained experimentally by considering suspensions of scatterers. An incident wave packet (a pulse) probes a static configuration of the scatterers. This results from the large ratio between the respective velocities of the light inside the medium and of the scatterers. Average quantities are thus obtained through time averaging. Hence, a pulse realizes an instantaneous picture of the disordered medium known as a speckle pattern. This pattern consists in a random distribution of bright and dark spots which signals large fluctuations of the relative intensity. This observation can be made more quantitative and the intensity distribution obeys a Rayleigh law which states that the intensity fluctuations are of the order of the average intensity.

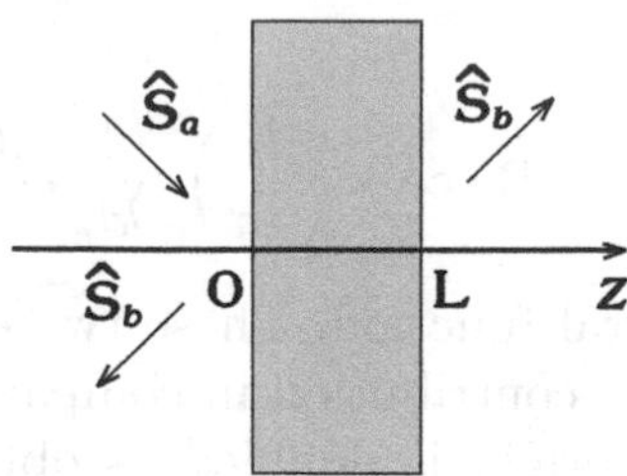

Figure 7. The geometry of a slab of width L and section S used for the measurement of the angular correlation functions both in reflection and in transmission.

There is a large variety of measurements that can be performed to study speckle patterns using electromagnetic waves. The first is given by the angular correlation function either in transmission or in reflection. To that purpose, we consider the slab geometry of Fig. 7. An incident beam along the direction $\hat{\mathbf{s}}_a$ is either reflected or transmitted along the direction $\hat{\mathbf{s}}_b$. We shall be mostly concerned with the transmission coefficient $\mathcal{T}_{ab}$ defined as the intensity $I(R, \hat{\mathbf{s}}_a, \hat{\mathbf{s}}_b)$ transmitted along the direction $\hat{\mathbf{s}}_b$ in the far field, *i.e.*, at a distance R much larger than any of the characteristic dimensions of the slab:

$$\mathcal{T}_{ab} = \frac{R^2}{S} \frac{I(R, \hat{\mathbf{s}}_a, \hat{\mathbf{s}}_b)}{I_0} \tag{32}$$

where I_0 is the intensity of the incoming wave and S is the section of the slab. It is important to notice that this definition differs from those used in the waveguide geometry. In this latter case, the incident and transmitted waves are plane waves with boundary conditions imposed by the waveguide. This gives rise to the quantization of the transverse channels. Here, instead, we have incident plane waves but transmitted spherical waves. This corresponds to different boundary conditions and to a continuous distribution of the transmitted angular directions.

We shall be interested in the normalized correlation function:

$$C_{aba'b'} = \frac{\overline{\delta\mathcal{T}_{ab}\delta\mathcal{T}_{a'b'}}}{\overline{\mathcal{T}}_{ab}\overline{\mathcal{T}}_{a'b'}} \tag{33}$$

where $\delta\mathcal{T}_{ab} = \mathcal{T}_{ab} - \overline{\mathcal{T}}_{ab}$.

By definition, this correlation function is build from the product of complex amplitudes which correspond to all possible multiple scattering sequences in the disordered medium. Like for the calculation of the average

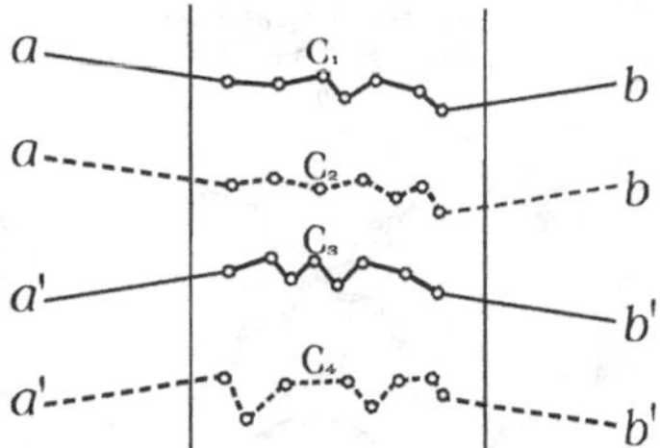

Figure 8. The transmission angular correlation function corresponding to four waves incident along the directions $\hat{\mathbf{s}}_a$ and $\hat{\mathbf{s}}_{a'}$ and outgoing along the directions $\hat{\mathbf{s}}_b$ and $\hat{\mathbf{s}}_{b'}$. A non zero contribution corresponds to the pairing of two amplitudes into a diffuson.

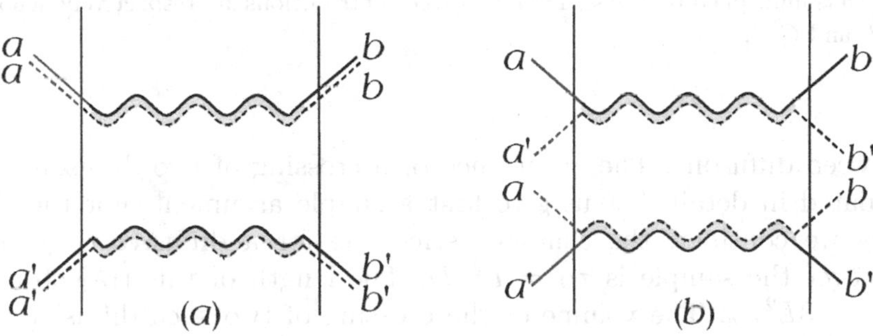

Figure 9. Two contributions to the product $\overline{\mathcal{T}_{ab}\mathcal{T}_{a'b'}}$ which correspond respectively to the pairing $\mathcal{C}_1 = \mathcal{C}_2, \mathcal{C}_3 = \mathcal{C}_4$ and $\mathcal{C}_1 = \mathcal{C}_4, \mathcal{C}_2 = \mathcal{C}_3$. The first gives $\overline{\mathcal{T}}_{ab}\overline{\mathcal{T}}_{a'b'}$. The second corresponds to the angular correlation function noted $C^{(1)}_{aba'b'}$ in the text.

probability, the non vanishing contributions correspond to cases where the amplitudes can be paired into diffusons (see Figs. 8 and 9).

In the case $a = a'$ and $b = b'$, we obtain,

$$\overline{\delta \mathcal{T}_{ab}^2} = \overline{\mathcal{T}_{ab}}^2 \tag{34}$$

This constitutes the Rayleigh law which accounts for the characteristic granular structure of a speckle pattern,*i.e.*, relative fluctuations of order unity. This is the most important and most "visible" property of a speckle pattern. It exists also in the single scattering regime.

In multiple scattering, there are additional contributions which result from the long range nature of the diffuson and from the existence of cross-

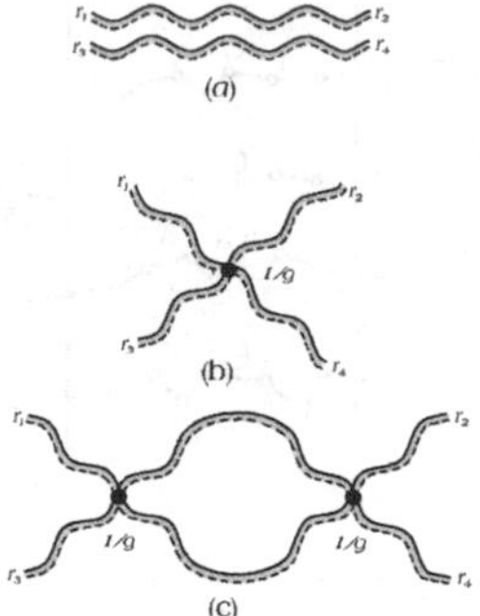

Figure 10. Classification of the contributions to the correlation function $C_{aba'b'}$ in terms of the number of crossings of two diffusons. At each crossing, the corresponding contribution is multiplied by $1/g \ll 1$. These three contributions are respectively denoted $C^{(1)}$, $C^{(2)}$, and $C^{(3)}$.

ings between diffusons. The occurrence of a crossing of two diffusons can be calculated in detail. Let us give first a simple argument. For the slab geometry we consider, the characteristic time for a diffusive trajectory to cross over the sample is $\tau_D = L^2/D$. The length of this trajectory is $\mathcal{L} = c\tau_D = 3L^2/l_e$. The volume of the crossing of two such diffusive trajectories namely the volume of a Hikami box is $\lambda^2 l_e$. We shall thus characterize a diffuson by its length $\mathcal{L}$ and its section λ^2. The occurrence of a crossing of two diffusons is therefore given by the ratio of the two volumes $\frac{\lambda^2 \mathcal{L}}{\Omega} = \frac{\lambda^2 L}{l_e S} \propto \frac{1}{g}$, where we have defined the dimensionless quantity

$$g = \frac{k^2 l_e S}{3\pi L} \tag{35}$$

with $k = 2\pi/\lambda$. This is the so-called dimensionless conductance of a wire of length L and section S. In the limit $kl_e \gg 1$ of a weak disorder, g can be very large, typically of order 10^2. Therefore, we may assume that the crossing events are uncorrelated, so that the probability of n crossings is given by $1/g^n$. This allows us to classify the contributions to the angular correlation functions in terms of the number of crossings as represented in Fig. 10.

In order to go further, we need to calculate the average transmission coefficient $\overline{T}_{ab}$. In the slab geometry, it can be written in terms of the structure factor Γ (taken at $\omega = 0$) defined by (17),

$$\overline{T}_{ab} = \frac{1}{(4\pi)^2 S} \int d\mathbf{r}_1 d\mathbf{r}_2 e^{-z_1/\mu_a l_e} e^{-|L-z_2|/\mu_b l_e} \Gamma(\mathbf{r}_1, \mathbf{r}_2) \tag{36}$$

where μ_a (μ_b) is the projection of $\hat{\mathbf{s}}_a$ ($\hat{\mathbf{s}}_b$) along the Oz axis. Moreover, Γ depends on the variables z_1, z_2 and $\rho = (\mathbf{r}_2 - \mathbf{r}_1)_\perp$. Thus, integrating over z_1, z_2 and using the relation (21) in the diffusion approximation, we obtain

$$\overline{T}_{ab} = \frac{c}{4\pi}\mu_a^2\mu_b^2 \int_S d\rho\, P(\rho, l_e, L - l_e) = \frac{c}{4\pi}\mu_a^2\mu_b^2 P(k_\perp = 0, l_e, L - l_e) \quad (37)$$

which depends on the two-dimensional Fourier transform of the diffuson P_d given by (16):

$$P(k_\perp, z, z') = \frac{1}{D}\frac{\sinh k_\perp z_m \sinh k_\perp (L - z_M)}{k_\perp \sinh k_\perp L} \quad (38)$$

with $z_m = \min(z, z')$ et $z_M = \max(z, z')$. For $k_\perp = 0$, we have

$$P(0, z, z') = \frac{z_m}{D}\left(1 - \frac{z_M}{L}\right) \quad (39)$$

Finally, by inserting this relation into (37) we obtain in the limit of small angles $\mu_a \simeq \mu_b \simeq 1$,

$$\overline{T}_{ab} = \frac{3}{4\pi}\frac{l_e}{L} \quad (40)$$

It should be emphasized here that this relation results from a given choice of boundary conditions for the diffusion equation, namely a vanishing of the probability at the boundaries $z = 0, L$ of the slab. A more precise calculation shows that there is an extrapolation length $z_0 \simeq l_e$ at which the probability vanishes outside the medium [1].

5.1. THE SHORT RANGE CORRELATION $C^{(1)}$

The main contribution (*i.e.* without crossing) to the angular correlation function is given by Fig. 9 (b). Its calculation is very similar to those of the average transmission coefficient apart from additional phase factors. We define the vectors $\Delta\hat{\mathbf{s}}_a = \hat{\mathbf{s}}_a - \hat{\mathbf{s}}_{a'}$ and $\Delta\hat{\mathbf{s}}_b = \hat{\mathbf{s}}_b - \hat{\mathbf{s}}_{b'}$ and we neglect their projection along the z-axis so that,

$$\overline{\delta T_{ab}\delta T_{a'b'}} = \left(\frac{1}{(4\pi)^2 S}\int d\mathbf{r}_1 d\mathbf{r}_2 e^{ik[\Delta\hat{\mathbf{s}}_a.\mathbf{r}_1 - \Delta\hat{\mathbf{s}}_b.\mathbf{r}_2]} e^{-z_1/l_e} e^{-|L-z_2|/l_e}\Gamma(\mathbf{r}_1, \mathbf{r}_2)\right)^2 \quad (41)$$

Performing the z-integrals and defining $q_a = k|\Delta\hat{\mathbf{s}}_a|$, we obtain in the limit $q_a l_e \ll 1$,

$$C^{(1)}_{aba'b'} = \delta_{\Delta\hat{\mathbf{s}}_a, \Delta\hat{\mathbf{s}}_b} F_1(q_a L) = \delta_{\Delta\hat{\mathbf{s}}_a, \Delta\hat{\mathbf{s}}_b}\left(\frac{q_a L}{\sinh q_a L}\right)^2 \quad (42)$$

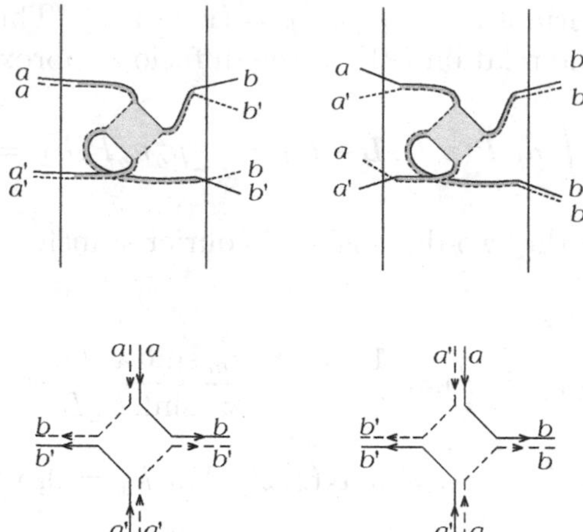

Figure 11. Contribution to $\overline{\delta \mathcal{T}_{ab} \delta \mathcal{T}_{a'b'}}$ with one crossing. The different cases correspond to configurations of incident plane waves along $\hat{\mathbf{s}}_a$ and $\hat{\mathbf{s}}_{a'}$ and outgoing along $\hat{\mathbf{s}}_b$ and $\hat{\mathbf{s}}_{b'}$. (a) depends on $\Delta\hat{\mathbf{s}}_b$ but not on $\Delta\hat{\mathbf{s}}_a$ and the opposite for (b).

with

$$F_1(x) = \left(\frac{x}{\sinh x}\right)^2 \tag{43}$$

This expression of the angular correlation function allows to understand, at least qualitatively, the *memory effect* [6, 7].

5.2. THE LONG-RANGE CORRELATION $C^{(2)}$

The next contribution to the correlation function arises from terms which involve one crossing of two diffusons. Due to the structure of the Hikami box within which the crossing occurs, the angular correlation function is different from $C^{(1)}$. This is represented in Fig. 11 and it gives rise only to the two possibilities

$$(aa)(a'a') \longrightarrow (bb')(bb') \tag{44}$$

and

$$(aa')(aa') \longrightarrow (bb)(b'b') \tag{45}$$

The corresponding expression for the first one is

$$\overline{\delta T_{ab}\delta T_{a'b'}}^{(2)} = \frac{1}{(4\pi)^4 S^2}\int\prod_{i=1}^{4} d\mathbf{r}_i[e^{ik\Delta\hat{\mathbf{s}}_b.(\mathbf{r}_2-\mathbf{r}_4)} + e^{ik\Delta\hat{\mathbf{s}}_a.(\mathbf{r}_1-\mathbf{r}_3)}]E(z_i)$$
$$\int\prod_{i=1}^{4} d\mathbf{R}_i H(\mathbf{R}_i)\Gamma(\mathbf{r}_1,\mathbf{R}_1)\Gamma(\mathbf{r}_3,\mathbf{R}_3)\Gamma(\mathbf{R}_2,\mathbf{r}_2)\Gamma(\mathbf{R}_4,\mathbf{r}_4) \quad (46)$$

where $H(\mathbf{R})$ is the expression of the Hikami box calculated in the diffusion approximation and given by [6, 8]

$$H(\mathbf{r}_1,\mathbf{r}_2,\mathbf{r}_3,\mathbf{r}_4) = \frac{l_e^5}{48\pi k_0^2}\int d\mathbf{r}\prod_{i=1}^{4}\delta(\mathbf{r}-\mathbf{r}_i)\left[2\nabla_\mathbf{2}.\nabla_\mathbf{4} - \nabla_\mathbf{1}{}^2 - \nabla_\mathbf{3}{}^2\right] \quad (47)$$

and where we have defined

$$E(z_i) = e^{-(z_1+z_3)/l_e} e^{-|L-z_2|/l_e} e^{-|L-z_4|/l_e} \quad (48)$$

In the diffusion approximation, and using the relation (21), the crossing of two diffusons rewrites

$$\int\prod_{i=1}^{4} d\mathbf{R}_i H(\mathbf{R}_i)\Gamma(\mathbf{r}_1,\mathbf{R}_1)\Gamma(\mathbf{r}_3,\mathbf{R}_3)\Gamma(\mathbf{R}_2,\mathbf{r}_2)\Gamma(\mathbf{R}_4,\mathbf{r}_4)$$
$$= \frac{32\pi^3 c^4}{3k^2 l_e^3}\int d\mathbf{R} P_d(\mathbf{r}_1,\mathbf{R})P_d(\mathbf{r}_3,\mathbf{R})[\nabla_\mathbf{R} P_d(\mathbf{R},\mathbf{r}_2)].[\nabla_\mathbf{R} P_d(\mathbf{R},\mathbf{r}_4)] \quad (49)$$

Using the relation (38) to evaluate the gradients and considering the limit $q_b l_e \ll 1$, it remains

$$\overline{\delta T_{ab}\delta T_{a'b'}}^{(2)} = \frac{81}{48\pi}\frac{l_e}{k^2 LS}F_2(kL\Delta\hat{\mathbf{s}}_b) \quad (50)$$

where we have defined

$$F_2(x) = \frac{1}{\sinh^2 x}\left(\frac{\sinh 2x}{2x} - 1\right) \quad (51)$$

Adding the contribution of the second diagram of Fig. 11, we finally obtain

$$C^{(2)}_{aba'b'} = \frac{\overline{\delta T_{ab}\delta T_{a,b'}}}{\overline{T_{ab}}^2} = \frac{1}{g}[F_2(kL\Delta\hat{\mathbf{s}}_a) + F_2(kL\Delta\hat{\mathbf{s}}_b)] \quad (52)$$

where the conductance g has been defined in (35). This contribution to the correlation function is thus smaller than $C^{(1)}$ by a factor $1/g$. But instead

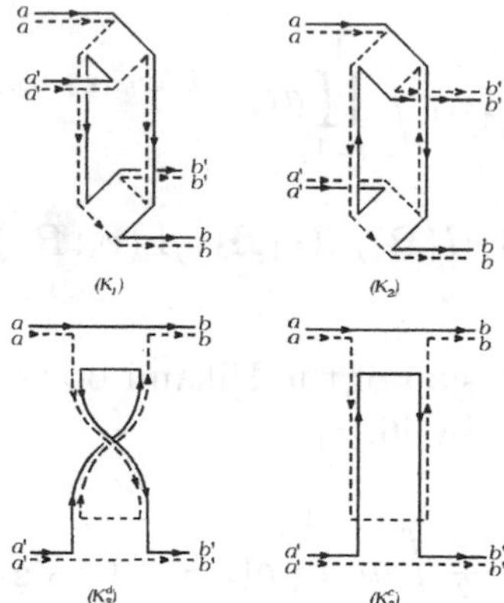

Figure 12. Contributions with two crossings. Those diagrams do not induce angular structure.

of decreasing exponentially, it behaves as a power law and vanishes when both $\Delta\hat{\mathbf{s}}_a$ and $\Delta\hat{\mathbf{s}}_b$ are large.

The next term in the expansion of the angular correlation function in powers of $1/g$ involves two crossings and is represented in Fig. 12. It is thus easy to see that it does not involve any angular structure due to the pairing of the trajectories. The corresponding correlation function thus corresponds to the angular structure

$$C^{(3)}: \qquad (aa)(a'a') \longrightarrow (bb)(b'b') \tag{53}$$

After a calculation similar to those of $C^{(2)}$ but which involves now two Hikami boxes, one finds

$$C^{(3)}_{aba'b'} = \frac{\overline{\delta \mathcal{T}_{ab} \delta \mathcal{T}_{a'b'}}^{(3)}}{\overline{\mathcal{T}}^2_{ab}} = \frac{2}{15}\frac{1}{g^2} \tag{54}$$

But, there exists other contributions to the two crossings term with the angular structure of either $C^{(1)}$ or $C^{(2)}$. They correspond to higher order terms in the $1/g$ expansion of the corresponding correlation functions.

While performing the integration over all the ingoing and outgoing angular directions, only the last term survives so that the relative fluctuations of the total transmission coefficient reduces to

$$\frac{\overline{\delta \mathcal{T}^2}}{\overline{\mathcal{T}}^2} = \frac{2}{15}\frac{1}{g^2} \tag{55}$$

This result is well known in the context of electronic systems as the so-called *universal conductance fluctuations.* There the conductance and the transmission coefficient in a waveguide geometry are related through the Landauer-Büttiker formalism. It is interesting to notice that we have obtained an identical result using a different definition of the transmission coefficient, *i.e.*, a different geometry where the conduction channels do not appear. This is true for the relative fluctuations but is not true anymore for the second moment.

These different contributions associated to the crossings of diffusons have been identified and measured [9]. We shall come back to it in the next section.

6. Diffusive Wave Spectroscopy

We have seen previously that a way to perform disorder averages is to consider suspensions in which the motion of the scatterers provides different realizations of the disorder. Hence the time averaging is equivalent to the averaging over realizations. The measurement of time correlation functions of speckle patterns provides also a very useful tool to investigate the dynamics of the scatterers in the multiple scattering regime. As such it corresponds to a generalization of the quasi-elastic light scattering, a well known experimental tool available in the single scattering regime [10].

We shall be interested in the time correlation functions of the intensity and of the electromagnetic field defined by

$$g_2(T) = \frac{\langle I(T)I(0)\rangle}{\langle I(0)\rangle^2} - 1 \tag{56}$$

and

$$g_1(T) = \frac{\langle E(T)E^*(0)\rangle}{\langle |E(0)|^2\rangle} \tag{57}$$

where the intensity and the field are related as usual by $I(T) = |E(T)|^2$. The notation $\langle\ldots\rangle$ denotes an average over all the multiple scattering sequences in the medium and over the dynamics of the scatterers. We shall assume here that this dynamics corresponds to a non deterministic Brownian motion characterized by a diffusion coefficient D_b (not to be mistaken with the diffusion coefficient $D = \frac{1}{d}vl_e$ obtained previously in the diffusion approximation). Then, the time correlation function of the electric field is expressed in terms of the probability P_d (*i.e.*, the contribution of the diffuson) and using the relation (21) so that

$$\langle E(\mathbf{r},T)E^*(\mathbf{r},0)\rangle = \int_0^\infty dt P_d(\mathbf{r},t)e^{-tT/2\tau_b\tau_e} \tag{58}$$

where the characteristic time $\tau_b = 1/4D_b k^2$ accounts for the motion of one scatterer.

To calculate the intensity correlation function $g_2(T)$, we notice that it involves the average of the product of four electric fields. Hence, using the pairing of these amplitudes as for the calculation of the angular correlation function (see Fig. 9), we obtain [11]

$$g_2(T) = |g_1(T)|^2 \tag{59}$$

In the limit $T = 0$, we recover the second moment of the Rayleigh law, namely $\langle I^2 \rangle = 2\langle I \rangle^2$. From the two expressions (58) and (59) we can deduce the expression of $g_2(T)$. For instance, for the slab geometry using the expression (38) and fixed ingoing and outgoing waves, we obtain a first contribution to $g_2(T)$, usually written as [12, 13]

$$g_2^{(1)}(T) = F_1(L/L_\gamma) = \left(\frac{L/L_\gamma}{\sinh L/L_\gamma} \right)^2 \tag{60}$$

with $L_\gamma = l_e \sqrt{\frac{2\tau_b}{3T}}$. It can be simply deduced from the relation (42) by replacing $1/q_a$ by L_γ.

The relation (59) is a consequence of the absence of crossings of diffusons. We now address the question of the effect of crossings on the time correlation function. The possible pairings of complex amplitudes is represented in Fig. 13. For the one crossing case, we notice that, unlike $g_2^{(1)}(T)$, the corresponding correlation function $g_2^{(2)}(T)$ involves two kind of diffusons, namely those with amplitudes taken at the same time and those taken at the two times 0 and T. The calculation can be done along the same lines as for the angular correlation function $C^{(2)}_{aba'b'}$ with the result

$$g_2^{(2)}(T) = \frac{2}{g} F_2(L/L_\gamma) \tag{61}$$

where the function $F_2(x)$ has been defined previously. We notice, that, unlike $g_2^{(1)}(T)$, this correlation function decreases at large times like a power law. It is nevertheless smaller than $g_2^{(1)}(t)$ by a factor $1/g$, so that its measurement requires to get rid of $g_2^{(1)}(T)$. This can be done by an angular integration over the outgoing directions or equivalently by averaging over a large number of speckle spots [14].

The case of the correlation function $g_2^{(3)}(T)$ which involves two diffusons crossings is more involved. We cannot anymore use the result of the angular correlation function and replace q_a by $1/L_\gamma$ since, as we have seen before,

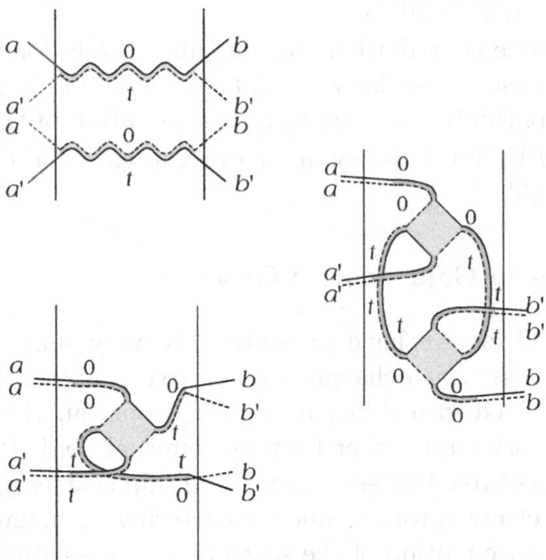

Figure 13. Diagrams contributing to the time correlation functions.

$g_2^{(3)}$ has no angular structure. Using the rules we have previously set for the Hikami boxes and the expression (38) for P_d for the slab geometry, we obtain [1, 15]

$$g_2^{(3)}(T) = \int_0^L \int_0^L dz dz' P_\gamma^2(z, z') = \frac{L^4}{8D^2} F_3(L/L_\gamma) \tag{62}$$

so that

$$g_2^{(3)}(T) = \frac{1}{g^2} F_3(L/L_\gamma) \tag{63}$$

where we have defined

$$F_3(x) = \frac{3}{2} \frac{2 + 2x^2 - 2\cosh 2x + x \sinh 2x}{x^4 \sinh^2 x} \tag{64}$$

We recover the expression $C^{(3)}(0) = 2/15$ for $L_\gamma \to \infty$. The expression of $g_2^{(3)}(t)$ given by (62) involves the integral of P_d^2. This originates from the closed loop which appear in the corresponding diagram of Fig. 13. This expression looks quite close to the one proposed in [9]. But it gives a much slower time dependence which should be sought in the distribution of all closed loops in the slab and not only those touching the boundaries.

The contribution of $g_2^{(3)}(T)$ is much more difficult to observe [9] since it is proportional to $1/g^2 \simeq 10^{-4}$.

In all the three contributions to the time correlation function, we have considered the case of a scalar wave in the absence of absorption. It is justified to treat separately the effect of the polarization of the electromagnetic wave for $g_2^{(1)}(T)$. But it is not anymore the case for the two remaining contributions [15].

7. Dephasing in Cold Atomic Gases

The coherent effects we have presented are very sensitive to dephasing, *i.e.*, to any process which changes the relative phase of the two interfering amplitudes involved in a diffuson or in a cooperon. Roughly speaking, a dephasing may originate either from an external field [16, 17] or from additional degrees of freedom such as the spin-flip scattering in metals where the spin of the electron rotates due to scattering by magnetic impurities or a non deterministic motion of the scatterers such as studied before for the diffusive wave spectroscopy [16, 1]. In the presence of dephasing, the probability of quantum diffusion can be written as $P(\mathbf{r}, \mathbf{r}', t) \left\langle e^{i\phi(t)} \right\rangle$, where the random variable $\phi(t)$ is the relative phase of the two interfering paths. Its distribution depends on the origin of the dephasing and we denote by $\langle ... \rangle$ the average over this distribution. In most cases we have $\left\langle e^{i\phi(t)} \right\rangle \simeq e^{-t/\tau_\phi}$ at least for long enough times t. The characteristic time τ_ϕ is the phase coherence or dephasing time. An exponential decrease of the probability of quantum diffusion does not necessarily describe a dephasing process. For instance, the intensity of an electromagnetic wave which propagates in an absorbing medium decreases exponentially. But this is not a dephasing process since it affects equally both the coherent and incoherent contributions by a decrease of the overall intensity. The propagation of photons in cold atomic gases addresses similar questions and provides new sources of dephasing like internal atomic degrees of freedom. Coherent multiple scattering effects have been observed in such gases and analyzed in great details [18]. The dephasing induced by the internal Zeeman atomic degrees of freedom can be obtained in a closed form [19] in terms of both the polarization state of the photons and of the Zeeman degeneracy.

Each of the N atoms of the gas is taken to be a two-level system of characteristic transition frequency ω_0 [20]. The ground state defines the zero of energy and has total angular momentum J. The excited state has a total angular momentum J_e and a natural width Γ due to coupling to the vacuum fluctuations. We shall assume, moreover, that the velocity v of the atoms is small compared to Γ/k (k is the light wave-vector) but large compared to

$\hbar k/M$ (M being the mass of the atom), so that it is possible to neglect the Doppler and recoil effects. The external degrees of freedom of the atoms are therefore the classical assigned positions $\mathbf{r}_\alpha$ $(\alpha = 1, \ldots, N)$ uncorrelated with one another. The atom-photon interaction is described within the dipole approximation [21], and the elastic scattering process between the two states $|\mathbf{k}\epsilon, Jm\rangle$ and $|\mathbf{k}'\epsilon', Jm'\rangle$, where $|\mathbf{k}\epsilon\rangle$ is a one-photon Fock state of the free transverse electromagnetic field in the mode $\mathbf{k}$ of polarization ϵ, is described by the single scattering transition amplitude $t_{ij}(m, m', \omega) = t(\omega)\langle Jm'|d_i d_j|Jm\rangle$. where the resonant scattering amplitude is given by

$$t(\omega) = \frac{3}{2\pi\rho_0(\omega)} \frac{\Gamma/2}{\delta + i\Gamma/2} \tag{65}$$

with $\delta = \omega - \omega_0$ being the detuning of the probe light from the atomic resonance and $\rho_0(\omega) = \mathcal{V}\omega^2/2\pi^2$ is the free photon spectral density. The average over the disorder is taken over both the uncorrelated positions $\mathbf{r}_\alpha$ of the atoms and over the magnetic quantum numbers m_α of the atoms. The first, standard average restores the translation invariance. The internal average, a trace with a scalar density matrix ρ assuming that the atoms are prepared independently and equally in their ground states, restores rotational invariance. The elastic mean free path is inversely proportional to $t(\omega)$ given in (65). Due to the tensorial structure of the transition amplitude, the diffuson and the cooperon have now three eigenmodes each. The dephasing associated to the internal degrees of freedom affects only the cooperon which is given in the diffusion approximation by

$$X_K(q) \propto \frac{1}{Dq^2 + \tau_d^{-1}(K) + \tau_\phi^{-1}(K)} \tag{66}$$

for $K = 0, 1, 2$. The characteristic times $\tau_d(K)$ are the depolarization times which affect the diffuson as well. They describe the depolarization of the initial light beam and can be calculated in the classical Rayleigh case [22]. The dephasing times associated to the intensity of the field are given by [19]

$$\frac{\tau_\phi(0)}{\tau_{\rm tr}} = \begin{cases} (J(2J+3))^{-1}, & J_e = J+1 \\ J^2 + J - 1, & J_e = J \\ (2J^2 + J - 1)^{-1}, & J_e = J-1 \end{cases} \tag{67}$$

where the transport time $\tau_{\rm tr} = l_e + \Gamma^{-1}$ (in units where $\hbar = c = 1$). An absence of dephasing only occurs for the classical dipole $J = 0$ (Rayleigh scattering) and in the semi-classical limit $J = J_e \to \infty$.

Acknowledgments

This work is supported in part by a grant from the Israel Academy of Sciences, by the fund for promotion of Research at the Technion and by the French-Israeli Arc-en-ciel program. E.A. acknowledges the very kind hospitality of the Laboratoire de Physique des Solides at the university of Paris (Orsay). G.M. acknowledges the very kind hospitality of the Technion.

References

1. E. Akkermans and G. Montambaux, *Propagation cohérente dans les milieux aléatoires: électrons et photons*, EDP Sciences-CNRS (2003) (to be published)
2. E. Akkermans and G. Montambaux, in *"Waves and Imaging through Complex Media"*, P. Sebbah (ed.), Kluwer, Dordrecht, 2001, pp. 29–52 (cond-mat/0104013)
3. S. F. Edwards, Phil. Mag. **3**, 1020 (1958)
4. Notice that we use a notation for $G(\mathbf{r}_i, \mathbf{r})$ where the initial point is on the left and the final point on the right.
5. S. Chakraverty and A. Schmid, Phys. Rep. **140**,193 (1986)
6. R. Berkovits and S. Feng, Phys. Rep. **238**, 135 (1994)
7. I. Freund, M. Rosenbluh and S. Feng, Phys. Rev. Lett. **61**, 2328 (1988)
8. M.C.W. van Rossum and T.M. Nieuwenhuizen, Rev. Mod. Phys. **71**, 313 (1999)
9. F. Scheffold and G. Maret, Phys. Rev. Lett. **81**, 5800 (1998)
10. B.J. Berne and R. Pecora, *Dynamic Light Scattering with Applications to Chemistry, Biology and Physics*, John Wiley, 1976
11. G. Maret and E. Wolf, Z. Phys. **B65**, 409 (1987)
12. P.E. Wolf and Maret, in *Scattering in Volumes and Surfaces,* M. Nieto-Vesperinas and C. Dainty (eds.), North-Holland, Amsterdam, 1990, pp. 37–52
13. D.J. Pine, D.A. Weitz, P.M. Chaikin and E. Herbolzheimer, Phys. Rev. Lett. **60**, 1134 (1988)
14. F. Scheffold, W. Hartl, G. Maret and E. Matijevic, Phys. Rev. **B 56**, 10942 (1997)
15. E. Akkermans and G. Montambaux, in preparation (2002)
16. G. Bergmann, Phys. Rep. **107**, 1 (1984); S. Hikami, A.I. Larkin and Y. Nagaoka, Prog. Theor. Phys. **63**, 707 (1980)
17. B.L. Altshuler, A.G. Aronov and D.E. Khmelnitsky, J. Phys. C **15**, 7367 (1982)
18. G. Labeyrie, F. de Tomasi, J.-C. Bernard, C.A. Müller, C. Miniatura and R. Kaiser, Phys. Rev. Lett. **83**, 5266 (1999), T. Jonckheere, C.A. Müller, R. Kaiser, C. Miniatura and D. Delande, Phys. Rev. Lett. **85**, 4269 (2000) and G. Labeyrie, C.A. Müller, D.S. Wiersma, C. Miniatura and R. Kaiser, J.Opt. **B2**, 672 (2000)
19. E. Akkermans, Ch. Miniatura and C.A. Müller, ArXiv:cond-mat/0206298, to be published
20. C.A. Müller, PhD thesis (Munich/Nice, 2001), unpublished, and C.A. Müller and C. Miniatura, submitted, preprint physics/0205029
21. C. Cohen-Tannoudji, J. Dupont-Roc and G. Grynberg, *Processus d'interaction entre photons et atomes*, Savoirs actuels, InterEditions, Editions du CNRS, (1988).
22. E. Akkermans, P.E. Wolf, R. Maynard and G. Maret, J. de Physique (France), **49**, 77 (1988)

MESOSCOPIC DYNAMICS

A Study of Phase

A.Z. GENACK AND A.A. CHABANOV
Physics Department
Queens College of the City University of New York
65-30 Kissena Boulevard, Flushing, NY 11367-0904, USA
P. SEBBAH
Laboratoire de la Physique de la Matière Condensée
CNRS/Université de Nice, Parc Valrose
F-06108, Nice Cedex 02, France
AND
B.A. VAN TIGGELEN
Laboratoire de Physique et Modélisation des Milieux Condensés
CNRS/Université Joseph Fourier, B.P. 166
38042 Grenoble Cedex 09, France

1. Introduction

Spatial coherence is rapidly lost as a wave penetrates a multiply scattering medium since the wave at any point is the sum of randomly scattered partial waves. Nonetheless, in samples in which the wave is not inelastically scattered, the wave is temporally coherent. As a result, the intensity is correlated over distances much greater than the wavelength scale in which the field loses coherence. In such mesoscopic samples, local fluctuations in intensity or particle flux do not self-average. When the flux is summed over the output surface for a single incident mode, giant fluctuations are found in the total transmission with variation of frequency or of sample configuration. Enhanced fluctuations are also found for the transmitted intensity as well as for the conductance, which is proportional to the sum of all transmission coefficients between incident and outgoing modes averaged over all relative phases for the incident modes. Because wave propagation is frequency dependent, the focus of statistical optics has been on characterizing a sample at a specific frequency at which the scattering characteristics of the sample are well-defined. This emphasis upon static characteristics is reinforced by

B.A. van Tiggelen and S.E. Skipetrov (eds.),
Wave Scattering in Complex Media: From Theory to Applications, 125–150.

the difficulties of performing time dependent electronic measurements on times small compared to the electron dephasing time. But in practice, all experiments are bounded in time. It is therefore natural to consider the statistics of dynamical aspects of propagation. We shall be primarily concerned with fluctuations in pulse characteristics in collections of statistically equivalent samples, which may include variations that occur with changes in position, frequency, or sample configuration. We will describe the key dynamical parameters and will follow their changing statistical character in the transition from diffusive propagation to localization.

The discussion of mesoscopic dynamics emphasizing fluctuation in pulse transmission will be placed in the context of the long-standing interest in the time-of-flight distribution of particles transmitted through random media. This distribution represents the ensemble average of the temporal variation of intensity following short pulse excitation. The connection of the photon transit time distribution to the underlying field is manifest in its equivalence to the Fourier transform of the correlation function of the field with frequency shift. The relationship of fluctuations in pulse characteristics to the underlying field is even more apparent in the statistics of individual pulses. In the limit of long pulses, the average photon dwell time of an individual pulse is equal to the spectral derivative of the phase of the field. The statistics of dwell time of such pulses is tightly bound to the statistics of intensity. Thus the phase and amplitude play distinctive, but correlated roles in wave dynamics.

2. Time of Flight and Delay Time in Random Media

In a homogeneous medium of thickness L and with group velocity v_g, the phase accumulated as the angular frequency is incremented by $\Delta\omega$ is $\Delta\phi = \Delta kL = \Delta\omega L/(\mathrm{d}\omega/\mathrm{d}k)$, so that the phase variation with frequency is $\Delta\phi/\Delta\omega = L/v_g$. In this simple situation, the "group delay" L/v_g is a measure of the transit time of a narrow-band pulse through a homogeneous, dispersive, and non-absorbing medium [1] and equals the phase derivative $\phi' \equiv \mathrm{d}\phi/\mathrm{d}\omega$.

In a random medium, however, the wave is multiply scattered and the output field at a point is the superposition of many partial waves. The total phase accumulated by a partial wave on its way through the medium, following some path of length s, is $ks = \omega s/c$ plus some random phase shift, $\Delta\phi(s)$, due to scattering within the sample and reflections at the boundaries. Even for constant s the phase derivative with frequency is a random variable. A major challenge is to quantify the whole distribution of the delay time of an arbitrary incident pulse in a random medium. This review represents the first step in this direction.

A logical way to study dynamic wave diffusion in disordered media would be to send in a pulse that is short compared to the diffuse transit time $t_D = L^2/6D$, where D is the diffusion coefficient, but still long compared to the wave period, $T = 2\pi/\omega$, and to characterize its statistics. In frequency space, this implies a relatively sharp frequency power spectrum with a width that is much smaller than its central frequency, but much broader than the Thouless linewidth $6D/L^2$, resulting from diffusion out of the sample. This regime can be modeled theoretically using the slowly-varying envelope approximation [2]. In typical studies with visible light, where $t_D \approx 3$ ps and $T \approx 1$ fs, this can only be accomplished with femto-second lasers that are now available [3]. In the microwave experiments discussed below, where $t_D \approx 30$ ns and $T \approx 30$ ps, this regime is satisfied for nanosecond pulses. The photon time-of-flight distribution can be computed from the frequency response function of the in- and out-of-phase components of the field transmission coefficient.

The variation with time of the transmission coefficient of intensity from channel a to channel b in a particular sample configuration, $I_{ab}(t)$, displays a modulation on the time scale of the incident pulse. When averaged over a random ensemble, for incident pulses with vanishing width, we obtain the time-of-flight distribution, $P_{ab}(t) = \langle I_{ab}(t)\rangle / \int \langle I_{ab}(t)\rangle \mathrm{d}t$. The time-of-flight distribution may also be obtained from spectral field measurements. It is given by the Fourier transform of the field correlation function,

$$P_{ab}(t) = \int \mathrm{d}\omega \int \mathrm{d}\Omega\, \mathrm{e}^{-i\Omega t} \left\langle t_{ab}(\omega - \frac{1}{2}\Omega)\, t^*_{ab}(\omega + \frac{1}{2}\Omega) \right\rangle, \tag{1}$$

where ω is the circular frequency of the waves, and Ω is the frequency shift. In most cases, measurements of $P_{ab}(t)$ have been in agreement with diffusion theory [4]. However, deviations from diffusion theory have been observed in thin samples [5], in which the transmitted pulse was found to rise earlier than predicted by diffusion theory, possibly as a result of incomplete randomization at short times. In addition, the diffusion coefficient determined from an exponential fit of the tail of $P_{ab}(t)$ gave a value of D that increased with sample thickness [6]. Recently, a nonexponential tail of pulsed transmission was observed [7]. This reflects the distribution of the decay rates of the quasi-normal modes of the sample. These measurements and earlier calculations [8, 9] have highlighted the increasing importance of long-lived modes with increasing delay time in mesoscopic media. Such long-lived modes may play an important role in the initiation of random lasing [10].

To obtain information on the statistics of dynamics, the narrow-band limit, $\Delta\omega \ll 6D/L^2$, must be considered, which is commonly taken in discussions of group velocity and phase delay in homogeneous media [1].

Key variables are the "single-channel delay time" t_{ab} and the energy transmission coefficient ϵ_{ab} for a pulse $I_a^{in}(t)$ incident in spatial mode a and scattered into outgoing mode b, defined in terms of the time-dependent transmission coefficient $I_{ab}(t)$ as $\tau_{ab} = \int \mathrm{d}t I_{ab}(t)t / \int \mathrm{d}t I_{ab}(t)$, and $\epsilon_{ab} = \int \mathrm{d}t I_{ab}(t)/\langle \int \mathrm{d}t I_a^{in}(t)\rangle$. The interplay between the single-channel delay time and the energy transmission coefficient is best expressed in terms of the joint probability distribution $P(\epsilon_{ab}, \tau_{ab})$ and the correlation functions with frequency of ϵ_{ab}, τ_{ab}, and $\epsilon_{ab}\tau_{ab}$. As we allow the incident pulse to broaden in time until its spectral width becomes less than the field correlation frequency, the delay time distribution $P(\tau_{ab})$ broadens and approaches a limiting distribution. In this limit, the single-channel delay time approaches the spectral derivative $\mathrm{d}\phi_{ab}/\mathrm{d}\omega \equiv \phi'_{ab}$ of the phase accumulated by the field as it propagates through the sample [11, 12], while the energy transmission coefficient ϵ_{ab} approaches the single-channel static transmission coefficient $I_{ab}(\omega)$.

2.1. PULSE MEASUREMENTS AND TIME-OF-FLIGHT DISTRIBUTION

It is worthwhile to begin the discussion of pulsed statistics by considering an individual pulse transmitted through a random medium. Measurements of pulsed propagation were made in samples composed of randomly positioned 1.27-cm-diameter polystyrene spheres at a volume filling fraction of 0.52 contained within a 1-m long, 7.6-cm-diameter copper tube. Wire antennas are used as the emitter and detector at the input and output surfaces of the sample. The response to a 1 ns pulse shaped by a pulse forming network and mixed in a local oscillator at 19 GHz is measured using high bandwidth B&H Electronics amplifiers. The results are collected using a digital sampling oscilloscope. The component of the transmitted field, which is in phase with the 19 GHz carrier frequency of the 1 ns incident pulse, is shown in Fig. 1. The squares are taken of the in- and out-of-phase field responses and added together to give the intensity variation with time. The average over 4096 configurations is taken and shown in Fig. 2. Because the incident pulse is much shorter than the typical traversal time through the medium this measurement gives the time-of-flight distribution of photons propagating through the sample at the carrier frequency of 19 GHz. The field correlation function with frequency shift is obtained in this sample using a Hewlett-Packard 8722C vector network analyzer. Measurements are made with frequency steps of 625 kHz at points displaced by 1 mm along a 4-cm line running symmetrically about the center of the output surface. After the spectrum is taken at a given position, the detector is translated by 1 mm. Measurements are made for two positions of the input antenna separated by 3 cm. Once spectra are taken at each

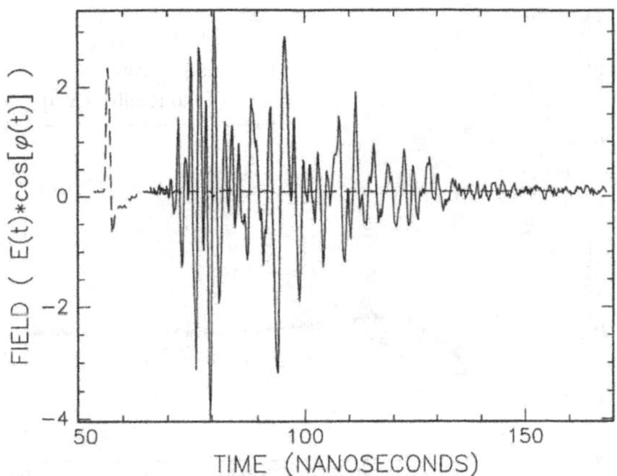

Figure 1. In-phase component of the field (full line) in response to a 1 ns microwave pulse (dashed line) at carrier frequency 19 GHz.

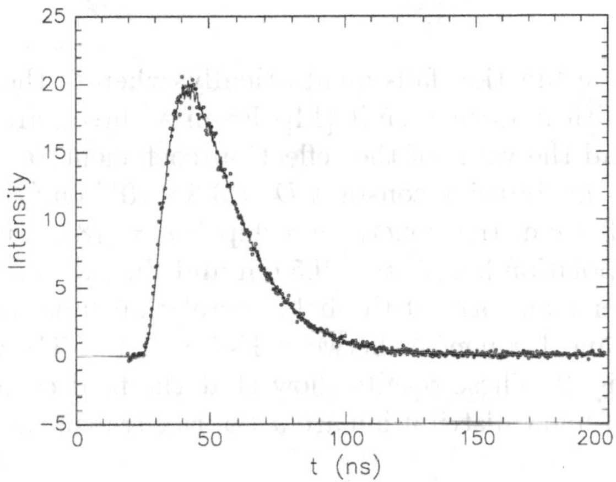

Figure 2. Time-of-flight distribution obtained by averaging the time response shown in Fig. 1 over 4096 configurations. The continuous line is the Fourier transform of the field correlation function shown in Fig. 3 in the range 18.5 to 19.5 GHz. (Sebbah *et al.* (2000) [13]).

point on the line, the copper tube is rotated briefly to create a new sample configuration. Measurements are made in 200 sample realizations and spectral correlation functions are computed from these data. The real and imaginary parts of the field correlation function with frequency shift over the frequency range 16.8 to 17.8 GHz are shown in Fig. 3. Averaging has been performed over sample configuration, position of the source and detector and over frequency [13]. For small frequency shifts, the real part

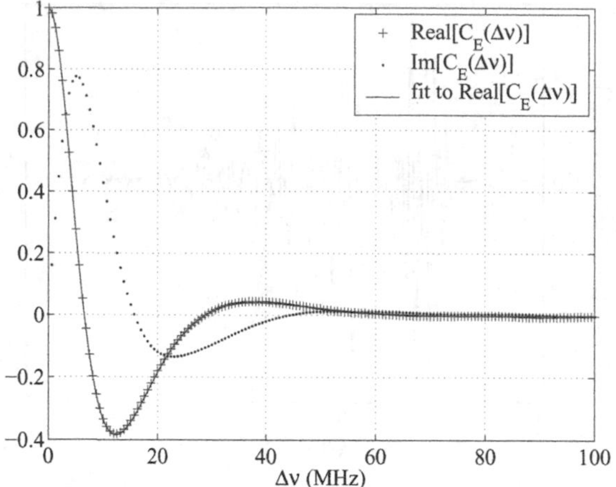

Figure 3. Real (crosses) and imaginary (dots) part of the field correlation function with frequency shift. The solid line is a theoretical fit. (Sebbah *et al.* (2000) [13]).

of the correlation function falls quadratically, whereas the imaginary part rises linearly with frequency shift [14]. From a three-parameter fit, using Eq. 2 of [13] and the value of the reflection coefficient $R = 0.13$ measured in [15], we find the diffusion constant $D = 3.3 \times 10^{10}$ cm^2/s, the absorption length $L_a = 33.3$ cm, the penetration depth $a = 18.3$ cm, which in turn gives the extrapolation length $z_b = 9.5$ cm and the mean free path $\ell = 11.0$ cm. The Fourier transform of the field correlation function averaged over configurations and frequencies between 18.5 and 19.5 GHz is presented as the curve in Fig. 2. These results show that the field correlation function and the time-of-flight distribution are a Fourier transform pair.

2.2. FLUCTUATIONS OF SINGLE-CHANNEL DELAY TIME

As we have seen above, direct dynamical microwave measurements are possible in principle, but generating precisely shaped pulses is not always practical. The complex response to an incident pulse with carrier frequency ω_0, $E(t) = |E(t)| \exp(\omega_0 t + \phi(t))$, can be obtained by measuring the frequency spectrum of a transmitted pulse and finding its Fourier transform. The spectrum of a transmitted pulse is simply the product of the spectra of the incident pulse and of the field transmission coefficient. An example of the amplitude $|E(t)|$ and phase $\phi(t)$ of the response to a pulse constructed by Fourier transforming the field spectrum in a particular sample configuration is presented in Fig. 4 for two pulses centered at 10 GHz with Gaussian envelopes $1/(\sqrt{2\pi}\sigma) \exp(-t^2/2\sigma^2)$, where $\sigma = 1$ ns and 100 ns. Use of a

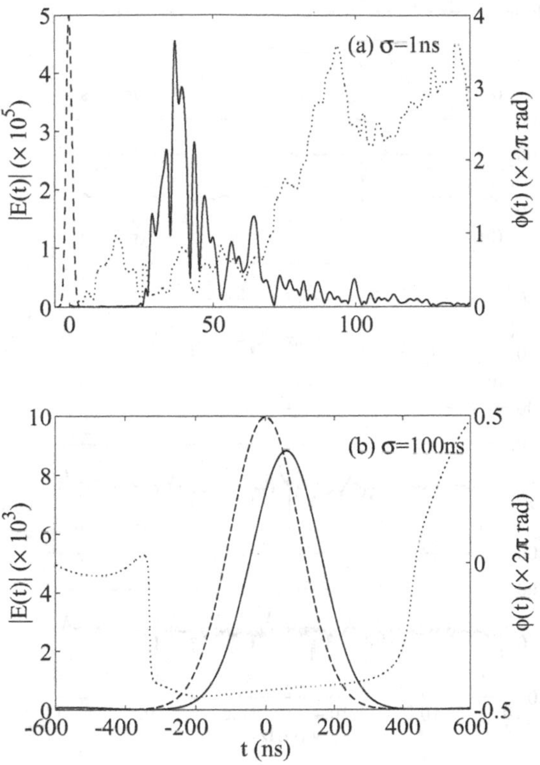

Figure 4. Amplitude of the time response (solid line) to a Gaussian pulse (dashed line) with (a) $\sigma = 1$ ns and (b) $\sigma = 100$ ns, with carrier frequency 10 GHz. Total energy of the input pulse, $\int |E|^2\, dt$, is normalized to unity. The slowly varying component of the phase $\phi(t)$ is shown as the dotted curve. (Sebbah *et al.* (1999) [12]).

Gaussian pulse is particularly convenient since its Fourier transform is a Gaussian and it therefore has rapidly falling tails in both the time and frequency domains. By changing the central frequency of the incident pulse, one can follow the variation with frequency of the single-channel delay time for a given sample configuration and pulse bandwidth. The frequency dependence of $\tau_{ab}(\omega, \Delta\omega)$ for an input pulse with decreasing bandwidths $\Delta\omega$ are plotted in Fig. 5a to 5d. We find that fluctuations in τ increase as the pulse bandwidth narrows [12]. For narrow bandwidth pulses, large positive and negative values of τ are found. These large values occur most commonly when a null in the speckle pattern passes near the detector, as the frequency is varied. The phase is undetermined at these nulls and jumps by π radians when a null passes through the detector, as the frequency is tuned. The

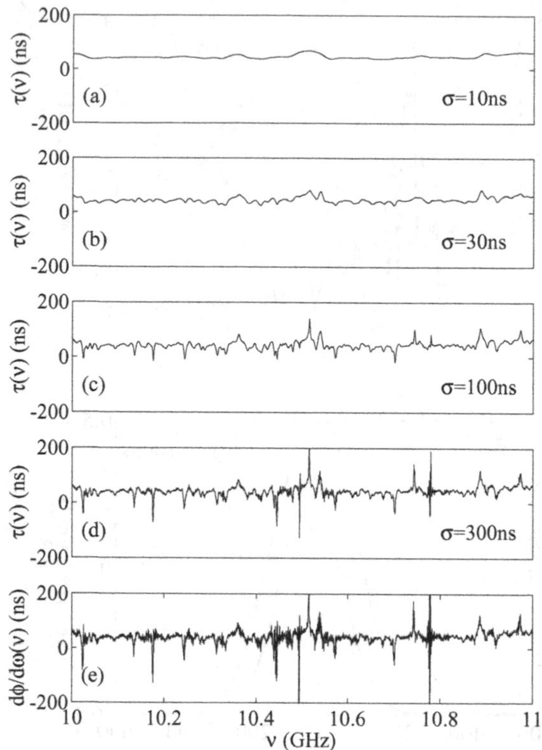

Figure 5. Time delay between 10 and 11 GHz for an input pulse with (a) $\sigma = 10$ ns, (b) $\sigma = 30$ ns, (c) $\sigma = 100$ ns, (d) $\sigma = 300$ ns. The phase derivative for the same frequency range is shown in (e). (Sebbah *et al.* (1999) [12]).

approach of the single-channel delay time τ in a random medium to the phase derivative is clear from Fig. 5 and can be shown formally in the limit of diverging pulse width and vanishing bandwidth.

The measured spectra of the amplitude and the phase modulus 2π, $\phi_{2\pi}$, of the transmitted field at a point for a particular configuration between 10 and 10.5 GHz are plotted in Fig. 6a and b. The intensity is obtained by squaring the field, and the phase derivative is taken once the cumulative phase is constructed (Fig. 6c and d). This is done by adding a phase $\pm 2\pi$ to the cumulative phase ϕ whenever $\phi_{2\pi}$ suffers a jump of magnitude larger than π, when the frequency is incremented, so that the jump in the cumulative phase is smaller than π. The cumulative phase can thus be expressed as $\phi = \phi_{2\pi} + 2n\pi$, where n is an integer. This integer can be determined by following the phase rollup starting from low frequencies where the phase

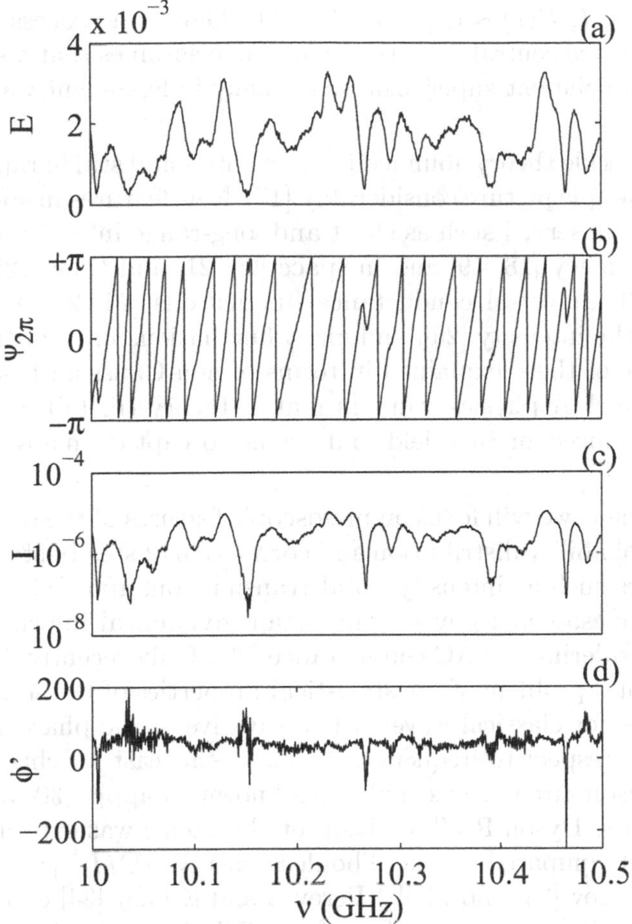

Figure 6. (a) Field amplitude, (b) phase modulus 2π, (c) logarithm of the transmitted intensity, and (d) phase derivative between 10 and 10.5 GHz. (Sebbah *et al.* (1999) [12]).

approaches zero. Large positive and negative peaks in the phase derivative are correlated with small values of intensity, when the detector is near a null in the intensity at which the in- and out-of-phase components of the field vanish so that the phase is undefined.

3. Mesoscopic Approach towards Dynamics

Fluctuations in wave propagation in disordered media are extensively treated in standard references [16]. Perhaps the best known feature is the apparently Gaussian statistics of the complex wave field, with real and imaginary parts as independent variables. This results in the familiar Rayleigh statistics of intensity I, $P(I) \sim \exp(-I/\langle I \rangle)$. The Gaussian process is a natural consequence of the central limit theorem, if one assumes that a speckle spot results from a coherent superposition of many independent wave trajectories.

Modern speckle theory, founded in the eighties and still in rapid progress, changes this simple picture considerably [17]. New features in speckle statistics have been observed such as short and long-range intensity correlations — e.g. in frequency [18, 19] and in space [20, 21] and time [22], the memory effect [23], universal conductance fluctuations [24, 25], non-Rayleigh statistics for the intensity [26], and giant fluctuations in total transmission [19, 27], many of these explained in terms of non-Gaussian field statistics. A tool from nuclear physics, random matrix theory (RMT), has been successfully introduced in this field and seems to capture many "universal" aspects [28].

In this review, we will focus on mesoscopic features of the dynamics, and not on the probability distribution and correlation of steady-state transmission quantities such as intensity, total transmission, and DC conductance. In electronic mesoscopic physics, important dynamical information is obtained by considering the AC conductance [29]. Only recently, RMT solved the fundamental problem of the statistical properties of the so-called phase delay times — for classical waves, the derivative of the phase-shifts of the S-matrix with respect to frequency, $\mathrm{d}\phi/\mathrm{d}\omega$, — at least for chaotic billiards where the Dyson circular ensembles are known to apply [30, 31]. The applicability of the Dyson RMT in disordered systems was expected only for energies small compared to the Thouless energy $\hbar D/L^2$ [32]. Pioneering work by Dorokhov [33] and Mello, Pereyra and Kumar [34] extended RMT to disordered wires. The extension of DMPK theory to dynamical problems is a subject of active research [35, 36]. Results have been obtained for one-dimensional [39] and quasi-one-dimensional disordered systems [40].

From the discussion of τ_{ab} in Sec. 2, the single-channel phase delay time $d\phi_{ab}/d\omega$ and the delay time weighted by the intensity, $W_{ab} = I_{ab} d\phi_{ab}/d\omega$, were put forward as important dynamical quantities. We carried out microwave experiments to study the statistics of the dynamics in transmission [41]. The sample consisted of the polystyrene spheres described above, contained in a 5-cm-diameter copper tube of length $L = 100$ cm. Measurements are taken between 18 and 19 GHz with frequency steps of 100 kHz. 10,000 spectra are taken. The transport mean free path is approximately 7 cm [15]. We will discuss the results of these experiments and compare these to a statistical theory for the dynamical matrix elements W_{ab} and $\mathrm{d}\phi_{ab}/\mathrm{d}\omega$ for diffuse waves [14, 41], using concepts developed for the static intensity I_{ab} [17]. Enhanced fluctuations in the intensity I_{ab}, the total transmission

$T_a = \sum_b I_{ab}$, and the transmittance $T = \sum_{ab} I_{ab}$, which is equivalent to the dimensionless conductance, as well as enhanced spectral correlation arise [21, 42]. Analogously, we find a long-range C_2 correlation term in the dynamics [41]. This is seen in both the frequency correlation function of W_{ab} as well as in its probability distribution. In contrast, the statistics of the single channel delay time $\phi'_{ab} = d\phi_{ab}/d\omega$ is found to be remarkably well described by the Gaussian or C_1 approximation for diffusive waves [14, 41].

When summed over all input and output modes, W_{ab} relates directly to a fundamental dynamic quantity in condensed matter, namely the number of microstates per frequency interval dω inside the scattering medium $N(\omega)$ [43],

$$\frac{1}{\pi}\sum_{ab}^{2M} I_{ab}\frac{\mathrm{d}\phi_{ab}}{\mathrm{d}\omega} = N(\omega), \tag{2}$$

where $\sqrt{I_{ab}}\exp(i\phi_{ab}) \equiv t_{ab}$ defines the complex transition amplitude from mode a to b. Unlike the summation in the Landauer formula for the conductance [44], the summation runs over M channels in both reflection *and* transmission. Eq. (2) is a manifestation of Friedel's theorem [45], originally devised for screening problems in the solid state, though with elegant applications to many scattering problems [46, 47, 48], including RMT [49]. Altshuler and Shklovskii [32] demonstrated the central role of the statistics of $N(\omega)$ in the understanding of the relation between level repulsion — a concept from RMT — and universal conductance fluctuations, a relation confirmed numerically [35]. Crucial is that the density of states in an open (scattering) medium is ill-defined due to the finite width of the levels. As shown in Ref. [43], the l.h.s. of Eq. (2) is well-defined and proportional to $\int \mathrm{d}\mathbf{r}\, [|\psi_\omega(\mathbf{r})|^2 - 1]$, which is — within the original Friedel argument — recognized as the "stored charge". For light, it equals the stored electromagnetic energy [48].

Eq. (2) calls for the interpretation of $W_{ab} \equiv I_{ab}\mathrm{d}\phi_{ab}/\mathrm{d}\omega$ as the weighted delay time for a transition from channel a to b [50], to be distinguished from the "proper" delay times defined as the eigenvalues of the Wigner-Smith matrix $\mathbf{Q} = -i\mathbf{S}^* \cdot \partial\mathbf{S}/\partial\omega$, and its trace $\mathrm{Tr}\,\mathbf{Q} = \sum_{ab} W_{ab} = \pi N(\omega)$, called the Heisenberg time τ_H; the channel-average $\tau_H/2M$ is associated with *the* Wigner-Smith phase delay time τ_W. Substantial theoretical effort in mesoscopic physics has focused on the probability distributions of the M "proper delay times" [53], whose measurement requires knowledge of the whole Wigner-Smith matrix.

The basic assumption in our theory is the validity of the so-called C_1 approximation, which is known to work particularly well for the static "one channel in — one channel out" matrix element I_{ab}, provided the dimensionless conductance $g = M\ell/L \gg 1$, but which has never been worked out for the dynamic element W_{ab}. The summation $\sum_b W_{ab}$ equals the *diagonal* element Q_{aa} of the Wigner-Smith matrix and may, like the transmittance $\sum_b T_{ab}$, be subject to C_2 correlations [18], and is left as a future challenge, since even a Gaussian process contains phase correlations between speckle spots [54].

3.1. GAUSSIAN STATISTICS OF SINGLE-CHANNEL PHASE DELAY

The C_1 approximation corresponds to the assumption of a circular complex Gaussian process [16] of the complex set t_{ab}. A circular process for K complex field amplitudes $E_i = t(i)E_i^{\rm in}$ requires that $\langle E_i \rangle = 0$ and that $\langle E_i\, E_j \rangle = 0$. The index i here labels K different frequencies for a given channel transition ab. The joint distribution is given by,

$$P(E_1, \cdots, E_K) = \frac{1}{\pi^K \det \mathbf{C}} \exp\left(- \sum_{ij=1}^{K} \bar{E}_i C_{ij}^{-1} E_j \right), \tag{3}$$

where $C_{ij} = \langle E_i \bar{E}_j \rangle$ is the Hermitian variance matrix. We shall normalize $\langle E_j \bar{E}_j \rangle = 1$ for all j, assuming $\langle I_j \rangle$ to be independent of j. For small frequency difference, $\omega_1 - \omega_2 = \omega$, we can expand $C_{12} = 1 + ia\omega + b\omega^2 + \mathcal{O}(\omega^3)$, where a and b can be calculated from diffusion theory [17, 55], involving D, L, L_a, and the transport velocity v_E. The latter contains the Wigner delay time of the scattering objects [56] and thus forms the crucial link between micro and macro dynamics. For the tube geometry in transmission, we use $C(\Omega) = z/\sinh z$ with $z = \sqrt{i\Omega L^2/D + (L/L_a)^2}$. In reflection, see Ref. [56].

Probability distributions can be derived for $K = 2$ using a change of variables $E_j = A_j \exp(i\phi_j)$. As $\omega \to 0$, the variables transform into $I = A^2$, ϕ', $R \equiv \partial \log A/\partial\omega$, and ϕ. Integrating out the latter yields the joint distribution,

$$P(I, \phi', R) = \frac{I}{\pi Q a^2} \exp(-I) \exp\left[-\frac{I}{Qa^2}(\phi' - a)^2 - \frac{I}{Qa^2} R^2 \right]. \tag{4}$$

The distribution functionally depends on a single parameter, $Q \equiv -2b/a^2 - 1 > 0$, which is shown in Fig. 7 as a function of absorption L/L_a. From the diffusion formula for $C(\Omega)$ in transmission used in Ref. [17], it follows that

$$Q = \frac{A^2 - 2\sinh^2 A + \frac{1}{2} A \sinh 2A}{(A\cosh A - \sinh A)^2}, \tag{5}$$

with $A = L/L_a$. In the absence of absorption, the mean delay time $\langle \phi' \rangle = a$ equals the diffuse transit time $L^2/6D$ for waves in transmission, and

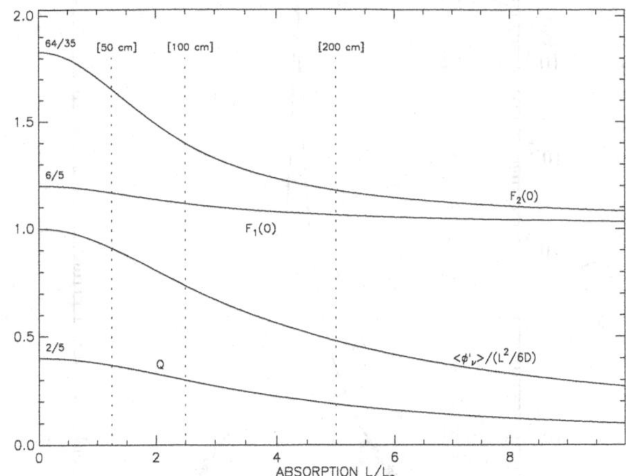

Figure 7. Several parameters showing up in the statistics of the delay time are shown here as a function of the absorption L/L_a in transmission through a thick slab of length L. Dashed vertical lines estimate their values in the experiments with different lengths of the tube. The function $F_1(0)$ denotes the short range "C_1" contribution to $\langle W^2 \rangle - \langle W \rangle^2$; $\frac{1}{g}F_2(0)$ is the non-Gaussian contribution calculated from the C_2 speckle theory. Q is the dimensionless parameter determining the probability distribution of the quantities W and ϕ'. The average delay time $\langle \phi'_\nu \rangle$ has been normalized to the diffuse traversal time $L^2/6D$. The relation $\langle (\Delta\phi_\nu)^2 \rangle \sim \langle \phi_\nu \rangle$ for the cumulative phase at frequency ν has been established in Eq. 13 and the proportionality factor equals the integral in this equation. (Genack *et al.* (1999) [41]).

$4L/3v_E$ in reflection. Even in the Gaussian approximation, I and ϕ' are dependent observables: for fixed values of the intensity I, the delay time is normally distributed with spread $\Delta\phi'/\langle\phi'\rangle = \sqrt{Q/2I}$, in agreement with microwave measurements [41]. The phase at nulls in the speckle pattern is ill-defined [54], giving rise to strong fluctuations in ϕ' at low intensities. Upon integration over I and R one finds,

$$P\left(\widehat{\phi'} \equiv \frac{\phi'}{\langle\phi'\rangle}\right) = \frac{1}{2}\frac{Q}{\left[Q + (\widehat{\phi'} - 1)^2\right]^{3/2}}. \tag{6}$$

This algebraic law agrees with microwave measurements in transmission [41] over *seven* orders of magnitude, as can be seen in Fig. 8. Figure 9 demonstrates the validity of this law in transmission of red light ($\lambda = 685nm$) from highly disordered semiconductor samples. The complex phase is much more difficult to measure for red light. In this case, this was facilitated by heterodyne detection [57]. Schomerus *et al.* [58] confirmed this formula using RMT and was able to extend it to the localized regime, about which we shall speak later.

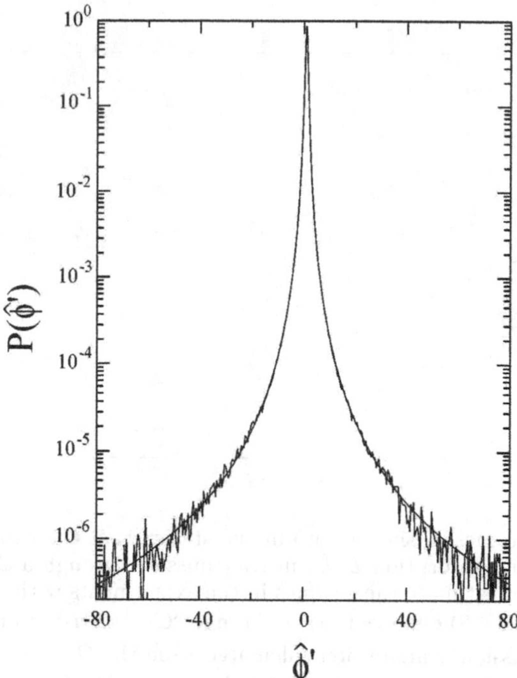

Figure 8. Distribution of the phase derivative with frequency, measured with microwaves in a quasi-1D sample with Thouless conductance $g \approx 5$. The solid line denotes the theoretical prediction (6) with a value $Q = 0.31$ that corresponds to the known amount of absorption $L/L_a = 5$. (Genack *et al.* (1999) [41]).

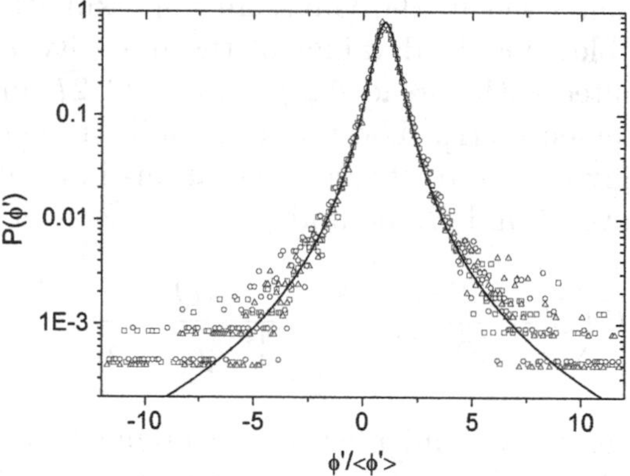

Figure 9. Distribution of the phase derivative with frequency, measured with red light ($\lambda = 685$ nm) in 3D highly disordered semiconductor powders Pa-GaP ($k\ell \approx 2$) and A-GaP ($k\ell \approx 5$) of different thicknesses ($L = 5$, 10, 18 μm). The solid line is the theoretical prediction (6) with $Q = 0.4$, that is the maximum possible value predicted by the diffusion approximation in the absence of absorption. Other measurements confirm that $L_a > 60$ μm, i.e. much longer than the sample lengths. (Lagendijk *et al.* (2001) [57], with kind permission of the authors).

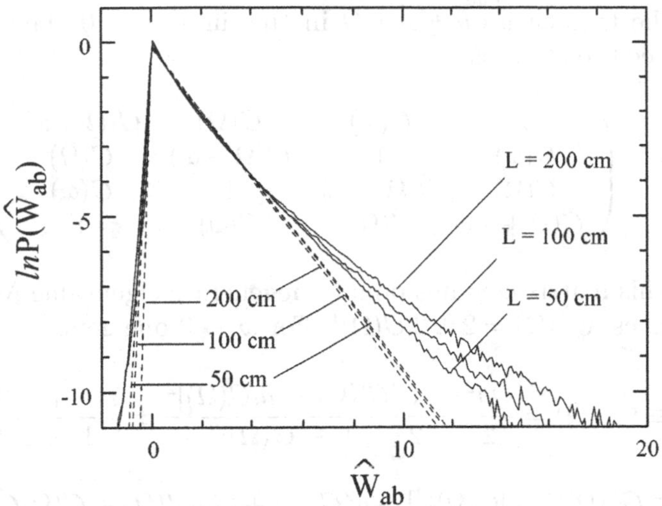

Figure 10. Probability distribution of the $W_{ab} = I_{ab} \mathrm{d}\phi_{ab}/\mathrm{d}\omega$, measured with microwaves, compared to Gaussian theory. Note the presence of significant non-Gaussian statistics (Genack *et al.* (1999) [41]).

Equation (6) has the property that $\langle (\phi')^2 \rangle = \infty$, though any finite frequency grid $\Delta\omega$ transforms this divergence into a finite result, $-\log \Delta\omega$. From Eq. (4), we find for the distribution of the dynamic matrix element W_{ab},

$$P\left(\widehat{W} \equiv \frac{W}{\langle W \rangle}\right) = \frac{1}{\sqrt{Q+1}} \exp\left(\frac{-2|\widehat{W}|}{\theta(\widehat{W}) + \sqrt{Q+1}}\right), \tag{7}$$

where $\theta(x) = x/|x|$. Here, $\langle W \rangle = \langle I_{ab} \rangle a \sim L/2Mv_E$ for both transmission and reflection. Just as for I, W has an exponential distribution. But unlike I, W can be negative. Though less probable, the existence of negative "delay" times is an interesting feature that is also observed in experiments [41] and is allowed by scattering theory [59]. Negative values of delay are most probable in "dark spots". In transmission through a thick slab without absorption ($Q = 2/5$), positive values for W are 12 times more probable than negative values. In reflection [56], $Q \approx (3L/7\ell)^2 \gg 1$, implying nearly equal probabilities for positive and negative W_{ab}.

3.2. FREQUENCY CORRELATIONS OF PHASE DELAY

Correlation functions at two close frequencies provide a sensitive test of the validity of the Gaussian approximation in the experiment. The frequency correlation function of ϕ'_{ab} and W_{ab} can be obtained from Eq. (3) with $K=4$ at the frequencies $\nu \pm \omega \pm \Omega$ in the limit $\omega \to 0$. The correlation matrix we need to study is,

$$\mathbf{C}(\omega,\Omega) = \begin{pmatrix} 1 & C(\omega) & C(\Omega) & C(\Omega+\omega) \\ \bar{C}(\omega) & 1 & C(\Omega-\omega) & C(\Omega) \\ \bar{C}(\Omega) & \bar{C}(\Omega-\omega) & 1 & C(\omega) \\ \bar{C}(\Omega+\omega) & \bar{C}(\Omega) & \bar{C}(\omega) & 1 \end{pmatrix}. \tag{8}$$

For $\omega = 0$, this matrix contains doubly degenerate eigenvalue $\lambda_{1,2} = 0$ and two eigenvalues $\lambda_{3,4}(\Omega) = 2 \pm 2|C(\Omega)|$. For $\omega \to 0$ one gets,

$$\frac{\lambda_{1,2}(\Omega)}{\omega^2} \equiv \xi_{1,2}(\Omega) = \frac{Qa^2}{2} + \frac{1}{2}\frac{|C'(\Omega) - iaC(\Omega)|^2}{1-|C(\Omega)|^2} \pm \frac{1}{2}\frac{|z(\Omega)|}{1-|C(\Omega)|^2}, \tag{9}$$

with $z(\Omega) \equiv \bar{C}''(\Omega)[1-|C(\Omega)|^2] - a^2\bar{C}(\Omega) + 2ia\bar{C}'(\Omega) + C(\Omega)\bar{C}'(\Omega)^2$. The corresponding four eigenfunctions can be derived straightforwardly, among which the first two will be required to order ω. A careful and tedious analysis for $\omega \to 0$ in the subspace spanned by the first two eigenvectors provides an expression for the joint distribution $P(A_1, A_3, A'_1, A'_3, \phi_1, \phi_3, \phi'_1, \phi'_3)$ for the 4 complex fields [14], containing many correlations of which we shall discuss only two. The calculation of the frequency correlation function of $W_{ab}(\nu)$ involves straightforward integrations that can all be done analytically,

$$\begin{aligned} \left\langle \widehat{W}_{ab}(\nu-\Omega/2)\widehat{W}_{ab}(\nu+\Omega/2)\right\rangle_c &= \frac{1}{2a^2}\left(|C'(\Omega)|^2 - \mathrm{Re}\, C(\Omega)\bar{C}''(\Omega)\right) \\ &\equiv F_1(\Omega)\,. \end{aligned} \tag{10}$$

The variance, $(\Delta\widehat{W}_{ab})^2 = F_1(0)$, has been plotted in Fig. 7 as a function of absorption. Also plotted is $F_2(0)$, which is the non-Gaussian (C_2) contribution calculated from the Hikami box published by Nieuwenhuizen and Van Rossum [60], whose technical details were reported in Ref. [14].

The derivation of the frequency correlation function for the single channel delay time $\phi'_{ab}(\nu)$ involves more work. With $R_\phi \equiv \mathrm{Re}\, C(\Omega)\mathrm{e}^{i\phi}$, we find

$$\begin{aligned} \left\langle \frac{\mathrm{d}\phi}{\mathrm{d}\nu}(\nu-\Omega/2)_{ab}\frac{\mathrm{d}\phi}{\mathrm{d}\nu}(\nu+\Omega/2)_{ab}\right\rangle &= \frac{1}{4\pi^2(1-|C|^2)}\int_{-2\pi}^{2\pi}\mathrm{d}\phi\,(2\pi-|\phi|) \\ &\left[-2\mathrm{Re}\left(z\mathrm{e}^{-i\phi}\right)\mathcal{H}_0\left(R_\phi\right)\right. \\ &\left. + \frac{\mathcal{H}_2\left(R_\phi\right)\left(\mathrm{Im}^2\gamma_1 + \mathrm{Im}^2\gamma_2\mathrm{e}^{i\phi}\right) - 2\mathrm{Im}\,\gamma_1\,\mathrm{Im}\,\gamma_2\,\mathcal{H}_1\left(R_\phi\right)}{1-R_\phi^2}\right], \end{aligned} \tag{11}$$

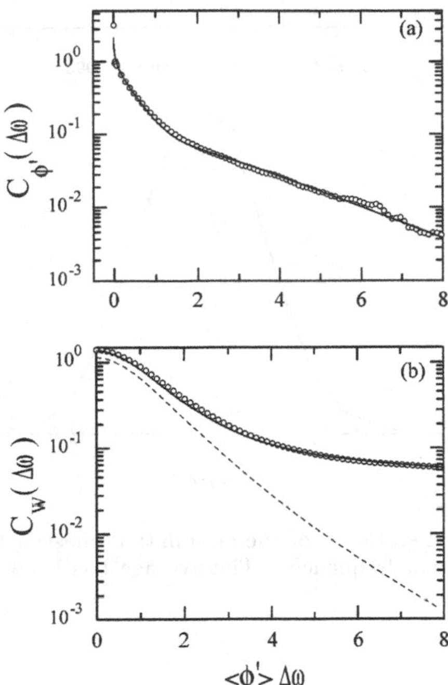

Figure 11. Frequency correlation functions of $d\phi_{ab}/d\omega$ (top) and $W_{ab} = I_{ab}d\phi_{ab}/d\omega$ (bottom), measured with microwaves, compared to theory. For W_{ab}, the C_2 contribution was added, assuming a Thouless conductance $g = 2.5$. The dashed line is the C_1 prediction, which clearly fails (Genack *et al.* (1999) [41]).

where $z(\Omega)$ and $C(\Omega)$ have been defined earlier, $\gamma_1(\Omega) \equiv C(\Omega)\bar{C}'(\Omega) + ia$, $\gamma_2(\Omega) \equiv iaC(\Omega) - C'(\Omega)$, $\mathcal{H}_0(x) = \arctan\left(\sqrt{(1+x)/(1-x)}\right)/\sqrt{1-x^2}$, $\mathcal{H}_1 = 2\mathcal{H}_0 + x$, and $\mathcal{H}_2 = 2x\mathcal{H}_0 + 1$.

The good fit of experimental data to both Eq. (6) and Eq. (12) is shown in Fig. 11. For W_{ab}, we require the C_2 contribution to achieve a good fit to the data, but for the phase derivative itself the Gaussian theory seems perfect. As $\Omega \to 0$, Eq. (12) predicts a logarithmic divergence, not encountered in intensity correlation functions, but which seems to exist in the experiment.

3.3. STATISTICS OF CUMULATIVE PHASE

The phase at a particular frequency and point in space is defined only modulo 2π, since only the real and imaginary parts (in- and out-of-phase components of the electromagnetic field) are direct observables. The cumulative phase φ is defined such that it becomes a continuous variable with

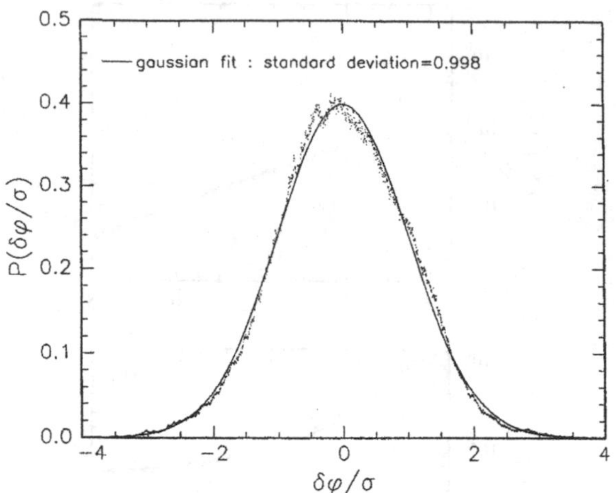

Figure 12. The observed statistics of the cumulative phase with frequency, measured with microwaves at different frequencies. The average has been subtracted. (Sebbah *et al.* (1997) [11]).

frequency or position: each time the jump in phase is greater than π, we add or subtract 2π. In the diffuse regime, we have $\langle E \rangle = 0$ so that $\langle \phi \rangle = 0$. But for the cumulative phase, this is not the case.

Mathematically, the cumulative phase with frequency can be defined as,

$$\varphi(\omega) = \int_{\omega_0}^{\omega} \mathrm{d}\omega \, \frac{\mathrm{d}\phi}{\mathrm{d}\omega}. \tag{12}$$

The integral is by construction a continuous function with frequency. In the previous section, we showed that the phase delay time $\mathrm{d}\phi/\mathrm{d}\omega$ has a short-range frequency correlation with range D/L^2. If $\omega - \omega_0$ is much larger than this Thouless frequency, the cumulative phase is — by Eq. (12) — basically a sum of many independent, identically distributed random variables. We thus expect it to be normally distributed, with the average $\langle \varphi \rangle$ and the variance $var(\varphi)$ as the only parameters.

This conjecture is confirmed in Fig. 12. It shows measurements taken at many different frequencies and then normalized to the standard deviation σ, yielding a universal Gaussian curve. We also observed a linear relation between the variance of the cumulative phase and its average [11]. This relation can easily be understood from Eq. (12). We have,

$$var(\varphi) \quad = \quad \int_{\omega_0}^{\omega} \mathrm{d}\omega \int_{\omega_0}^{\omega} \mathrm{d}\omega' \left\langle \frac{\mathrm{d}\phi}{\mathrm{d}\omega} \frac{\mathrm{d}\phi}{\mathrm{d}\omega'} \right\rangle$$

$$
\begin{aligned}
&\approx \int_{\omega_0}^{\omega} \mathrm{d}\omega \left\langle \frac{\mathrm{d}\phi}{\mathrm{d}\omega} \right\rangle^2 \int_{-\infty}^{\infty} \mathrm{d}\Delta\omega\, C_\phi \left(\Delta\omega \times \frac{\mathrm{d}\phi}{\mathrm{d}\omega} \right) \\
&= \int_{\omega_0}^{\omega} \mathrm{d}\omega \left\langle \frac{\mathrm{d}\phi}{\mathrm{d}\omega} \right\rangle \times \int_{-\infty}^{\infty} \mathrm{d}x\, C_\phi(x) \\
&= \langle \varphi \rangle \times \int_{-\infty}^{\infty} \mathrm{d}x\, C_\phi(x). \qquad (13)
\end{aligned}
$$

This relation confirms the linearity between the variance of the phase and its average. The constant of proportionality is given by the integral of the phase-delay correlation (12), and is shown in Fig. 7 as a function of absorption. The value $1.0\,\langle \phi'_\nu \rangle^2$ for this integral near $L/L_a \approx 5$ coincides with the experimental value reported in Ref. [11].

4. Statistics of Dynamics for Localized Waves

As stated above, the mesoscopic wave dynamics reflects the interaction between the single-channel delay time ϕ' and the intensity I, which is evident in the joint probability distribution $P(\phi', I)$ or the conditional probability distribution $P_I(\phi')$ for a fixed intensity. For diffusive waves, the distribution $P_I(\phi')$ is a Gaussian centered at $\phi' = \langle \phi' \rangle_I$ independent of I, i.e. $\langle \phi' \rangle_I / \langle \phi' \rangle \equiv \langle \widehat{\phi}' \rangle_I = 1$, whereas the variance of ϕ' depends upon I according to $var(\phi')_I / \langle \phi' \rangle^2 = var(\widehat{\phi}')_I = Q/2I$ [41]. But the statistics of dynamics for localized waves differ fundamentally from those for diffusive waves since the transmission spectrum appears as a series of narrow spikes, reflecting the condition $\delta\nu < \Delta\nu$, where $\Delta\nu$ and $\delta\nu$ are the linewidth and spacing between quasi-states in the random medium. Unlike diffusive waves, for which $\langle W \rangle \equiv \langle \phi' I \rangle = \langle \phi' \rangle \langle I \rangle$, long delay times for localized waves are associated with peaks in the transmitted intensity, associated with resonant tunneling through localized states so that $\langle W \rangle / \langle \phi' \rangle \langle I \rangle > 1$. Thus the correlation between the delay time and intensity also provides a dynamical test of photon localization, as will be discussed below. The interaction between ϕ' and I in diffusive samples with a considerable degree of long-range intensity correlation, was observed to lead to a departure from Gaussian statistics of $P_I(\phi')$ [61, 62] and $P(W)$ (see Fig. 10) for large ϕ' and I.

The focus of theoretical work on the statistics of dynamics in localized samples has mostly been on reflected waves [63]. Recent theoretical work focussed on the distribution of proper delay times and resonance widths near the mobility edge, reporting universal powerlaw behavior [40]. The possible application to random lasers was recently discussed by Patra [64]. The delay time statistics for localized waves transmitted through a disordered waveguide was studied by Schomerus [65]. The first measurements of the statistics of dynamics for localized waves were performed with microwaves in quasi-one-dimensional random samples of dielectric spheres [61].

In these experiments, we measured the microwave field transmission coefficient $\sqrt{I_{ab}}\exp(i\phi_{ab})$ in ensembles of randomly positioned alumina spheres. The amplitude $\sqrt{I_{ab}}$ and the phase ϕ_{ab} of the field at the output surface, referenced to the field at the input surface, were obtained using a Hewlett-Packard HP8722C vector network analyzer. Alumina spheres of diameter 0.95 cm and dielectric constant 9.86 were contained in a 7.3-cm-diameter copper tube at a volume fraction of 0.068. This low density was achieved by embedding the alumina spheres in 1.9-cm-diameter Styrofoam spheres with dielectric constant 1.08. The measurements were carried out in samples with lengths ranging from 25 to 90 cm.

The degree of localization in these samples is given by the variance of the total transmission normalized to its ensemble average value, $var(s_a)$, where $s_a = \Sigma_b I_{ab}/\langle \Sigma_b I_{ab}\rangle$. At a threshold of order unity, $var(s_a)$ crosses over from a nearly linear increase for extended waves to an exponential for localized waves [66, 68, 67]. Calculated values of $var(s_a)$ as well as measurements of scaling of $var(s_a)$ and $\langle I_{ab}\rangle$ in identical alumina samples have allowed us to establish that, as the sample length increases, the wave becomes localized in a narrow frequency range centered at $f \approx 10$ GHz, slightly above the first Mie resonance of the alumina spheres (see [69], in this volume).

The probability distribution $P(\widehat{\phi'} \equiv \phi'/\langle\phi'\rangle)$ of the normalized single-channel delay time in the frequency interval of 9.94–10.1 GHz within the localization window, in samples with $L = 25$ and 90 cm, are shown in Fig. 13. The values of $var(s_a)$ in these samples are 1.0 and 7.1, and the ensemble-averaged values of ϕ' are $\langle\phi'\rangle = 21$ and 122 ns, respectively. As Fig. 13 suggests, $P(\widehat{\phi'})$ becomes more asymmetrical as L increases, and reaches its peak at values of $\widehat{\phi'}$ further below its average value of unity. The measured distributions $P(\widehat{\phi'})$ of Fig. 13 are similar to the calculated distributions of Ref. [65], shown in Fig. 14, though a closer comparison reveals a discrepancy in the tails that may be due to the presence of absorption in alumina samples.

The conditional probability distribution $P_I(\widehat{\phi'})$ is of particular interest. In contrast to the Gaussian distribution for diffusive waves, the distribution $P_I(\widehat{\phi'})$ for localized waves was found to exhibit an exponential fall-off, with an asymmetry in the distribution, which increases with decreasing I [61]. The variation of $\langle\widehat{\phi'}\rangle_I$ and $var(\widehat{\phi'})_I$ with $\widehat{I} \equiv I/\langle I\rangle$ for different values of L are presented in Fig. 15. As seen in Fig. 15a, $\langle\widehat{\phi'}\rangle_I$ increases markedly with $\widehat{I}$ to an extent, which increases with L. This deviation from diffusive behavior is consistent with the expected trend for resonant transmission through localized modes. The variation of $var(\widehat{\phi'})_I$, seen in Fig. 15b, is

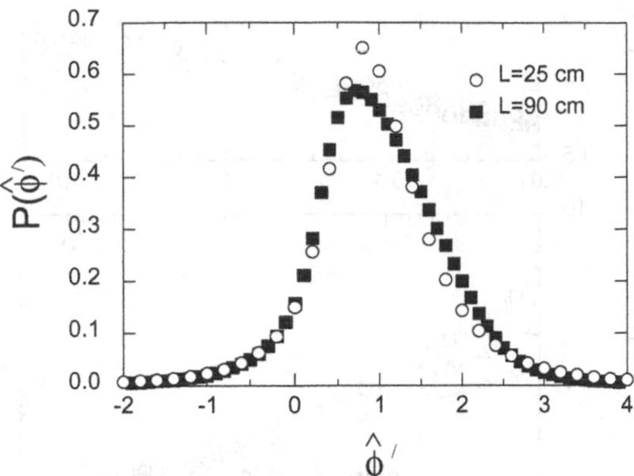

Figure 13. Probability distribution of the normalized delay time, $P(\widehat{\phi'})$, for localized waves in alumina samples with $L = 25$ (circles) and 90 cm (squares).

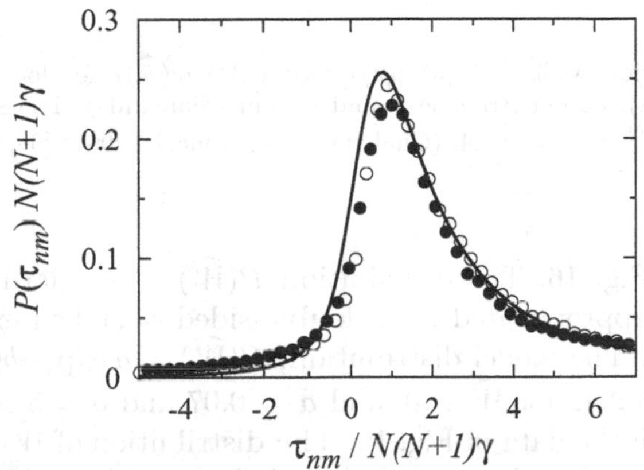

Figure 14. Distribution of transmission delay time τ_{mn} obtained from a numerical simulation of random scattering in planar waveguides with $N = 2$ (open dots) and $N = 30$ (filled dots) channels. The solid line is the analytical prediction (from Eq. (29) of Ref. [65]), independent of N. (Schomerus (2001) [65], with kind permission of the author).

even more striking. For all sample lengths, we find that for $\widehat{I} > 0.5$, $var(\widehat{\phi'})_I$ converges to $q/(\widehat{I})^{1/4}$, with $q = 0.4$, shown as the solid line in Fig. 15b. For smaller values of $\widehat{I}$, $var(\widehat{\phi'})_I$ becomes smaller and, at the same time, is falling more slowly with $\widehat{I}$, as the sample length increases. For $L = 90$ cm, $var(\widehat{\phi'})_I = 0.4/(\widehat{I})^{1/4}$ for any measured $\widehat{I}$.

The probability distributions of the normalized weighted delay time, $P(\widehat{W} \equiv W/\langle W\rangle)$, for localized waves in alumina samples with $L = 90$

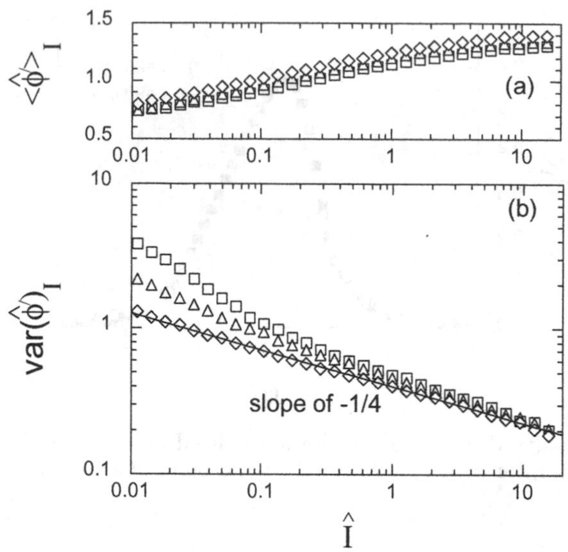

Figure 15. Variation with $\widehat{I}$ of (a) $\langle\widehat{\phi'}\rangle_I$ and (b) $var(\widehat{\phi'})_I$ for localized waves for $L = 49$ cm (squares), 65 cm (triangles), and 90 cm (diamonds). The solid line in (b) is $var(\widehat{\phi'})_I = q/(\widehat{I})^{1/4}$, with $q = 0.4$. (Chabanov and Genack (2001) [68]).

cm is shown in Fig. 16. The distribution $P(\widehat{W})$ is extraordinarily broad and can be well approximated by a double-sided stretched exponential to the power of 1/3. The model distribution, $P(\widehat{W}) = a\exp(-b|\widehat{W}|^{1/3})$, with $a = 0.44$ and $b = 2.42$ for $\widehat{W} > 0$, and $a = 0.07$ and $b = 5.50$ for $\widehat{W} < 0$, is plotted through the data in Fig. 16. The distribution of $\widehat{W}$ is wider than that of $\widehat{I}$ [66], reflecting the enhanced probability of long dwell times and the increased variance of dwell times at large values of the intensity for localized waves.

We have previously shown that $var(\widehat{I})$ as well as $var(s_a)$ serve as indicators of localization, even in the presence of absorption [66, 67, 68]. We find that the interaction between $\widehat{\phi'}$ and $\widehat{I}$ may also be used to identify the range of localization. The correlation between $\widehat{I}$ and $\widehat{\phi'}$ can be expressed as the dimensionless ratio, $\langle\widehat{I}\,\widehat{\phi'}\rangle \equiv \langle I\phi'\rangle/\langle I\rangle\langle\phi'\rangle$. The frequency variation of this ratio in a sample with $L = 80$ cm is plotted in Fig. 17. It is unity in the diffusive limit, since $P_I(\widehat{\phi'})$ is then a Gaussian centered at $\langle\widehat{\phi'}\rangle_I = 1$, and rises above unity, as localization is approached. $\langle\widehat{I\phi'}\rangle$ and $var(\widehat{I})$ both rise appreciably only within the localization window (see Fig. 2c of [69]). The localization threshold, at which $var(\widehat{I}) = 7/3$, corresponds to the condition $\langle\widehat{I}\,\widehat{\phi'}\rangle = 1.1$, shown as the dotted line in Fig. 17. Thus dynamical as well as

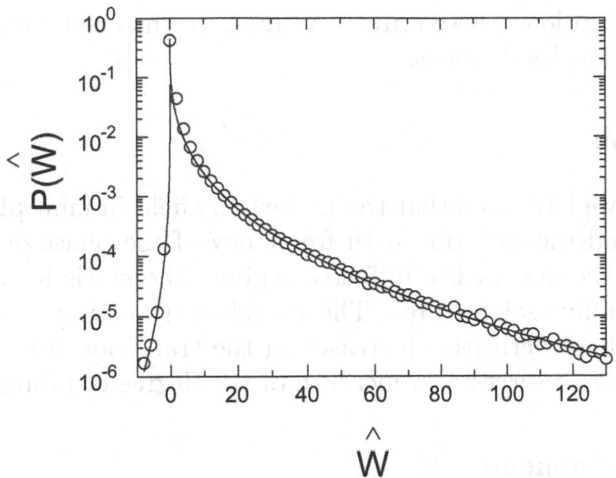

Figure 16. Probability distribution of the normalized weighted delay time, $P(\widehat{W})$, for localized waves (circles) in alumina samples with $L = 90$ cm. The solid line is the model distribution, $P(\widehat{W}) = a\exp(-b|\widehat{W}|^{1/3})$, with $a = 0.44$ and $b = 2.42$ for $\widehat{W} > 0$, and $a = 0.07$ and $b = 5.50$ for $\widehat{W} < 0$.

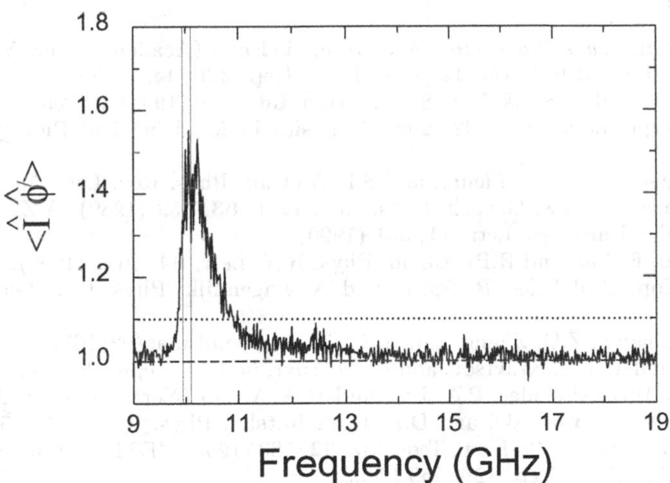

Figure 17. Dimensionless ratio, $\langle \widehat{I}\,\widehat{\phi'} \rangle \equiv \langle I\phi' \rangle / \langle I \rangle \langle \phi' \rangle$, versus frequency for $L = 80$ cm. The dashed line corresponds to the value of unity of this ratio in the diffusive limit [41]. The dotted line represents the condition $\langle \widehat{I}\,\widehat{\phi'} \rangle = 1.1$, which corresponds to the localization criterion $var(\widehat{I}) = 7/3$ [69]. The peak above this line indicates the window of localization. The frequency interval used in computing the statistics of dynamics for localized waves in Figs. 13, 15, and 16 is marked by the vertical lines.

steady state criteria for localization can be formulated, which are robust in the face of absorption. These criteria adumbrate the emergence of discrete and long-lived localized modes.

5. Conclusion

In conclusion, we have seen that the underlying field in multiply scattering static media and the variation with frequency of the phase determine the statistics of dynamics. In the diffusive regime, the statistics of dynamics exhibit many universal features. The correlation between the amplitude and spectral phase derivative increases in the transition from diffusive to localized waves and serves as a measure of the degree of localization.

6. Acknowledgements

We thank O. Legrand and M. Stoytchev for stimulating discussions and contribution to this work. This work was supported by the National Science Foundation, the US Army Research Office, and the Groupements de Recherches POAN and PRIMA of the CNRS.

References

1. L. Brillouin, *Wave Propagation and Group Velocity* (Academic, New York, 1960).
2. A. Lagendijk and B.A. van Tiggelen, Phys. Rep. **270**, 143 (1996).
3. R.H.J. Kop and R. Sprik, Rev. Sci. Instrum. **66**, 5459 (1995); J. Pearce and D. Mittleman, Opt. Lett. **26**, 2002 (2001); Physics in Medicine and Biology, **47**, 3823 (2002).
4. G.H. Watson, Jr., P.A. Fleury, and S.L. McCall, Phys. Rev. Lett. **58**, 945 (1987); J.M. Drake and A.Z. Genack, Phys. Rev. Lett. **63**, 259 (1989); A.Z. Genack and J.M. Drake, Europhys. Lett. **11**, 331 (1990).
5. K.M. Yoo, F. Liu, and R.R. Alfano, Phys. Rev. Lett. **64**, 2647 (1990).
6. R.H.J. Kop, P. deVries, R. Sprik, and A. Lagendijk, Phys. Rev. Lett. **79**, 4369 (1997).
7. A.A. Chabanov, Z.Q. Zhang, and A.Z. Genack, cond-mat/0211651.
8. B.L. Altshuler, V.E. Kravtsov, and I.V. Lerner, in *Mesoscopic Phenomena in Solids*, edited by B.L. Altshuler, P.A. Lee, and R.A. Webb (North Holland, Amsterdam, 1991); B.A. Muzykantskii and D.E. Khmelnitskii, Phys. Rev. B **51**, 5480 (1995); A.D. Mirlin, Pis'ma Zh. Eksp. Teor. Fiz. **62**, 583 (1995) [JETP Lett. **62**, 603 (1995)].
9. A.D. Mirlin, Phys. Rep. **326**, 259 (2000).
10. H. Cao, Y.G. Zhao, S.T. Ho, E.W. Seelig, Q.H. Wang, and R.P.H. Chang, Phys. Rev. Lett. **82**, 2278 (1999); S.V. Frolov, Z.V. Vardeny, and K. Yoshino, Phys. Rev. B **57**, 9141 (1998).
11. P. Sebbah, O. Legrand, B.A. van Tiggelen, and A.Z. Genack, Phys. Rev. E **56**, 3619 (1997).
12. P. Sebbah, O. Legrand, and A.Z. Genack, Phys. Rev. E **59**, 2406 (1999).
13. P. Sebbah, R. Pnini, and A.Z. Genack, Phys. Rev. E **62**, 7348 (2000).
14. B.A. van Tiggelen, P. Sebbah, M. Stoytchev, and A.Z. Genack, Phys. Rev. E **59**, 7166 (1999).

15. A. Z. Genack, J. H. Li, N. Garcia, and A. A. Lisyansky, in *Photonic Band Gaps and Localization*, edited by C. M. Soukoulis (Plenum Press, New York, 1993).
16. J.W. Goodman, *Statistical Optics* (John Wiley, New York, 1985).
17. R. Berkovits and S. Feng, Phys. Rep. **238**, 135 (1994).
18. A.Z. Genack, Phys. Rev. Lett. **58**, 2043 (1987); N. Garcia and A.Z. Genack, Phys. Rev. Lett. **63**, 1678 (1989); A.Z. Genack, N. Garcia, and W. Polkosnik, Phys. Rev. Lett. **65**, 2129 (1990).
19. M.P. van Albada, J.F. de Boer and A. Lagendijk, Phys. Rev. Lett. **64**, 2787 (1990).
20. B. Shapiro, Phys. Rev. Lett. **57**, 2168 (1986).
21. G. Cwilich and M. Stephen, Phys. Rev. Lett. **59**, 285 (1987).
22. F. Sheffold, W. Härtl, G. Maret, and E. Matijević, Phys. Rev. E **56**, 10942 (1997).
23. I. Freund, M. Rosenbluh, and S. Feng, Phys. Rev. Lett. **61**, 459 (1988).
24. P.A. Lee, Physica **140A**, 169 (1986).
25. R.A. Webb, S. Washburn, C.P. Umbach, and R.B. Laibowitz, Phys. Rev. Lett. **54**, 2696 (1985).
26. E. Kogan, M. Kaveh, R. Baumgartner, and R. Berkovits, Phys. Rev. B **48**, 9404 (1993).
27. M. Stoytchev and A.Z. Genack, Phys. Rev. Lett. **79**, 309 (1997).
28. See: C.W.J. Beenakker, Rev. Mod. Phys. **69**, 731 (1997).
29. M.H. Pedersen, S.A. van Langen, and M. Büttiker, Phys. Rev. B **57**, 1838 (1998); M. Büttiker, A. Prêtre, and H. Thomas, Phys. Rev. Lett. **70**, 4114 (1993).
30. Y.V. Fyodorov and H. J. Sommers, Phys. Rev. Lett.**76**, 4709 (1996).
31. P.W. Brouwer, K.M. Frahm, and C.W.J. Beenakker, Phys. Rev. Lett. **78**, 4737 (1997).
32. B.L. Altshuler and B.I. Shklovskii, Sov. Phys. JETP **64**, 127 (1986).
33. O.N. Dorokhov, JETP Lett. **36**, 318 (1982).
34. P.A. Mello, P. Pereyra, and N. Kumar, Ann. Phys. (N.Y.) **181**, 290 (1988).
35. E. Mucciolo, R.A. Jalabert, and J.L. Pichard, J. Phys. I (France) **10**, 1267 (1997).
36. T. Guhr, A. Müller-Groeling, and H.A. Weidemüller, Phys. Rep. **299**, 189 (1998).
37. M. Titov and Y.V. Fyodorov, Phys. Rev. B **61**, R2444 (2000).
38. C.J. Bolton-Heaton *et al.*, Phys. Rev. B **60**, 10569 (1999).
39. C. Texier and A. Comtet, Phys. Rev. Lett. **82**, 4220 (1999).
40. T. Kottos and M. Wiess, Phys. Rev. Lett. **89**, 056401 (2002).
41. A.Z. Genack, P. Sebbah, M. Stoytchev, and B.A. van Tiggelen, Phys. Rev. Lett. **82**, 412 (1999).
42. S. Feng, C. Kane, P. Lee, and A.D. Stone, Phys. Rev. Lett. **61**, 834 (1988).
43. J.M. Jauch, K.B. Sinha, and B.N. Misra, Helv. Ph. Acta **45**, 398 (1972); R.G. Newton, *Scattering Theory of Waves and Particles* (Springer-Verlag, New York, 1982), section 11.13.
44. R. Landauer, Philos. Mag. **21**, 863 (1970); E.N. Economou and C.M. Soukoulis, Phys. Rev. Lett. **46**, 618 (1981); D.S. Fisher and P.A. Lee, Phys. Rev. B **23**, 6851 (1981).
45. see G.D. Mahan, *Many Particle Physics* (Plenum, New York, 1981), section 4.1C.
46. V.L. Lyuboshits, Phys. Lett. B **72**, 41 (1977).
47. E. Akkermans, A. Auerbach, J.E. Avron, and B. Shapiro, Phys. Rev. Lett. **66**, 76 (1991).
48. B.A. van Tiggelen and E. Kogan, Phys. Rev. A **49**, 708 (1994).
49. R.A. Jalabert and J.L. Pichard, J. Phys. France **5**, 287 (1995).
50. W. Jaworski and D.M. Wardlaw, Phys.Rev.A **37**, 2843 (1988); A.M. Steinberg, Phys. Rev. A **52**, 32 (1995).
51. M. Büttiker, in: *Electronic Properties of Multilayers and Low-Dimensional Semiconductor Structures*, edited by J.M. Chamberlain *et al.* (Plenum, New York, 1990), p. 297.
52. B.A. van Tiggelen, A. Tip, and A. Lagendijk, J. Phys. A (math. physics) **26**, 1731 (1993).

53. H.J. Sommers, D. Savin, and V.V. Sokolov, Phys. Rev. Lett. **87**, 094101 (2001).
54. I. Freund, Waves in Random Media **8**, 119 (1998).
55. A.Z. Genack, in: *Scattering and Localization of Classical Waves in Random Media*, edited by Ping Sheng (World Scientific, Singapore, 1990).
56. M.P. van Albada, B.A. van Tiggelen, A. Tip, and A. Lagendijk, Phys. Rev. Lett. **66**, 3132 (1991).
57. A. Lagendijk, J. Gomez Rivas, A. Imhof, F.J.P. Schuurmans, and R. Sprik, in: *Photonic Crystals and Light Localization in the 21st Century*, edited by C.M. Soukoulis (Kluwer Academic, 2001), p. 447.
58. H. Schomerus, K.J.H. van Bemmel, and C.W.J. Beenakker, Phys. Rev. E, 026605 (2001).
59. E.P. Wigner, Phys. Rev. **98**, 145 (1955); F.T. Smith, Phys. Rev. **118**, 349 (1960).
60. Th.M. Nieuwenhuizen and M.C.W. van Rossum, Phys. Lett. A **177**, 102 (1993).
61. A.A. Chabanov and A.Z. Genack, Phys. Rev. Lett. **87**, 233903 (2001).
62. J. Pearce, Z. Jian, and D. Mittleman, Statistics of multiply scattered broadband terahertz pulses, manuscript in preparation (2003).
63. C.W.L. Beenakker, in: *Photonic Crystals and Light Localization in the 21st Century*, edited by C.M. Soukoulis (Kluwer Academic, 2001), pp. 489–08.
64. M. Patra, Phys. Rev. E **67**, 016603 (2003).
65. H. Schomerus, Phys. Rev. E **64**, 026606 (2001).
66. A.A. Chabanov, M. Stoytchev, and A.Z. Genack, Nature (London), **404**, 850 (2000).
67. A.Z. Genack, and A.A. Chabanov, in *Waves and Imaging through Complex Media*, edited by P. Sebbah (Kluwer Academic Publishers, 2001), p. 53.
68. A.A. Chabanov and A.Z. Genack, Phys. Rev. Lett. **87**, 153901 (2001).
69. A.A. Chabanov and A.Z. Genack, Photon Localization in Resonant Media, in this volume.

DIFFUSING ACOUSTIC WAVE SPECTROSCOPY: FIELD FLUCTUATION SPECTROSCOPY WITH MULTIPLY SCATTERED ULTRASONIC WAVES

Ultrasonic Wave Transport and Spectroscopy in Complex Media I

J.H. PAGE[1] AND M.L. COWAN[2]
Department of Physics and Astronomy, University of Manitoba, Winnipeg MB R3T 2N2 Canada

D.A. WEITZ
Department of Physics and DEAS, Harvard University, Cambridge, Massachusetts 02138 U.S.A.

AND

B.A.VAN TIGGELEN
Laboratoire de Physique et Modélisation des Milieux Condensés, Maison des Magistères CNRS, 25 Avenue des Martyrs, B.P. 166, 38042 Grenoble Cedex 9, France

Abstract. Diffusing Acoustic Wave Spectroscopy (DAWS) is a powerful technique in field fluctuation spectroscopy for investigating the dynamics of strongly scattering media and studying mesoscopic wave phenomena. The principles underlying DAWS are described and illustrated with measurements of the particle velocity fluctuations in fluidized suspensions. Two examples of the potential of DAWS for elucidating mesoscopic wave physics are presented: understanding the phase statistics of temporally fluctuating multiply scattered fields, and investigating the breakdown of the Siegert relation for multiply scattered waves when correlations exist in the scattering medium.

[1]Author to whom correspondence should be addressed. E-mail: jhpage@cc.umanitoba.ca. Website: http://www.physics.umanitoba.ca/~jhpage.

[2]Present address: Dept. of Physics, University of Toronto, 60 St. George St., Toronto, Ontario M5S 1A7 Canada.

B.A. van Tiggelen and S.E. Skipetrov (eds.),
Wave Scattering in Complex Media: From Theory to Applications, 151–174.

1. Introduction

The role that acoustic waves, and ultrasonic waves in particular, are playing in understanding the rich diversity of wave phenomena in complex media [1–4] is becoming increasing appreciated. In large part, this role reflects the fact that ultrasonic techniques measure the wave field directly, without the need for complicated interferometry, and are normally easier to perform with pulses than with continuous waves (cw). The advantage of the first is that it gives direct experimental access to the wave function and/or Green's function, while the second allows the dynamics of the wave fields to be explored, as well as the time-of-flight distribution or path-length dependence of multiply scattered waves. Furthermore, scattering contrast is governed by differences in both density and velocity, offering versatile control of the scattering strength. Thus, experiments with acoustic and elastic waves can make important contributions to both fundamental studies and practical applications of wave scattering in complex media, and are complementary to optical and microwave methods for investigating these phenomena.

This is the first of two papers that summarize recent progress in using ultrasonic waves to explore two different aspects of wave transport and spectroscopy in strongly scattering media. In this paper we focus on random systems in which the scatterers are moving, and describe a technique in field fluctuation spectroscopy, called Diffusing Acoustic Wave Spectroscopy (DAWS), that uses multiply scattered waves to measure their dynamics [5–8]. We review the application of this technique to investigating the dynamics of particles in fluidized suspensions, where a detailed understanding of the complex particulate flows has remained elusive, despite the fact that the dynamics are mediated by one of the simplest many-body interactions, the hydrodynamic interactions between particles in a liquid. We also show how DAWS provides an opportunity to study fundamental properties of multiply scattered waves through the measurement of phase and amplitude fluctuations. In the second paper, we turn to ordered systems and examine ultrasonic wave propagation and tunnelling in three-dimensional phononic crystals, where the character of waves is strongly modified by the existence of band gaps and anisotropy of the wave speeds.

The organization of this paper is as follows. In section 2, we present measurements and theory of the pulsed DAWS field autocorrelation function, highlighting the dynamic quantities that can be measured in fluidized suspensions of particles. Section 3 deals with measurements of the phase and amplitude of multiply scattered waves, focusing on the phase statistics of time-varying fields and what can be learned from the comparison of field and intensity correlation functions. The paper ends with some concluding remarks and prospects for future work.

2. The Pulsed DAWS Field Autocorrelation Function and Scatterer Dynamics

When the scatterers in a multiple scattering material move, the speckle pattern fluctuates, reflecting the changes that occur in the interference of waves travelling different scattering paths through the sample. In Diffusing Acoustic Wave Spectroscopy, these fluctuations of the multiply scattered wave field $\psi(T)$ are measured in one (or more) speckle spots and analyzed to provide a sensitive technique for probing the dynamics of the scatterers. The most direct way of determining the movement of the scatterers is through the autocorrelation function of the scattered acoustic field,

$$g_1(\tau) = \frac{\int \psi(T)\, \psi^*(T+\tau)\, dT}{\int |\psi(T)|^2 dT}. \tag{1}$$

The relationship between $g_1(\tau)$ and the scatterers' dynamics can be illustrated qualitatively by noting that $g_1(\tau)$ decays to approximately 1/2 when the total rms change in phase of the scattered field due to the scatterers' motion is about a radian. Here we use T to denote the time scale on which the fluctuations in the scattered field are measured, as distinct from the propagation time t of acoustic waves in the sample ($t \ll T$). To facilitate the calculation of $g_1(\tau)$ when the waves are multiply scattered, and hence link the phase change directly to the dynamics of the scatterers, we model the propagation of sound though the material using the diffusion approximation. Thus, we take advantage of the simple physical picture of wave transport in a multiple scattering medium as a random walk process along paths characterized by a step length equal to the transport mean free path l^*, which is related to the energy velocity v_e and the diffusion coefficient D by $D = v_e l^*/3$. In this paper we will illustrate the technique by reviewing the application of DAWS to investigating the motion of particles in a fluidized bed, where all the scatterers are moving in locally correlated flow patterns that evolve rapidly in time. In this case, DAWS measures the relative mean square displacement of the scattering particles that are separated, on average, by a distance equal to the transport mean free path. The method can obviously be generalized to many other scenarios, such as those where only a fraction of the scatters may move or where the type of motion or system dynamics is entirely different.

Pulsed DAWS, in which the incident signal is a short pulse or tone burst, provides an important advantage over a cw (continuous wave) approach in that it allows the field fluctuations to be measured for multiple scattering paths of fixed length $s = (n+1)l^*$, where n is the number of steps. If we isolate one such path, the phase change resulting from the motion of all the scatterers p along the path is

$$\begin{aligned}\Delta\phi^{(n)}(\tau) &= \sum_{p=0}^{n}\Delta\phi_p(\tau)\\ &= \sum_{p=0}^{n}\left[\mathbf{k}_p\cdot\left(\Delta\mathbf{r}_{p+1}(\tau)-\Delta\mathbf{r}_p(\tau)\right)\right]\\ &\simeq \sum_{p=1}^{n-1}\mathbf{k}_p\cdot\Delta\mathbf{r}_{rel,p}(\tau,l^*),\end{aligned} \tag{2}$$

where we have neglected (for the time being) the phase change due to motion of the first and last scatterer relative to the source and detector, as this is a small contribution for large n. Here $\mathbf{k}_p$ is the wave vector of the wave scattered from the pth to the $(p+1)^{th}$ particle, and $\Delta\mathbf{r}_{rel,p}(\tau) = \Delta\mathbf{r}_{p+1}(\tau) - \Delta\mathbf{r}_p(\tau)$ is their relative displacement during the time interval τ. By averaging over all paths with n steps, assuming that the successive phase shifts $\Delta\phi_p(\tau)$ are uncorrelated, and using a cumulant expansion, the pulsed field correlation function can be written in terms of the variance of the phase change per scattering event p as

$$\begin{aligned}g_1(\tau) = \left\langle\exp\left[-i\Delta\phi^{(n)}(\tau)\right]\right\rangle &\simeq \left\langle\exp\left[-i\Delta\phi_p(\tau)\right]\right\rangle^n\\ &\simeq \exp\left[-\left\langle\Delta\phi_p^2(\tau)\right\rangle n/2\right]\end{aligned} \tag{3}$$

Hence $g_1(\tau)$ can be written in terms of the relative mean square displacement of the scatterers using Eq. (2), from which we obtain $\left\langle\Delta\phi_p^2(\tau)\right\rangle = \left\langle(\mathbf{k}_p\cdot\Delta\mathbf{r}_{rel,p})^2\right\rangle = \left\langle\left(k_p\,\Delta r_{rel,p}(\theta_p)\,\cos\theta_p\right)^2\right\rangle$, where θ_p is the angle between $\Delta\mathbf{r}_{rel,p}$ and $\mathbf{k}_p$. In particular, when there are no correlations between the directions of $\Delta\mathbf{r}_{rel,p}$ and $\mathbf{k}_p$, the phase change can be written as $\left\langle\Delta\phi_p^2(\tau)\right\rangle = k^2\left\langle\Delta r_{rel}^2\right\rangle/3$. Thus, the autocorrelation function takes on the simple form:

$$g_1(\tau) \approx \exp\left[-\frac{nk^2}{6}\left\langle\Delta r_{rel}^2(\tau,l^*)\right\rangle\right]. \tag{4}$$

This equation shows clearly that the decay of $g_1(\tau)$ is governed by the relative mean square displacement $\left\langle\Delta r_{rel}^2(\tau,l^*)\right\rangle$ of particles that are separated by the average step length l^* of the diffusing sound's random walk paths through the sample.

For fluidized suspensions where the spatial correlations decay rapidly, Eq. (4) is a very good approximation for $g_1(\tau)$. However, for other types of flows, it may be necessary to account for correlations in θ_p. For example, in uniform shear flow, $\left\langle\Delta\phi_p^2(\tau)\right\rangle = 0.6k^2\left\langle\Delta r_{rel}^2\right\rangle/3$, while in pure rotational flow $\langle \Delta\phi_p^2(\tau)\rangle = 0$ [8]. In general, these correlations can be accounted for formally by expressing $g_1(\tau)$ in terms of the average local strain as

$$g_1(\tau) \approx \exp\left[-\frac{nk^2l^{*2}}{6}\bar{\varepsilon}^2(\tau, l^*)\right], \tag{5}$$

where

$$\bar{\varepsilon}^2 \equiv \tfrac{2}{5}\left[\left\langle\left(\sum \varepsilon_{ii}\right)^2\right\rangle + 2\sum_{i,j}\langle\varepsilon_{ij}^2\rangle\right], \tag{6}$$

$\varepsilon_{i,j}$ is the strain tensor

$$\varepsilon_{ij}(\tau) = \tfrac{1}{2}\left(\frac{\partial u_i(\tau)}{\partial r_j} + \frac{\partial u_j(\tau)}{\partial r_i}\right), \tag{7}$$

and $u_i(\tau)$ are the components of $\mathbf{\Delta r}(\tau)$ [8, 9, 10]. For uncorrelated motion, $\bar{\varepsilon}^2 = \langle \Delta r_{rel}^2\rangle / l^*$, and Eq. (5) reduces to Eq. (4) as expected.

To illustrate the application of pulsed DAWS to investigating the dynamics of particulate suspensions, experiments were performed on a fluidized bed containing 0.88-mm-diameter glass beads in a liquid mixture of water and glycerol. By flowing the liquid vertically upward to counterbalance gravitational sedimentation, stable suspensions could be achieved over a wide range of particle volume fractions ϕ by varying the flow velocity V_f. Even under quiescent conditions at low flow velocities, the fluctuations in the particle velocities are remarkably large, and DAWS offers a novel approach for addressing some of the outstanding scientific challenges in understanding their complex behaviour. As discussed elsewhere in more detail [5, 11], some of the important questions that can be investigated using DAWS include how the dynamics of the particles are influenced by volume fraction, system size and Reynolds number, which measures the relative importance of viscous and inertial effects in the hydrodynamic interactions between the particles.

The majority of our experiments [8] to investigate these effects have been performed by sending a train of short ultrasonic pulses, with a central carrier frequency of typically 2 MHz, towards the sample and measuring the scattered field at a particular transit time t_s after each pulse has entered the sample. The experiments described here were performed in transmission mode, with the fluidized bed immersed in a large water tank, using a small hydrophone to detect the scattered field within a single speckle spot on the far side of the sample from the incident beam. In this pulsed realization of DAWS, the pulse repetition frequency sets the rate at which the field fluctuations are measured, and the sampling time t_s sets the average path length of diffusing sound in the sample and hence the number of scattering events

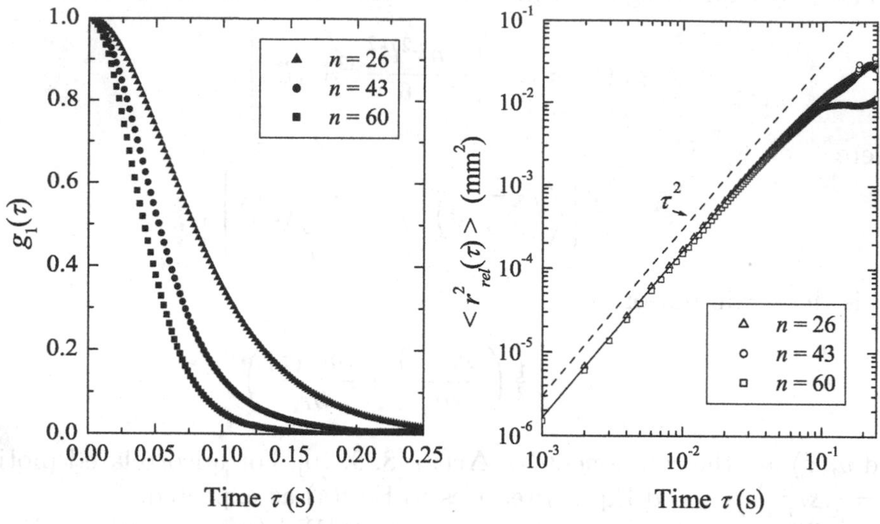

Figure 1. Field autocorrelation functions for several values of n in pulsed DAWS (left) and corresponding values of the mean square relative displacements as a function of time for particles in a fluidized bed (right).

n since $n \approx s/l^* - 1 \approx s/l^* = v_e\, t_s/l^*$. From digitized records of temporal field fluctuations at a particular value of n, acquired using a digital oscilloscope PC card, the field correlation function was calculated numerically using fast Fourier transforms and the correlation theorem, which allows the correlation function to be efficiently determined. An example of pulsed DAWS field autocorrelation functions measured for different values of n is shown in Fig. 1. As t_s and hence n increases, the fluctuations become more rapid due to accumulation of larger total phase changes along the longer paths involved, and the correlation function decays more quickly. Making use of independent measurements of the transport mean free path l^*, the energy velocity v_e and the wave vector $k = \omega/v_p$ of diffusing acoustic waves in the suspension [8, 12–16], Eq. (4) can be inverted to determine the relative mean square displacements of the particles, shown on the right side of Fig. 1. Despite the large differences in the correlation functions measured at the different sampling times, essentially identical values of the relative mean free paths are found, indicating that the dependence on path length in Eq. (4) correctly describes the data for this range of values of n. It is also worth noting that measurements at large values of n (longer t_s) give better sensitivity to small displacements at early times (down to nearly a micron, or $\approx \lambda/500$ for these data), because of the more rapid decay of the correlation function noted above. By contrast, smaller n (shorter t_s) can be used to increase the sensitivity to larger displacements of the particles, as the correlation function can be measured for longer times before it reaches the noise level; this is evident in Fig. 1 where it can be seen that the measurements for $n = 60$ saturate at $\langle \Delta r_{rel}^2(\tau) \rangle < 0.01$, while the maximum measurable value of $\langle \Delta r_{rel}^2(\tau) \rangle$ for $n = 26$ is a factor of 3-4 times bigger.

Figure 1 also shows that $\langle \Delta r_{rel}^2(\tau) \rangle$ varies quadratically with time, at least for $\tau < 0.1$ s, indicating that the particles follow ballistic trajectories for short times. Eventually, these ballistic trajectories become altered by interactions with neighbouring particles in the suspension, and the rate at which $\langle \Delta r_{rel}^2(\tau) \rangle$ increases slows down. We represent this behaviour with the empirical expression

$$\langle \Delta r_{rel}^2(\tau) \rangle = \frac{\langle \Delta V_{rel}^2 \rangle \tau^2}{1 + (\tau/\tau_{cl})^2} \tag{8}$$

where $\langle \Delta V_{rel}^2 \rangle$ is the variance in the relative velocities of the particles and τ_{cl} is the local crossover time, or the average time interval during which the local relative motion of the particles is not impeded by interparticle"collisions". The solid curve in Fig. 1 shows a fit of this expression to the data, and gives an excellent description of the time dependence over four orders of magnitude in $\langle \Delta r_{rel}^2(\tau) \rangle$ and more than two orders of magnitude in τ. From this fit, an accurate measurement of the rms relative velocity fluctuations of the particles $\Delta V_{rel} = \sqrt{\langle \Delta V_{rel}^2 \rangle}$ is obtained, as well as a good indication of the average change in particle separation $\Delta d_{sep} = \Delta V_{rel} \tau_{cl}$ before interactions modify the particle trajectories.

The path length dependence of the measurements in pulsed DAWS is examined in more detail in Fig. 2, where data at shorter path lengths are also included. In order to correctly analyze data for short paths (small n), it is necessary to consider the contributions to g_1 due to the motion of the first and last scatterer relative to the source and detector, respectively. As shown by Cowan *et al.* [8], this can be accomplished by adding a correction term to $\langle \Delta r_{rel}^2(\tau, l^*) \rangle$ in Eq. (4) given by $[\langle \Delta r_{rel}^2(\tau, R) \rangle - \langle \Delta r_{rel}^2(\tau, l^*) \rangle]/n$. Here R is the linear distance between the first and last scatterer. This correction term is zero if the motion of the scatterers is uncorrelated for $R \geq l^*$, but its magnitude is otherwise of order $1/n$. For fluidized suspensions, where there are significant local correlations in the velocities of the particles, this term may not be negligible if the path length is short enough. The experimental data indicated by the symbols in Fig. 2 show that this is indeed the case. In this figure, we plot ΔV_{rel} determined from Eq. (4) without this correction, divided by an extrapolation of ΔV_{rel} as $n \to \infty$. For $n < 20$, this ratio

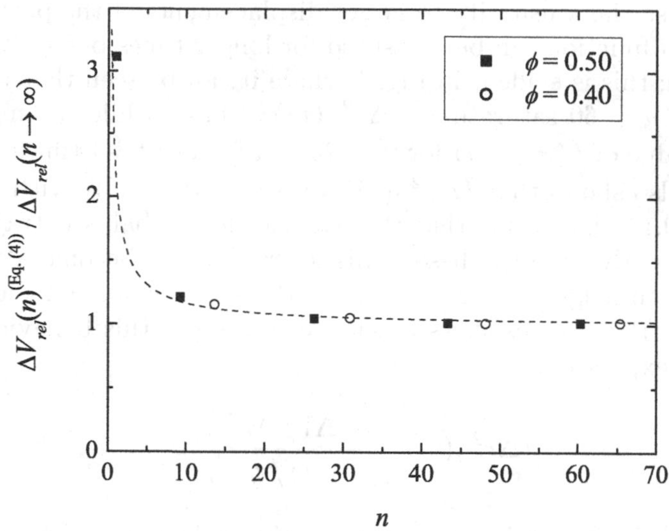

Figure 2. Values of ΔV_{rel} as a function of the number of scattering events n determined from Eq. (4), without accounting from the motion of the first and last scatterer with respect to the source and detector. The data are normalized by the limiting values of ΔV_{rel} at large n, and compared with a fit to the correction term described in the text (dashed curve).

becomes significantly larger than unity, showing that appreciable errors in the measurement of ΔV_{rel} will result if the correction is not taken into account. Fig. 2 also compares a fit of the correction term to the data for $\phi = 0.50$, normalized in the same way, which is shown by the dashed curve. In this fit, there is one unknown: $\langle \Delta V_{rel}^2 (R) \rangle = \langle \Delta r_{rel}^2 (R) \rangle / \tau^2$. It is clear from Fig. 2 that the correction term does indeed give a satisfactory fit to the data; furthermore it yields a value of $\Delta V_{rel}(R)$ that is physically reasonable, lying between the minimum possible value at $R = L$, where L is the sample thickness, and the maximum possible value at $R = \xi$, where ξ is the velocity correlation length (see below). These results show that this correction term can be used to make measurements for short paths and hence extend the range of paths lengths over which data can be collected. They also show that the simple expression for $g_1(\tau)$ given by Eq. (4), neglecting the motion of the first and last scatterers relative to the source and detector, accurately accounts for the path length dependence in the correlation function for n greater than about 20.

Most of our DAWS measurements have been performed using quasi-

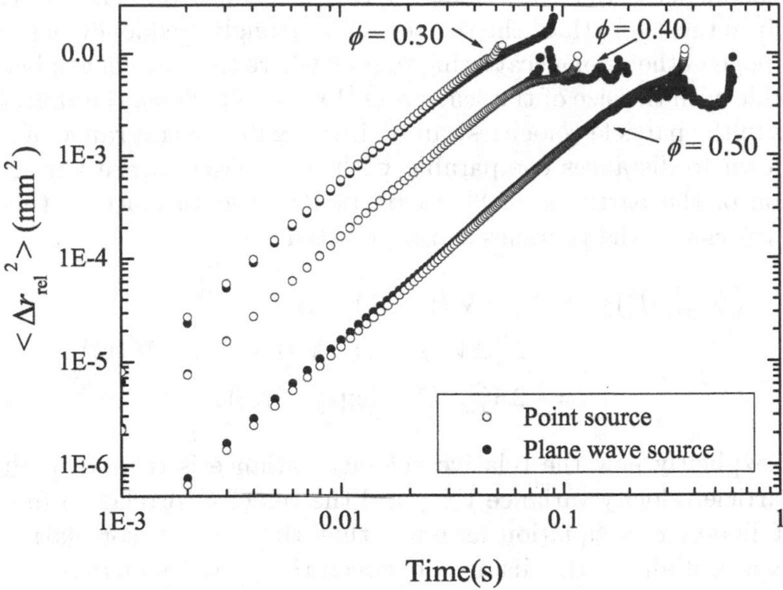

Figure 3. Comparison of DAWS experiments performed with a point source (open symbols) and plane wave source (closed symbols).

plane wave input pulses generated in the far field of planar immersion transducers. To improve signal to noise, a focusing transducer can be used to focus the input pulse on the sample face, producing a quasi-point source. The advantage of a point source is that there is more signal, since all of the input energy is focused down on a small spot, instead of being spread out in a quasi-plane wave. Since extensive signal averaging cannot be used to measure time varying fields, other ways of optimizing signal to noise can be especially important in DAWS. While in principle, the source geometry should not matter, as what counts is only the the average length of the scattering paths that are being measured, it is still important to check that this holds in practice. Fig. 3 shows the results of such a test, where data for $\langle \Delta r_{rel}^2(\tau) \rangle$ using a point source are compared with data obtained with a planar source for three volume fractions ϕ of scatterers. The very good agreement between the data taken with these two different input geometries demonstrates that focusing transducers can indeed be used reliably to increase the single-to-noise ratio in DAWS, and provides a useful check of the robustness of the DAWS technique itself.

As emphasized above, DAWS measures the relative motion of the scatterers $\Delta V_{rel}(R)$ on a length scale determined by the transport mean free path of the ultrasound, l^*. By changing the ultrasonic frequency, l^* can be varied, since the strength of the scattering is strongly frequency dependent near the edge of the strong scattering regime where the wavelength becomes comparable with the size of the scatterers [13, 14, 16]. Thus, the spatial correlations in the particle velocities can be investigated over a range of length scales, down to distances comparable with the average nearest neighbour separation of the particles [5, 8]. In particular, the variance in the local relative velocity of the particles can be written

$$\begin{aligned} \langle \Delta V_{rel}^2 (l^*) \rangle &= \langle [\Delta \mathbf{V} (\mathbf{r} + l^*) - \Delta \mathbf{V} (\mathbf{r})]^2 \rangle \\ &= 2 \langle \Delta V^2 \rangle - 2 \langle \Delta \mathbf{V} (\mathbf{r} + l^*) \cdot \Delta \mathbf{V} (\mathbf{r}) \rangle \\ &= 2\, V_{rms}^2 \left(1 - \exp\left[-l^*/\xi\right]\right), \end{aligned} \tag{9}$$

showing explicitly how the relative velocity variance is related to the absolute particle velocity variance V_{rms}^2 and the velocity correlation function. The last line of this equation assumes that the velocity correlations decrease exponentially as the distance between the particles increases, with a characteristic decay length given be the velocity correlation length ξ. Since V_{rms} can be measured using Dynamic Sound Scattering (DSS) in the single scattering regime at low frequencies [5, 11], the instantaneous velocity correlation length can be measured from the dependence of the relative velocity fluctuations on l^*. Experimental data over a wide range of volume fractions from 0.08 to 0.50 are shown in Fig. 4, where the relative velocity is normalized by the asymptotic value at large distances $\sqrt{2}V_{rms}$, and the measurement length scale l^* is normalized by the correlation length ξ. Fig. 4 shows that for small inter-particle separations, the relative velocity increases as the square root of distance, but levels off as the correlation length is approached. The predictions of Eq. (9) are represented by the solid curve and are in very good agreement with the data, confirming that the velocity correlation function decreases exponentially with distance and indicating the velocity correlation length of the particles can be reliably measured using this technique.

We end this section with a couple of examples of typical data, taken as part of an extensive investigation of the velocity fluctuations in fluidized suspensions [11]. For these data, the particle Reynolds number Re_p is 0.9, a value at which we find very similar behaviour to that in creeping flow conditions at much lower Re_p, even though it is close to the range where inertial effects are expected to become important. Fig. 5(a) shows the volume fraction dependence of the relative velocity fluctuations at a distance given by the nearest neighbour particle separation, the shortest distance in the suspension over which the local relative velocity can be defined. In

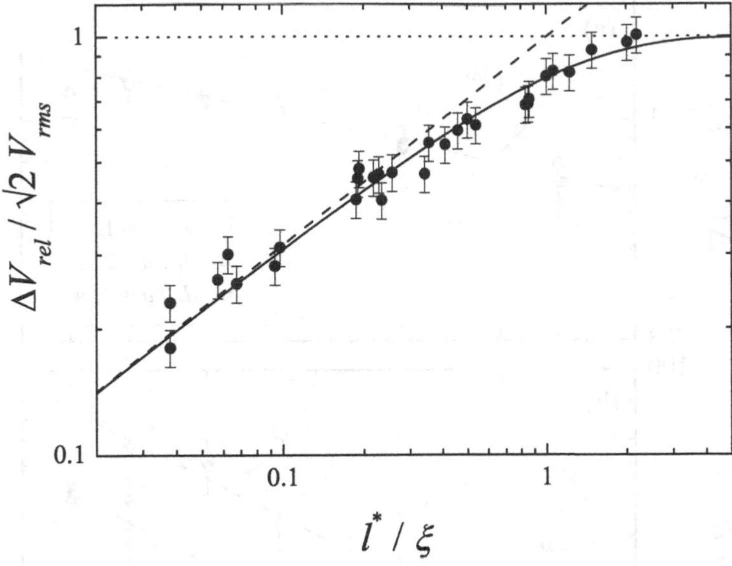

Figure 4. Scaling plot showing the length scale dependence of ΔV_{rel}. Here $\Delta V_{rel}(l^*)$ is normalized by the value reached at large distances where the motion is uncorrelated ($\sqrt{2}V_{rms}$), and the average separation between the particles at each measurement of ΔV_{rel} is normalized by the measured correlation length ξ. The solid curve represents the predictions of Eq. (9), and the dashed line represents the $\sqrt{l^*/\xi}$ dependence seen at short length scales.

this figure, the velocity is normalized by the average fluid velocity V_f. Data from three rectangular cells with different thicknesses L_z are included, the other dimensions of the cells being larger and constant at 1030 and 407 bead radii a. The relative velocity is remarkably large even at these short length scales, being similar in magnitude to V_f, and increases with volume fraction ϕ as $\phi^{1/3}$ up to $\phi \sim 0.4$. At the highest ϕ, the relative velocity drops as the particles move more in step with each other. Note that throughout this range of volume fractions, the local relative velocity is unaffected by the thickness of the cell, even when the shortest distance between the walls is as small as 10 particle diameters. By contrast, the velocity correlation length, shown in Fig. 5(b), shows a larger variation with volume fraction and decreases significantly as the thickness of the cell is reduced. The volume fraction dependence is similar for the different cells, despite the fact that the magnitude of the correlation length may be either substantially larger or smaller than the smallest cell dimension. For $\phi \lesssim 0.2$, the vol-

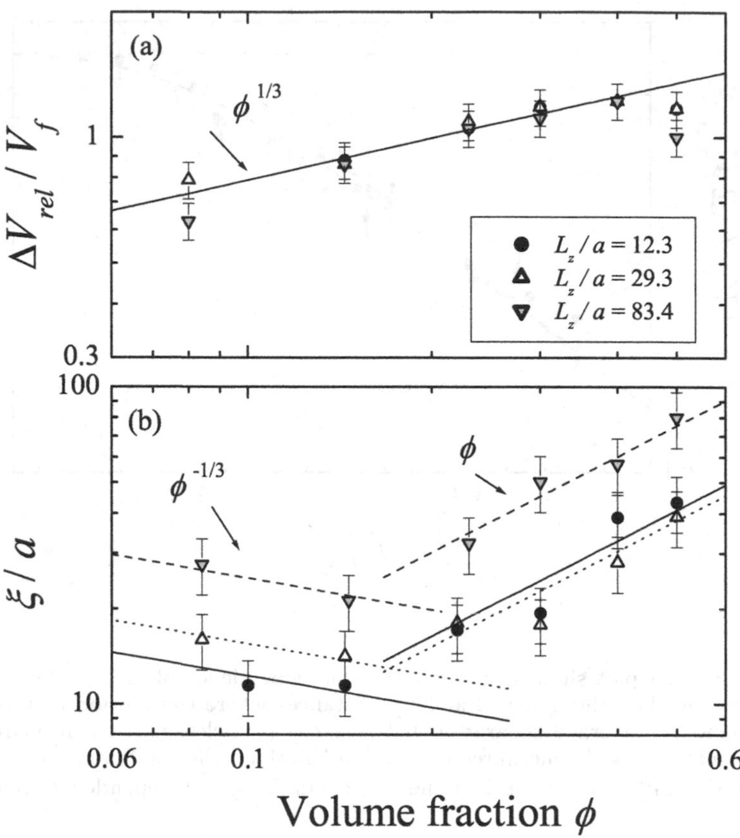

Figure 5. (a) The root mean square relative velocity at the nearest-neighbour particle separation, normalized by the fluid velocity V_f, as a function of volume fraction. Data for three sample thicknesses are shown. The solid line shows the $\phi^{1/3}$ dependence that is consistent with the data up to $\phi \sim 0.4$. (b) The volume fraction dependence of the instantaneous velocity correlation length, normalized by the particle radius, for the same three values of sample thickness. The data are consistent with a $\phi^{-1/3}$ dependence at low ϕ and a linear dependence at high ϕ, as indicated by the solid, dotted and dashed lines which represent fits to the data for the three thicknesses.

ume fraction dependence is consistent with $\phi^{-1/3}$ behaviour expected if the number of particles in the correlation volume is independent of volume fraction and is in agreement with the extrapolation of measurements using particle imaging velocimetry at low ϕ and Re_p [17]. However, at higher ϕ the correlation length increases quite rapidly, approximately linearly in ϕ, indicating that the number of particles in the correlation volume grows rapidly as ϕ^4. The fact that the thickness of the cells influences the magnitude of the correlation length indicates that cell walls play a major role in determining the spatial extent of the velocity correlations, at least in small cells. These data suggest the following simple physical picture: the magnitude of the velocity fluctuations in fluidized suspensions may be set initially at the local level by the relative motion of neighbouring particles in the suspension, the relative motion then growing with the square root of distance until the fluctuations are cut off at the correlation length by cell walls or other screening effects [18, 19, 20]. However, the development of a microscopic model of the particle dynamics observed in our experiments remains an intriguing theoretical challenge.

3. Phase and Amplitude Fluctuations of Multiply Scattered Waves

In addition to being a powerful technique for probing the dynamic behaviour of multiple scattering systems, DAWS provides an almost unique opportunity to investigate fundamental properties of multiply scattered waves. This opportunity arises naturally from the ability of ultrasonic techniques to directly access the multiply scattered wave field, as opposed to the scattered intensity, and the relative ease with which this can be accomplished compared to optical techniques. Thus, the temporal evolution of both the phase and amplitude of the wave field can be measured independently, and their statistical behaviour analyzed. Furthermore, using a pulsed technique, this information can be obtained as a function of the path length of the multiply scattering waves in the sample, potentially adding additional insight into the nature of the wave fields. Here we address the following question: what can we learn using DAWS, both about wave physics and the dynamic properties of complex media, from the amplitude and phase fluctuations of multiply scattered ultrasonic waves?

The measurement of the phase and amplitude of the scattered field was performed as follows (see Fig. 6). Again we use moving spherical glass particles in a fluidized bed as an archetypical dynamic system to illustrate the method. In order to maximize the rate at which the data could be recorded, a short segment of the entire transmitted waveform (top panel of Fig. 6) was recorded for each repetition of the input pulse, each segment being about 4.5 periods long. The segments were centered around a fixed sampling time $t_s = 18$ μs after the input pulse arrived at the sample, each time window corresponding to an average of 34 ± 2 scattering events. Typical data for the scattered field segments, recorded at time intervals separated by 150 repetitions of the input pulse (pulse repetition frequency = 500 Hz), are shown in the left column of the bottom panel of Fig. 6. We

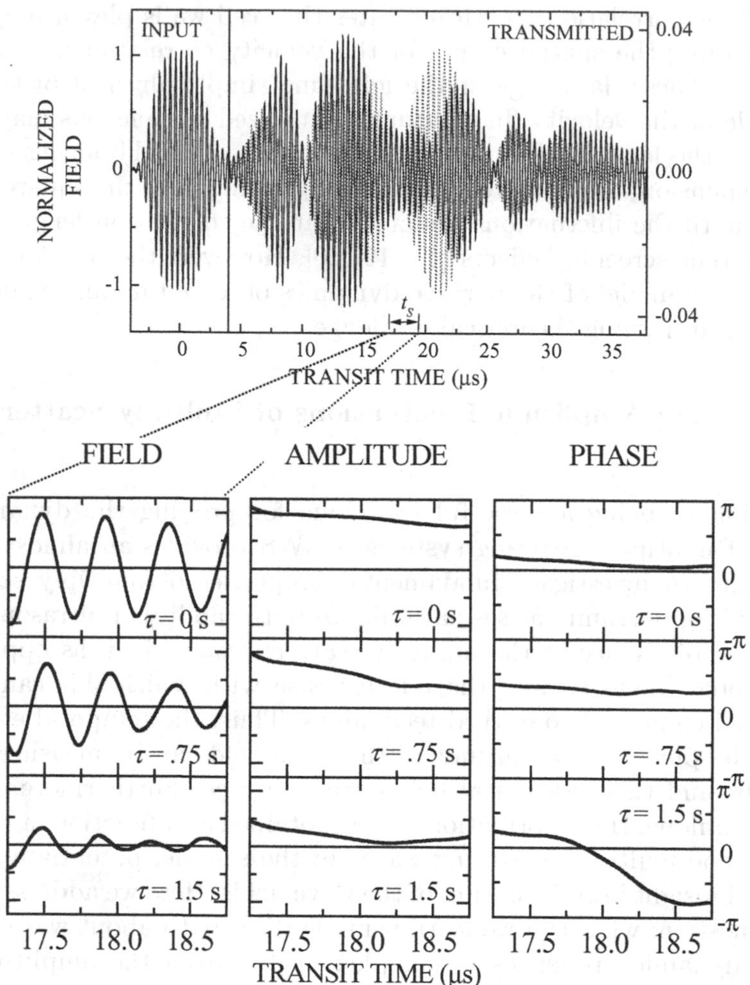

Figure 6. Top panel: The input and transmitted pulses in a typical transmission DAWS experiment. Two transmitted pulses, measured after a sufficiently long time interval that the field has evolved considerably, are shown by the solid and dotted traces. Bottom panel: Three snapshots of the scattered wave field, amplitude and phase (left, middle and right columns) in a 1.5 μs long window after the scatterers have moved during time intervals of $\tau = 0.75$ s.

used a simple numerical technique to determine the phase and amplitude, in which the digitized field data are first multiplied by a sine and cosine wave at the central frequency of the pulse and then low-pass filtered to extract the dc components $S(t)$ and $C(t)$; the amplitude and phase are then given

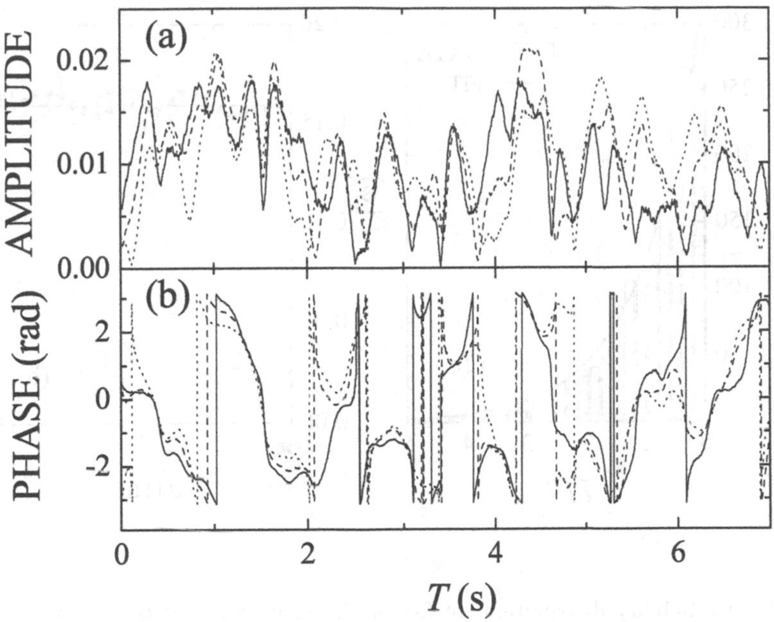

Figure 7. Evolution of (a) the amplitude and (b) the wrapped phase at three different sampling times, corresponding to times near the beginning, at the middle and near the end of the segment windows shown in Fig. 6. The amplitude is expressed in arbitrary units corresponding to the voltage measured on the digital oscilloscope.

by $A(t) = 2\sqrt{S^2(t) + C^2(t)}$ and $\varphi(t) = \tan^{-1}[-S(t)/C(t)]$. The frames in the middle and right columns of the bottom panel illustrate snapshots of the amplitude and wrapped phase $[-\pi : \pi]$ for different positions of the scatterers after they have moved during 0.75 s intervals. To measure the amplitude and phase fluctuations with good statistical accuracy, 10 sets of 8300 consecutive pulses were recorded.

An example of the evolution of the measured amplitude and phase fluctuations at three different sampling times is plotted in Fig. 7. The three times correspond to times which are 0.15, 0.5, and 0.85 of the way across the segment windows shown in Fig. 6. These sampling times are far enough away from each other that the amplitudes and phases are not the same; however, they are still very strongly correlated. The amplitude gets extremely close to zero several times in the 7 seconds shown, which is typical for this data set; on average this occurs about once every 2 seconds. The variations for the same three sampling times are shown in Fig. 7(b). The

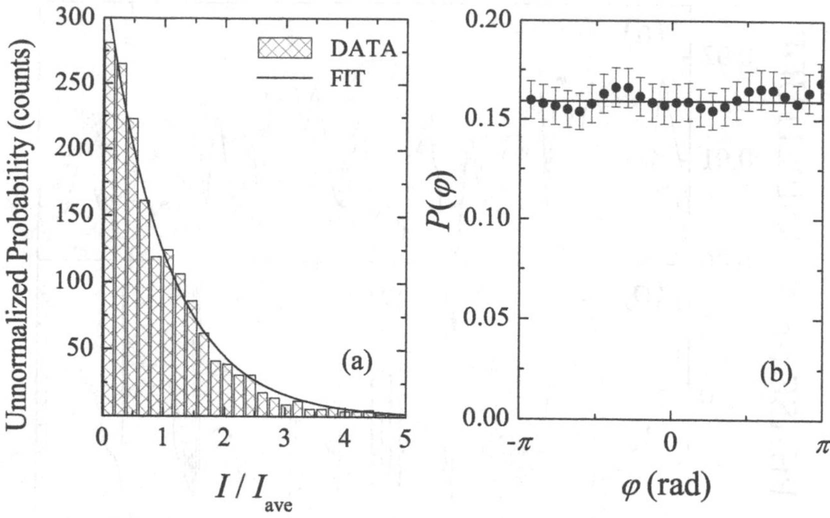

Figure 8. Probability distributions of (a) the intensity and (b) the wrapped phase of multiply scattered acoustic waves in fluidized suspensions.

behavior of the wrapped phase at the three sampling times is very similar, except for differences during the rapid variations of π in the phase that occur in the dark speckles when the amplitude is close to zero.

The first-order statistics of the scattered intensity ($I(T) \propto |A(T)|^2$) and phase ($\varphi(T)$) are shown in Fig. 8. The probability density function for the intensity (Fig. 8(a)) shows the usual negative exponential form characteristic of Rayleigh statistics of uncorrelated speckles, for which $P(I) = (1/I_{ave})\exp(-I/I_{ave})$. The solid curve in Fig. 8(a) is a fit of this distribution to the data, and confirms that the standard deviation is equal to the mean, as expected [21]. Figure 8(b) shows that the corresponding phase distribution function is flat between -π and $+\pi$, i.e. the phases are distributed randomly with $P(\varphi) = 1/(2\pi)$. Together, these results indicate that the multiply scattered wave field is a complex Gaussian random variable, despite the correlations in the motion of the scatterers discussed in the previous section.

3.1. PHASE INFORMATION: THE WRAPPED PHASE DIFFERENCE PROBABILITY DISTRIBUTION

To gain more insight into the phase statistics of multiply scattered waves and to see how information on the scatterer dynamics can be uncovered, we examine the phase difference $\Delta\varphi(\tau) = \varphi(T+\tau) - \varphi(T)$. Because of the random character of the wave field, we are forced to take a statistical approach. The statistics of the phase difference can be obtained from the joint probability distribution $P(\psi_T, \psi_{T'})$ of the fields at times T and $T' = T+\tau$. For a complex Gaussian process, $P(\psi_T, \psi_{T'})$ can be written [21, 22]

$$P(\psi_T, \psi_{T'}) = \frac{1}{\pi^2 \det \mathbf{C}} \exp\left(- \sum_{i,j=T,T'} \psi_i^* C_{ij}^{-1} \psi_j \right), \tag{10}$$

where $C_{ij} = \langle \psi_i \psi_j^* \rangle$ is the covariance matrix. We normalize the fields so that $\langle \psi_i \psi_i^* \rangle = \langle |\psi(T)|^2 \rangle = 1$ and $\langle \psi_i \psi_j^* \rangle = g_1(\tau)$, the field autocorrelation function. Eq. (10) for $P(\psi_T, \psi_{T'})$ can then be rewritten in terms of the amplitude and phase to give

$$\begin{aligned} P(A_T, A_{T'}, \varphi_T, \varphi_{T'}) = & \frac{A_T A_{T'}}{\pi^2 (1-g_1^2)} \\ & \times \exp\left\{ - \frac{A_T^2 + A_{T'}^2 - 2 g_1 A_T A_{T'} \cos(\varphi_T - \varphi_{T'})}{1-g_1^2} \right\}. \end{aligned} \tag{11}$$

The wrapped phase difference probability distribution can then be obtained from Eq. (11) by integrating out the dependence on A_T, A_T' and φ_T' at constant $\Delta\varphi$, giving

$$\begin{aligned} P(\Delta\varphi) = & \frac{2\pi - |\Delta\varphi|}{4\pi^2} \left[\frac{1-g_1^2}{1 - g_1^2 \cos^2(\Delta\varphi)} \right] \\ & \times \left[1 + \frac{g_1 \cos(\Delta\varphi) \arccos\{-g_1 \cos(\Delta\varphi)\}}{\sqrt{1 - g_1^2 \cos^2(\Delta\varphi)}} \right]. \end{aligned} \tag{12}$$

Eq. (12) shows that the wrapped phase difference probability distribution contains information on the dynamics of the scatterers through its dependence on $g_1(\tau)$. At short times and small $\Delta\varphi$, $P(\Delta\varphi)$ can be written as a simple algebraic law

$$P(\Delta\varphi) \cong \frac{\frac{1}{2}\langle \Delta\phi^2(\tau) \rangle}{[\langle \Delta\phi^2(\tau) \rangle + \Delta\varphi^2]^{3/2}}, \tag{13}$$

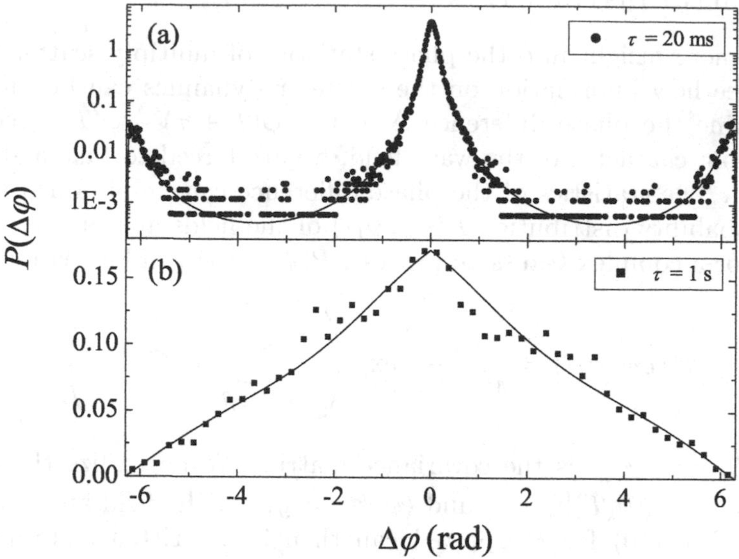

Figure 9. The wrapped phase difference probability distribution for two different time intervals τ. The solid symbols represent the experimental data, and the curves are the single-parameter fits of Eq. (12).

whose width $\langle \Delta\phi^2(\tau) \rangle \simeq n \langle \Delta\phi_p^2(\tau) \rangle$ is the variance of the change in phase for all paths containing n scattering events (c.f. Eq. (3)) and is related to the relative motion of the scatterers through Eq. (4). This equation, which is equivalent to the probability distribution for the phase derivative with time $P(d\varphi/d\tau)$, has the same form as the distribution for the phase derivative with frequency investigated previously with microwaves [23, 22]. At long times, the phase difference probability distribution approaches a triangular function

$$P(\Delta\phi) = \frac{2\pi - |\Delta\phi|}{4\pi^2}, \tag{14}$$

reflecting the complete lack of correlations in the phase difference as $\tau \to \infty$ and the underlying phase distribution is flat in $[-\pi : \pi]$.

These theoretical predictions for the phase derivative probability distribution are compared with experiment in Fig. 9, where our measurements of $P(\Delta\varphi)$ at two different times τ (representative of the behaviour at small and large τ) are plotted along with fits of Eq. (12) to the data. At the relatively short time of $\tau = 20$ ms, the central part of the distribution shows the algebraic form predicted by Eq. (13), but rises again near $\pm 2\pi$

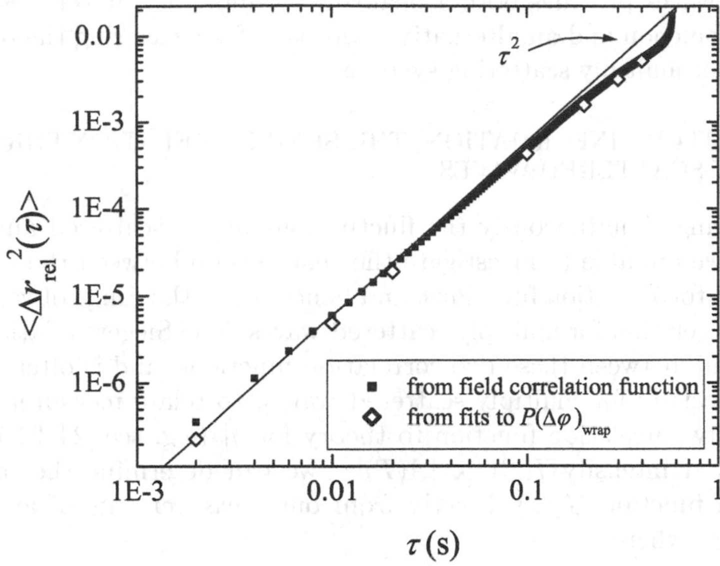

Figure 10. The mean square relative displacement of particles in a fluidized suspension measured from the width of the wrapped phase difference probability distribution (open diamonds) and directly from the field autocorrelation function (solid squares).

due to the effect of the 2π jumps in the phase due to phase wrapping. At the longer time of $\tau = 1$ s, the shape of $P(\Delta\varphi)$ has changed completely and is starting to approach the triangular function predicted by Eq. (14). Thus, excellent agreement between theory and experiment is found over many orders of magnitude in $P(\Delta\varphi)$ and over the entire range of $\Delta\varphi$ from -2π to 2π, providing a very convincing demonstration of the validity of this theoretical model for the phase difference statistics.

Our theory for $P(\Delta\varphi)$ also provides a framework for using phase fluctuations of multiply scattered waves to measure the dynamics of complex media, thereby showing how phase information can be used in a regime where more traditional methods such as Doppler imaging break down. The relative mean square displacement of the scattering particles determined from fitting the full theory for $P(\Delta\varphi)$ (Eq. (12)) to our phase data is plotted as open symbols in Fig. 10 for a range of different times τ. In this figure, the results from the phase measurements are also compared with data obtained from the field correlation function, measured as described in section 2. Again excellent agreement is found, further confirming the validity of our model for the phase difference probability distribution. Thus, we can conclude that this investigation of the phase statistics of multiply scattered ultrasonic waves provides both a beautiful example of a novel mesoscopic wave phenomenon and an alternative approach for measuring the dynamic behaviour of multiply scattering systems.

3.2. AMPLITUDE INFORMATION: THE SIEGERT RELATION FOR MULTIPLY SCATTERED WAVES

By measuring simultaneously the fluctuations in the scattered amplitude and field, we are able to investigate the relationship between the intensity and field autocorrelation functions, and hence probe the range of validity of the Siegert relation for multiply scattered waves. The Siegert relation gives a simple link between these two correlation functions, and is often used in optics, especially for multiply scattered waves, to relate measurements of the intensity correlation function to theory for g_1 (e.g. see [24–27]). Since the scattered intensity $I(T) \propto |A(T)|^2$, we can determine the intensity correlation function $G_2(\tau)$ directly from our measurements of amplitude fluctuations, where

$$G_2(\tau) = \frac{\langle I(T)I(T+\tau)\rangle}{\langle I(T)\rangle^2} = 1 + \beta g_2(\tau). \tag{15}$$

Here β is the coherence factor, which is determined by the detector geometry and is normally close to unity. For complex random Gaussian fields, the correlation function $g_2(\tau)$ is equal to the square of the field correlation function $g_1(\tau)$, as expressed by the Siegert relation:

$$g_2(\tau) = |g_1(\tau)|^2 \tag{16}$$

Since we have shown above that the scattered wave field can be represented as a complex Gaussian random variable under the conditions for which the phase measurements were performed, it would seem natural to assume that this is always the case, and that the Siegert relation is generally obeyed for multiply scattered waves. In this section we study this question critically by examining the validity of the Siegert relation as a function of the path length of multiply scattered waves in fluidized suspensions, where the motions of the scatterers are at least locally correlated.

Pulsed DAWS measurements of the correlation functions $g_2(\tau)$ and $g_1(\tau)$ are shown in Fig. 11 for the same fluidized suspensions (glass beads in a water/glycerol mixture) used as examples of dynamic strongly scattering media throughout this paper. To ensure very efficient data collection, the amplitude of the scattered field was measured using a diode detector, and the field and amplitude sampled at the same point t_s in the scattered wave form using two separate boxcars. (The gate width in the boxcars was set to

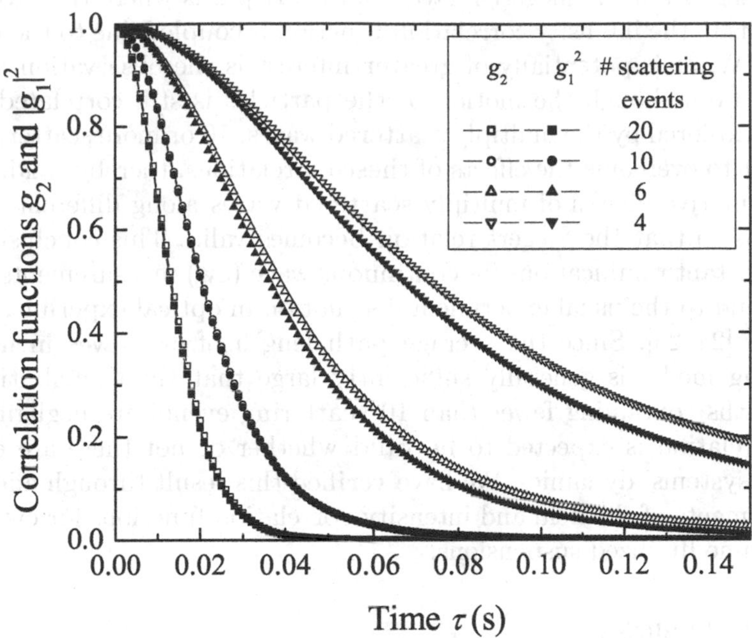

Figure 11. Comparison of the intensity (open symbols) and field (closed symbols) autocorrelation functions measured using pulsed DAWS for different path lengths of diffusing sound. The number of scattering events for each data set is shown in the insert.

be much less than the wave period to enable the field to be accurately captured at a single point in the wave form.) Figure 11 shows that $g_2(\tau)$ and $|g_1(\tau)|^2$ are indistinguishable for long paths. However, for $n < 10$, $g_2(\tau)$ is significantly larger than $|g_1(\tau)|^2$ at all times τ, indicating clearly that the Siegert relation breaks down for short paths in this system. We have investigated this effect for different volume fractions of scatterers ranging from 0.245 to 0.5, and find similar behaviour at all volume fractions. To gain insight into this phenomenon, we compare the volume Ω_ξ over which the particle motions are correlated with the volume Ω_D probed by diffusing acoustic waves for 10 scattering events. From our DAWS and DSS measurements of the velocity correlation length ξ in the suspensions, we find that the particle correlation volume Ω_ξ ranges from about 500 mm^3 at $\phi = 0.245$ to 5000 mm^3 at $\phi = 0.5$. By contrast, simple estimates of Ω_D for $n = 10$ scattering events using the diffusion approximation are much smaller, ranging form 130 mm^3 at $\phi = 0.245$ to 17 mm^3 at $\phi = 0.5$. Thus, the motion of the particles in a path containing 10 steps is strongly correlated, leading to the breakdown of the Siegert relation for short paths where the product of four fields in the intensity correlation function is coupled due to the relative motion. What is potentially of greater interest is the observation that for $n \geq 10$, even though the motion of the particles is still correlated in the volume explored by the multiply scattered waves, 10 or more scatterings are sufficient to overcome the effects of these correlations, thereby randomizing the cumulative phases of multiply scattered waves along different scattering paths so that the Siegert relation becomes valid. This conclusion also has important ramifications for continuous wave (cw) measurements, which correspond to the usual experimental situation in optical experiments such as DWS [24–26]. Since the average path length of cw waves in multiply scattering media is generally sufficiently large that the contributions for short paths containing fewer than 10 scattering events are negligible, the Siegert relation is expected to be valid whether or not there are correlations in systems' dynamics. We have verified this result through additional measurements of the field and intensity correlation functions for cw DAWS in the same fluidized suspensions.

3.3. CONCLUSIONS

Diffusing Acoustic Wave Spectroscopy (DAWS) is a powerful approach for investigating the dynamics of strongly scattering media where direct imaging fails. Because the wavelengths of seismic waves, sound and ultrasound are very much larger that light, this technique and its many potential extensions are applicable to different classes of materials and systems than those that can be studied with the optical technique of DWS. DAWS is based on measurements of the scattered wave field, not intensity, so that it is an example of field fluctuation spectroscopy. This has important consequences not only for investigating the dynamics of complex media but also for studying wave phenomena in the presence of strong multiple scattering. Two examples of the latter have been presented in this paper: understanding the phase statistics of temporally fluctuating multiply scattered fields, and investigating the breakdown of the Siegert relation for multiply scattered waves due to particle velocity correlations.

In this paper, the potential applications of DAWS have been illustrated with experiments on fluidized suspensions of non-Brownian particles. In this case, DAWS provides a very sensitive probe of the local relative velocity of the particles as well as the correlations in the velocities over a wide range of length and time scales. Recently a number of other applications and analogous techniques have been or are being developed, including the monitoring of fish in a reverberant aquarium [28], the sensitive measurement of ultrasonic velocity changes with temperature [29] and the monitoring seismic and laboratory-scale events in geophysics, where the relevance of wave diffusion has recently been convincingly demonstrated [30] and field fluctuation spectroscopy with multiply scattering acoustic waves has been given the seismically relevant name of Coda Wave Interferometry [31]. Many other applications in the nondestructive evaluation of complex media will likely emerge in the future.

4. Acknowledgements

Support from Natural Sciences and Engineering Research Council of Canada is gratefully acknowledged.

References

1. *Scattering and Localization of Classical Waves in Random Media*, edited by P. Sheng (World Scientific, Singapore, 1990).
2. P. Sheng, *Introduction to Wave Scattering, Localization, and Mesoscopic Phenomena*, (Academic Press, San Diego, 1995).
3. *Diffuse Waves in Complex Media*, edited by J.-P. Fouque, (Kluwer, Dordrecht, 1999)
4. *Photonic Crystals and Light Localization in the 21st Century*, edited by C. M. Soukoulis, (Kluwer, Dordrecht, 2001)
5. M. L. Cowan, J. H. Page and D. A. Weitz, *Phys. Rev. Lett.* **85**, 453 (2000).
6. J. H. Page, M. L. Cowan and D. A. Weitz, *Physica B* **279**, 130 (2000).
7. J. H. Page, M. L. Cowan, P. Sheng and D. A. Weitz, in *IUTAM Symposium 99/4: Mechanical and Electromagnetic Waves in Structured Media*, edited by R. C. McPhedran, L. C. Botten and N. A. Nicorovici (Kluwer Academic, 2001), p. 121.
8. M. L. Cowan, I. P. Jones, J. H. Page, and D. A. Weitz, *Phys. Rev. E.* **65**, 066605 (2002).
9. D. Bicout and R. Maynard, *Physica A* **199**, 387 (1993).
10. D. Bicout and G. Maret, *Physica A* **210**, 87 (1994).
11. M. L. Cowan, *Ph.D. Thesis* (University of Manitoba, 2001).
12. J. H. Page, H. P. Schriemer, A. E. Bailey and D. A. Weitz, *Phys. Rev. E* **52**, 3106 (1995).
13. J. H. Page, P. Sheng, H. P. Schriemer, I. Jones, X. Jing and D. A. Weitz, *Science* **271**, 634 (1996).
14. H. P. Schriemer, M. L. Cowan, J. H. Page, P. Sheng, Z. Liu, and D. A. Weitz, *Phys. Rev. Lett.* **79**, 3166 (1997).
15. A. Tourin, A. Derode, P. Roux, B. A. van Tiggelen and M. Fink, *Phys. Rev. Lett.* **79**, 3637 (1997).
16. M. L. Cowan, K. Beaty, J. H. Page, Z. Liu and P. Sheng, *Phys. Rev. E* **58**, 6626 (1998).
17. P. N. Segrè, E. Herbolzheimer and P. M. Chaikin, *Phys. Rev. Lett.* **79**, 2574 (1997).
18. M. P. Brenner, *Phys. Fluids* **11**, 754 (1999).
19. D. L. Koch and E. S. G. Shaqfeh, *J. Fluid Mech.* **224**, 275 (1991).
20. A. Levine, S. Ramaswamy, E. Frey, and R. Bruinsma, *Phys. Rev. Lett.* **81**, 5944 (1998).
21. J. W. Goodman, *Statistical Optics* (Wiley, New York, 1985).
22. B. A. van Tiggelen, P. Sebbah, M. Stoytchev, and A. Z. Genack, *Phys. Rev. E* **59**, 7166 (1999).
23. A. Z. Genack, P. Sebbah, M. Stoytchev, and B. A. van Tiggelen, *Phys. Rev. Lett.*

82, 715 (1999).
24. G. Maret and P. E. Wolf, *Z. Phys. B* **65**, 409 (1987).
25. D. J. Pine, D. A. Weitz, P. M. Chaikin and E. Herbolzheimer, *Phys. Rev. Lett.* **60**, 1134 (1988).
26. D. J. Pine, D. A. Weitz, G. Maret, P. E. Wolf, E. Herbolzheimer, and P. M. Chaikin, in *Scattering and Localization of Classical Waves in Random Media*, edited by P. Sheng (World Scientific, Singapore, 1990) p. 312.
27. J. X. Zhu, D. J. Pine and D. A. Weitz, *Phys. Rev. A* **44**, 3948 (1991).
28. J. de Rosny and P. Roux, *J. Acoust. Soc. Am.* **109**, 2587 (2001); J. de Rosny, P. Roux, M. Fink and J. H. Page, *Phys. Rev. Lett.*, in press (2003).
29. R. L. Weaver and O. I. Lobkis, *Ultrasonics* **38**, 491 (2000).
30. M. Campillo, L. Margerin and N.M. Shapiro, in *Diffuse Waves in Complex Media*, edited by J.-P. Fouque, (Kluwer, Dordrecht, 1999) p. 383.
31. R. Snieder, A. Gret, H. Douma, and J. Scales, *Science* **295**, 2253, (2002).

INTENSITY CORRELATIONS AND FLUCTUATIONS OF WAVES TRANSMITTED THROUGH RANDOM MEDIA

J.J. SÁENZ
*Laboratoire d'Energétique Moléculaire et Macroscopique, Combustion, Ecole Centrale Paris, Centre National de la Recherche Scientifique
92295 Châtenay-Malabry Cedex, France.*†

L.S. FROUFE-PÉREZ
*Departamento de Física de la Materia Condensada, Universidad Autónoma de Madrid
E-28049 Madrid, Spain.*

AND

A. GARCÍA-MARTÍN
*Institut für Theorie der Kondensierten Materie, Universität Karlsruhe
76128 Karlsruhe, Germany*

Abstract. The correlations between waves transmitted through random media are analyzed by use of a random matrix approach. Although the intensity and conductance fluctuations are practically independent of the sample length, the correlations present an unexpected strong dependence on the length of the disordered region. Interestingly, we show that the long range correlations $C^{(2)}$ and $C^{(3)}$, usually associated with intensity and conductance fluctuations, respectively, become negative as the length of the system decreases. Flux conservation is shown to be the main source of negative correlations in the quasi-ballistic regime.

†Permanent address: Departamento de Física de la Materia Condensada and Instituto "Nicolás Cabrera", Universidad Autónoma de Madrid, 28049 Madrid, Spain. E-mail: juanjo.saenz@uam.es.

*B.A. van Tiggelen and S.E. Skipetrov (eds.),
Wave Scattering in Complex Media: From Theory to Applications*, 175–187.

1. Introduction

The propagation of waves through random media produces complex, irregular intensity distributions (*speckle* patterns) [1] which are not nearly as random as intuition suggests. The statistical behavior of such patterns is dominated by the underlying correlations between transport coefficients. In the last decades much attention has been focused on this subject in connection with the discovery of interesting multiple scattering effects [2]. Different phenomena like the enhanced coherent backscattering [3] and intensity correlations in transmitted and reflected electromagnetic and other classical waves [4–13] are closely connected to weak localization and universal conductance fluctuations (UCF) [14–18] in electron transport. These interesting phenomena are not restricted to the case of volume scattering. Enhanced backscattering [19] and correlations of the speckle pattern intensity [20] arising in the light scattering from randomly rough surfaces have been intensively investigated. Recently, much attention has been paid to the reported experimental observation of strong localization [21] and universal conductance fluctuations [22, 23] of light.

Most of the previous work have been devoted to the diffusive regime, where the length of the sample L is much larger than the mean free path ℓ, but still smaller than the localization length ξ. Various types of correlations (usually referred to as $C^{(1)}$, $C^{(2)}$ and $C^{(3)}$) have been identified [4, 5, 9, 24]. Recently, the existence of a new type of long-range intensity correlation has been pointed out [13]. It was suggested that in quasi-one-dimensional geometries (e.g. disordered waveguides), these new correlations could even dominate the "standard" long-range correlations for small waveguide lengths. The standard description of the above-mentioned problems, usually based on a perturbative treatment or on numerical simulations, is of a *microscopic* nature [5]. In contrast with these approaches, random matrix theory (RMT) [25] provides a *macroscopic* approach since it deals with the statistical distribution of the scattering matrix for the full system [16, 26–28]. The purpose of this work is to discuss the behavior of the different correlations as a function of the length of the disordered region within the macroscopic random matrix theory (RMT) approach.

Very recently, it has been shown [29] that, although the intensity and conductance fluctuations are practically independent of the sample length, the correlations present an unexpected strong dependence on the length of the disordered region. Within the approach developed by Dorokhov, Mello, Pereyra and Kumar (DMPK) [27, 28], it is remarkable that both $C^{(2)}$ and $C^{(3)}$ may be even negative as the length of the system decreases [29]. Moreover, the results of numerical simulations of wave transport through surface randomly corrugated waveguides [29, 30] were found to be in full

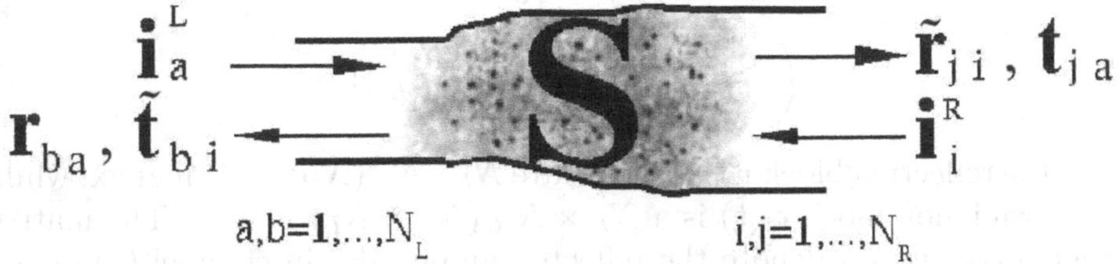

Figure 1. Sketch of the system.

qualitative agreement with the predictions based on the DMPK approach.

The macroscopic theory of multichannel disordered conductors (DMPK) [27, 28] is based on two general properties of the scattering system: flux conservation and time-reversal invariance together with an additional *maximum entropy* hypothesis (concerning the appropriate combination law when two conductors are put together). Our aim here is to explicitly discuss the length dependence of the statistical correlations based only on the above-mentioned general conditions.

First, we will introduce the definition of different transport coefficients as well as some general properties of the scattering matrix (section 2). The definition of the different correlation and fluctuation functions is given in section 3. Some recent results for both chaotic cavities and disordered waveguides are presented in sections 4 and 5. Finally, we discuss the physical origin of correlations in section 6 where we present a simple argument to explain the intriguing behavior of correlations as a function of the length of the disordered wire. As we will see, in the quasi-ballistic regime, flux conservation is the main source of long-range (C_2 and C_3) correlations. In the diffusive regime, the correlations are dominated by the intensity (total transmission) and conductance fluctuations.

2. Scattering Matrix and Transport Coefficients

To simplify the problem, we consider scalar waves instead of vector waves, thus neglecting polarization effects. Our geometry has also been chosen as simple as possible. It consists of a central section of disordered media connected to two perfect leads (waveguides) where the transverse confinement define N_L and N_R channels or propagating modes (see Fig. 1).

The scattering matrix $\mathbf{S}$ relates the asymptotic propagating outgoing waves $(\mathbf{o}^R, \mathbf{o}^L)$ with the incoming ones $(\mathbf{i}^R, \mathbf{i}^L)$[1],

$$\begin{pmatrix} \mathbf{o}^R \\ \mathbf{o}^L \end{pmatrix} = \begin{pmatrix} \mathbf{r} & \tilde{\mathbf{t}} \\ \mathbf{t} & \tilde{\mathbf{r}} \end{pmatrix} \begin{pmatrix} \mathbf{i}^R \\ \mathbf{i}^L \end{pmatrix} \tag{1}$$

where the reflection block $\mathbf{r}$ ($\tilde{\mathbf{r}}$) is a square $N_L \times N_L$ ($N_R \times N_R$) matrix, while the transmission block $\mathbf{t}$ ($\tilde{\mathbf{t}}$) is a $N_L \times N_R$ ($N_R \times N_L$) matrix. The matrix elements r_{ba} and t_{ja} denote the reflected amplitudes in channel b and the transmitted amplitudes in channel j when there is a unit flux incident from the left in channel a; $\tilde{r}_{ji}$ and $\tilde{t}_{bi}$ have an analogous meaning, except that the incident flux comes from the right. Flux conservation implies $\mathbf{S}$ unitary ($\mathbf{SS}^+ = \mathbf{S}^+\mathbf{S} = \mathbf{I}$), while reciprocity requires $\mathbf{S}$ to be symmetric ($\mathbf{S} = \mathbf{S}^T$).

For a given incoming mode a [for example, in a light scattering experiment a laser beam is incident in a given direction (channel) a] the transmission (T_{aj}) and reflection (R_{ab}) coefficients are defined as:

$$\begin{aligned} R_{ab} &\equiv |r_{ba}|^2 \\ T_{aj} &\equiv |t_{ja}|^2 \end{aligned} \tag{2}$$

The speckle pattern observed in disordered samples is just the complex interference pattern T_{aj} as a function of the outgoing direction j.

The total transmission (reflection) for the incoming mode a is given by:

$$\begin{aligned} T_a &= \sum_{j=1}^{N_L} T_{aj} \\ (R_a &= \sum_{b=1}^{N_R} R_{ab}) \end{aligned} \tag{3}$$

3. Fluctuations and Correlations in Speckle Patterns

The seemingly unrelated problems of the correlation effects in laser speckle patterns and fluctuation phenomena in mesoscopic conductors are tied together by the Landauer formula [18]: The dimensionless conductance G of a waveguide can be defined, both for classical and quantum waves, as a sum over transmission coefficients T_{aj} connecting all input modes a and output modes j, $G = \sum_{aj} T_{aj}$. Since $\mathbf{S}$ is unitary, we also have

$$G = \sum_{aj} T_{aj} = \sum_{ja} \tilde{T}_{ja}$$

[1]For a discussion of the general properties of the S-matrix, including evanescent channels, see [31].

$$G = N_L - \sum_{ab} R_{ab} = N_R - \sum_{ij} \tilde{R}_{ij} \tag{4}$$

For disordered mesoscopic systems, the transmission coefficients vary from sample to sample (that is, for different realizations of the locations of defects), so that the variance of conductance fluctuations $\mathrm{var}\{G\} = \langle G^2 \rangle - \langle G \rangle^2$ is given by

$$\mathrm{var}\{G\} \equiv \sum_{aj} \sum_{a'j'} \langle \delta T_{aj} \delta T_{a'j'} \rangle \tag{5}$$

where $\langle \ldots \rangle$ denotes an average over an ensemble of samples with different disordered configurations, and $\delta T_{ab} \equiv T_{ab} - \langle T_{ab} \rangle$.

$$\langle \delta T_{aj} \delta T_{a'j'} \rangle \equiv C^T_{aja'j'} \tag{6}$$

is the correlation function among individual transmission channels. In quantum electronic transport, multiple scattering leads to anomalously large conductance fluctuations (UCF) of the order of $G_0 = 2e^2/h$ (i.e. $\delta G \approx 1$) which, to a large extent, do not depend on the size of the sample nor on the degree of disorder. The physical origin of the UCF is directly related to the statistical properties of the channel (angular) intensity-intensity correlation function $C_{aja'j'}$.

3.1. INTENSITY-INTENSITY CORRELATION FUNCTION

The statistical analysis of $C_{aja'j'}$ can be achieved in a light scattering experiment: a laser beam is incident in a given direction (channel) a and the transmitted light intensity T_{aj} can be measured in any direction j. $C_{aja'j'}$ can then be constructed from the experimental data by collecting T_{aj} for different samples.

The correlation function of the transmission coefficients for scalar wave propagation through disordered media was first calculated by Feng *et al.* [4] by using a diagrammatic technique. Since then, the correlation function is usually written as

$$C^T_{aja'j'} = C^{(1)}_{aja'j'} + C^{(2)}_{aja'j'} + C^{(3)}_{aja'j'}. \tag{7}$$

This is a rather general result that does not depend on the details of the scattering mechanism. As a matter of fact, essentially the same results are obtained from a RMT approach [5, 11]: Mello and coworkers [5] showed that for transmitted waves

$$C^T_{aja'j'} = \langle T_{aj} \rangle \langle T_{a'j'} \rangle \left\{ C^T_1 \delta_{aa'} \delta_{jj'} + C^T_2 \left(\delta_{aa'} + \delta_{jj'} \right) + C^T_3 \right\}, \tag{8}$$

which has the same structure as that found by Feng *et al.* It is worth noticing that an important hypothesis of the RMT approach is that the phase factors of the scattering matrix have uniform statistical distributions. One of the consequences of this *isotropy* hypothesis is that the results are independent of the particular modes involved. This implies that

$$g \equiv \langle G \rangle = N_L \langle T_a \rangle = N_L N_R \langle T_{aj} \rangle. \tag{9}$$

The same holds for reflection $[g = N_R(1 - \langle R_a \rangle)]$ where all coefficients are equivalent except for the enhanced backscattering factor, $\langle R_{ab} \rangle = (1 + \delta_{ab}) \langle R_a \rangle / (N_R + 1)$, associated to reciprocity [5, 11, 25].

We can construct equivalent expressions for correlations among reflection coefficients or reflection and transmission coefficients both for right and left incoming beams [32]. For example, the correlation function for reflected waves can be written as [5, 32]

$$C^R_{aba'b'} = \langle R_{ab} \rangle \langle R_{a'b'} \rangle \begin{cases} C^R_1 \left(\delta_{aa'} \delta_{bb'} + \delta_{ab'} \delta_{a'b} - \delta_{aba'b'} \right) \\ + C^R_2 \left(\delta_{aa'} + \delta_{bb'} + \delta_{ab'} + \delta_{ba'} \right) \\ + C^R_3 \end{cases} \tag{10}$$

3.2. FLUCTUATIONS

The fluctuations of the different transport coefficients (T_{aj}, T_a and g) are directly related to the different correlations C_1, C_2 and C_3. From equation (8) we can define three fluctuation coefficients C_{1F}, C_{2F} and C_{3F}:

- C_{1F}. The normalized variance of the intensity of a given speckle pattern [sometimes referred to as (far field) *speckle contrast*]:

$$C^T_{1F} \equiv \frac{\mathrm{var}\{T_{aj}\}}{\langle T_{aj} \rangle^2} = C_{ajaj} = C^T_1 + 2C^T_2 + C^T_3 \tag{11}$$

  It is well known [1] that, under rather general conditions, the speckle contrast is of the order of 1 for both transmitted and reflected speckles, i.e. $C^T_{1F} = C^R_{1F} \approx 1$.
- C_{2F}. The normalized variance of the total transmitted intensity:

$$C^T_{2F} \equiv \frac{\mathrm{var}\{T_a\}}{\langle T_a \rangle^2} = \sum_{jj'} C_{ajaj'} = \frac{1}{N_L} C^T_1 + (1 + 1/N_L) C^T_2 + C^T_3 \tag{12}$$

- C_{3F}. The normalized variance of the conductance:

$$C_{3F} \equiv \frac{\mathrm{var}\{g\}}{g^2} = \sum_{aja'j'} C_{aja'j'} = \frac{1}{N_L N_R} C^T_1 + (1/N_L + 1/N_R) C^T_2 + C^T_3 \tag{13}$$

3.3. FLUCTUATIONS AND CORRELATIONS IN THE DIFFUSIVE REGIME

For a disordered waveguide with $N = N_L = N_R$ in the diffusive regime (where $1 \ll g \ll N$), the amplitude of the different correlation coefficients was found to scale as $C_1 \propto 1$, $C_2 \propto 1/g$ and $C_3 \propto 1/g^2$. The fluctuation of different transport parameters is dominated by the different correlation processes:

- C_1: *Speckle Contrast*
  C_1 is the leading term in the correlation function when $a = a'$ (only one incoming channel) and $j = j'$ (only one outgoing channel). Therefore C_1 is the normalized variance of the intensity of a given speckle pattern:
  $C_1^{T,R} \approx C_{1F}^{T,R} \approx 1$

- C_2: *Intensity correlations in a speckle pattern*
  C_2 is the leading term in the correlation function when $a = a'$ (only one incoming channel) and $j \neq j'$ (i.e. the correlation between two well separated speckle spots). In the diffusive regime, C_2^T correlations are responsible for the fluctuations of the total transmitted intensity:
  $C_2^T \approx C_{2F} \approx 2/(3g)$

- C_3: *Intensity correlations between spots of different speckle patterns*
  C_3 is the leading term in the correlation function when $a \neq a'$ (two incoming channels) and $j \neq j'$ (two outgoing channels). Therefore C_3 is the correlation between two different spots of two different speckle patterns. Deep in the diffusive regime, the conductance fluctuations are dominated by C_3^T. As a matter of fact, C_3^T is numerically equal to the normalized variance of the conductance:
  $C_3^T \approx C_{3F} \approx 2/(15g^2)$

Although physically different, the results summarized above may lead to consider an implicit equivalence between correlations (C_2, C_3) and fluctuations $(C_{2F}, \text{var}\{G\}/g^2)$. However, as we will discuss below, this equivalence holds only deep in the diffusive regime.

4. Fluctuations and Correlations in a Chaotic Cavity

The statistical properties of transport through a chaotic cavity connected to two leads with N_R and N_L channels are usually well described assuming that the elements of the scattering matrix $\mathbf{S}$ have a random phase, while their moduli are constrained by the conditions of unitarity and reciprocity.

In a chaotic cavity, except for the enhanced backscattering factor, reflection and transmission coefficients are statistically equivalent.

The correlations in a chaotic cavity, which have been discussed only in connection with reflected waves [11], have the same structure as equations (8) and (10). Following reference [11], we found (for $N_L, N_R \gg 1$)

$$C_1 \approx 1, \; C_{2F} \approx 1/(4g) \approx -C_2 \text{ and } \mathrm{var}\{G\} \approx 2/16 \approx g^2 C_3 \tag{14}$$

$[g \approx N_L N_R/(N_L + N_R)]$ for both transmission and reflection. These results are strikingly different to those obtained in the diffusive regime. C_2^T *is negative*, i.e. just the opposite to the diffusive regime. The fluctuations of the transmitted intensity [C_{2F}, equation (12)] are essentially dominated by C_1 and C_2, while C_3 is the only relevant correlation behind the conductance fluctuations, because for this last quantity the contribution of C_1 and C_2 exactly cancel each other [see equation (13)].

5. Fluctuations and Correlations in a Disordered Wire

Let us know consider a wire (waveguide) with $N = N_L = N_R$ channels, a disordered region of length L and an elastic mean free path ℓ [33], with the definition $s \equiv L/\ell$, in a regime where $1 \ll g \lesssim N$. The results for a chaotic cavity would roughly correspond to $s = L/\ell \approx 1$ ($g \approx N/2$) since a mean free path is the characteristic length for phase randomization. The chaotic cavity correlations discussed above together with the well known results in the diffusive regime, suggest a delicate interplay among the different correlations as the length of the sample increases. This can be analytically investigated within the DMPK approach.

The basic ingredient of the DMPK approach (together with the above mentioned unitarity and reciprocity conditions) is a *maximum entropy* hypothesis concerning the appropriate combination law when two conductors are put together. By using the method of moments of Mello and Stone [5, 17, 25] one can compute the variance of both T_a and g. To leading order in $N \gg 1$, the results are [5, 17] $g \approx N/(1+s)$, $C_{1F}^T \approx 1$,

$$C_{2F}^T \approx \frac{s^2}{g(1+s)^3}\left(\frac{2}{3}s + 1\right) \tag{15}$$

$$\mathrm{var}\{G\} \approx \frac{2}{15}\left(1 - \frac{1+6s}{(1+s)^6}\right) \tag{16}$$

From equations (11), (12) and (13) we obtain the s dependence of C_1^T, C_2^T and C_3^T: $C_1^T \approx 1$,

$$C_2^T \approx \frac{2}{3g}\frac{1}{(1+s)^3}\left(s^3 - 3s - \frac{3}{2}\right) \tag{17}$$

$$C_3^T \approx \frac{2}{15g^2}\left(1-\frac{1+6s}{(1+s)^6}\right)-\frac{1}{g^2(1+s)^4}\left[\frac{4}{3}s^3+s^2-2s-1\right] \quad (18)$$

The same s dependence of the correlations was obtained previously [29] by using directly the method of moments.

The interplay between fluctuations and correlations as a function of the sample length is summarized in Fig. 2 where we have plotted the results obtained from the DMPK approach and from numerical simulations of randomly corrugated waveguides (after García-Martín *et al.* [29]). In Fig. 2a we represent gC_{2F}^T together with gC_2^T versus $s = L/\ell$. For small s values C_2^T is negative and, only when the system length is larger than a few transport mean free paths, it approaches to C_{2F}^T. Dashed line in Fig. 2a corresponds to the well known asymptotic result $gC_{2F} \approx gC_2^T \approx 2/3$. In Fig. 2b we show the DMPK results for g^2C_3 and var$\{G\}$ versus s. Dashed line in Fig. 2b corresponds to the well known asymptotic result $g^2C_3^T \approx \text{var}\{G\} \approx 2/15$. C_3 is positive both for small ($s \lesssim 1$) and large ($s \gg 1$) values of s but, interestingly, there is a region in which it may become negative, i.e. within this region two different speckle patterns are not correlated but *anticorrelated.* It is worth noticing that, while var$\{G\}$ (and, to some extent, gC_{2F}^T) rapidly becomes constant and equal to its asymptotic value (already for $s \approx 1$), the relative weight of the underlying correlations strongly changes with sample length. *This is in striking contrast with the standard interpretation of UCF in terms of long range C_3 correlations.*

6. Flux Conservation and Correlations

The asymptotic results for both fluctuations and correlations can be obtained from perturbative diagrammatic expansions as well as from a random matrix DMPK approach. The intriguing dependence of the correlations with the length of the system described above, based only on the DMPK results, was not clearly understood. Here we offer a simple general argument which may provide new insights in the problem. Most of the behavior described above can be understood from the simple condition $T_a + R_a = 1$ together with the definition of the fluctuation coefficients.

The interesting interplay between conductance and total transmission/reflection fluctuations and the correlations is mainly imposed by the unitarity (flux conservation) condition. The first and simplest consequence of the unitarity condition is

$$\text{var}\{T_a\} = \text{var}\{R_a\} = -\langle \delta R_a \delta T_a \rangle \quad (19)$$

i.e., for a given incoming angle, the intensities of the transmitted and reflected speckles are anticorrelated. In other words, when one reflected spot

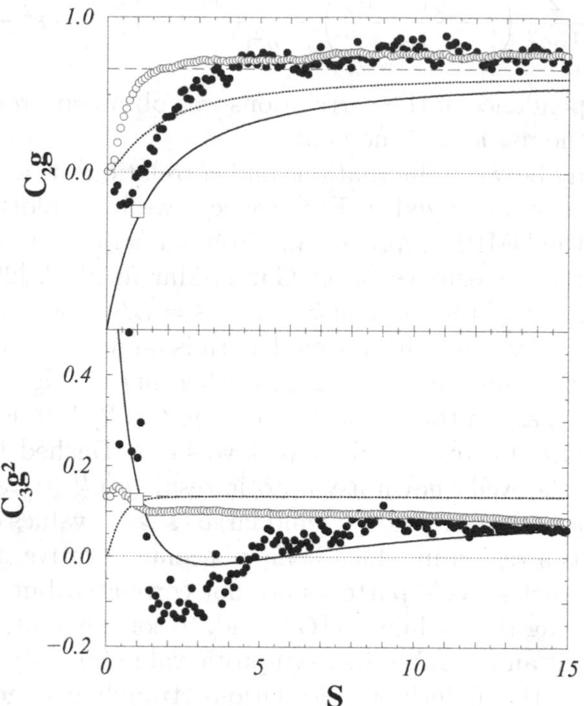

Figure 2. Dependence of correlations and fluctuations of the transmission coefficients on the waveguide length ($s = L/\ell$) obtained from the DMPK approach (lines) and from numerical simulations of randomly corrugated waveguides (symbols). (a) C_2^T correlations (solid lines and full symbols) and C_{2F}^T intensity fluctuations (dotted line and open symbols). (b) C_3^T correlations (solid lines and full symbols) and conductance fluctuations (var$\{G\}$) (dotted line and open symbols). Horizontal lines indicate the asymptotic values: (a) $gC_2^T \approx gC_{2F}^T \approx 2/3$ and (b) $g^2C_3^T \approx \text{var}\{G\} \approx 2/15$. As expected, the results for a chaotic cavity (open squares) are very close to those obtained from the DMPK approach for $s \approx 1$. (After García-Martín *et al.* [29]).

is brighter than the average, a transmitted spot is likely to be darker than the average.

From equations (10)–(13)[2] we obtain

$$C_2^T \approx [C_{2F}^T - 1/N] \tag{20}$$

$$C_2^R \approx [C_{2F}^T/\tilde{s}^2 - 1/N] \tag{21}$$

$$C_3^T \approx [C_{3F} - 2C_{2F}^T/N + 1/N^2] \tag{22}$$

$$C_3^R \approx [C_{3F}/\tilde{s}^2 - 4C_{2F}^T/(N\tilde{s}^2) + 2/N^2] \tag{23}$$

where $1/\tilde{s} \equiv \langle T_{aj}\rangle / \langle R_{ab}\rangle_{a\neq b}$ ($\tilde{s} = s$ except in the localization regime).

[2]Equations (10)–(13) are general results based only on the properties of symmetry of the **S**-matrix together with the *isotropy* assumption.

When most of the intensity is transmitted, flux conservation leads to negative C_2^T correlations: In the quasi-ballistic regime ($s \ll 1, g \approx N$), most of the intensity is transmitted and we have $T_a \approx 1$ and $\mathrm{var}\{T_a\} \approx 0$, i.e. $C_{2F}^T \approx 0$. From (20) we then have $C_2^T \approx -1/g$, i.e. for a given incoming angle a, the intensities of two well separated spots (j and j') in a transmitted speckle pattern are anticorrelated. The same arguments lead to $C_{3F} \approx 0$ and, from (22), $C_3^T \approx 1/g^2$. In the quasi-ballistic regime, the fluctuations of the transmission coefficients are negligible and flux conservation is the only origin of correlations.

As the length of the disordered region increases, the fluctuations increases and $\mathrm{var}\{G\}$ and gC_{2F}^T rapidly become constant and equal to their asymptotic values. Substituting the well known results $C_{3F} = \mathrm{var}\{G\}/g^2 \approx 2/(15g^2)$, $C_{2F} \approx 2/(3g), g \approx N/s$ in equations (20) and (22), we have

$$C_2^T \approx \frac{1}{3N}(2s - 3) \tag{24}$$

$$C_3^T \approx \frac{2}{15N^2}(s^2 - 10s + 15/2) \tag{25}$$

These approximate results present exactly the same qualitative behavior as obtained with the DMPK equation. In particular, they show that, *in the diffusive regime, C_3 correlations may be negative*, this being a simple and direct consequence of the fluctuations of the total transmission and conductance. Finally, deep in the diffusive regime, most of the intensity has been reflected and correlations become dominated by the transmittance of the waveguide (if, for a given realization one speckle spot is darker than the average, the others would also be darker than average).

In analogy, correlations between reflected waves show the opposite behavior, when most of the intensity is reflected, flux conservation leads to negative C_2^R correlations: In the localization regime ($s \gg 1$, $g \approx 0$), most of the intensity is reflected and we have $R_a \approx 1$, $\mathrm{var}\{R_a\} \approx 0$ (i.e. $C_{2F}^R = C_{2F}^T/\bar{s}^2 \approx 0$) and, from (21), $C_2^R \approx -1/N$ [from (23) we also have $C_3^R \approx 2/N^2$], i.e. for a given incoming angle a, the intensities of two well separated spots (b and b') in a reflected speckle pattern are anticorrelated. In the diffusive regime we have

$$C_2^R \approx \frac{1}{3N}\left(\frac{2}{s} - 3\right) \tag{26}$$

$$C_3^R \approx \frac{32}{15N^2}\left(1 - \frac{5}{4s}\right) \tag{27}$$

i.e. C_2^R correlations are negative in agreement with diagrammatic calculations [12].

7. Conclusions

We have discussed the physical origin of the intensity-intensity angular correlations in the wave transport through disordered systems.

In the quasi-ballistic regime, the fluctuations of the transmission coefficients are negligible and flux conservation is the only origin of correlations. In this regime, C_2 correlations of transmitted waves are negative (when one spot in the transmitted speckle pattern is brighter than the average, another spot is likely to be darker than the average).

In the diffusive regime, C_3 correlations may be negative, this being a simple and direct consequence of the fluctuations of the total transmission and conductance. This is in contrast with the standard interpretation of UCF and fluctuations of total transmitted intensity in terms of correlations.

Finally, deep in the diffusive regime, most of the intensity has been reflected and correlations become dominated by the transmittance of the waveguide (if, for a given realization, one speckle spot is darker than the average, the others would also be darker than average). Only deep in the diffusive regime, fluctuations and correlations become equivalent.

8. Acknowledgements

We gratefully acknowledge stimulating discussions with R. Carminati, G. Cwilich, J.J. Greffet, P.A. Mello, M. Nieto-Vesperinas and F. Scheffold. This work has been supported by the Spanish MCyT (Refs. No. /PB98-0464/ and BFM2000-1470-C02-02). J.J.S. acknowledge the support of the Spanish MECyD (Ref. No. PR2002-0161) during his stay in Paris.

References

1. J.C. Dainty (Ed.), *Laser Speckle Patterns and Related Phenomena* (Springer-Verlag, Berlin, 1984).
2. P. Sebbah (Ed.), *Waves and Imaging through Complex Media* (Kluwer, Dordretch, 2001).
3. Y. Kuga and J. Ishimaru, J. Opt. Soc. Am. A **1**, 831 (1984); M.P. van Albada and A. Lagendijk, Phys. Rev. Lett. **55**, 2692 (1985); P.E. Wolf and G. Maret, Phys. Rev. Lett. **55**, 1696 (1985). E. Akkermans, P.E. Wolf and R. Maynard, Phys. Rev. Lett. **56**, 1471 (1986); M. Kaveh, M. Rosenbulth, M. Edrei and I. Freud, Phys. Rev. Lett. **57**, 2949 (1986)
4. S. Feng, C. Kane, P.A. Lee and A.D. Stone, Phys. Rev. Lett. **61**, 834 (1988); J. Freund, M. Rosenbluh and S. Feng, Phys. Rev. Lett. **61**, 2328 (1988)
5. P.A. Mello, E. Akkermans and B. Shapiro, Phys. Rev. Lett. **61**, 459 (1988); P.A. Mello, J. Phys. A **23**, 4061 (1990); P.A. Mello and A.D. Stone, Phys. Rev. B **44**, 3559 (1991).
6. R. Pnini and B. Shapiro, Phys. Rev. B **39**, 6986 (1989); R. Pnini, pp. 391–412 in Ref. [2] .
7. N. García and A.Z. Genack, Phys. Rev. Lett. **63**, 1678 (1989); A.Z. Genack, N. García, J.H. Li and W. Polkosnik, Phys. Rev. Lett. **65**, 2129 (1990); P. Sebbah, R.

Pnini and A.Z. Genack, Phys. Rev. E **62**, 7348 (2000).
8. M.P. van Albada, J.F. de Boer and A. Lagendijk, Phys. Rev. Lett. **64**, 2787 (1990); J.F. de Boer, M.P. van Albada and A. Lagendijk, Phys. Rev. B **45**, 658 (1992).
9. M.C.W. van Rossum and T.M. Nieuwenhuizen, Rev. Mod. Phys. **71**, 313 (1999); R. Berkovits and S. Feng, Phys. Rep. **238**, 135 (1994).
10. F. Scheffold, W. Härtl, G. Maret and E. Matijević, Phys. Rev. B **56**, 10942 (1997).
11. E. Bascones, M.J. Calderón, D. Castelo, T. López and J.J. Sáenz, Phys. Rev. B **55**, R11911 (1997).
12. D.B. Rogozkin, JETP Lett. **69**, 117 (1999).
13. B. Shapiro, Phys. Rev. Lett. **83**, 4733 (1999); S.E. Skipetrov and R. Maynard, Phys. Rev. B **62**, 886 (2000).
14. C.P. Umbach, S. Washburn, R.B. Laibowitz and R.A. Webb, Phys. Rev. B **30**, 4048 (1984); R.A. Webb, S. Washburn, C.P. Umbach and R.B. Laibowitz, Phys. Rev. Lett. **54**, 2696 (1995).
15. P.A. Lee and A.D. Stone, Phys. Rev. Lett. **55**, 1622 (1985); B.L. Alt'shuler, JETP Lett. **41**, 648 (1985); P.A. Lee, A.D. Stone and H. Fukuyama, Phys. Rev. B **35**, 1039 (1987).
16. K.A. Muttalib, J.-L. Pichard and A.D. Stone, Phys. Rev. Lett. **59**, 2475 (1987).
17. P.A. Mello, Phys. Rev. Lett. **60**, 1089 (1988).
18. S. Feng and P.A. Lee, Science **251**, 633 (1991).
19. A.R. McGurn, A.A. Maradudin and V. Celli, Phys. Rev. B **31**, 4866 (1985); V. Celli, A.A. Maradudin, A.M. Marvin and A.R. McGurn, J. Opt. Soc. Am. A **2**, 2225 (1985); E.R. Mendez and K.A. O'Donnell, Opt. Commun. **61**, 91 (1987); K.A. O'Donnell and E.R. Mendez, J. Opt. Soc. Am. A **4**, 1194 (1987); C.S. West and K.A. O'Donnel, J. Opt. Soc. Am. A **12**, 390 (1995); M. Nieto-Vesperinas and J.M. Soto-Crespo, Opt. Lett. **12**, 979 (1987); J.M Soto-Crespo and M. Nieto-Vesperinas, J. Opt. Soc. Am. A **6**, 3 (1989).
20. D. Léger, E. Mathieu and J.C. Perrin, Appl. Opt. **14**, 872 (1975); D. Léger and J. Perrin, J. Opt. Soc. Am. **66**, 1210 (1976); M. Nieto-Vesperinas and J.A. Sanchez-Gil, Phys. Rev. B **46**, 3112 (1992); J. Opt. Soc. Am. A **10**, 150 (1993); A. Arsenieva and S. Feng, Phys. Rev. B **47**, 13047 (1993).
21. D.S. Wiersma *et al.*, Nature **390**, 671 (1997); *ibid.* **398**, 207 (1999); F. Scheffold and G. Maret, *ibid.* **398**, 206 (1999) ; A. A. Chabanov, M. Stoytchev, and A. Z. Genack, *ibid.* **404**, 850 (2000); J. Gomez-Rivas *et al.*, Phys. Rev. E **63**, 046613 (2001).
22. F. Scheffold and G. Maret, Phys. Rev. Lett. **81**, 5800 (1998).
23. F. Scheffold and G. Maret, p. 413–434 in Ref. [2].
24. P. Sebbah, B. Hu, A. Z. Genack, R. Pnini and B. Shapiro, Phys. Rev. Lett. **88**, 123901 (2002).
25. C.W.J. Beennakker, Rev. Mod. Phys. **69**, 731 (1997).
26. Y. Imry, Europhys. Lett. **1**, 249 (1986).
27. O.N. Dorokhov, Solid State Commun. **51**, 381 (1984).
28. P.A. Mello, P. Pereyra and N. Kumar, Ann. Phys. (N.Y.) **181**, 290 (1988).
29. A. García-Martín, F. Scheffold, M. Nieto-Vesperinas and J.J. Sáenz, Phys. Rev. Lett. **88**, 143901 (2002).
30. A. García-Martín *et al.*, Appl. Phys. Lett. **71**, 1912 (1997); Phys. Rev. Lett. **80**, 4165 (1998); *ibid.* **81**, 329 (1998); *ibid.* **84**, 3578 (2000).
31. R. Carminati, J.J. Sáenz, J.-J. Greffet and M. Nieto-Vesperinas, Phys. Rev. A **62**, 012712 (2000).
32. L.S. Froufe-Pérez, A. García-Martín and J.J. Sáenz, unpublished (2002).
33. The definition of ℓ within RMT differs by a numerical coefficient from that of the transport mean free path ℓ^* of kinetic theory [25]. In three-dimensional systems $\ell = (4/3)\ell^*$.

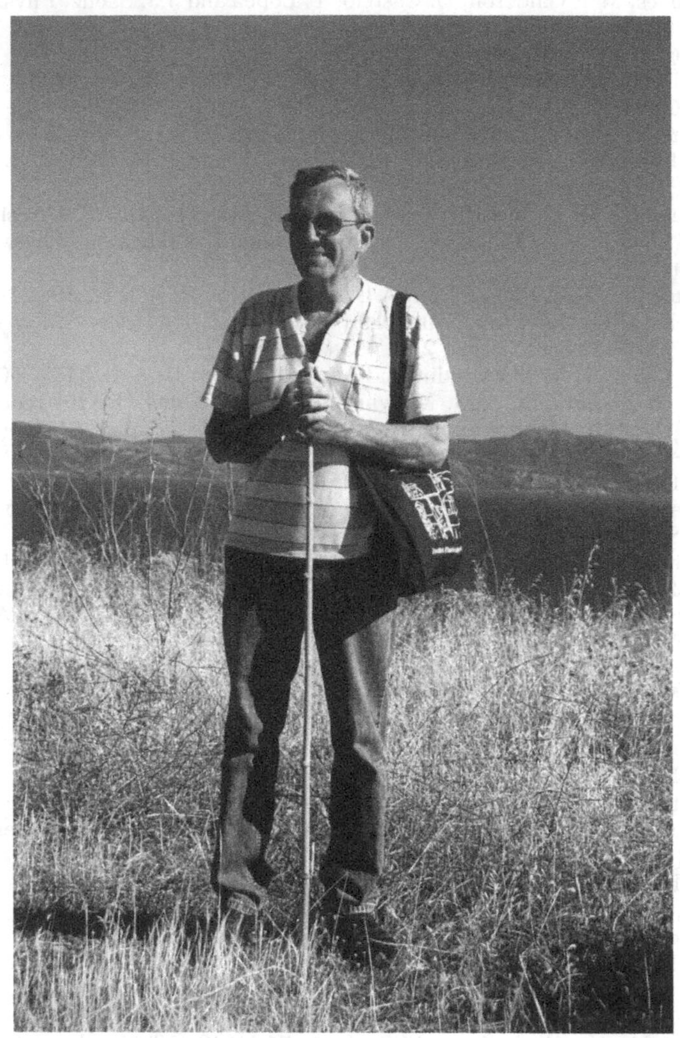

Valentin Freilikher. Photo by Konstantin Bliokh

SPECTRAL COHERENCE OF WAVE FIELDS IN RANDOM MEDIA

GREGORY SAMELSOHN AND VALENTIN FREILIKHER

The Jack and Pearl Resnick Institute of Advanced Technology,
Department of Physics,
Bar-Ilan University, Ramat-Gan 52900, Israel

Abstract. An original approach to the analysis of classical wave localization and diffusion in random media is developed. This approach is based on a cumulant path integral technique and, being universal with respect to the dimensionality of the system, accounts explicitly for the correlation properties of the disorder. The general theory is applied to the evaluation of a two-frequency mutual coherence function, which is an important quantity in itself, and also determines the evolution of transient signals in the time domain. Our results describe a ballistic to diffusive transition in the wave transport, and, for not too large distances, are consistent, in general, with a classical diffusion paradigm. In particular, the coherence function is shown to possess a well-pronounced two-scale structure, especially, when the absorption is taken into account. However, the final results are related to a power spectrum of the disorder, rather than to a phenomenological diffusion constant. Also, the propagation in random media is described explicitly as a process of wave scattering by resonant Bragg lattices hidden in a disordered structure, and an additional dwell time due to local resonances is taken into account. Since the coherence function is expressed via an arbitrary form power spectrum, the results obtained in the work open a new avenue in studying wave transport in anisotropic and/or fractally correlated systems.

1. Introduction

For several decades, the propagation and scattering of waves in random media has been the subject of intensive investigations in various areas of

B.A. van Tiggelen and S.E. Skipetrov (eds.),
Wave Scattering in Complex Media: From Theory to Applications, 189–201.

both theoretical physics and applied research. Typical examples include electromagnetic, acoustic, and elastic waves propagating in natural media and artificial structures, and also electron waves in disordered solids [1–4]. While considering the propagation of wave fields in random media, one usually deals with two extreme regimes. The first one is a near ballistic propagation when the wave is scattered mostly in the forward direction, and the problem can be reduced to that of studying a well established parabolic wave equation, in which the backscattering is fully neglected, and the lateral diffusion of the wave energy is described in a paraxial approximation. In the other extreme regime, a so-called diffuse wave scattering, the direction of initial wave propagation is completely lost. A conventional way to describe wave behavior in this regime is to use a phenomenological radiative transfer equation, which can be reduced, under some conditions, to the diffusion equation [1,4]. The weak point of this approach is that, being based on the energy balance principle, it ignores completely phase coherence and related interference effects, and, therefore, fails to predict many wave nature phenomena, such as strong, Anderson type, localization [5].

Despite a huge number of related investigations, there is still a need in a universal model describing on an equal footing diffusion and localization for both time-harmonic and pulsed waves propagating in random media of arbitrary dimensionality. In particular, it is important to formulate the relation between the localization and its characteristics, say, localization length, on the one hand, and the correlation properties of the scattering potential, on the other [6,7]. In order to achieve this goal, a perturbative path integral analysis of the wave intensity, supplemented also by a filtering in the spectral domain, has been developed in Ref. [7]. It has been shown there that the localization in multidimensional systems is determined by the power spectrum of the scattering potential taken at the frequencies of Bragg resonances , in a close analogy to what was known in one dimension.

In the present work we extend the above approach to studying a two-frequency mutual coherence function for wave fields propagating in random media, the problem which is important in itself, and also because the spectral correlator determines the behavior of transient signals in the time domain. Our theory describes a ballistic to diffusive transition in the wave transport, and, for not too large distances, is consistent, in general, with a classical diffusion paradigm. In particular, the coherence function is shown to possess a well-pronounced two-scale structure, especially, when the absorption is taken into account. However, the shape of the pulse, which can be reconstructed by using an appropriate Fourier transformation, is related to a power spectrum of the disorder, rather than to a phenomenological diffusion constant. The propagation in random media is described explicitly as a process of wave scattering by resonant Bragg lattices hidden in a disordered structure, and an additional dwell time due to local resonances is taken into account. Since the coherence function is expressed via an arbitrary form power spectrum, the results obtained in the work open a new avenue in studying wave transport in anisotropic and/or fractally correlated systems.

The outline of the paper is as follows. In Sec. 2 we present general results for the coherence function given in the form of a weighted integral over the power spectrum of the disorder. Section 3 is devoted to a spectral analysis of different scattering phenomena and their relation to both localization and diffusion of classical waves. The results of the calculations of spectral coherence for three-dimensional systems with isotropic disorder are shown in Sec. 4. Finally, Sec. 5 contains a summary and concluding remarks.

2. General Results

Our starting point is the Helmholtz equation, describing the propagation of a scalar time-harmonic wave in an m-dimensional medium ($m = 1, 2$, or 3) characterized by a random distribution of a scattering potential $\tilde{\varepsilon}(\mathbf{R})$. For a point source excitation, this equation takes the form

$$\nabla^2 G(\mathbf{R}|\mathbf{R}_0) + k^2 [1 + \tilde{\varepsilon}(\mathbf{R})] G(\mathbf{R}|\mathbf{R}_0) = -\delta(\mathbf{R} - \mathbf{R}_0), \tag{1}$$

where $k = \omega/c$ is the wave number in a homogeneous (reference) medium. In order to evaluate the two-frequency mutual coherence function,

$$\Gamma(\omega, \Omega) = \left\langle G_{\omega+\Omega/2}(\mathbf{R}|\mathbf{R}_0) G^*_{\omega-\Omega/2}(\mathbf{R}|\mathbf{R}_0) \right\rangle, \tag{2}$$

we adopt the method of proper time originally proposed by Fock for the integration of quantum mechanical equations [8]. Specifically, we consider an auxiliary problem for a function $g(\mathbf{R}, \tau|\mathbf{R}_0, \tau_0)$ satisfying the generalized parabolic equation,

$$2ik\partial_\tau g + \nabla^2 g + k^2 \tilde{\varepsilon}(\mathbf{R}) g(\mathbf{R}, \tau|\mathbf{R}_0, \tau_0) = 0, \quad \tau > \tau_0, \tag{3}$$

which is supplemented by the initial condition,

$$g(\mathbf{R}, \tau_0|\mathbf{R}_0, \tau_0) = \delta(\mathbf{R} - \mathbf{R}_0). \tag{4}$$

Hence, the Green's function $G(\mathbf{R}|\mathbf{R}_0)$ is defined through the solution of the latter equations as

$$G(\mathbf{R}|\mathbf{R}_0) = \frac{i}{2k} \int_{\tau_0}^{\infty} d\tau \exp\left[i\frac{k}{2}(\tau - \tau_0) \right] g(\mathbf{R}, \tau|\mathbf{R}_0, \tau_0). \tag{5}$$

Here, it is supposed that while $\tilde{\varepsilon}$ is a real function, k contains an infinitesimally small positive imaginary part that enforces the radiation condition at infinity and provides the convergence of the corresponding integral. Then, by using a stationary phase approximation and a cumulant path integral technique (a comprehensive picture of all the related details may be found in Refs. [7,9–11]), we arrive at

$$\Gamma(\omega,\Omega) = \Gamma_0(\omega,\Omega)\exp\left[-\chi(\omega,\Omega)\right], \tag{6}$$

where $\Gamma_0(\omega,\Omega)$ is the coherence function in a homogeneous medium. The cumulant series $\chi(\omega,\Omega)$ entering Eq. (6), is then approximated by a (linearized) first cumulant which, for non-dispersive media, reads

$$\chi(\omega,\Omega) = \frac{\pi}{2}k^3 L \int d\mathbf{K} f(\mathbf{K},\omega,\Omega)\,\Phi_\epsilon(\mathbf{K}), \tag{7}$$

where $L = |\mathbf{R}-\mathbf{R}_0|$ is the distance between the source and the observation point, and $\Phi_\epsilon(\mathbf{K})$ is the power spectrum (Fourier transformed correlation function) of the disorder. The filtering function $f(\mathbf{K},\omega,\Omega)$ has the form

$$f(\mathbf{K},\omega,\Omega) = K^{-1}\frac{2}{\pi}\int_0^\infty dx\cos(2x\mathbf{k}\cdot\mathbf{K}/K)$$

$$\times\left[\cos(xK) - \mathrm{sinc}(xK)\exp\left(i\Omega LK^2\big/8ck^2 - i\Omega L\mathbf{k}\cdot\mathbf{K}\big/4ck^2\right)\right], \tag{8}$$

where $\mathrm{sinc}(x) \equiv \sin(x)/x$. In terms of generalized functions the latter expression becomes

$$f(\mathbf{K},\omega,\Omega) = K^{-1}\delta(K - |2\mathbf{k}\cdot\mathbf{K}/K|)$$

$$-K^{-2}\exp\left[i\Omega L\left(K\big/8ck^2\right)(K - 2\mathbf{k}\cdot\mathbf{K}/K)\right]\vartheta(K - |2\mathbf{k}\cdot\mathbf{K}/K|), \tag{9}$$

where, as previously, $\delta(x)$ is the Dirac δ function, and $\vartheta(x)$ is the Heaviside step function. The latter equations are valid for any dimensionality of the system and for an arbitrary spectrum of the disorder. However, the approximations leading to Eq. (9) can be justified only if the dimensionless parameter $LK^2/4k$ is large enough. This condition is obviously violated in the high-frequency limit, where $K/2k \to 0$. On the other hand, in the situation we are mainly interested in, namely, when the radiation wavelength is comparable with a correlation scale of the disorder, the approximations used do not disturb essentially the accuracy of the final result, because in this case not too much energy of the spectrum $\Phi_\varepsilon(\mathbf{K})$ is concentrated near the origin of the Fourier space.

For a given spectrum $\Phi_\varepsilon(\mathbf{K})$, the final result depends on both the modulus and the direction of the wave vector $\mathbf{k}$. In what follows, we will concentrate on the analysis of wave propagation in isotropic structureless media, where the scattering is independent of the direction. Integration in Eq. (7) over angular variables for isotropic spectra $\Phi_\varepsilon(\mathbf{K})$ yields

$$\chi(\omega,\Omega) = a_m k^2 L \int_0^\infty dK K^{m-2} f(K,\omega,\Omega)\,\Phi_\epsilon(K)\,, \tag{10}$$

where a_m is the numerical coefficient depending on the dimensionality of the system:

$$a_1 = \pi/2, \qquad a_2 = \pi, \qquad a_3 = \pi^2. \tag{11}$$

Note, that for convenience, we have extracted the general factor K^{m-2} from the filtering function.

In the one-dimensional case, we obtain

$$f(K,\omega,\Omega) = 2k\delta(K-2k)$$
$$-(2k/K)\exp\left(i\Omega LK^2\big/8ck^2\right)\cos(\Omega LK/4ck)\,\vartheta(K-2k)\,. \tag{12}$$

In two dimensions, the integration over angular variable gives

$$f(K,\omega,\Omega) = \left[\left(1-K^2\big/4k^2\right)^{-1/2}\right.$$
$$\left.-(2k/K)\exp\left(i\Omega LK^2\big/8ck^2\right)\mathcal{A}(K/2k,\Omega LK/4ck)\right]\vartheta(2k-K)$$
$$-(2k/K)\exp\left(i\Omega LK^2\big/8ck^2\right)\mathcal{A}(1,\Omega LK/4ck)\,\vartheta(K-2k)\,, \tag{13}$$

where we have defined the function

$$\mathcal{A}(x,z) = \int_0^\infty d\xi\left(1-\xi^2\right)^{-1/2}\cos(z\xi)\,. \tag{14}$$

Analogous procedure performed in three dimensional case leads to

$$f(K,\omega,\Omega) = \left[1-\exp\left(i\Omega LK^2\big/8ck^2\right)\mathrm{sinc}\left(\Omega LK^2\big/8ck^2\right)\right]\vartheta(2k-K)$$
$$-(2k/K)\exp\left(i\Omega LK^2\big/8ck^2\right)\mathrm{sinc}(\Omega LK/4ck)\,\vartheta(K-2k)\,. \tag{15}$$

In a degenerate case of zero frequency shift ($\Omega = 0$), coherence function $\Gamma(\omega,0)$ represents an average intensity of the wave, $\langle I\rangle$, and Eqs. (6)–(15) can be simplified to the corresponding expressions obtained previously in Ref. [7].

3. Spectral Filtering and Wave Localization

Before proceeding with the analysis of spectral coherence, it is useful to consider how the general results obtained above reflect the existence of wave localization, at least in low-dimensional systems. When the localization is present, the intensity of the wave is known to behave typically as $I(L) \sim \exp(-L/\xi)$, where ξ is the localization length, even without any dissipative mechanism [3,4]. The inverse localization length (Lyapunov exponent), ξ^{-1}, is a self-averaging quantity, i.e., its value obtained for a single realization tends to a non-random limit if the size of the system increases. A classical perturbative result for the inverse localization length in 1D systems reads

$$\xi^{-1}(k) = \frac{\pi}{2} k^2 \Phi_\epsilon (2k), \tag{16}$$

see, e.g., Ref. [12]. Relation (16) shows that in the lowest-order approximation, only $\pm 2k$ components of the spectrum are responsible for the localization of a time-harmonic wave with wave number k.

At the same time, the coherence function calculated above corresponds (when $\Omega = 0$) to a mean intensity of the wave, $\langle I \rangle$, which is obtained by averaging over the ensemble of all possible realizations of the scattering potential, including the low-probable though representative realizations with high-Q *local resonances.* Despite the rarity of such realizations, their contribution to the mean intensity may be significant, and the behavior of the mean intensity differs from that observed in typical realizations. If we now inspect Eq. (12) in the one-frequency case, we see that the filtering function $f(K)$ is split into two terms. One is discrete and extracts the Bragg harmonics with $K = \pm 2k$. The other is continuous and, accounting for a high-frequency tail of the spectrum $\Phi_\epsilon(K)$, is obviously related to local resonances, because only these high-frequency harmonics could generate spatially localized, resonant spikes of the wave intensity observed in the configuration space.

Since the mixing of different spectral components is absent in the first cumulant, the procedure of extracting the Bragg resonances, which are responsible for the wave localization, corresponds to a filtering in the Fourier space. By keeping in χ the discrete term only, we arrive at nothing else than Eq. (16) obtained by using an alternative approach. In some sense, such spectral filtering is reminiscent of the so-called *resonance approximation,* where the scattering potential is *a priori* replaced by Fourier components with wave numbers close to 0 and $\pm 2k$, see, e.g., Refs. [12,13] and references therein. The difference is in respective order of the operations. In contrast to the resonant approximation, we first calculate the mean value for a complete form of the spectrum and only then extract the relevant spectral components.

In multidimensional systems, the components of the spectrum participating in the Bragg scattering and, hence, determining the *global (Bragg) resonances,* are located within the limiting sphere of radius $2k$ in the Ewald construction (this is a generalization of a simple three-point diagram in one

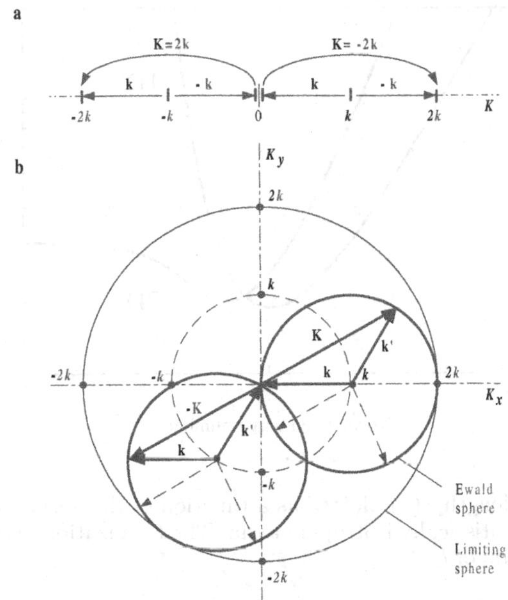

Figure 1. Momentum diagram of resonant scattering in a weakly disordered medium. a) One-dimensional case. The wave vectors of two counter-propagating waves, $\mathbf{k}$ and $-\mathbf{k}$, are mapped onto the K-space in such a way that their endpoints lie at the origin, $K = 0$. The exchange of the momenta between these two waves is governed in the lowest-order approximation by the relevant Bragg components ($\mathbf{K} = \pm 2\mathbf{k}$) of the scattering potential. b) Multidimensional case (2D example is shown). The points of the Ewald sphere for a given wave vector $\mathbf{k}$ determine all spectral components that transform the incident wave into a scattered one. The limiting sphere encircles all spectral components coupling any two wave vectors in the process of elastic scattering.

dimension), see Fig. 1. The Ewald construction provides a mapping of the wave vectors onto the K space of a medium (reciprocal lattice, in the case of a periodic system like crystal). Indeed, the lowest-order resonant component $\mathbf{K}$ of the spectrum should satisfy the Bragg law $\mathbf{k}' = \mathbf{k} + \mathbf{K}$, where $\mathbf{k}$ and $\mathbf{k}'$ are the wave vectors of the incident and resonantly scattered waves, respectively. Wave vector $\mathbf{k}$ of the incident wave is drawn such that its endpoint is at the origin of the Fourier space. Since the initial point of the wave vector $\mathbf{k}'$ coincides with that of vector $\mathbf{k}$, the endpoint of $\mathbf{k}'$ lies, therefore, on a sphere of radius k due to the energy conservation in elastic scattering, $|\mathbf{k}'| = |\mathbf{k}|$. The points of this sphere which is referred to as the Ewald sphere of reflection (Ewald circle in the 2D case), include all possible spectral components of the medium which elastically transform the incident wave into a resonantly scattered one.

In periodic systems, a strong scattering occurs when the Ewald sphere

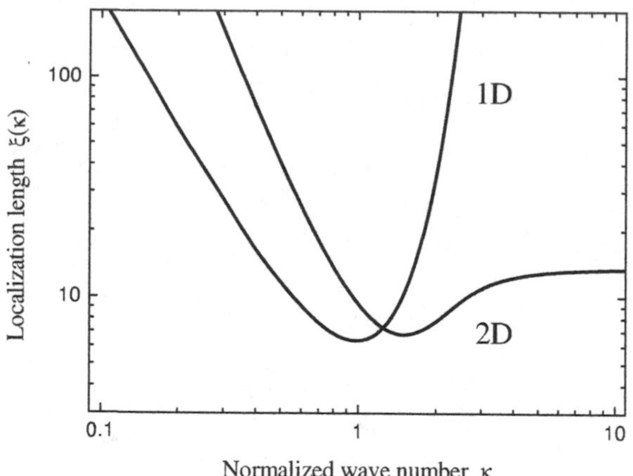

Figure 2. Localization length $\xi(k)$ plotted as a function of the normalized wave number $\kappa = kl_\varepsilon$ for 1D and 2D statistically isotropic media. The localization length is normalized to σ_ε^2 and is given in units of l_ε.

passes through a point of a discrete reciprocal lattice. In structureless random media, the spectrum is continuous and resonant scattering is produced by all points of the Ewald sphere. These points, therefore, define those propagation channels that are coupled to a given one with wave vector $\mathbf{k}$. One of these channels, say, that is defined by the wave vector $\mathbf{k}'$ shown in Fig. 1b, determines another Ewald sphere and, consequently, a set of possible spectral components $\mathbf{K}$ that provide a scattering into all other coupled channels. Rotation of vector $\mathbf{k}$ with its endpoint being fixed at the origin covers all possible directions of the incident wave and demonstrates that the components of the spectrum participating in the Bragg scattering and, hence, determining the global resonances, are located within the limiting sphere of radius $2k$.

The idea of spectral filtering proposed above for the 1D case, can be extended to multidimensional systems as well. Namely, in order to estimate the behavior of the intensity in typical (most probable) realizations, we use the expressions obtained for the mean value, but keeping in Eq. (7) only the spectral components lying within the limiting sphere.

To exemplify the results, we use a Gaussian model for the correlation function,

$$B_\varepsilon(R) = \sigma_\varepsilon^2 \exp\left(-R^2 \big/ l_\varepsilon^2\right), \tag{17}$$

where σ_ε^2 and l_ε characterize the strength and correlation scale of the disorder, respectively. Figure 2 shows the localization length $\xi(k)$ plotted as a function of the normalized wave number $\kappa = kl_\varepsilon$. Although a well-defined minimal value of ξ at some intermediate frequency band is found in two dimensions, in the high-frequency limit the localization length is unexpectedly constant, independent of the wave number k. The same effect has recently been observed in numerical simulations [14] where the localization length for 2D strongly disordered systems was shown to saturate at relatively high frequencies. In three dimensions, $\chi_{\mathrm{f}}(k) \equiv 0$ for isotropic systems, which means that our perturbative approach cannot describe the diffusion-to-localization transition, and other techniques have to be used in this case.

As will be shown in the next section, by using the approximation obtained for the coherence function $\Gamma(\omega, \Omega)$ we can study not only the transport of time-harmonic waves but also the propagation of pulsed radiation in random media. Mapping the spectral expansion of the cumulant $\chi(\omega, \Omega)$ onto the Ewald construction may help in attributing various propagation mechanisms to different domains in the Fourier space and analyze their relative role in wave scattering.

4. Waves in 3D Media with Isotropic Disorder

Here we will concentrate on the analysis of wave behavior in 3D isotropic media, where a diffusion like spread of the wave energy takes place. In this case, our goal is to evaluate the normalized correlator,

$$\tilde{\Gamma}(\omega, \Omega) = \Gamma(\omega, \Omega)/\Gamma(\omega, 0), \tag{18}$$

which should be much more robust with respect to our perturbative procedure than the coherence function itself. Actually, the normalized correlator remains meaningful even when $\Gamma(\omega, \Omega)$ diverges exponentially with the distance L, the behavior prescribed by the *ansatz* when the cumulant $\chi(\omega, \Omega)$ becomes negative. Also, it can be shown, that for a very short but still narrowband wave packet, the function $\tilde{\Gamma}(\omega, \Omega)$ and the normalized impulse response function $J_{\omega}(t)$ constitute a Fourier transform pair [1,15],

$$J_{\omega}(t) = \frac{1}{2\pi} \int_{-\infty}^{\infty} d\Omega \exp(-i\Omega t)\, \tilde{\Gamma}(\omega, \Omega), \tag{19}$$

and, therefore, the calculations of $\tilde{\Gamma}(\omega, \Omega)$ are intimately related to the propagation of transient waves in random media.

Thus, by substituting Eq. (6) into (18) we obtain

$$\tilde{\Gamma}(\omega, \Omega) = \tilde{\Gamma}_0(\omega, \Omega) \exp[-\tilde{\chi}(\omega, \Omega)], \tag{20}$$

where $\tilde{\Gamma}_0(\omega, \Omega)$ is the normalized coherence function in a homogeneous medium, and the tilded cumulant $\tilde{\chi}$ is calculated by using equation like

(7) with an appropriate filtering function $\tilde{f}(K,\omega,\Omega)$. The first factor in Eq. (20), $\tilde{\Gamma}_0(\omega,\Omega)$, describes a ballistic wave propagation, while the second factor, $\exp[-\tilde{\chi}(\omega,\Omega)]$, corresponds to a scattering of the wave by Bragg lattices of all possible directions and periodicities hidden in a disordered structure, and includes also an additional time delay due to local resonances. Taken as a whole, this can be interpreted as a random walk in the momentum space, but, unlike classical diffusion approximation, the result accounts for the wave interference effects and also for the dwell time cased by the scattering on resonant clusters, cf. Ref. [16]. Moreover, the shape of the pulse, which can be reconstructed by using Fourier transformation, Eq. (19), is related to real parameters of the scattering medium (correlation properties of its microstructure) rather than to a phenomenological diffusion constant.

When the distance L is large enough, the behavior of the cumulant is determined by small values of Ω, and we can expand the cumulant $\tilde{\chi}(\omega,\Omega)$ into a series,

$$\tilde{\chi}(\omega,\Omega)=\tilde{\chi}(\omega,\Omega)\,\Omega+\tilde{\chi}(\omega,\Omega)\,\Omega^2\big/2+\ldots. \tag{21}$$

As was shown in Ref. [10], the first, linear term (proportional to L^2) is related to a pulse delay time, while the quadratic term (increasing with distance as L^3) corresponds to a pulse width squared. The higher order terms specify an asymmetry and other details of the pulse shape. For extremely large values of L, the higher order terms may be neglected, which leads to a symmetric Gaussian form of the impulse response function $J_\omega(t)$. This asymptotic behavior of the pulse shape contradicts a power law (in the absence of absorption) form of the tail predicted by the diffusion approximation. Since the diffusion theory also relies on a perturbative procedure and its applicability to describing wave nature effects is an open question, the higher order terms in the cumulant expansion of the path integral should be taken into account in order to shed more light on the origin of this discrepancy. However, as will be shown below, even in the regime of rather strong multiple scattering, where the distance is much larger than a properly defined transport (scattering) mean-free-path, the coherence function differs essentially from that of a Gaussian form.

In three dimensions, the filtering function $\tilde{f}(K,\omega,\Omega)$ is given by

$$\tilde{f}(K,\omega,\Omega)=\left[1-\exp\left(i\Omega LK^2\big/8ck^2\right)\mathrm{sinc}\left(\Omega LK^2\big/8ck^2\right)\right]\vartheta(2k-K)$$
$$+(2k/K)\left[1-\exp\left(i\Omega LK^2\big/8ck^2\right)\mathrm{sinc}\left(\Omega LK/4ck\right)\right]\vartheta(K-2k). \tag{22}$$

According to the analysis performed in Sec. 3, the first term of the filtering function is attributed to the effective Bragg lattices, whereas the second

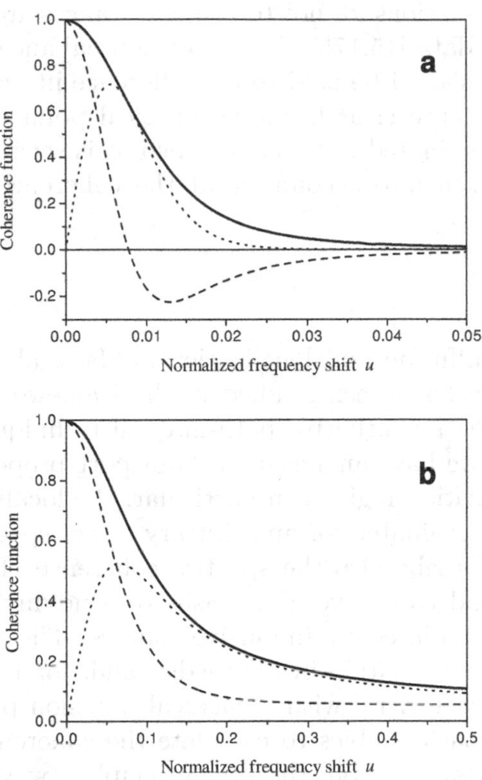

Figure 3. Normalized coherence function $\tilde{\Gamma}(\omega, \Omega)$ vs. normalized frequency shift $u = \Omega L/c$, for $L/l_{\rm s} = 50$ and $\kappa = 0.1$; (a) infinite absorption length, $L/l_{\rm a} = 0.0$, (b) finite absorption length, $L/l_{\rm a} = 0.1$. The absolute value of $\tilde{\Gamma}(\omega, \Omega)$ is shown by solid line, while the dashed and dotted lines correspond, respectively, to the real and imaginary parts of $\tilde{\Gamma}(\omega, \Omega)$.

term accounts for the contribution of local resonances. It can be easily verified that our approximation for the coherence function obeys the symmetry relation $\Gamma(\omega, -\Omega) = \Gamma^*(\omega, \Omega)$, and, therefore, the impulse response function is everywhere real. Note that if the dissipation is present, the real valued wave number k should be replaced by $k + i\gamma$, where γ is the decrement of the field (so that the absorption length is $l_{\rm a} = 1/2\gamma$). Hence, the absorption can be taken into account by obvious substitution $\Omega \to \Omega + 2ic\gamma$ performed in the final expression.

Figure 3 shows two typical examples obtained for the Gaussian model, Eq. (17). The correlator $\tilde{\Gamma}(\omega, \Omega)$ is plotted as a function of the normalized frequency shift, $u = \Omega L/c$. The distance is measured in terms of the scattering mean-free path $l_{\rm s}$ which is given by $l_{\rm s}^{-1} = (\sqrt{\pi}/4)\, k^2 l_\varepsilon \sigma_\varepsilon^2$. In general, the results of the calculations for not too large distances are consistent with known experimental data [15,17]. The absorption enhances essentially the coherence (attention should be paid to the difference in scales between the two plots), which is quite clear from the physical point of view, because the long paths are eliminated in this case. Also, it is seen, especially when the absorption is taken into account, that the coherence function has a two-scale structure.

5. Summary

In this work, both diffusion and localization of classical waves propagating in random media have been studied in the framework of a universal model. Combining the perturbative path-integral technique with the idea of spectral filtering, we have analyzed the transport properties of random systems in the localization regime. In particular, the localization length in 2D systems has been evaluated for an arbitrary power spectrum of the disorder. Our results describe also the spectral coherence of the wave fields in random media, and can serve as a basis for studying a diffusion like spread of the waves radiated by impulsive sources. The theory describes a ballistic to diffusive transition in 3D media, and, for not too large distances, is consistent in general with a classical diffusion paradigm. At the same time, our approach enables to calculate the coherence function and the pulse shape by starting from the first principles (wave equation) and to express these quantities through microscopic parameters of the disorder (power spectrum of the scattering potential). A similar analysis of spectral correlation and pulse evolution can be easily performed also for 1D and 2D media. Since the coherence function has been expressed via an arbitrary form power spectrum, not just isotropic Gaussian one, the results obtained in the work open a new avenue in studying wave transport in anisotropic and/or fractally correlated systems.

Acknowledgement

This work has been supported, in part, by the Office of Naval Research, under Grant No. N00014-00-1-0672.

References

1. A. Ishimaru, *Wave Propagation and Scattering in Random Media* (Academic, New York, 1978).
2. I. M. Lifshits, S. A. Gredeskul, and L. A. Pastur, *Introduction to the Theory of Disordered Systems* (Wiley, New York, 1988).
3. *Scattering and Localization of Classical Waves in Random Media*, edited by P. Sheng (World Scientific, Singapore, 1990).

4. *Diffuse Waves in Complex Media*, edited by J.-P. Fouque (Kluwer, Dordrecht, 1999).
5. P. W. Anderson, Absence of diffusion in certain random lattices, Phys. Rev. **109**, 1492 (1958).
6. S. John, The localization of light and other classical waves in disordered media, Comments Cond. Matt. Phys. **14**, 193 (1988).
7. G. Samelsohn, S. A. Gredeskul, and R. Mazar, Resonances and localization of classical waves in random systems with correlated disorder, Phys. Rev. E **60**, 6081 (1999).
8. V. Fock, Die Eigenzeit in der klassischen und in der Quantenmechanik, Phys. Z. Sow. **12**, 404 (1937).
9. G. Samelsohn and R. Mazar, Path-integral analysis of scalar wave propagation in multiple-scattering random media, Phys. Rev. E **54**, 5697 (1996).
10. G. Samelsohn and V. Freilikher, Two-frequency mutual coherence function and pulse propagation in random media, Phys. Rev. E **65**, 046617 (2002).
11. G. Samelsohn and V. Freilikher, submitted.
12. S. A. Gredeskul, A. V. Marchenko, and L. A. Pastur, Particle and wave transmission in one-dimensional disordered systems, in *Surveys in Applied Mathematics*, vol. 2. (Plenum, New York, 1995).
13. V. D. Freilikher and Yu. V. Tarasov, Propagation of wave packets in randomly stratified media, Phys. Rev. E **64**, 056620 (2001).
14. M. M. Sigalas, C. T. Chan, and C. M. Soukoulis, Propagation of electromagnetic waves in two-dimensional disordered systems, in *Wave Propagation in Complex Media*, edited by G. Papanicolaou (Springer, New York, 1998).
15. P. Sebbah, R. Pnini, and A. Z. Genack, Field and intensity correlation in random media, Phys. Rev. E **62**, 7348 (2000).
16. M. P. van Albada, B. A. van Tiggelen, A. Lagendijk, and A. Tip, Speed of propagation of classical waves in strongly scattering media, Phys. Rev. Lett. **66**, 3132 (1991).
17. Z. Q. Zhang, I. P. Jones, H. P. Schriemer, J. H. Page, D. A. Weitz, and P. Sheng, Wave transport in random media: The ballistic to diffusive transition, Phys. Rev. E **60**, 4843 (1999).

Andrey Chabanov. Photo by Gabriel Cwilich

PHOTON LOCALIZATION IN RESONANT MEDIA

A.A. CHABANOV
Department of Chemical Engineering and Materials Science,
University of Minnesota,
421 Washington Avenue SE, Minneapolis, MN 55455, USA

AND

A.Z. GENACK
Physics Department,
Queens College of the City University of New York,
65-30 Kissena Boulevard, Flushing, NY 11367, USA

In analogy with electron localization in insulators [1], it has been anticipated that photons may be trapped by the constructive interference of waves returning to a point within a strongly scattering medium [2]. Since the closeness to the transition between diffusive and localized waves determines the statistics of multiply scattered waves, charting this transition is of fundamental interest for both statistical optics and electronic mesoscopic physics. However, the very possibility of observing photon localization in random systems has been called into question by the difficulties of achieving strong scattering and of unambiguously detecting electromagnetic localization. Unlike electrons that can be trapped by the Coulomb interaction at atomic sites, photons are not bound by individual particles. They are not strongly scattered by particles either, except at Mie resonances where the scattering cross section can considerably exceed the geometric cross section. Measurements of the exponential scaling of transmission [3–5] have not definitively established photon localization since such scaling may also be due to the presence of absorption. However, recent measurements of coherent backscattering in macroporous GaP networks [6] along with theoretical predictions [7] suggest that the approach to localization can be observed in the rounded backscattering peak from weakly absorbing samples. Also recently, the variance of relative fluctuations has been shown to provide a decisive test for localization, even in the presence of strong absorption [8]. This provides a sure guide in the search for photon localization, which can be used to sort out the precise material and structural

B.A. van Tiggelen and S.E. Skipetrov (eds.),
Wave Scattering in Complex Media: From Theory to Applications, 203–212.

characteristics that may edge samples towards and potentially across the localization threshold.

In the present study of microwave propagation and localization in low-density collections of resonant alumina spheres, we show that localization can be fostered by striking a balance between competing considerations with regard to each of a number of sample characteristics. These include the scatterer size, concentration, and structural correlation. Though precipitous drops in transmission and sharp peaks in the photon transit time are observed near the first five Mie resonances of the alumina spheres, localization occurs only in a narrow frequency window above the first resonance.

We begin with a discussion of key localization parameters in a random medium. These are the Thouless number, which is the ratio of the width to the spacing between quasi-states of a random medium, $\delta = \delta\nu/\Delta\nu$ [9], the dimensionless conductance g [10], and a parameter g' [8], which is proportional to the inverse of the variance of transmission normalized to its ensemble average value. These parameters capture, respectively, the relation to localization of average dynamics, average static transmission, and fluctuations in static transmission.

The Thouless criterion identifies the onset of localization as the point at which the level width $\delta\nu$ becomes smaller than the level spacing $\Delta\nu$ [9]. Beyond this point, transport is inhibited since modes in different blocks of a sample do not overlap. The level width $\delta\nu$ may be identified with the field correlation frequency [11]. This is inversely related to the width of the time-of-flight distribution of the transmitted wave and to the average transit time through the medium, τ [12]. The level spacing $\Delta\nu$ is the inverse of the density of states $N(\nu)$, $\Delta\nu = 1/N(\nu)$, thus giving $\delta = \delta\nu N(\nu)$. The hurdles to localizing radiation, even in resonant media, can be seen from a dynamical perspective. The precipitous drop in $\delta\nu$ expected at resonance, reflecting the sharp peak in τ, does not necessarily result in a diminished value of δ because it is countered by an increasing value of $N(\nu)$.

The dimensionless conductance g is given by the sum $g = \Sigma_{ab}T_{ab}$, where T_{ab} is the transmission coefficient between incident mode a and output mode b. In electronic systems, g can be obtained from the measurement of the conductance G, since $G = (e^2/h)g$, whereas for classical waves it can be found from the measurement of the total transmission $T_a = \Sigma_b T_{ab}$, since $g = N\langle T_a\rangle$, where N is the number of transverse modes given by $N = Ak^2/2\pi$, where A is the area of the cross section of a sample and k is the wave number, and the angle brackets denote the average over an ensemble of statistically equivalent random samples. For diffusive waves in nonabsorbing samples of length L, $g = N\ell/L$, where ℓ is the transport mean free path. The parameter g is essentially the number of diffusive channels in a sample [13], and thus the localization threshold is reached when its value falls below unity. This is consistent with the Thouless criterion for localization since for diffusive waves, $g = \delta > 1$ [10].

In the diffusive limit in the absence of absorption, the variance of the total transmission normalized to its ensemble average is given by $var(s_a) \simeq 2/3g$, where $s_a = T_a/\langle T_a \rangle$ [14, 15]. This allows us to construct a measure of fluctuations, $g' \equiv 2/3var(s_a)$ [8, 16], which reduces to g in the above limit. The localization transition should occur when $g' \simeq 1$ [8]. Since $var(s_a)$ and $var(s_{ab})$, are related by $var(s_a) = \frac{1}{2}(var(s_{ab}) - 1)$ [8, 15], where $s_{ab} = T_{ab}/\langle T_{ab} \rangle$, we are able to evaluate g' from measurements of fluctuation in either total transmission, obtained with use of an integrating sphere [16], or intensity, measured with a probe on the sample surface [8].

Measurements of microwave field transmission spectra have been carried out in low-density collections of 0.95-cm-diameter alumina spheres (99.95% Al_2O_3) with use of a Hewlett-Packard 8772C network vector analyzer. Low densities are produced by embedding the alumina spheres in Styrofoam spheres with low refractive index. The refractive index of the alumina spheres, n, is inferred from a comparison of extinction of the coherent intensity I_c in a thin three-dimensional alumina sample with an alumina volume fraction $f \simeq 0.01$ with the calculated scattering cross section σ_{sc} for isolated alumina spheres. In the independent scatterer approximation, $I_c \propto \exp(-n_{sc}\sigma_{sc}L)$, where n_{sc} is the density of the alumina spheres. The natural logarithm of I_c is compared to calculations of σ_{sc} with $n = 3.14$ in Fig. 1. This choice of n gives the closest correspondence between the positions of the Mie resonances and those of the dips in $\ln I_c$.

Measurements of microwave propagation and localization have been made in quasi-one-dimensional samples with an alumina volume fraction of $f \simeq 0.068$, contained in a 7.3-cm-diameter copper tube of $L = 12$ and $L = 80$ cm. The alumina spheres are put at the centers of 1.9-cm-diameter Styrofoam spheres. An ensemble of 5,000 random sample configurations is created by rotating the tube between subsequent field measurements. A linearly polarized microwave field is launched from a horn positioned 60 cm in front of the sample. The transmitted field is detected with a probe at a point on the output surface of the sample. For each sample configuration, the field spectra yield the corresponding frequency variation of the transmitted intensity I_{ab}. The average intensity $\langle I_{ab} \rangle$ for samples with $L = 80$ cm, shown in Fig. 2a, exhibits distinct drops near Mie resonances. The resonant character of scattering is further indicated by the sharp peaks in the average photon transit time τ, seen in Fig. 2b. The transit time is given by $\tau = \langle s_{ab} d\phi_{ab}/d\omega \rangle$, where ϕ_{ab} is the phase accumulated by the field as it propagates through the sample and ω is the angular frequency [17]. Since low transmission can be due to absorption and long dwell time can be associated with microstructure resonances [18, 19], these indicators do not

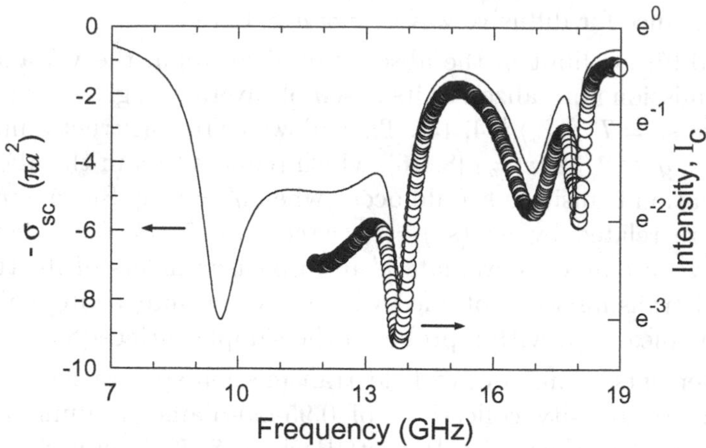

Figure 1. Comparison of the scattering cross section σ_{sc} of a 0.95-cm-diameter alumina sphere (solid) and extinction of the coherent intensity I_c in a thin three-dimensional alumina sample with $f \simeq 0.010$ (circles). In calculations, the alumina sphere is assumed to be lossless and the refractive index is adjusted so that the positions of the Mie resonances are located at the dips in the experimental data. The curve shown is for $n = 3.14$.

provide a definitive measure of the closeness to the localization threshold. This can be obtained, however, from the measurement of $var(s_{ab})$, shown in Fig. 2c. For diffusive waves obeying Rayleigh statistics, $var(s_{ab}) = 1$. Lower values of $var(s_{ab})$, seen below 8.5 GHz, indicate a significant ballistic component in the transmitted field, whereas higher values indicate the presence of substantial long-range correlation. The horizontal dashed line in Fig. 2c represents the localization threshold, $var(s_{ab}) = 7/3$, which corresponds to $var(s_a) = 2/3$ and $g' = 1$. In $L = 80$ cm alumina samples, the localization threshold is crossed in a narrow frequency range above the first Mie resonance. Large fluctuations in measurements of $var(s_{ab})$ above the localization threshold, seen in Fig. 2c, are a result of the greatly enhanced tail of the intensity distribution for localized waves [8].

In order to more carefully examine the nature of transport in the region of localization, the data of Fig. 2 are superimposed and displayed using an expanded scale in Fig. 3a. The position of the resonance is indicated by the peak in the photon transit time. The shift of extrema in $\langle I_{ab} \rangle$ and $var(s_{ab})$ to frequencies above the maximum in τ indicates that localization occurs above the resonance. Since localization occurs within a trough of δ in a frequency range in which τ is smaller than its peak value by a factor of three, we expect $N(\nu)$ drops from its peak value on resonance by an even greater factor. We also expect that, in the absence of absorption, $g \simeq g'$

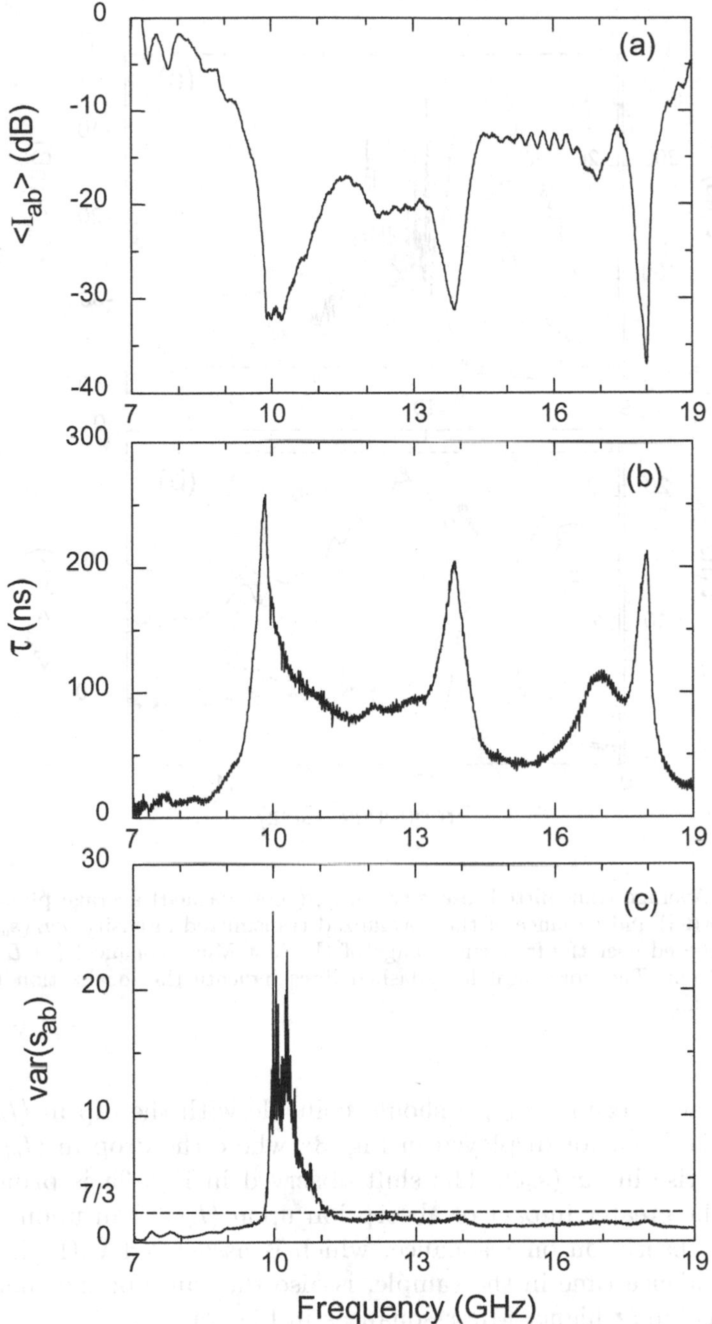

Figure 2. (a) Average transmitted intensity, $\langle I_{ab} \rangle$, (b) average photon transit time, τ, (c) variance of the normalized transmitted intensity, $var(s_{ab})$, in quasi-one-dimensional alumina samples with length $L = 80$ cm and alumina volume fraction $f \simeq 0.068$. The long-dashed line indicates the localization threshold.

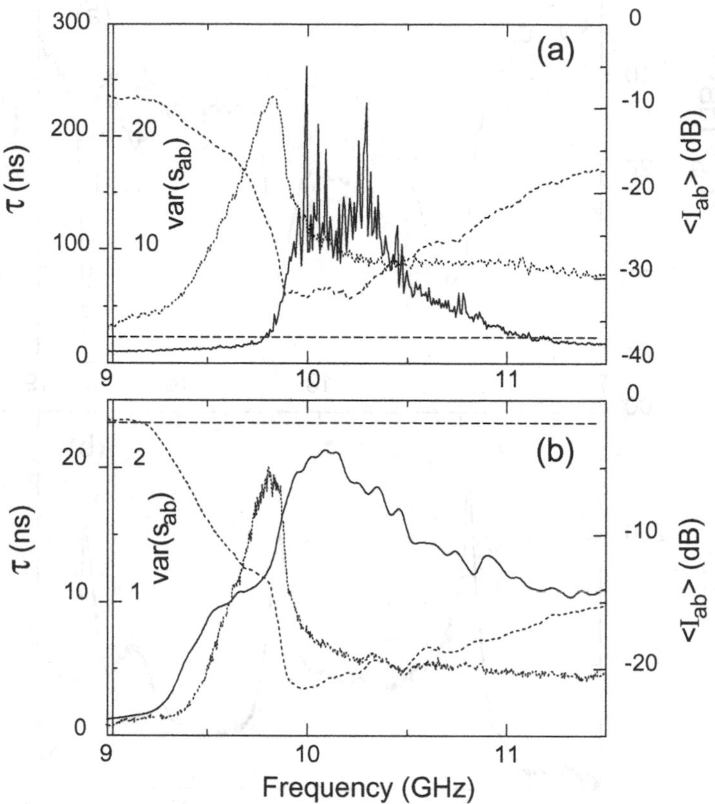

Figure 3. Average transmitted intensity, $\langle I_{ab} \rangle$, (short-dashed) average photon transit time, τ, (dotted) and variance of the normalized transmitted intensity, $var(s_{ab})$, (solid) are superimposed over the frequency range of the first Mie resonance: (a) $L = 80$ cm; (b) $L = 12$ cm. The horizontal long-dashed lines indicate the localization threshold, $var(s_{ab}) = 7/3$.

and that the peak in $var(s_{ab})$ should coincide with the dip in $\langle I_{ab} \rangle$. This is unlike the behavior displayed in Fig. 3a where the drop in $\langle I_{ab} \rangle$ occurs before the rise in $var(s_{ab})$. The shift observed in Fig. 3a is principally a result of the greater impact of absorption upon $\langle I_{ab} \rangle$ than upon $var(s_{ab})$. Enhanced absorption on resonance, which is associated with the greater photon residence time in the sample, is also the cause of the sharp drops in transmission at higher Mie resonances in Fig. 2a.

To investigate the factors those incline a sample towards localization, we carry out the same measurements as presented in Fig. 3a, but in a shorter sample with $L = 12$ cm (see Fig. 3b), in which $\ell < L < \xi, L_a$, where $\xi = N\ell$

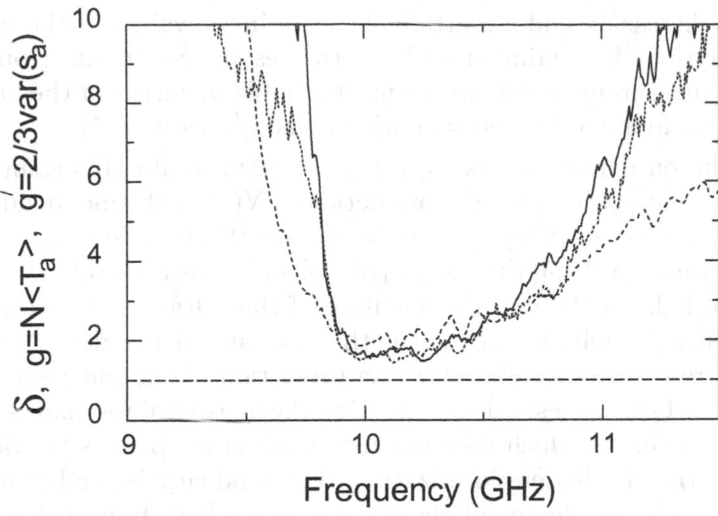

Figure 4. Localization parameters δ (short-dashed), g (dotted), and g' (solid) over the frequency range of the first Mie resonance in alumina samples with $L = 12$ cm. The horizontal long-dashed line indicates the localization threshold.

is the localization length and L_a is the exponential absorption length. In this weakly absorbing sample δ, g, and g' can be associated unambiguously with the proximity to the localization threshold. We find that transport is diffusive above 9.6 GHz, where $1 < var(s_{ab}) < 7/3$. The quantities g and g' are obtained from measurements of the total transmission T_a, via the expressions $g = N\langle T_a \rangle$ and $g' = 2/3var(s_a)$, respectively. Good agreement between g and g' is found, as seen in Fig. 4. Also notice that, since the minimum of g occurs above the resonance, the minimum of ℓ must also lie above that of the scattering length, ℓ_{sc}.

To examine more closely the balance of competing effects near the resonance, we measure $\delta\nu$, $\Delta\nu$, and δ in samples with $L = 12$ cm. The level width $\delta\nu$ is determined from the measurement of the field correlation function with frequency shift. We find that $\delta\nu$ is related to the average transit time τ of Fig. 3b by $\delta\nu \simeq D/L^2 \simeq 1/6\tau$. The level spacing $\Delta\nu$ is found by measuring transmission between a source and a probe in each of the copper end caps placed on the sample to lengthen the dwell time in the sample and thereby narrow the linewidth. In this case, the mode spacing can exceed the linewidth and can be measured (see Fig. 5a). When determining $\Delta\nu$, both positive and negative peaks in spectra of $d\phi_{ab}/d\omega$ are accounted. The negative peaks appear because of interference between two closely spaced modes. A comparison of the level spacing $\Delta\nu$ calculated in an ensemble of

360 sample configurations with the level width $\delta\nu$ is shown in Fig. 5b. We find that whereas $\delta\nu$ and $\Delta\nu$ attain their minimum values at the resonance peak their ratio is a minimum above the resonance. In the trough of δ, where the measurements of $\Delta\nu$ are most reliable by virtue of the low values of $N(\nu)$, δ is in a good agreement with g and g' (see Fig. 4).

The photon density of states, $N(\nu) = 1/\Delta\nu$, in alumina samples with $L = 12$ cm is compared with the prediction of $N(\nu)$ in the independent scatterer approximation, $N(\nu) = (8\pi\nu^2/v_0^3)(v_p/v_E)V$ [18]. Here $v_p = \omega/k$ and v_E are the phase [18] and the energy [19] velocities, respectively, $v_0 = c/n_{eff}$ is speed of light in the effective medium of the refractive index n_{eff}, and V is the sample volume. In Fig. 6, the decrease in the density of states above the resonance is well below the prediction of the independent scatterer model. This suggests that increasing the density of resonant scatterers beyond the point at which they begin to interact suppresses the density of states and thereby fosters localization. The tendency is further enhanced when there is some degree of short-range order [20]. Indeed, for periodic samples exhibiting a complete photonic band gap, the vanishing of $N(\nu)$ automatically results in localization within a pseudogap, when disorder is introduced. We find that, when the degree of order in the sample is reduced by displacing the alumina spheres by half of their radius from the centers of the embedding Styrofoam spheres, $var(s_{ab})$ is only slightly suppressed. For alumna samples with $L = 55$ cm, $var(s_{ab})$ drops from a maximum of 6.0 to 4.5. This small reduction, for a considerable broadening of the radial correlation function, suggests that collective scattering from randomly positioned alumina spheres and not their residual positional order is the chief reason for the sharp drop in $N(\nu)$ above the first Mie resonance.

In conclusion, we have measured the key localization parameters – the Thouless number, the dimensionless conductance, and the variance of transmission fluctuations – in low-density collections of alumina spheres in a quasi-one-dimensional geometry. We find that these parameters are equivalent measures of localization for $L < L_a, \xi$. For samples longer than L_a, only $var(s_a)$ serves as a reliable guide to localization. Photon localization is found in a narrow frequency range above the first Mie resonance in longer samples. Localization does not occur on resonance because the narrowing of the level width is offset by the enhancement of the density of states so that the value of δ does not fall below unity. However, a sharp drop in $N(\nu)$ above resonance leads to a minimum in δ. The drop in $N(\nu)$ and the consequent appearance of localization are the result of collective scattering.

We are grateful to A. Geyfman, E. Kuhner and Z. Ozimkowski for their help in constructing the experimental apparatus. We thank V. Kopp and O. Pudeyev for assistance in analysis of spectra of $d\phi_{ab}/d\omega$, and L.I. Deych, A.A. Lisyansky, M. Stoytchev, B.A. van Tiggelen, and A. Yamilov for stimu-

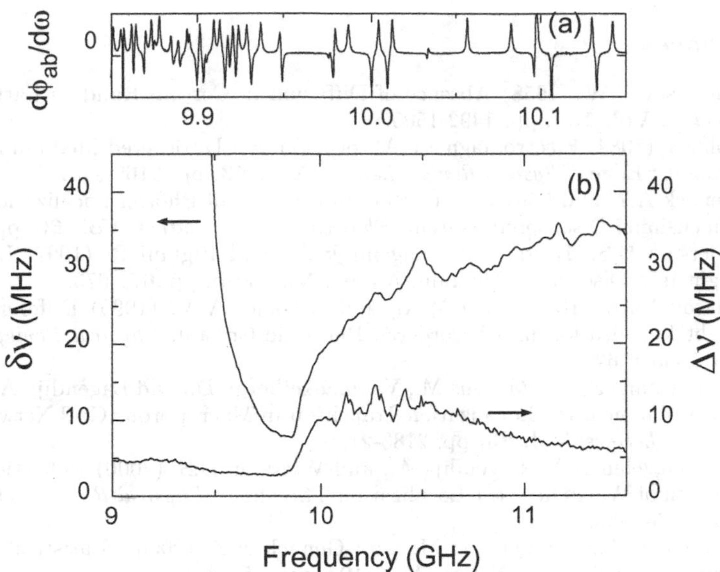

Figure 5. (a) Part of a spectrum of the spectral derivative, $d\phi_{ab}/d\omega$, for a single sample configuration of $L = 12$ cm with the peaks marking the mode frequencies; (b) level width, $\delta\nu$, and level spacing, $\Delta\nu$, in alumina samples of $L = 12$ cm. For all frequencies, $\delta\nu > \Delta\nu$ and thus $\delta > 1$.

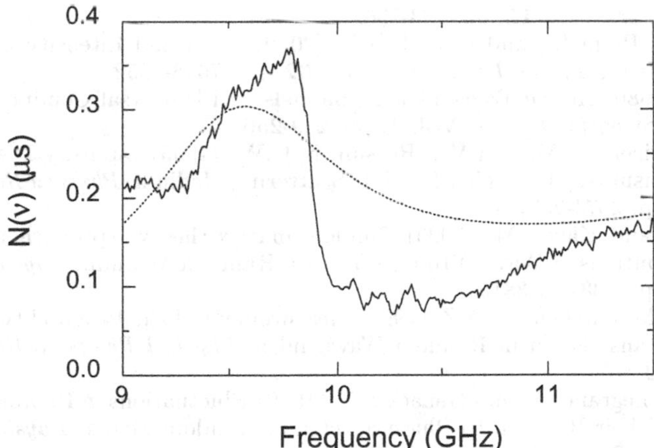

Figure 6. Photon density of states, $N(\nu)$, in alumina samples with $L = 12$ cm; the measured values of $N(\nu) = 1/\Delta\nu$ (solid), and the prediction of independent scatterer model [18] (dotted).

lating discussions. This research was supported by grants from the National Science Foundation, the U.S. Army Research Office, and PSC-CUNY.

References

1. Anderson P.W. (1958) Absence of Diffusion in Certain Random Lattices, *Physical Review*, **Vol. 109**, pp. 1492-1505.
2. John S. (1984) Electromagnetic Absorption in a Disordered Medium near a Photon Mobility Edge, *Physical Review Letters*, **Vol. 53**, pp. 2169-2172.
3. Genack A.Z. and Garcia N. (1991) Observation of Photon Localization in a Three-dimensional Disordered System, *Physical Review Letters*, **Vol. 66**, pp. 2064-2067.
4. Wiersma D.S., Bartolini P., Lagendijk A., and Righini R. (1997) Localization of Light in a Disordered Medium, *Nature*, **Vol. 390**, pp. 671-673.
5. Vlasov Yu.A., Kaliteevski M.A., and Nikolaev V.V. (1999) Different Regimes of Light Localization in a Disordered Photonic Crystal, *Physical Review B*, **Vol. 60**, pp. 1555-1562.
6. Schuurmans F.J.P., Megens M., Vanmaekelbergh D., and Lagendijk A. (1999) Light Scattering near the Localization Transition in Macroporous GaP Networks, *Physical Review Letters*, **Vol. 83**, pp. 2183-2186.
7. Van Tiggelen B.A., Lagendijk A., and Wiersma D.S. (2000) Reflection and Transmission of Waves near the Localization Threshold, *Physical Review Letters*, **Vol. 84**, pp. 4333-4336.
8. Chabanov A.A., Stoytchev M., and Genack A.Z. (2000) Statistical Signatures of Photon Localization, *Nature*, **Vol. 404**, pp. 850-853.
9. Thouless D.J. (1977) Maximum Metallic Resistance in Thin Wires, *Physical Review Letters*, **Vol. 39**, pp. 1167-1169.
10. Abrahams E., Anderson P.W., Licciardello D.C., and Ramakrishnan T.V. (1979) Scaling Theory of Localization: Absence of Quantum Diffusion in Two Dimensions, *Physical Review Letters*, **Vol. 42**, pp. 673-676.
11. Genack A.Z. (1990) Universality of Wave Propagation in Random Media, *Europhysics Letters*, **Vol. 11**, pp. 733-736.
12. Sebbah P., Pnini R., and Genack A.Z. (2000) Field and Intensity Correlation in Random Media, *Physical Review E*, **Vol. 62**, pp. 7348-7352.
13. Imry Y. (1986) Active Transmission Channels and Universal Conductance Fluctuations, *Europhysics Letters*, **Vol. 1**, pp. 249-256.
14. Nieuwenhuizen Th.M. and Van Rossum M.C.W. (1995) Intensity Distributions of Waves Transmitted through a Multiple Scattering Medium, *Physical Review Letters*, **Vol. 74**, pp. 2674-2677.
15. Kogan E. and Kaveh M. (1995) Random-matrix-theory Approach to the Intensity Distributions of Waves Propagating in a Random Medium, *Physical Review B*, **Vol. 52**, pp. R3813-R3815.
16. Stoytchev M. and Genack A.Z. (1997) Measurement of the Probability Distribution of Total Transmission in Random Waveguides, *Physical Review Letters*, **Vol. 79**, pp. 309-312.
17. Sebbah P., Legrand O., and Genack A.Z. (1999) Fluctuations in Photon Local Delay Time and Their Relation to Phase Spectra in Random Media, *Physical Review E*, **Vol. 59**, pp. 2406-2411.
18. Lagendijk A. and Van Tiggelen B.A. (1996) Resonant Multiple Scattering of Light, *Physics Reports*, **Vol. 270**, pp. 143-215.
19. Van Albada M.P., Van Tiggelen B.A., Lagendijk A., and Tip A. (1991) Speed of Propagation of Classical Waves in Strongly Scattering Media, *Physical Review Letters*, **Vol. 66**, pp. 3132-3135.
20. John S. (1987) Strong Localization of Photons in Certain Disordered Dielectric Superlattices, *Physical Review Letters*, **Vol. 58**, pp. 2486-2489.

ANDERSON LOCALIZATION OF ELECTROMAGNETIC WAVES IN CONFINED DISORDERED DIELECTRIC MEDIA

MARIAN RUSEK AND ARKADIUSZ ORŁOWSKI
Institute of Physics, Polish Academy of Sciences
Aleja Lotników 32/46, 02–668 Warszawa, Poland

1. Introduction

Scattering of electromagnetic waves from various kinds of obstacles is rich of interesting and sometimes unexpected phenomena. Already for two scatterers placed together well within a wavelength an extremely narrow proximity resonance can appear [10]. For many randomly distributed scatters we may expect that, for same range of parameters, Anderson localization can show up.

A convincing experimental demonstration that Anderson localization of electromagnetic waves is possible in three-dimensional disordered dielectric structures has been given recently [12]. The strongly scattering medium has been provided by a semiconductor powder with a very large refractive index. By decreasing the average particle size it was possible to observe a clear transition from linear scaling of transmission ($T \propto L^{-1}$) to an exponential decay ($T \propto e^{-L/\xi}$). Some localization effects have been also reported in previous experiments on microwave localization in copper tubes filled with metallic and dielectric spheres [4]. However, the latter experiments were plagued by large absorption, which makes the interpretation of the data quite complicated.

Another experiment on microwave localization has been performed in a two-dimensional medium [1]. The scattering chamber was set up as a collection of dielectric cylinders randomly placed between two parallel aluminum plates on half the sites of a square lattice. These authors attributed the observed sharp peaks of transmission to the existence of localized modes and measured the energy density of the electromagnetic field localized by their random structures.

B.A. van Tiggelen and S.E. Skipetrov (eds.),
Wave Scattering in Complex Media: From Theory to Applications, 213–227.

As shown in our previous paper [7], Anderson localization of classical waves can be studied theoretically by investigating a striking phase transition in the spectra of certain random matrices. The matrices whose elements are equal to the Green's function calculated for the differences between positions of any pair of scatterers. The Breit-Wigner's model of a single scatterer allows one to give a sound physical interpretation to the universal properties of these matrices. In this case the spectrum can be considered as an approximation to the resonance poles of the system. In the limit of an infinite random medium all eigenvalues condense to a smooth line instead of redistributing themselves over the complex plane. Physically speaking this corresponds to the formation of an entire frequency band of spatially localized electromagnetic waves.

The Green's function is one of the fundamental basic building blocks for constructing a self-consistent description of multiple scattering processes. In the free-space case the Green's function is something very simple. It describes a spherical (in 3D), cylindrical (in 2D) or a pair of two plane (in 1D) outgoing wave(s) centered at the scatterers position. Boundary conditions may change the Green's function in a complicated way. It is therefore interesting to extend the random-matrix description of localization to encompass the case of nontrivial boundary conditions. For this purpose in this paper we investigate Anderson localization of electromagnetic waves in a disordered dielectric medium confined within a metallic waveguide. The results of our previous paper [9] are extended to the case of a multimode waveguide and new physical interpretation based on the Breit-Wigner's model of a single scatterer is presented.

2. Basic Assumptions

In the following we study the properties of the stationary solutions of the Maxwell's equations in two-dimensional media consisting of randomly placed parallel dielectric cylinders of infinite height (i.e., very long as compared to the wavelength of the electromagnetic field). This means that one, say (y), out of three dimensions is translationally invariant and only the remaining two (x, z) are random. In the present model we place the disordered dielectric medium between two infinite, perfectly conducting mirrors described by the equations $x = 0$ and $x = d$. Thus we will consider the case where the cylinders are oriented parallel to the mirrors. For simplicity our discussion will be restricted to TE modes polarized along the y axis only. The main advantage of this two-dimensional approximation is that we can use the scalar theory of electromagnetic waves [8]:

$$\mathbf{E}(\mathbf{r}, t) = \mathrm{Re}\left\{\mathbf{e}_y\, \mathcal{E}(x, z)\, e^{-i\omega t}\right\}. \tag{1}$$

Consequently, the polarization of the medium takes the form:

$$\mathbf{P}(\mathbf{r},t) = \mathrm{Re}\left\{\mathbf{e}_y\,\mathcal{P}(x,z)\,e^{-i\omega t}\right\}. \tag{2}$$

Localization of electromagnetic waves in disordered 2D media is usually studied experimentally in microstructures consisting of dielectric cylinders with diameters and mutual distances being comparable to the wavelength [1]. However it seems to be a reasonable assumption that what really counts for the basic features of localization is the scattering cross-section and not the real geometrical size of the scatterer itself. Therefore we will represent the dielectric cylinders located at the points (x_a, z_a) by single 2D electric dipoles:

$$\mathcal{P}(x,z) = \sum_{a=1}^{N} p_a\,\delta^{(2)}(x - x_a, z - z_a). \tag{3}$$

It should be stressed, that the boundary conditions considered in this paper are different from those encountered in the experiment of Ref. [1]. To minimize the effect of the waves reflected off the edges of the scattering chamber, its perimeter was lined with a layer of microwave absorber. Therefore, to model that particular experiment, it is appropriate to use the free space boundary conditions (as we did in our previous papers [8, 11]).

3. Planar Waveguide

Several particular cases may be considered ($k\,d = \pi$ is the cut-off thickness of the waveguide). For $N = 0$ and $k\,d < \pi$ there are no guided modes in the waveguide as well as there are no localized waves. This case is analogous to the electronic band gap in a solid. If $N > 0$ and $k\,d < \pi$ there are still no guided modes in the waveguide but localized waves can appear for any distribution of the cylinders. It is again analogous to the solid state physics situation where isolated perturbations of the periodicity of crystals (like impurities or lattice defects) can lead to the formation of localized electronic states with energies within the forbidden band. Another possibility corresponds to $N = 0$ and $k\,d > \pi$. In this case there are guided modes but the system supports no localized waves. This is very similar to the conductance band in solids. Guided modes correspond to extended electronic states described by Bloch functions. In this paper we perform a detailed study of the regime where $N > 0$ and $k\,d > \pi$. For this range of parameters there are both the guided modes and the resonances of transmission. Isolated localized waves can be seen for certain distributions of the cylinders. The signs of Anderson localization emerging in the limit of an infinite medium can be observed both in analysis of transmission and in the properties of the spectra of certain random matrices. Eventually we will consider a limiting case of $N \to \infty$ and $k\,d > \pi$. It turns out that in this case the guided modes no longer exist in the waveguide. Instead a band of localized waves will be formed for any distribution of the cylinders. It is an interesting analog of the Anderson localization in noncrystalline solids such as amorphous semiconductors or disordered metals.

4. Field Expansion

The electric field of the electromagnetic wave incident on the cylinders

$$\mathcal{E}^{(0)}(x,z) = \sum_{m=1}^{M} \iota_m\, \mathcal{E}^{(m)}(x,z), \tag{4}$$

may be expanded into the guided modes of the waveguide [5]:

$$\mathcal{E}^{(m)}(x,z) = \frac{2}{\sqrt{\beta_m\, d}}\, \sin(\alpha_m\, x)\, e^{i\beta_m\, z}, \tag{5}$$

where the propagation constants are given by:

$$\alpha_m = \frac{\pi}{d}\, n, \quad \beta_m = \sqrt{k^2 - \alpha_m^2}. \tag{6}$$

The total field that can be measured *far* from the cylinders is fully described by the reflection ρ_m and transmission τ_m coefficients of all guided modes:

$$\mathcal{E}(x,z) = \sum_{m=1}^{M} \iota_m\, \mathcal{E}^{(m)}(x,z) + \sum_{m=1}^{M} \rho_m\, \mathcal{E}^{(m)*}(x,z) \quad \text{for} \quad z \to -\infty \tag{7}$$

$$\mathcal{E}(x,z) = \sum_{m=1}^{M} \tau_m\, \mathcal{E}^{(m)}(x,z) \quad \text{for} \quad z \to +\infty \tag{8}$$

Using the Lorentz theorem and repeating the straightforward but lengthly calculations (see, e.g., [5]) we easily arrive at the following expressions determining the transmission coefficients

$$\tau_m = \iota_m + i\pi\, k^2 \sum_{a=1}^{N} p_a\, \mathcal{E}^{(m)*}(x_a, z_a), \tag{9}$$

and the reflection coefficients

$$\rho_m = i\pi\, k^2 \sum_{a=1}^{N} p_a\, \mathcal{E}^{(m)}(x_a, z_a), \tag{10}$$

for given dipole moments p_a. In the following sections we will relate p_a to the values of the incident field calculated at the positions of the cylinders $\mathcal{E}^{(0)}(x_a, z_a)$.

5. Method of Images

A simple way to take into account the boundary conditions of parallel mirrors and their influence on the electromagnetic field is to use the method of images. This technique has been used, i.e., in QED calculations of spontaneous emission in cavities [2, 3]. To reproduce the correct boundary conditions on the radiation field of each cylinder (3) the mirrors are replaced by an array of image cylinders whose phases alternate in sign:

$$\mathcal{P}(x,z) = \sum_{a=1}^{N} \sum_{j=-\infty}^{\infty} (-1)^j p_a \, \delta^{(2)}(x - x_a^{(j)}, z - z_a), \tag{11}$$

where

$$x_a^{(j)} = (-1)^j x_a + jd. \tag{12}$$

Thus a finite system of dielectric cylinders (3) placed within a metallic waveguide is fully equivalent to an infinite system of cylinders (11) forming a slab in a free space. This fact allows us to utilize some results from our previous paper concerning dielectric cylinders in free space [8].

6. Elastic Scattering

It is now a well-established fact that to use safely the point-scatterer approximation it is essential to use a representation for the cylinders that conserves energy in the scattering processes. In the case of a system of cylinders in free space this requirement means that for each cylinder the optical theorem holds [8]:

$$\pi \, k^2 \, |p_a|^2 = \operatorname{Im} \left\{ p_a^* \, \mathcal{E}'(x_a, z_a) \right\}. \tag{13}$$

Eq. (13) gives the following form of the coupling between the dipole moment p_a and the electric field incident on the cylinder $\mathcal{E}'(x_a, z_a)$:

$$i\pi \, k^2 \, p_a = \frac{e^{i\phi} - 1}{2} \, \mathcal{E}'(x_a, z_a), \tag{14}$$

The same result holds also for a system of cylinders placed in a metallic waveguide (11).

7. Multiple Scattering

In the case of a confined medium the field acting on the ath cylinder $\mathcal{E}'(x_a, z_a)$ from Eq. (14) is the sum of the incident guided mode $\mathcal{E}^{(0)}$, which obeys the Maxwell's equations in an empty waveguide, and waves scattered by all other cylinders *and* by all images:

$$\mathcal{E}'(x_a, z_a) = \mathcal{E}^{(0)}(x_a, z_a) + \frac{e^{i\phi} - 1}{2} \sum_{b=1}^{N} G_{ab}\, \mathcal{E}'(x_b, z_b), \quad a = 1, \ldots N. \quad (15)$$

Thus, in the present model the G-matrix from Eq. (15) needs to be defined differently than in Ref. [8]:

$$i\pi\, G_{ab} = 2 \sum_{\rho_{ab}^{(j)} \neq 0} (-1)^j\, K_0(-ik\rho_{ab}^{(j)}), \quad (16)$$

where

$$\rho_{ab}^{(j)} = \sqrt{(x_a - x_a^{(j)})^2 + (z_a - z_b)^2} \quad (17)$$

denotes the distance between the ath cylinder and the jth image of the bth cylinder and K_0 is the modified Bessel function of the second kind. Note that summation in Eq. (16) is performed over all j, for which $\rho_{ab}^{(j)} \neq 0$.

The system of linear equations (15) fully determines the field acting on each cylinder $\mathcal{E}'(x_a, z_a)$ for a given field of the guided mode $\mathcal{E}^{(0)}(x_a, z_a)$ incident on the system. Analogous relationships between the stationary outgoing wave and the stationary incoming wave are known in the general scattering theory as the Lippmann-Schwinger equations [6]. If we solve Eqs. (15) and use Eqs. (14) to find p_a, then we are able to find the transmission and reflection coefficients given by Eqs. (9) and (10).

8. Reflection Coefficient

Substituting Eq. (14) into Eq. (10) we get the following expression for the reflection coefficient of a mth guided mode from a system of N identical cylinders:

$$|\rho_m|^2 = \sin^2 \frac{\phi}{2} \left| \sum_{a=1}^{N} \mathcal{E}'(x_a) \cdot \mathcal{E}^{(m)}(x_a, y_a) \right|^2 . \quad (18)$$

If $\mathcal{E}'(x_a)$ is an eigenvector of the G-matrix (16) corresponding to the eigenvalue λ,

$$\sum_{b=1}^{N} G_{ab}\, \mathcal{E}'(x_b, y_b) = \lambda\, \mathcal{E}'(x_a, y_a), \quad a = 1, \ldots, N, \quad (19)$$

then the reflection coefficient takes the form of

$$|\rho_m|^2 = \frac{1}{(\cot\phi + \operatorname{Im}\lambda)^2 + (1 + \operatorname{Re}\lambda)^2} \times \left| \sum_{a=1}^{N} \mathcal{E}^{(0)}(x_a, y_a) \cdot \mathcal{E}^{(m)}(x_a, y_a) \right|^2 . \quad (20)$$

Only the first term in Eq. (20) depends on the model of the scatterer (through the phase shift ϕ). The incident field appears only in the second term. Both terms depend on the geometry of the system (the first one through the eigenvalue λ) and frequency.

9. Breit-Wigner Scatterers

The cylinders necessarily have an internal structure. Thus in general the phase shift ϕ should be regarded as a function of frequency ω. For example to model a simple scattering process with one internal Breit-Wigner type resonance one can write:

$$\cot \frac{\phi(\omega)}{2} = -\frac{\omega - \omega_0}{\gamma_0}. \tag{21}$$

The total scattering cross-section $k\,\sigma = 4\,\sin^2 \phi/2$ [8] takes then the familiar Lorentzian form:

$$k\,\sigma = \frac{4\,\gamma_0^2}{(\omega - \omega_0)^2 + \gamma_0^2}. \tag{22}$$

Substituting the Breit-Wigner model of scattering from Eq. (21) into Eq. (20) we get the following expressions for the reflection coefficients of the guided modes:

$$\begin{aligned} |\rho_m(\omega)|^2 &= \overbrace{\frac{\gamma_0^2}{[\omega - (\omega_0 + \gamma_0 \operatorname{Im} \lambda)]^2 + [\gamma_0(1 + \operatorname{Re} \lambda)]^2}}^{\text{rapidly varying}} \\ &\times \underbrace{\left| \sum_{a=1}^{N} \mathcal{E}^{(0)}(x_a, y_a) \cdot \mathcal{E}^{(m)}(x_a, y_a) \right|^2}_{\text{slowly varying}}. \end{aligned} \tag{23}$$

Most localization experiments are performed in the range of optical or microwave frequencies. In this case usually $\gamma_0/\omega_0 \ll 1$. It is therefore reasonable to assume that the first term in Eq. (23) varies with frequency much faster than the second one. Thus we get a Lorentzian-type resonance of width

$$\gamma \simeq \gamma_0(1 + \operatorname{Re} \lambda) \tag{24}$$

centered around frequency

$$\omega \simeq \omega_0 + \gamma_0 \operatorname{Im} \lambda. \tag{25}$$

10. Single Scattering

Let us begin by considering a single dielectric cylinder placed in a metallic waveguide. In the following we will assume that only *one* guided mode exists. The scattering is elastic. Thus on average the energy scattered by the cylinder must be equal to the energy given to the cylinder by the incident wave. Therefore:

$$|\tau_1|^2 + |\rho_1|^2 = 1. \tag{26}$$

Substituting into Eq. (26) the formulas for the transmission and reflection coefficients Eqs. (9) and (10) we get the following form of the coupling between the dipole moment p_1 of the cylinder and the electric field of the incident wave calculated at the cylinder $\mathcal{E}^{(0)}(x_1, z_1)$:

$$i\pi\, k^2\, p_1 = \frac{e^{i\phi'} - 1}{2}\, \mathcal{E}^{(0)}(x_1, z_1). \tag{27}$$

Thus all the scattering properties of a cylinder are perfectly described by a phase shift ϕ'. The explicit form of the transmission and reflection coefficients from a single cylinder then reads

$$\tau_1 = \frac{e^{i\phi'} + 1}{2}, \quad \rho_1 = \frac{e^{i\phi'} - 1}{2}. \tag{28}$$

Notice that the waveguide may introduce its own phase shift depending on the position of the cylinder with respect to its walls. Therefore the free space phase shift ϕ from Eq. (21) is different from the phase shift ϕ' from Eq. (27). This results in changing the width and position of the transmission and reflection resonance.

Indeed, as an example let us consider a cylinder placed between the mirrors separated by a distance $k\,d = 3\pi/2$. We have calculated numerically the corresponding 1×1 G-matrix (16). The images of the cylinder from Eq. (11) were summed from $j = -50000$ to $j = 50000$. The resulting "spectrum" ($\lambda = G_{11}$) of the G-matrix is plotted on the left panel of Fig. 1. We see, that as opposed to the free-space case ($G_{11} = 0$) it is nonzero.

According to Eqs. (24) and (25) the eigenvalues λ can be considered as a first-order approximation (in γ_0/ω_0) to the positions of the resonance poles in the system. Thus if $\lambda \neq 0$ then $\omega \neq \omega_0$, $\gamma \neq \gamma_0$. To check if the shape of the resonance has really changed on the right panel of Fig. 1 we have plotted as a solid line the reflection coefficient $R = |\rho_1|^2$. The dashed line on the same plot corresponds to the free-space case $\lambda = 0$.

11. Collective Resonances

In the next step in Figs. 2 and 3 we plot the reflection R of the systems of $N = 10, 100$ cylinders as a function of the frequency ω. In the same plots

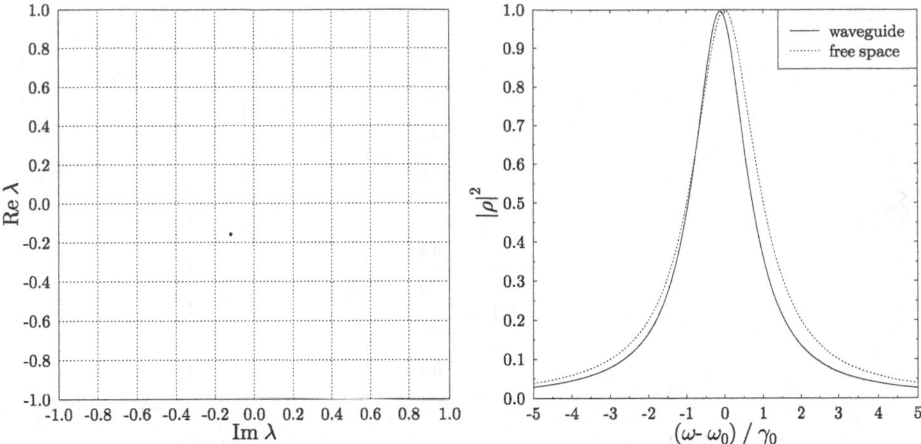

Figure 1. Reflection R of a single dielectric cylinder placed in a planar metallic waveguide plotted as a function of the frequency ω (right) and the corresponding spectrum of the G-matrix (left). The dotted line corresponds to the transmission in a free-space case.

we have also the corresponding approximate values of the resonance poles given by the spectrum of the G-matrix. The cylinders were distributed randomly with constant uniform density $n = 1$ cylinder per wavelength squared. Therefore for each N the size of the system was proportional to the number of cylinders $L \propto N$.

We have seen that in the case $N = 1$ the incident wave was totally reflected for a *single* value of $\omega = \tilde{\omega}$. Note that not necessarily $\tilde{\omega} = \omega_0$, and therefore for this value of ω the total scattering cross-section σ of an individual dielectric cylinder (22) does not approach its maximal value. However, for systems containing $N = 10$ and $N = 100$ scatterers the entire *regions* of the values of frequencies ω exist for which $R \simeq 1$. They are separated by narrow maxima of transmission. Moreover, inspection of Figs. 2 and 3 suggests that in the limit $N \to \infty$ the number of these maxima increases and simultaneously they became narrower and sharper. Therefore we may expect that for sufficiently large N the incident waves will be totally reflected for almost any ω except the discrete set $\omega = \omega_l$ for which the reflection is close to zero. Physically speaking this means that different realizations of sufficiently large system of randomly placed cylinders are hardly distinguishable from each other by a transmission experiment.

It follows from inspection of Eqs. (9) and (10) that the maximum of transmission $T = \sum_m |\tau_m|^2 = 1$ (and minimum of reflection $R =$

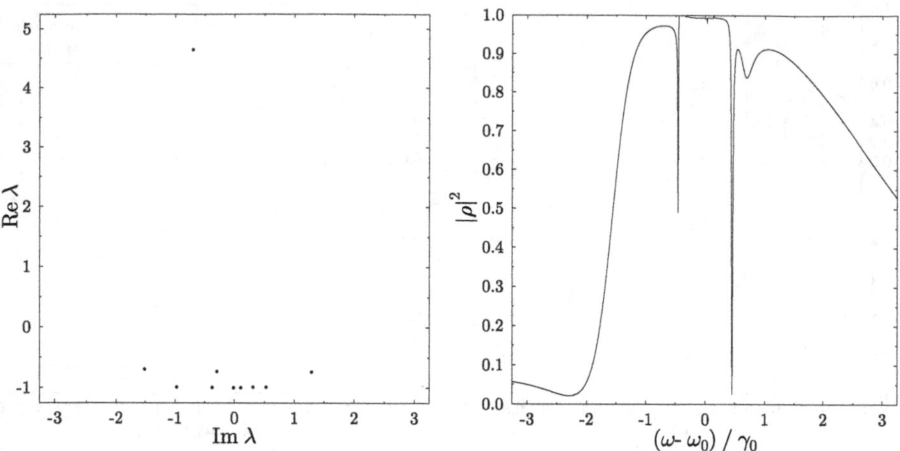

Figure 2. On the right: reflection R of a system of $N = 10$ dielectric cylinders placed randomly in a planar metallic waveguide plotted as a function of the frequency ω. On the left: first-order approximation to the resonance poles in the system given by the spectrum λ of the corresponding G-matrix.

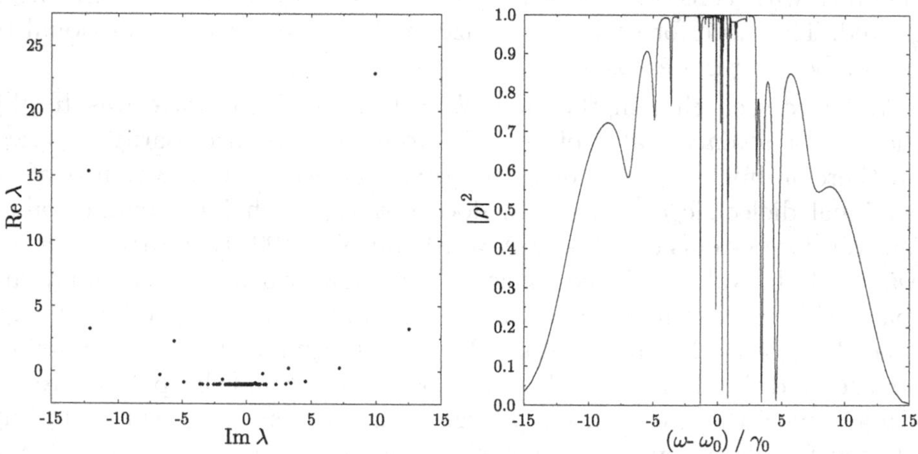

Figure 3. On the right: reflection R of a system of $N = 100$ dielectric cylinders placed randomly in a planar metallic waveguide plotted as a function of the frequency ω. On the left: first-order approximation to the resonance poles in the system given by the spectrum λ of the corresponding G-matrix.

$\sum_m |\rho_m|^2 = 0$, because the medium is non-dissipative) corresponds to the case when the polarization of the medium fulfills the following condition:

$$\sum_{a=1}^{N} p_a\, \mathcal{E}^{(m)*}(x_a, z_a) = \sum_{a=1}^{N} p_a\, \mathcal{E}^{(m)}(x_a, z_a) = 0 \quad \text{for} \quad m = 1, \dots M. \qquad (29)$$

This means that in the expansion of the field radiated by the medium into waveguide modes the coefficients on nearly *all* guided modes Eq. (5) vanish. Therefore the radiated field consists only of evanescent modes with imaginary propagation constants β_m and thus it is exponentially localized in the vicinity of the medium. In the next section we will show that such a field can exist also without any incident wave and therefore represents a truly *localized* wave.

12. Localized Waves

By definition, an electromagnetic wave is localized in a certain region of space if its magnitude is (at least) exponentially decaying in any direction from this region. We will show now that electromagnetic waves localized in the system of dielectric cylinders placed in a planar metallic waveguide correspond to nonzero solutions $\mathcal{E}_l'(x_a, z_a) \neq 0$ of Eqs. (15) for the incoming wave equal to zero, i.e., $\mathcal{E}^{(0)}(x, z) \equiv 0$. Note that we added an index l which labels the localized waves.

Indeed, let us suppose that the field is exponentially localized. This means that there are no guided modes in the radiation field. Therefore (as shown in the previous section) Eq. (29) holds. Using Eq. (14) we see that the vector formed by the values of the field acting on the cylinders is *orthogonal* to the vector formed by the values of incident field calculated at the positions of the cylinders:

$$\sum_{a=1}^{N} \mathcal{E}_l'(x_a, z_a)\, \mathcal{E}^{(0)*}(x_a, z_a) = 0. \qquad (30)$$

But simultaneously $\mathcal{E}_l'(x_a, z_a)$ is a solution of a system of linear Eqs. (15) where $\mathcal{E}^{(0)}(x_a, z_a)$ is the right-hand-side. Therefore $\mathcal{E}_l'(x_a, z_a)$ is also a solution of Eqs. (15) with $\mathcal{E}^{(0)}(x_a, z_a) \equiv 0$.

The proof works also the other way round. Suppose that $\mathcal{E}_l'(x_a, z_a)$ is a solution of Eqs. (15) for $\mathcal{E}^{(0)}(x_a, z_a) \equiv 0$. As the considered medium is non-dissipative, the time average energy stream integrated over a closed surface surrounding it must vanish. This means that there are again no guided modes in the radiation field (which in the case $\mathcal{E}^{(0)}(x, z) \equiv 0$ is equal to the total field). Therefore Eq. (29) holds and the wave is localized.

13. Eigenproblem

Note that for the incoming wave equal to zero, i.e., $\mathcal{E}^{(0)}(x,z) \equiv 0$, the system of equations (15) is equivalent to the *eigenproblem* for the G-matrix:

$$\sum_{b=1}^{N} G_{ab}\, \mathcal{E}'_l(x_b, z_b) = \lambda_l\, \mathcal{E}'_l(x_a, z_a), \tag{31}$$

where

$$\lambda_l = -1 - i \cot\frac{\phi}{2}. \tag{32}$$

Let us stress that Eq. (32) can be fulfilled *only* if the real part of an eigenvalue satisfies

$$\mathrm{Re}\,\lambda_l = -1. \tag{33}$$

Substituting the Breit-Wigner model of scattering (21) we see that the imaginary part of the eigenvalue and the frequency of the localized wave are then related by

$$\mathrm{Im}\,\lambda_l = \frac{\omega_l - \omega_0}{\gamma_0}. \tag{34}$$

Therefore only those eigenvectors $\mathcal{E}'_l(x_a, z_a)$ of the G-matrix which correspond to the eigenvalues λ_l satisfying the condition (33) may be related to localized waves. Moreover, those waves can exist only for discrete frequencies ω_l given by Eq. (34). As discussed in the previous sections they are the same values of ω for which the transmission is equal to unity.

14. Anderson Localization

Note that in finite dielectric media no localized states are supported by Maxwell's equations in two dimensions [8]. However, this is not the case with confined media, where localized waves do exist even in finite media. Therefore a clear distinction between localized waves with isolated frequencies and the dense band of localized waves (due to Anderson localization) is needed. We show now that this distinction may be provided by investigation of a phase transition which occurs in the limit of $N \to \infty$ in the spectra of G_{ab} matrices corresponding to systems of randomly distributed dielectric cylinders.

To support this statement let us recall Figs. 2 and 3. We have plotted there the spectra λ of a G-matrix (diagonalized numerically) corresponding to certain specific configurations of $N = 10$ and $N = 100$ cylinders placed randomly with the uniform density $n = 1$ cylinder per wavelength squared. We see that already in the case of $N = 10$ quite a lot of eigenvalues are located near the $\mathrm{Re}\,\lambda = -1$ axis. This tendency is more and

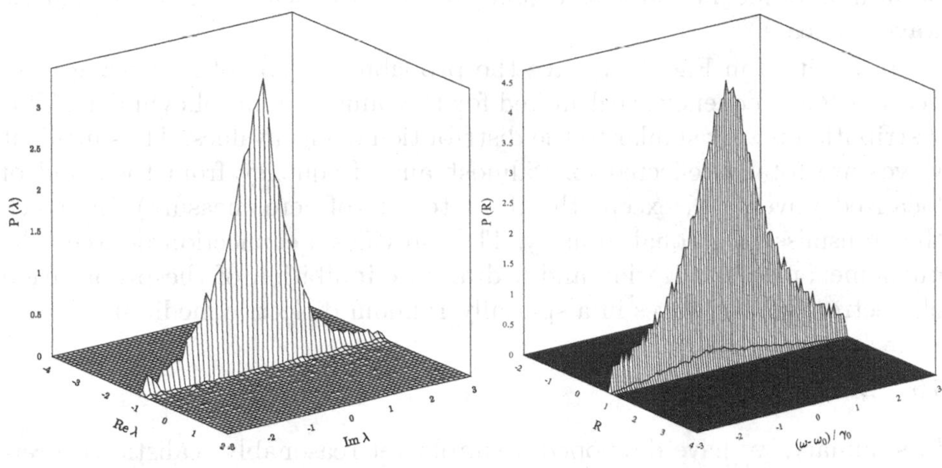

Figure 4. On the left: density of eigenvalues $P(\lambda)$ of the matrix G calculated from 10^2 distributions of $N = 100$ cylinders. On the right: probability $P(R)$ of measuring a reflection R at frequency ω calculated for the same systems. The similarity is striking.

more pronounced with increasing size of the system measured by N. This is a universal property of 2D G matrices, not restricted to this specific realization of the system only. To prove these statements we diagonalize numerically the G-matrix (16) for 10^2 different distributions of $N = 100$ cylinders. Then we construct two-dimensional histogram of eigenvalues λ from all distributions. It approximates the corresponding probability distribution $P(\lambda)$ which is normalized in the standard way $\int d^2\lambda\, P(\lambda) = 1$. In Fig. 4 we have the surface plot of the function $P(\lambda)$. It clearly shows that for all configurations (without, may be, a set of zero measure) most eigenvalues are located near the $\mathrm{Re}\,\lambda = -1$ axis.

Our numerical investigations indicate that in the limit of an infinite medium, the probability distribution under consideration will tend to the delta function in the real part:

$$\lim_{N\to\infty} P(\lambda) = \delta(\mathrm{Re}\,\lambda + 1)\, f(\mathrm{Im}\,\lambda). \tag{35}$$

This means that in this limit for almost any random distribution of the cylinders, an infinite number of eigenvalues satisfies the condition Eq. (33). It is therefore reasonable to expect that in the case of a random and infinite system a countable set of frequencies ω_l corresponding to localized waves becomes *dense* in some finite interval. But it is always difficult to separate such frequencies from frequencies which may be arbitrarily near and physically the spectrum is always a coarse-grained object. Therefore in the limit of an infinite medium an entire *band* of spatially localized electromagnetic waves appears.

In addition in Fig. 4 we have the probability $P(R)$ of measuring a reflection R at frequency ω calculated for the same systems of cylinders. This distribution is very similar to the distribution of eigenvalues. Thus incident waves are totally reflected for "almost any" frequency from the band of localized waves, i.e., except the discrete set (of zero measure) for which the transmission is equal to unity. This provides a connection between the phenomenon of localization and a dramatic inhibition of the propagation of electromagnetic waves in a spatially random dielectric medium.

15. Summary

In summary, we have developed a simple yet reasonably realistic theoretical approach to Anderson localization of electromagnetic waves in two-dimensional dielectric media confined within a metallic waveguide. The results of our previous papers dealing with the free space configurations are now extended to encompass the case of nontrivial boundary conditions. A sound physical interpretation in terms of transmission experiment is also proposed. By confining a system of randomly distributed dielectric cylinders into a planar metallic waveguide we are able to observe clear signs of Anderson localization already for $N = 100$ scatterers. One of the indicators of localization is the phase transition in the spectra of certain random matrices. This property of random Green's functions is now generalized to the case of a confined dielectric medium (where the Green's function is different). It may be interpreted as an appearance of the band of localized electromagnetic waves emerging in the limit of the infinite medium. A connection between this phenomenon and a dramatic inhibition of the propagation of electromagnetic waves in a spatially random dielectric medium was provided. A clear distinction between isolated localized waves (which do exist in finite confined media) and the band of localized waves (which appears in the limit of an infinite random medium) was also presented.

Acknowledgements

The authors are grateful to Sergey Skipetrov and Bart van Tiggelen for the kind hospitality in Cargese. This work was supported in part by Polish KBN grant No. 2 P03B 044 19.

References

1. Dalichaouch, R., Armstrong, J.P., Schultz, S., Platzman, P.M., and McCall, S.L. (1991) Microwave localization by two-dimensional random scattering, *Nature* **354**,

53.
2. Dowling, J.P. (1993) Spontaneous emission in cavities: How much more classical can you get? *Found. Phys.* **23**, 895.
3. Dowling, J.P., Scully, M.O., and DeMartini, F. (1991) Radiation pattern of a classical dipole in a cavity, *Opt. Comm.* **82**, 415.
4. Genack, A.Z., and Garcia, N. (1991) Observation of photon localization in a three-dimensional disordered system, *Phys. Rev. Lett.* **66**, 2064.
5. Jackson, J.D. (1962) Classical Electrodynamics, John Wiley & Sons, New York.
6. Lippmann, B.A., and Schwinger, J. (1950) Variational principles for scattering processes, *Phys. Rev.* **79**, 469.
7. Rusek, M., Mostowski, J., and Orłowski, A. (2000) Random Green matrices: From proximity resonances to Anderson localization, *Phys. Rev. A* **61**, 022704.
8. Rusek, M., and Orłowski, A. (1995) Analytical approach to localization of electromagnetic waves in two-dimensional random media, *Phys. Rev. E* **51**, R2763.
9. Rusek, M., and Orłowski, A. (1999) Anderson localization of electromagnetic waves in confined dielectric media, *Phys. Rev. E* **59**, 3655.
10. Rusek, M., and Orłowski, A. (2001) Localization of light in three-dimensional disordered dielectrics, in: Optics of Nanostructured Materials, Markel, V.A., and George, T.F. (eds.), John Wiley & Sons, New York, pp. 201–226.
11. Rusek, M., Orłowski, A., and Mostowski, J. (1997) Band of localized electromagnetic waves in random arrays of dielectric cylinders, *Phys. Rev. E* **56**, 4892.
12. Wiersma, D.S., Bartolini, P., Lagendijk, A., and Righini, R. (1997) Localization of light in a disordered medium, *Nature* **390**, 671.

During a lunch. In the lower right corner: Arkadiusz Orłowski and Marian Rusek. Photo by Gabriel Cwilich

LASER ACTION IN THE REGIME OF STRONG LOCALIZATION

M. PACILLI, P. SEBBAH AND C. VANNESTE
Laboratoire de Physique de la Matière Condensée
CNRS UMR 6622
Université de Nice - Sophia Antipolis
Parc Valrose, 06108, Nice Cedex 02, France

Abstract. We present numerical results of laser action in an active random medium in the localized regime. It is shown that the localized modes play the same role as the cavity modes of a conventional laser. We demonstrate that the introduction of a local gain enables to activate the localized modes individually. The origin of coherent feedback in disordered media is then discussed.

1. Introduction

It has been now well established experimentally that introducing light amplification in disordered materials can lead to two different kinds of lasing action. Lasing with incoherent feedback, which is expected to occur in weakly disordered media, describes amplified spontaneous emission (ASE). In this regime, which was proposed by Letokhov [1] in 1967 and was first observed by Gouedart *et al.* [2] in 1993 and Lawandy [3] in 1994, scattering simply increases the path length of light in the active region. As a result, amplification of light is more efficient than in a homogeneous medium. In contrast, lasing with coherent feedback, which is expected in strongly disordered media, describes true laser oscillation. In this regime, which has been observed recently by several groups [4–7], the photon mean free path is supposed to be sufficiently small for the scattering medium to provide coherent feedback. If the origin of coherent feedback in actual experiments has been a matter a debate [3, 8, 9], there is one situation where coherent feedback should naturally occur, namely the case when light is localized. In this case, localized modes are expected to play the same role as the

B.A. van Tiggelen and S.E. Skipetrov (eds.),
Wave Scattering in Complex Media: From Theory to Applications, 229–239.

cavity modes in a conventional laser. This work addresses this issue. Most analysis to date have described wave propagation in disordered media in the diffusion approximation. However, this approximation, which is well adapted to describe the ASE regime, does not take into account the interference effects, which underlie the regime of Anderson localization. To handle properly the interference effects, direct resolution of Maxwell's equations must be used as first performed by Soukoulis *et al.* [10]. In this article, we start from Maxwell's equations to investigate numerically laser action in a two-dimensional (2D) medium [11, 12]. First, the system is chosen to be sufficiently scattering to supports modes, which are spatially localized. Next, by using a four-level atomic system to describe the amplifying medium, the lasing modes are proven to be identical to the localized modes of the passive random medium. Eventually, by locally pumping the system, it is shown that it is possible to select individually the excited mode of the random laser. This result is similar to recent experimental observations [4].

2. Description of the System

We consider a 2D system of size L^2, which consists of circular particles with radius r, optical index n_2, randomly distributed in a background medium of optical index n_1. This system is equivalent to a random collection of cylinders oriented along the z-axis. The filling fraction of the particles is ϕ (Fig. 1). The background medium of index n_1 is chosen as the active part of

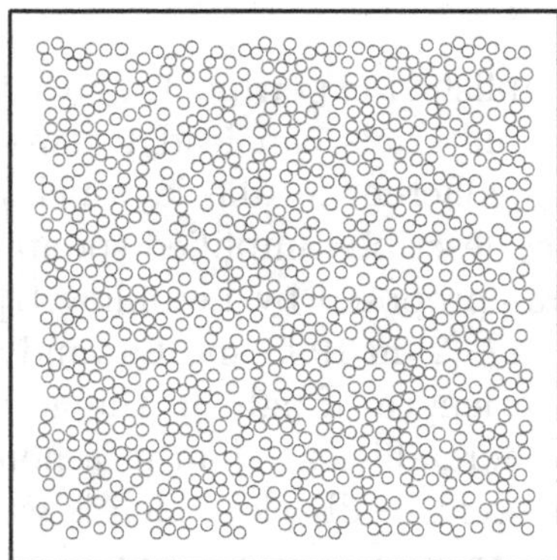

Figure 1. Random realization of particles: $L = 5.5$ μm, $r = 60$ nm and $\Phi = 40\%$.

the system, in order to control the randomness and the gain independently. This active medium induces a polarization term in Maxwell's equations, which are written

$$\mu_0 \partial H_x/\partial t = -\partial E_z/\partial y \tag{1a}$$
$$\mu_0 \partial H_y/\partial t = \partial E_z/\partial x \tag{1b}$$
$$\varepsilon_i \varepsilon_0 \partial E_z/\partial t + \partial P/\partial t = \partial H_y/\partial x - \partial H_x/\partial y \ , \tag{1c}$$

where P is the polarization density.

Here, we have considered a 2D transverse magnetic (TM) field with components H_x, H_y and E_z. ε_0 and μ_0 are the electric permittivity and the magnetic permeability of vacuum respectively.

The time evolution of the four-level atomic system is described by conventional rate equations. The electrons in the ground level 1 are transferred to the upper level 4 by an external pump at a fixed rate W_p. Electrons in level 4 flow downward to level 3 by means of fast decay processes with a characteristic time τ_{43} so that they quickly populate level 3. The intermediate levels 3 and 2 are the upper and lower levels respectively of the laser transition. The spontaneous decay rates downward from these two levels are $1/\tau_{32}$ and $1/\tau_{21}$ respectively. Moreover, stimulated transitions due to the electromagnetic field take place between these two levels. The corresponding equations have the following form [13]

$$dN_1/dt = N_2/\tau_{21} - W_p N_1 \tag{2a}$$
$$dN_2/dt = N_3/\tau_{32} - N_2/\tau_{21} - (E_z/\hbar\omega_l)\, dP/dt \tag{2b}$$
$$dN_3/dt = N_4/\tau_{43} - N_3/\tau_{32} + (E_z/\hbar\omega_l)\, dP/dt \tag{2c}$$
$$dN_4/dt = -N_4/\tau_{43} + W_p N_1 \ , \tag{2d}$$

where N_i is the population density in level i, i=1 to 4. The stimulated transition rate is given by the term $(E_z/\hbar\omega_l) dP/dt$ where $\omega_l (E_3 - E_2)/\hbar$ is the transition frequency between levels 2 and 3. The fields, the polarization and the atomic populations E_z, P and N_i, i=1 to 4, not only depend on the time t but also on the position $\vec{r}$ in the system. The evolution of the polarization is described by the following equation [13]

$$d^2P/dt^2 + \Delta\omega_l dP/dt + \omega_l^2 P = \kappa \cdot \Delta N \cdot E_z \ , \tag{3}$$

where $\Delta N = N_2 - N_3$ is the difference between the populations densities in the lower and upper levels of the atomic transition. Amplification takes place when the external pumping mechanism produces inverted population differences $\Delta N < 0$. The linewidth of the atomic transition is $\Delta\omega_l = 1/\tau_{32} + 2/T_2$ where the dephasing time T_2 is usually much shorter than the lifetime τ_{32}. The constant κ is given by $\kappa = 6\pi\epsilon_0 c^3/\omega_l^2 \tau_{32}$.

We have used the following values of the different parameters:

TABLE 1. Random system of particles

System size	$L = 5.5\ \mu$m
Volume fraction of particles	$\Phi = 40\%$
Radius of the particles	$r = 60$ nm
Optical index of the particles	$n_2 = 2$
Optical index of the background medium	$n_1 = 1$

TABLE 2. Four-level atomic system

Total atomic density	$N_T = 3.313 \times 10^{24}\text{m}^{-3}$
Frequency of the atomic transition	$\nu_l = \omega_l/2\pi = 6.71 \times 10^{14}$ Hz ($\lambda = 446.9$ nm)
Lifetime of level 4	$\tau_{43} = 10^{-13}$ s
Lifetime of level 3	$\tau_{32} = 10^{-10}$ s
Lifetime of level 2	$\tau_{21} = 5 \times 10^{-12}$ s
Collision time	$T_2 = 2 \times 10^{-14}$ s

Note that in order to approach the localized regime, a large optical index contrast has been chosen, namely $n_1 = 1$ and $n_2 = 2$.

The above values are close to those of dye molecules such as rhodamine 640, which has been used in several experiments [3, 6, 14, 15, 16].

We have used the finite-difference time-domain (FDTD) method [17] to solve the Maxwell's equations and PML (perfectly matched layer) absorbing conditions [18] in order to model an open system. These absorbing conditions efficiently absorb the waves that leave the system thus preventing from spurious reflections at the boundaries. Otherwise, the boundaries would play the role of an artificial numerical cavity and would alter the lasing action.

The space increment $\Delta x = \Delta y = 10$ nm. The time increment has been chosen to be $\Delta t = \Delta x / c\sqrt{2} \approx 2.8 \times 10^{-17}$ s, where c is the velocity of light in vacuum, to ensure the stability of the FDTD algorithm. These values are sufficiently small compared to the typical optical wavelength $\lambda \approx 500$ nm and optical period $T \approx 1.5 \times 10^{-15}$ s considered in the following. The total size of the system $L = 5.5\ \mu$m corresponds to $L = 550 \times \Delta x$.

3. Modes of the System without Gain

To study the passive modes of the system, the pumping rate W_p and the populations N_1, N_2 and N_3 are set to zero in Eqs. 2. Two different sources have been used. The first source is a short electromagnetic pulse, which is launched inside the system. The response is recorded at several locations and Fourier transformed in order to get the power spectrum. The second source is monochromatic at an eigenfrequency ν selected in the power spectrum in order to excite the corresponding eigenmode alone.

3.1. SPECTRUM

Since the system is open, most of the energy of the short pulse escapes rapidly through the boundaries. However, after this fast transient, which lasts a fraction of picosecond due to the small size of the system, a small part of the energy remains trapped for much longer times. The field is then recorded at several locations and Fourier transformed. An example of the resulting power spectrum is displayed in Fig. 2. It exhibits several peaks,

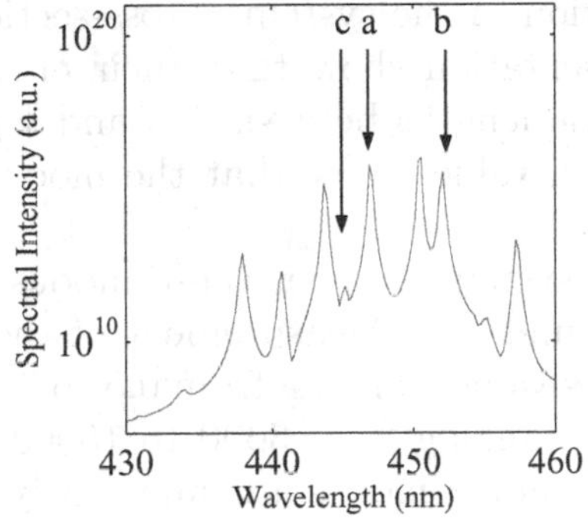

Figure 2. Power spectrum after excitation of the system by a short pulse.

which correspond to long-lived modes. The number and position of these peaks depend on the realization of the disorder. However, we have shown [12] that the wavelength range where they appear (430 nm to 460 nm in Fig. 2), does not depend on the realization of the disorder. It only depends on the indices n_1, n_2, the size r and the filling fraction ϕ of the particles. Other frequency windows exhibiting similar peaks also show up at smaller wavelengths.

3.2. EIGENMODES

In order to study the modes associated to each peak of the previous spectrum, the system is then excited by a monochromatic source at the corresponding frequencies. The source has a Gaussian envelope of duration larger than the reciprocal of the level spacing between two adjacent peaks so that each mode is excited separately. When the emission of the monochromatic source has stopped, the spatial map of the field is recorded. Fig. 3 displays the maps corresponding to a few peaks in Fig. 2. They display a

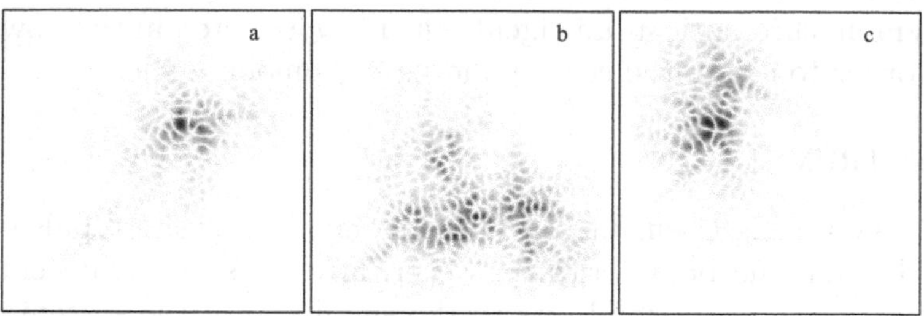

Figure 3. Spatial distribution of the field amplitude of 3 localized modes corresponding to peaks a to c in Fig. 2. (a) mode a ($\lambda = 446.9$ nm), (b) mode b ($\lambda = 452.1$ nm) and (c) mode c ($\lambda = 445.0$ nm).

strong spatial localization in the system. Cross-sections of their shapes in semilogarithmic representation show that their envelope decay exponentially with characteristic lengths between 0.5 and 1 μm. Since the system size is $L = 5.5$ μm, such values show that the modes are well localized in the system.

However, since the system is open, these modes also decay by energy leaking through the boundaries. Measurement of the escape times for the modes in Fig. 3 provides values ranging from 0.5 to 3.5 ps. They correspond to quality factors $\nu/\delta\nu$ ranging from 3000 to 15000. Such values indicate that indeed the corresponding modes are sufficiently long-lived to be comparable to the cavity modes of a conventional laser. This remark will be corroborated in the next section when gain is introduced in the background medium.

4. Modes of the Random Laser

In this section, gain is introduced in the background medium by setting the pumping rate W_p at a non-zero value. Above the threshold value, laser action takes place. First, we shall describe the situation where the pumping rate is uniform over the whole system. Just above threshold, a single mode is excited. This mode is the most favored one in the system. This is the mode for which the laser gain counterbalances first the losses through the boundaries of the system. It will be shown that this mode is one of the localized modes that were found in the passive system without gain. Next, when the pumping rate increases further, laser action enters the multimode regime where other modes are excited. The power spectrum of the laser shows that the excited frequencies are identical to the eigenfrequencies of the passive system without gain, indicating that the laser field is a superposition of modes of the passive system. Eventually, we shall describe the situation where the laser gain is spatially localized. By adjusting the rate and the position of the pump, we show that it is possible to select any of the localized modes of the passive system.

4.1. UNIFORM GAIN

When the pumping rate is adjusted just above threshold for laser action, a single mode is excited. The single mode regime is corroborated by the presence of a single peak in the power spectrum of the laser field (not shown). Figure 4 displays the spatial map of the field amplitude of this mode. We see that this map is identical to the map of mode a displayed in

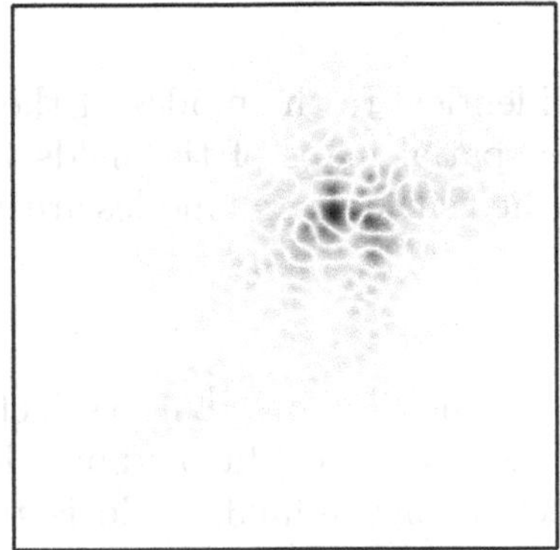

Figure 4. Spatial distribution of the field amplitude of the laser emission just above threshold.

Fig. 3. The reason why mode a has been selected rather than another mode is first due to the fact that mode a has the longest decay time $\tau = 3.49$ ps. Moreover, we have chosen on purpose the center frequency of the atomic transition equal to the frequency of mode a. Hence, the gain is maximum at the frequency of mode a. Thus, it is not surprising that mode a is excited first just above threshold. However, the main result in this example is the fact that the laser mode is identical to one of the localized modes of the passive system without gain.

When the pumping rate is further increased, new modes are progressively excited as indicated by the emergence of new peaks in the power spectrum. Figure 5 shows the power spectrum at a high pumping rate where many modes are simultaneously excited. The important result here is that the wavelength positions of the highest peaks of this spectrum can be identified with a good precision with the peaks of the power spectrum displayed in Fig. 2. This result indicates that in this multimode regime, the

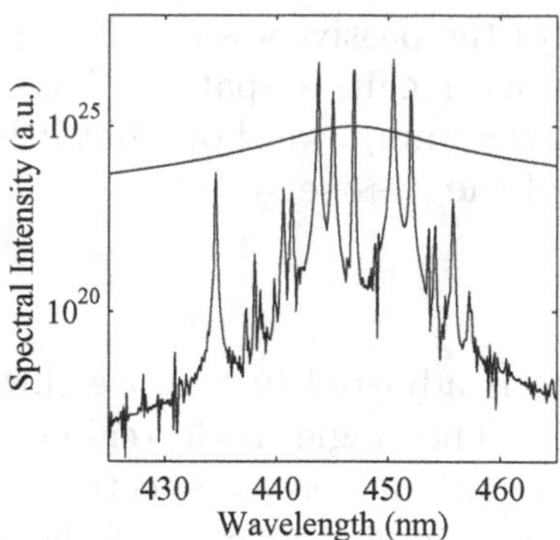

Figure 5. Power spectrum of the laser field at high pumping rate. The Lorentzian lineshape of the gain has also been displayed.

excited modes are still identical to the modes of the passive system. However, comparison of the spatial maps of the fields with and without gain cannot be readily performed since several modes are excited simultaneously.

4.2. LOCALIZED GAIN

Hence, one can wonder whether it is possible to select individually localized modes different from mode a, the most favored one. In a conventional laser, a common way to select a longitudinal mode is to introduce an etalon into the cavity. Obviously, this method is not possible for a random laser. Another possibility would be to adjust the center frequency of the atomic transition to the frequency of the mode to be selected. Though possible numerically, it is not usually easy to adjust at will the gain curve in actual experiments. Moreover, since the selection of the excited mode is due to the competition between the gain and the losses, it is not obvious that this method will work for all the modes, especially those, which are strongly damped. Even by adjusting the maximum of the gain curve to the frequency of such a mode, it is quite possible that a long-lived mode, not too far from the center frequency of the atomic transition, will emerge first. However, there is another way to select the random laser modes individually, which take advantage of the spatial properties of the localized modes. Since the localized modes are randomly located in the system, it is possible to favor one of them by optimizing the spatial overlap of the external pump with this mode. For this purpose, local instead of uniform pumping of the system can be used, the position and the rate of the pump being adjusted to select the desired mode.

We have used an external pump with a Gaussian spatial profile. The width σ has been chosen of the order of the localization length $\sigma \approx \xi \approx 0.5$

μm. Its position has been fixed at different locations in the system. For each position of the pump, the pumping rate is adjusted just above threshold for laser action in order to excite a single mode. As expected, we found that this threshold value depends on the position of the external pump. By scanning the position of the pump across the system, we have been able to excite each of the modes in Fig. 3 individually.

Figure 6 displays a first example of such local pumping. In this ex-

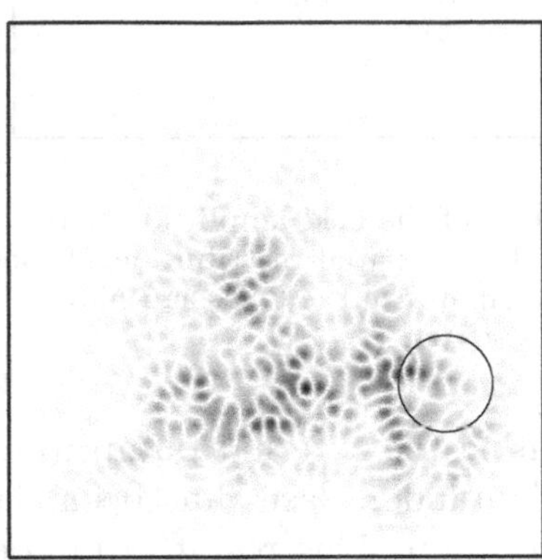

Figure 6. Spatial distribution of the field amplitude of the laser emission for spatially localized gain. Compare with the spatial distribution of mode b in Fig. 3. The circle represents the spatial extension at σ of the Gaussian pump.

ample, we see that the excited mode is mode b in Fig. 3. First, this result demonstrates again that the laser mode is identical to one of the localized modes of the system without gain. Moreover, it confirms that it is possible to select a mode less favored than mode a since mode b has a shorter decay time $\tau = 2.25$ ps and is frequency-shifted from the maximum of the gain curve.

A second example of local pumping is displayed in Fig. 7. Once again, the excited mode is a localized mode of the system without gain, namely mode c in Fig. 3. In Fig. 3, mode c is the mode, which has the shortest decay time, $\tau = 0.72$ ps. Nevertheless, even in this very unfavorable case, proper adjustment of the location of the pump allows to excite this mode alone.

5. Discussion

The main result of this work is that the excited modes of a random laser in the localized regime are identical to the localized modes of the same medium without gain [11, 12, 19]. Hence, the localized modes of a strongly scattering system are comparable to the cavity modes of a conventional

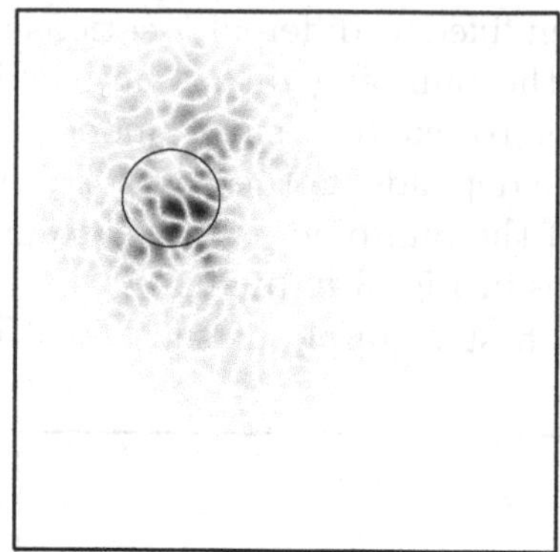

Figure 7. Spatial distribution of the field amplitude of the laser emission for spatially localized gain. Compare with the spatial distribution of mode c in Fig. 3. The circle represents the spatial extension at σ of the Gaussian pump.

laser. Moreover, by limiting the external pump to a small part of the system, the spectrum of the random laser has been shown to depend on the location of the external pump. This result is in good agreement with recent experimental observations [4]. In particular, we have shown that local pumping of the system enables to excite individually any localized mode by properly adjusting the position and the rate of the pump. Hence, introducing laser amplification in a random system can be used as a efficient tool to study Anderson localization as achieved in a recent experimental work by Cao *et al.* [20].

Although random lasing in the localized regime is well understood, it is experimentally difficult to achieve. As a matter of fact, all experimental observations of laser action in disordered systems have been reported for systems that seem to be far from the localized regime. When available, measurements of the mean free path ℓ indicate values of the product $k\ell$ much larger than 1. Hence, such systems are still in the diffusive regime. Nevertheless, laser oscillation with coherent feedback rather than ASE is now well established in those systems. Hence, current investigation is largely devoted to the study of lasing in systems, which are not supposed to support localized modes providing the mechanism for coherent feedback. Recently, Apalkov *et al.* [21] have put forward the idea that resonators with high quality factors are likely to appear in random systems even if on average light propagation is diffusive. Those resonators are analogous to the prelocalized states that have been investigated in electronic systems for about ten years [22]. The presence of such resonators would be sufficient to explain the coherent feedback responsible for lasing. Yet, such a scenario, which is probably pertinent in appropriate random media, is to be confirmed experimentally.

Actually, we have performed preliminary numerical analysis of 2D random systems, which seem to indicate that lasing with coherent feedback is possible in the diffusive regime without any favorable configuration of scattering particles acting as high quality resonators. The first main difference with the localized regime is that the threshold pumping rate is much higher. Future investigation is needed to understand the origin of laser oscillation in such systems. Clearly, in spite of much progress in recent years, laser action in random media appears to be more complex than the simple division between incoherent and coherent feedback and will still be matter of discussion in the next future.

This work is currently supported by the Centre National de la Recherche Scientifique, the Research Group PRIMA and the Direction Générale de l'Armement under contract No. 0034041.

References

1. V. S. Letokhov, Sov. Phys. JETP **26**, 835 (1968).
2. C. Gouedart, D. Husson, C. Sauteret, F. Auzel, and A. Migus, J. Opt. Soc. Am. **B 10**, 2358 (1993).
3. N.M. Lawandy, R.M. Balachandra, A.S.L. Gomes, and E. Sauvain, Nature **368**, 436 (March 1994).
4. H. Cao, Y.G. Zhao, S.T. Ho, E.W. Seelig, Q. H. Wang, and R. P. H. Chang, Phys. Rev. Lett. **82**, 2278 (1999).
5. H. Cao, J. Y. Xu, D.Z. Zhang, S.-H. Chang, S.T. Ho, E.W. Seelig, X. Liu and R.P.H. Chang, Phys. Rev. Lett. **84**, 5584 (2000).
6. S.V. Frolov, Z.V. Vardeny, K. Yoshino, A. Zakhidov, and R.H. Baughman, Phys. Rev **B 59**, 5284 (1999).
7. R.C. Polson, J.D. Huang and Z.V. Vardeny, Synth. Metals **119**, 7 (2001).
8. D. S. Wiersma, M. P. van Albada and A. Lagendijk, Nature **373**, 203 (1995).
9. H. Cao, J. Y. Xu, S.-H. Chang, S.T. Ho, Phys. Rev. **E61**, 1985 (2000).
10. X. Jiang and C.M. Soukoulis, Phys. Rev. Lett. **85**, 70 (2000).
11. C. Vanneste and P. Sebbah, Phys. Rev. Lett. **87**, 183903 (2001).
12. P. Sebbah and C. Vanneste, Phys. Rev. **B66**, 144202 (2002).
13. A. E. Siegman, *Lasers* (University Science Books, Mill Valley, 1986).
14. M. Siddique, R. R. Alfano, G. A. Berger, M. Kempe and A. Z. Genack, Opt. Lett. **21**, 450 (1996).
15. G. van Soest, F. J. Poelwijk, R. Sprik and A. Lagendijk, Phys. Rev. Lett. **86**, 1522 (2001).
16. Y. Ling, H. Cao, A.L. Burin, M.A. Ratner, X. Liu and R.P.H. Chang, Phys. Rev. **A64**, 063808 (2001).
17. A. Taflove, *Computational Electrodynamics: The Finite-Difference Time-Domain Method* (Artech House, Norwood,1995).
18. J.P. Berenger, J. Comput. Phys. **114**, 185 (1995).
19. X. Jiang and C. M. Soukoulis, Phys. Rev. **E65**, 025601 (2002).
20. H. Cao, Y. Ling, J.Y. Xu and A.L. Burin, Phys. Rev. **E66**, 025601 (2002).
21. V.M. Apalkov, M.E. Raikh and B. Shapiro, Phys. Rev. Lett. **89**, 016802 (2002)
22. V.G. Karpov, Phys. Rev. **B48**, 12539 (1993).

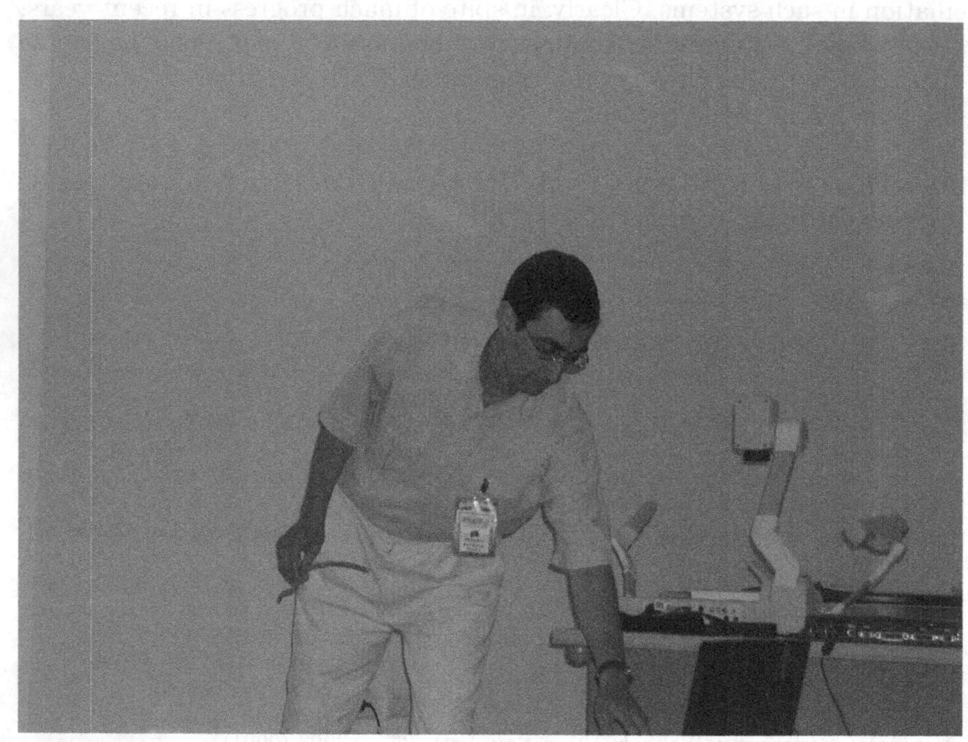

Patrick Sebbah. Photo by Valentin Freilikher

STATISTICS OF REFLECTED SPECKLE INTENSITIES ARISING FROM LOCALIZED STATES INSIDE THE GAP OF DISORDERED PHOTONIC CRYSTALS

ZHAO-QING ZHANG AND XIANGDONG ZHANG
Department of Physics
and Institute of Nano Science and Technology
The Hong Kong University of Science and Technology,
Clear Water Bay, Kowloon, Hong Kong, China

Abstract. The field distributions of reflected speckles arising from localized states inside the gap of disordered photonic crystals in two dimensions were studied through numerical simulations using the multiple-scattering method. By separating the field into the coherent and diffuse parts, we have studied the statistics of field and phase distributions for both diffuse and total fields as well as their speckle contrasts as a function of the amount of disorder. For the non-Bragg angles, it is found that the intensity distribution crosses over from non-Rayleigh to Rayleigh statistics when disorder is increased. This is similar to the crossover from ballistic to diffusive wave propagation for the transmitted waves and can be described by the random-phasor-sum model (RPS). For the Bragg angle, only non-Rayleigh statistics were found. Both the RPS and K distribution have limited range of validity in this case.

1. Introduction

In the past few years, the scattering of both classical and quantum waves from random media has been under intensive study [1]. A wave undergoes multiple scatterings due to the presence of inhomogeneities/impurities as it propagates through a random medium. As a result, the scattered intensities become highly irregular. Those speckle pattern should be described in statistical terms. There has been considerable study of the statistics of transmitted speckle intensities in random media [1–9]. It is well-known that, in the absence of interference, wave propagation is diffusive and the

B.A. van Tiggelen and S.E. Skipetrov (eds.),
Wave Scattering in Complex Media: From Theory to Applications, 241–254.

intensities of the transmitted speckles follow Rayleigh statistics (a negative exponential law). The presence of interference always produces an anomalous exponential tail [3–5]. In a strong scattering medium, the distribution can cross over to log-normal behavior when waves become localized [8,9]. Thus, a strong deviation from Rayleigh statistics is an indication of localization. Deviation from Rayleigh statistics also appears in thin samples when the number of multiple scatterings is insufficient to randomize the phase of the incident wave. The field distributions exhibit interesting crossover from ballistic to diffusive behavior with increasing sample thickness [10,11].

In contrast to the transmitted speckles, in the reflection geometry, random matrix theory has predicted that, in a waveguide with large number of channels, the intensity of reflected speckles follows Rayleigh statistics even in the localized regime [12]. This is because for the reflected waves, the dominant contributions come from short and non intersecting paths, which produce Rayleigh statistics [13]. However, this may not be true if the reflected waves experience only a small number of impurity scatterings before exiting the sample, similar to the case of thin samples in the transmission geometry. This can occur when the frequency of the incident wave lies inside the gap of a weakly disordered photonic crystal sample. Photonic crystals themselves have attracted a lot of attention in recent years [14]. For frequencies inside a band gap, waves are evanescent and cannot propagate in those crystals. When disorder is introduced, strongly localized states can appear inside a gap [15–17]. The presence of these localized states, in turn, suppresses the evanescent nature of the wave propagation and, therefore, facilitates the propagation of waves deeper into the sample. However, when disorder is small, the wave propagation is predominantly evanescent. Due to the small number of multiple scatterings, a non-Rayleigh distribution is expected for the intensity distribution of the reflected speckles. With the increase of the degree of disorder, more localized states appear in the gap. When the density of states is sufficiently large, the gap may be destroyed completely and wave propagation may lose its evanescent nature for the entire gap. In this case, the number of multiple scatterings occurred inside the sample can be sufficiently large to recover the Rayleigh statistics. In this work, we study the statistics of reflected speckles for disordered photonic crystals with strongly localized states inside the gap, as a function of the amount of disorder. For the non-Bragg angles, our results indeed show a crossover from non-Rayleigh to Rayleigh statistics, which is similar to the crossover from ballistic to diffusive behavior found in the transmission geometry.

2. Numerical Simulation of Reflected Waves from Strongly Localized States inside the Gap of Disordered PBG

The 2D photonic crystals considered here consist of a square array of dielectric cylinders in an air background. The cylinders have the same radius, R, and dielectric constant, $\epsilon = 11.4$. The Maxwell equation for the s-polarized waves takes the form $[c^{-2}\omega^2\epsilon(\mathbf{r}) + \nabla^2]E(\mathbf{r}) = 0$, where $E(\mathbf{r})$ is the electric field along the cylinder axis and $\epsilon(\mathbf{r})$ is the position-dependent dielectric constant. It has been shown that complete gaps exist for a certain range of R/a, where a is the lattice constant [14].

Disordered photonic crystals were produced by randomizing an ordered one. The degree of randomness was controlled. Consider two cases: small randomness and complete randomness. For the case of complete randomness, the sample was produced by randomly altering the position of each cylinder within a distance $(a - 2R)/2$. The procedure was repeated 1000 times to ensure complete randomization. A move was forbidden if two cylinders overlapped. For the case of small randomness, each cylinder was moved randomly, but only once within a range $[-dr, dr]$, where $dr < (a - 2R)/2$.

The transmission, reflection, and scattering properties of the disordered photonic crystals were calculated using the multiple-scattering method [18]. The multiple scattering method is best suited for a finite collection of cylinders with a continuous incident wave of fixed frequency. The source is prepared by passing a plane wave through an open slit in front of the sample. The width of the slit is about 20% smaller than the sample width to avoid the scattering at the sample edges. The detailed description of this method has been given in Ref. [18].

In our calculations, a sample was chosen of width W and thickness L, denoted as $W \times L$. In order to ensure sufficient angular resolution for the study of statistics of reflected waves, W should be sufficiently large. In all the calculations discussed below, W was chosen to be $101a$. In an ordered system, the transmission coefficient, T, as a function of the renormalized frequency, f $(=\omega a/2\pi c)$, for a sample of size $101a \times 7a$ along the $\Gamma - X$ direction is plotted in Fig. 1 as a solid line for $R = 0.3a$. There exist two gaps for frequencies below $f = 0.6$. Inside these two gaps, the wave propagation is evanescent in nature. For any given frequency inside these gaps, the reflected intensity in the far field always exhibits a sharp peak at the Bragg reflection angle. There also exist many small satellite peaks in the non-Bragg angles due to the diffraction effect arising from the finite width of the incident beam [13]. In the absence of disorder, the reflected intensity is considered as coherent as will be discussed later.

When randomness is introduced, localized states may appear inside the gap and the gap size reduces accordingly. In the case of small randomness with $dr = 0.2$, the transmission spectrum of one particular random con-

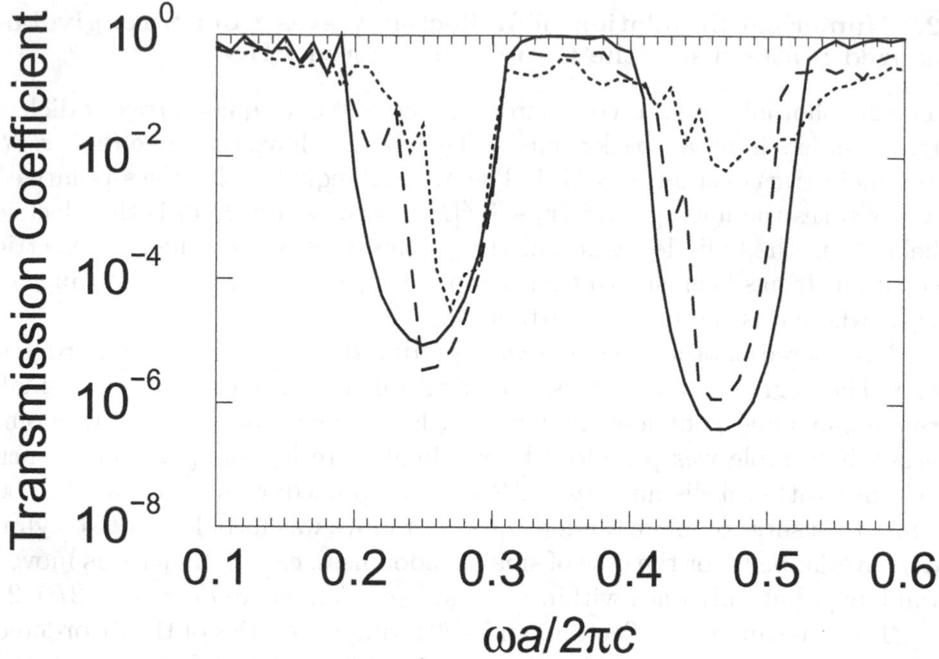

Figure 1. The solid line represents the transmission coefficient of an ordered photonic crystal consisting of a square array of dielectric cylinders of radius $R = 0.3a$ and $\epsilon = 11.4$. The dashed line is the case of small randomness with $dr = 0.2$ and the dotted line is the case of complete randomness.

figuration is shown as a dashed line in Fig. 1. Inside each gap, there is a peak near the lower band edge. This indicates the existence of some localized states at the peak frequencies. The presence of localized states reduces the size of a gap and, therefore, suppresses the evanescent nature of the waves inside the gap. In the case of complete randomness, the transmission spectrum of one particular random configuration is shown as a dotted line in Fig. 1. In this case, more localized states appear deeper inside the gap. As a result, the first gap is further reduced, whereas the second gap is now completely destroyed and the wave propagation loses its evanescent nature. Here we will focus on those frequencies near the center of the second gap. In the calculations below, we choose a fixed frequency $f = 0.44$.

In order to analyze our simulation data, we apply the theoretical tools used in Refs. [10] and [11]. Thus, we write the complex field, $E(\theta)$, as the sum of the average and the residual fields,

$$E(\theta) = \langle E(\theta) \rangle + \delta E(\theta), \tag{1}$$

where the angle brackets represent the ensemble average over different disorder realizations. The coherent and residual intensities can be written as $I_c = |\langle E(\theta)\rangle|^2$ and $I_{res} = |\delta E(\theta)|^2$, and the averaged total intensity is $\langle I\rangle = \langle I_{res}\rangle + I_c$.

Another interesting statistical quantity to study is speckle contrast (SC) as a function of the degree of disorder. The SC is defined as $SC(I) = \sigma_I / \langle I\rangle$ (with $\sigma_I^2 \equiv var(I)$) [11,19]. When SC is one, the intensity distribution is of the Rayleigh type. Deviation from one implies a non-Rayleigh distribution. It has been pointed out that in the case of the residual field, a deviation from $SC = 1$ indicates that the phase randomization is not complete and speckles are not completely developed. In this case, the distribution function of the residual field should be governed by K distributions [11]. It should be pointed out that since the coherent intensity is much larger in the case of the Bragg angle reflection than at non-Bragg angles, their statistical distributions are also very different.

For the non-Bragg angles, we calculate the scattered intensities at 20 different angles between 155° and 175° for each configuration. The process is repeated for 8000 configurations. The sample width in all the calculations presented below is taken as $8a$. For both the residual intensity ($SC(I_{res})$) and the total intensity ($SC(I)$), our simulation results show very similar behavior as a function of the degree of disorder for different angles. However, as a function of angle, there exist some oscillations in the values of $SC(I_{res})$ and $SC(I)$ due to the presence of oscillations in the coherent part. Here, we only show the behavior of $SC(I_{res})$ and $SC(I)$ for the case of $\theta = 174°$ in Fig. 2 by open squares and open triangles, respectively. It should be pointed out that the filled square and filled triangle represent the case of complete randomness. They do not represent the case of $dr = 0.25$. It is interesting to point out that the monotonic increase of $SC(I)$ as a function of the degree of disorder is similar to what has been found in the SC of the speckle pattern as a function of surface roughness [19]. Also in Fig. 2, we show the corresponding results at the Bragg angle, i.e., $\theta = 180°$, by circles and diamonds. The large differences in SC between the Bragg and non-Bragg angles are obvious. We have also performed the simulations for an oblique incidence (angle of incidence 45° with respect to the normal to the sample surface). Similar results have been found. We also plot these results as crosses and pluses in Fig. 2. They completely overlap with the results of normal incidence. Thus, the statistical properties of the reflected speckle intensities do not depend on the angle of incidence. Since the behaviors of $SC(I)$ and $SC(I_{res})$ are very different for the Bragg angle and the non-Bragg angles, the statistical properties in these two cases will also be very different. Thus, we will discuss these two cases separately.

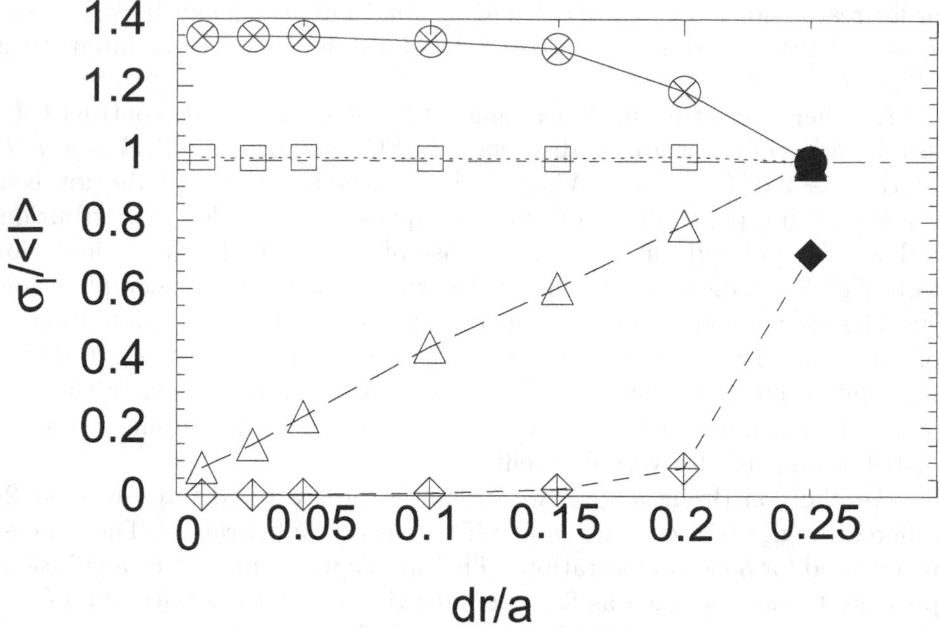

Figure 2. Speckle contrast (SC), $\sigma_I / \langle I \rangle$, as a function of the amount of disorder. Solid line (circles) and dashed line (diamonds) are for the residual field and the total field in the Bragg direction; Dotted line (squares) and long-dashed line (triangles) for the residual field and the total field in the non-Bragg direction. The corresponding filled points are for the case of complete randomness.

2.1. STATISTICS OF THE REFLECTED WAVES AT NON-BRAGG ANGLES

In the non-Bragg directions, we again find that the statistical properties of $E(\theta)$ and $\delta E(\theta)$ are very similar at different angles. Here, we take $\theta = 174°$ as an example. For the case of complete disorder, the distribution functions of the normalized amplitude $A = E(\theta)/\sigma$ and the phase of the reflected field are presented in Figs. 3(a) and 3(b), respectively. Here $2\sigma^2 = \langle |E(\theta)|^2 \rangle - |\langle E(\theta) \rangle|^2$. The symbols "squares" and "circles" denote the results for the total and residual fields. The same symbols will be used in all other cases discussed below. Very little difference in the amplitude distributions of the residual and total fields indicates a negligible coherent component in this case. If we use the parameter $k = |\langle E(\theta) \rangle|/\sigma$ (or $k^2 = 2a/(1-a)$ with $a = I_c / \langle I \rangle$) to represent the amount of the coherent component in the reflected field [19], the value of k in this case is 0.242 according to the results of simulations. Since the phase distribution of the residual field (circles in Fig. 3(b)) is constant, indicating a complete randomization of the phase, the random-phasor-sum (RPS) should give a good description of these data. The analytical distribution functions for the amplitude and phase obtained from the RPS are, respectively [10,19]

$$P(A,k) = A\exp(-\frac{A^2+k^2}{2})I_0(Ak) \tag{2}$$

and

$$P(\varphi,k) = \frac{1}{2\pi}\exp(-\frac{k^2}{2}) + \frac{k\cos\varphi}{\sqrt{2\pi}}\exp(-\frac{(k\sin\varphi)^2}{2})\mathrm{erf}(k\cos\varphi), \tag{3}$$

where $I_0(x)$ is the modified Bessel function of the first kind of zeroth order, and $\mathrm{erf}(x)$ is the error function. By using the simulated value of $k = 0.24$ in the above equations we obtain two solid lines in Figs. 3(a) and 3(b). Indeed, they overlap well with the simulated data of the total field. By taking $k = 0$ in the above equations, we obtain two dashed lines. They also agree well with the simulated data for the residual field. The dashed line in Fig. 3(a) represents Rayleigh statistics. The presence of a small coherent component makes the total intensity distribution very close to the Rayleigh distribution. Although the intensity distribution of the total field is close to the Rayleigh distribution, its phase distribution is not completely random due to the presence of a small coherent component. Such a sensitive dependence of the phase distribution of the total field on the presence of a small coherent part has also been observed for transmitted waves [10,11].

The results corresponding to the case of small randomness ($dr = 0.1$) are shown in Figs. 4(a) and 4(b). In this case, the value of k is 4.17. Again, the RPS gives a very good description of the simulated data. Due to the presence of a larger coherent part (or higher value of k), the distribution of the total field (squares and the solid line in Fig. 4(a)) is very different from that of the residual field (circles and the dashed line in Fig. 4(a)). The phase of the residual field is not completely randomized and appear angular dependent.

Thus, in the case of non-Bragg angles, we indeed find a crossover from non-Rayleigh to Rayleigh distribution as a function of the degree of disorder, which is very similar to that of ballistic to diffusive wave propagation in the transmission geometry. In fact, our Figs. 3 and 4 are very similar to Figs. 2 and 3 of Ref. [10]. When the randomness is further reduced below $dr = 0.1$, the value of $SC(I)$ is reduced. However, the value of $SC(I_{res})$ remains close to one. This seems to indicate that the RPS is still valid and that the amplitude distribution of the residual field is of Rayleigh type. Indeed, this is what we have found [13].

2.2. STATISTICS OF THE REFLECTED WAVES AT THE BRAGG ANGLE

In contrast to the case of the non-Bragg angles, the statistical distributions at the Bragg angle are more complicated. For the case of complete ran-

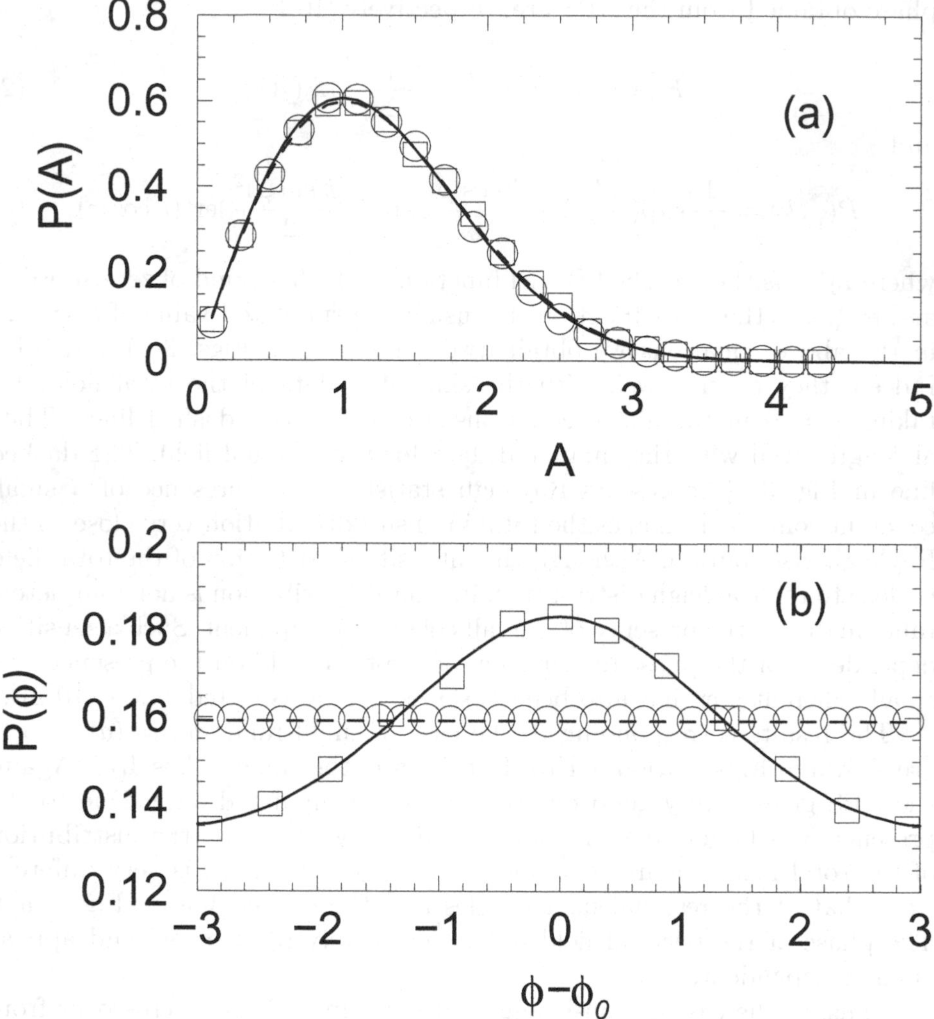

Figure 3. Distributions of the normalized amplitude (a) and phase (b) of the total field (squares) and the residual field (circles) in the non-Bragg direction for the case of complete randomness with $R = 0.3$ and $f = 0.44$. Solid lines and dotted lines are the theoretical predictions using the random-phasor-sum model.

domness, the results are shown in Fig. 5. It is clear that the RPS is still valid in this case. This is not surprising for the small values of $k = 2.07$ and $SC(I_{res}) \approx 1$ (filled circles in Fig. 2). It is also clear that, even for the case of complete randomness, the amplitude distribution of the total field is not Rayleigh. This is due to the presence of a large coherent component in the total field, which also makes the value of $SC(I)$ smaller than one as

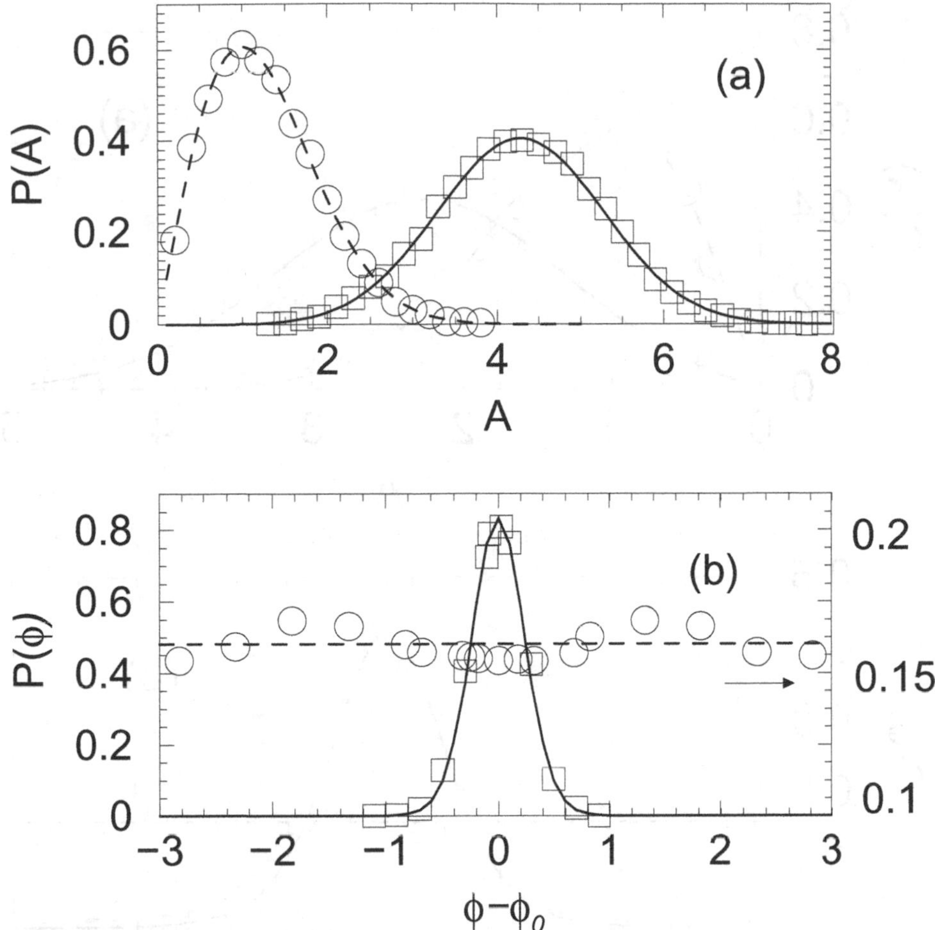

Figure 4. Distributions of the normalized amplitude (a) and phase (b) of the total field (squares) and the residual field (circles) in the non-Bragg direction for a system with small randomness $dr = 0.1$, $R = 0.3$ and $f = 0.44$. Solid lines and dotted lines are the theoretical predictions using the random-phasor-sum model.

shown by the filled diamonds in Fig. 2.

When the disorder is reduced, we see that the value of $SC(I_{res})$ increases monotonically, whereas the value of $SC(I)$ decreases rapidly (Fig. 2). Since $SC(I_{res}) > 1$, it is expected that the amplitude distribution of the residual field will deviate from the Rayleigh distribution. This is indeed what we have found. In Fig. 6, we show the results for the case of $dr = 0.2$. A clear deviation of the circles from the dashed line in Fig. 6(a) can be seen. This deviation is also reflected in the much larger non-uniformity in the phase distribution of the residual field as indicated by the circles in

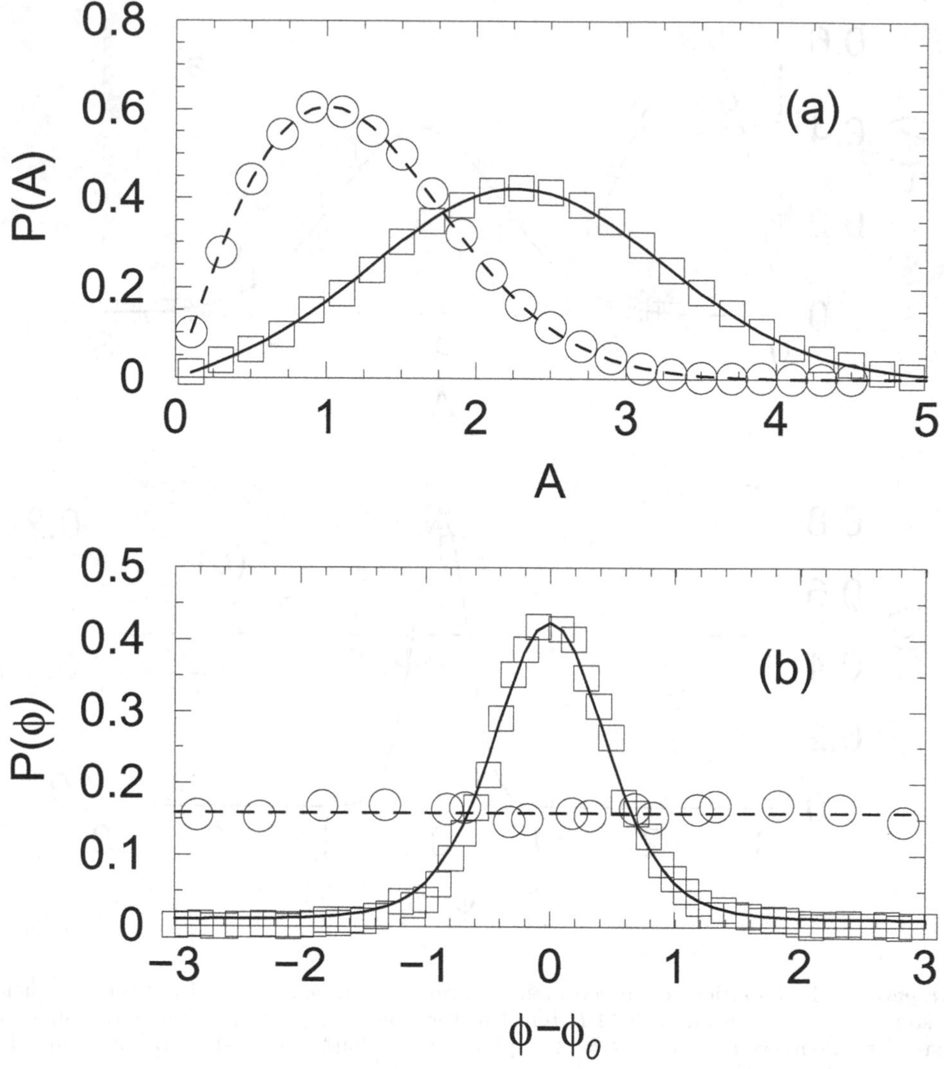

Figure 5. Distributions of the normalized amplitude (a) and phase (b) of the total field (squares) and the residual field (circles) in the Bragg direction for the case of complete randomness with $R = 0.3$ and $f = 0.44$. Solid lines and dotted lines are the theoretical predictions using the random-phasor-sum model.

Fig. 6(b). Similar behavior has been seen for the transmitted waves and is attributed to the incomplete randomization of the phase of the residual field, so that the speckle pattern is not completely developed [11]. In this case, the RPS cannot describe the statistical distribution of the total field. Indeed, in the inset of Fig. 6(a), we see large deviations of the sim-

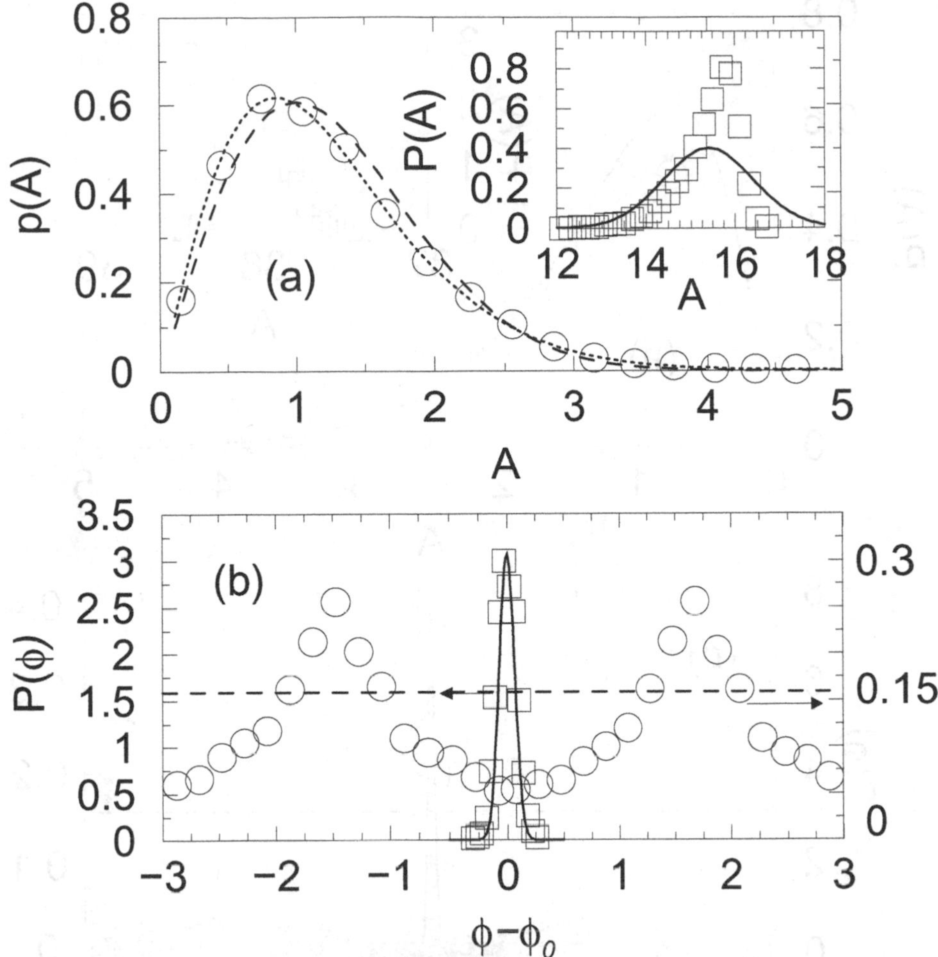

Figure 6. Distributions of the normalized amplitude (a) and phase (b) of the total field (squares) and the residual field (circles) in the Bragg direction for a system with small randomness $dr = 0.2$, $R = 0.3$ and $f = 0.44$. Solid lines and dotted lines are the theoretical predictions using the random-phasor-sum model.

ulated data (squares) from the prediction of Eq. (3) (solid line), although certain agreement between the two in the phase distribution remains, i.e., the agreement between the squares and the solid line in Fig. 6(b). The value of k in this case is 15.36, indicating a large coherent component in the total field. It is interesting to point out that the value of $k(= 15.36)$ in this case is not very different from that in the case of non-Bragg angles with $dr = 0.1$, i.e., $k = 14.03$. Both cases show a large coherent part in the scattered intensities. However, their values of $SC(I_{res})$ are very different.

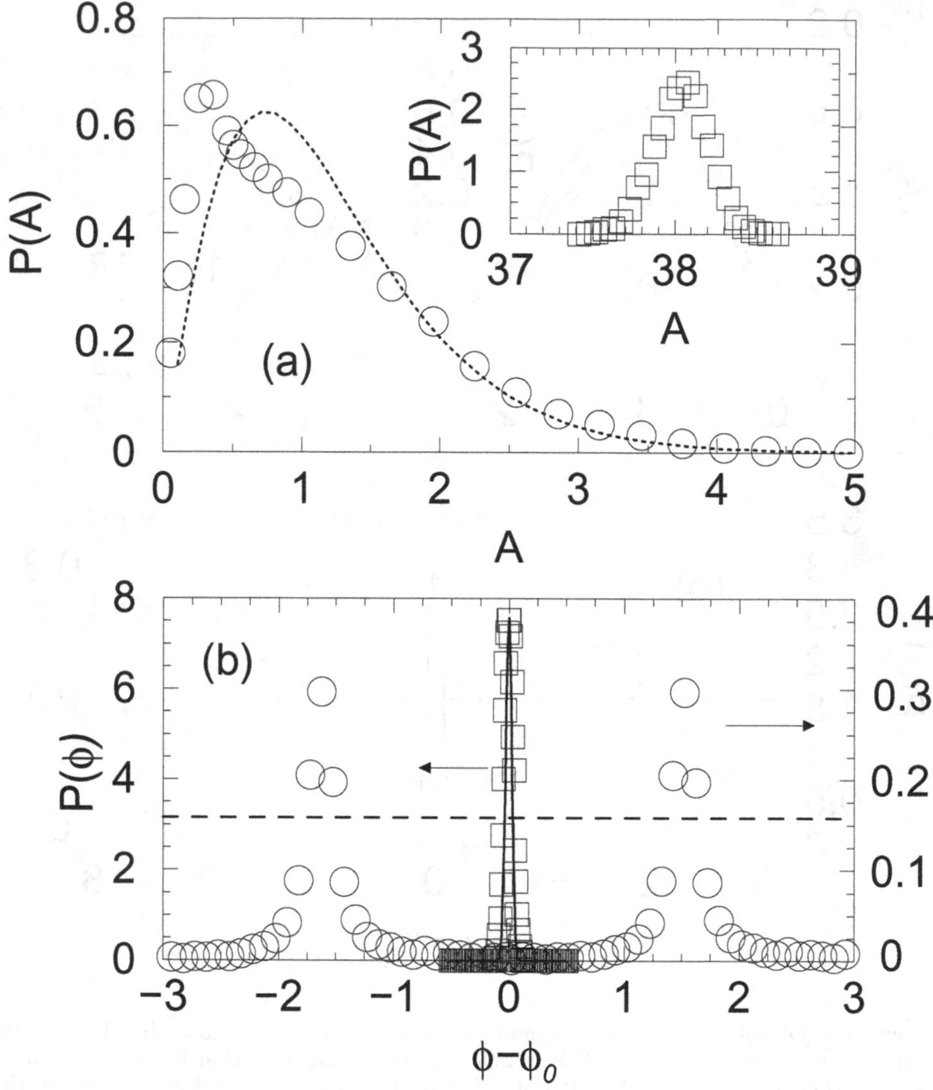

Figure 7. Distributions of the normalized amplitude (a) and phase (b) of the total field (squares) and the residual field (circles) in the Bragg direction for a system with small randomness $dr = 0.1$, $R = 0.3$ and $f = 0.44$. Solid lines and dotted lines are the theoretical predictions using the random-phasor-sum model.

It has been suggested that when $SC > 1$, one can use K distribution to describe the amplitude distribution of the residual field. When SC approaches one, the Rayleigh distribution is recovered. The K distribution was derived to describe the statistics of the partially developed speckle intensities and has the form [11,19]:

$$P(A) = \frac{2\sqrt{2M}}{\Gamma(M)} \left(\frac{MA^2}{2} \right)^{M/2} K_{M-1}(A\sqrt{2M}), \qquad (4)$$

where $K_\nu(x)$ is the modified Bessel function of the second kind of order ν and $2/M = SC(I_{res})^2 - 1$. Using the value of $SC(I_{res}) = 1.18$ in Eq. (4), we obtain the K distribution and plot it as the dotted line in Fig. 6(a). The excellent agreement between the dotted line and the circles indicates the validity of the K distribution.

In Fig. 7, we show the results for the case of $dr = 0.1$. In this case, $SC(I_{res}) = 1.34$. A clear deviation from the K distribution is seen in Fig. 7(a), where the dotted line does not agree with the circles. In this case, the phase distribution of the residual field, i.e, the circles in Fig. 7(b), is far from randomized and has two sharp peaks on each side of the phase of the coherent field. It should be pointed out that the total field is predominated by the coherent wave and only 0.1% of the total intensity comes from the scattered field. Finally, we would like to point out we have repeated the calculations for the case of oblique incidence. The results are similar to the case of normal incidence shown in Figs. 3–7.

3. Conclusions

By using the multiple-scattering method, we have studied in details the statistics of reflected speckle intensities arising from localized states inside the gap of disordered photonic crystals in two dimensions. For the non-Bragg angles, we found that the intensity distribution crosses over from non-Rayleigh to Rayleigh when the amount of disorder is increased. Such a crossover behavior is very similar to that from ballistic to diffusive wave propagation for the transmitted waves and can be described by the RPS. For the Bragg angle, due to the presence of large coherent intensity only the non-Rayleigh statistics were found. However, the statistics varies sensitively with the degree of disorder. It should be mentioned that the randomness we considered here is the position randomness only. In reality, the radius of each cylinder can also vary, i.e., size randomness. It has been shown that the size randomness is more efficient in producing localized states inside the gap [20]. Thus, if we also include the size randomness in the above calculations, it is possible to observe Rayleigh statistics for the Bragg angle case when the size randomness is sufficiently large.

This work was supported by Hong Kong RGC Grant Nos. HKUST 6137/97P and 6160/99P.

References

1. See, for example, *Scattering and Localization of Classical Waves in Random Media*, edited by P. Sheng (World Scientific, Singapore, 1990); P. Sheng, *Introduction to Wave Scattering, Localization and Mesoscopic Phenomena* (Academic Press, New York, 1995); M. C. W. van Rossum and T. M. Nieuwenhuizen, Rev. Mod. Phys. **71**, 313 (1999).
2. E. Wolf, Phys. Rev. Lett. **56**, 1370 (1986); B. Shapiro, *ibid.* **57**, 2168 (1986); M. Stephen and G. Cwilich, *ibid.* **59**, 285 (1987); S. Feng, C. Kane, P. A. Lee and A. D. Stone, *ibid.* **61**, 834 (1988); I. Freund, M. Rosenbluh and S. Feng, *ibid.* **61**, 2328 (1988); N. Garcia and A. Z. Genack, *ibid.* **63**, 1678 (1989); R. Berkovits, M. Kaveh and S. Feng, Phys. Rev. B **40**, 737 (1989); M. P. van Albada, J. F. de Boer and A. Lagendijk, Phys. Rev. Lett. **64**, 2787 (1990).
3. A. Z. Genack and N. Garcia, Europhys. Lett. **21**, 753 (1993); J. F. de Boer, M. C. W. van Rossum, M. P. van Albada, Th. M. Nieuwenhuizen and A. Lagendijk, Phys. Rev. Lett. **73**, 2567 (1994); M. Stoytchev and A. Z. Genack, *ibid.* **79**, 309 (1997).
4. Th. M. Nieuwenhuizen and M. C. W. van Rossum, Phys. Rev. Lett. **74**, 2674 (1995); E. Kogan and M. Kaveh, Phys. Rev. B **52**, R3813 (1995).
5. I. Edrei et al., Phys. Rev. Lett. **62**, 2120 (1989); N. Shnerb and M. Kaveh, Phys. Rev. B **43**, 1279 (1991); E. Kogan et al., *ibid.* **48**, 9404 (1993); E. Kogan and M. Kaveh, *ibid.* **51**, 16400 (1995).
6. P. A. Mello, E. Akkermans and B. Shapiro, Phys. Rev. Lett. **61**, 459 (1988).
7. C. W. J. Beenakker, Rev. Mod. Phys. **69**, 731 (1997).
8. S. A. van Langen, P. W. Brouwer and C. W. J. Beenakker, Phys. Rev. E **53**, R1344 (1996).
9. A. Garcia-Martin, J. A. Torres, J. J. Saenz and M. Nieto-Vesperinas, Phys. Rev. Lett. **80**, 4165 (1998).
10. A. A. Chabanov and A. Z. Genack, Phys. Rev. E **56**, R1338 (1997).
11. A. Garcia, J. J. Saenz and M. Nieto-Vesperinas, Phys. Rev. Lett. **84**, 3578 (2000).
12. A. Garcia-Martin, T. Lopez-Ciudad, and J. J. Saenz, Phys. Rev. Lett. **81**, 329 (1998).
13. X. Zhang and Z. Q. Zhang, Phys. Rev. B **65**, 245115 (2002).
14. See, for example, *Photonic Band Gaps and Localization*, edited by C. M. Soukoulis (Plenum, New York, 1993); J. Opt. Soc. Am. B **10**, 208–408 (1993); *Photonic Band Gap Materials*, edited by C. M. Soukoulis (Kluwer, Dordrecht, 1996); J. D. Joannopolous, R. D. Meade and J. N. Winn, *Photonic Crystals* (Princeton University, Princeton, 1995).
15. S. John, Phys. Rev. Lett. **58**, 2486 (1987).
16. R. Dalichaouch et al., Nature (*London*) **354**, 53 (1991).
17. Z. Zhang et al., Phys. Rev. Lett. **81**, 5540 (1998).
18. L. M. Li, and Z. Q. Zhang, Phys. Rev. B **58**, 9587 (1998).
19. J. W. Goodman, in *Laser Speckle and Related Phenomena*, edited by J. C. Dainty (Springer-Verlag, Berlin, 1984); J. W. Goodman, *Statistical Optics* (Wiley, New York, 1985); J. Ohtsubo and T. Asakura, Opt. Comm. **14**, 30 (1975); H. Fujii and T. Asakura, Opt. Comm. **11**, 35 (1974); H. M. Pederson, Opt. Comm. **12**, 156 (1974).
20. For instance, see, Z. Y. Li, X. Zhang, and Z. Q. Zhang, Phys. Rev. B **61**, 15738 (2000).

CHAPTER III

ACOUSTIC TIME REVERSAL

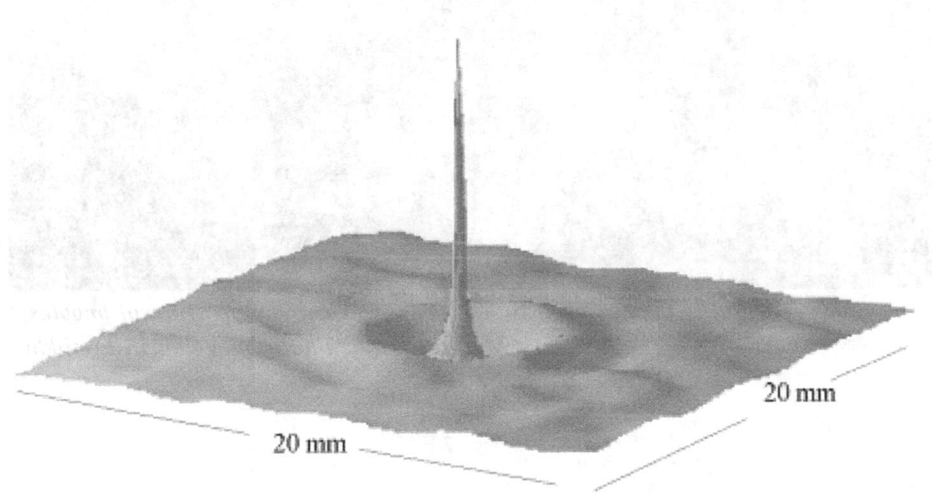

Focusing pattern obtained in a time reversal experiment for elastic waves propagating in a chaotic cavity. The initial source was replaced by its time-reversed counterpart, i.e. an acoustic sink, which makes it possible to overcome the classical diffraction limit. Image provided by Mathias Fink, ESPCI, Paris, France.

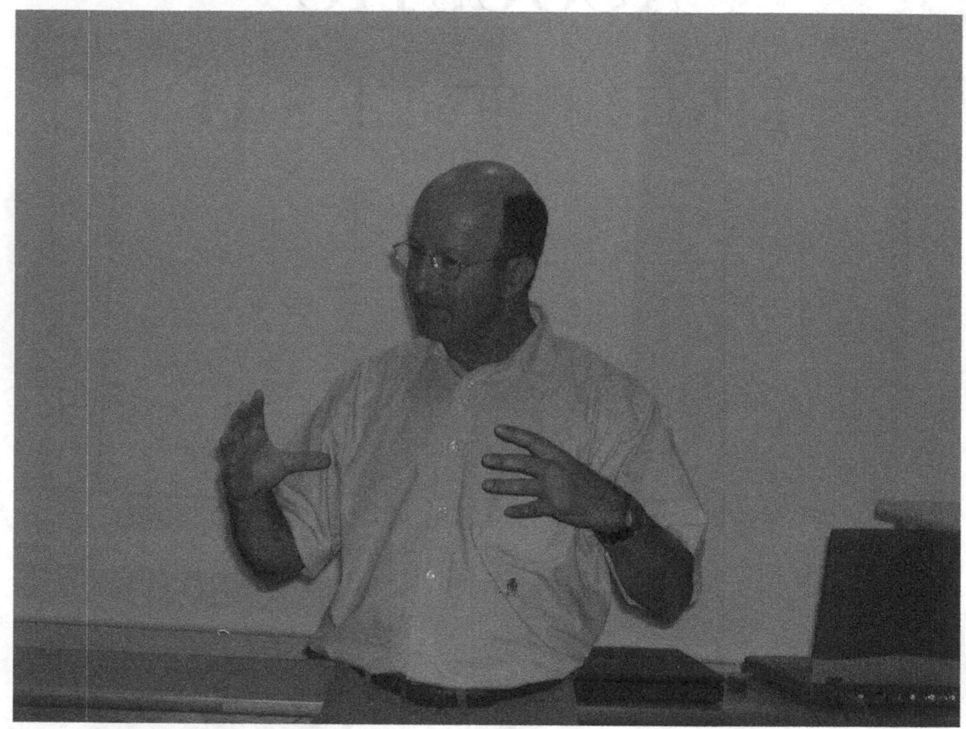

"Time reversal is the best way to achieve spherical compression of bubbles," says Mathias Fink. Photo by Valentin Freilikher

AUTO-FOCALISATION, COMMUNICATION AND SONOLUMINESCENCE WITH ACOUSTIC TIME REVERSAL

M. FINK, A. TOURIN AND J. DE ROSNY
Laboratoire Ondes et Acoustique, ESPCI, Universite Paris VII
UMR 7587, 10 rue Vauquelin 75005 Paris, France

Abstract. The basic limitations of Time-Reversal Mirrors and the means to overcome them are first discussed: especially the concept of acoustic sink is introduced. Then, time-reversal experiments in chaotic cavities and multiple scattering media are presented. In the case of time reversal in chaotic cavities, we show an experimental realisation of the acoustic sink. As to time reversal in multiple scattering media, it is interpreted as a robust estimator of both the time and space correlation function of the multiply scattered signals. Using this approach, the difference between broadband time-reversed acoustics and monochromatic phase conjugation is highlighted. Finally, two recent applications of adaptive focusing to communication on the one hand and to boosting sonoluminescence on the other hand are presented.

1. Introduction

It is well known that the acoustic wave equation in a non-dissipative heterogeneous medium is invariant under a time-reversal operation. Indeed, it contains only a second-order time-derivative operator. Therefore, for every burst of sound $p(\mathbf{r}, t)$ diverging from a source (and possibly reflected, refracted or scattered by any heterogeneous media) there exists in theory a set of waves $p(\mathbf{r}, -t)$ that precisely retraces all of these complex paths and converges in synchrony, at the original source as if time were going backwards. This idea gives the basis of time-reversal acoustics.

In practice, to synthetise the time-reversed field $p(\mathbf{r}, -t)$, the acoustic field $p(\mathbf{r}, t)$ has to be measured in every point of a closed surface surrounding the medium, what is called the time-reversal cavity (TRC), and retransmitted through the medium in a time-reversed chronology; then the wave will

B.A. van Tiggelen and S.E. Skipetrov (eds.),
Wave Scattering in Complex Media: From Theory to Applications, 257–279.

travel back to its source and recover its original shape. It must be pointed out that both time-reversal invariance and spatial reciprocity (Jackson and Dowling, 1991) are required to reconstruct the exact time-reversed wave in the whole volume by means of a two-dimensional time-reversal operation.

From an experimental point of view, a TRC would consist of a two-dimensional piezoelectric transducer array that would sample the wavefield over a closed surface. The important point is that a transducer is able to record, in a large bandwidth, the field itself, and not only the intensity, as in optics. An array pitch of the order of $\lambda/2$ where λ is the smallest wavelength of the pressure field is needed to ensure the recording of all the information on the wavefield. Each transducer is connected to its own electronic circuitry that consists of a receiving amplifier, an A/D converter, a storage memory and a programmable transmitter able to synthesize a time-reversed version of the stored signal that it will send back in the medium. In practice, TRC is difficult to realize and the TR operation is usually performed on a limited angular area, that we call a Time-Reversal Mirror (TRM), thus limiting reversal and focusing quality. A TRM consists typically of some hundred elements, or time-reversal channels.

Based on this principle, we have developed a lot of applications in therapy and medical imaging on the one hand and in non-destructive testing of materials on the other hand. From a more theoretical point of view, our works on multiple scattering in random media and reverberation in chaotic cavities have shown that the complexity of the medium can be taken advantage of to reduce the number of reversible transducers used for a time-reversal experiment and to enhance the spatial resolution of the focal spot.

Here, we first present the fundamentals of time-reversal acoustics. Especially, the basic limitations of time-reversal mirrors are discussed as well as the methods to overcome them. Then, two types of time-reversal experiments performed in complex media with TRMs are discussed. The first one is conducted in a chaotic cavity. We show that for waves propagating in such a closed reverberant medium, multiple reflections on the boundaries significantly increase the apparent aperture of the TRM. Furthermore in a reflecting cavity with chaotic boundaries, even a one-channel time-reversal mirror is sufficient to ensure reversibility and optimal focusing. In this context, we present an experimental realisation of the acoustic sink. In the second experiment, it is shown that the wave reversibility is improved if the wave traverses a random multiply scattering medium before arriving on the transducer array. Indeed, the multiple scattering processes allow to redirect one part of the initial wave towards the TRM, that normally misses the transducer array. After the time-reversal operation, the whole multiply scattering medium behaves as a coherent focusing source, with a large angular aperture for enhanced resolution. As a consequence, in multiply scattering media, one is able to reduce the size and the complexity of the TRM.

In the context of communication, another way of interpreting a time-reversal experiment is that the complexity of the medium adds new independent channels of communication, this number being related to the number of focal spots that a transmitting antenna can create in a given region of space. Based on this idea, we briefly discuss in the fourth section the application of time-reversal to communications in random media. Finally, we present another recent application of time-reversal focusing for sonoluminescence boosting.

2. Basic Limitations of Time-Reversal Mirrors

The basic theory employs a scalar wave formulation and, hence, is strictly applicable to acoustic or ultrasound propagation in fluid. However, the basic ingredients and conclusions apply equally well to elastic waves in solid and to electromagnetic fields. In any propagation experiment, the acoustic sources and the boundary conditions determine a unique solution $\phi(\mathbf{r}, t)$ in the fluid. The goal, in time-reversal experiments, is to modify the initial conditions in order to generate the dual solution $\phi(\mathbf{r}, T-t)$ where T is a delay due to causality requirements. D. Dowling, D. Jackson and D. Cassereau (Jackson and Dowling, 1991; Cassereau and Fink, 1992) have studied theoretically the conditions necessary to ensure the generation of $\phi(\mathbf{r}, T-t)$ in the entire volume of interest.

2.1. THE CLOSED TIME-REVERSAL MIRROR

Although reversible acoustic retinas usually consist of discrete elements, it is convenient to examine the behavior of idealized retinas, defined by continuous surfaces. In the case of a time-reversal cavity, we assume that the retina completely surrounds the source. The basic time-reversal experiment can be described in the following way. In a first step, a point-like source located at $\mathbf{r}_0$ inside a volume V surrounded by the retina surface, emits a pulse at $t = t_0$. The wave equation in a medium of density $\rho(\mathbf{r})$ and compressibility $\chi(\mathbf{r})$ reads

$$(L_r + L_t)\phi(\mathbf{r}, t) = -A\delta(\mathbf{r} - \mathbf{r}_0)\delta(t - t_0) \tag{1}$$

with $L_r = \nabla(\frac{1}{\rho(\mathbf{r})}\nabla)$ and $L_t = -\chi(\mathbf{r})\partial_{tt}$. A is a dimensional constant that ensures the compatibility of physical units between the two sides of the equation; for simplicity reasons, this constant will be omitted in the following. The solution to Eq. (1) reduces to the Green's function $G(\mathbf{r}, t|\mathbf{r}_0, t_0)$.

Classically, $G(\mathbf{r}, t|\mathbf{r}_0, t_0)$ is written as a diverging spherical wave (homogeneous and free space case) and additional terms that describe the interaction of the field itself with the inhomogeneities (multiple scattering) and the boundaries. We assume that we are able to measure the pressure field and its normal derivative at any point on the surface S during the interval $[0, T]$. In all the following, we suppose that the contribution of multiple scattering decreases with time, and that T is chosen such that the information loss can be considered as negligible inside the volume V. During the second step of the time-reversal process, the initial source $\mathbf{r}_0$ is removed and we create on the surface of the cavity monopole and dipole sources that correspond to the time reversal of those same components measured during the first step. The time-reversal operation is described by the transform $t \to T - t$ and the secondary sources are

$$\phi_s(\mathbf{r}, t) = G(\mathbf{r}, T - t|\mathbf{r}_0, t_0), \quad \partial_n \phi_s(\mathbf{r}, t) = \partial_n G(\mathbf{r}, T - t|\mathbf{r}_0, t_0) \qquad (2)$$

In this equation, ∂_n is the normal derivative operator with respect to the normal direction $\mathbf{n}$ to S, oriented outward. Due to these secondary sources on S, a time-reversed pressure field $\phi_{tr}(\mathbf{r}_1, t_1)$ propagates inside the cavity. It can be calculated using a modified version of the Helmholtz-Kirchhoff integral:

$$\begin{aligned} \phi_{tr}(\mathbf{r}_1, t_1) &= \int\!\!\int dS \int_{-\infty}^{+\infty} dt \, [G(\mathbf{r}_1, t_1|\mathbf{r}, t)\, \partial_n \phi_s(\mathbf{r}, t) \\ &- \phi_s(\mathbf{r}, t)\, \partial_n G(\mathbf{r}_1, t_1|\mathbf{r}, t)] \frac{1}{\rho(\mathbf{r})} \end{aligned} \qquad (3)$$

Spatial reciprocity and time-reversal invariance of the wave equation (1) yield the following expression:

$$\phi_{tr}(\mathbf{r}_1, t_1) = G(\mathbf{r}_1, T - t_1|\mathbf{r}_0, t_0) - G(\mathbf{r}_1, t_1|\mathbf{r}_0, T - t_0) \qquad (4)$$

This equation can be interpreted as the superposition of incoming and outgoing spherical waves, centered on the initial source position. The incoming wave collapses at the origin and is always followed by a diverging wave. Thus the time-reversed field, observed as a function of time, from any location in the cavity, shows two wavefronts, where the second one is the exact replica of the first one, multiplied by (-1). If we assume that the retina does not perturb the propagation of the field (free-space assumption) and that the acoustic field propagates in an homogeneous fluid, the free-space Green's function G reduces to a diverging spherical impulse wave that propagates with a sound speed c. Introducing its expression in (4) yields the following formulation of the time-reversed field:

$$\phi_{tr}(\mathbf{r}_1, t_1) = K(\mathbf{r}_1 - \mathbf{r}_0, t_1 - T + t_0) \qquad (5)$$

where the kernel distribution is given by

$$K(\mathbf{r},t) = \frac{1}{4\pi|\mathbf{r}|}\delta\left(t+\frac{|\mathbf{r}|}{c}\right) - \frac{1}{4\pi|\mathbf{r}|}\delta\left(t-\frac{|\mathbf{r}|}{c}\right) \tag{6}$$

The kernel distribution $K(\mathbf{r},t)$ (the propagator) corresponds to the difference between two impulse spherical waves that respectively converge to and diverge from the origin of the spatial coordinate system, i.e. the location of the initial source. It results from this superposition that the pressure field remains finite for all time throughout the cavity, although the converging and diverging spherical waves show a singularity at the origin. If we consider a wide-band excitation function instead of a Dirac distribution $\delta(t)$, the two wavefronts overlap near the focal point, therefore resulting in a temporal distortion of the acoustic signal. It can be shown that this distortion yields a temporal derivation of the initial excitation function at the focal point. If we now calculate the Fourier transform of (6) over the time variable t, we obtain

$$\tilde{K}(\mathbf{r},\omega) = \frac{1}{2j\pi}\frac{\sin(\omega|\mathbf{r}|/c)}{|\mathbf{r}|} = \frac{1}{j\lambda}\frac{\sin(k|\mathbf{r}|)}{k|\mathbf{r}|} \tag{7}$$

where λ and k are the wavelength and wavenumber respectively. As a consequence, for a monochromatic excitation, the time-reversed field is effectively focused on the initial source position, but with a focal spot size limited to one half-wavelength. A similar interpretation can be given in the case of an inhomogeneous fluid, but the Green's function G now takes into account the interaction of the pressure field with the inhomogeneities of the medium.

If we were able to create a movie of the propagation of the acoustic field during the first step of the process, the final result could be interpreted as a projection of this movie in the reverse order, immediately followed by a re-projection in the initial order.

2.2. THE ACOUSTIC SINK: FOCUSING BELOW THE DIFFRACTION LIMIT

The apparent failure of the time-reversed operation that leads to diffraction limitation can be interpreted in the following way. The second step described above is not strictly the time reversal of the first step. During the second step of an ideal time-reversal experiment, the initial active source (that injects some energy into the system) must be replaced by a sink (the time reversal of a source). An acoustic sink is a device that absorbs all arriving energy without reflecting it. Taking into account the source term in the wave equation, reversing time leads to the transformation of the source into a sink. For an initial point source transmitting a waveform $s(t)$, the wavefield obeys the wave equation with a source term:

$$(L_r + L_t)\,\phi\,(\mathbf{r}, t) = -A\delta\,(\mathbf{r} - \mathbf{r}_0)\, s(t) \tag{8}$$

and is transformed by the time-reversal operation into

$$(L_r + L_t)\,\phi\,(\mathbf{r}, -t) = -A\delta\,(\mathbf{r} - \mathbf{r}_0)\, s(-t) \tag{9}$$

To achieve a perfect time reversal experimentally, the field on the surface of the cavity has to be time-reversed, and the source has to be transformed into a sink. Therefore one may achieve time-reversed focusing below the diffraction limit. The role of the new source term $\delta\,(\mathbf{r} - \mathbf{r}_0)\, s(-t)$ is to transmit a diverging wave that exactly cancels the usual outgoing spherical wave. Taking into account the evanescent waves concept, the necessity of replacing a source by a sink in the complete time-reversed operation can be interpreted as follows. In the first step a point-like source of size quite smaller than the transmitted wavelengths radiates a field whose angular spectrum contains both propagating waves and evanescent waves. The evanescent wave components (that contains details of the source smaller than the wavelength) are lost after propagating a few wavelengths (Goodman, 1996). In the time-reversed step, the time-reversed field retransmitted by the surface of the cavity does not contain evanescent components. The role of the sink is to be a source modulated by $s(-t)$ that radiates exactly, with the good timing, the evanescent waves that have been lost during the first step. Therefore the resulting field contains the evanescent part that is needed to focus below diffraction limits. With J. de Rosny, we have recently built such a sink in our laboratory and we have observed focal spot size quite below diffraction limits (typically with dimension $\lambda/20$) (De Rosny and Fink, 2002) as we discuss in the third section.

2.3. THE TIME-REVERSAL MIRROR

The above theoretical model of a closed time-reversal cavity is interesting since it affords an understanding of the basic limitations of the time-reversed self-focusing process; but it has several limitations, particularly compared to classical experimental setup:

1. Classical experimental setup usually works without any acoustic sink.

2. From a practical point of view, it is not possible to measure both the field and its normal derivative. Fortunately, it can be proven that it is not necessary to measure and time-reverse both the scalar field (acoustic pressure) and its normal derivative on the cavity surface:

measuring the pressure field and re-emitting the time-reversed field in the backward direction yields the same results, under the condition that the evanescent parts of the acoustic fields have vanished (propagation along several wavelengths) (Cassereau and Fink, 1994). This comes from the fact that each transducer element of the cavity records the incoming field from the forward direction, and retransmits it (after the time-reversal operation) in the backward direction (and not in the forward direction). The change between the forward and backward directions replaces the measurement and the time reversal of the field normal derivative.

3. From an experimental point of view, it is not possible to measure and re-emit the pressure field at all points of a 2D surface: experiments are carried out with transducer arrays that spatially sample the receiving and emitting surface. The spatial sampling of the TRC by a set of transducers may introduce grating lobes. These lobes can be avoided by using an array pitch smaller than $\lambda_{min}/2$, where λ_{min} is the smallest wavelength of the transient pressure field. In this latter case, each transducer is sensitive to all the wave vectors of the incident field.

4. The temporal sampling of the data recorded and transmitted by the TRC has to be at least of the order of $T_{min}/8$ (T_{min} the minimum period) to avoid secondary lobes (Kino, 1987).

5. It is generally difficult to use acoustic arrays that surround completely the area of interest, and the closed cavity is usually replaced by a TRM of finite angular aperture, which limits the spatial resolution of time-reversal refocusing.

3. Time-Reversal Experiments

3.1. TIME-REVERSAL IN CHAOTIC CAVITIES

In the time-reversal cavity approach, the transducer array samples a closed surface surrounding the acoustic source. Since experimentally, we have not disposal of such TRC, the basic idea is to replace one part of the TRC transducers by reflecting boundaries that redirect one part of the incident wave towards the TRM aperture. Thus spatial information is converted into the time domain and the reversal quality depends crucially on the duration of the time-reversal window, i.e. the length of the recording to be reversed. Experiments performed by P. Roux in rectangular ultrasonic waveguides showed the effectiveness of the TR processing to compensate for multipath effects (Roux *et al.*, 1997). Impressive time recompression has been observed, that compensated for reverberation and dispersion. Besides, the TR beam is focused on a spot which is much thinner than the one observed in free water. This can be interpreted by the theory of images in a medium bounded by two mirrors. For an observer, located at the source point, the TRM seems to be escorted by an infinite set of virtual images related to multipath propagation and effective aperture 10 times larger than the real aperture has been observed. Acoustic waveguides are also currently found in underwater acoustic, especially in shallow water, and TRMs can compensate for the multipath propagation in oceans that limits the capacity of underwater communication systems. The problem arises because acoustic transmission in shallow water bounces off the ocean surface and floor, so that a transmitted pulse gives rise to multiple copies of it arriving at the receiver. Recently, underwater acoustic experiments have been leaded by W. Kuperman and his group from San Diego University in a sea water channel of 120 m depth, with a 24-element TRM working at 500 Hz and 3.5 kHz. They observed focusing and multipath compensation at a distance up to 30 kms (Kuperman *et al.*, 1998).

In this paragraph, we are interested in another aspect of multiply reflected waves: waves confined in closed reflecting cavities such as elastic waves propagating in a silicon wafer. With such boundary conditions, no information can escape from the system and a reverberant acoustic field is created. If, moreover, the cavity shows ergodic and mixing properties and negligible absorption, one may hope to collect all information at only one point. C. Draeger *et al.* (Draeger and Fink, 1997; Draeger and Fink, 1999) have shown experimentally and theoretically that in this particular case the time reversal can be obtained using only one TR channel operating in a closed cavity. The field is measured at one point over a long period of time and the time-reversed signal is re-emitted at the same position. The experiment is 2-dimensional and has been carried out by using elastic surface waves propagating along monocrystalline silicon wafer with a D-shape. This geometry is chosen to avoid quasi-periodic orbits. Silicon was selected for its weak absorption. The elastic waves which propagate in such a plate are Lamb waves. An aluminum cone coupled to a longitudinal transducer generates these waves at one point of the cavity. A second transducer is used as a receiver. The central frequency of the transducers is 1 MHz and their bandwidth is 100%. At this frequency, only three Lamb modes are possible (one flexural, two extensional). The source is isotropic and considered point-like because the cone tip is much smaller than the central wavelength. A heterodyne laser interferometer measures the displacement field as a function of time at different points on the cavity. Assuming that there are nearly no mode conversion between the flexural mode and other modes at the boundaries, we have only to deal with one field, the flexural

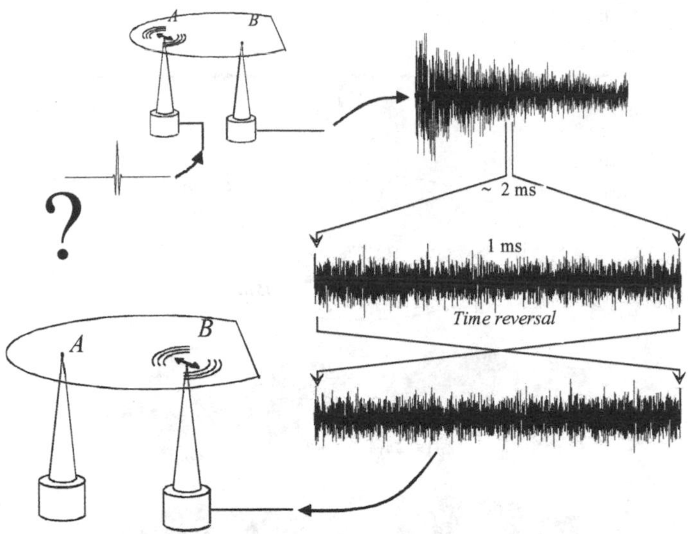

Figure 1. Time-reversal experiment performed in a chaotic cavity with flexural waves. In a first step, a point transducer located at point A transmits a 1 μs long signal. The signal is recorded at point B by a second transducer. The signal spreads on more than 30 ms due to reverberation. In the second step of the experiment, a 2 ms portion of the recorded signal is time-reversed and retransmitted back into the cavity.

scalar field. The experiment is a two-step process as described above. In the first step, one of the transducers, located at point A, transmits a short omnidirectional signal of duration 1 μs into the wafer. Another transducer, located at B, observes a very long chaotic signal, that results from multiple reflections of the incident pulse along the edges of the cavity, and which continues for more than 30 milliseconds corresponding to some hundred reflections on the boundaries. Then, a portion of 2 milliseconds of the signal is selected, time-reversed and re-emitted by point B. As the time-reversed wave is a flexural wave that induces vertical displacements of the silicon surface, it can be observed using the optical interferometer that scan the surface around point A (Fig. 1).

One observes both an impressive time recompression at point A and a refocusing of the time-reversed wave around the origin [Figs. 2(a) and 2(b)], with a focal spot whose radial dimension is equal to half the wavelength of the flexural wave.

The success of this time-reversal experiment is particularly interesting with respect to two aspects. Firstly, it proves again the feasibility of time reversal in wave systems with chaotic ray dynamics. Paradoxically, in the case of one-channel time reversal, chaotic dynamics is not only harmless but also even useful, as it guarantees ergodicity and mixing. Secondly, using a

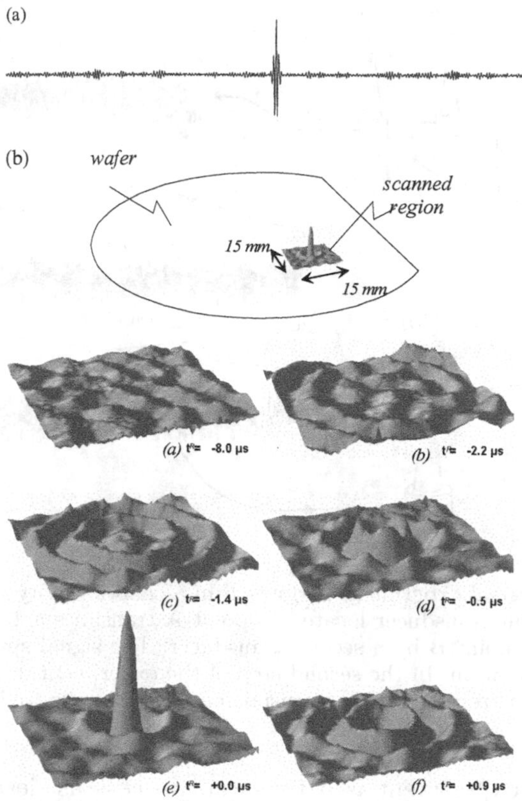

Figure 2. (a) Time-reversed signal observed at point A during 210 μs. The duration of the recompressed peak is about 1 μs. (b) Time-reversed wavefield observed at different times around point A on a square of 15 mm $\times$ 15 mm.

source of vanishing aperture, we obtain an almost perfect focusing quality. The procedure approaches the performance of a closed TRC, which has an aperture of 360°. Hence, a one-point time reversal in a chaotic cavity produces better results than a TRM in an open system. Using reflections at the edge, focusing quality is not aperture limited, and in addition, the time-reversed collapsing wavefront approaches the focal spot from all directions. Although one obtains excellent focusing, a one-channel time reversal is not perfect, as a weak noise level throughout the system can be observed. Residual temporal and spatial sidelobes persist even for time-reversal windows of infinite size. They are due to multiple reflections passing over the locations of the TR transducers and they have been expressed in closed form by C. Draeger (Draeger and Fink, 1999). Using an eigenmode analysis of the wavefield, he showed that, for long time-reversal windows, there is a saturation regime that limits the SNR. The reason is the following. Once the saturation regime is reached, all the eigenmodes are well resolved. However, A and B are always located at the nodes of some eigenmodes and these eigenfrequencies cannot be transmitted into nor received from the cavity.

3.1.1. *The acoustic sink*

As discussed above, to achieve a perfect time reversal experimentally, the field on the surface of the cavity has to be time-reversed but also the source has to be transformed into a sink. With J. de Rosny, we have imagined the following experiment to realize an acoustic sink.

From a practical point of view, we have used a slightly different experimental setup than the one previously described in order to be able to record the field close to the time-reversed source. The cavity is a 1.9 mm thick transparent glass plate whose shape is a 80 mm by 80 mm quarter stadium. This chaotic geometry is chosen to obtain quasi-isotropic focusing. Elastic waves that propagate in such a plate are Lamb waves. A brass cone coupled to a longitudinal transducer generates the field at one point of the cavity. The dimension of the contact zone is less than 100 μm, which corresponds to $\lambda/14$ at 500 kHz central working frequency. The -3 dB bandwidth of this transducer device is equal to 200 kHz. In the previous experiment, a second identical transducer was used as the TR device. In the present setup the same transducer is used to generate both the forward field and the time-reversed field. A heterodyne laser interferometer records the field time dependence. The optical beam that scans the field has a lateral dimension of 6 μm. A thin aluminum layer (1 μm thick) is evaporated on the far face of the plate. The laser beam is thus reflected back from the face of the plate that is in contact with the source. During the first step of the experiment, a 5 μs pulse is emitted by the point-like transducer. This 5 μs initial pulse generates a field that reverberates for more than 2 ms inside the cavity. With the laser spot directed towards the position of the source, the normal displacement is recorded and sampled until the field completely vanishes. In the second step, a 1.5 ms long window (between times 250 μs and 1750 μs) is time-reversed and re-emitted by the transducer. The interferometer is placed on a 2-axis stepping motor bench which allows us to record the time evolution of the time-reversed wave over a 20 mm by 20 mm square centered on the source. On Fig. 3(a), we clearly observe that a diverging wave follows the converging one as explained in the second section.

In a second experiment, a short pulse equal to the time-reversed initial waveform [function $f(-t)$] is added at time 1750 μs to the signal emitted by the transducer. Now, the time reversed source cancels the diverging wave [Fig. 3(b)] and only a converging wave remains, i.e. the TR image of the source emission. A more striking difference lies in the two focus patterns.

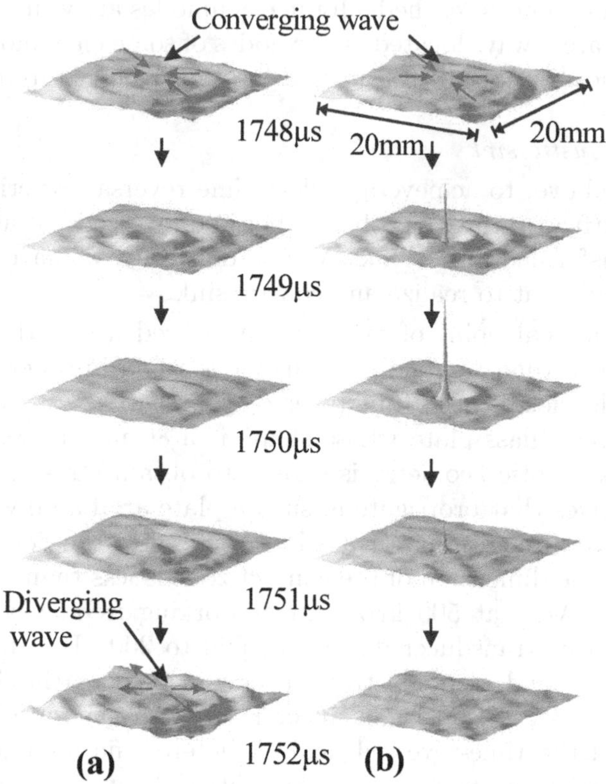

Figure 3. Time-reversal experiment performed in a chaotic cavity with flexural waves. In a first step, a point transducer located at point A transmits a 1 μs long signal. The signal is recorded at point B by a second transducer. The signal spreads on more than 30 ms due to reverberation. In the second step of the experiment, a 2 ms portion of the recorded signal is time-reversed and retransmitted back into the cavity.

Indeed, when the time-reversed source is off, the focusing pattern is smooth with a focal spot size equal to 2 mm, corresponding to half the wavelength. When the time-reversed source is switched on, a sharp and strong peak appears. In this case the focal spot is narrower and equal to 0.3 mm ($\lambda/14$). Therefore, a perfect time-reversal process yields a converging wave with a focal spot size much smaller than $\lambda/2$. The sub-wavelength focus implies that evanescent waves (near field) are involved in the complete time reversal as explained in the second section.

3.2. TIME REVERSAL THROUGH RANDOM MEDIA

A few years ago, A. Derode *et al.* (Derode *et al.*, 1995) carried out the first experimental demonstration of the reversibility of an acoustic wave

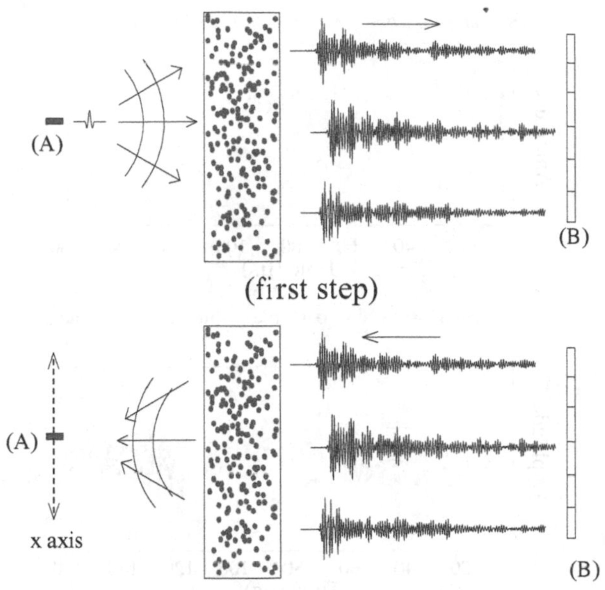

Figure 4. Sketch of the time-reversal experiment in a random medium experiment.

propagating through a 2-dimensional random collection of scatterers with strong multiple scattering contributions. In an experiment such as the one depicted on Fig. 4, a multiple scattering sample is placed between the source and an array made of 128 elements. The whole setup is in a water tank. The scattering medium consists of a set of 2000 randomly distributed parallel steel rods (diameter 0.8 mm). The sample thickness is $L = 40$ mm, and the average distance between rods is 2.3 mm. The source is 30 cm away from the TRM and transmits a short 1 μs ultrasonic pulse (3 cycles at a 3.5 MHz central frequency).

Figure 5(b) shows one part of the waveform received on the TRM by one of the elements. It spreads over more than 200 μs, i.e 200 times the initial pulse duration. After the arrival of a first wavefront corresponding to the ballistic wave, a long incoherent wave is observed, which results from the multiply scattered contribution. In the second step of the experiment, the 128 signals are time-reversed and transmitted and an hydrophone measures the time-reversed wave around the source location. Two different aspects of this problem have been studied: the property of the signal recreated at the source location (time compression) and the spatial property of the time-reversed wave around the source location (spatial focusing). The time-reversed wave traverses the rods back to the source, and the signal received on the source is represented on Fig. 5(c): an impressive compression is ob-

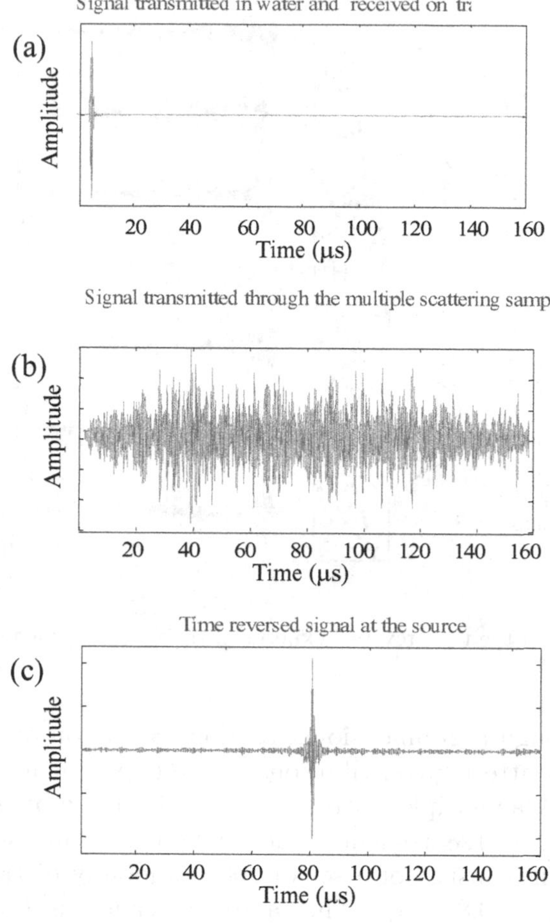

Figure 5. (a) Signal transmitted in water. (b) Signal transmitted through the sample (c) Time-reversed signal recorded at the source location.

served, since the received signal lasts about 1 μs, against over 300 μs for the scattered signals. The pressure field is also measured around the source, in order to get the directivity pattern of the beam emerging from the rods after time reversal and the results are plotted on Fig. 6. Surprisingly, multiple scattering has not degraded the resolution of the system: indeed, the resolution is found to be six times finer (solid line) than the classical diffraction limit (dotted line)! This effect does not contradict the laws of diffraction. The intersection of the incoming wavefront with the sample has a typical size D. After time reversal, the waves travel on the same scattering paths and focus back on the source as if they were passing through a converging lens with size D. The angular aperture of this pseudo-lens is much wider

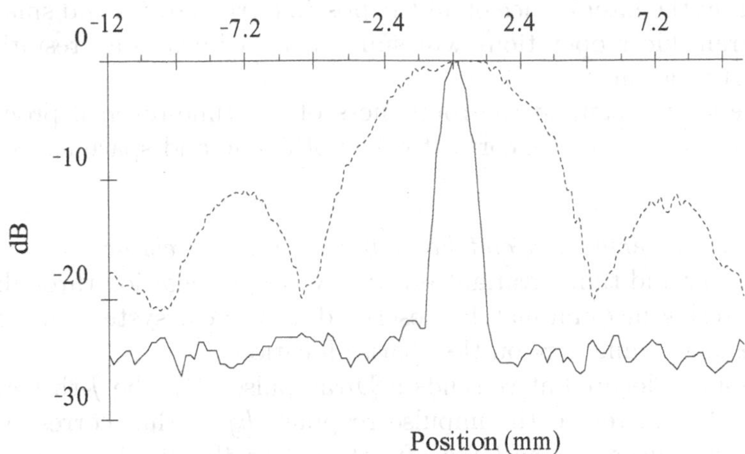

Figure 6. Directivity patterns of the time-reversed wave in water (dotted line) and through the random medium (solid line).

than that of the array alone, hence an improvement in resolution. In other words, because of the scattering sample, the array is able to detect higher spatial frequencies than in a purely homogeneous medium. High spatial frequencies that would have been lost otherwise are redirected towards the array, due to the presence of the scatterers in a large area.

This experiment shows also that the acoustic time-reversal experiments are surprisingly stable. The recorded signals have been sampled with 8-bit analog-to-digital converters that introduce discretization errors and the focusing process still works. This has to be compared to time-reversal experiments involving particles moving like balls on a elastic billiard of the same geometry. Computation of the direct and reversed particle trajectory moving in a plane among a fixed array of some thousand convex obstacles (Lorentz gas) shows that the complete trajectory is irreversible. Indeed, such a Lorentz gas is a well-known example of chaotic system that is highly sensitive to initial conditions. The finite precision that occurs in the computer leads to an error in the trajectory of the time-reversed particle that grows exponentially with the number of scattering encounters.

This difference highlights the fact that waves and particles react in fundamentally different ways to perturbations of the initial conditions. The physical reason for this is that each particle follows a well-defined trajectory whereas waves travel along all possible trajectories, visiting all the scatterers in all possible combinations. While a small error on the initial velocity or position makes the particle miss one obstacle and completely changes its future trajectory, the wave amplitude is much more stable because it results from the interference of all the possible trajectories and small errors on the transducer operations will sum up in a linear way resulting in a small perturbation.

To better understand the robustness of the time-reversal process it is worth interpreting it as a correlator in both time and space as we discuss now.

3.2.1. *Time reversal as a matched filter or time correlator*

As any linear and time-invariant process, wave propagation through a multiple scattering medium may be described as a linear system with impulse responses depending here on the spatial location.

If a source, located at $\mathbf{r}_0$ sends a Dirac pulse $\delta(t)$, the j-th transducer of the TRM will record the impulse response $h_j(t)$ that corresponds, for a point transducer, to the Green function $G(\mathbf{r}, t|\mathbf{r}_0, 0)$. Moreover, due to reciprocity, $h_j(t)$ is also the impulse response describing the propagation of a pulse from the j-th transducer to the source. Thus, neglecting the causal time delay T, the time-reversed signal at the source is equal to the convolution product $h_j(t) * h_j(-t)$. This convolution product, in terms of signal analysis, is typical of a matched filter. Given a signal as input, a matched filter is a linear filter whose output is optimal in some sense. Whatever the impulse response $h_j(t)$, the convolution is maximum at time $t = 0$. This maximum is always positive and equals $\int h_j^2(t)\,dt$, i.e the energy of the signal $h_j(t)$.

This has an important consequence. Indeed, with an N-element array, the time-reversed signal recreated on the source can be written as a sum

$$\phi_{tr}(\mathbf{r}_0, t) = \sum_{j=1}^{N} h_j(t) * h_j(-t) \tag{10}$$

Even if the $h_j(t)$ are completely random and apparently uncorrelated signals, each term in this sum reaches its maximum at time $t = 0$. So all contributions add constructively around $t = 0$, whereas at earlier or later times uncorrelated contributions tend to destroy one another. Thus the re-creation of a sharp peak after time-reversal on a N-element array can be viewed as an interference process between the N outputs of N matched filters.

The robustness of the TRM can also be accounted for through the matched filter approach. If for some reason, the TRM does not exactly retransmits $h_j(t)$ but rather $h_j(-t) + n_j(t)$ where $n_j(t)$ is an additional noise on channel j, then the re-created signal is

$$\sum_{j=1}^{N} h_j(t) * h_j(-t) + \sum_{j=1}^{N} h_j(t) * n(t)$$

The time-reversed signals are tailored to exactly match the medium impulse response, which results in a sharp peak. Whereas an additional small noise is not matched to the medium and, given the extremely long duration involved, it generates a low-level long-lasting background noise instead of a sharp peak.

3.2.2. *Time reversal as a spatial correlator*

Another way to consider the focusing properties of the time-reversed wave is to follow the impulse response approach and treat the time-reversal process as a spatial correlator. If we denote by $h'_j(t)$ the propagation impulse response from the j-th element of the array to an observation point different from the source location, the signal recreated at $\mathbf{r}_1$ at time $t_1 = 0$ writes

$$\phi_{tr}(\mathbf{r}_1, 0) = \int h_j(t)\, h'_j(t)\, dt \tag{11}$$

Notice that this expression can be used as a way to define the directivity pattern of the time-reversed waves around the source. Now, due to reciprocity, the source S and the receiver can be exchanged, i.e $h'_j(t)$ is also the signal that would be received at $\mathbf{r}_1$ if the source were the j-th element of the array. Therefore, we can imagine this array element is the source, and the transmitted field is observed at two points $\mathbf{r}_0$ and $\mathbf{r}_1$. The spatial correlation function of this wavefield would be $\left\langle h_j(t)\, h'_j(t) \right\rangle$ where the impulse responses product is averaged over different realizations of disorder. Therefore Eq. (10) can be viewed as an estimator of this spatial correlation function. Note that in one time-reversal experiment we have only access to a single realization of disorder. However, the ensemble average can be replaced by a time average, a frequency average or by a spatial average on a set of transducers. In that sense, the spatial resolution of the time-reversal mirror (i.e. the -6 dB width of the directivity pattern) is simply an estimate of the correlation length of the scattered wavefield (Derode *et al.*, 2001). This has an important consequence. Indeed, if the resolution of the system essentially depends on correlation properties of the scattered wavefield, it should become independent from the array's aperture. This is confirmed by the experimental results. Fig. 7 presents the directivity patterns obtained through a 40 mm thick multiple scattering sample, using either 1 array element or the whole array (122 elements) as a time-reversal mirror.

In both cases, the spatial resolution at -6 dB is the same: ~ 0.85 mm. In total contradiction with what happens in a homogeneous medium, enlarging the aperture of the array does not change the -6 dB spatial resolution. However, even though the number N of active array elements does not influence the typical width of the focal spot, it has a strong impact on the

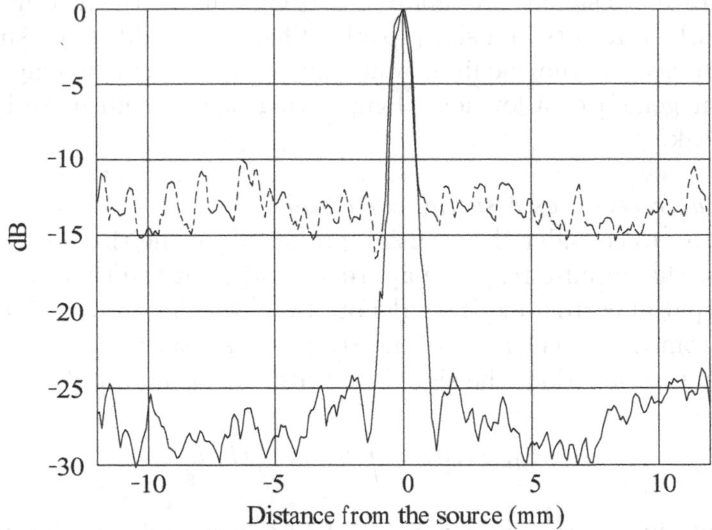

Figure 7. Directivity patterns through a 40 mm thick sample using 122 time-reversal channels (full line) or 1 time-reversal channel (dotted line).

background level of the directivity pattern [$\sim(-12)$ dB for $N = 1$, $\sim(-28)$ dB for $N = 122$], as can be seen on Fig. 7. However, it is important to point out that the signal-to-noise ratio is appreciable even if a single element is used, which would be impossible in a monochromatic phase conjugation experiment as we discuss now.

3.2.3. *Relation between the signal-to-noise ratio of the focal spot and the typical width of the frequency correlation function*

Let us consider a situation where only one transmitter is used, instead of an array, to focus on the receiver. In this case, TR focusing can be still achieved as we have just seen. Here the importance of frequency decorrelation must be emphasized. Imagine a single element trying to focus on the same receiver but in a narrow frequency bandwidth; the phase-conjugated wave has no reason at all to be focused on this receiver since the element only sends back a sinusoidal spherical wave through the medium. But if the frequency bandwidth $\Delta\omega$ is much larger than the correlation frequency $\delta\omega$, then the spectral components of the scattered field at two frequencies apart by more than $\delta\omega$ are decorrelated and there are roughly $\Delta\omega/\delta\omega$ decorrelated frequencies in the scattered signals. When we time-reverse (i.e. phase-conjugate coherently all along the bandwidth, and not just at one frequency) all these components, they add up in phase at the receiver position, because all the phases have been set back to 0 all along the bandwidth. Thus, the amplitude at this position increases as $\Delta\omega/\delta\omega$ whereas outside the receiver position, the various frequency components add up incoherently and their sum rises as $\sqrt{\Delta\omega/\delta\omega}$. On the whole, the peak-to-noise ratio increases as $\sqrt{\Delta\omega/\delta\omega}$ as the bandwidth is enlarged. Through a 40 mm thick forest of rods, we found $\delta\omega = 10$ kHz; the total bandwidth at half-maximum is 1.5 MHz, the frequency ratio is therefore ~ 150, thus an improvement of more than 20 dB compared to a monochromatic phase conjugation technique.

Finally, the fundamental properties of time reversal in a random medium rely on the fact that it is both a space and time correlator, and the time-reversed waves can be viewed as an estimate of the space and time auto-correlation functions of the waves scattered by a random medium. The estimate becomes better with a large number of transducers in the mirror and a larger bandwidth. Moreover, the system is not sensitive to small perturbations since adding a small noise to the scattered signals (e.g. by digitizing them on a reduced number of bits) may alter the noise level but does not drastically change the correlation time or the correlation length of the scattered waves. Even in the extreme case where the scattered signals are digitized on a single bit, A. Derode *et al.* have shown recently (Derode *et al.*, 1999) that the time and space resolution of the TRM were practically unchanged, which is a striking evidence for the robustness of wave time reversal in a random medium.

4. Two Applications of Time-Reversal Focusing

4.1. DIGITAL COMMUNICATION WITH TIME REVERSAL IN A MULTIPLE-SCATTERING MEDIUM

Based on the previous analysis, we present in the same book (cf. Tourin *et al.*) an experimental demonstration showing that contrary to first intuition the more scattering a mesoscopic medium is, the more information can be conveyed through it. We used a MIMO (Multiple Input Multiple Output) configuration: a multi-channel ultrasonic time-reversal antenna is used to transmit random series of bits simultaneously to different receivers which were only a few wavelengths apart. Whereas the transmission is free of error when multiple scattering occurs in the propagation medium, the error rate is huge in a homogeneous medium. This is discussed in relation with the number of eigenvalues of the time-reversal operator in a homogeneous or multiple scattering medium. This number is strongly related to the number of focal spots that a transmitting antenna can create in a given region of space and thus to the spatial correlation length of the field as discussed in the previous section. For communication purposes, the number of eigen-values of the time-reversal operator in the medium gives the number of independent receivers an antenna can talk to.

4.2. SONOLUMINESCENCE

Time-reversal focusing techniques may be used in various situations where one has to focus adaptively on a target whose position is unknown. For example, large amplitude sound waves can be adaptively focused in liquid to destruct kidney stones. Another interesting application of this concept is to upscale single bubble sonoluminescence from a bubble trapped in liquid by a low frequency sound field.

Sonoluminescence is observed as a transient phenomenon, for example, in converging fluid flows. As a quasi-continuous glow, it may be seen either as single bubble or multiple bubble sonoluminescence. Until roughly ten years ago, sonoluminescence was synonymous with the latter: a liquid is insonated, for example by mounting a sonic transducer in a water tank. At sufficiently high sound intensities a beam of light is observed which originates from a cloud of small, luminescing bubbles. In more recent years, Gaitan and Crum (Gaitan *et al.*, 1992) discovered single bubble sonoluminescence (SBSL), an interesting modification that makes use of a standing sound wave with one pressure antinode in the container. A single, pulsating bubble may be levitated stably for hours and emits light.

Single bubble cavitation may be observed in an acoustic resonator, e.g. a flask filled with a liquid driven at resonance by piezoelectric transducers (Gaitan *et al.*, 1992; Brenner *et al.*, 2002). If the amplitude of the sound pressure is too low, a bubble may be injected but it will drift to the water surface and disappear (buoyancy region). With slightly increased sound amplitudes, at the lowest trapping pressures, a seeded bubble oscillates linearly with the sound pressure, but it dissolves slowly in time. If the sound amplitude is further increased, the bubble begins to jump back and forth irregularly ('dancing region'), up to sound pressures of typically 1.2 atm, the threshold of stable single bubble sonoluminescence. At this amplitude a bubble is trapped at the antinode of the sound field. Depending on the liquid, its temperature, and on other parameters of the experiment, a sonoluminescent region of great stability exists, up to acoustic pressure amplitudes of roughly 1.4 atm. In this region bubble oscillations are highly nonlinear but synchronous with the sound field. The bubble undergoes large periodic nonlinear oscillations characterized by an expansion stage followed by a violent collapse. At the end of the collapse, one part of the acoustic energy stored by the bubble is converted into a short flash of light. When the amplitude of the stationary sound field exceeds 1.4 atm, the bubble disappears due to bubble shape instabilities and diffusive processes. Single bubble sonoluminescence is only stable in a very narrow range of experimental parameters. State-of-the-art models predict that the temperature and pressure inside the bubble reach several tens of thousands of degrees and atmospheres, respectively, at the end of the collapse.

Several teams investigated the possibility to further increase the energy concentration by modifying one or several parameters of the experiment: pressure, temperature, and the nature of the gas or the liquid. One goal was to reach the conditions required to initiate a nuclear fusion reaction. Most of these trials have rapidly reached the boundaries of the narrow stability domain discussed above, and water remains the best liquid to achieve sonoluminescence. Simultaneously, several studies have tried to control the pressure field in order to improve the concentration of energy. Moss proposed to add to the low frequency monochromatic sound field, that trapped the bubble, a brief intense pressure pulse during the collapse of the bubble. Several teams have attempted to enhance the flash of light according to this technique. However, their technique is limited by asymmetrical effects since they used only one transducer. Indeed, one limitation of a pulse technique is that it may induce instabilities of the spherical shape of the bubble. These instabilities are smoothed during the growth of the bubble, whereas they will develop during the collapse of the bubble. Thanks to time-reversed acoustics, with J.-L. Thomas, we have proposed a new way that leads to the transmission of a much higher incident sound field on the bubble and overcomes the classical stability domain of sonoluminescence (Thomas *et al.*, 2002). The principle is to apply a strong acoustic pressure pulse adaptively focused on the bubble at the time of collapse, in order to increase the speed of the bubble collapse. To achieve this goal, we have modified the classical setup used by Gaitan and Crum (Fig. 8).

We used a spherical glass cavity, filled with degassed water and driven at resonance by two piezoelectric transducers glued opposite each other along a circumference. This now classical setting is adapted to incorporate eight high-frequency piezoelectric transducers (central frequency of 700 kHz) operating in pulse mode. The eight transducers are oriented perpendicularly to the surface of the spherical cavity and paired opposite each other along planes that cut the sphere into quarters. The glass cavity has been cut at the eight locations of the transducers to optimize the acoustic coupling to the water. A time-reversal electronics drives the high-frequency transducers. From the high frequency part of the bubble echo, we can built a time-reversed wave that exactly focuses back on the bubble. In our first experiments, the acoustic pulse produced by the high-frequency transducers allows to obtain on the bubble location a spherical pressure pulse of 7 atm and 400 ns. The time-reversed pulse has to be sent with a good timing to reach the bubble at the collapse. Compared to the classical regime with 1.4 atm on the bubble, we obtain a gain of brightness of a factor 2 with this

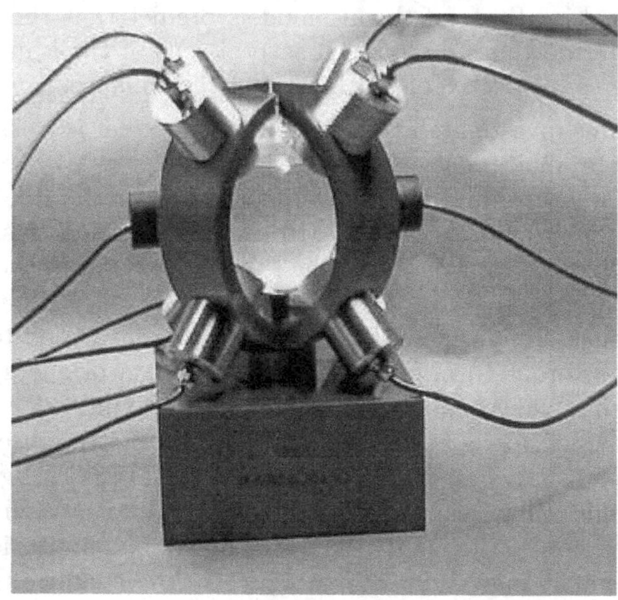

Figure 8. Experimental setup for boosting sonoluminescence by time reversal.

transient boosting without creating shape instability of the bubble. Since this transient boosting is no longer limited by the classical stability domain of SBSL, it could lead in the future to much higher temperatures inside the bubble. This experimental setup also provides a new domain of parameters for SBSL and, hence, a new test for numerous theories about light emission mechanisms.

5. Conclusion

In this paper, we have shown how chaotic ray dynamics in random media and in cavities enhances resolution in time-reversal experiments. Multipathing makes the effective size of the TRM much larger than its physical size. This study proves the feasibility of acoustic time reversal in media with chaotic ray dynamics. Paradoxically, chaotic dynamics is useful, as it reduces the number of channels needed to insure an accurate time-reversal experiment. These experiments are mainly 2-dimensional, however these results can be extended to 3D configurations. These results can also be easily extended to electromagnetic propagation through random and chaotic media. Perhaps one of the most spectacular applications of time-reversed technologies will be, in the future, wireless communications, where the multipathing may enhance the amount of information that can be transmitted between antenna arrays in complex scattering environments like cities. From quantum chaos to wireless communications, time-reversal symmetry plays an important role and, contrary to long-held beliefs, chaotic scattering of acoustic or microwave signals may strongly enhance the amount of independent information that can be transmitted from an antenna array to a volume of arbitrary shape.

References

Brenner, M., Hilgenfeldt, S., and Lohse, D. (2002) *Rev. Mod. Phys.* **74**, 425.

Cassereau, D., and Fink, M. (1992) *IEEE Trans. Ultrason. Ferroelec. Freq. Contr.* **39**(5), 579–592.

Cassereau, D., and Fink, M. (1994) *J. Acoust. Soc. Am.* **96**(5), 3145–3154.

De Rosny, J., and Fink, M. (2002) *Phys. Rev. Lett.* **89**(12), 124301.

Derode, A., Tourin, A., and Fink, M. (1999) *J. Appl. Phys.* **85**(9), 6343–6352.

Derode, A., Roux, P., and Fink, M. (1995) *Phys. Rev. Lett.* **75**(23), 4206–4209.

Derode, A., Tourin, A., and Fink, M. (2001) *Phys. Rev. E* **64**(3), 1063.

Draeger, C., and Fink, M. (1997) *Phys. Rev. Lett.* **79**(3), 407–410.

Draeger, C., and Fink, M. (1999) *J. Acoust. Soc. Am.* **105**(2), 611–617.

Fink, M. (1992) *IEEE Trans. Ultrason. Ferroelec. Freq. Contr.* **39**(5), 555–566.

Fink, M. (1997) *Physics Today* **50**(3), 34–40.

Fink, M., Prada, C., Wu, F., and Cassereau, D. (1989) *IEEE Ultrasonics Symposium Proceedings* **1**, 681–686.

Gaitan, D.F., Crum, L.A., Church, C., and Roy, R.A. (1992) *J. Acoust. Soc. Am.* **91**, 3166.

Goodman, J. (1996) Introduction to Fourier Optics. Vol. 1, second edition, McGraw-Hill, New York.

Jackson, D. and Dowling, D. (1991) *J. Acoust. Soc. Am.* **89**(1), 171.

Kino, G. (1987) Acoustics waves, In: Signal Processing Series, Prentice Hall, New York.

Kuperman, W., Hodgkiss, W., Song, H., Akal, T., Ferla, T., and Jackson, D. (1998) *J. Acoust. Soc. Am.* **103**, 25–40.

Roux, P., Roman, B., and Fink, M. (1997) *Appl. Phys. Lett.* **70**, 1811–1813.

Thomas, J.L., Forterre, Y., and Fink, M. (2002) *Phys. Rev. Lett.* **88**, 074302.

Negative refraction, does it exist? Manuel Nieto-Vesperinas, Mathias Fink and Roger Maynard (from left to right). Photo by Gabriel Cwilich

CHAPTER IV

WAVES IN METAMATERIALS

Metamaterials are artificially structured composite materials that can be engineered to have desired electromagnetic or acoustic properties, while having other advantageous material properties. The material shown in the photograph is a structure composed of copper rings and wires that appears as a continuous material to electromagnetic waves over a certain range of frequencies. This composite material, or metamaterial, exhibits a simultaneously negative permittivity and permeability, and can thus be thought of as an example of left-handed material. Text and photo are due to Prof. David R. Smith, University of California, San Diego, USA (http://physics.ucsd.edu/~ drs/left_home.htm, with kind permission of the author).

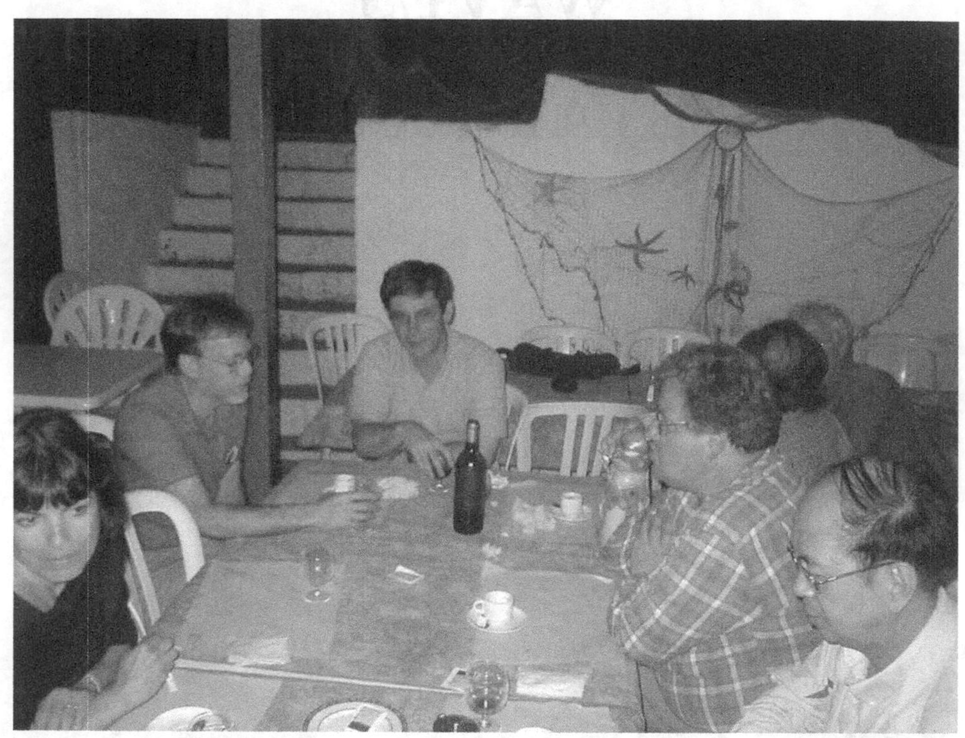

After-dinner discussions... Sylvia Cwilich, John Page, Richard Weaver, Ad Lagendijk and Costas Soukoulis, Rudolf Sprik, Zhao-Qing Zhang (from left to right). Photo by Gabriel Cwilich

3D PHONONIC CRYSTALS

Ultrasonic Wave Transport and Spectroscopy in Complex Media II

J. H. PAGE[1], SUXIA YANG[2] AND M. L. COWAN[3]
Department of Physics and Astronomy, University of Manitoba, Winnipeg MB R3T 2N2 Canada

ZHENGYOU LIU
Department of Physics, Wuhan University, Wuhan 430072 China

AND

C. T. CHAN AND PING SHENG
Department of Physics, Hong Kong University of Science and Technology, Clear Water Bay, Kowloon, Hong Kong

Abstract. Recent progress in the use of ultrasonic experiments and Multiple Scattering Theory to investigate wave transport in three-dimensional phononic crystals is summarized. Through appropriate choice of material properties, complete band gaps can be realized for acoustic or elastic waves in such structures. This has allowed us to demonstrate the tunnelling of ultrasound through the band gap and to explore the unexpected effect of absorption on evanescent waves in crystals. Wave propagation above the gap has also been investigated, where we have shown that anisotropy of the wave speeds leads to the focusing of ultrasound without the curved surfaces usually employed in lenses. These ultrasonic experiments and their interpretation using Multiple Scattering Theory illustrate the important contribution that the study of phononic crystals can make to learning about wave scattering and transport in ordered mesoscopic materials.

[1]Author to whom correspondence should be addressed. E-mail: jhpage@cc.umanitoba.ca. Website: www.physics.umanitoba.ca/~jhpage.

[2]Current address: Energenius Centre for Advanced Nanotechnology, University of Toronto, 170 College St. Toronto, Ontario, Canada, M5S 3E3. Suxia Yang was also a student at the Hong Kong University of Science and Technology while this research was performed.

[3]Current address: Dept. of Physics, University of Toronto, 60 St. George St., Toronto, Ontario Canada M5S 1A7.

B.A. van Tiggelen and S.E. Skipetrov (eds.),
Wave Scattering in Complex Media: From Theory to Applications, 283–307.

1. Introduction

Phononic crystals are periodic composite materials with variations of velocity and density on length scales comparable with the wavelength of sound (or ultrasound) [1–14]. Because the scattering contrast between the component materials depends on differences in both density and phase velocity, phononic crystals with very strong scattering can be realized experimentally, making such materials interesting candidates for studying the profound effects of lattice structure on wave propagation. This is one of the main reasons for the considerable growth of interest in phononic crystals that has occurred during the last decade. Much of this interest has focussed on phononic bandgaps, which correspond to ranges of frequency in which acoustic or elastic waves cannot propagate due to Bragg scattering, and are analogous to photonic band gaps [15–17] for electromagnetic waves. Through appropriate choice of materials, it has been demonstrated by a combination of theory and experiment that complete band gaps can be readily achieved for acoustic and elastic systems, so that wave propagation is forbidden in all directions. This is in sharp contrast with photonic materials, where engineering complete spectral gaps in three dimensions has been a difficult experimental challenge.

In this paper we describe a combination of experimental and theoretical results on ultrasonic wave transport in three-dimensional (3D) phononic crystals [10–14], which have been less studied experimentally than 2D structures [4–7,13], possibly because 3D structures have been considered more difficult to fabricate. After a brief description of the crystals used in our experiments and the expected band structures, we summarize recent pulse propagation experiments in which both amplitude and phase information is measured, allowing the transmission coefficient, the dispersion relation and the wave propagation dynamics to be investigated. These experiments are compared with the predictions of Multiple Scattering Theory (MST), which is ideally suited to the spherical scattering geometry of our crystals. This combination of experiment and theory is used to investigate two quite different types of wave phenomena that result from the underlying crystal structure: ultrasound tunnelling through the band gap and the focusing of ultrasound without the usual curved surfaces employed in traditional lenses. Some results of near field imaging experiments are also reported.

2. Our 3D Phononic Crystals and their Band Structures

The crystals used in our experiments consist of close-packed periodic arrays of spherical beads surrounded by a liquid or solid matrix. Both hcp and fcc arrays were assembled by placing the beads carefully by hand in a hexagonal template that was precisely machined to force the beads into

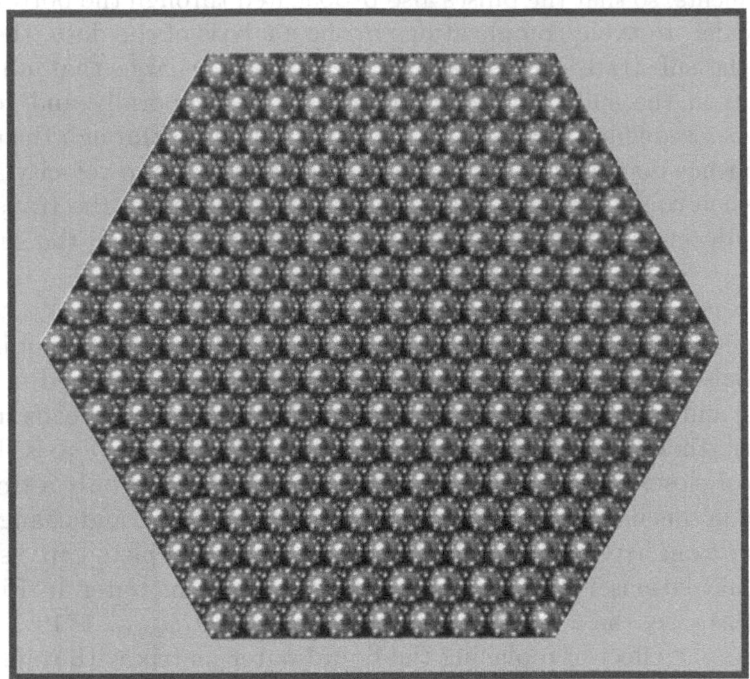

Figure 1. Picture of the top surface of a hcp phononic crystal consisting of monodisperse 0.8-mm-diameter stainless steel beads in water.

triangular layers. The layers were stacked vertically in either an ABAB.. or ABCABC... sequence to form slabs with a hexagonal (c-axis $\perp$ layers) or face-centred cubic ([111]-axis $\perp$ layers) structure, respectively. For the fcc structures, the template was designed with sloping sides to ensure that the layers were arranged in the correct sequence [18]. The beads were made of either stainless steel or tungsten carbide, and the matrix was either water or epoxy. For all these crystals, the scattering contrast was very high (the acoustic impedance ratios of bead to matrix ranged from 30 to 60) and the beads were very monodisperse (e.g. for the tungsten carbide beads, the sphere diameter d was 0.8000 ± 0.0006 mm), so that very high quality crystals could be prepared with patience and good manual dexterity. An example of one of the crystals is shown in Fig. 1, illustrating the excellent regularity of the structure that was achieved.

Ultrasonic pulse propagation through the crystals was measured by placing the crystals in a large water tank in between two immersion transducers, which acted as generator and detector [12, 14]. For the water-matrix crystals, it was necessary to keep the crystals in the template during the measurements, so that the pulses also propagated through the bottom plate of the holder; to avoid complications in the analysis of the data, the thickness of the substrate was chosen to be sufficiently large that ultrasonic reflections in the substrate could be separated temporally and removed from the subsequent analysis of the signal transmitted through the crystal. The frequency dependence of the phase velocity, the group velocity and the transmission coefficient was then measured by comparing the transmitted pulses with reference pulses that had travelled once through the substrate alone.

Before considering the experimental results, it is instructive to consider the theoretical band structures for three of the eight possible combinations of materials. These band structure calculations were performed using the MST [10] and are shown in Fig. 2. For a hcp crystal of steel beads in water [Fig. 2(a)], there is a reasonably large stop band along the *c*-axis, but the gap almost closes between the M and K points, so there is only a tiny complete gap in this material. Increasing the density contrast, and changing the symmetry from hcp to fcc, results in a much larger complete gap, as shown by the calculations for fcc tungsten carbide beads in water in Fig. 2(b) [19]. In this case, the width of the complete gap, $\Delta\omega/\omega_{\text{centre}}$ is 19%. Figure 2(c) shows the effect of replacing the liquid water matrix with solid epoxy. This changes the scattering, as the matrix can support both longitudinal and transverse polarizations, and results in an even bigger compete gap, with $\Delta\omega/\omega_{\text{centre}} = 90\%$, even though the longitudinal impedance contrast is reduced. However, the epoxy is quite lossy, and the absorption in the crystal becomes significant, so we will focus instead on the second system, fcc tungsten carbide beads in water, in the remainder of this paper.

To compare the theory more directly with the results of our pulsed transmission experiments, we use a layer MST to calculate the transmitted field though crystals consisting of a finite number of layers [10]. For a sample of thickness L, the transmitted field as a function of frequency ω can be written as

$$T(L,\omega) = A(L,\omega)\exp[i\phi(L,\omega)], \tag{1}$$

where A and ϕ are the amplitude and cumulative phase relative to the input field. Thus, the amplitude transmission coefficient is simply $|T(L,\omega)| = A(L,\omega)$, and the phase and group velocities, v_p and v_g, can be determined from cumulative phase in the usual way:

$$v_p(\omega) = \frac{\omega}{k} = L\frac{\omega}{\phi} = \frac{L}{t_p} \tag{2}$$

$$v_g(\omega) = \frac{d\omega}{dk} = L\frac{d\omega}{d\phi} = \frac{L}{t_g} \tag{3}$$

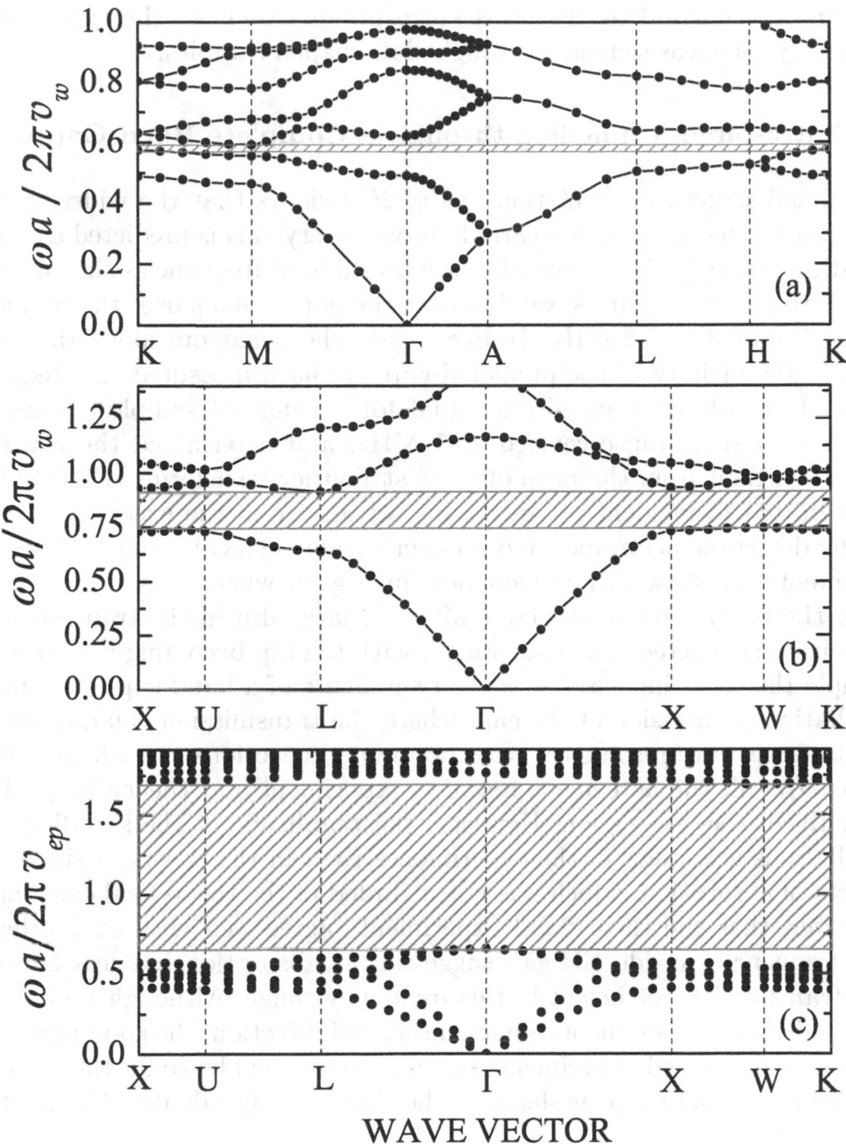

Figure 2. Band structure calculated using the Multiple Scattering Theory for (a) a hcp phononic crystal of stainless steel beads immersed in water, (b) a fcc phononic crystal of tungsten carbide beads in water, and (c) a fcc phononic crystal of tungsten carbide beads in epoxy. Here a is the lattice constant, and v_w and v_{ep} are the (longitudinal) sound velocities in water and epoxy, respectively. The shaded regions indicate the locations of the bandgaps in these materials. The letters denote the high symmetry points of the Brillouin zone.

Here k is the wave vector, and t_p and t_g are the phase delay and group delay times. This layer MST is expected to give an accurate description of our experimental results, as it corresponds closely to the experimental geometry for measurements through slab-shaped samples.

3. Ultrasound Tunnelling through a Complete Band Gap

The band structure calculation in Fig 2(b) shows that the widest part of the gap for the fcc tungsten-carbide-in-water crystals is predicted along the ΓL direction (|| [111]). Converting the normalized frequencies shown in this figure to laboratory units, we find that the gap extends over the frequency range from 0.8 to 1.2 MHz. To investigate the behaviour along this direction in the vicinity of the predicted gap, we have measured the frequency dependence of the transmitted signal for a range of sample thicknesses. We used a short pulse centred at 1 MHz, and determined the transmission coefficient from the ratio of the fast Fourier transforms (FFTs) of the transmitted to incident pulses, taking advantage of the excellent linearity of the detection electronics in our experiments. Typical results of these experiments are shown by the symbols in Fig. 3, where they are compared with the predictions of the layer MST. A large dip in the transmission is seen in the expected frequency range, with the dip becoming deeper as the sample thickness increases — a clear signature of a band gap. The smaller oscillations either side of the gap, where the transmission is large, are due to standing wave resonances from boundary reflections at each side of the slab. Overall, the agreement between experiment and theory is good, especially in the gap region. However, the amplitude of the standing wave oscillations is smaller in the experiments than in the theory, a sign of the effects of absorption, which was not included in the theoretical calculation. The fact that the theory and experiments agree well with each other at the frequencies inside the gap suggest that absorption has less effect on the transmission coefficient in this frequency range. Although we only did transmission measurements along one crystal direction, the good agreement between theory and experiments serves as strong evidence for the existence of a wide complete gap, as shown in the theoretically calculated band structure.

One important parameter that characterizes the band gap is the gap width. The band structure calculation only gives the theoretical size and position of a band gap for an infinite sample. Experimentally, samples are finite, and the width and position of the band gap may be thickness dependent. This is borne out by our experiments, which show that the gap width (defined here as the frequency interval between the positions of the peaks in the transmission coefficient at the band edges) decreases as the

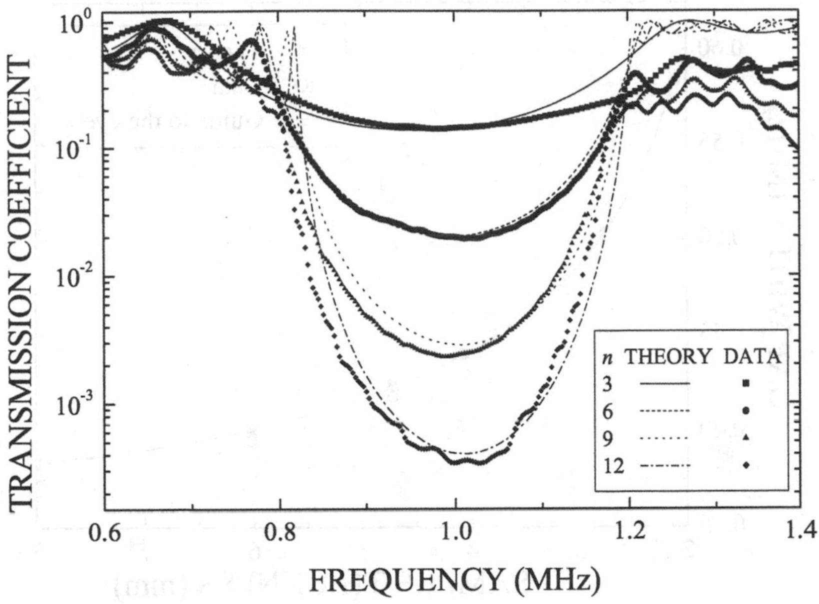

Figure 3. Frequency dependence of the transmission coefficient in the vicinity of the band gap in phononic crystals made from tungsten carbide beads in water. Experimental data for four sample thicknesses (symbols), consisting of $n = 3$, 6, 9 and 12 layers, are compared with the predictions of the layer MST (solid and broken curves).

sample thickness increases, tending towards a constant for our thickest samples (Figure 4). Again, the theory and experiments are in good agreement, especially for the thinner samples.

To look in more detail at the decrease in the transmission as a function of the number of layers in the crystal, we plot the thickness dependence of the transmission coefficient at the gap central frequency of 0.945 MHz in Fig. 5. From this figure, it can be seen that the transmission coefficient decays exponentially as the sample becomes thicker, as expected for evanescent waves in a band gap. From a fit to $T = \exp[-L/(2l)]$, we determine the value of the decay length l of the evanescent modes to be 0.54 mm at this frequency, very close to half the lattice constant ($a = \sqrt{2}d = 1.13$ mm). The value of l corresponds to an imaginary wave number κ of 0.92 mm^{-1}. This exponential decay with such a small value of the decay length, or large κ, is strongly suggestive that the modes in the gap are evanescent, not propagating, implying that the small signal that does get through thick samples does so by tunnelling. Note that the transmitted intensity for the

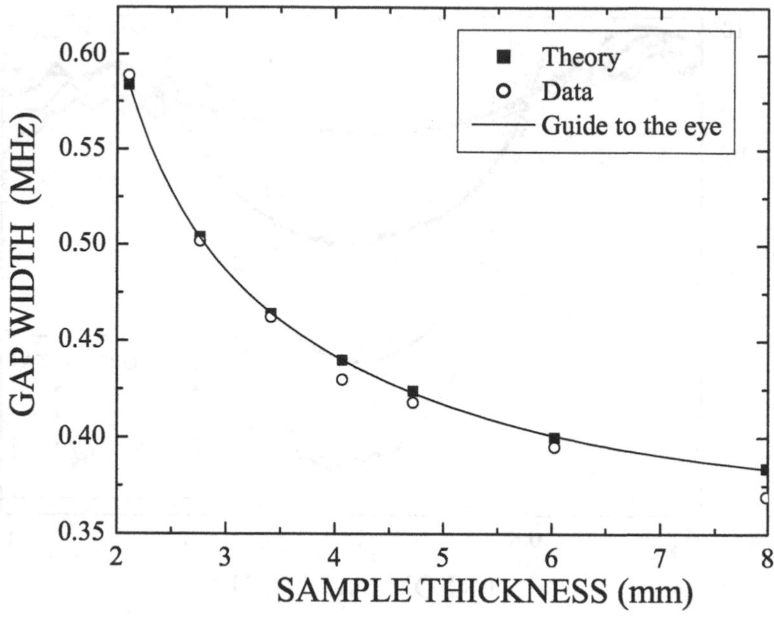

Figure 4. Variation of the gap width with crystal thickness.

thickest sample, consisting of only 12 layers, is nearly 7 orders of magnitude smaller than the incident intensity.

The frequency dependence of the cumulative phase ϕ and phase velocity $v_p(\omega) = L\omega/\phi$ in the vicinity of the gap are shown in Fig. 6. Here, data for the 12-layer sample are plotted as solid symbols. The experimental data for ϕ were obtained from the phase difference of the complex FFTs of the transmitted and input pulses, while resolving any possible ambiguity of 2π in the phase by making use of the condition that the cumulative phase must extrapolate to zero at zero frequency. The cumulative phase increases approximately linearly with frequency at low frequencies, with small oscillations due to the standing wave resonances mentioned above. However, in the gap between 0.8 and 1.2 MHz, there is a plateau in the cumulative phase; in the plateau, there is only a very small linear increase in the phase with frequency, implying a linear increase in the phase velocity and a large group velocity. The behaviour of the phase velocity, shown in Fig. 6(b), confirms this result, and also shows that the phase velocity decreases with frequency on both sides of the gap, reaching values substantially less

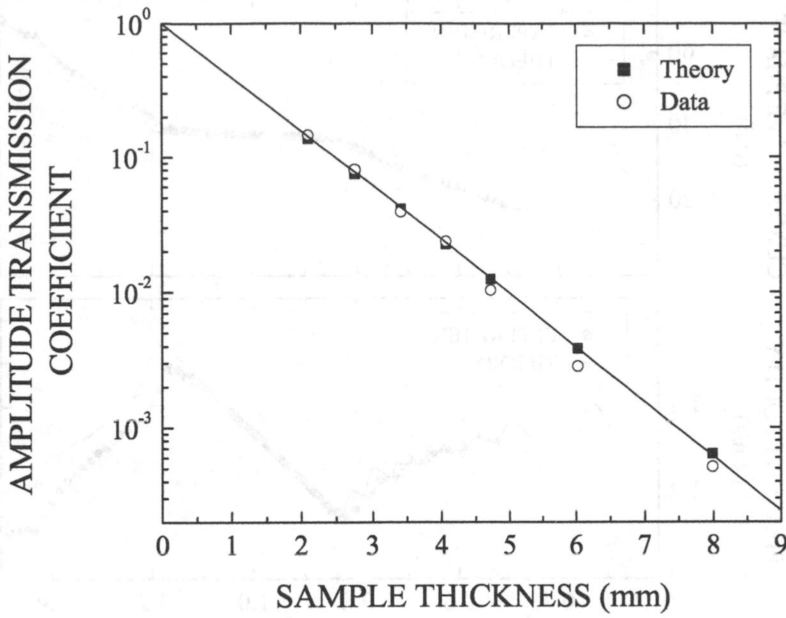

Figure 5. Amplitude transmission coefficient as a function of sample thickness, showing the strong exponential decay. Both the experimental data (solid symbols) and the predictions of the MST (open symbols) are in good agreement with the exponential fit (solid line), which gives a tunnelling decay length of $l = 0.54$ mm.

than the velocity of sound in the matrix material at the highest frequencies shown. The curves show that the behaviour for ϕ and v_p calculated from the transmission MST are in good overall agreement with the experimental data. The phase velocity data can also be used to compute the dispersion curve. Good agreement for the data from the 12-layer crystal is found with the band structure calculation, as shown in [14]. Here we compare the experimental dispersion curves in the extended zone scheme (left panel in Fig. 7) with theoretical predictions from the transmission MST theory (right panel in Fig. 7) for three different sample thicknesses. It can be seen that as the sample becomes thicker, the dispersion curve becomes steeper around the boundary of the first Brillouin zone. The overall structure is well captured by the theory, although slight differences can be noticed. The steep slope of the dispersion curve indicates that the group velocity is large in the gap, as the group velocity is equal to the slope of the dispersion curve.

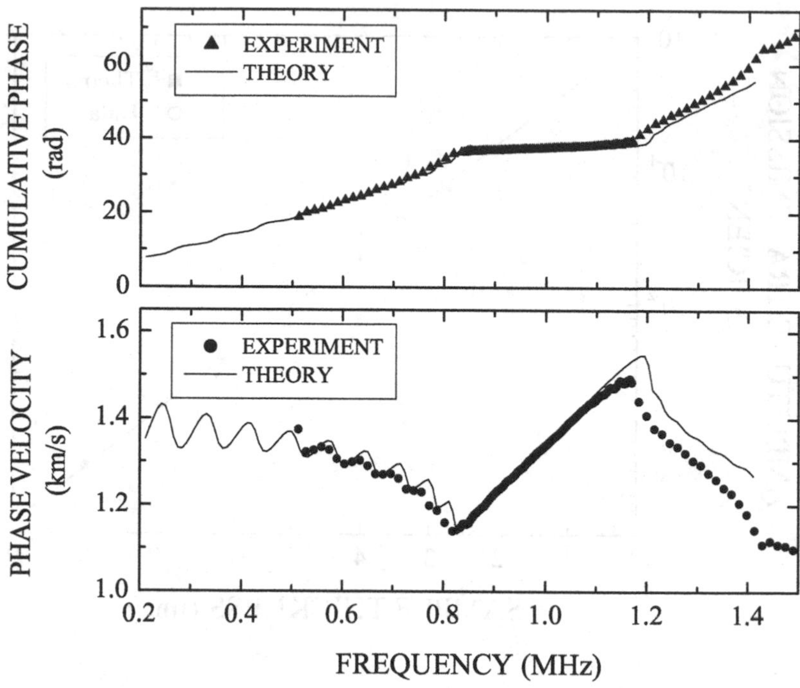

Figure 6. Cumulative phase and phase velocity as a function of frequency for a 12-layer tungsten carbide/water crystal. The solid symbols represent experimental data and the solid curves represent the predictions of the layer MST.

To examine the behaviour of the group velocity in more detail, and to obtain more definitive evidence that tunnelling of ultrasound is occurring in the gap, we have measured the group velocity directly from pulse transmission experiments [14]. These measurements were performed by digitally filtering the input and transmitted pulses using a narrow Gaussian bandwidth, as shown in Fig. 8 for a 12-layer phononic crystal. In this example, the bandwidth was 0.05 MHz and the central frequency was 0.945 MHz, the same frequency in the middle of the gap as in Fig. 5. It is clear from this figure that the pulse travelling through the crystal (bottom panel) arrives very soon after the input pulse (top panel), and travels much more quickly than an identical pulse transmitted through the same thickness of water (middle panel). Note also that the shape of the pulse that has travelled through the crystal is identical with the input pulse, although much reduced in amplitude; this confirms that despite the considerable variation

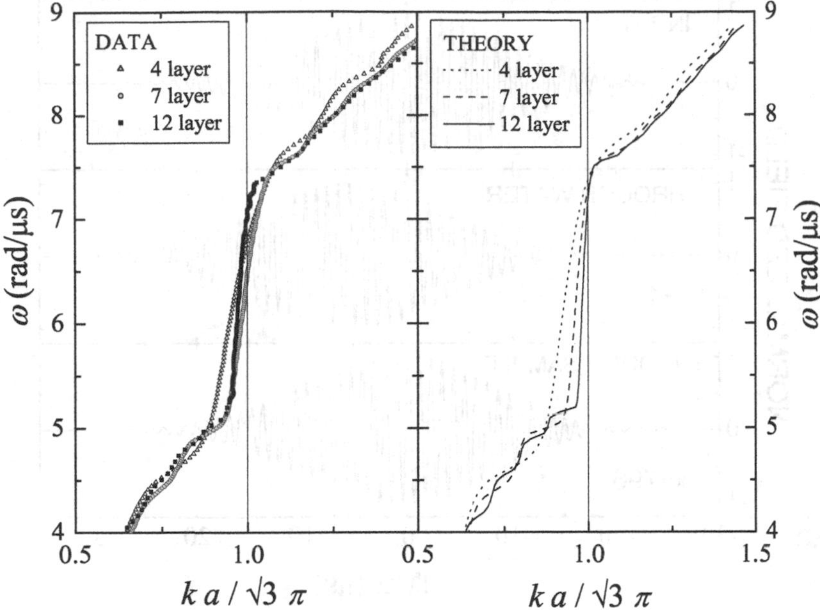

Figure 7. Comparison of the measured dispersion curves along the direction from Γ to L (left panel) with the predictions of the layer MST for the transmission (right panel). The wave vector k is divided by the value at the Brillouin zone boundary for this direction ($\sqrt{3}\pi/a$, where a is the lattice constant). Data and theory for three thickness, corresponding to 4, 7 and 12 layers, are shown.

of the phase velocity in the gap region, there is negligible pulse distortion, and the pulse transit time is well defined. The origin of the negligible pulse distortion lies in the linearity of the phase variation with frequency within the gap (see Fig. 6), implying that the group velocity dispersion $dk^2/d^2\omega$ is essentially zero, as we have verified directly in other measurements [18]. The delay between the peak arrival times of the sample and input pulses gives the group delay time t_g, from which we determine the group velocity experimentally ($v_g = L/t_g$).

The frequency dependence of the group delay time in the vicinity of the gap is shown in Fig. 9 for four crystals containing 3, 6, 9 and 12 layers. The behaviour is very striking. Below and above the gap, the group delay time undergoes large oscillations due to the standing wave resonances of the crystal slabs, and the delay time averaged over the oscillations increases in proportion to the sample thickness, as expected. Even the average delay

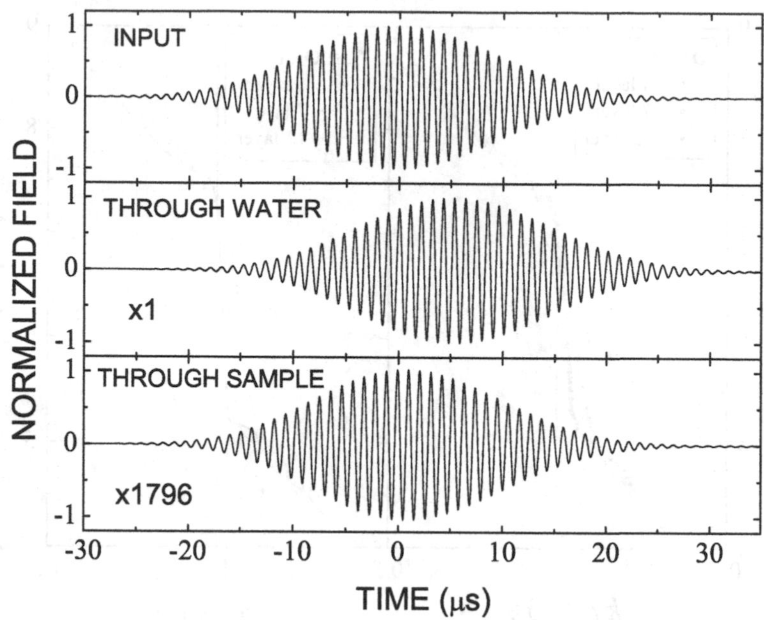

Figure 8. Digitally filtered input and transmitted pulses (top and bottom panels; bandwidth 0.05 MHz) for a 12-layer phononic crystal in the middle of the gap, compared with the pulse transmitted through the same thickness of water (middle panel).

time is quite large at these frequencies, and corresponds to a group velocity less than that of water. By contrast, in the gap, the delay time is very short and *essentially independent of sample thickness,* implying that the group velocity increases linearly with L as the sample becomes thicker. This behaviour shows convincingly that tunnelling is occurring, since one of the remarkable features of tunnelling is that the tunnelling time is independent of thickness [20]. This feature of tunnelling holds quite generally unless the sample or barrier is extremely thin.

To examine the tunnelling behaviour in more detail, we plot the group velocity near the centre of the gap at 0.945 MHz as a function of sample thickness L (Fig. 10). This figure shows that the group velocity in the gap increases monotonically with L, both in experiment (open circles) and in MST theory (solid line and squares). Note the large values of the group velocity for thick samples, where experimental and theoretical values can be larger than the longitudinal velocity in both water and tungsten carbide (horizontal dotted lines). However, there is a substantial difference

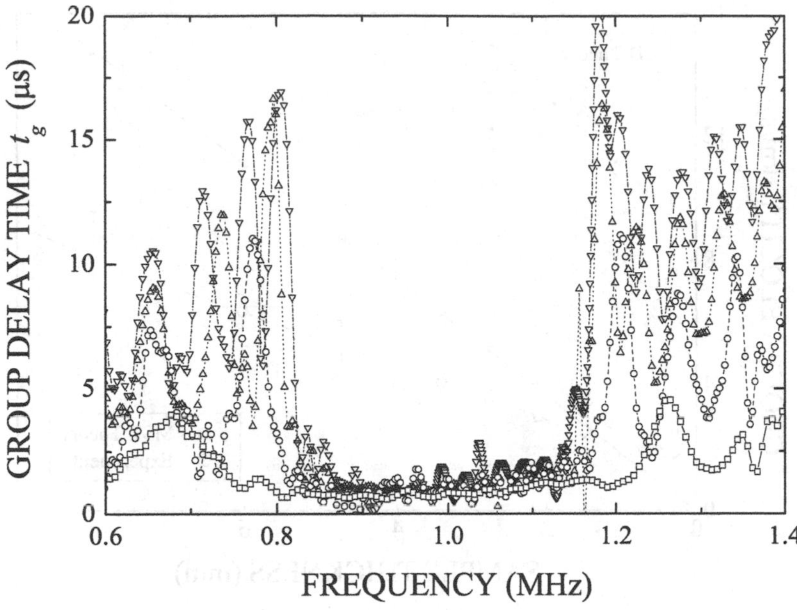

Figure 9. Measurements of the group delay time as a function of frequency for 4 phononic crystals containing 3, 6, 9 and 12 layers (squares, circles, up triangles and down triangles, respectively), corresponding to 1, 2, 3 and 4 complete unit cells along the [111] direction.

between the theoretical predictions and the experimental results, which we interpret as a consequence of absorption which was not included in the MST calculation [21]. For evanescent waves in the band gap of a phononic crystal, the consequences of absorption are quite interesting. As we have discussed previously [14], the main effect of absorption is to cut off long multiple scattering paths, with the result that the destructive interference of Bragg-scattered waves that give rise to the band gap becomes incomplete. Consequently, in addition to the dominant tunnelling mode, a small propagating component is 'created' by absorption, with an effective wave vector determined by the incomplete cancellation of the Bragg scattered waves. This has the effect of increasing the transit time and hence reducing the group velocity. We can understand the effects of absorption on the group velocity by a 'two-modes' model, based on the simple approximation that pulse transport can be viewed as a dominant tunnelling process in parallel with a small propagating component. In this model, the group

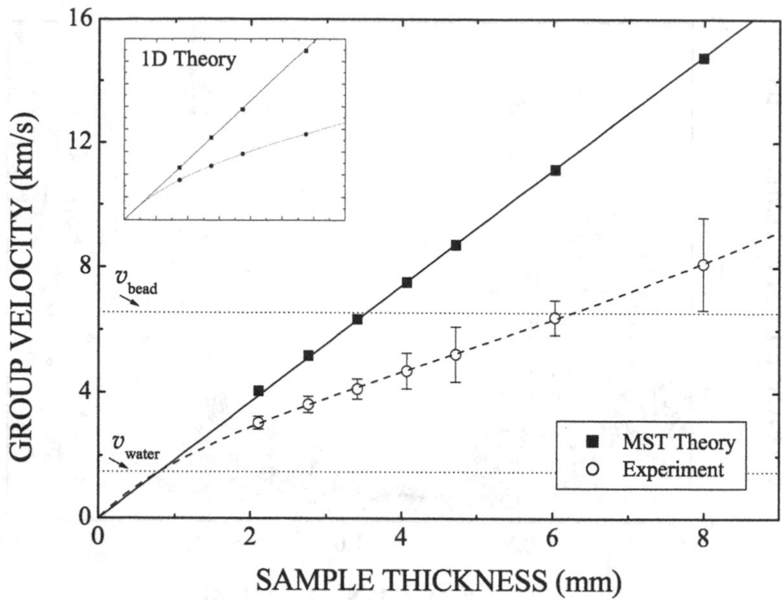

Figure 10. The group velocity as a function of sample thickness at 0.945 MHz. The horizontal dotted lines indicate the longitudinal velocities in water and tungsten carbide. The inset shows theoretical predictions in 1D for the group velocity with (solid circles) and without (solid squares) absorption. The dashed curves are fits of the two-modes model to the experimental data.

velocity can be estimated from the weighted average of the tunnelling time, $t_{\rm tun}$ and the propagation time $L/v_{\rm prop}$ [18]:

$$\bar{v}_g = \frac{L}{w_t t_{\rm tun} + w_p (L/v_{\rm prop})} \tag{4}$$

Note that $t_{\rm tun}$ and $v_{\rm prop}$, the group velocity of the propagating mode, are constant, independent of L. The weighting factors w_t and w_p are given by

$$w_t = \frac{c\ \exp[-L/(2l_{\rm tun})]}{c\ \exp[-L/(2l_{\rm tun})] + \sqrt{1-c^2}\ \exp[-L/(2l_{\rm prop})]} \tag{5}$$

$$w_p = \frac{\sqrt{1-c^2}\ \exp[-L/(2l_{\rm prop})]}{c\ \exp[-L/(2l_{\rm tun})] + \sqrt{1-c^2}\ \exp[-L/(2l_{\rm prop})]} \tag{6}$$

Here c is the coupling coefficient, and $l_{\rm tun}$ and $l_{\rm prop}$ are the extinction lengths of the tunnelling and propagating modes. Using the MST to calculate $t_{\rm tun} = 0.54\ \mu$s and $l_{\rm tun} = 0.54$ mm, and taking $v_{\rm prop} = 1.5$ km/s (the group velocity in the water matrix), we fit Eq. (4) to the experimental data in Fig. 10 with $l_{\rm prop}$ and c as the only free parameters. We find that the empirical parameter describing the decay of the propagating mode is very similar to the tunnelling mode ($l_{\rm prop} = 0.47$ mm), so that the weighting factors are almost independent of thickness, and that $c = 0.96$, confirming that tunnelling is the dominant component. It can be seen from Fig. 10 that this simple phenomenological model gives an excellent fit to the data over the entire range of thicknesses. To check the validity of this approach, we have also calculated the group velocity for a 1D phononic crystal in which absorption can be included rigorously. As shown in the inset to Fig. 10, absorption reduces the group velocity in the gap; furthermore, we can explain the reduction in terms of the two-modes model by fitting Eq. (4) to the calculation with absorption, giving the dotted curve shown in the inset. Thus, the two-modes model can successfully account for the effect of absorption in both cases. Note that the effects of absorption on wave transport in the gap are somewhat paradoxical, in the sense that absorption appears to modify tunnelling to produce a small-amplitude propagating component with a real wave vector, even though both absorption and tunnelling are themselves characterized by imaginary wave vectors.

It is important to recognize that the tunnelling time for ultrasonic waves in phononic crystals, both measured in these experiments and calculated from the MST theory, corresponds to the group delay time of a pulse. We have verified this directly by comparing the measured pulse delay time, defined as above by the time interval between the peaks of Gaussian input and transmitted pulses, and the group time $t_g = d\phi/d\omega$ determined by numerically differentiating the measured cumulative phase with respect to frequency. We find that the same values of $t_{\rm tun}$ are measured in both cases. Thus, some of the theoretical models of the tunnelling time [22–27], such as the 'dwell time' or the Büttiker-Landauer 'semi-classical' time [22, 26], do not apply here. Our results for the tunnelling time in the middle of the gap are summarized in Fig. 11, where we plot the tunnelling time as a function of the thickness of the crystals. Both theory and experiment approach an asymptotic limit for the thicker samples that is independent of L. Moreover, even though the times are short by ultrasonic standards for samples up to 8 mm thick, being about 1 μs or less, they are long enough to be easy to measure compared with the tunnelling times in optical band gap experiments, which are about 9 orders of magnitude shorter [28–30]. This large difference in the magnitude of the tunnelling times is a consequence of the relation that we find between the tunnelling time and the gap width $\Delta\omega_{\rm gap}$, a relationship that also holds for the group delay of light and electrons (where for electrons $\Delta\omega_{\rm gap}$ corresponds to the barrier height): $t_{\rm tun} \sim 1/\Delta\omega_{\rm gap}$ in the middle of the gap.

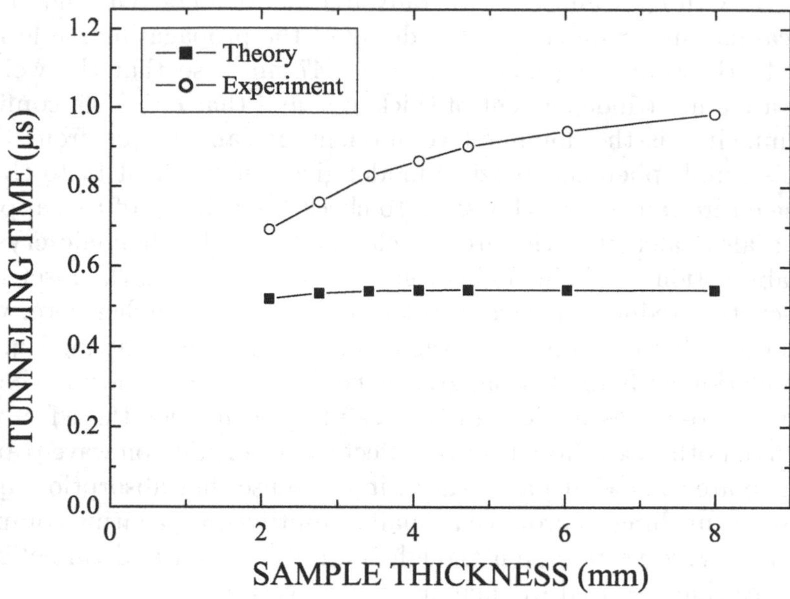

Figure 11. The tunnelling time as a function of sample thickness. For ultrasound tunnelling though a phononic crystal, the tunnelling time is equal to the group delay time.

The sample-thickness-independent tunnelling time observed in these and other experiments may be interpreted to imply that the group velocity can be greater than the speed of light for a sufficiently thick crystal (superluminal velocity). Does it violate causality? The answer is no. This superluminal phenomenon can be understood by a pulse reshaping process in which interference causes the later part of the pulse to be attenuated in such a way that the peak of the transmitted pulse shifts to an earlier time than that of the incident pulse. The energy transmitted at any time is much less than it would have been without the crystal in place; in fact for the values of the tunnelling time measured in our experiments, the thickness required for the group velocity to become superluminal is so great that the signal would be undetectably small (the relevant distance in everyday units amounts to the length of a football field!). Nonetheless, it may still be worth noting that the speed is 'supersonic' in the sense that it greatly exceeds the velocity of sound in water, which plays a role in these acoustic experiments that is analogous to the vacuum speed of light in optics experiments.

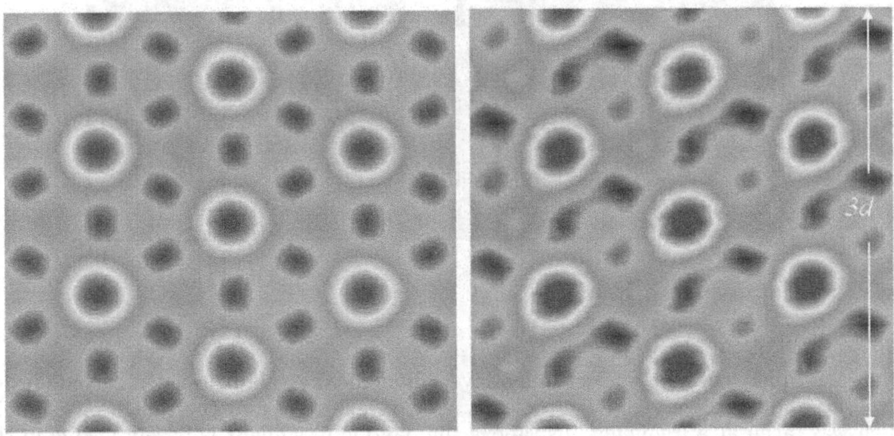

Figure 12. Theoretical (left) and experimental (right) near-field patterns for a 7-layer fcc crystal. The plane of the figure is perpendicular to the [111] direction. The height and width of the pictures are both equal to 3 bead diameters.

4. Near-field Ultrasonic Imaging of the Wave Field at the Surface of Phononic Crystals

In this section, we give some examples of the wave field close to the surface of the phononic crystals consisting of tungsten carbide beads in water. We use an ultrasound field pattern imaging technique, in which a plane-wave pulse is incident on one face of the crystal and the transmitted field is measured near the opposite face using a small hydrophone detector. By studying these wave field patterns, a clear picture can be obtained of the behaviour of the displacement field and the distribution of energy in perfect crystals, as well as monitoring the effects of defects in imperfect crystals.

The wave amplitude just above the surface of a phononic crystal is not uniform, but varies in periodic patterns that reveal the underlying structure of the crystal. Figure 12 shows our theoretical and experimental near-field wave patterns 3 mm away from the crystal surface for a 7-layer crystal. These field patterns were measured and calculated using the MST at 2.5 MHz, well above the complete band gap. Despite the fact that the measured pattern is slightly tilted, due to a very small drift in the 2D scanner

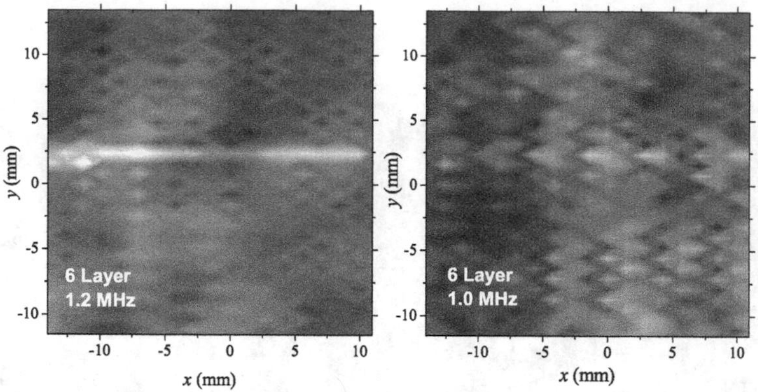

Figure 13. Image of a line defect in a 6-layer phononic crystal of tungsten carbide beads in water. The line defect shows up clearly as a bright streak in the image on the left, which was taken at a frequency just above the band gap, while it is barely visible in the image on the right, which was taken at a frequency in the gap.

used to position the hydrophone, good agreement between the theory and experiment is observed. The pattern shows the 3-fold symmetry expected in a plane perpendicular to the [111] direction, and suggests that measurements of such periodic near-field diffraction patterns could provide a novel way of determining crystal structures. The pattern varies rapidly as the frequency is varied, while always preserving the underlying symmetry, giving additional information on the interference of the multiply scattered wave field inside the crystal. For example, peaks in the field pattern at 2.4 MHz were found to become valleys at 2.8 MHz, reflecting the very different spatial distribution of the wave energy that that results from only a modest change in the frequency ($\sim 15\%$).

Most crystals are usually not perfect, and imaging the location of defects can be important for understanding and controlling their physical properties. An example of a subsurface defect, whose presence could not detected by visual inspection of the crystal, is shown in Fig. 13. Here the presence of a line defect in a 6-layer phononic crystal shows up clearly when imaged at frequencies above the gap (left), but is barely detectable in the gap (right), illustrating the increased sensitivity to this type of defect in the pass band. As another example of imaging defects, we also studied a point defect (vacancy) in the middle of the top layer of the 7 layer phononic crystal. The measured wave field patterns both before and after removing one of the beads are shown in Fig. 14. It can be seen that the field pattern of the

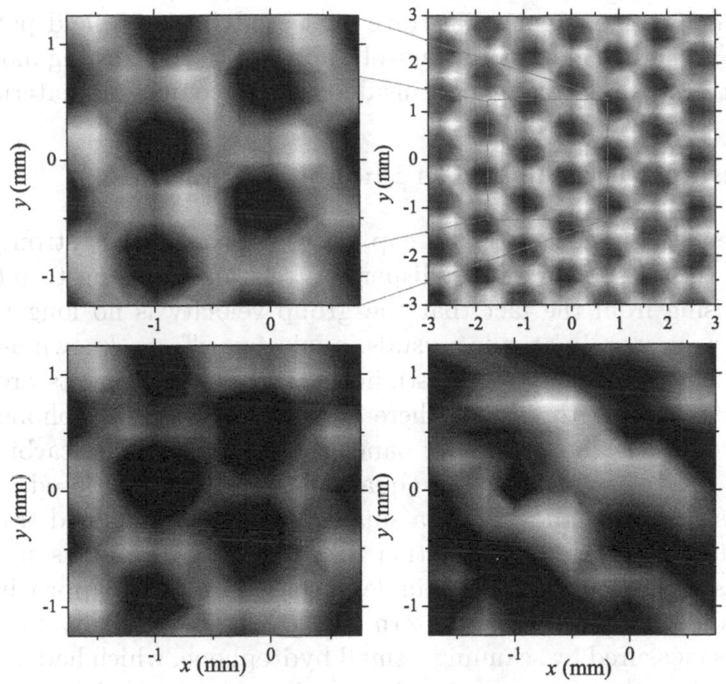

Figure 14. Observation of a point defect or vacancy created by removing one bead from the top layer of a 7-layer phononic crystal. The perfect lattice is shown in the top two images (the left picture being a magnified portion of the right picture). The lower left image shows the change in the field pattern as a result of removing the bead, while the difference (perfect crystal image minus the defect crystal image) is shown at the lower right. White corresponds to maximum amplitude and black to minimum amplitude.

'perfect' crystal has almost perfect periodicity (top right). For the crystal with the defect, the pattern changes around the vacancy, as can be see by comparing the magnified images before and after the defect was removed, (top and bottom left images in the figure). The field of view in the magnified images is approximately three bead diameters across, and the vacancy is located close to the centre. This figure shows that the effect of creating the vacancy is not to change the local symmetry of the field pattern, or to remove any of the features, but to substantially modify the amplitude of the field pattern in the vicinity of the defect. Rather surprisingly, the wave amplitude above the defect is smaller than for the perfect crystal, which is counterintuitive since one would naively imagine that removing a scatterer from the surface would enhance the field locally. To show this effect more clearly, the bottom right picture shows the difference between these two field patterns, providing a clearer indication of the location of the defect. These examples illustrate that studying such field patterns in phononic crystals may provide a novel opportunity for learning more about wave scattering and propagation inside periodic composite materials.

5. Focusing Ultrasound with Phononic Crystals

At frequencies outside the band gap, wave propagation is strongly influenced by the anisotropy of the dispersion relations, leading to interesting effects arising from the fact that the group velocity is no longer parallel to the wave vector. In atomic crystals, analogous effects, known as phonon focusing, have been extensively studied [31], but experiments are limited to the long wavelength regime where $\lambda \gg a$. To investigate phonon focusing at frequencies above the first band gap in the tungsten carbide/water phononic crystal, we replaced the quasi plane-wave source used in the experiments described above with a small-diameter disk-shaped transducer, which acts as a good approximation to a point source. The source transducer was placed close to the sample surface (3 mm, or approximately 2 wavelengths, away). The field pattern on the far side of the crystal and substrate was measured by scanning a small hydrophone, which had a diameter much less than the ultrasonic wavelength, in a plane parallel to the crystal surface. The experiments were performed with pulses. By taking FFTs of each pulse, the amplitude of the transmitted field at any frequency in the bandwidth of the pulse could then be measured as a function of position in the detecting plane.

As the frequency is varied, the detected field pattern varies widely, a result of the rapid changes in the anisotropy of the dispersion relations with frequency that take place in the higher pass bands (see Fig. 2 and Refs. [32]). One example of the measured field patterns for a 12-layer tungsten-carbide/water crystal is shown in Fig. 15(a); this pattern was measured at 1.57 MHz, which corresponds to a reduced frequency of 1.2 near the bottom of the third pass band along the ΓL direction [Fig. 2(b)]. The figure shows that the diverging beam from the source transducer, which has a FWHM in the detection plane of 65 mm (more than three quarters of the width of the region imaged in the figure), is sharply focussed to a tight spot with a FWHM of only 5 mm. Thus, it is clear from this example that a phononic crystal with flat parallel planar faces can be used to focus ultrasound. One remarkable feature of these data is that such a sharp focal spot is observed in a plane quite far from the crystal; the distance from the crystal face to the detection plane is approximately 130 wavelengths, while the distance from the source to the crystal is only 2 wavelengths.

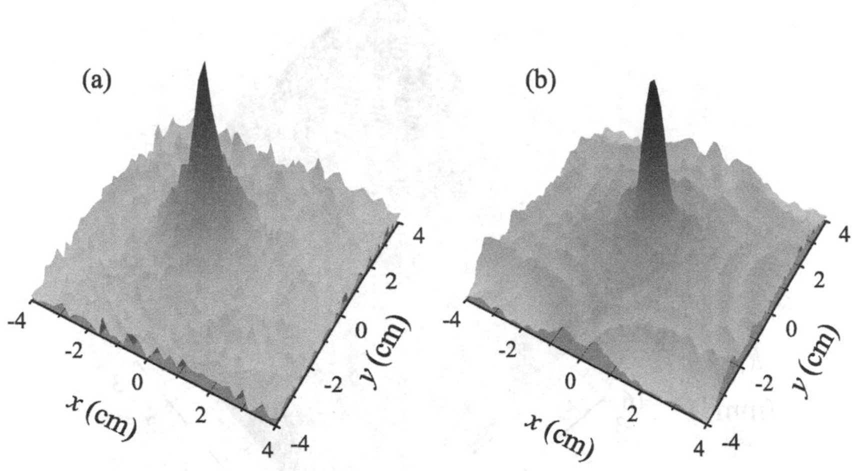

Figure 15. The spatial variation of the wave amplitude in a plane parallel to the surface of a tungsten-carbide/water phononic crystal when an incident point-like source is placed on the opposite side of the crystal. The frequency is 1.57 MHz. Instead of the diverging beam seen without the crystal in place, a tightly focused spot is observed. Fig. (a) shows the experimental data and (b) shows the theoretical prediction using a Fourier imaging technique, in which wave propagation through the crystal is described by the dispersion surface calculated using the MST.

The origin of the focusing effect can be understood from the dispersion (or slowness) surface, which, for constant frequency, represents the variation in the magnitude of the wave vector with direction. To interpret our data, we calculated the dispersion surface using the MST, solving for the magnitudes of the wave vectors along different wave vector directions at a particular frequency, in the same way as the band structure is calculated for different eigenfrequencies along particular wave vector directions. To represent forward propagation through the crystal at frequencies near 1.6 MHz, the dispersion surface calculated in the reduced Brillouin zone was translated to the extended zone, the correct translation being from $\vec{k}$ to $\vec{k}-2\vec{G}_{111}$ in this case. Figure 16 shows a 3D plot of the dispersion surface for wave vectors near the ΓL (or [111]) direction. It can be seen that the dispersion surface has a pronounced minimum for $\vec{k}$ parallel to the [111] direction.

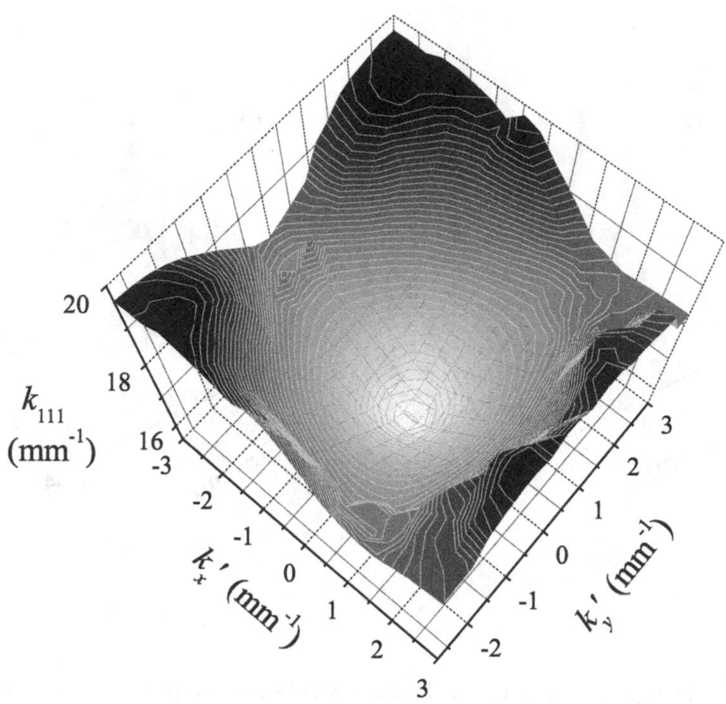

Figure 16. A 3D plot of the dispersion surface for wave vectors near the ΓL ([111]) direction. The dispersion surface is shown in the extended zone scheme, which also corresponds to part of the periodic zone scheme representation centred at $2G_{\Gamma L}$. The directions of k_x' and k_y' are parallel to the LK and LW directions, respectively. The frequency used in these calculations corresponds to the same frequency relative to the band edge as in the experimental data shown in Fig. 15.

Consequently, the normal to the dispersion surface, which represents the direction of the group velocity and hence the direction in which energy is transported inside the crystal, points back towards the [111] direction as the direction of the wave vector moves away from the [111] direction. This large anisotropy results in 'negative refraction' without a negative refractive index; this occurs since the net wave transport inside the crystal follows the directions of the group velocity, which points in directions that correspond to negative angles of refraction. However, it is important to remember that this is not a refraction phenomenon in the usual sense, since its origin is the direction of the group velocity and not the wave vector. The dip in the dispersion surface shown in Fig. 16 is quite deep, resulting in large effective 'negative refraction'; as a result, it is possible for the waves passing through the crystal to form a tight focal spot quite far away from the crystal even when the crystal is only 12 layers thick.

To interpret our data quantitatively, we calculated the field pattern corresponding to the data shown in Fig. 15(a) using a Fourier imaging technique. After the input beam is Fourier transformed spatially into plane waves, each plane wave is allowed to propagate through the crystal according to the wave vectors determined by its dispersion surface, accounting correctly for refraction at the front and back interfaces (i.e. the components of the wave vector parallel to the surface remain the same inside and outside the crystal). The field pattern in the detecting plane is then reconstructed by taking an inverse Fourier transform back into real space, giving the results shown in Fig. 15(b). Excellent agreement with experiment is seen, confirming that this model correctly incorporates the essential physics of this wave focusing phenomenon.

6. Conclusions

Ultrasonic pulse propagation in phononic crystals is making an important contribution to understanding wave transport in ordered structures. Because the wave field is measured directly in ultrasonic experiments, a complete picture of wave propagation is accessible, allowing the transmission coefficient, the dispersion relations and the dynamics of the wave fields to be investigated. In this paper, we have concentrated on two different types of wave phenomena in 3D phononic crystals: ultrasound tunnelling and focusing by 'negative refraction'. Some examples of near-field imaging experiments have also been presented.

We have summarized recent progress in demonstrating the tunnelling of ultrasonic wave pulses through a complete band gap, showing that remarkably large values of the group velocity can be measured, as the group velocity is proportional to sample thickness. The classic signature of tunnelling, that the tunnelling time is independent of sample thickness for thick samples, has been demonstrated, and its magnitude found to be equal to the reciprocal of the gap width in the middle of the gap. The experimental results have been interpreted using the Multiple Scattering Theory, which was found to give good overall agreement. The counterintuitive effects of absorption on the tunnelling dynamics were also investigated and interpreted using a simple 'two-modes' model; this model shows that the effect of absorption on evanescent waves in band gap materials is to modify their character by introducing a small propagating component.

The concave character of the dispersion surfaces in phononic crystals (as viewed looking back towards the origin) has been shown to lead to the focusing of ultrasonic waves from a diverging source in the absence of the usual curved surfaces employed in traditional lenses. Again the MST theory provides an excellent basis for understanding the experimental results, and gives predictions that are in good agreement with experiment. We are currently examining the behaviour at other frequencies and extending this work to solid 3D phononic crystals, which are easier to handle and more useful for practical applications.

7. Acknowledgements

Support from the Natural Sciences and Engineering Research Council of Canada and the RGC (HKUST6143/00P) of Hong Kong is gratefully acknowledged.

References

1. M. Sigalas and E. N. Economou, *Solid State Communications* **86**, 141 (1993).
2. E. N. Economou and M. Sigalas, *J. Acoust. Soc. Am.* **95**, 1734 (1994).
3. M. S. Kushwaha, P. Halevi, and G. Martinez, *Phys. Rev. B* **49**, 2313 (1994).
4. R. Mártinez-Sala, J. Sancho, J. V. Sánchez, J. Linres, and F. Mesegure, *Nature* **378**, 241 (1995).
5. F. R. Montero de Espinosa, E. Jiménez, and M. Torres, *Phys. Rev. Lett.* **80**, 1208 (1998).
6. J. O. Vasseur, P. A. Deymier, G. Frantziskonis, G. Hong, B. Dijafari-Rouhani, and L. Dobrzynski, *J.Phys.: Condens. Matter* **10**, 6051 (1998).
7. M. Torres, F. R. Montero de Espinosa, D. García-Pablos, and N. García, *Phys. Rev. Lett.* **82**, 3054 (1999).
8. M. Kafesaki and E. N. Economou, *Phys. Rev. B* **60**, 11993 (1999).
9. I. E. Psarobas, N. Stefanou, and A. Modinos, *Phys. Rev. B* **62**, 278 (2000).
10. Z. Liu, C. T. Chan, P. Sheng, A. L. Goertzen, and J. H. Page, *Phys. Rev. B* **62**, 2446 (2000).
11. Z. Liu, X. Zhang, Y. Mao, Y. Y. Zhu, Z. Yang, C. T. Chan, and P. Sheng, *Science* **289**, 1734 (2000).
12. J. H. Page, A. L. Goertzen, S. Yang, Z. Liu, C. T. Chan, and P. Sheng, in *Photonic Crystals and Light Localization in the 21st Century*, edited by C. M. Soukoulis (Kluwer Academic Publishers, Amsterdam, 2001), p. 59.
13. M. Torres, F. R. Montero de Espinosa, and J. L. Aragón, *Phys. Rev. Lett.* **86**, 4282 (2001).
14. Suxia Yang, J. H. Page, Zhengyou Liu, M. L. Cowan, C. T. Chan, and Ping Sheng, *Phys. Rev. Lett.* **88**, 104301 (2002).
15. E. Yablonovitch, *Phys. Rev. Lett.* **58**, 2059 (1987).
16. S. John, *Phys. Rev. Lett.* **58**, 2486 (1987).
17. *Photonic Crystals and Light Localization in the 21st Century*, edited by C. M. Soukoulis (Kluwer Academic Publishers, Amsterdam, 2001)
18. Suxia Yang, *Ph.D Thesis*, (The Hong Kong University of Science and Techology, 2002)
19. As is well known, the first Brillouin zone for the fcc structure is closer to a spherical shape than for hcp, and this facilitates the formation of a complete gap.
20. For a review of tunneling times with particular reference to photon tunnelling, see R. Y. Chiao and A. M. Steinberg, in *Progress in Optics*, edited by E. Wolf, Vol. 37, (Elsevier, Amsterdam, 1997), p. 347.
21. A rigorous first-principles calculation of ultrasonic absorption in the crystal remains

a challenging problem. Mechanisms involving interfacial effects are difficult to incorporate in the MST model because of problems with an ill-conditioned matrix when the effects of viscosity in the liquid are included.

22. M. Büttiker and R. Landauer, *Phys. Rev. Lett.* **49**, 1739 (1982).
23. E.H. Hauge, J.P. Falck and T.A. Fjeldly, *Phys. Rev. B* **36**, 4203 (1987).
24. E.H. Hauge and J.A.Støvneng, *Rev. Mod. Phys.* **61**, 917 (1989).
25. R. Landauer, *Nature* **341**, 567 (1989).
26. M. Büttiker, in *Electronic Properties of Multilayers and Low-Dimensional Semiconductor Structures*, edited by J. M. Chamberlain, *et al.* (Plenum Press, New York, 1990) p. 297
27. R. Y. Chiao and A. M. Steinberg, in *Progress in Optics*, edited by E. Wolf, Vol. 37, (Elsevier, Amsterdam, 1997), p. 347.
28. A. M. Steinberg, P. G. Kwist, and R. Y. Chiao, *Phys. Rev. Lett.* **71**, 708 (1993).
29. C. Spielmann, R. Szipocs, A. Stingl, and F. Krausz, *Phys. Rev. Lett.* **73**, 2308 (1994).
30. M. Mojahedi, E. Schamiloglu, F. Hegeler, and K. J. Malloy, *Phys. Rev. E* **62**, 5758 (2000).
31. J. P. Wolfe, *Imaging Phonons: Acoustic Wave Propagation in Solids* (Cambridge University Press, Cambridge, 1998).
32. H. Kosaka, T. Kawashima, A. Tomita, M. Notomi, T. Tamamura, T. Sato and S. Kawakami, *Phys. Rev. B* **58**, 10096 (1998); *Appl., Phys. Lett.* **74**, 1212 (1999); *Appl., Phys. Lett.* **74**, 1370 (1999).

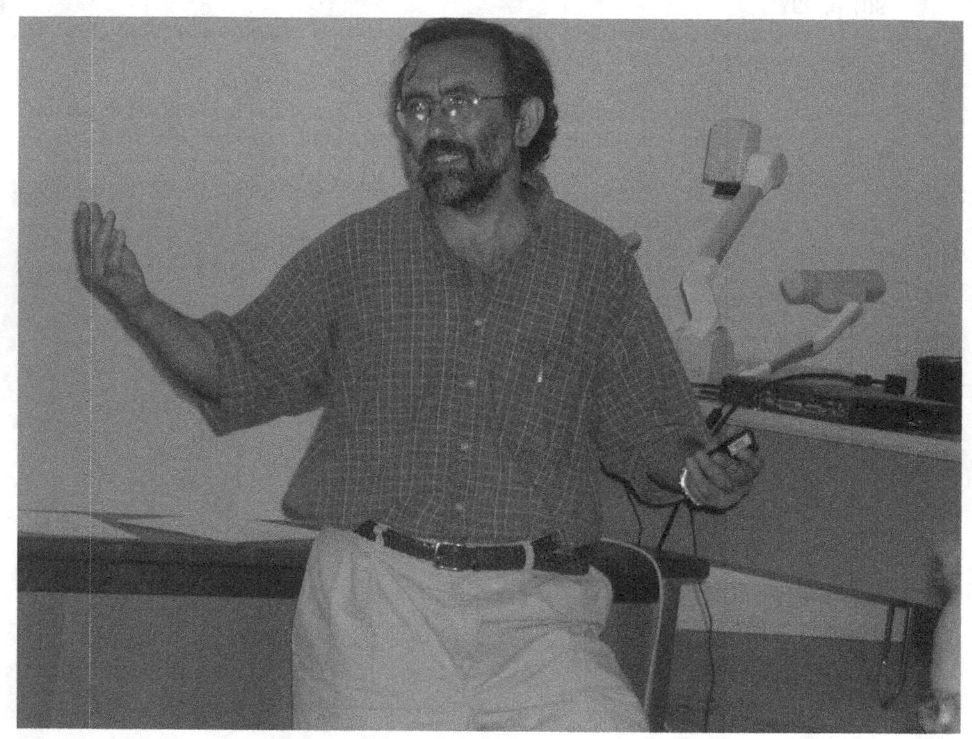

Costas Soukoulis. Photo by Valentin Freilikher

LEFT-HANDED MATERIALS

P. MARKOŠ[1,2] AND C.M. SOUKOULIS[1,3]
[1] *Ames Laboratory and Department of Physics and Astronomy Iowa State University, Ames, Iowa 50011*
[2] *Institute of Physics, Slovak Academy of Sciences Dúbravská cesta 9, 842 28 Bratislava, Slovakia*
[3] *Research Center of Crete 71110 Heraklion, Crete, Greece*

1. Introduction

Rapidly increasing interest in the left-handed materials (LHM) started after Pendry *et al.* predicted that certain man-made composite structure could possess, in a given frequency interval, a negative *effective* magnetic permeability μ_{eff} [1]. Combination of such a structure with negative effective permittivity medium — for instance the regular array of thin metallic wires [2–7] — enabled the construction of meta-materials with both *effective* permittivity and permittivity *negative*. This was confirmed by experiments [8, 9].

Structures with negative permittivity and permittivity were named "left-handed" by Veselago [10] over 30 years ago to emphasize the fact that the intensity of the electric field **E**, the magnetic intensity **H** and the wave vector **k** are related by a left-handed rule.[1] This can be easily seen by writing Maxwell's equation for a plane monochromatic wave: $\mathbf{k}\times\mathbf{E} = \frac{\omega\mu}{c}\mathbf{H}$ and $\mathbf{k}\times\mathbf{H} = -\frac{\omega\epsilon}{c}\mathbf{E}$ Once ϵ and μ are both positive, then **E**, **H** and **k** form a right set of vectors. In the case of negative ϵ and μ, however, these three vectors form a left set of vectors.

In his pioneering work, Veselago described the physical properties of LH systems: Firstly, the direction of the energy flow, which is given by the Poynting vector

$$\mathbf{S} = \frac{c}{4\pi}\,\mathbf{E}\times\mathbf{H} \tag{1}$$

[1]Besides the name "Left handed materials" (metamaterials), this new structures are also called "Negative index materials" (NIM), "Double negative metamaterials" (DNM) or "Backward-wave medium" in the literature.

B.A. van Tiggelen and S.E. Skipetrov (eds.),
Wave Scattering in Complex Media: From Theory to Applications, 309–329.

does not depend on the sign of the permittivity and permeability of the medium. Then, the vectors **S** and **k** are parallel (anti-parallel) in the right-handed (left-handed) medium, respectively. Consequently, the phase and group velocity of an electromagnetic wave propagate in *opposite* directions in the left-handed material. This gives rise to a number of novel physical phenomena, as were discussed already by Veselago. For instance, the Doppler effect and the Cherenkov effect are reversed in the LHM [10].

If both ϵ and μ are negative, then also the refraction index n is negative [10, 11]. This means the negative refraction of the electro magnetic wave passing through the boundary of two materials, one with positive and the second with negative n (negative Snell's law). Observation of negative Snell's law, reported experimentally [12] and later in [13], is today a subject of rather controversially debate [14–17]. Analytical arguments of the sign of the refraction index were presented in [11]. Numerically, negative phase velocity was observed in FDTD simulations [18]. Negative refraction index was calculated from the transmission and reflection data [19]. Finally, negative refraction on the wedge experiment was demonstrated also by FDTD simulations [20].

Negative refraction allows the fabrication of flat lens [10]. Maybe the most challenging property of the left-handed medium is its ability to enhance the evanescent modes [21]. Therefore flat lens, constructed from left-handed material with $\epsilon = \mu = -1$ could in principle work as perfect lens [21] in the sense that it can reconstruct an object without any diffraction error.

The existence of the perfect lens seems to be in contradiction with fundamental physical laws, as was discussed in a series of papers [14, 22, 23]. Nevertheless, more detailed physical considerations [15, 24, 25, 27, 28] not only showed that the construction of "almost perfect" lens is indeed possible, but brought some more insight into this phenomena [29-35].

As Veselago also discussed in his pioneering paper, the permittivity and the permeability of the left-handed material must depend on the frequency of the EM field, otherwise the energy density [36]

$$U = \frac{1}{2\pi} \int d\omega \left[\frac{\partial(\omega\epsilon')}{\partial\omega} |E|^2 + \frac{\partial(\omega\mu')}{\partial\omega} |H|^2 \right] \tag{2}$$

would be *negative* for *negative* ϵ' and μ' (real part of the permittivity and permeability). Then, according to Kramers-Kronig relations, the imaginary part of the permittivity (ϵ'') and of the permeability (μ'') are non-zero in the LH materials. Transmission losses are therefore unavoidable in any LH structure. Theoretical estimation of losses is rather difficult problem, and led even to the conclusion that LH materials are not transparent [16]. Fortunately, recent experiments [13, 37] confirmed the more optimistic theo-

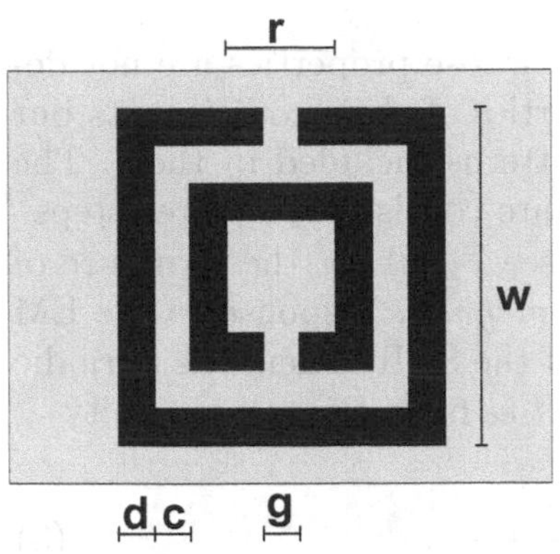

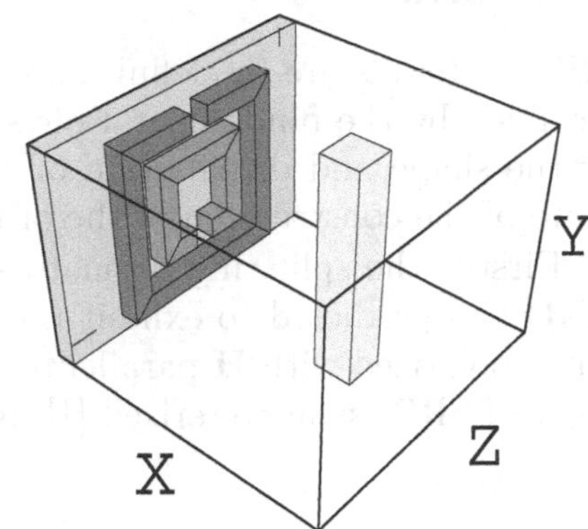

Figure 1. Left: Structure of the split ring resonator (SRR). The SRR consists of two splitted metallic "rings". The SRR is characterized by the size w, width of the rings d, and two gaps: g and c. The external magnetic field induces an electric current in both rings [1]. The shape of the SRR (square or circular) is not crucial for the existence of the magnetic resonance. Right: Structure of the unit cell of the left-handed material. Each unit cell contains the split ring resonator located on the dielectric board, and one wire. Left-handed structure is created by regular lattices of unit cells. The EM wave propagates along the z direction. Periodic boundary conditions are considered in the x and the y direction, which assures the periodic distribution of the EM field.

retical expectation [38], that the losses in the LH structures might be as small as in conventional RH materials.

The number of papers about left-handed materials increased dramatically last year. The present paper is not the first review about left-handed materials. For recent reviews, see [39] or the papers of Pendry [40]. Here we present typical structures of the left-handed materials (Sect. 2), discuss a numerical method of simulation of the propagation of EM waves based on the transfer matrix (Sect. 3), and present some recent results obtained by this method (Sects. 4, 5). We discuss in Sect. 5 how the transmission depends on various structural and material parameters of LH structure. The method of calculation of the refractive index and of the effective permittivity and permeability is presented and applied to the LH structure. An unambiguous proof of the negative refraction index is given and the *effective* permittivity and permeability are calculated in Sects. 6, and 7. The obtained data for the permittivity are rather counter-intuitive and require some physical interpretation. Finally, in the Section 8 we discuss some new directions of the development of both theory and experiments.

2. Structure

LHM materials are by definition composites, whose properties are not determined by the fundamental physical properties of their constituents but by the shape and distribution of specific patterns included in them. The route of the construction of the of LH structure consists from three steps.

Firstly, the split ring resonators (SRR) (see Fig. 1 for the structure of SRR) was predicted to exhibit the resonant *magnetic* response to the EM wave, polarized with **H** parallel to the axis of the SRR. Then, the periodic array of SRR is characterized [1] by the *effective* magnetic permeability

$$\mu_{\rm eff}(f) = 1 - \frac{F\nu^2}{\nu^2 - \nu_m^2 + i\nu\gamma} \,. \tag{3}$$

In (3), ν_m is the resonance frequency which depends on the structure of the SRR (Fig. 1) as $(2\pi\nu_m)^2 = 3L_x c_{\rm light}^2/[\pi \ln(2c/d)r^3]$. F is the filling factor of the SRR within one unit cell and γ is the damping factor $2\pi\gamma = 2L_x\rho/r$, where ρ is the resistivity of the metal.

Formula (3) assures that the *real* part of $\mu_{\rm eff}$ is *negative* at an interval $\Delta\nu$ around the resonance frequency. If an array of SRR is combined with a medium with negative *real* part of the permittivity, the resulting structure would possess *negative effective* refraction index in the resonance frequency interval $\Delta\nu$ [11].

The best candidate for the negative permittivity medium is a regular lattice of thin metallic wires, which acts as a high pass filter for the EM wave polarized with **E** parallel to the wires. Such an array exhibits negative effective permittivity

$$\epsilon_{\rm eff}(\nu) = 1 - \frac{\nu_p^2}{\nu^2 + i\nu\gamma} \tag{4}$$

[2, 4, 6] with the plasma frequency $\nu_p^2 = c_{\rm light}^2/(2\pi a^2 \ln(a/r))$ [2]. Sarychev and Shalaev derived another expression for the plasma frequency, $\nu_p = c_{\rm light}^2/(\pi a^2 \mathcal{L})$ with $\mathcal{L} = 2\ln(a/\sqrt{2}r) + \pi/2 - 3$ [6]. Apart from tiny differences in both formulas, the two theories are equivalent [7] and predict that effective permittivity is *negative* for $\nu < \nu_p$.

By combining both the above structures, a left-handed structure can be created. This was done for the first time in the experiments of Smith *et al.* [8]. Left-handed material is a periodic structure. A typical unit cell of the left-handed structure is shown in Fig. 1. Each unit cell contains a metallic wire and one split ring resonator (SRR), deposited on the dielectric board.

Fig. 2 shows the transmission of the EM waves through the left-handed structure discussed above. The transmission through the array of the SRR

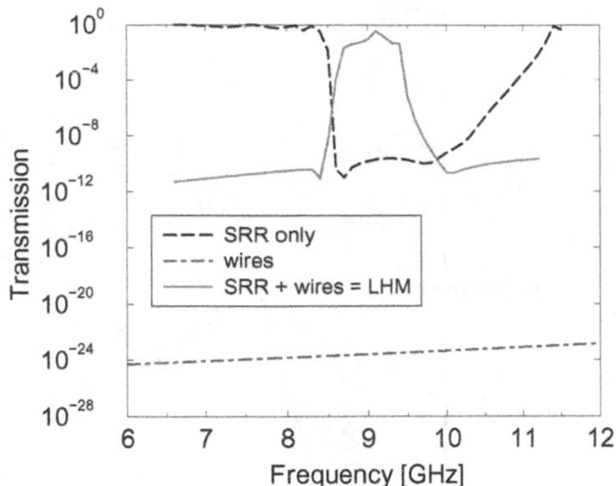

Figure 2. Transmission of the EM wave, polarized with $\mathbf{E} \parallel y$ and $\mathbf{H} \parallel x$, through a periodic array of split ring resonators, wires, and of both SRR and wires.

is close to unity for all the frequencies outside the resonance interval (8.5-11 GHz in this particular case) and decreases to -120 dB in this interval, because μ_{eff} is negative [Eq. (3)]. The transmission of the array of metallic wires is very small for all frequencies below the plasma frequency (which is ~ 20 GHz in this case), because ϵ_{eff} is negative [Eq. (4)]. The structure created by the combination of an array of SRR and wires exhibits high transmission $T \sim 1$ within the resonance interval, where both ϵ_{eff} and μ_{eff} are negative. For frequencies outside the resonance interval, the product $\mu_{\text{eff}}\epsilon_{\text{eff}}$ is negative. The transmission decays with the system length, and is only ~ -120 dB in the example of Fig. 2. Experimental analysis of the transmission of all the three structures was performed by Smith *et al.* [8].

We want to obtain a resonance frequency $\nu_m \approx 10$ GHz. This requires the size of the unit cell to be 3-5 mm. The wavelength of the EM wave with frequency ~ 10 GHz is ≈ 4 cm, and exceeds by a factor of 10 the structural details of the left-handed materials. We can therefore consider the left-handed material as macroscopically homogeneous. This is the main difference between the left-handed structures and the "classical" photonic band gap (PBG) materials, in which the wave length is comparable with the lattice period.

It is important to note that the structure described in Fig. 1 is strongly anisotropic. For frequencies inside the resonance interval, the effective ϵ_{eff} and μ_{eff} are negative only for EM field with $\mathbf{H} \parallel x$ and $\mathbf{E} \parallel y$. The left-handed properties appear only when a properly polarized EM wave propagates in the z direction. The structure in Fig. 1 is therefore *effectively* one-dimensional. Any EM waves, attempting to propagate either along the x or along the y direction would decay exponentially since the correspond-

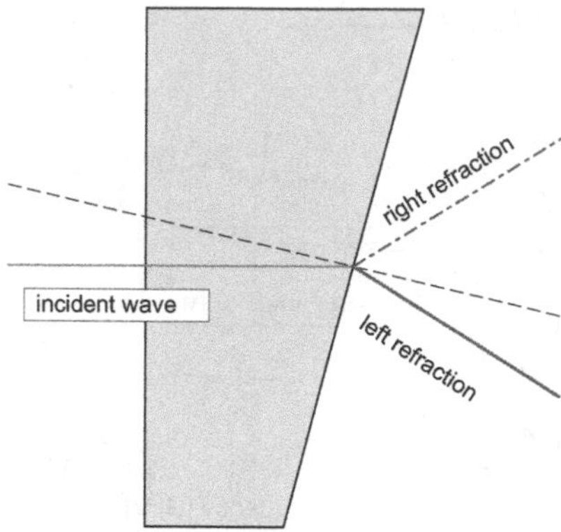

Figure 3. Refraction experiment on the left-handed material [12, 37]. The incident EM wave propagates from the left and hits perpendicularly the left boundary of the wedge. The angle of refraction is measured when the EM wave passes the right boundary of the inspected material and propagates for some time in the air. Two possible directions of the propagation of the refracted wave are shown: the right refraction for the conventional right-handed (RH) material, and left refraction for the left-handed material. This experimental design enables to use also strongly anisotropic one-dimensional LH samples, since the angle of refraction is measured *outside* the sample.

Figure 4. Two positions of the metallic wire in the unit cell. Left: wire is deposited on the *opposite* side of the dielectric board [9, 13]. Right: wire is located along the split-ring resonator [41].

ing product $\epsilon_{\text{eff}}\mu_{\text{eff}}$ is negative. This structure is therefore not suitable for the realization of the perfect lens. To test the negative Snell's law experimentally, a wedge type of experiment must be considered [12] (Fig. 3) in which the angle of refraction is measured *outside* the left-handed medium (in air) [12, 13].

Two dimensional structures have also been constructed. For instance, the anisotropy in the $x - z$ plane is removed if each unit cell contains two SRR located in two perpendicular planes [9, 12]. No three dimensional structures have been experimentally prepared yet.

It is also worth mentioning that some other one dimensional structures were prepared, in which the wires were deposited on the same side of the dielectric board with the SRRs. The wires could be located either on the opposite side of the dielectric board, as it was done in [9, 12, 37], or put next to the SRRs on the same side of board [41] as shown in the right panel of Fig. 4. More complicated one-dimensional structure was suggested by Ziolkowski [42]. Recently, Marques *et al.* found left-handed behavior in an array of SRR located inside a metallic wave guide [43].

3. Numerical Simulation

Various numerical algorithms were used to simulate the propagation of EM waves through the LH structure. We concentrate on the transfer matrix algorithm, developed in a series of papers by Pendry and co-workers [44]. The transfer matrix algorithm enables us to calculate the transmission, reflection, and absorption as a function of frequency [45, 46]. Others use commercial software: either Microwave studio [13, 47, 37] or MAFIA [8, 39], to estimate the position of the resonance frequency interval. Time-dependent analysis, using various forms of FDTD algorithms are also used [13, 18, 42, 48, 49].

In the transfer matrix algorithm, we attach in the z direction, along which EM wave propagates, two semi-infinite ideal leads with $\epsilon = 1$ and $\mu = 1$. The length of the system varies from 1 to 300 unit cells. Periodic boundary conditions along the x and y directions are used. This makes the system effectively infinite in the transverse directions, and enables us to restrict the simulated structure to only one unit cell in the transverse directions.

A typical size of the unit cell is 3.66 mm. Because of numerical problems, we are not able to treat very thin metallic structures. While in experiments the thickness of the SRRs is usually 17 μm, the thickness used in the numerical calculations is determined by the minimal mesh discretization, which is usually $\approx$ 0.33 mm. In spite of this constrain, the numerical data are in qualitative agreement with the experimental results. This indicates that the thickness of the SRR is not a crutial parameter, unless it decreases below or is comparable with the skin depth δ. As $\delta \approx 0.7\mu$m at GHz frequencies of interest, we are far from this limitation. We will discuss the role of discretization in Section 5.

4. Transmission

As discussed in Section 2 the polarization of the EM waves is crucial for the observation of the LH properties. The electric field E must be parallel to the wires, and the magnetic field H must be parallel to the axis of the SRRs.

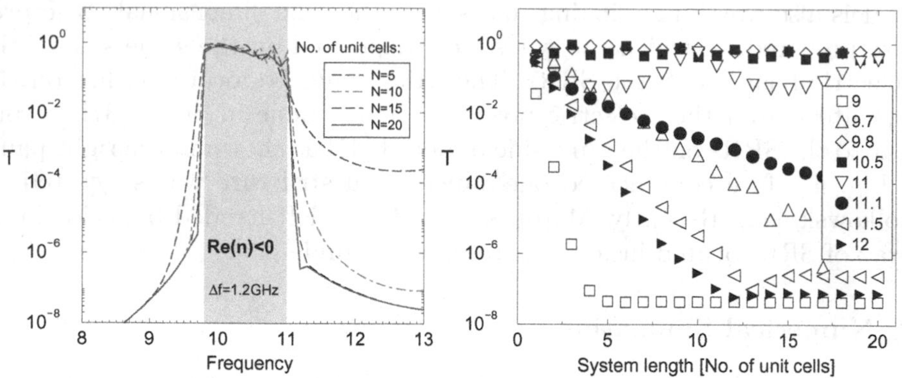

Figure 5. Transmission power T through the left-handed metamaterial of various lengths. Left: frequency dependence, right: length dependence.

In the numerical simulations, we treat simultaneously both polarizations, $\mathbf{E} \parallel x$ as well as $\mathbf{E} \parallel y$. Due to the non-homogeneity of the structure, these polarizations are not separated: there is always non-zero transmission t_{xy} from the x to y polarized wave. As we will see later, this effect is responsible for some unexpected phenomena. At present, we keep in mind that they must be included into the formula for absorption

$$A_x = 1 - |t_{xx}|^2 - |t_{xy}|^2 - |r_{xx}|^2 - |r_{xy}|^2 \tag{5}$$

and in the equivalent relation for A_y.

Figure 5 shows typical data for the transmission in the resonance frequency region. A resonance frequency interval, in which the transmission increases by many orders of magnitude is clearly visible. Of course, high transmission does not guarantee negative refraction index. The sign of n must be obtained by other methods, which will be described in Sect. 6.

In contrast to the original experimental data, numerical data show very high transmission, indicating that LH structures could be as transparent as the "classical" right-handed ones. This is surprising, because due to the dispersion, high losses are expected. Fig. 6 shows the transmission as the function of the system length, for three different frequencies inside the resonance frequency interval.

Figure 5 shows also that the transmission never decreases below a certain limit. Due to the non-homogeneity of the structure there is a non-zero probability t_{yx} that the EM wave, polarized with $\mathbf{E} \parallel y$, is converted into the polarization $\mathbf{E} \parallel x$. The total transmission t_{yy} consists therefore not only from the "unperturbed" contribution $t_{yy}^{(0)}$, but also from additional terms, which describe the conversion of the y-polarized wave into x-polarized and

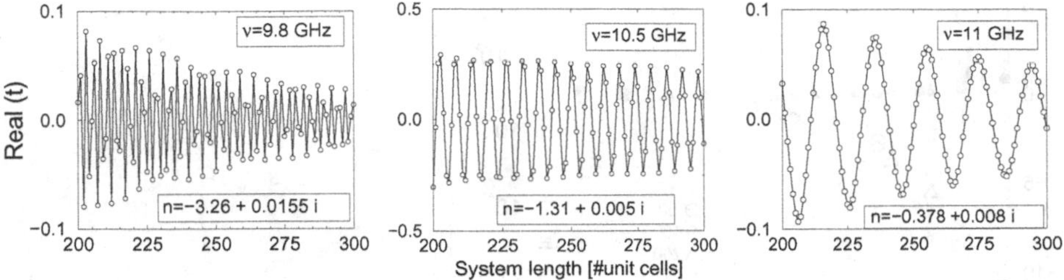

Figure 6. System length dependence of the real part of the transmission through the left-handed structure for three various frequencies within the left-handed band. The system length is given as the number of unit cells in the propagation direction. Note the different scale on the y axis for the three cases.

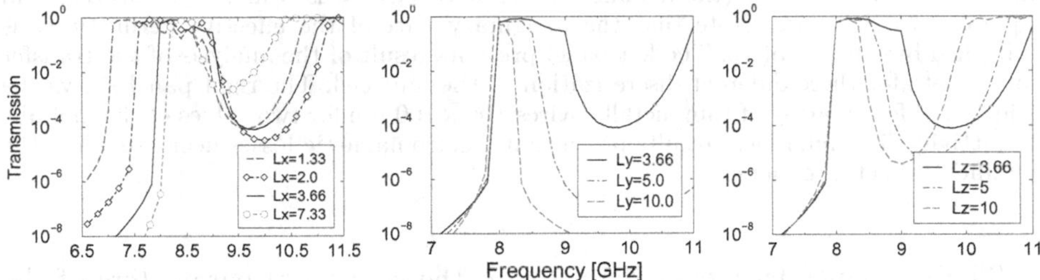

Figure 7. The transmission peak for various sizes of the unit cell. In contrast to the structure shown in Fig. 1, the metallic wire is located on the opposite side of the dielectric board. This enables to compress the width of the unit cell as shown in the far left panel.

back to y-polarized:

$$t_{yy}(0,L) = t_{yy}^{(0)}(0,L) + \sum_{z,z'} t_{xy}(0,z)t_{xx}(z,z')t_{yx}(z'L) + \dots \quad (6)$$

$t_{yy}^{(0)}(0,L)$ decreases exponentially with the system length L, while the second term, which represents the conversion of the y-polarized wave into x-polarized wave and back, remains system-length independent, because $t_{xx}(z,z') \sim 1$ for any distance $|z-z'|$.

5. Structure and Transmission

Numerical simulations confirm the existence of the resonance left-handed frequency interval. Before we proceed in the calculation of the effective system parameters, let us briefly discuss how the structure of the unit cell influences the position and the width of the resonance interval.

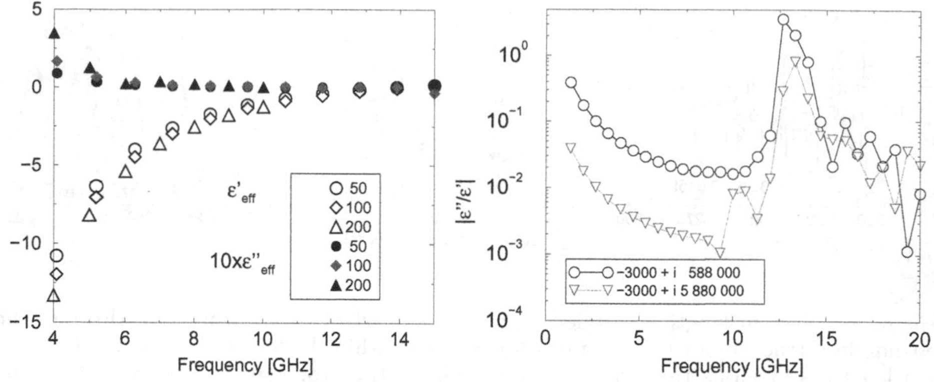

Figure 8. Left: effective permittivity $\epsilon_{\rm eff}$ for an array of metallic wires with radius 0.1 mm. The lattice constant (the distance between the wires) is 5 mm. The electric field is parallel to the wires. Note that the imaginary part of the effective permittivity is multiplied by a factor of 10. The left panel presents result of the analysis of the transfer matrix data for three different discretization of the unit cell. The right panel shows the ratio $|\epsilon''/\epsilon'|$ for an array of thin metallic wires 100×100 μm for two values of the metallic permittivity. The numerical results prove that electromagnetic losses decrease when the metallic conductance increases.

The resonance frequency depends on the structural parameters of the SRR and on the parameters of the unit cell [1]. Qualitative agreement between the theoretical formulas and the numerical results was obtained [47, 46]. Another problem is the dependence of the resonance interval on the size and shape of the unit cell. As an example, we show in Fig. 7 how the transmission through the various LH structures depends on the width/high/length of the unit cell. Notice that the width of the LH frequency interval increases substantially as shown in the far left panel of Fig. 7.

Another important parameter is the permittivity $\epsilon_{\rm m}$ of the metallic components. Within the first approximation, we can consider both the SRR and the wires made from a perfect metal. Then both the conductivity σ and the imaginary part of the metallic permittivity $\epsilon''_{\rm m}$ are infinite [50]. This option is often used in the simulations of the commercial software [47]. In the transfer matrix algorithm, however, $\epsilon''_{\rm m}$ is finite, of the order of 10^5. For copper, which is currently used in the experimentally fabricate LH structures, $\epsilon''_{\rm m} \approx 10^7$ [50]. Some test calculations with higher values of $\epsilon''_{\rm m}$ gave us almost the same result, indicating that $\epsilon''_{\rm m} \sim 10^5$ is already sufficient to simulate realistic materials [38].

As was mentioned in Sect. 1, the left-handed structure must be dispersive. Dispersion requires a non-zero imaginary part for the permittivity and permeability. Therefore, transmission losses can not be avoided in LH

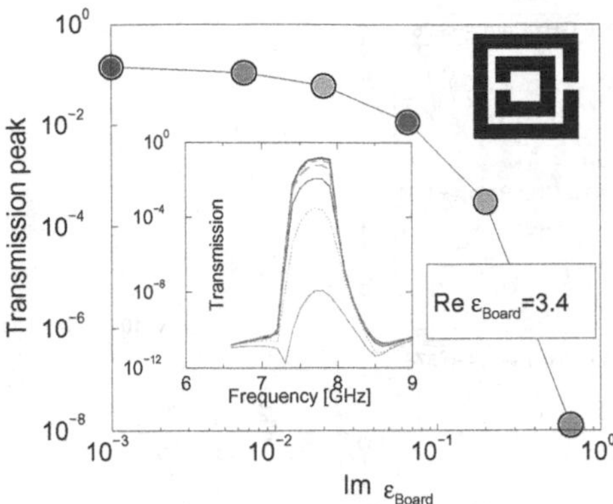

Figure 9. The dependence of the transmission peak on the imaginary part of the dielectric permittivity of the dielectric board. Note that standard dielectrics have $\varepsilon'' \approx 10^{-2}$.

systems. This seems to be in agreement with the first experimental data: the transmission measured in the experiments [8, 9], was only of the order 10^{-3}. Although recent experiments reported the transmission very close to unity, there are still serious doubts in the literature [16, 48] about the possibility to create highly transmitted LH structures. Results of numerical simulations, as shown in Fig. 5, however give that the transmission through the LH structure could be very high, of order of unity.

The main argument against the expectation of high transmission in LH structures [16] is that the effective permittivity of the array of thin metallic wires is mostly imaginary than real and negative [48]. This is a serious objection, because the thickness of the wires in Fig. 1 is 1 mm, which exceeds more than 10 times the realistic parameters of the experimentally analyzed structures. There is no chance to simulate very tiny metallic structures in the transfer matrix algorithm. Fortunately, Pendry *et al.* (formula (36) of [3]) had shown that the problem of the thickness of the wires could be avoided by simultaneously re-scaling the metallic permittivity. An increase of the wire radius by factor α could be compensated by a *decrease* of the permittivity of the metal by factor of α^2. As the metallic permittivity used in [19] is ≈ 100 times *lower* than the realistic permittivity of Cu, we assume that the size of the wires (1 mm) corresponds to Cu wires of thickness 0.1 mm, which is close to size used in experiments (0.2×0.017 mm).

In Fig. 8 we present results for the *effective* permittivity of an array of thin metallic wires. The left panel confirms that the transfer matrix algorithm provides us with realistic data for the transmission and reflection. Three different discretizations of the unit cells were used. For each of them the effective permittivity was calculated by the method explained in Sects.

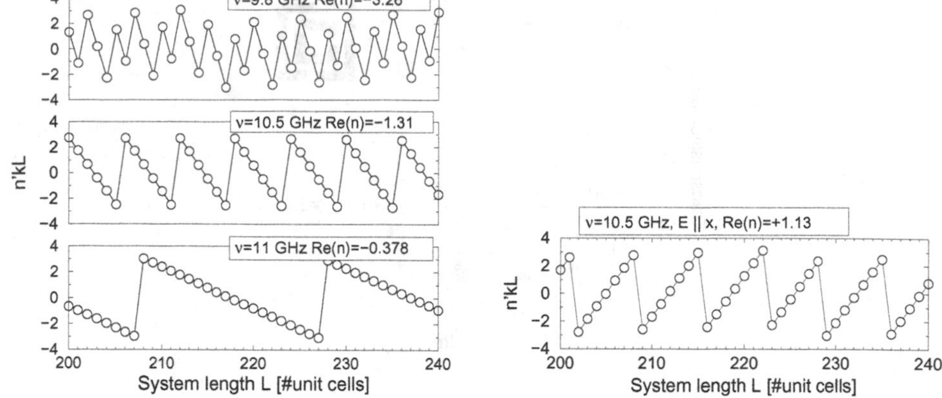

Figure 10. Left: $n'kL$ as a function of the system length L for the EM wave polarized with $\mathbf{E}$ parallel to the wires. The slope is negative which confirms that n' is negative. Right: $n'kL$ as a function of the system length L for the EM wave polarized $\mathbf{E} \parallel x$. The slope is positive which confirms that Re n' is positive. A linear fit gives that $n' = 1.1$. There is almost no interaction of the LH structure with the EM wave.

6, 7. The right panel of figure 8 shows that transmission losses are small in realistic LH structures [51]. Losses in the metallic components of the left-handed structure are therefore not responsible for the low transmission, measured in the experiments [8, 9]. The role of the material properties of the dielectric board on which SRR is deposited was also studied [38]. As is shown in Fig. 9, very small imaginary part of ϵ_{Board} causes a rapid decrease of the transmission. This surprising result seems to be in agreement with experiments [37]. Much higher transmission was obtained in cases with extremely small losses in the dielectric board [37].

6. Effective Index of Refraction

As was discussed above, the structural inhomogeneities of the LH materials are approximately ten times smaller than the wavelength of the EM wave. It is therefore possible, within a first approximation, to consider the slab of the LH material as an homogeneous material. Then we can and use the textbook formulas for the transmission te^{ikL} and reflection r for the homogeneous slab:

$$t^{-1} = \left[\cos(nkL) - \frac{i}{2}\left(z + \frac{1}{z}\right)\sin(nkL)\right] \tag{7}$$

and

$$\frac{r}{t} = -\frac{i}{2}\left(z - \frac{1}{z}\right)\sin(nkL). \tag{8}$$

Here, z and n are the *effective* impedance and the refraction index, respectively, k is the wave vector of the incident EM wave in vacuum, and L is the length of the LH slab. We only consider perpendicular incident waves, so that only the z components of the effective parameters are important. To simplify the calculations, we neglect also the conversion of the polarized EM wave into another polarization, discussed in the Section 4. More accurate analysis should treat both t and r as 2×2 matrices. Here we assume that the off-diagonal elements of these matrices are negligible:

$$|t_{xy}| \ll |t_{yy}|. \tag{9}$$

In the present analysis, we use the numerical data for the transmission and the reflection obtained by the transfer matrix simulation. The expressions for the transmission and the reflection can be inverted as

$$z = \pm\sqrt{\frac{(1+r)^2 - t^2}{(1-r)^2 - t^2}}, \tag{10}$$

$$\cos(nkL) = X = \frac{1}{2t}\left(1 - r^2 + t^2\right). \tag{11}$$

The sign of z is determined by the condition

$$z' > 0 \tag{12}$$

which determines z unambiguously. The obtained data for z enables us also to check the assumption of the homogeneity of the system. We indeed found that z is independent of the length of the system L.

The second relation, (11), is more difficult to invert since $\cos^{-1}$ is not an unambiguous function. One set of physically acceptable solutions is determined by the requirement

$$n'' > 0 \tag{13}$$

which assures that the material is passive. The real part of the refraction index, n', however, suffers from the unambiguity of $2\pi m/(kL)$ (m is an integer). To avoid this ambiguity, data for various system length L were used. As n characterizes the material property of the system, it is L independent. Using also the requirement that n should be a continuous function of the frequency, the proper solution of (11) was found and the resonance frequency interval, in which n' is *negative* was identified [19]. Here, we use another method for the calculation of n and z: Equation (11) can be written as

$$e^{-n''kL}\left[\cos(n'kL) + i\ \sin(n'kL)\right] = Y = X \pm \sqrt{1 - X^2}. \tag{14}$$

Relation (14) enables us to find unambiguously both the real and the imaginary part of the refraction index from the linear fit of $n''kL$ and $n'kL$ *vs* the system length L. The requirement (13) determines the sign in the r.h.s. of Eq. (14), because $|Y| < 1$. Then, the linear fit of $n'kL$ *vs* L determines unambiguously the real part of n'.

Figure 10 shows the L dependence of $n'kL$ for three frequencies inside the resonance interval (transmission data for these frequencies are presented in Fig. 6). The numerical data proves that the real part of the refraction index, n', is indeed *negative* in the resonance interval. For comparison, we present also $n'kL$ *vs* L for the x polarized wave outside the resonance interval. As expected, the slope is positive and gives that $n' = 1.13$, which is close to unity.

Besides the sign of the real part of the refraction index, the value of the imaginary part of n, n'' is important, since it determines the absorption of the EM waves inside the sample. Fortunately, n'' is very small, it is only of the order of 10^{-2} inside the resonance interval. As is shown in Fig. 6, quite good transmission was numerically obtained also for samples with length of 300 unit cells (which corresponds to a length of the system 1.1 m). This result is very encouraging and indicates that left-handed structures could be as transparent as right-handed materials.

There are two constrains in the present method. (i) While the above method works very well in the right side of the resonance interval, we had problems to estimate n in the neighborhood of the left border of the resonance region, where n' is very large and negative. This is, however, not surprising since in this frequency region the wavelength of the propagating wave becomes comparable with the size of the unit cell, so that the effective parameters have no physical meaning. (ii) Outside the resonance interval, we have serious problems to recover proper values of n. This is due to the conversion of the x polarized wave into a y polarized. As a result, we do not have enough numerical data for obtaining the L-dependence of the transmission t. As it is shown in Fig. 5, only data for 2-3 unit cells are representative for the $\mathbf{E} \parallel y$ wave. The condition (9) is not any more fulfilled, and the present theory must be generalized as discussed above.

7. Effective Permittivity and Permeability

Once n and z are obtained

$$\epsilon_{\text{eff}} = n/z \qquad \text{and} \quad \mu_{\text{eff}} = nz, \tag{15}$$

one can estimate unambiguously the effective permittivity and permeability of the investigated structure. Results are shown in Fig. 11. Frequency dependence of the effective refraction index and ϵ'_{eff}, μ'_{eff} is also presented in [19]. Our numerical results confirm the resonance behavior of the μ'_{eff}, in agreement with theoretical predictions of Pendry *et al.* [1]. In addition to

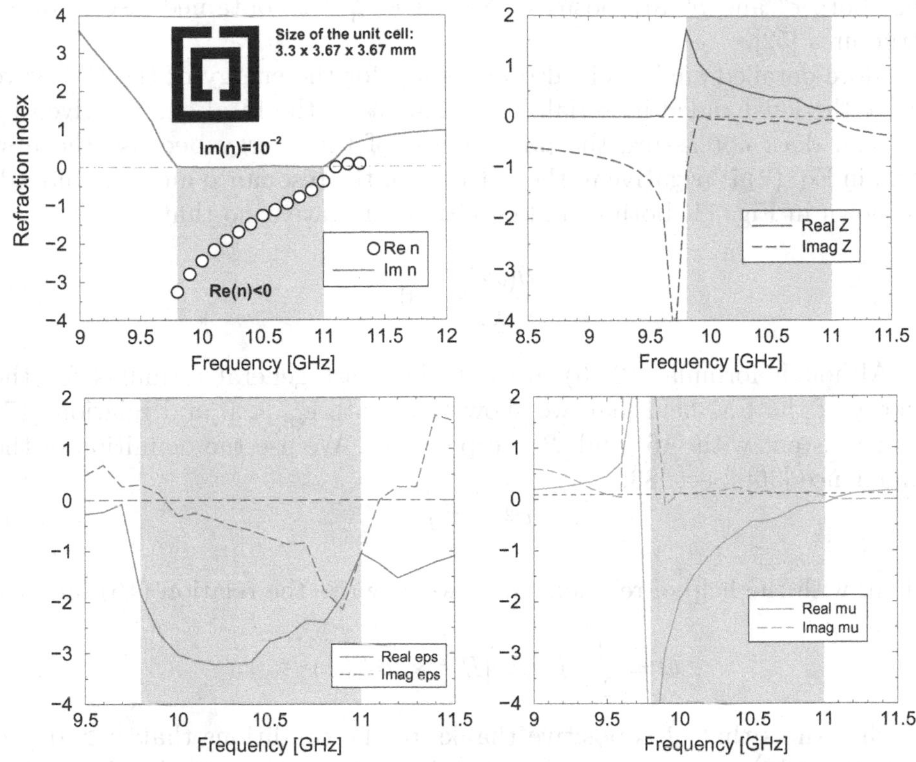

Figure 11. Refraction index n, impedance z, permittivity ϵ and permeability μ as a function of frequency in the resonance interval. Close to the resonance frequency $\nu = 9.8$ GHz, there are problems to estimate n and z, because the wave length of the propagated EM wave becomes comparable with the size of the unit cell of the left-handed structure. Dashed area shows the resonance interval.

these expected results, we also found rather strong electric response of the SRR, which manifests itself as the "anti-resonant" behavior of $\epsilon'_{\rm eff}$ in the resonance frequency interval and by the decrease of the plasma frequency (Fig. 4 of Ref. [19], see also [52]).

The real parts of the permittivity and permeability, $\epsilon'_{\rm eff}$ and $\mu'_{\rm eff}$ are *negative* in the resonance interval, as expected. Surprisingly, the imaginary part of the effective permittivity, $\epsilon''_{\rm eff}$ was also found *negative.* This seems to contradict our physical intuition, since, following [36], Sect. 80, the electromagnetic losses are given by

$$Q = \frac{1}{2\pi}\int d\omega\omega\left[\epsilon''|E|^2 + \mu''|H|^2\right]. \tag{16}$$

In passive materials, we require $Q > 0$. This is trivially satisfied if we require that both ϵ'' and μ'' are positive. Negative $\epsilon''_{\rm eff}$ was obtained also in other structures [52].

More detailed analysis is also necessary for the energy of the EM wave inside the left-handed material. The formula for the total energy, given by Eq. (2), does not assure the positiveness of the energy because the first term in Eq. (2) is negative in the left part of the resonance interval. Indeed, as shown in Fig. 11, both ϵ' and $\partial\epsilon'/\partial\omega$ are negative, so that

$$\frac{\partial(\omega\epsilon')}{\partial\omega} < 0. \tag{17}$$

Although formulas (2,16) are not the most general formulas for the energy of the EM field [53], we show that both $\epsilon''_{\rm eff} < 0$ and relation (17) are consistent with (16) and (2), respectively. We use the definition of the impedance [36], Sect. 83,

$$E^2 = \frac{\mu}{\epsilon}H^2. \tag{18}$$

Then, with the help of relations (15), we re-write the relation (16) into the form

$$Q = \frac{1}{2\pi}\int d\omega\omega|H|^2 \times 2n''(\omega)z'(\omega) \tag{19}$$

which assures that Q is positive thanks to the conditions that $z' > 0$ and $n'' > 0$ (12,13). There is therefore no physical constrain to the sign of the imaginary part of the permittivity and permeability.

With the help of the relation (18) Eq. (2) can be rewritten into the form

$$U = \frac{1}{2\pi}\int d\omega|H|^2\left[|z|^2\frac{\partial(\omega\epsilon')}{\partial\omega} + \frac{\partial(\omega\mu')}{\partial\omega}\right]. \tag{20}$$

Using the numerical data for the impedance and for real part of the permittivity, we checked that indeed the expression in the bracket of the r.h.s. of (20) is always positive.

8. Further Development, Unsolved Problems and Open Questions

We reviewed some recent experiments and numerical simulations on the transmission of the electromagnetic waves through left handed structures. For completeness, we note that recently, Notomi [54] has studied the light propagation in strongly modulated two dimensional photonic crystals (PC). In these PC structures the permittivity is periodically modulated in space and is *positive*. The permeability is equal to unity. Such PC behaves as a material having an effective refractive index controllable by the band structure. For a certain frequency range it was found by FDTD simulations [54–56] that n_{eff} is negative. It is important to examine if left-handed behavior can be observed in photonic crystals at optical frequencies.

Negative refraction on the interface of a three dimensional PC structure has been observed experimentally by Kosaka *et al.* [58] and a negative refractive index associated to that was reported. Large beam steaming has been observed in [58], that authors called "the superprism phenomena". Similar unusual light propagation has been observed in one-dimensional and two dimensional refraction gratings. Finally, a theoretical work [59] has predicted a negative refraction index in photonic crystals.

Studies of the left handed structures open a series of new challenging problems for theoreticians as well as for experimentalists. The complete understanding of the properties of left-handed structures requires the reevaluation of some "well known" facts of the electromagnetic theory. There is no formulas with *negative* permeability in classical textbooks of electromagnetism [36, 50]. Application of the existing formulas to the analysis of left handed structures may lead to some strange results. The theory of EM field has to be reexamined assuming negative μ and ϵ. We need to understand completely the relationship between the real and the imaginary parts of the permittivity and the permeability. Kramers-Kronig relations should be valid, but nobody have verified them yet in the case of the left-handed structures. The main problem is that we need ϵ_{eff} and μ_{eff} in the entire range of frequencies, which is difficult to obtain numerically. Then, due to the anisotropy of the structure as well as the nonzero transmission t_{xy} in Eq. (7), Kramers-Kronig relations should be generalized. We do not believe that today's numerical data enables their verifications with sufficient accuracy.

Both in the photonic crystals and LHM literature there is a lot of confusion about what is the correct definitions of the phase and group refractive index and what is their relations to negative refraction. In additions, it is instructive to see how the LH behavior is related with the sign of the phase and group refractive indices for the PC system. The conditions of obtaining LH behavior in PC were recently examined in [55]. It was demonstrated that the existence of negative refraction is neither a prerequisite nor guarantees a negative effective refraction index and so LH behavior. Contrary, *LH behavior can be seen only if phase refractive index* n_{phase} *is negative.* Once n_{phase} is negative, the product $\mathbf{S} \cdot \mathbf{k}$ is also negative, and the vectors $\mathbf{k}$, $\mathbf{E}$ and $\mathbf{H}$ form a left handed set, as discussed in the Introduction.

Problems of causality arises also in connection with "well known" and accepted relations like $\partial(n'(\omega)\omega)/\partial\omega > 0$. It is evident, that this relation can not be valid in the vicinity of the left border of left-handed frequency interval, since both n' and $\partial n'/\partial\omega$ are negative. The same problem arises for the real part of $\epsilon_{\rm eff}$. Also relation (17) requires more exact and complete treatment. We need more general relations for the energy of the EM field [53] which incorporates all the allowed signs of the real and imaginary parts of the permittivity and permeability.

Following problems that are currently discussed in literature. The negative Snell's law requires the understanding in more detail of the relationship between the Poynting vector, the group velocity, and the phase velocity. We believe, that there is no controversial in this phenomena [15]. Anisotropy of real left-handed materials inspires further development of super-focusing [31].

We believe that further analysis, of what happens when EM wave crosses the boundary of the left-handed and right-handed systems, will bring more understanding of the negative refraction as well as perfect lensing. Numerical FDTD simulations of the transmission of the EM wave through the interface of the positive and negative refraction index [56] showed that the wave is trapped temporarily at the interface and after a long time the wave front moves eventually in the direction of negative refraction. Computer simulations of the transmission through LH wedge [20] also confirm that EM wave spends some time on the boundary before the formation of the left-handed wave front. Formation of surface waves [29, 30, 28] can explain "perfect lensing" without violating causality. Recent development, both numerical [49] and experimental [57] indicate that perfect (although not absolutely perfect) lensing might be possible. We need also to understand some peculiar properties of the left handed structures due to its anisotropy [60] and bi-anisotropy [61].

Last but not least, let us mention an attempt to find new left-handed structures, both in the traditional metallic left-handed structures [62] and in the photonic crystals [52, 63].

Acknowledgments We want to thank D.R. Smith, E.N. Economou, S. Foteinopoulou and I. Rousochatzakis for fruitful discussions. This work was supported by Ames Laboratory (contract. n. W-7405-Eng-82). Financial support of DARPA, NATO (Grant No. PST.CLG.978088) and APVT (Project n. APVT-51-021602) and EU project DALHM are also acknowledged.

References

1. Pendry J.B., Holden A.J., Robbins D.J. and Stewart W.J. (1999) Magnetism from conductors and enhanced nonlinear phenomena *IEEE Trans. on Microwave Theory and Techn.* **47** 2075
2. Pendry J.B., Holden A.J., Stewart W.J. and Youngs I. (1996) Extremely Low Frequency Plasmons in Metallic Mesostructures *Phys. Rev. Lett.* **76**, 4773
3. Pendry J.B., Holden A.J., Stewart W.J. and Youngs I. (1998) *J. Phys.: Condens. Matt.* **10**, 4785

4. Sigalas M., Chan C.T., Ho K.M., and Soukoulis C.M. (1995) *Phys. Rev. B* **52**, 11 744
5. Smith D.R., Schultz S., Kroll N., Sigalas M., Ho K. M. and Soukoulis C.M. (1994) Defect studies in a two-dimensional periodic photonic lattice *Appl. Phys. Lett.* **65**, 645
6. Sarychev A.K. and Shalaev V.M. (2001) Comment on [2], *e-print* cond-mat/0103145
7. Pokrovsky A.L., and Efros A.L. (2002) Electrodynamic of metallic photonic crystals and the problem of left-handed materials *Phys. Rev. Lett.* **89**, 093901
8. Smith D.R., Padilla W.J., Vier D.C., Nemat-Nasser S.C. and Schultz S. (2000) A Composite medium with simultaneously negative permeability and permittivity *Phys. Rev. Lett.* **84**, 4184
9. Shelby R.A., Smith D.R., Nemat-Nasser S.C. and Schultz S. (2001) Microwave transmission through a two-dimensional, isotropic, left-handed meta material *Appl. Phys. Lett.* **78**, 489
10. Veselago V.G. (1968) The electrodynamics of substances with simultaneously negative values of permittivity and permeability *Sov. Phys. Usp.* **10**, 509
11. Smith D.R. and Kroll N. (2000) Negative Refractive Index in Left-Handed Materials *Phys. Rev. Lett.* **85** 2933
12. Shelby R.A., Smith D.R. and Schultz S. (2001) Experimental verification of a negative index of refraction *Science* **292**, 77
13. Parazzoli C.G., Gregor R.B., Li K., Koltenbah B.E.C., Tanielian M. (2002) Experimental verification and simulation of negative index of refraction using Snell's law, *preprint*
14. Valanju P.M., Walser R.M., and Valanju A.O., Wave refraction in negative-index materials: Always positive and very inhomogeneous *Phys. Rev. Lett.* **88**, 187401 (2002)
15. Pendry J.B. and Smith. D.R. (2002) reply to [14] *e-print* cond-mat/0206563
16. Garcia N. and Nieto-Vesperinas M. (2002) Is there an Experimental Verification of a Negative Index of Refraction yet? *Optics Lett.* **27**, 885
17. Sanz V., Papageorgopoulos A.C., Egelhoff Jr. W.F., Nieto-Vesperinas M., and Garcia N. (2002) Wedge-shaped absorbing samples look left handed: The problem of interpreting negative refraction, and its solution *e-print* cond-mat/0206464
18. Ziolkowski R.W. and Heyman E. (2001) Wave propagation in media having negative permittivity and permeability *Phys. Rev. E* **64** 056625
19. Smith D. R., Shultz S., Markoš P. and Soukoulis C.M. (2002) Determination of Effective Permittivity and Permeability of Metamaterials from reflection and Transmission Coefficient *Phys. Rev. B* **65** 195104
20. Parazzoli C.G. (2002) *private communication*
21. Pendry J.B. (2000) Negative refraction makes a perfect lens *Phys. Rev. Lett.* **85**, 3966
22. t'Hooft G. W. (2001) Comment on [21] *Phys. Rev. Lett.* **87** 249701; Williams J. M. (2001) Comment on [21] *Phys. Rev. Lett.* **87** 249703;
23. Garcia N. and Nieto-Vesperinas M. (2002) Negative refraction does not makes perfect lens *Phys. Rev. Lett.* **88**, 122501 (2002); Garcia N. and Nieto-Vesperinas M. (2002) Answer to [25], *e-print* cond-mat/0207413; Garcia N. and Nieto-Vesperinas M. (2002) Answer to [30], *e-print* cond-mat/0207489
24. Pendry J. B. (2001) Replies to [22] *Phys. Rev. Lett* **87** 249702; *ibid* 249704
25. Pendry J. B. (2002) Comment on [23] *e-print* cond-mat/020561
26. Pendry J.B. and Smith D.R. (2002) Comment on [14] *e-print* cond-mat/0206563
27. Lu W.T., Sokoloff J.B. and Sridhar S (2002) Comment on [14] *e-print* cond-mat/0207689
28. Gómez-Santos G. (2002) Universal features of time dynamics in a left-handed perfect lens *e-print cond-mat/0210283*
29. Ruppin R. (2000) Surface polaritons of a left-handed medium *Phys. Lett. A* **277** 61; Ruppin R. (2001) Surface polaritons of a left-handed material slab *J. Phys.:*

Condens. Matt. **13** 1811

30. Haldane F. D. M. (2002) Electromagnetic surface modes at interfaces with negative refractive index make a "Not-quite-perfect" lens *e-print* cond-mat/0206420
31. Smith D.R. and Schurig D. (2002) Electromagnetic wave propagation in media with indefinite permittivity and permeability tensors *e-print* cond-mat/0210625
32. Feise M. W., Bevelacqua P. J., Schneider J. B. (2002) Effects of surface waves on the behavior of perfect lenses *Phys. Rev. B* **66** 035113
33. Ramakrishna S. A. Pendry J. B., Schurig D., Smith D.R., and Shultz S. (2002) The asymmetry lossy near-perfect lens *e-print* cond-mat/0206564; Pendry J. B. and Ramakrishna S. A. (2002) Near-field lenses in two dimensions *J. Phys.: Condens. matt.* **14** 8463
34. Nefedov I. S. and Tretyakov S. A. (2002) Photonic band gap structure containing meta material with negative permittivity and permeability *Phys. Rev. B* **66** 036611
35. Zhang Z.M., Fu C.J. (2002) Unusual photon tunneling in the presence of a layer with negative refractive index *Appl. Phys. Lett.* **80** 1097
36. Landau L.D., Lifshitz E.M. and Pitaevskiĭ L.P. Electrodynamics of Continuous Media, Pergamon Press 1984
37. Li K., McLean S.J., Gregor R.B., Parazzoli, C.G., Tanielian M.H. (2002) Free-space focused-beam characterization of left handed materials, *preprint*
38. Markoš P., Rousohatzakis I. and Soukoulis C.M. (2002) Transmission Losses in Left-handed materials *Phys. Rev. E* **66** 045601
39. Smith D.R., Padilla W.J., Vier D.C., Shelby R., Nemat-Nasser S.C., Kroll N. and Shultz S. (2000) Left-handed metamaterials, in *Photonic Crystals and Light Localization*, ed. Soukoulis C. M. (Kluwer, Netherlands)
40. Pendry J.B. (2001) Electromagnetic materials enter the negative age *Phys. World* **14** 47; Pendry J.B. (2000) Light runs backward in time *Phys. Word* **13** 27; Pendry J.B. (2000) *Physics Today* **53**(5) 17
41. Bayindir M., Aydin K., Ozbay E., Markoš P. and Soukoulis C.M. (2002) Transmission Properties of Composite Metamaterials in Free Space *Appl. Phys. Lett.* **81** 120
42. Ziolkowski R.W. (2002) Design, fabrication and testing of double negative metamaterials *preprint*
43. Marqués R., Martel J., Mesa F. and Medina F. (2002) Left-handed simulation and transmission of EM waves in sub-wavelength split-ring-resonator-loaded metallic waveguides *Phys. Rev. Lett.* **89** 183901
44. Pendry J.B., and MacKinnon A. (1992) *Phys. Rev. Lett.* **69** 2772; Pendry J.B., MacKinnon A., and Roberts P.J., (1992) *Proc. Roy. Soc. London Ser. A* **437**, 67; Pendry J.B. (1994) Photonic band gap structures *J. Modern Optics* **41** 209; Ward A.J. and Pendry J.B. (1996) Refraction and geometry in Maxwell's equations *J. Modern Optics* **43** 773; Pendry J.B. and Bell P.M. 1996 in *Photonic Band Gap Materials* vol. **315** of *NATO ASI Ser. E: Applied Sciences*, ed. by C.M. Soukoulis (plenum, NY) p. 203
45. Markoš P. and Soukoulis C.M. (2002) Transmission Studies of the Left-handed materials *Phys. Rev. B* **65** 033401
46. Markoš P. and Soukoulis C.M. (2002) Numerical Studies of Left-handed materials and Arrays of Split Ring Resonators *Phys. Rev. E* **65** 036622
47. Weiland T. *et al.* (2001) *J. Appl. Phys* **90** 5419
48. Garcia N., and Ponizovskaya E.V. (2002) Calculation of the effective permittivity of a array of wires and the left-handed materials *e-print cond-mat/0206460* Ponizovskaya E.V., Nieto-Vesperinas M., and Garcia N. (2002) Losses for microwave transmission metamaterials for producing left-handed materials: The strip wires *e-print cond-mat/0206429*
49. Contributions to PIERS' 2002 Conference, Boston, July 1.-5.: Zhang Y., Grzegorczyk T.M., and Kong J.A. Propagation of electromagnetic waves in a left-handed medium; Kik P.G., Maier S.A., and Atwater H.A. (2002) The perfect Lens in a

non-perfect world; Moss C.D., Zhang Y., Grzegorczyk T.M., and Kong J.A. FTDT simulation of propagation through LHM

50. Jackson J.D. Classical Electrodynamic (3rd edition), J.Willey and Sons, 1999, p. 312
51. Markoš P. and Soukoulis C.M. (2002) unpublished
52. O'Brien S. and Pendry J.B. (2002) Photonic band gap effects and magnetic activity of dielectric composites *J. Phys.: Condens. Matter* **14** 4035; O'Brien S. and Pendry J.B. (2002) Magnetic activity at infrared frequencies in structured metallic photonic crystals *J. Phys.: Condens. Matter* **14** 6389
53. Ruppin R. (2002) Electromagnetic energy density in a dispersive and absorptive material *Phys. Lett. A* **299** 309
54. Notomi M. (2000) Theory of light propagation in strongly modulated photonic crystals: Refraction behavior in the vicinity of the photonic band gap *Phys. Rev. B* **62** 10696
55. Foteinopoulou S., and Soukoulis C.M. (2002) Negative refraction and left-handed behavior in two-dimensional photonic crystals *Phys. Rev. Lett.* (submitted);
56. Foteinopoulou S., Economou E.N., and Soukoulis C.M. (2002) Refraction at media with negative refractive index *Phys. Rev. Lett.* (submitted)
57. Cubukcu E., Aydin K., Ozbay E., Foteinopoulou S., and Soukoulis C.M. (2002) Experimental demonstration of superlensing in two dimensional photonic crystals *Nature* submitted
58. Kosaka H., Kanashima T., Tomita A., Notomi, M, Tamamura T., Sato T., and Kawakami S. (1998) Superprism phenomena in photonic crystals *Phys. Rev. B* **58** 10096
59. Gralak B., Enoch S. and Tayeb G. (2000) Anomalous refractive properties of photonic crystals *J. Opt. Soc. Am. A* **17** 1012
60. Hu L., and Chui S.T. (2002) Characteristics of electromagnetic wave propagation in uniaxially anisotropic left-handed materials *Phys. Rev. B* **66** 085108
61. Marqués R., Medina F., and Rafii-Idrissi R. (2002) Role of bi-anisotropy in negative permeability and left-handed metamaterials *Phys. Rev. B* **65** 144440
62. Gorkunov M., Lapine M., Shamonina E. and Ringhofer K. H. (2002) Effective magnetic properties of a composite material with circular conductive elements *Eur. Phys. J. B* **28** 263
63. Pendry J.B. and O'Brien S (2002) Very low frequency magnetic plasma *J. Phys.: Condens. Matter* **14** 7404

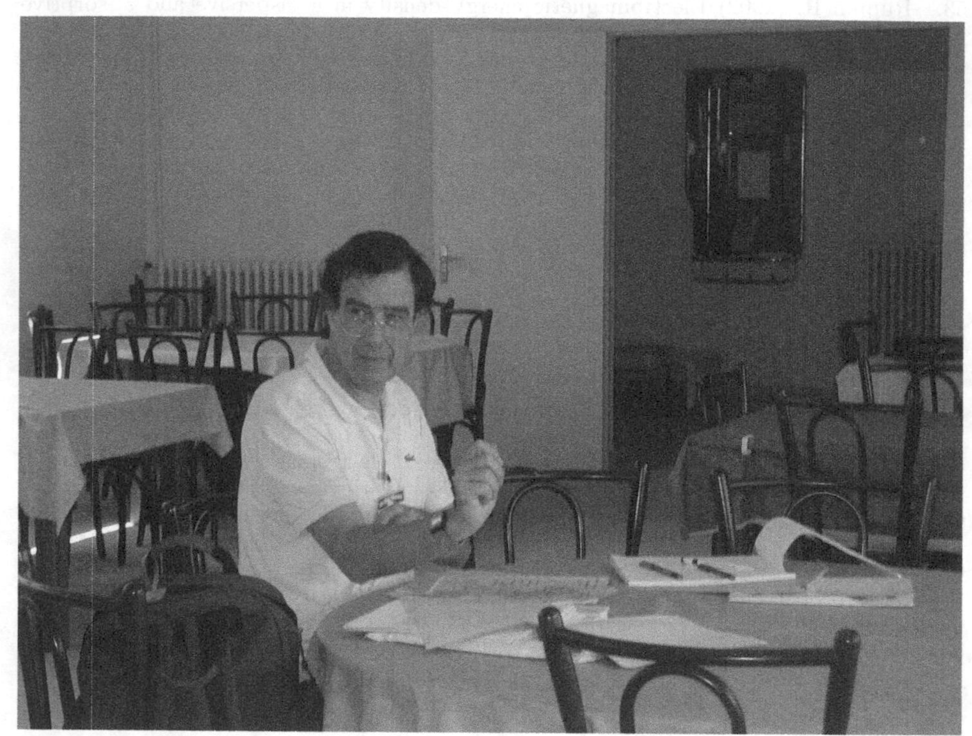

Manuel Nieto-Vesperinas. Photo by Valentin Freilikher

NEGATIVE INDEX COMPLEX METAMATERIALS

MANUEL NIETO-VESPERINAS
Instituto de Ciencia de Materiales de Madrid, Consejo Superior de Investigaciones Cientificas, Campus de Cantoblanco, Madrid 28049, Spain. E-mail: mnieto@icmm.csic.es

AND

NICOLAS GARCIA
Laboratorio de Fisica de sistemas Pequeños, Consejo superior de Investigaciones Cientificas, Serrano 144, Madrid 28006, Spain.

Abstract. A discussion is presented on electromagnetic wave propagation in a new kind of complex media that has recently received attention: metamaterials with overall negative dielectric permittivity and magnetic permeability. This study is made in the light of recent experiments reporting the observation of negative refraction in these media, and also as regards proposals of employing slabs of these so-called negative index media to obtain resolution beyond the Rayleigh half-wavelength diffraction limit.

1. Introduction

In 1968 the Russian physicist V.G. Veselago [1] speculated on the electrodynamic properties of media that he postulated had negative dielectric permittivity, magnetic permeability, and refractive index. The set of vectors corresponding to the electric and magnetic fields and the wave vector of a plane electromagnetic wave in such a system, would be left-handed, namely, the wave vector would point opposite to the direction of energy transport. Thus originating the terminology "left-handed" for such materials. Among other properties, refraction at the surface of such medium would occur at negative angles (negative refraction), the Doppler and the Vavilov-Cherenkov effects would be reversed. Also, the radiation pressure would become light tension towards the source. Since no such medium was

B.A. van Tiggelen and S.E. Skipetrov (eds.),
Wave Scattering in Complex Media: From Theory to Applications, 331–342.

known to exist naturally, and given the difficulties to devise a material whose permeability behaved in analogy to the permittivity, thus involving magnetic plasmons, the idea did not progress until recently [2] when it was suggested that a metamaterial formed by a cubic lattice of copper split ring resonators and strip wires, immersed in a dielectric, would possess an effective dielectric permittivity and magnetic permeability both being negative in the microwave region. It was also proposed [3] that a slab of such a material would amplify evanescent waves, thus focusing them with superresolution and acting as a perfect lens. After recent work [4]–[6], we next discuss the possibility of negative refraction and propagation in such complex media, and the consequences of evanescent wave amplification.

2. Negative Refraction

Let us consider a metamaterial composed of a dielectric medium, (e.g., fiber glass or air), hosting a cubic lattice of metallic (Cu) split-ring resonators and wires [7, 8]. The size of the unit cell is $a_x \times a_y \times a_z = 3.6 \times 3.6 \times 5$ mm^3. We shall consider microwave frequencies near 10 GHz. Both the effective permittivity ϵ_t and the effective permeability μ_t have negative real parts, and of course their imaginary parts are positive. For the energy to be conserved, the refractive index n_t should have a positive imaginary part, which requires it to be defined as a negative square root: $n_t = -\sqrt{\epsilon_t \mu_t}$. Such a metamaterial is named left-handed (LHM) because the set formed by the electric **E**, magnetic **H**, and propagation **k** vectors is such that the vector product **E** × **H** (and thus the Poynting vector **S** and the group velocity) have a direction opposite to that of **k** in normal, right handed materials (RHM), where ϵ_t and μ_t are positive. Negative refraction at the interface of the LHM is met when the angle of refraction θ_t for light incident from the exterior RHM with index $n_i > 0$ is

$$\sin\theta_t = \frac{n_i \sin\theta_i}{-n_t}, \tag{1}$$

where θ_i is the angle of incidence. Thus the angle of refraction is negative and the energy flow goes into the LHM with its normal component being positive, but with negative component parallel to the interface.

At these microwave frequencies an effective medium model can be constructed with "plasmonic" frequencies for the overall ϵ_t and μ_t. These quantities are

$$\epsilon_t = 1 - \frac{f_{ep}^2 - f_{eo}^2}{f^2 - f_{eo}^2 + i\gamma f}, \tag{2}$$

$$\mu_t = 1 - \frac{f_{mp}^2 - f_{mo}^2}{f^2 - f_{mo}^2 + i\gamma f}, \tag{3}$$

where $f_{ep} = 12.8$ GHz, $f_{mp} = 10.95$ GHz, $f_{eo} = 10.3$ GHz and $f_{mo} = 10.05$ GHz are the electronic plasma, magnetic plasma, electronic resonance and magnetic resonance frequencies, respectively; γ represents the losses. Eqs. (2) and (3) show a region $1.03 < f < 1.095$ where the refractive index becomes negative (from now on, all frequencies are expressed in units of 10^{10} Hz). On the other hand, for $1.095 < f < 1.28$ the medium is metallic with imaginary dielectric constant, and for $f > 1.28$ it becomes transparent. By contrast, in the regions $1.005 < f < 1.03$ and $f < 1.005$, the medium has an imaginary refractive index and is a right handed dielectric, respectively. For $f < \gamma = 0.001$, losses dominate, but in the region of parameters of the experiment of [7, 8] losses may be neglected.

Let an electromagnetic wave be incident from air into a complex medium slab of thickness d with effective parameters ϵ_t and μ_t. Then, the trasnmittivity of this system can be obtained from continuity conditions at the interfaces $z = 0$ and $z = d$. For s polarization (electric field parallel to the cylinder axis) the transmitted amplitude is

$$t = \frac{\frac{4q_i q_t}{\mu_t} e^{-iq_i d}}{\left(q_i + \frac{q_t}{\mu_t}\right)^2 e^{-iq_t d} - \left(q_i - \frac{q_t}{\mu_t}\right)^2 e^{iq_t d}}, \tag{4}$$

with $q_i = 2\pi n_i \frac{cos\theta_i}{\lambda}$ and $q_t = 2\pi n_t \frac{\cos\theta_t}{\lambda}$, λ being the wavelength of the incident radiation. The transmittivity is $T = |t|^2$. We address normal incidence. Then, when for any value of μ_t the dielectric permittivity is dominated by its imaginary part: $\epsilon_t = i\epsilon$, absorption would dominate. Like in metals at microwave frequencies [10], $n_t = -\alpha + i\beta$ with $\alpha = \beta$, and one has

$$T = \frac{4\left|\frac{\epsilon_t}{\mu_t}\right|}{4\left|\frac{\epsilon_t}{\mu_t}\right| + \left(\left|\frac{\epsilon_t}{\mu_t}\right| + 1\right)^2 \left[(\sin\kappa d \cosh\kappa d)^2 + (\cos\kappa d \sinh\kappa d)^2\right]}. \tag{5}$$

In Eq. (5), $\kappa = \frac{\sqrt{2}\pi f |n_t|}{2c}$, c being the speed of light in vacuum. This case is of interest to interpret the experimental data of Refs. [7] and [8]. In Fig. 1, T is plotted in the region $\mu_t = 0$. Notice that T has a peak only in the region $\mu_t = 0$, T being negligible, below detection limit, otherwise. When $\mu_t = 0$, Eq. (5) yields $T = \frac{4}{4+\left(\frac{\pi\epsilon f d}{c}\right)^2}$, but when the frequency departs from the region where $\mu_t = 0$, then T decreases exponentially.

If Eq. (3) describes the magnetic permeability for non-magnetic wires ($\mu_t = 1$), as is the case for copper (Cu), and it is given by the response of the circuit elements, then for thin wires of 0.15 mm radius the value of γ is 0.2, and not 0.001 as claimed in [8]. The dielectric permittivity for such small section wires at microwave frequencies may be imaginary (losses

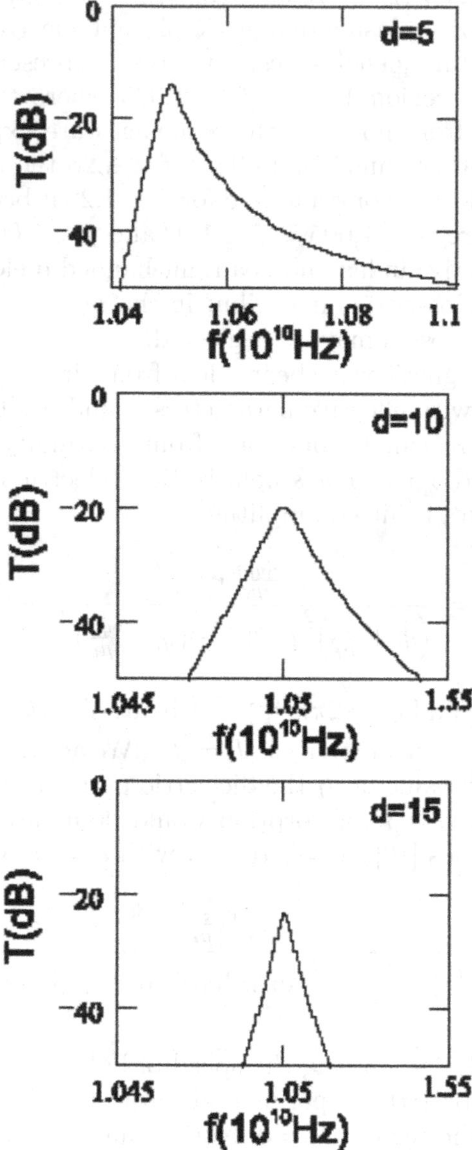

Figure 1. Plot of T for a slab of thickness $d = 5$, 10, and 15 cm for losses as described in the text and for $f_{mp} = 1.05$.

dominate [11]) below the cut-off frequency $f_c = \frac{c}{2a} = 3$ for the structure lattice constant equal to $a_z = 5$ mm. Therefore, for $f_c < 3$ the effective medium is no longer dielectric and $\epsilon_t = i\epsilon$, the transmittivity T then being very low. Then the maximum of T is obtained at $f = 1.095$ where the permeability is exactly zero ($\mu_t = 0$). Notice that the wire thickness is larger than the wave penetration length, the cases of experiments of wire radii between 1.5 mm and 0.10 mm correspond to penetration lengths of 10^{-4}–10^{-5} cm. Losses dominate in the wires. This is confirmed by a finite-difference time-domain calculation of T for the structure with the cylinders only. For the effective medium theory we have an imaginary permittivity $\epsilon_t = \eta\epsilon_{\rm Cu}$, where $\epsilon_{\rm Cu}(f) = 1 - \frac{f_p^2({\rm Cu})}{f(f+i\gamma({\rm Cu}))}$ is the frequency-dependent dielectric constant of Cu and η is an effective "filling factor" of Cu wires that takes into account the permittivity of Cu in vacuum. This is estimated from the skin depth of the wires of radius r: $\eta = p\frac{r\lambda}{4a^2|\epsilon_{\rm Cu}|^{\frac{1}{2}}}$, where losses and plasma frequency for Cu are $\gamma({\rm Cu}) = 0.15$ eV and $f_p({\rm Cu}) = 8.17$, respectively. The values obtained for the skin depth are 10^{-4}–10^{-5} cm, in agreement with previous results [11]. Notice that Re $\epsilon_{\rm Cu}(10^{10}\ {\rm Hz}) \simeq 2000$, while Im $\epsilon_{\rm Cu}(10^{10}\ {\rm Hz}) \simeq 10^7$. The wire radii are assumed to be 1.5 mm. In the above expression, p is a factor that can be used to fit $T(\mu_t = 0)$ and can be roughly estimated to be $p \simeq \left(\frac{\pi^{\frac{1}{2}} r}{a}\right)$. By using the permeability given by Eq. (3) and Eq. (5), we have the transmittivity T depicted in Fig. 1. Only a peak is observed around $\mu_t = 0$, like in the experiments reported in [7]. Otherwise, the values of T are below the detection threshold. Fig. 1 also shows plots for $d = 5$, 10 and 15 cm. When μ_t is around zero, the value of n_t is complex, then the angle of refraction $-\theta_t(1 - i)$ is a small complex number and hence the electromagnetic wave is inhomogeneous.

To conclude this section, we should say that, as shown, absorption dominates microwave transmission through a complex medium of such thin metallic elements at which lossy transmission is predominant, the wave inside the metamaterial then being inhomogeneous. Therefore, until additional data are provided in experiments like that of [8], namely: the correct damping constant of the wires [5], the appropriate functional form for the permittivity and permeability of the metallic circuit elements contained in the complex material, as well as the sample dimensions and distance to the detector, there is no clear evidence that one can infer a negative refractive index. However, there is the possibility that thicker metallic wires and split-ring resonators, at which reflection with small losses dominates, should not involve such strong absorption [9]. Also, we show that transmission is peaked at zero values of the effective magnetic permeability. This effect constitutes a sharp frequency transmission filter in this kind of structures.

3. Evanescent Waves Inside Negative Index Media

It is well-known that optical resolution outside the near-field is limited to a half-wavelength $\lambda/2$ of the electromagnetic wave [11]. This is due to the loss of evanescent components of the wave emitted by the object. It is however proposed [3] that a slab of left-handed metamaterial, can amplify and restore such evanescent wavefield components, thus acting as a superlens. This has been controversial [4, 6, 13]. Here we analyze the possibilities and consequences of such an amplification of evanescent waves.

The electric vector $\mathcal{E}$ of a wavefield from an object, propagating in a source-free half-space, is well known to be represented by its angular spectrum $\mathbf{A}(k_x, k_y)$ of plane wave components with wavevectors $\mathbf{k} = (k_x, k_y, k_z)$, $k_x^2 + k_y^2 + k_z^2 = (2\pi/\lambda)^2$ [12, 14]:

$$\mathcal{E}(\mathbf{r}) = \int_{-\infty}^{\infty} \mathbf{A}(k_x, k_y) \exp(i\mathbf{k} \cdot \mathbf{r}) dk_x dk_y. \tag{6}$$

This expression contains both propagating and evanescent plane wave components. In what follows we shall focus our attention on a given evanescent wave $\mathbf{E}(k_x, k_y, \mathbf{r}) = \mathbf{A}(k_x, k_y) \exp(i\mathbf{k} \cdot \mathbf{r})$. (Unless explicitly stated, the k_x, k_y-dependence of $\mathbf{E}$ will not be included in the notation). Let $\mathbf{E}^{(i)}$ be the electric field of an evanescent component of an s-polarized electromagnetic wave in the half-space $z < 0$ occupied by vacuum, incident on the plane $z = 0$ that limits a LHM, filling the region $z > 0$ of dielectric permittivity $-\epsilon$, magnetic permeability $-\mu$, and refractive index $-n$ ($\epsilon > 0$, $\mu > 0$, and $n > 0$). For brevity, we drop the subscript t of ϵ_t and μ_t from here on. A time harmonic dependence $\exp(-i\omega t)$ will be assumed throughout. We then choose the geometry such that

$$\mathbf{E}^{(i)}(z \leq 0) = (A^{(i)}, 0, 0) \exp(i\mathbf{k}^i \cdot \mathbf{r}) \tag{7}$$

with the wavevector $\mathbf{k}^i = (0, k_y^i, k_z^i)$, $k_z^i = \pm i K_i$, $K_i = \sqrt{k_y^{i2} - k_0^2}$, $k_0 = \omega/c = 2\pi/\lambda$. The sign of the square root is discussed next.

The corresponding reflected and transmitted waves, $\mathbf{E}^{(r)}$ and $\mathbf{E}^{(t)}$, respectively, are

$$\begin{aligned} \mathbf{E}^{(r)}(z \leq 0) &= (A^{(r)}, 0, 0) \exp(i\mathbf{k}^r \cdot \mathbf{r}), \\ \mathbf{E}^{(t)}(z \geq 0) &= (A^{(t)}, 0, 0) \exp(i\mathbf{k}^t \cdot \mathbf{r}), \end{aligned} \tag{8}$$

with wavevectors $\mathbf{k}^r = (0, k_y^i, k_z^r)$, $\mathbf{k}^t = (0, -nk_y^t, -nk_z^t)$, and $k_z^r = \pm i K_i$, $K_i = \sqrt{k_y^{i2} - k_0^2}$, $k_z^t = \pm i K_t$, $K_t = \sqrt{k_y^{t2} - k_0^2}$. Of course, $A^{(i)}$, $A^{(r)}$, and $A^{(t)}$ are functions of k_y. Continuity conditions at $z = 0$: $[E_x^{(i)} + E_x^{(r)}]_{z=0} = [E_x^{(t)}]_{z=0}$ and $\partial_z [E_x^{(i)} + E_x^{(r)}]_{z=0} = (-1/\mu)\partial_z [E_x^{(t)}]_{z=0}$, impose that $k_z^i = -nk_z^t$, which characterizes negative refraction.

As regards the signs of k_z^i, k_z^r, and k_z^t, let us consider the case in which $\mathbf{E}^{(i)}$ is evanescent and grows as $z \to -\infty$, $\mathbf{E}^{(r)}$ is evanescent and decays as $z \to -\infty$, $\mathbf{E}^{(t)}$ is evanescent and grows as $z \to \infty$. Now, the correct normalization of $\mathbf{E}^{(i)}$ imposes it to be restricted to the strip $-z_0 \leq z \leq 0$, and has the amplitude $A^{(i)} \exp(-K_i z_0)$. Then the matching conditions at $z = 0$ yield

$$r = \frac{-\mu K_i + K_t}{-\mu K_i - K_t} \exp(-K_i z_0), \qquad t = \frac{-2\mu K_i}{-\mu K_i - K_t} \exp(-K_i z_0). \tag{9}$$

As shown in [1], focusing both inside the LHM slab and, consequently, behind it, is obtained when both its permittivity and permeability equal those of the surrounding medium. Considering this last to be vacuum, let us assume therefore that at a certain frequency ω, $\epsilon = \mu = n = 1$, then $K_t = K_i$ and the coefficients of Eqs. (9) become $r = 0$ and $t = \exp(-K_i z_0)$. Hence, the wave transmitted into the LHM is

$$\mathbf{E}^{(t)}(k_y, y, z \geq 0) = (A^{(i)}(k_y^i), 0, 0) \exp[ik_y^i y + K_i(z - z_0)]. \tag{10}$$

Eq. (10) shows that the evanescent wave components transmitted into the LHM are now amplified as z increases, but since all objects are limited in space, the angular spectrum A decreases as $k_y^i \to \pm\infty$ as an inverse power of k_y^i. (Obviously, this is also true for the k_x dependence of A if any other polarization were chosen) [12]. Also, as energy should be conserved, $\mathcal{E}^{(t)}(x, y, z \geq 0)$ and $\mathbf{E}^{(t)}(k_y^i, y, z \geq 0)$ must be square integrable in y and k_y^i, respectively [12, 15]. This imposes that $z \leq z_0$. Namely, the waves given by (10) can only exist within a strip $0 \leq z < z_0$ in the LHM half-space. Therefore, such a mode cannot exist in the LHM if $z > z_0$. Its physical meaning is that the formation of a surface state requires a transient time since the light source is switched on, that is infinite and hence involves infinite energy density when $z > z_0$.

One may also wonder what is special in the plane $z = z_0$. The answer is that this is precisely where focusing inside the slab occurs, namely where the wavefront changes from being incoming to being outgoing. Such a change in the wavefield would involve a transformation in its evanescent components from being amplified towards increasing z for $z \leq z_0$ to becoming decaying towards increasing z for $z > z_0$. However, such a transformation can only take place by the presence of sources at $z = z_0$, which would violate the continuity conditions of the field and its derivative through $z = z_0$. Observe that this problem does not occur for the propagating part of the wavefield which has the well-known phase anomaly at the focus [11, 12].

Since, as shown above, $r = 0$ when $\epsilon = \mu = n = 1$, the transmitted mode (10) of a semi-infinite LHM also applies inside a LHM slab of width $d \leq z_0$ that at a certain frequency has $\epsilon = \mu = n = 1$. Let this layer be embedded in vacuum, limited by the planes $z = 0$ and $z = d$. And let an evanescent wave, decaying as $z \to 0$ in $z < 0$ (i.e., amplifying as $z \to -\infty$), be incident on the interface $z = 0$. One then has in each of the three regions:

$$\begin{aligned} \mathbf{E}(-z_0 \leq z \leq 0) &= (A^{(i)}, 0, 0) \exp[ik_y^i y - K_i(z + z_0)] \\ &+ (rA^{(i)}, 0, 0) \exp[ik_y^i y + K_i z], \quad (11) \\ \mathbf{E}(0 \leq z \leq d) &= (A, 0, 0) \exp[ik_y^i y - K_i z] \\ &+ (B, 0, 0) \exp[ik_y^i y + K_i z], \quad (12) \\ \mathbf{E}(z \geq d) &= (tA^{(i)}, 0, 0) \exp[ik_y^i y - K_i z], \quad (13) \end{aligned}$$

where r and t are the reflection and transmission coefficients at $z = 0$ and $z = d$, respectively, and A and B depend on k_y. The incident evanescent wave in $z < 0$ has been normalized again to $\exp[-K_i z_0]$ and, as before, it is restricted to the strip $0 \leq z \leq d$ in order to ensure that it is square integrable. The matching conditions at $z = 0$ and $z = d$ give: $r = A = 0$, $B = A^{(i)} \exp[-K_i z_0]$, and $t = A^{(i)} \exp[K_i(2d - z_0)]$. Hence, the resulting waves in each of the three regions are $\mathbf{E}(-z_0 \leq z \leq 0) = (A^{(i)}, 0, 0) \exp[ik_y^i y - K_i(z + z_0)]$, $\mathbf{E}(0 \leq z \leq d) = (A^{(i)}, 0, 0) \exp[ik_y^i y + K_i(z - z_0)]$, and $\mathbf{E}(z \geq d) = (A^{(i)}, 0, 0) \exp[ik_y^i y - K_i(z + z_0 - 2d)]$. This, as mentioned, shows that the wave $\mathbf{E}(0 \leq z \leq d)$ inside the LHM is again like in Eq. (10). Its proper normalization imposes once again that $d \leq z_0$, otherwise this wavefunction will be zero, as required by its normalization. Hence there is no transmitted evanescent component inside the LHM slab when $d > z_0$. Notice that the current density that characterizes the energy transport along OY for each evanescent component inside the slab is [16]: $J_y = Ck_y^i \exp[2K_i(z - z_0)]$, C being a constant.

As before, a width $d = 2z_0$ would imply a wavefunction $\mathbf{E}(k_y^i, 0 \leq z \leq d)$ inside the slab to contain a factor $\exp[K_i(z - z_0)]$ that prevents it from being square integrable in k_y^i at $z > z_0$. As a consequence, *the evanescent components cannot be restored by the slab of LHM* beyond $z = z_0$. Hence, the evanescent components at the exit of the slab $z = d$ are, at best, like those at the plane $z = -z_0$, where they were originally created, but cannot be amplified any further.

We illustrate the above with the effect of a LHM slab on evanescent waves in a tunneling barrier between two dielectric semi-infinite media. This situation is relevant since ultimately, in the detection process, evanescent waves have to couple into propagating waves at a certain interface. The system to study is then composed of five regions as depicted in Fig. 2: the space $0 < z \leq a$ and $a + d < z \leq 2a + d$ are air gaps, and $a < z \leq a + d$ is occupied by the LHM that at a certain frequency ω has $\epsilon = \mu = n = 1$. The dielectric semi-infinite regions are $z < 0$ and $z > 2a + d$, respectively. Let us consider one of the plane propagating components of a wavefield $\mathcal{E}$

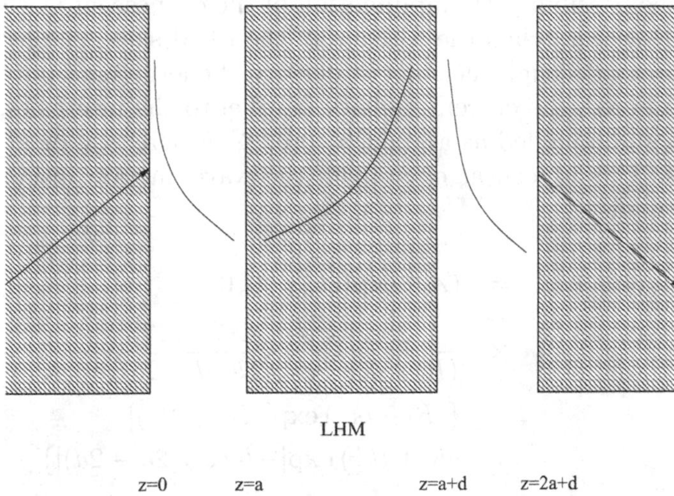

Figure 2. Geometry for a tunnel barrier with a slab of left-handed material (LHM). An incident plane wave component of a propagating wavefield is totally reflected at $z = 0$, thus creating an evanescent wave, which after traversing the two air gaps and the LHM slab couples again to a propagating wave at $z = 2a + d$.

that is incident on $z = 0$ from the first dielectric medium of refractive index $n_d > 0$ in $z < 0$ beyond the critical angle. In this medium, the wavefunction is

$$\mathbf{E}(z \leq 0) = (A^{(i)}, 0, 0)\exp(i\mathbf{k}^i \cdot \mathbf{r}) + (rA^{(i)}, 0, 0)\exp(i\mathbf{k}^r \cdot \mathbf{r}), \tag{14}$$

where $\mathbf{k}^i = (0, k_y^i, k_z^i)$, $k_z^i = \sqrt{k_0^2 n_d^2 - k_y^{i2}}$. The transmitted wave into the second dielectric medium is

$$\mathbf{E}(z \geq 2a + d) = (tA^{(i)}, 0, 0)\exp(i\mathbf{k}^i \cdot \mathbf{r}). \tag{15}$$

Continuity conditions at the interfaces lead to the transmission and reflection coefficients t and r:

$$t = \frac{4iKk_z^i \exp[-ik_z^i(2a + d)]\exp[K(d - 2a)]}{(K + ik_z^i)^2 \exp[2K(d - 2a)] - (K - ik_z^i)^2}, \tag{16}$$

$$r = \frac{(k_z^{i2} + K^2)(1 - \exp[2K(d - 2a)])}{(K + ik_z^i)^2 \exp[2K(d - 2a)] - (K - ik_z^i)^2}. \tag{17}$$

where $K = \sqrt{k_y^{i2} - k_0^2}$. Of course, when $d = 0$, Eqs. (16) and (17) become those of t and r for the usual tunnel effect in an air gap of width $2a$. Also, notice that when $d = 2a$ then $t = 1$ and $r = 0$. At first sight, this seems to suggest the possibility of amplification of the evanescent wave inside the LHM slab. However, when the width d of the LHM slab is larger than that of the air gap a, although the current density of energy transport along OZ remains conserved, the current density parallel to the interfaces, i.e. along OY, becomes unbounded as argued before, as d increases, or at fixed d as K increases. This is seen at once from the wavefunction inside the LHM slab, which reads

$$\begin{aligned}
\mathbf{E}(a \leq z \leq a+d) &= (A^{(i)} \exp(ik_y^i y), 0, 0) \\
&\times \frac{2ik_z^i}{(K+ik_z^i)^2 \exp[2K(d-2a)] - (K-ik_z^i)^2} \\
&\times \{(K-ik_z^i) \exp[K(z-2a)] \\
&+ (K+ik_z^i) \exp[-K(z+2a-2d)]\}. \qquad (18)
\end{aligned}$$

Again, the wavefunction given by Eq. (18) is not square integrable in k_y when $d > a$. In fact, when $d = 2a$, the energy density inside the LHM slab grows as $\cosh[2K(z-2a)]$ when either z or K increase. We see, therefore, that these divergences prevent the LHM slab from restoring the evanescent waves. Then, since $d \leq a$, according to Eq. (11), $t \neq 1$ and $r \neq 0$. Hence, the coupling of the evanescent waves with the propagating waves at $z > d+2a$, which is necessarily involved in the detection process, unavoidly leads to an amplitude of the transmitted propagating wave different from the one of the corresponding component of the incident field. Similar results are obtained for p polarization by exchanging μ by ϵ in the matching conditions.

4. Role of Absorption

So far, ideal lossless media have been addressed. However, the positivity of the electromagnetic energy imposes that the LHM exhibits frequency dispersion [1]. This, implies that the LHM is absorbing [10]. We shall next see how the presence of absorption, even if small, drastically changes the nature of waves in the LHM, as it produces an evanescent wavefield, decaying as z increases, rather than an amplifying wave. We once again consider a slab of LHM surrounded by vacuum and limited by the planes $z = 0$ and $z = d$. We address an evanescent component of a wavefield incident on the slab from $z < 0$. This evanescent wave decays as z increases. Let the refractive index of the LHM at some frequency ω be $n = -1 + in_2$ with $0 < n_2 \ll 1$. Here we have made use of the relation $n = -\sqrt{(-\epsilon_r + i\epsilon_i)(-\mu)}$, where $-\epsilon_r$ and ϵ_i are the real and imaginary parts of the dielectric permittivity. When $\mu = 1$ and $\epsilon_r = 1$, one has $n = -\sqrt{1 - i\epsilon_i} \simeq -(1 - i\epsilon_i/2)$, so that $n_2 = \epsilon_i/2$. Then $K = \sqrt{k_y^2 - k_0^2}$ becomes $K_i(1 - in_2)$ in the LHM.

Normalizing the incident evanescent wave to $\exp(-K_i z_0)$, as before, so that at some plane $z = -z_0$ the wave amplitude is one, one can make use of Eqs. (11), (12), and (13) by writing $k_z^i = iK_i$ (incident evanescent wave on the slab), substituting K by $K_i(1 - in_2)$, and setting $a = 0$ (no air gaps). Then, provided that $n_2 \exp(K_i d) \gg 2$, the second term in the denominator of Eqs. (16), (17), and (18) can be neglected. Thus the fields read

$$\begin{aligned}
\mathbf{E}(z_0 \leq z \leq 0) &= (A^{(i)} \exp(-K_i z_0), 0, 0)[\exp(ik_y^i y - K_i z) \\
&- \frac{n_2 + 2i}{n_2} \exp(ik_y^i y + K_i z)], \qquad (19) \\
\mathbf{E}(0 \leq z \leq d) &= (A^{(i)} \exp(-K_i z_0), 0, 0) \\
&\times \frac{2}{n_2^2}[(2 - in_2) \exp[ik_y^i y + K_i(z - 2d) + iK_i n_2(2d - z)] \\
&- in_2 \exp(ik_y^i y - K_i z + iK_i n_2 z)]], \qquad (20) \\
\mathbf{E}(z \geq d) &= (A^{(i)} \exp(-K_i z_0), 0, 0) \\
&\times \frac{4(1 - in_2)}{n_2^2} \exp(ik_y^i y - K_i z) \exp(iK_i n_2 d). \qquad (21)
\end{aligned}$$

Notice the remarkable fact that the existence of some absorption, which should be accounted for when $n_2 \exp(K_i d) \gg 2$, gives rise to a decaying evanescent wave in the LHM slab due to the second term of Eq. (20). This is in striking contrast with the amplifying wave transmitted into the LHM in the absence of such an absorption. Also, there is now a wave reflected at $z = d$, represented by the first term of Eq. (20). In fact, when $K_i d \gg 1$, the second term of Eq. (20) dominates near $z = 0$, whereas the first term contributes near $z = d$.

As an example, for $n_2 = 0.1$, $K_i d = 10$, and $d = 2.8\lambda$, the intensity at the exit of the slab is $|\mathbf{E}(z = d)|^2 = 32 \times 10^{-5} \exp(-2K_i z_0)$, whereas if $n_2 = 0$, $|\mathbf{E}(z = d)|^2 = e^{20} \exp(-2K_i z_0) \simeq 5 \times 10^8 \exp(-2K_i z_0)$. On the other hand, the intensity at the entrance of the slab is $|\mathbf{E}(z = 0)|^2 = |1 - (n_2 + 2i)/n_2|^2 = 4 \times 10^2 \exp(-2K_i z_0)$. This quantity is remarkably larger than the value $\exp(-2K_i z_0)$ of the incident field intensity at $z = 0$ (which also coincides with the total field at $z = 0$ when $n_2 = 0$). This increase is due to the contribution $(n_2 + 2i)/n_2$ of the reflected wave at $z = 0$ [namely, the second term of Eq. (19)].

5. Conclusions

We conclude that an unambiguous proof of effective negative refractive index in complex materials must be able to overcome the strong absorption of the metallic tiny elements. If thicker wires and split ring resonators provide a means to this end, then one must be able to associate to this system a local permeability and permittivity with negative values that keep an analogy with those of an effective medium, in other words, both the effective ϵ and μ should depend on the frequency ω only, and not on the components of the wavevector. This is not at all obvious from the combination of the metallic circuit elements [17].

Concerning superlenses with complex media constituting negative index slabs, we prove that amplified waves inside ideally lossless, dispersiveless negative index media are limited to a penetration depth equal to the focusing distance inside such slabs, and hence no superfocusing including evanescent components can be obtained. In addition, the presence of absorption is very difficult to overcome, it easily transforms amplified components of the wavefield into decaying ones.

Acknowledgements

Work supported by the Spanish DGICYT.

References

1. Veselago, V.G. (1968) Sov. Phys. Usp. 10, 509.
2. Pendry, J.B., Holden, W., Stewart, W., and Youngs, I. (1996) Phys. Rev. Lett. **76**, 4773.
3. Pendry, J.B. (2000) Phys. Rev. Lett. **85**, 3966.
4. Valanju, P.M., Walser, R.M., and Valanju, A.P. (2002) Phys. Rev. Lett. **88**, 187401-1.
5. Garcia, N. and Nieto-Vesperinas, M. (2002) Opt. Lett.**27**, 885.
6. Garcia, N. and Nieto-Vesperinas, M. (2002) Phys. Rev. Lett. **88**, 207403.
7. Shelby, R.A., Smith, D.R., Nemat-Nasser, S.C., and Scultz, S. (2001) Appl. Phys. Lett. **78**, 489.
8. Shelby, R.A., Smith, D.R., and Schultz, S. (2001) Science **292**, 77.
9. Ponizovskaya, E.V., Nieto-Vesperinas, M., and Garcia, N., Appl. Phys. Lett. **81**, 4470 (2002).
10. Landau, L.D. and Lifshitz, E.M. (1963) *Electrodynamics of Continuous Media*, Pergamon Press, Oxford.
11. Born, M. and Wolf, E. (1999) *Principles of Optics*, 7th Edition, Cambrige University Press, Cambridge.
12. Boas, R.P. (1954) *Entire Functions*, Academic Press, New York; Nieto-Vesperinas, M. (1991), *Scattering and Diffraction in Physical Optics*, J. Wiley, New York; Wolf, E. and Nieto-Vesperinas, M. (1985) J. Opt. Soc. Am. **A 2**, 886.
13. t'Hooft, G.V. (2001) Phys. Rev. Lett. **87**, 249701; Williams, J.M. (2001) *Phys. Rev. Lett.* **87**, 249703.
14. Mandel, L. and Wolf, E. (1995) *Optical Coherence and Quantum Optics*, Cambridge University Press, Cambridge.
15. Sherman, G.C. and Bremermann, H.J. (1969) J. Opt. Soc. Am. **59**, 146.
16. Cohen-Tannoudji, C., Diu, B., and Laloe, F. (1977) *Quantum Mechanics*, J. Wiley, New York.
17. Simovski, C.R., Below, P.A., and He, S., IEEE Trans. Antennas Propag., in press (see also arXiv: cond-mat/0211205).

LOCAL FIELD STATISTIC AND PLASMON LOCALIZATION IN RANDOM METAL-DIELECTRIC FILMS

DENTCHO A. GENOV, ANDREY K. SARYCHEV
AND VLADIMIR M. SHALAEV
School of Electrical and Computer Engineering,
Purdue University, West Lafayette, Indiana 47907-1285

Abstract. A new, exact and efficient numerical method for calculating the effective conductivity and local-field distributions in random R–L–C networks is developed. Using this method, the local field properties of random metal dielectric films are investigated in a wide spectral range and for a variety of metal concentrations p. It is shown that for metal concentrations close to the percolation threshold ($p = p_c$) and frequencies close to the resonance, the local field intensity is characterized by a non-Gaussian, exponentially broad distribution. For low and high metal concentrations a scaling region is formed that is due to the increasing number of non-interacting dipoles. The local electric fields are studied in terms of characteristic length parameters. Properties of both localized and extended eigenmodes in the Kirchhoff's Hamiltonian are investigated.

1. Introduction

The last two decades was a time of immense improvement in our understanding of the optical properties of inhomogeneous media [1]. One of the important representatives of such media is a metal-dielectric composite near the percolation threshold. This type of nanostructured materials has attracted recently lots of attention because of their unique electromagnetic properties. Many fundamental phenomena, such as localization and delocalization of electrons and optical excitations, play an important role in random media. The light-induced plasmon modes in metal-dielectric composites can result in dramatic enhancement of optical responses in a broad spectral range. In particular, percolation metal-dielectric films can be employed for surface-enhanced spectroscopy with unsurpassed sensitivity and

B.A. van Tiggelen and S.E. Skipetrov (eds.),
Wave Scattering in Complex Media: From Theory to Applications, 343–364.

for development of novel optical elements, such as optical switches and efficient optical filters, with transparency windows induced by local photomodification in the composite films.

In the optical and infrared spectral ranges, the metal dielectric permittivity has, typically, a negative real part, so that metal particles can be viewed as inductance elements with small losses (R–L elements). In accordance with this assumption, a metal-dielectric composite can be treated as an R–L–C network, where the C elements stand for dielectric grains, which have a positive dielectric permittivity. Many different approaches based on the effective-medium theories and various numerical models have been suggested to describe the optical nonlinearities of such systems [2]. In particular, a number of numerical simulations have been carried out by using the real space renormalization group [3–8]. A recently developed scaling theory [4–8] for the field fluctuations and high-order field moments predicts localization of the surface plasmons in percolation composites and strong enhancement for the local field, resulting from the localization. Experimental observations [7, 9] in accord with the theoretical predictions show the existence of giant local fields, which can be enhanced by a factor of 10^5 for the linear response and 10^{20} and greater for the nonlinear response. A recent study [10] of the plasmon modes in metal-dielectric films gives more insights into the problem. Thus, in Ref. [10] it was found that for all studied systems the local fields are concentrated in nanometer size areas, while some of the eigenstates are not localized.

Despite the progress, computer modeling of the electric field distribution in metal-dielectric nanocomposites was restricted so far to mainly approximate methods, such as the real space renormalization group (RSRG). To some extent, this was justified since the focus of those calculations was on the effective properties, such as the macroscopic conductivity and dielectric permittivity. Many fast algorithms were suggested for determining the effective conductivities; those include very efficient models, such as Frank and Lobb $Y - \nabla$ transformation [11], the exact numerical renormalization in a vicinity of the percolation threshold [12–14], and the transfer matrix method [15]. Unfortunately, all these methods cannot be used for precise calculation of the local-field distribution and a new approach is needed. The relaxation method (RM) was one of the first algorithms to give some insight into the field distributions [16]. This method has the advantage of using the minimum possible memory, which is proportional to the number of the sites, L^d, where L is the size of the system and d is the space dimensionality. The fast Fourier acceleration [17] allows one to perform calculations for both $2D$ and $3D$ percolation systems. However, the "critical slowing down" effect and the problem of stability (occurring when the imaginary part of the local conductivity takes both positive and negative values) restricts the use of this approach. Thus, the local-field statistics for percolation composites in the optical and infrared spectral ranges was not investigated until very recently, with direct numerical methods not involving any *a priori* assumptions. In their work, Zekri, Bouamrane, and Zekri [18] suggested a substitution method, which allows one to calculate the local field distributions in percolation metal-dielectric composites in the optical range. However, results obtained for the local-field intensity $I = |E|^2$ distribution function $P(I)$ appear to be rather surprising. Specifically, instead of the predicted theoretically and observed experimentally enhancement for the local field, the authors of Ref. [18] obtained average local field intensities far less than the applied field. We note that the high local fields play a crucial role in enhancement for nonlinear optical effects and thus it is important to verify this prediction by exact calculations.

In this work we apply a new numerical method, which we refer to as block elimination (BE) [19]. The BE method allows calculations of effective parameters (such as the conductivity, dielectric permittivity, etc.) and, most importantly, the local field distribution in inhomogeneous media. We focus our attention on the local-field distribution $P(I)$ and compare results obtained by the BE with those following from the RSRG, the relaxation method, and the Zekri-Bouamrane-Zekri (ZBZ) method. Specifically, we investigate the properties of two-dimensional random metal-dielectric composites by modeling them as a square lattice with the lattice size L comprised of dielectric and metal bonds, with conductivity σ_d and σ_m, respectively. The probability of a bond to have the metal conductivity σ_m is equal to p (where p is the metal concentration) while the probability to have the dielectric conductivity σ_d is equal to $1 - p$. We obtain that the local electric field is characterized by sharp peaks that can exceed the applied field by several orders of magnitude, in agreement with earlier theoretical predictions and experimental observations [4–9]. The field maxima are associated with the localized surface plasmons. For the first time a full set of field distribution functions $P(I)$ that gradually transform from the "one-dipole" field distribution to the log-normal distribution are calculated by using the newly developed BE method. Relying on an approach based on the inverse participation ratio, we find important relations for the field correlation length ξ_e, average field localization length ξ_f, and the average distance between the metal particles ξ_a. The eigenvalue problem is solved directly and effects due to the existence of extended states, predicted in [10] are investigated.

2. Block Elimination (BE) Method

In the explanation of the Block Elimination method we will follow the outline introduced in our previous work [19]. We will consider the problem of a local field distribution in nanoscale metal-dielectric films at and away from the percolation threshold. When the wavelength λ of the incident light is much larger than the metal grain size a we can introduce local potential $\varphi(\mathbf{r})$ and local current $\mathbf{j}(\mathbf{r}) = \sigma(\mathbf{r}) \cdot (-\nabla\varphi(\mathbf{r}) + \mathbf{E}_0)$, where $\mathbf{E}_0$ is the applied field and $\sigma(\mathbf{r})$ is the local conductivity. In the quasistatic approximation, the problem of the potential distribution is reduced to solving the current conservation law $\nabla \cdot \mathbf{j}(\mathbf{r}) = 0$, which leads to the Laplace equation $\nabla \cdot [\sigma(\mathbf{r}) \cdot (-\nabla\varphi(\mathbf{r}) + \mathbf{E}_0)] = 0$ for determining the potentials. Now we use the discretization procedure based on the tight-binding model. The film is described as a binary composite of metal and dielectric particles, which are represented by metal and dielectric bonds in the square lattice. The current conservation for lattice site i acquires the following form

$$\sum_j \sigma_{ij}(\varphi_i - \varphi_j + E_{ij}) = 0, \tag{1}$$

where φ_i is the field potential of site i. The summation is over the nearest (to i) neighbor sites j; $\sigma_{ij} = \sigma_{ji}$ are the conductivities of bonds connecting neighbor sites i and j and E_{ij} are the electromotive forces. The electromotive forces E_{ij} are defined so that $E_{ij} = aE_0$, for the bond leaving site i in the "$+y$" direction, and $E_{ij} = -aE_0$, for the bond in the "$-y$" direction; E_{ij} is zero for the "x" bonds. Note that $E_{ij} = -E_{ji}$.

Numerical solution to the Kirchhoff's equations (1) in the case of large lattice sizes encounters immense difficulties and requires very large memory storage and high operational speed. A full set of the Kirchhoff equations for a square lattice of size L is comprised of L^2 separate equations. This system of equations can be written in the matrix form

$$\hat{\mathbf{H}} \cdot \mathbf{\Phi} = \mathbf{F}, \tag{2}$$

where $\hat{\mathbf{H}}$ is a symmetric, $L^2 \times L^2$, matrix that depends on the structure and composition of the lattice, $\mathbf{\Phi} = \{\varphi_i\}$, and $\mathbf{F} = \left\{-\sum_j \sigma_{ij}E_{ij}\right\}$ are vectors of size L^2, which represent the potentials and applied field at each site and bond. In the literature, the matrix $\hat{\mathbf{H}}$ is called the Kirchhoff Hamiltonian (KH) and it is shown to be similar to the Hamiltonian for the Anderson transition problem in quantum mechanics [5, 7–9]. The Kirchhoff Hamiltonian is a sparse random matrix with diagonal elements $H_{ii} = \sum_j \sigma_{ij}$ (where the summation is over all bond conductivities σ_{ij} that connect the i-th site with it neighbors) and nonzero off-diagonal elements $H_{ij} = -\sigma_{ij}$. For detailed description of the KH see the Appendix.

In principle, Eq. (2) can be solved directly by applying the standard Gaussian elimination to the matrix $\hat{\mathbf{H}}$ [20]. This procedure has a run time proportional to $\sim L^6$ and requires a memory space of the order of L^4. Simple estimations show that the direct Gaussian elimination cannot be applied for large lattice sizes, $L > 40$, because of the memory restrictions and long run times for all contemporary personal computers. Fortunately, the KH matrix $\hat{\mathbf{H}}$ has a simple symmetrical structure that allows implementation of block elimination procedure that can reduce significantly the operational time and memory.

In calculations, we can apply the periodic boundary conditions for the "x" and "y" directions; alternatively, we can also impose parallel or "L"-electrode-type boundaries. In the case of the periodic boundary conditions, we suppose that the sites in the first row of the $L \times L$ lattice are connected to the L-th row, whereas the sites of the first column are connected to the last column. Then the Kirchhoff's equations for the first site in the first row, for example, have the following form

$$\sigma_{1,L}\left(\varphi_1 - \varphi_L\right) + \sigma_{1,2}\left(\varphi_1 - \varphi_2\right) + \sigma_{1,L^2-L+1}\left(\varphi_1 - \varphi_{L^2-L+1} - aE_0\right) + \sigma_{1,L+1}\left(\varphi_1 - \varphi_{L+1} + aE_0\right) = 0, \tag{3}$$

where $\sigma_{1,L}$ is the conductivity of the bond connecting the first and the last sites in the first row. The $\sigma_{1,2}$ conductivity connects the first and second sites in the first row, σ_{1,L^2-L+1} connects the first site of the first row and the first site of the L-th row, $\sigma_{1,L+1}$ connects the first sites of the first and the second rows, and the external field E_0 is applied in the "$+y$" direction. Note that the $\sigma_{1,L}$ and σ_{1,L^2-L+1} connections are due to the periodic boundary conditions in the "x" and "y" directions, respectively.

In Eq. (3) we numerate the sites of the $L \times L$ lattice "row by row", from 1 (for the first site in the first row) to L^2 (for the last site in the L-th row). Then, the KH matrix $\hat{\mathbf{H}}$ acquires a block-type structure. As an example, for a system with size $L = 5$, the matrix $\hat{\mathbf{H}}$ takes the following block form:

$$\hat{\mathbf{H}} = \begin{pmatrix} h^{(11)} & h^{(12)} & 0 & 0 & h^{(15)} \\ h^{(21)} & h^{(22)} & h^{(23)} & 0 & 0 \\ 0 & h^{(32)} & h^{(33)} & h^{(34)} & 0 \\ 0 & 0 & h^{(43)} & h^{(44)} & h^{(45)} \\ h^{(51)} & 0 & 0 & h^{(54)} & h^{(55)} \end{pmatrix}, \tag{4}$$

where $h^{(jj)}$ are $L \times L$ tridiagonal matrices with diagonal elements $h_{ii}^{(jj)} = \sum_k \sigma_{i+(j-1)L,\,k}$ (the summation is over the nearest neighbors of the site $i + (j-1)L$, which are located in the j-th row, $1 \le i \le L$), while the diagonal matrices $h^{(kl)} = h^{(lk)}(k \neq l)$ connect the k-th row with the l-th row and vice versa. The matrices in the right upper and in the left bottom corners of the KH matrix $\hat{\mathbf{H}}$ are due to the periodical boundary conditions: they connect the top and the bottom rows and the first and the last columns. The explicit forms for the matrices $h^{(jj)}$ and $h^{(kl)}$ are given in the Appendix.

For large sizes L, the majority of the blocks $h^{(ij)}$ are zero matrices and thus Gaussian elimination will be a very inefficient way to solve the system (2). In fact, in a process of elimination of all block elements below $h^{(11)}$ in matrix (4), the only matrix elements that will change are $h^{(11)}$, $h^{(12)}, h^{(22)}, h^{(15)}$ and $h^{(55)}$ with two more elements appearing in the second and last rows. Thus to eliminate the first block column of the KH we can instead of $\hat{\mathbf{H}}$ work with the following $3L \times 3L$ block matrix (recall that in the considered example we choose, for simplicity, $L = 5$):

$$\hat{\mathbf{h}}^{(1)} = \begin{pmatrix} h^{(11)} & h^{(12)} & h^{(15)} \\ h^{(21)} & h^{(22)} & 0 \\ h^{(51)} & 0 & h^{(55)} \end{pmatrix}. \tag{5}$$

Now to eliminate the first block column of matrix $\hat{\mathbf{h}}^{(1)}$ we apply a standard procedure [19], where by using the diagonal elements of block matrix $h^{(11)}$ as pivots we transform $h^{(11)}$ in a triangle matrix $h^{*(11)}$and simultaneously eliminate $h^{(21)}$ and $h^{(51)}$. The elimination of the first column of $\hat{\mathbf{h}}^{(1)}$and respectively $\hat{\mathbf{H}}$ thus requires only L^3 simple arithmetical operations which is to be compared with L^5 operations needed if we work directly with the whole matrix $\hat{\mathbf{H}}$. After the first step of this block elimination is completed the matrix $\hat{\mathbf{H}}$ has the following form:

$$\hat{\mathbf{H}}^{(1)} = \begin{pmatrix} h^{*(11)} & h^{*(12)} & 0 & 0 & h^{*(15)} \\ 0 & h^{*(22)} & h^{(23)} & 0 & h^{(25)} \\ 0 & h^{(32)} & h^{(33)} & h^{(34)} & 0 \\ 0 & 0 & h^{(43)} & h^{(44)} & h^{(45)} \\ 0 & h^{(52)} & 0 & h^{(54)} & h^{*(55)} \end{pmatrix}, \tag{6}$$

where we denote all blocks that have changed in the elimination process by the "$*$" superscript. The two new block elements $h^{(25)}$ and $h^{(52)}$ appeared due to the interactions of the first row with the second and the fifth rows.

As a second step, we apply the above procedure for the minor $\hat{\mathbf{H}}_{11}^{(1)}$ of the matrix $\hat{\mathbf{H}}^{(1)}$ (which now plays the role of $\hat{\mathbf{H}}$), therefore we work again with $3L \times 3L$ matrix:

$$\hat{\mathbf{h}}^{(2)} = \begin{pmatrix} h^{*(22)} & h^{(23)} & h^{(25)} \\ h^{(32)} & h^{(33)} & 0 \\ h^{(52)} & 0 & h^{*(55)} \end{pmatrix}. \tag{7}$$

Repeating with $\hat{\mathbf{h}}^{(2)}$ all operations we performed on $\hat{\mathbf{h}}^{(1)}$ we put $h^{*(22)}$ in the triangular form and eliminate $h^{(32)}$ and $h^{(52)}$. We continue this procedure until the whole matrix $\hat{\mathbf{H}}$ is converted into the triangular form with all elements below the diagonal being zero. The backward substitution for a triangular matrix is straightforward, namely we obtain first the site potentials in the L-th row (the fifth row, in our example) and then, by calculating the potentials, in the $L-1$ row and so on, until the potentials in all rows are obtained. The total number of operations needed is estimated as $\sim L^4$, for the described block elimination (BE) method, which is less than the number L^6 needed for Gaussian or **LU** (for symmetric matrixes) elimination [20]. The BE has operational speed on the same order of magnitude as in the transfer-matrix method [15] and the Zekri-Bouamrane-Zekri (ZBZ) method [18]. However, BE allows the calculation of the local fields, as opposed to the Franck-Lobb method, and we believe that it is much easier in numerical coding when compared to the ZBZ method.

For a Pentium II 450 MHz processor, the run time we observed is given by the formula $T(L) \simeq 3.2 \cdot 10^{-7} \cdot L^4\,\mathrm{s}$, which for $L = 250$ is less than 23 min. For each step of the BE procedure, we need to keep only L^2 (the matrix $\hat{\mathbf{h}}^{(k)}$) complex numbers in the operational memory and L^3 on a hard disk. By using the hard drive we do not decrease the speed performance significantly because only L loadings of L^2 numbers are required, i.e., L^3 additional operations in total. Note that the BE, similar to the Gaussian elimination, is well suited for parallel computing.

We performed various tests to check the accuracy of the BE algorithm described above. First, the sum of the currents at each site was calculated and the average value $\sim 10^{-14}$ was found; this is low enough to claim that the current conservation holds in the method. Our calculations, using the standard Gaussian elimination (for small lattice sizes) and the relaxation method (for the case of all positive conductivities), for the effective conductivity and the local field distribution show full agreement with results obtained using the developed block elimination procedure.

3. Results for 2D Parallel and L-type Lattices

In inhomogeneous media, such as metal-dielectric composites, both dielectric permittivity $\varepsilon(\mathbf{r})$ and conductivity $\sigma(\mathbf{r}) = -i\omega\varepsilon(\mathbf{r})/4\pi$ depend on the position $\mathbf{r}$. When the size of the composite is much larger than the size of inhomogeneities, the effective conductivity σ_e can be introduced. As discussed above, we model the composite by an R–L–C network and then apply the BE method to find the field potentials at all sites of the square lattice. When the potential distribution is known we can calculate the effective conductivity:

$$\sigma_e \left|\mathbf{E}_0\right|^2 = \frac{1}{S}\int \sigma(\mathbf{r}) \left|\mathbf{E}(\mathbf{r})\right|^2 d\mathbf{r} \tag{8}$$

where $\mathbf{E}(\mathbf{r})$ and $\mathbf{E}_0$ are the local and the applied fields, respectively (see, e.g., [2]). The integration is performed over the film surface S.

It is well known that the effective DC conductivity for a two component random mixture ($\sigma_m \gg \sigma_d$) should vanish as a power law, when the metal concentration p approaches the percolation threshold p_c, i.e.,

$$\sigma_e \sim \sigma_m (p - p_c)^t, \tag{9}$$

where t is the critical exponent, which was calculated and measured by many authors. In the 2D case, the critical exponent is given by $t = 1.28 \pm 0.03$, according to Derrida and Vannimenus [15], and $t = 1.29 \pm 0.02$, according to Frank and Lob [11]. The value $t = 1.33 \pm 0.03$ was found by Sarychev and Vinogradov [13], who used the exact renormalization group procedure and reached the lattice size $L = 500$ in their simulations. In all cases, the critical exponent t was calculated using the finite-size scaling theory [21]. When the volume fraction p of the conducting elements reaches the percolation threshold p_c, the correlation length increases as $\xi \sim (p - p_c)^{-\nu}$, where $\nu = 4/3$ is the critical exponent for the correlation length [2]. Because the correlation length ξ determines the minimum size of the network, for which it can be viewed as homogeneous, one expects that for $L \ll \xi$, the effective conductivity depends on the system size L. The finite-size scaling theory [22, 23] predicts the following dependence:

$$\sigma_e(L) \sim L^{-t/\nu} f(\eta), \tag{10}$$

where the argument $\eta = L^{1/\nu}(p - p_c)$ depends on the system size L and on the proximity to the percolation threshold p_c. For a self-dual lattice, such as the square lattice considered here, the percolation threshold is known exactly: $p_c = 0.5$. When calculations are carried out for $p = p_c$ there is no need for knowledge of the specific form of the function f in Eq. (10).

We calculate the effective conductivity $\sigma_e(L)$ for different sizes L. In order to improve the statistics for each size L, a number of distinct realizations were performed. Specifically we used $40,000$ realizations for $L = 10$; $5,000$ realizations for $L = 20$; $1,000$ realizations for $L = 60$; and 100 realizations for $L = 150$. The data from our calculations was fit to Eq. (10) and the χ^2 analysis was applied to determine the critical exponents. Thus we found that $t/\nu = 0.96 \pm 0.03$ and $t = 1.28 \pm 0.04$. This result is in good agreement with the estimates of Derrida-Vannimenus and Frank-Lobb, but somewhat lower than the $t/\nu = 1.0$ obtained by Sarychev and Vinogradov. Note that the value $t/\nu = 1.0$ is expected for the sizes $L > 300$ that are greater than those we used in our estimates.

4. Local-Field Distribution Function

To further verify the accuracy of the block elimination method, we tested explicitly the field distribution function, for the case when the conductivities are positive and real numbers (i.e., the dielectric permittivity is purely imaginary in this case). The local field distribution $P(I)$ we sampled in terms of $\log I$, where $I = (|\mathbf{E} - \mathbf{E}_0|/|\mathbf{E}_0|)^2$ is the local field intensity fluctuation with $|\mathbf{E}_0|^2$ being the intensity of the applied field. If the bond conductivities σ_d and σ_m are positive (resistor network), we can also apply the relaxation method [17] and compare the results with those obtained with the BE procedure. Such a comparison is presented in Fig. 1, where the metal concentration is chosen to be equal to the percolation threshold $p = p_c$. The distributions obtained with the two exact methods, the Block Elimination (BE) and the Relaxation Method (RM), are nearly the same. The minor deviations are due to the differences in the calculation procedures resulting in different round-off errors, and also because of non-sufficient relaxation times. The local-field intensities are distributed in very wide range that extends from low filed intensities of the order of $I \sim 10^{-8}$ to very high values reaching $I \sim 10^4$.

In the same figure, the field distribution obtained with the real space renormalization group (RSRG) method is also shown. Among most important results obtained with this non-exact method is the extension of the distribution function toward small values of the field intensity I. Such a distortion is obtained for all distributions calculated with this method; however, this does not considerably affect the method's applicability for processes depending on the local field moments. The n-th moment of the field $M_n = \langle |E|^n \rangle / |E_0|^n$ is given by the spatial average over the film surface and thus depends mainly on the local fields with the largest intensities. Because the RSRG calculations differ from the exact values only for low-intensity fields, the method can be used for estimation of the field moments.

Although the case of real positive values for the conductivities is of considerable interest, more important physical problems arise when the metal conductivity is complex. One special case corresponds to the surface plasmon resonance, which plays a crucial role in the optical and infrared spectral ranges for metal-dielectric composites. For the two-dimensional case, this resonance for individual particles occurs when $\sigma_d = -\sigma_m$, and

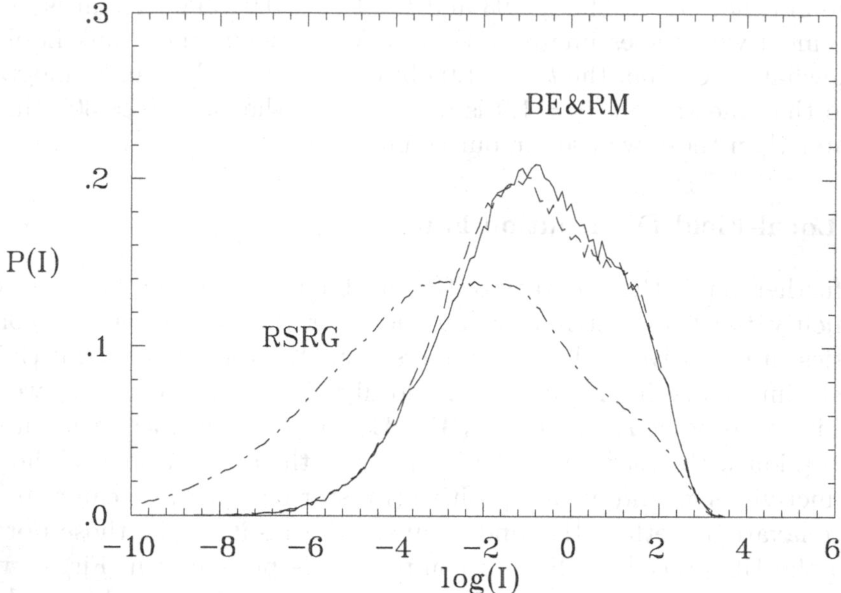

Figure 1. The local-field distribution $P(I)$ calculated with two exact methods, Relaxation Method (RM) and Block Elimination (BE). Results of calculations with approximate, real-space renormalization group (RSRG) are also shown. The ratio of the (real) conductivities for metal and dielectric bonds is chosen as $\sigma_m/\sigma_d = 10^{-3}$.

it can be investigated using a dimensionless set of conductivities $\sigma_d = -i$, and $\sigma_m = i + \kappa$, where $i = \sqrt{-1}$ and κ is a small real conductivity that represents the loses in the system. Recall that in metal-dielectric films the conductivity $\sigma_m = -i\omega\varepsilon_m/4\pi$ is predominantly imaginary with very small real part [23].

In Fig. 2, we show the local-field distributions calculated for three different values of κ, using both the block elimination (BE) and the real space renormalization group (RSRG) procedures. All functions obtained by these two methods differ in shape and peak positions; however, taking into account that the RSRG is indeed an approximate procedure, we can conclude that qualitatively it performs relatively well. All the three local field distributions, which are calculated with the exact BE method, can be approximated by the log-normal function:

$$P(I) = \frac{1}{\Delta I\sqrt{2\pi}} \exp\left[-\frac{(\log I - \langle \log I \rangle)^2}{2\Delta^2}\right], \tag{11}$$

where $\langle \log I \rangle$ is the average value for the logarithm of the local field intensity I and Δ is the standard deviation in terms of $\log I$ ($\log x \equiv \log_{10} x$).

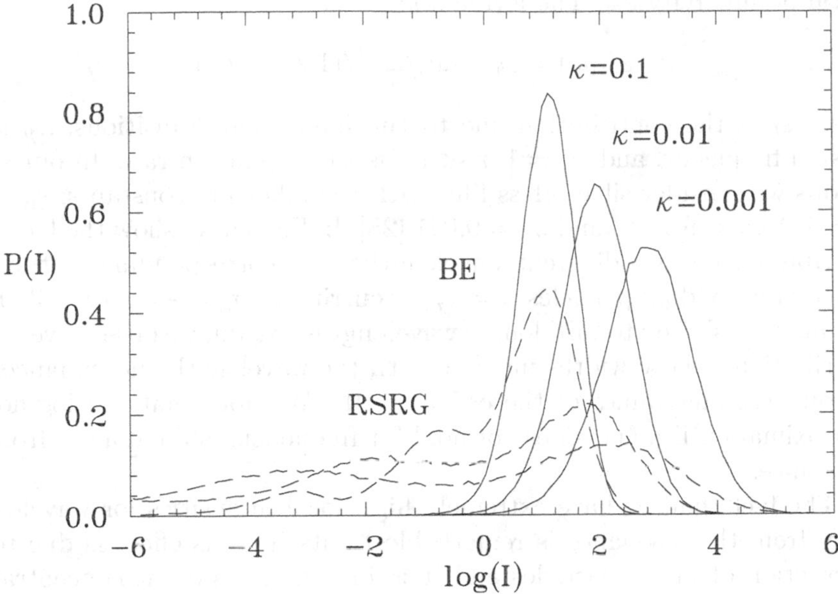

Figure 2. Local-field distributions $P(I)$ calculated for three different loss factors $\kappa = 0.1, 0.01$, and 0.001, using BE and RSRG methods. All distributions are obtained for $p = p_c$.

This approximation for the field distribution seems to work sufficiently well around the average value $\overline{s} = \langle \log I \rangle$. We note, however, that according to Ref. [25], where the current distribution was studied, Eq. (11) probably will fail for the intensities I far from the logarithmic average $\overline{s}$. The occurrence of log-normal distribution in a disordered system is related to localization of plasmon modes. A similar type of dependence was found for the conductance in the Anderson transition problem [26]. In Fig. 2 we can also see that $\langle \log I \rangle$ and Δ both increase when κ decreases. The increase in the logarithm of the average of the local field can be explained by correlations between the loss parameter κ and the quality-factor, which leads to relation $\left\langle |E|^2 \right\rangle \sim \kappa^{-1}$ [4]: the smaller the losses, the higher the local fields.

The reference system with $\sigma_d = -i$ and $\sigma_m = i + \kappa$ is an important case for studying some fundamental properties of metal-dielectric films, but it can not be applied for real metals, where σ_m depends on the wavelength. In order to extend our studies to arbitrary materials we can use available experimental data and theoretical models. For the case of metals, the Drude formula can be used that describes well important characteristics of the metal permittivity ε_m. The formula is

$$\varepsilon_m(\omega) = \varepsilon_b - (\omega_p/\omega)^2/(1 + i\omega_\tau/\omega), \tag{12}$$

where ε_b is the contribution due to the inter-band transitions, ω_p is the plasma frequency, and $\omega_\tau = 1/\tau \ll \omega_p$ is the relaxation rate. In our calculations we consider silver-glass film with the following constants: $\varepsilon_d = 2.2$, $\varepsilon_b = 5.0$, $\omega_p = 9.1$ eV, and $\omega_\tau = 0.021$ [28]. In Fig. 3a we show the local field distribution for two different wavelengths: one corresponding to the resonance of individual particles $\omega = \omega_r$, occurring at $\sigma_d = -\sigma_m$ ($\lambda \sim 370$ nm) and another shifted toward longer wavelengths. Again, we observe very wide distributions whose widths increase with the wavelength and enhancement factors reaching values of the order of 10^5. We note that the log-normal approximation Eq. (11) does not hold for frequencies shifted away from the resonance.

The fact that we have extremely high local intensities for wavelengths away from the resonance is remarkable by itself. This effect is due to the interaction of metal particles and it is best manifested at concentrations close to the percolation threshold. A similar long-wavelength spectral behavior was observed in fractal aggregates and is quantitatively explained by the long range character of the dipole-dipole interaction [29]. Because the dipole-dipole interactions are relatively weak, it is expected that for low metal concentrations there should be considerable change in the field distribution. To investigate thoroughly this dependence, we calculated the local field distribution function $P(I)$ for surface metal coverages that deviate from the percolation threshold value. Our results are shown in Fig. 3b where we plot the field distribution for three different metal concentrations: $p = 0.5$, 0.01, and 0.001 at the resonant wavelength $\lambda = 370$ nm. We also include the case of a single metal bond (dipole) positioned in the center of the lattice. The graph shows that there is an apparent transition from the log-normal ($p = p_c$) distribution to a distributions with a "scaling" (power-law) dependence. The appearance of such scaling regions is due to the transformation of the composite film from a strongly coupled dipole system at the percolation threshold into a randomly distributed, sparse configuration of non-interacting dipoles at lower metal concentrations. The range of the scaling interval increases gradually with the decrease of the metal concentration until it "consumes" the entire distribution for the case of a single dipole. In two dimensions, a single dipole placed in the center of the coordinate system induces an electric field with intensity $I_{dip}(r, \theta) = \gamma \cos^2\theta/r^4$, where $r = |\mathbf{r}|$ is the modulus of the radius-vector $\mathbf{r} = \{x, y\}$ and θ is the angle between the field polarization and $\mathbf{r}$. To find the actual one-dipole field distribution $P_{dip}(I)$ we consider the above one-dipole intensity $I_{dip}(r, \theta)$ over the square lattice and then we count the "identical" magnitudes of the logarithm of

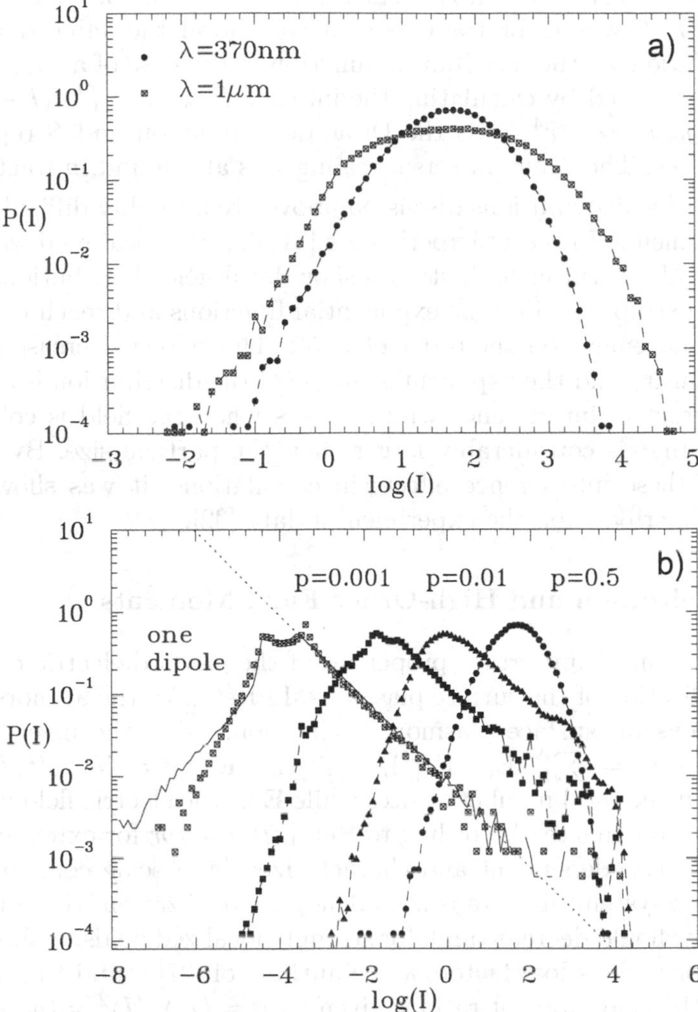

Figure 3. Local-field distributions $P(I)$ for silver-glass films, (a) for $\lambda = 370$ nm and $\lambda = 1\mu m$ at $p = p_c$; (b) for different metal filling factors p at $\lambda = 370$ nm (corresponding to $\sigma_m = -\sigma_d$).

the field-intensity I. The resultant curve for the one-dipole field distribution (the solid line in Fig. 3b) should be compared with the field distribution obtained by BE calculations for one metal bond positioned in the center of the film. Both distributions match extremely well; it can be seen that our method captures even the smallest effects on the distribution caused by the cosine term and the square geometry of the lattice. The slope of the scaling region is preserved for all concentrations p and can be fit to the relation $P(I) \sim I^{-\alpha}$, where for the exponent we obtain the value $\alpha = 3/2$. The same relation for the distribution function in the case of a single dipole can be easily derived by calculating the integral $P_{dip}(I) = \iint \delta(I - I_{dip}(r))dS$, where $I_{dip}(r) \sim 1/r^4$, δ is the Dirac delta-function and S represents the film surface. The same universal scaling was also found in fractals [30].

The wide distributions discussed above are probably difficult to observe in experiments. In recent experiments [31, 32], the local filed was measured over the film surface; both studies show local-field distributions $P(I)$ that are well extrapolated by the exponential functions and reach the maximum field enhancement on the order of ~ 50. The strong decrease in the local field intensity and the exponential shape of the distribution is explained by the destructive interference which occurs when the field is collected from an area that is considerably larger then the particle size. By taking into account these interference effects in calculations, it was shown that the theory describes well the experimental data [32].

5. Localization and High-Order Field Moments

One of the most important properties of the metal-dielectric composites is the localization of the surface plasmons. In Ref. [27], the authors performed estimations for surface plasmon localization, using the inverse participation ratio $ipr = (\sum_i^N |\mathbf{E}_i - \mathbf{E}_0|^4)/(\sum_i^N |\mathbf{E}_i - \mathbf{E}_0|^2)^2 = N^{-1}\langle I^2\rangle/\langle I\rangle^2$, where $N = L^d$ it the total number of sites while $\mathbf{E}_i$ is the electric field vector corresponding to i-th site. According to Ref. [27], the ipr for extended plasmons should be size-dependent and characterized by a scale comparable to the size of the system; if there is a tendency to localization, the corresponding exponent should decrease and, for strongly localized fields, it should become unity. For various loss factors κ the authors of [27] found that $ipr \sim L^{-1.3}$ so that the field moment ratio is given as $R = \langle I^2\rangle/\langle I\rangle^2 = ipr \times L^d \sim L^{0.7}$. This result leads to size-dependent field moments which for large L should not be the case. Below we show that the earlier theory [4–8], which is based on Eq. (1), is indeed size-independent and supports the conclusion on plasmon localization with the exact BE method. We will also extract some important relationships that describe statistical properties of the local fields in semicontinuous metal films.

We first focus on the most simple case when there is only one dipole in the entire space. For a single dipole it is easy to obtain the relation $R = \langle I^2\rangle/\langle I\rangle^2 \simeq \frac{1}{3}\varkappa^{\frac{1}{2}}$, where $\varkappa = I_{\max}/I_{\min}$ is the ratio of the maximum (close to the particle) and minimum (away from the particle) in the field intensities. Because of the power-law dependence $I_{dip} \sim r^{-4}$, there is a size dependence, $R \simeq \frac{1}{3}(l/a)^2 = \frac{1}{3}L^2$, where l is the length scale of space that is under consideration and a is the average particle size. The size dependence for the one-dipole local-field moments is an expected result since the weight of the low-magnitude fields becomes progressively larger with the increase of the film surface. However, for practical applications, we are interested in systems with large numbers of particles so that they can be viewed as macroscopically homogeneous. We can write this condition as $n_a = (l/\xi_a)^d \gg 1$, or $(aL/\xi_a)^d = pL^d \gg 1$, where p is the volume fraction and ξ_a is the average distance between the metal particles. Now for the theory to be size-independent ($R(L) \sim$ const) the condition $L \gg p^{-1/d}$ has to be enforced.

By investigating the dynamics of the field moments ratio R we can also determine relationships between important statistical quantities, such as field correlation length ξ_e and field localization length ξ_f. By the field correlation length ξ_e, we understand the average distance between the field peaks, while we characterize their spatial extension by the field localization length ξ_f [23]. For non-overlapping peaks, one can find that $R = N/(N_e N_F) = (\xi_e/\xi_f)^d$, where $N = (l/a)^d = L^d$ is the total number of sites, $N_e = (l/\xi_e)^d$ is the total number of the field peaks, each one occupying $N_F = (\xi_f/a)^d$ sites. In general, for $L \gg p^{-1/d}$, we expect R to be a function of p (but not of L) and κ; the same is true for the statistical length ξ_e.

To determine this dependence we run calculations for two loss factors, $\kappa = 0.1$ and $\kappa = 0.01$. As illustrated in Fig. 4, for both cases, R can be approximated as:

$$R(\kappa, p) = \eta(\kappa) \left\{ \left[\theta(p) - \theta(p - \frac{1}{2}) \right] p^{-\tau} + \theta(p - \frac{1}{2})(1-p)^{-\tau} \right\}, \qquad (13)$$

where θ is the step-function. For the exponent τ, we obtain the value which is close to the ratio 2/3. For $p \leq 0.5$ and $d = 2$ this value yields the following relationship for the field correlation length: $\xi_e \simeq \xi_f p^{-1/3}\sqrt{\eta(\kappa)} = \xi_f(\xi_e/a)^{2/3}\sqrt{\eta(\kappa)}$, where the function $\eta(\kappa)$ increases when κ decreases. The analysis of the ratio ξ_e/ξ_f shows that we should expect an increase of the localization strength with a decrease of both surface coverage p and loss factor κ. In the special case of a single dipole we have $R = (\xi_e/\xi_f)^2 = \frac{1}{3}L^2$, which, combined with $\xi_e = aL$, yields for the field localization length $\xi_f = a\sqrt{3}$.

The localization of the electric filed into "hot" spots can be easily seen in Fig. 5, where we show (for different wavelengths) the spatial distribution of the local intensity $I(\mathbf{r})$, and the fourth moment of the local fields, $I^2(\mathbf{r})$. Note, that $I^2(\mathbf{r})$ is proportional to the local Raman scattering provided that Raman-active molecules are covering the film [7]. As mentioned, res-

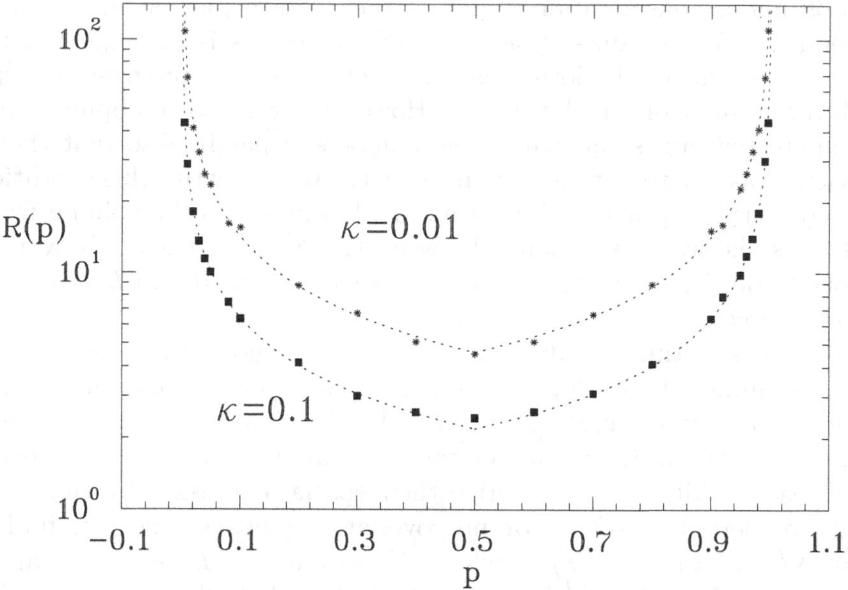

Figure 4. The ratio of the local-field moments, $R = M_4/(M_2)^2$ as a function of the metal coverage p, for two different values of the loss-factor κ (both values satisfy the inequality $\xi_e \ll aL$); the dashed lines represent fits based on Eq. (13).

onance condition for isolated silver particles is fulfilled at the wavelength $\lambda \approx 370$ nm. In Fig. 5, we see that the fluctuating local fields are well localized and enhanced with the enhancement of the order of 10^4 for $I(\mathbf{r})$, and 10^9 for $I^2(\mathbf{r})$. The spatial separation of the local peaks has a minimum when the wavelength of the applied filed corresponds to the single particle resonance. In this case, most of particles resonate and the local filed is enhanced randomly all over the film surface (Fig. 5a,b). With the increase of the wavelength, only few spatial regions can support propagation of plasmon modes which in turn leads to very high fields, significantly larger than those observed in the single particle resonance case. All these results support the assumption of plasmon localization in random metal-dielectric films and they are in qualitative agreement with the previously developed theory [4–8].

Based on similarities between the Kirchhoff's Hamiltonian $\hat{\mathbf{H}}$ and the quantum-mechanical Hamiltonian for the Anderson transition problem, the scaling theory predicts that there should be a power-law dependence for the

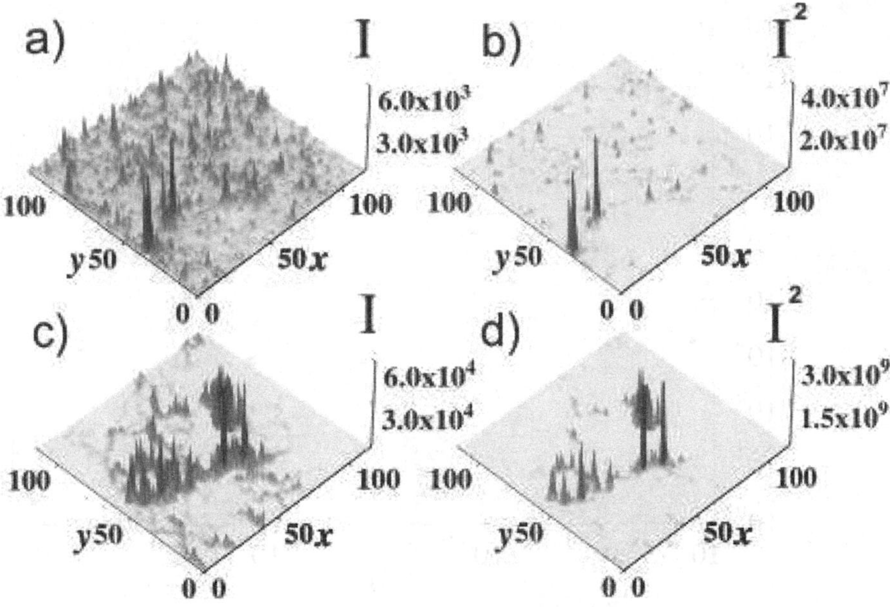

Figure 5. The spatial distributions of the normalized local intensity, $I(x, y)$, and of the "local Raman enhancement factor", $I^2(x, y)$. The distributions are calculated at two different wavelengths: $\lambda = 0.370\ \mu$m (a, b) and and $\lambda = 10\ \mu$m (c, d). The metal filling factor is chosen as $p = p_c$ for all cases.

higher-order field moments:

$$M_n = \langle |\mathbf{E}|^n \rangle / |\mathbf{E}_0|^n \sim \int d\Lambda \frac{\rho(\Lambda)[a/\xi_f(\Lambda)]^{2n-d}}{[\Lambda^2 + \kappa^2]^{n/2}} \sim \kappa^{-n+1}, \tag{14}$$

where $n = 2, 3, 4, \ldots$, $\rho(\Lambda)$ is the density of states, $\xi_f(\Lambda)$ represents the average single mode localization length which corresponds to eigenvalue Λ and κ is the loss factor [7]. This functional dependence was checked earlier, using the approximate real space renormalization group (RSRG) method, where qualitative agreement was accomplished with Eq. (14). However, since the renormalization procedure is not exact, it is worth estimating the field moments with the exact BE method. To determine the field moments M_n, we used the BE procedure for surface filling fraction $p = p_c$ and $\kappa \in 1 \div 10^{-3}$. Our results are shown on the log-log scale in Fig. 6. The data points represent a fit to a power law with each field moment having different exponents.

For M_2 we obtained the exponent $x_2 = 1.0 \pm 0.1$, which is close to the one predicted by the scaling theory. For the third and the fourth moments, we obtain that $M_{3,4} \sim \kappa^{-x_{3,4}}$, where the exponents $x_{3,4}$ are estimated as

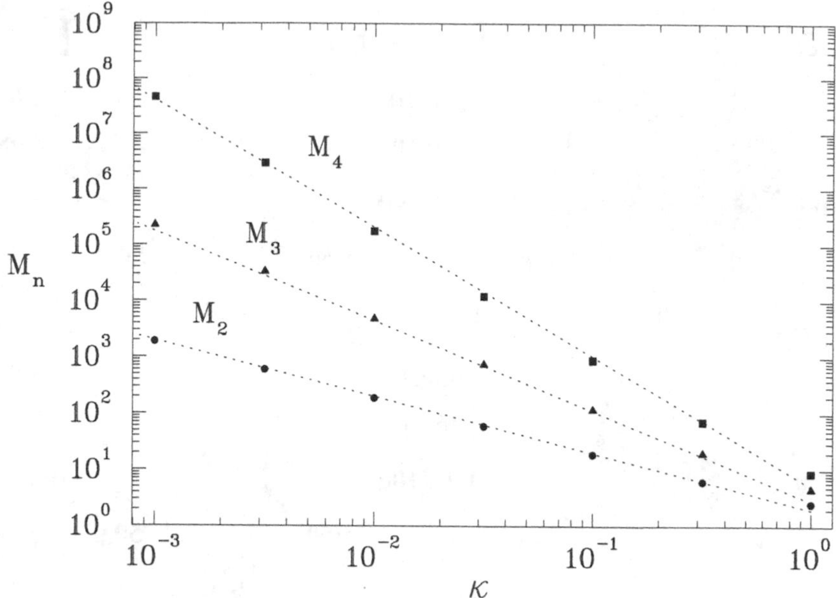

Figure 6. High-order field moments M_2, M_3, and M_4, as functions of the loss parameter κ; the calculations were performed for 100 different realizations in each case, for a lattice with size $L = 150$.

$x_3 = 1.7 \pm 0.1$, and $x_4 = 2.4 \pm 0.2$ which are somewhat different from the values predicted by Eq. (14): $x_3 = 2$ and $x_4 = 3$. This slight difference between the predicted and calculated values of the field moments exponents suggests the possibility of existence of non-localized or extended eigenmodes in the bond percolation model.

As we have mentioned above the scaling solution (14) is based on the assumption that the localization length $\xi_f(\Lambda)$ is finite for all Λ and it does not scale with the size of the system. If the function $\xi_f(\Lambda)$ has a pole, for example, at $\Lambda = 0$ (note that in the previous publications, we used the notation ξ_A for this case), this can lead to a change in the scaling indices, which is responsible for the difference above in the indices in Eq. (14) and those found from the exact numerical method. The minimum of the correlation length ξ_e at the percolation threshold and the log-normal distribution resulting from the strong coupling between the dipoles also suggest that at $\Lambda = 0$ we can expect localization-delocalization transition [10, 33]. To explicitly determine $\xi_f(\Lambda)$ we solve the eigenvalue problem for the real part $\hat{\mathbf{H}}'$ of the Kirchhoff's Hamiltonian $\hat{\mathbf{H}} = \hat{\mathbf{H}}' + i\kappa\hat{\mathbf{H}}''$ in $2D$. The eigenvalue problem was solved with *Mathematica* software for lattice sizes up to $L = 50$. In our calculations of the localization length we used the

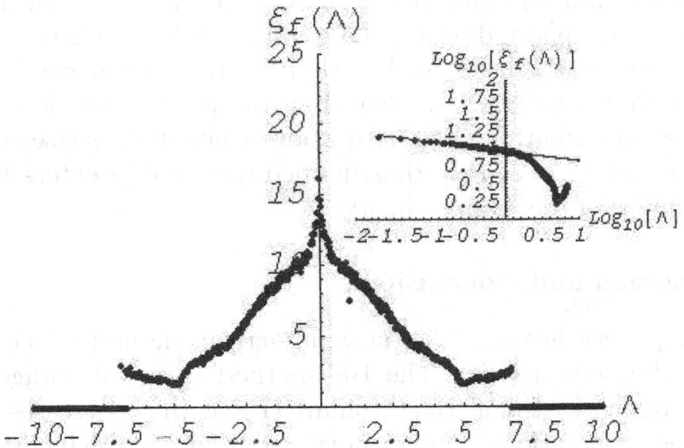

Figure 7. Localization length $\xi_f(\Lambda)$ as a function of the eigenvalues Λ calculated for metal concentration equal to the percolation threshold p_c. The log-log inset depicts the scaling region with an exponent $\chi \approx 0.14$.

inverse participation ratio so that for each eigenmode $\mathbf{\Psi}_n$ that satisfies the equation $\hat{\mathbf{H}}'\mathbf{\Psi}_n = \Lambda_n \mathbf{\Psi}_n$, the localization length for n-th mode $\xi_f^{(n)}$ is given by $\xi_f^{(n)} = [(\sum_{i,j}^N |\mathbf{E}_n(i,j)|^4)/(\sum_{i,j}^N |\mathbf{E}_n(i,j)|^2)^2]^{-1/2}$, where $\mathbf{E}_n = -\nabla\mathbf{\Psi}_n$. Results for the average localization length $\xi_f(\Lambda)$ are shown in Fig. 7.

This figure illustrates that all states but $\Lambda = 0$ are localized as predicted by the theory. Localization lengths are symmetrically distributed with respect to the zero eigenvalue and scale as a power law $\xi_f(\Lambda) \sim \Lambda^{-\chi}$ (that can be seen from the log-log inset). For the delocalization exponent we obtain a value $\chi = 0.14 \pm 0.02$. By substituting the power law dependence for the field localization length in Eq. (14) and performing the integration we arrive at a new modified expression for the field moments in the form $M_n \sim \kappa^{-n(1-\chi)+1}$. Using this equation we easily obtain new exponents x_3 and x_4 that have the values $x_3 = 1.58 \pm 0.06$ and $x_4 = 2.44 \pm 0.08$. These exponents are in much better agreement with those found in the simulations. We note that although the presence of delocalized states at $\Lambda = 0$ results in a slight change of the critical exponents in Eq. (14), all basic conclusions of the previously developed scaling theory still hold because the relative weight of the delocalized states is small.

The presence of non-localized states in random metal-dielectric films was also investigated by Stockman, Faleev and Bergman [10]. While the results we have presented are in qualitative agreement with Ref. [10], it is difficult to compare them quantitatively. This difficulty arises from the fact that in the calculations of the localization length, the authors of [10] rely on the gyration radius. However, for eigenstates consisting of two (or more) spatially separated peaks, the gyration radius is characterized by the distance between the peaks rather than by the spatial sizes of individual peaks, which can be much smaller than the peak separation. In contrast, the inverse participation ratio used above characterizes sizes of individual peaks. We note that namely thus defined quantity ξ_f enters Eq. (14) and other formulas of the scaling theory.

6. Discussion and Conclusions

In this paper we introduced a new numerical method which we refer to as block elimination (BE). The BE method takes advantage of a block structure of the Kirchhoff Hamiltonian $\hat{\mathbf{H}}$ and thus decreases the amount of numerical operations and memory required for solving the Kirchhoff equations for square networks. Note that this method is exact as opposed to previously used numerical methods, most of which are approximate. The results obtained show that the BE method reproduces well the known critical exponents and distribution functions obtained by other methods. The BE verifies the large enhancement of the local electric field predicted by the earlier theory [4–8]. Specifically, the BE results are in good accord with the estimates following from the real space renormalization group.

Besides suggesting a new efficient numerical method, we thoroughly examined the local field distribution function $P(I)$ for different metal filling factors p and loss factors κ. The important result here is that in the optical and infrared spectral range, the local electric field intensity is distributed over an exponentially broad range; specifically, the function $P(I)$ can be characterized by the log-normal function. The latter result, however, holds only in a close vicinity of the percolation threshold and for the light frequencies close or equal to the surface plasmon resonance of individual metal particles. For metal concentrations far away from the percolation region, a power-law behavior was found for $P(I)$. This "scaling" tail in the local field distribution can be related to the one-dipole distribution function. The BE method also verifies the localization of plasmons predicted earlier by the scaling theory. The ensemble average high-order moments for the local field have also been calculated. We found a power law exponents that are in qualitative accord with the scaling theory. With the introduction of corrections due to the presence of extended eigenmodes in the KH we obtained very good agreement between theory and simulations.

Acknowledgement

This work was supported by Battelle under contract DAAD19-02-D-0001, NASA (NCC-1-01049), ARO (DAAD19-01-1-0682), and NSF (E SC-02104

45).

Appendix

In this Appendix we outline the construction of the KH in terms of the bond conductivities. As we show in Section 2, the Kirchhoff equations in the quasistatic approximation provide solutions for the field distribution in a composite medium. We consider the construction of the matrix Eq. (4) for the two-dimensional case (the three-dimension procedure is analogous) and treat a metal-dielectric film as a square lattice of size L. The field potentials at the sites of the lattice are described by vector $\{\varphi_i\}$, where $i = 1, 2, \ldots, L^2$. All sites are connected by conducting bonds $\sigma_{i,j}$, where index $j = \{i-1, i+1, i+L, i-L\}$ includes all the nearest neighbors for site i. Then, we can re-write Eq. (1) in the following form:

$$
\begin{aligned}
&-\frac{1}{\Delta}[\sigma_{i,i+1}(\varphi_{i+1}-\varphi_i)-\sigma_{i,i-1}(\varphi_i-\varphi_{i-1})]+E_{0x}(\sigma_{i,i+1}-\sigma_{i,i-1}) \qquad (15)\\
&-\frac{1}{\Delta}[\sigma_{i,i+L}(\varphi_{i+L}-\varphi_i)-\sigma_{i,j-L}(\varphi_i-\varphi_{i-L})]+E_{0y}(\sigma_{i,i+L}-\sigma_{i,i-L})=0,
\end{aligned}
$$

where $\Delta = a = 1/L$ is the bond length and the pair (E_{0x}, E_{0y}) represents the components of the applied electric field. We can rewrite Eq. (15) in a slightly different way:

$$
\begin{aligned}
h_{i,i}^{(jj)}\varphi_{i+(j-1)L}&+h_{i,i+1}^{(jj)}\varphi_{i+(j-1)L+1}+h_{i,i-1}^{(jj)}\varphi_{i+(j-1)L-1}\\
&+h_{i,i}^{(j,j+1)}\varphi_{i+jL}+h_{i,i}^{(j-1,j)}\varphi_{i+(j-2)L}=F_i^{(j)}, \qquad (16)
\end{aligned}
$$

where $i' = i + (j-1)L$. If $L < i' < L^2 - L$, the components of matrices $h^{(ij)}$ and vectors $F^{(j)}$ can be written as $h_{i,i}^{(jj)} = \sigma_{i',i'+1} + \sigma_{i',i'-1} + \sigma_{i',i'+L} + \sigma_{i',i'-L}$, $h_{i,i+1}^{(jj)} = -\sigma_{i',i'+1}$, $h_{i,i-1}^{(jj)} = -\sigma_{i',i'-1}$, $h_{i,i}^{(j,j+1)} = -\sigma_{i',i'+L}$, $h_{i,i}^{(j-1,j)} = -\sigma_{i',i'-L}$, and $F_i^{(j)} = -\Delta\left[E_{0x}(\sigma_{i',i'+1} - \sigma_{i',i'-1}) + E_{0y}(\sigma_{i',i'+L} - \sigma_{i',i'-L})\right]$. The elements in the first and the last rows of matrix $\hat{\mathbf{H}}$ in Eq. (4), however, must be described in accordance with the boundary conditions. If the parallel boundaries (zero on the bottom and unity on the top) are used, then $h_{i,j}^{(11)} = h_{i,j}^{(LL)} = \delta_{ij}$; for the periodic boundary conditions, the matrix elements are described by relations similar to Eq. (3). We should point out that for the periodic boundaries, the matrix $\hat{\mathbf{H}}$ has rank $L^2 - 1$, and in order for the system to have a solution, one of the site potentials must be grounded.

References

1. W.L. Mochan and R.G. Barrera, *Physica A* **241** (1997), 1–452 .
2. D.J. Bergman and D. Stroud, *Solid State Physics* **46** (1992), 147–269.
3. A.K. Sarychev, *Zh. Eksp. Teor. Fiz.* **72** (1977), 1001–1006.
4. V.M. Shalaev and A.K. Sarychev, *Phys. Rev. B* **57** (1998), 13265–13288.
5. A.K. Sarychev, V.A. Shubin, and V.M. Shalaev, *Phys. Rev. B* **60** (1999), 16389–16409.
6. A.K. Sarychev, V.A. Shubin and V.M. Shalaev, *Phys. Rev. E* **59** (1999), 7239–7242.
7. A.K. Sarychev and V.M. Shalaev, *Physics Reports* **335** (2000), 275–371.
8. V. M. Shalaev, *Nonlinear Optics of Random Media: Fractal Composites and Metal-Dielectric Films*, Springer Tracts in Modern Physics v. 158, Springer, Berlin, Germany, 2000.
9. S. Gresillon, L. Aigouy , A.C. Boccara, J.C. Rivoal, X. Quelin, C. Desmarest, P. Gadenne, V.A. Shubin, A.K. Sarychev and V.M. Shalaev, *Phys. Rev. Lett.* **82** (1999), 4520–4523.
10. M.I. Stockman, S.V. Faleev and D.J. Bergman, *Phys. Rev. Lett.* **87** (2001), 167401–167404.
11. D.J. Frank and C.J. Lobb, *Phys. Rev. B* **37** (1988), 302–307.
12. L. Tortet, J.R. Gavarri, J. Musso, G. Nihoul, J.P. Clerc , A.N. Lagarkov and A.K. Sarychev, *Phys. Rev. B* **58** (1998), 5390.
13. A.K. Sarychev and A.P. Vinogradov, *J. Phys. C: Solid State Phys.* **14** (1981), L487–L490 .
14. J.P. Clerc, V.A. Podolskiy, A.K. Sarychev, *Europhys. J. B* **15** (2000), 507–516.
15. B. Derrida and J. Vannimenus, *J. Phys. A: Math. Gen.* **15** (1982), L557–L564.
16. S. Kirkpatrick, *Phys. Rev. Lett.* **27** (1971), 1722–1741.
17. G.G. Bartrouni, A. Hansen, and M. Nelkin, *Phys. Rev. Lett.* **57** (1986), 1336–1339.
18. L. Zekri, R. Bouamrane and N. Zekri, *J. Phys. A: Math. Gen.* **33** (2000), 649–656.
19. D.A. Genov, A.K. Sarychev and V.M. Shalaev, *Phys. Rev. E* (2002) (submited).
20. R. Coult et al., *Computational Methods in Linear Algebra,* John Wiley & Sons, New York, 1975.
21. V. Privman, *Finite-Size Scaling and Numerical Simulations of Statistical Systems,* World Scientific, Singapore, 1988.
22. D. Stauffer and A. Aharony, *Introduction to Percolation Theory,* 2nd Edition, Taylor and Francis, London, 1992.
23. F. Brouers, S. Blacher and A. K. Sarychev, *Phys. Rev. B* **58** (1998), 15897–15907.
24. J. Cardy, *Finite-Size Scaling,* North Holland, Amsterdam, 1981.
25. A. Aharony, R. Blumenfeld and A.B. Harris, *Phys. Rev. B* **47** (1993), 5756–5769.
26. B. Kramer and A. MacKinnon, *Rep. Prog. Phys.* **56** (1993), 1469–1564.
27. L. Zekri, R. Bouamrane, N. Zekri and F. Brouers, *J. Phys.: Cond. Matter* **12** (2000), 283–291.
28. E.D. Palik (Ed.), *Hand book of Oprical Constants of Solids,* Academic Press, New York, 1985.
29. V.A. Markel, V.M. Shalaev, E.B. Stechel, W. Kim and R.L. Armstrong, *Phys. Rev. B* **53** (1996), 2425,-2436.
30. M.I. Stockman, N.L. Pandey, L.S. Muratov and T.F. George, *Phys. Rev. Lett.* **72** (1994), 2486–2489.
31. S. Bozhevolnyi and V. Coello, *Phys. Rev. B* **64** (2001), 115414–115421.
32. Katayayani Seal, M.A. Nelson, Z.C. Ying, D.A. Genov, A.K. Sarychev and V. M. Shalaev, *Phys. Rev. B* (2002) (submited).
33. K. Müller, B. Mehling, F. Milde and M. Schreiber, *Phys. Rev. Lett.* **78** (1997), 215–218.

CHAPTER V

RADIATIVE TRANSFER

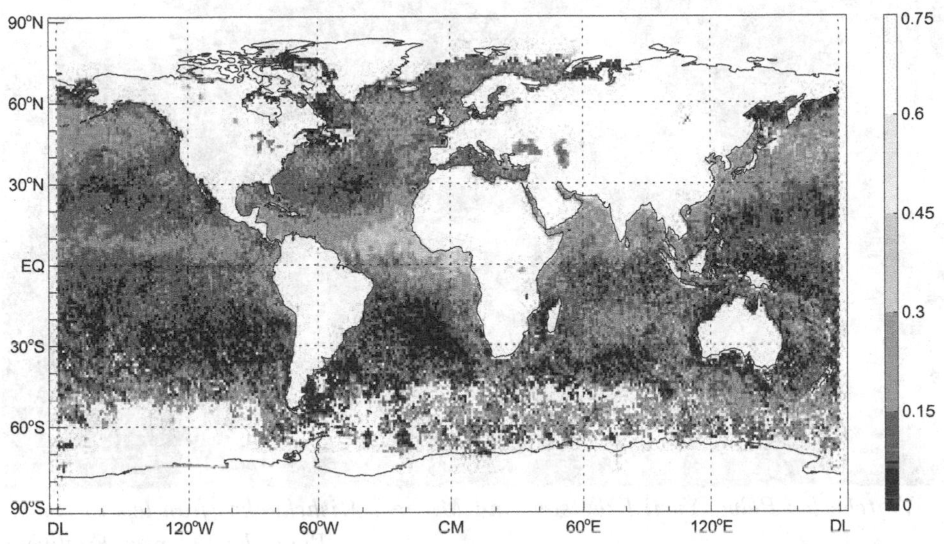

Global map of aerosol optical thickness averaged over the months of February–July 2000. The results are retrieved from channel 1 and 2 radiances measured by the Advanced Very High Resolution Radiometer on board of NOAA weather satellites. The data are not available over the land as well as over areas with extensive cloud coverage (these areas are left white). The details of the retrieval process are described in the paper M.I. Mishchenko, I.V. Geogdzhayev, B. Cairns, W.B. Rossow, and A.A. Lacis (1999) Aerosol retrievals over the ocean by use of channels 1 and 2 AVHRR data: sensitivity analysis and preliminary results, Appl. Opt. 38, 7325-7341. Image courtesy of Igor Geogdzhayev and Michael Mishchenko, Goddard Institute for Space Studies, New York, USA.

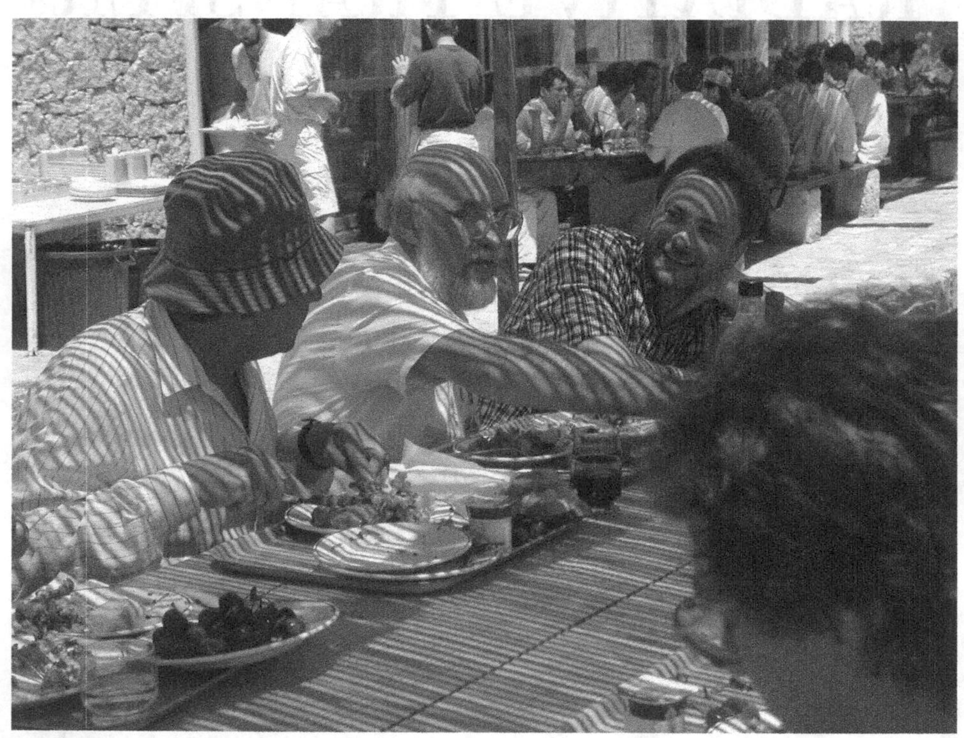

Viatcheslav Belyi, Yurii Kravtsov, and Michael Mishchenko (fron left to right).
Photo by Valentin Freilikher

RADIATIVE TRANSFER THEORY: FROM MAXWELL'S EQUATIONS TO PRACTICAL APPLICATIONS

MICHAEL I. MISHCHENKO
NASA Goddard Institute for Space Studies, 2880 Broadway, New York, NY 10025, USA

1. Introduction

Since the pioneering papers by Khvolson [1] and Schuster [2], the radiative transfer theory (RTT) has been a basic working tool in astrophysics, atmospheric physics, and remote sensing [3–11], while the radiative transfer equation (RTE) has become a classical equation of mathematical physics [12–15]. However, the RTT has been often criticized for its phenomenological character, lack of solid physical background, and unknown range of applicability [e.g., 16]. The past three decades have demonstrated substantial progress in studies of the statistical wave content of the RTT (e.g., [17–24] and references therein). This research has resulted in a much better understanding of the physical foundation of the RTT and has ultimately made the RTE a corollary of the statistical electromagnetics [25].

The aim of this chapter is to demonstrate how the RTE follows from the Maxwell equations when the latter are applied to the problem of multiple electromagnetic scattering in discrete random media and to discuss how this equation can be solved in practice. The following section contains a brief summary of those principles of classical electromagnetics that form the basis of the theory of single light scattering by a small particle. Section 3 outlines the derivation of the general RTE starting from the vector form of the Foldy-Lax equations for a fixed N-particle system and their far-field version. Based on the assumption that particle positions are completely random, the RTE is derived by applying the Twersky approximation to the coherent electric field and the Twersky and ladder approximations to the coherency dyad of the diffuse field in the limit $N \to \infty$. We then discuss in detail the assumptions leading to the RTE and the physical meaning of the quantities entering this equation. The final section describes a general technique for solving the RTE that allows efficient software implementation and leads to physically based practical applications.

2. Single scattering

Many quantities used in the derivation of the RTE and finally entering it originate in the electromagnetic theory of scattering by a single particle. Therefore, we will introduce in this section the necessary single-scattering concepts and definitions and briefly recapitulate the

B.A. van Tiggelen and S.E. Skipetrov (eds.),
Wave Scattering in Complex Media: From Theory to Applications, 367–414.

results that will be necessary for understanding the material presented in the following sections. A comprehensive treatment of the subject of single scattering, including explicit derivations of all formulas, can be found in [26].

2.1. COHERENCY MATRIX, COHERENCY VECTOR, AND STOKES VECTOR

In order to introduce the basic radiometric and polarimetric characteristics of a transverse electromagnetic wave, we use a local Cartesian coordinate system with origin at the observation point (Fig. 1) and specify the direction of propagation of the wave by a unit vector $\hat{\mathbf{n}} = \{\theta, \varphi\}$, where $\theta \in [0, \pi]$ is the zenith angle and $\varphi \in [0, 2\pi)$ is the azimuth angle measured from the positive x-axis in the clockwise direction when looking in the direction of the positive z-axis. Because the wave is assumed to be transverse, the electric field at the observation point can be expressed as $\mathbf{E} = \mathbf{E}_\theta + \mathbf{E}_\varphi = E_\theta \hat{\boldsymbol{\theta}} + E_\varphi \hat{\boldsymbol{\varphi}}$, where $\mathbf{E}_\theta$ and $\mathbf{E}_\varphi$ are the θ- and φ-components of the electric field vector.

Consider a time-harmonic plane electromagnetic wave propagating in a homogeneous, linear, isotropic, and nonabsorbing medium with a real electric permittivity ε and a real magnetic susceptibility μ:

$$\mathbf{E}(\mathbf{r}) = \mathbf{E}_0 \exp(ik\hat{\mathbf{n}} \cdot \mathbf{r}), \qquad \mathbf{E}_0 \cdot \hat{\mathbf{n}} = 0, \tag{1}$$

where the time factor $\exp(-i\omega t)$ is omitted, $k = \omega\sqrt{\varepsilon\mu}$ is the wave number, and ω is the angular frequency. The 2×2 coherency matrix $\boldsymbol{\rho}$ is defined by

$$\boldsymbol{\rho} = \begin{bmatrix} \rho_{11} & \rho_{12} \\ \rho_{21} & \rho_{22} \end{bmatrix} = \frac{1}{2}\sqrt{\frac{\varepsilon}{\mu}} \begin{bmatrix} E_{0\theta}E_{0\theta}^* & E_{0\theta}E_{0\varphi}^* \\ E_{0\varphi}E_{0\theta}^* & E_{0\varphi}E_{0\varphi}^* \end{bmatrix}, \tag{2}$$

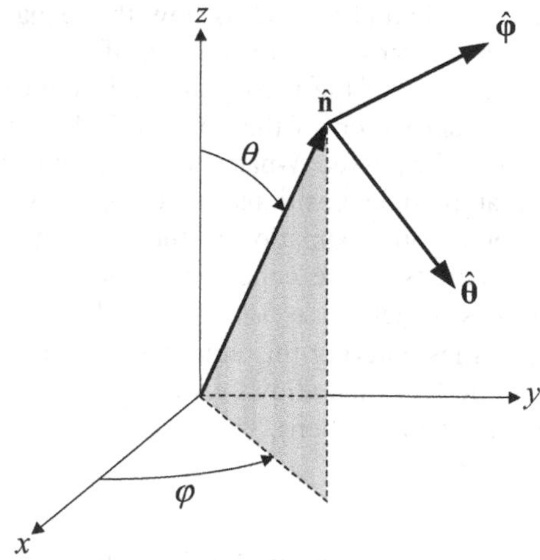

Fig. 1. Local coordinate system used to describe the direction of propagation and the polarization state of a transverse electromagnetic wave.

where the asterisk denotes a complex-conjugate value. The elements of $\boldsymbol{\rho}$ have the dimension of monochromatic energy flux (Wm^{-2}) and can be also grouped into a 4×1 coherency column vector:

$$\mathbf{J}=\begin{bmatrix}\rho_{11}\\ \rho_{12}\\ \rho_{21}\\ \rho_{22}\end{bmatrix}=\frac{1}{2}\sqrt{\frac{\varepsilon}{\mu}}\begin{bmatrix}E_{0\theta}E_{0\theta}^{*}\\ E_{0\theta}E_{0\varphi}^{*}\\ E_{0\varphi}E_{0\theta}^{*}\\ E_{0\varphi}E_{0\varphi}^{*}\end{bmatrix}. \tag{3}$$

The Stokes parameters I, Q, U, and V are then defined as the elements of a 4×1 column Stokes vector $\mathbf{I}$:

$$\mathbf{I}=\mathbf{DJ}=\frac{1}{2}\sqrt{\frac{\varepsilon}{\mu}}\begin{bmatrix}E_{0\theta}E_{0\theta}^{*}+E_{0\varphi}E_{0\varphi}^{*}\\ E_{0\theta}E_{0\theta}^{*}-E_{0\varphi}E_{0\varphi}^{*}\\ -E_{0\theta}E_{0\varphi}^{*}-E_{0\varphi}E_{0\theta}^{*}\\ \mathrm{i}(E_{0\varphi}E_{0\theta}^{*}-E_{0\theta}E_{0\varphi}^{*})\end{bmatrix}=\begin{bmatrix}I\\ Q\\ U\\ V\end{bmatrix}, \tag{4}$$

where

$$\mathbf{D}=\begin{bmatrix}1 & 0 & 0 & 1\\ 1 & 0 & 0 & -1\\ 0 & -1 & -1 & 0\\ 0 & -\mathrm{i} & \mathrm{i} & 0\end{bmatrix}. \tag{5}$$

2.2. VOLUME INTEGRAL EQUATION AND LIPPMANN-SCHWINGER EQUATION

Consider a scattering object that occupies a finite interior region V_{INT} and is surrounded by the infinite exterior region V_{EXT}. The interior region is filled with an isotropic, linear, and possibly inhomogeneous material.

The monochromatic Maxwell curl equations describing the scattering of a time-harmonic electromagnetic field are as follows:

$$\left.\begin{aligned}\nabla\times\mathbf{E}(\mathbf{r})&=\mathrm{i}\omega\mu_1\mathbf{H}(\mathbf{r})\\ \nabla\times\mathbf{H}(\mathbf{r})&=-\mathrm{i}\omega\varepsilon_1\mathbf{E}(\mathbf{r})\end{aligned}\right\}\ \text{for } \mathbf{r}\in V_{\mathrm{EXT}}, \tag{6}$$

$$\left.\begin{aligned}\nabla\times\mathbf{E}(\mathbf{r})&=\mathrm{i}\omega\mu_2(\mathbf{r})\mathbf{H}(\mathbf{r})\\ \nabla\times\mathbf{H}(\mathbf{r})&=-\mathrm{i}\omega\varepsilon_2(\mathbf{r})\mathbf{E}(\mathbf{r})\end{aligned}\right\}\ \text{for } \mathbf{r}\in V_{\mathrm{INT}}, \tag{7}$$

where subscripts 1 and 2 refer to the exterior and interior regions, respectively. Since the first relations in Eqs. (6) and (7) yield the magnetic field provided that the electric field is known everywhere, we will look for the solution of these equations in terms of only the electric field. Assuming that the host medium and the scattering object are nonmagnetic, i.e., $\mu_2(\mathbf{r})\equiv\mu_1=\mu_0$, where μ_0 is the permeability of a vacuum, and following the approach described in [26], one can reduce Eqs. (6) and (7) to the following volume integral equation:

$$\mathbf{E}(\mathbf{r}) = \mathbf{E}^{\text{inc}}(\mathbf{r}) + k_1^2 \int_{V_{\text{INT}}} d^3\mathbf{r}' \overleftrightarrow{G}(\mathbf{r},\mathbf{r}') \cdot \mathbf{E}(\mathbf{r}')[m^2(\mathbf{r}')-1], \qquad \mathbf{r} \in \mathbf{R}^3, \tag{8}$$

where $\overleftrightarrow{G}(\mathbf{r},\mathbf{r}')$ is the free space dyadic Green's function, $m(\mathbf{r}) = k_2(\mathbf{r})/k_1$ is the refractive index of the interior relative to that of the exterior, and $k_1 = \omega\sqrt{\varepsilon_1 \mu_0}$ and $k_2(\mathbf{r}) = \omega\sqrt{\varepsilon_2(\mathbf{r})\mu_0}$ are the wave numbers in the exterior and interior regions, respectively. Alternatively, the scattered field $\mathbf{E}^{\text{sca}}(\mathbf{r}) = \mathbf{E}(\mathbf{r}) - \mathbf{E}^{\text{inc}}(\mathbf{r})$ can be expressed in terms of the incident field by means of the dyad transition operator $\overleftrightarrow{T}$:

$$\mathbf{E}^{\text{sca}}(\mathbf{r}) = \int_{V_{\text{INT}}} d^3\mathbf{r}' \overleftrightarrow{G}(\mathbf{r},\mathbf{r}') \cdot \int_{V_{\text{INT}}} d^3\mathbf{r}'' \overleftrightarrow{T}(\mathbf{r}',\mathbf{r}'') \cdot \mathbf{E}^{\text{inc}}(\mathbf{r}''), \qquad \mathbf{r} \in \mathbf{R}^3. \tag{9}$$

Substituting Eq. (9) in Eq. (8) yields the Lippmann-Schwinger equation for $\overleftrightarrow{T}$:

$$\overleftrightarrow{T}(\mathbf{r},\mathbf{r}') = k_1^2 [m^2(\mathbf{r})-1]\delta(\mathbf{r}-\mathbf{r}')\overleftrightarrow{I} + k_1^2 [m^2(\mathbf{r})-1] \int_{V_{\text{INT}}} d^3\mathbf{r}'' \overleftrightarrow{G}(\mathbf{r},\mathbf{r}'') \cdot \overleftrightarrow{T}(\mathbf{r}'',\mathbf{r}'), \qquad \mathbf{r},\mathbf{r}' \in V_{\text{INT}}, \tag{10}$$

where $\overleftrightarrow{I}$ is the identity dyad.

2.3. FAR-FIELD SCATTERING

We now choose a point O at the geometrical center of the scatterer as the common origin of all position vectors (Fig. 2) and make the standard far-field-zone assumptions that $k_1 r \gg 1$ and that r is much larger than any linear dimension of the scatterer. Then Eq. (8) becomes

$$\mathbf{E}^{\text{sca}}(\mathbf{r}) \underset{r\to\infty}{=} \frac{\exp(ik_1 r)}{r} \frac{k_1^2}{4\pi} (\overleftrightarrow{I} - \hat{\mathbf{r}} \otimes \hat{\mathbf{r}}) \cdot \int_{V_{\text{INT}}} d^3\mathbf{r}' [m^2(\mathbf{r}')-1]\mathbf{E}(\mathbf{r}')\exp(-ik_1\hat{\mathbf{r}}\cdot\mathbf{r}'), \tag{11}$$

where $\otimes$ denotes a dyadic product of two vectors and $\hat{\mathbf{r}} = \mathbf{r}/r$ is a unit vector in the direction of $\mathbf{r}$. The factor $\overleftrightarrow{I} - \hat{\mathbf{r}} \otimes \hat{\mathbf{r}} = \hat{\boldsymbol{\theta}} \otimes \hat{\boldsymbol{\theta}} + \hat{\boldsymbol{\varphi}} \otimes \hat{\boldsymbol{\varphi}}$ ensures that the scattered spherical wave in the far-field zone is transverse so that

$$\mathbf{E}^{\text{sca}}(\mathbf{r}) \underset{r\to\infty}{=} \frac{\exp(ik_1 r)}{r} \mathbf{E}_1^{\text{sca}}(\hat{\mathbf{r}}), \qquad \hat{\mathbf{r}} \cdot \mathbf{E}_1^{\text{sca}}(\hat{\mathbf{r}}) = 0, \tag{12}$$

where the scattering amplitude $\mathbf{E}_1^{\text{sca}}(\hat{\mathbf{r}})$ is independent of r and describes the angular distribution of the scattered radiation.

Assuming that the incident field is a plane electromagnetic wave $\mathbf{E}^{\text{inc}}(\mathbf{r}) = \mathbf{E}_0^{\text{inc}} \exp(ik_1\hat{\mathbf{n}}^{\text{inc}} \cdot \mathbf{r})$ yields

$$\mathbf{E}_1^{\text{sca}}(\hat{\mathbf{n}}^{\text{sca}}) = \overleftrightarrow{A}(\hat{\mathbf{n}}^{\text{sca}}, \hat{\mathbf{n}}^{\text{inc}}) \cdot \mathbf{E}_0^{\text{inc}}, \tag{13}$$

where $\hat{\mathbf{n}}^{\text{sca}} = \hat{\mathbf{r}}$ (Fig. 2). The elements of the so-called scattering dyad $\overleftrightarrow{A}(\hat{\mathbf{n}}^{\text{sca}}, \hat{\mathbf{n}}^{\text{inc}})$ have the dimension of length.

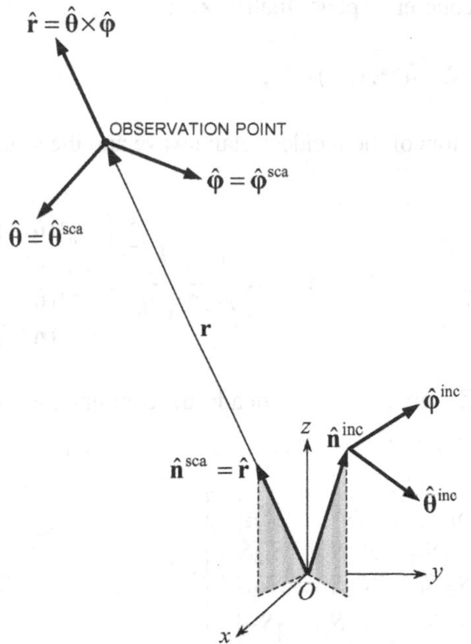

Fig. 2. Scattering in the far-field zone.

It follows from Eq. (12) that $\hat{\mathbf{n}}^{\text{sca}} \cdot \overleftrightarrow{A}(\hat{\mathbf{n}}^{\text{sca}}, \hat{\mathbf{n}}^{\text{inc}}) = 0$. Since $\mathbf{E}_0^{\text{inc}} \cdot \hat{\mathbf{n}}^{\text{inc}} = 0$, the dot product $\overleftrightarrow{A}(\hat{\mathbf{n}}^{\text{sca}}, \hat{\mathbf{n}}^{\text{inc}}) \cdot \hat{\mathbf{n}}^{\text{inc}}$ is not defined by Eq. (13). To complete the definition, we take this product to be zero. As a consequence, only four out of nine components of the scattering dyad are independent. It is therefore convenient to introduce a 2×2 amplitude matrix S, which describes the transformation of the θ- and φ-components of the incident plane wave into the θ- and φ-components of the scattered spherical wave (Fig. 2):

$$\mathsf{E}^{\text{sca}}(r\hat{\mathbf{n}}^{\text{sca}}) \underset{r \to \infty}{=} \frac{\exp(\mathrm{i}k_1 r)}{r} \mathsf{S}(\hat{\mathbf{n}}^{\text{sca}}, \hat{\mathbf{n}}^{\text{inc}}) \mathsf{E}_0^{\text{inc}}, \tag{14}$$

where E denotes a two-component column formed by the θ- and φ-components of the electric vector. The elements of the amplitude matrix are expressed in terms of the scattering dyad as follows:

$$\begin{aligned} S_{11} &= \hat{\boldsymbol{\theta}}^{\text{sca}} \cdot \overleftrightarrow{A} \cdot \hat{\boldsymbol{\theta}}^{\text{inc}}, \quad & S_{12} &= \hat{\boldsymbol{\theta}}^{\text{sca}} \cdot \overleftrightarrow{A} \cdot \hat{\boldsymbol{\varphi}}^{\text{inc}}, \\ S_{21} &= \hat{\boldsymbol{\varphi}}^{\text{sca}} \cdot \overleftrightarrow{A} \cdot \hat{\boldsymbol{\theta}}^{\text{inc}}, \quad & S_{22} &= \hat{\boldsymbol{\varphi}}^{\text{sca}} \cdot \overleftrightarrow{A} \cdot \hat{\boldsymbol{\varphi}}^{\text{inc}}. \end{aligned} \tag{15}$$

2.4. PHASE AND EXTINCTION MATRICES

The relationship between the coherency vectors of the incident and scattered light for scattering directions away from the incidence direction ($\hat{\mathbf{r}} \neq \hat{\mathbf{n}}^{\text{inc}}$) in the far-field zone is described by the 4×4 coherency phase matrix $\mathbf{Z}^J$:

$$\mathbf{J}^{\rm sca}(r\hat{\mathbf{n}}^{\rm sca}) = \frac{1}{r^2}\mathbf{Z}^J(\hat{\mathbf{n}}^{\rm sca},\hat{\mathbf{n}}^{\rm inc})\,\mathbf{J}^{\rm inc}, \tag{16}$$

where the coherency vectors of the incident plane wave and the scattered spherical wave are given by

$$\mathbf{J}^{\rm inc} = \frac{1}{2}\sqrt{\frac{\varepsilon_1}{\mu_0}}\begin{bmatrix} E_{0\theta}^{\rm inc}E_{0\theta}^{\rm inc*} \\ E_{0\theta}^{\rm inc}E_{0\varphi}^{\rm inc*} \\ E_{0\varphi}^{\rm inc}E_{0\theta}^{\rm inc*} \\ E_{0\varphi}^{\rm inc}E_{0\varphi}^{\rm inc*} \end{bmatrix}, \qquad \mathbf{J}^{\rm sca}(r\hat{\mathbf{n}}^{\rm sca}) = \frac{1}{r^2}\frac{1}{2}\sqrt{\frac{\varepsilon_1}{\mu_0}}\begin{bmatrix} E_{1\theta}^{\rm sca}(\hat{\mathbf{n}}^{\rm sca})[E_{1\theta}^{\rm sca}(\hat{\mathbf{n}}^{\rm sca})]^* \\ E_{1\theta}^{\rm sca}(\hat{\mathbf{n}}^{\rm sca})[E_{1\varphi}^{\rm sca}(\hat{\mathbf{n}}^{\rm sca})]^* \\ E_{1\varphi}^{\rm sca}(\hat{\mathbf{n}}^{\rm sca})[E_{1\theta}^{\rm sca}(\hat{\mathbf{n}}^{\rm sca})]^* \\ E_{1\varphi}^{\rm sca}(\hat{\mathbf{n}}^{\rm sca})[E_{1\varphi}^{\rm sca}(\hat{\mathbf{n}}^{\rm sca})]^* \end{bmatrix}, \tag{17}$$

and the elements of $\mathbf{Z}^J(\hat{\mathbf{n}}^{\rm sca},\hat{\mathbf{n}}^{\rm inc})$ are quadratic combinations of the elements of the amplitude matrix $\mathbf{S}(\hat{\mathbf{n}}^{\rm sca},\hat{\mathbf{n}}^{\rm inc})$:

$$\mathbf{Z}^J = \begin{bmatrix} |S_{11}|^2 & S_{11}S_{12}^* & S_{12}S_{11}^* & |S_{12}|^2 \\ S_{11}S_{21}^* & S_{11}S_{22}^* & S_{12}S_{21}^* & S_{12}S_{22}^* \\ S_{21}S_{11}^* & S_{21}S_{12}^* & S_{22}S_{11}^* & S_{22}S_{12}^* \\ |S_{21}|^2 & S_{21}S_{22}^* & S_{22}S_{21}^* & |S_{22}|^2 \end{bmatrix}. \tag{18}$$

The corresponding Stokes transformation law is

$$\mathbf{I}^{\rm sca}(r\hat{\mathbf{n}}^{\rm sca}) = \frac{1}{r^2}\mathbf{Z}(\hat{\mathbf{n}}^{\rm sca},\hat{\mathbf{n}}^{\rm inc})\,\mathbf{I}^{\rm inc}, \tag{19}$$

where $\mathbf{I}^{\rm inc} = \mathbf{D}\mathbf{J}^{\rm inc}$ and $\mathbf{I}^{\rm sca} = \mathbf{D}\mathbf{J}^{\rm sca}$. The explicit expressions for the elements of the Stokes phase matrix $\mathbf{Z}$ follow from Eq. (18) and the obvious formula

$$\mathbf{Z}(\hat{\mathbf{n}}^{\rm sca},\hat{\mathbf{n}}^{\rm inc}) = \mathbf{D}\mathbf{Z}^J(\hat{\mathbf{n}}^{\rm sca},\hat{\mathbf{n}}^{\rm inc})\mathbf{D}^{-1}. \tag{20}$$

The coherency vector of the total field for directions $\hat{\mathbf{r}}$ very close to $\hat{\mathbf{n}}^{\rm inc}$ is defined as

$$\mathbf{J}(r\hat{\mathbf{r}}) = \frac{1}{2}\sqrt{\frac{\varepsilon_1}{\mu_0}}\begin{bmatrix} E_\theta(r\hat{\mathbf{r}})[E_\theta(r\hat{\mathbf{r}})]^* \\ E_\theta(r\hat{\mathbf{r}})[E_\varphi(r\hat{\mathbf{r}})]^* \\ E_\varphi(r\hat{\mathbf{r}})[E_\theta(r\hat{\mathbf{r}})]^* \\ E_\varphi(r\hat{\mathbf{r}})[E_\varphi(r\hat{\mathbf{r}})]^* \end{bmatrix}, \tag{21}$$

where $\mathbf{E}(r\hat{\mathbf{r}}) = \mathbf{E}^{\rm inc}(r\hat{\mathbf{r}}) + \mathbf{E}^{\rm sca}(r\hat{\mathbf{r}})$ is the total electric field. Integrating $\mathbf{J}(r\hat{\mathbf{r}})$ over the surface ΔS of a collimated detector facing the incident wave, one can obtain for the total polarized signal:

$$\mathbf{J}(r\hat{\mathbf{n}}^{\rm inc})\Delta S = \mathbf{J}^{\rm inc}\Delta S - \mathbf{K}^J(\hat{\mathbf{n}}^{\rm inc})\mathbf{J}^{\rm inc} + \mathbf{O}(r^{-2}), \tag{22}$$

where the elements of the 4×4 coherency extinction matrix $\mathbf{K}^J(\hat{\mathbf{n}}^{\rm inc})$ are expressed in terms of the elements of the forward-scattering amplitude matrix $\mathbf{S}(\hat{\mathbf{n}}^{\rm inc},\hat{\mathbf{n}}^{\rm inc})$ as follows:

$$\mathbf{K}^J = \frac{\mathrm{i}2\pi}{k_1}\begin{bmatrix} S_{11}^* - S_{11} & S_{12}^* & -S_{12} & 0 \\ S_{21}^* & S_{22}^* - S_{11} & 0 & -S_{12} \\ -S_{21} & 0 & S_{11}^* - S_{22} & S_{12}^* \\ 0 & -S_{21} & S_{21}^* & S_{22}^* - S_{22} \end{bmatrix}. \tag{23}$$

In the Stokes-vector representation,

$$\mathbf{I}(r\hat{\mathbf{n}}^{\mathrm{inc}})\Delta S = \mathbf{I}^{\mathrm{inc}}\Delta S - \mathbf{K}(\hat{\mathbf{n}}^{\mathrm{inc}})\,\mathbf{I}^{\mathrm{inc}} + \mathbf{O}(r^{-2}), \tag{24}$$

where $\mathbf{I}(r\hat{\mathbf{n}}^{\mathrm{inc}}) = \mathbf{DJ}(r\hat{\mathbf{n}}^{\mathrm{inc}})$. Expressions for the elements of the 4×4 Stokes extinction matrix $\mathbf{K}(\hat{\mathbf{n}}^{\mathrm{inc}})$ follow from Eq. (23) and the formula

$$\mathbf{K}(\hat{\mathbf{n}}^{\mathrm{inc}}) = \mathbf{DK}^J(\hat{\mathbf{n}}^{\mathrm{inc}})\mathbf{D}^{-1}. \tag{25}$$

Equations (22) and (24) represent the most general form of the optical theorem and show that the presence of the scattering particle changes not only the total power of the electromagnetic radiation received by the detector facing the incident wave, but also, perhaps, its state of polarization. The latter phenomenon is called dichroism and results from different attenuation rates for different polarization components of the incident wave.

2.5. OPTICAL CROSS SECTIONS

Important optical characteristics of the scattering object are the total scattering, absorption, and extinction cross sections. The scattering cross section C_{sca} is defined such that the product of C_{sca} and the incident monochromatic energy flux gives the total monochromatic power removed from the incident wave owing to scattering of the incident radiation in all directions. Similarly, the product of the absorption cross section C_{abs} and the incident monochromatic energy flux is equal to the total monochromatic power removed from the incident wave as a result of absorption of light by the object. Finally, the extinction cross section C_{ext} is the sum of the scattering and absorption cross sections and characterizes the total monochromatic power removed from the incident light due to the combined effect of scattering and absorption.

Explicit formulas for the extinction and scattering cross sections are as follows:

$$C_{\mathrm{ext}} = \frac{1}{I^{\mathrm{inc}}}[K_{11}(\hat{\mathbf{n}}^{\mathrm{inc}})\,I^{\mathrm{inc}} + K_{12}(\hat{\mathbf{n}}^{\mathrm{inc}})Q^{\mathrm{inc}} + K_{13}(\hat{\mathbf{n}}^{\mathrm{inc}})U^{\mathrm{inc}} + K_{14}(\hat{\mathbf{n}}^{\mathrm{inc}})V^{\mathrm{inc}}], \tag{26}$$

$$C_{\mathrm{sca}} = \frac{1}{I^{\mathrm{inc}}}\int_{4\pi}\mathrm{d}\hat{\mathbf{r}}\,[Z_{11}(\hat{\mathbf{r}},\hat{\mathbf{n}}^{\mathrm{inc}})I^{\mathrm{inc}} + Z_{12}(\hat{\mathbf{r}},\hat{\mathbf{n}}^{\mathrm{inc}})Q^{\mathrm{inc}} + Z_{13}(\hat{\mathbf{r}},\hat{\mathbf{n}}^{\mathrm{inc}})U^{\mathrm{inc}} + Z_{14}(\hat{\mathbf{r}},\hat{\mathbf{n}}^{\mathrm{inc}})V^{\mathrm{inc}}]. \tag{27}$$

We then have $C_{\mathrm{abs}} = C_{\mathrm{ext}} - C_{\mathrm{sca}} \geq 0$. The single-scattering albedo is defined as the ratio of the scattering and extinction cross sections,

$$\varpi = C_{\mathrm{sca}}/C_{\mathrm{ext}} \le 1, \tag{28}$$

and is equal to unity for nonabsorbing particles.

2.6 SINGLE SCATTERING BY A SMALL COLLECTION OF RANDOMLY POSITIONED PARTICLES

The formalism described above can also be applied to *single scattering* by tenuous particle collections under certain simplifying assumptions. Consider a volume element having a linear dimension l and comprising a number N of randomly positioned particles. We assume that N is sufficiently small and that the mean distance between the particles is large enough that the contribution of light scattered by the particles to the total field exciting each particle is much weaker than the external incident field and can be neglected. We also assume that the positions of the particles are sufficiently random that there are no systematic phase relations between individual waves scattered by different particles. Consider now far-field scattering by the entire volume element by assuming that the observation point is located at a distance much greater than both l and the wavelength of the incident light. It can then be shown [26] that the cumulative optical characteristics of the entire volume element are obtained by incoherently adding the respective optical characteristics of the individual particles:

$$C_{\mathrm{ext}} = \sum_{i=1}^{N} (C_{\mathrm{ext}})_i = N\langle C_{\mathrm{ext}} \rangle, \tag{29}$$

$$C_{\mathrm{sca}} = \sum (C_{\mathrm{sca}})_i = N\langle C_{\mathrm{sca}} \rangle, \tag{30}$$

$$C_{\mathrm{abs}} = \sum (C_{\mathrm{abs}})_i = N\langle C_{\mathrm{abs}} \rangle, \tag{31}$$

$$\mathbf{K} = \sum \mathbf{K}_i = N\langle \mathbf{K} \rangle, \tag{32}$$

$$\mathbf{Z} = \sum \mathbf{Z}_i = N\langle \mathbf{Z} \rangle, \tag{33}$$

where the index i numbers the particles and $\langle C_{\mathrm{ext}} \rangle$, $\langle C_{\mathrm{sca}} \rangle$, $\langle C_{\mathrm{abs}} \rangle$, $\langle \mathbf{K} \rangle$, and $\langle \mathbf{Z} \rangle$ are the average extinction, scattering, and absorption cross sections and the extinction and phase matrices per particle, respectively.

2.7 MACROSCOPICALLY ISOTROPIC AND MIRROR-SYMMETRIC SCATTERING MEDIA

By definition, the phase matrix relates the Stokes parameters of the incident and the scattered beam defined relative to their respective meridional planes. In contrast, the scattering matrix $\mathbf{F}$ relates the Stokes parameters of the incident and the scattered beam defined with respect to the scattering plane, i.e., the plane through the $\hat{\mathbf{n}}^{\mathrm{inc}}$ and $\hat{\mathbf{n}}^{\mathrm{sca}}$. A simple way to introduce the scattering matrix is to direct the z-axis of the laboratory reference frame along the incident beam and superpose the meridional plane with $\varphi = 0$ and the scattering plane:

$$\mathbf{F}(\theta^{\mathrm{sca}}) = \mathbf{Z}(\theta^{\mathrm{sca}}, \varphi^{\mathrm{sca}} = 0;\ \theta^{\mathrm{inc}} = 0, \varphi^{\mathrm{inc}} = 0). \tag{34}$$

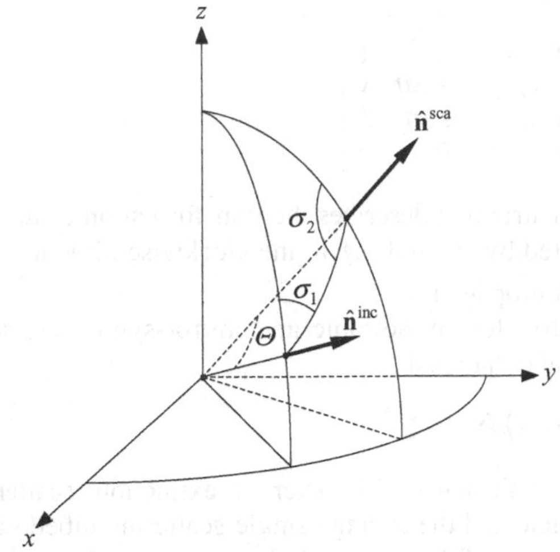

Fig. 3. Relationship between the scattering and phase matrices.

The concept of scattering matrix is especially useful in application to so-called *macroscopically isotropic and mirror-symmetric* scattering media composed of randomly oriented particles with a plane of symmetry and/or equal numbers of randomly oriented particles and their mirror-symmetric counterparts. Indeed, in this case the scattering matrix of a particle collection is independent of incidence direction and orientation of the scattering plane, is functionally dependent only on the scattering angle $\Theta = \arccos(\hat{\mathbf{n}}^{\text{inc}} \cdot \hat{\mathbf{n}}^{\text{sca}})$, and has a simple structure:

$$\mathbf{F}(\Theta) = \begin{bmatrix} F_{11}(\Theta) & F_{12}(\Theta) & 0 & 0 \\ F_{12}(\Theta) & F_{22}(\Theta) & 0 & 0 \\ 0 & 0 & F_{33}(\Theta) & F_{34}(\Theta) \\ 0 & 0 & -F_{34}(\Theta) & F_{44}(\Theta) \end{bmatrix} = N\langle \mathbf{F}(\Theta) \rangle, \tag{35}$$

where $\langle \mathbf{F}(\Theta) \rangle$ is the ensemble-averaged scattering matrix per particle.

Knowledge of the scattering matrix can be used to calculate the Stokes phase matrix for an isotropic and mirror-symmetric scattering medium (Fig. 3). Specifically, to compute the Stokes vector of the scattered beam with respect to its meridional plane, one must:

- calculate the Stokes vector of the incident beam with respect to the scattering plane;
- multiply it by the scattering matrix, thereby obtaining the Stokes vector of the scattered beam with respect to the scattering plane; and finally
- compute the Stokes vector of the scattered beam with respect to its meridional plane.

This procedure yields:

$$\mathbf{Z}(\theta^{\text{sca}}, \varphi^{\text{sca}}; \theta^{\text{inc}}, \varphi^{\text{inc}}) = \mathbf{L}(-\sigma_2)\,\mathbf{F}(\Theta)\,\mathbf{L}(\pi - \sigma_1), \tag{36}$$

where

$$\mathbf{L}(\eta) = \begin{bmatrix} 1 & 0 & 0 & 0 \\ 0 & \cos 2\eta & -\sin 2\eta & 0 \\ 0 & \sin 2\eta & \cos 2\eta & 0 \\ 0 & 0 & 0 & 1 \end{bmatrix} \tag{37}$$

is the Stokes rotation matrix that describes the transformation of the Stokes vector as the reference plane is rotated by an angle η in the clockwise direction when one is looking in the direction of light propagation.

The extinction matrix for an isotropic and mirror-symmetric scattering medium is direction independent and diagonal:

$$\mathbf{K}(\hat{\mathbf{n}}) \equiv \mathbf{K} = N\langle C_{\text{ext}}\rangle \mathbf{\Delta}, \tag{38}$$

where $\mathbf{\Delta}$ is the 4×4 unit matrix. The average extinction, scattering, and absorption cross sections per particle and the average single-scattering albedo are also independent of the propagation direction of the incident light as well as of its polarization state.

It is convenient in the RTT to use the so-called normalized scattering matrix

$$\widetilde{\mathbf{F}}(\Theta) = \frac{4\pi}{\langle C_{\text{sca}}\rangle}\langle \mathbf{F}(\Theta)\rangle = \begin{bmatrix} a_1(\Theta) & b_1(\Theta) & 0 & 0 \\ b_1(\Theta) & a_2(\Theta) & 0 & 0 \\ 0 & 0 & a_3(\Theta) & b_2(\Theta) \\ 0 & 0 & -b_2(\Theta) & a_4(\Theta) \end{bmatrix} \tag{39}$$

with dimensionless elements. The (1, 1)-element of this matrix, traditionally called the phase function, is normalized to unity according to

$$\tfrac{1}{2}\int_0^{\pi} d\Theta \sin\Theta \, a_1(\Theta) = 1. \tag{40}$$

Similarly, the normalized phase matrix can be defined as

$$\widetilde{\mathbf{Z}}(\vartheta^{\text{sca}}, \varphi^{\text{sca}};\, \vartheta^{\text{inc}}, \varphi^{\text{inc}}) = \frac{4\pi}{\langle C_{\text{sca}}\rangle}\langle \mathbf{Z}(\vartheta^{\text{sca}}, \varphi^{\text{sca}};\, \vartheta^{\text{inc}}, \varphi^{\text{inc}})\rangle. \tag{41}$$

3. Multiple Scattering

3.1. FOLDY-LAX EQUATIONS

We will now study *multiple scattering* by large particle collections and eventually derive the RTE. We begin by considering electromagnetic scattering by a *fixed* group of N particles collectively occupying the interior region $V_{\text{INT}} = \bigcup_{i=1}^{N} V_i$, where V_i is the volume occupied by the ith particle. Equation (8) now reads

$$\mathbf{E}(\mathbf{r}) = \mathbf{E}^{\text{inc}}(\mathbf{r}) + \int_{\mathbf{R}^3} \mathrm{d}^3\mathbf{r}' \, U(\mathbf{r}') \overleftrightarrow{G}(\mathbf{r},\mathbf{r}') \cdot \mathbf{E}(\mathbf{r}'), \qquad \mathbf{r} \in \mathbf{R}^3, \tag{42}$$

where the total potential function $U(\mathbf{r})$ is given by

$$U(\mathbf{r}) = \sum_{i=1}^{N} U_i(\mathbf{r}), \qquad \mathbf{r} \in \mathbf{R}^3, \tag{43}$$

and $U_i(\mathbf{r})$ is the ith-particle potential function. The latter is defined by

$$U_i(\mathbf{r}) = \begin{cases} 0, & \mathbf{r} \notin V_i, \\ k_1^2\,[m_i^2(\mathbf{r}) - 1], & \mathbf{r} \in V_i, \end{cases} \tag{44}$$

where $m_i(\mathbf{r}) = k_{2i}(\mathbf{r})/k_1$ is the relative refractive index of particle i. All position vectors originate at the origin O of an arbitrarily chosen laboratory coordinate system. It can then be shown [25] that the total electric field everywhere in space can be expressed as

$$\mathbf{E}(\mathbf{r}) = \mathbf{E}^{\mathrm{inc}}(\mathbf{r}) + \sum_{i=1}^{N} \int_{V_i} \mathrm{d}^3\mathbf{r}'\, \overleftrightarrow{G}(\mathbf{r},\mathbf{r}') \cdot \int_{V_i} \mathrm{d}^3\mathbf{r}''\, \overleftrightarrow{T}_i(\mathbf{r}',\mathbf{r}'') \cdot \mathbf{E}_i(\mathbf{r}''), \quad \mathbf{r} \in \mathbf{R}^3, \tag{45}$$

where the field $\mathbf{E}_i$ exciting particle i is given by

$$\mathbf{E}_i(\mathbf{r}) = \mathbf{E}^{\mathrm{inc}}(\mathbf{r}) + \sum_{j(\neq i)=1}^{N} \mathbf{E}_{ij}^{\mathrm{exc}}(\mathbf{r}), \tag{46}$$

and the $\mathbf{E}_{ij}^{\mathrm{exc}}$ are partial exciting fields given by

$$\mathbf{E}_{ij}^{\mathrm{exc}}(\mathbf{r}) = \int_{V_j} \mathrm{d}^3\mathbf{r}'\, \overleftrightarrow{G}(\mathbf{r},\mathbf{r}') \cdot \int_{V_j} \mathrm{d}^3\mathbf{r}''\, \overleftrightarrow{T}_j(\mathbf{r}',\mathbf{r}'') \cdot \mathbf{E}_j(\mathbf{r}''), \qquad \mathbf{r} \in V_i. \tag{47}$$

The $\overleftrightarrow{T}_i$ satisfies the Lippmann-Schwinger equation

$$\overleftrightarrow{T}_i(\mathbf{r},\mathbf{r}') = U_i(\mathbf{r})\,\delta(\mathbf{r}-\mathbf{r}')\,\overleftrightarrow{I} + U_i(\mathbf{r}) \int_{V_i} \mathrm{d}^3\mathbf{r}''\, \overleftrightarrow{G}(\mathbf{r},\mathbf{r}'') \cdot \overleftrightarrow{T}_i(\mathbf{r}'',\mathbf{r}'), \quad \mathbf{r},\mathbf{r}' \in V_i, \tag{48}$$

and is the dyad transition operator of particle i in the absence of all other particles.

The Foldy-Lax equations (45)–(47) directly follow from Maxwell's equations and describe the process of multiple scattering by a fixed group of N particles. Specifically, Eq. (45) decomposes the total field into the vector sum of the incident field and the partial fields generated by each particle in response to the corresponding exciting fields, whereas Eqs. (46) and (47) show that the field exciting each particle consists of the incident field and the fields generated by all other particles.

3.2. FAR-FIELD ZONE APPROXIMATION

Assume now that the distance between any two particles in the group is much greater than the wavelength and much greater than the particle sizes, which means that each particle is located in the far-field zones of all other particles. This assumption allows us to considerably simplify the Foldy-Lax equations. Indeed, the contribution of the jth particle to the field exciting the ith particle in Eq. (46) can now be represented as a simple outgoing spherical wave centered at the origin of particle j:

$$\mathbf{E}_{ij}^{\text{exc}}(\mathbf{r}) \approx G(r_j)\,\mathbf{E}_{1ij}(\hat{\mathbf{r}}_j) \approx \exp(-\mathrm{i}k_1\hat{\mathbf{R}}_{ij}\cdot\mathbf{R}_i)\,\mathbf{E}_{ij}\,\exp(\mathrm{i}k_1\hat{\mathbf{R}}_{ij}\cdot\mathbf{r}), \qquad \mathbf{r}\in V_i, \tag{49}$$

where

$$G(r) = \frac{\exp(\mathrm{i}k_1 r)}{r}, \tag{50}$$

$$\mathbf{E}_{ij} = G(R_{ij})\,\mathbf{E}_{1ij}(\hat{\mathbf{R}}_{ij}), \qquad \mathbf{E}_{ij}\cdot\hat{\mathbf{R}}_{ij} = 0, \tag{51}$$

$\hat{\mathbf{r}}_j = \mathbf{r}_j/r_j$, $\hat{\mathbf{R}}_{ij} = \mathbf{R}_{ij}/R_{ij}$, $r_j = |\mathbf{R}_{ij} + \mathbf{r} - \mathbf{R}_i| \underset{R_{ij}\to\infty}{=} R_{ij} + \hat{\mathbf{R}}_{ij}\cdot(\mathbf{r}-\mathbf{R}_i)$, and the vectors $\mathbf{r}$, $\mathbf{r}_j$, $\mathbf{R}_i$, $\mathbf{R}_j$, and $\mathbf{R}_{ij}$ are shown in Fig. 4(a). Obviously, $\mathbf{E}_{ij}$ is the partial exciting field at the origin of the *i*th particle caused by the *j*th particle. Thus, Eqs. (46) and (49) show that each particle is excited by the external field and the superposition of *locally* plane waves from all other particles with amplitudes $\exp(-\mathrm{i}k_1\hat{\mathbf{R}}_{ij}\cdot\mathbf{R}_i)\,\mathbf{E}_{ij}$ and propagation

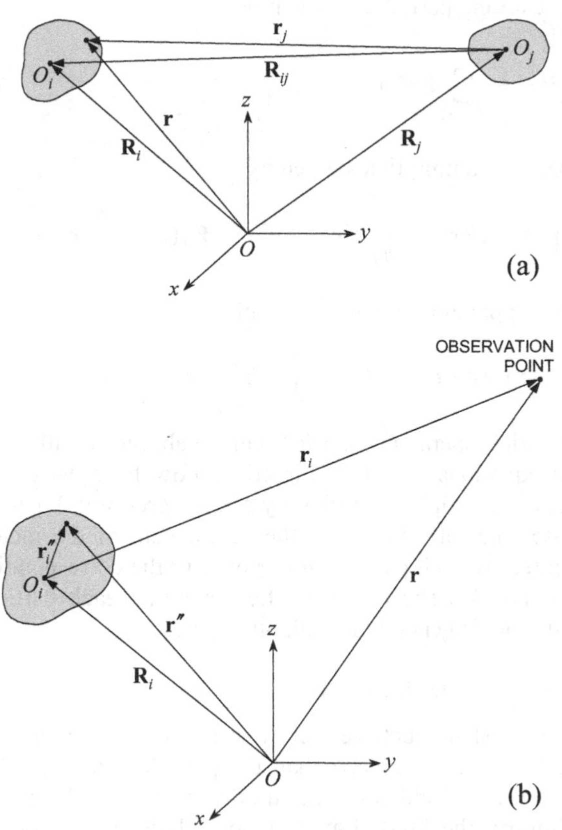

Fig. 4. Scattering by widely separated particles. The local origins O_i and O_j are chosen arbitrarily inside particles *i* and *j*, respectively.

directions $\hat{\mathbf{R}}_{ij}$:

$$\mathbf{E}_i(\mathbf{r}) \approx \mathbf{E}_0^{\rm inc} \exp(\mathrm{i}k_1\hat{\mathbf{s}}\cdot\mathbf{r}) + \sum_{j(\neq i)=1}^{N} \exp(-\mathrm{i}k_1\hat{\mathbf{R}}_{ij}\cdot\mathbf{R}_i)\,\mathbf{E}_{ij}\exp(\mathrm{i}k_1\hat{\mathbf{R}}_{ij}\cdot\mathbf{r}), \qquad \mathbf{r}\in V_i, \tag{52}$$

where we have assumed that the external incident field is a plane electromagnetic wave $\mathbf{E}^{\rm inc}(\mathbf{r}) = \mathbf{E}_0^{\rm inc}\exp(\mathrm{i}k_1\hat{\mathbf{s}}\cdot\mathbf{r})$.

According to Eqs. (12) and (13), the outgoing spherical wave generated by the jth particle in response to a plane-wave excitation of the form $\mathbf{E}\exp(\mathrm{i}k_1\hat{\mathbf{n}}\cdot\mathbf{r}_j)$ is given by $G(r_j)\,\vec{A}_j(\hat{\mathbf{r}}_j,\hat{\mathbf{n}})\cdot\mathbf{E}$, where $\mathbf{r}_j$ originates at O_j and $\vec{A}_j(\hat{\mathbf{r}}_j,\hat{\mathbf{n}})$ is the jth particle scattering dyad centered at O_j. To make use of this fact, we must rewrite Eq. (52) for particle j with respect to the jth-particle coordinate system centered at O_j, Fig. 4(a). Taking into account that $\mathbf{r} = \mathbf{r}_j + \mathbf{R}_j$ yields

$$\mathbf{E}_j(\mathbf{r}) \approx \mathbf{E}^{\rm inc}(\mathbf{R}_j)\exp(\mathrm{i}k_1\hat{\mathbf{s}}\cdot\mathbf{r}_j) + \sum_{l(\neq j)=1}^{N}\mathbf{E}_{jl}\exp(\mathrm{i}k_1\hat{\mathbf{R}}_{jl}\cdot\mathbf{r}_j), \qquad \mathbf{r}\in V_j. \tag{53}$$

The electric field at O_i generated in response to this excitation is simply

$$G(R_{ij})\left(\vec{A}_j(\hat{\mathbf{R}}_{ij},\hat{\mathbf{s}})\cdot\mathbf{E}^{\rm inc}(\mathbf{R}_j) + \sum_{l(\neq j)=1}^{N}\vec{A}_j(\hat{\mathbf{R}}_{ij},\hat{\mathbf{R}}_{jl})\cdot\mathbf{E}_{jl}\right). \tag{54}$$

Equating this expression with the right-hand side of Eq. (49) evaluated for $\mathbf{r} = \mathbf{R}_i$ finally yields a system of linear algebraic equations for the partial exciting fields $\mathbf{E}_{ij}$:

$$\mathbf{E}_{ij} = G(R_{ij})\left(\vec{A}_j(\hat{\mathbf{R}}_{ij},\hat{\mathbf{s}})\cdot\mathbf{E}^{\rm inc}(\mathbf{R}_j) + \sum_{l(\neq j)=1}^{N}\vec{A}_j(\hat{\mathbf{R}}_{ij},\hat{\mathbf{R}}_{jl})\cdot\mathbf{E}_{jl}\right), \qquad i,j=1,\ldots,N, \quad j\neq i. \tag{55}$$

After the system (55) is solved, one can find the electric field exciting each particle and the total field. Indeed, Eq. (53) gives for a point $\mathbf{r}''\in V_i$:

$$\mathbf{E}_i(\mathbf{r}'') \approx \mathbf{E}^{\rm inc}(\mathbf{R}_i)\exp(\mathrm{i}k_1\hat{\mathbf{s}}\cdot\mathbf{r}_i'') + \sum_{j(\neq i)=1}^{N}\mathbf{E}_{ij}\exp(\mathrm{i}k_1\hat{\mathbf{R}}_{ij}\cdot\mathbf{r}_i''), \quad \mathbf{r}''\in V_i \tag{56}$$

[see Fig. 4(b)], which is a vector superposition of locally plane waves. Substituting $\mathbf{r}_i''=0$ in Eq. (56) yields

$$\mathbf{E}_i(\mathbf{R}_i) = \mathbf{E}^{\rm inc}(\mathbf{R}_i) + \sum_{j(\neq i)=1}^{N}\mathbf{E}_{ij}. \tag{57}$$

Finally, substituting Eq. (56) in Eq. (45), we derive for the total electric field:

$$\mathbf{E}(\mathbf{r}) = \mathbf{E}^{\rm inc}(\mathbf{r}) + \sum_{i=1}^{N}G(r_i)\,\vec{A}_i(\hat{\mathbf{r}}_i,\hat{\mathbf{s}})\cdot\mathbf{E}^{\rm inc}(\mathbf{R}_i) + \sum_{i=1}^{N}G(r_i)\sum_{j(\neq i)=1}^{N}\vec{A}_i(\hat{\mathbf{r}}_i,\hat{\mathbf{R}}_{ij})\cdot\mathbf{E}_{ij}\,, \tag{58}$$

where the observation point $\mathbf{r}$, Fig. 4(b), is assumed to be in the far-field zone of any particle forming the group.

3.3. TWERSKY APPROXIMATION

We will now rewrite Eqs. (58) and (55) in a compact form:

$$\mathbf{E} = \mathbf{E}^{\rm inc} + \sum_{i=1}^{N} \vec{\vec{B}}_{ri0} \cdot \mathbf{E}_i^{\rm inc} + \sum_{i=1}^{N} \sum_{j(\neq i)=1}^{N} \vec{\vec{B}}_{rij} \cdot \mathbf{E}_{ij}, \tag{59}$$

$$\mathbf{E}_{ij} = \vec{\vec{B}}_{ij0} \cdot \mathbf{E}_j^{\rm inc} + \sum_{l(\neq j)=1}^{N} \vec{\vec{B}}_{ijl} \cdot \mathbf{E}_{jl}, \tag{60}$$

where $\mathbf{E} = \mathbf{E}(\mathbf{r})$, $\mathbf{E}^{\rm inc} = \mathbf{E}^{\rm inc}(\mathbf{r})$, $\mathbf{E}_i^{\rm inc} = \mathbf{E}^{\rm inc}(\mathbf{R}_i)$,

$$\begin{aligned} \vec{\vec{B}}_{ri0} &= G(r_i)\, \vec{\vec{A}}_i(\hat{\mathbf{r}}_i, \hat{\mathbf{s}}), \qquad & \vec{\vec{B}}_{rij} &= G(r_i)\, \vec{\vec{A}}_i(\hat{\mathbf{r}}_i, \hat{\mathbf{R}}_{ij}), \\ \vec{\vec{B}}_{ij0} &= G(R_{ij})\, \vec{\vec{A}}_j(\hat{\mathbf{R}}_{ij}, \hat{\mathbf{s}}), \qquad & \vec{\vec{B}}_{ijl} &= G(R_{ij})\, \vec{\vec{A}}_j(\hat{\mathbf{R}}_{ij}, \hat{\mathbf{R}}_{jl}). \end{aligned} \tag{61}$$

Iterating Eq. (60) yields

$$\mathbf{E}_{ij} = \vec{\vec{B}}_{ij0} \cdot \mathbf{E}_j^{\rm inc} + \sum_{\substack{l=1 \\ l\neq j}}^{N} \vec{\vec{B}}_{ijl} \cdot \vec{\vec{B}}_{jl0} \cdot \mathbf{E}_l^{\rm inc} + \sum_{\substack{l=1 \\ l\neq j}}^{N} \sum_{\substack{m=1 \\ m\neq l}}^{N} \vec{\vec{B}}_{ijl} \cdot \vec{\vec{B}}_{jlm} \cdot \vec{\vec{B}}_{lm0} \cdot \mathbf{E}_m^{\rm inc} + \cdots, \tag{62}$$

whereas substituting Eq. (62) in Eq. (59) gives an order-of-scattering expansion of the total electric field:

$$\begin{aligned} \mathbf{E} = {} & \mathbf{E}^{\rm inc} + \sum_{i=1}^{N} \vec{\vec{B}}_{ri0} \cdot \mathbf{E}_i^{\rm inc} + \sum_{i=1}^{N} \sum_{\substack{j=1 \\ j\neq i}}^{N} \vec{\vec{B}}_{rij} \cdot \vec{\vec{B}}_{ij0} \cdot \mathbf{E}_j^{\rm inc} \\ & + \sum_{i=1}^{N} \sum_{\substack{j=1 \\ j\neq i}}^{N} \sum_{\substack{l=1 \\ l\neq j}}^{N} \vec{\vec{B}}_{rij} \cdot \vec{\vec{B}}_{ijl} \cdot \vec{\vec{B}}_{jl0} \cdot \mathbf{E}_l^{\rm inc} \\ & + \sum_{i=1}^{N} \sum_{\substack{j=1 \\ j\neq i}}^{N} \sum_{\substack{l=1 \\ l\neq j}}^{N} \sum_{\substack{m=1 \\ m\neq l}}^{N} \vec{\vec{B}}_{rij} \cdot \vec{\vec{B}}_{ijl} \cdot \vec{\vec{B}}_{jlm} \cdot \vec{\vec{B}}_{lm0} \cdot \mathbf{E}_m^{\rm inc} + \cdots. \end{aligned} \tag{63}$$

The terms with $j = i$ and $l = j$ in the triple summation on the right-hand side of Eq. (63) are excluded, but the terms with $l = i$ are not. Therefore, we can decompose this summation as follows:

$$\sum_{i=1}^{N} \sum_{\substack{j=1 \\ j\neq i}}^{N} \sum_{\substack{l=1 \\ l\neq i \\ l\neq j}}^{N} \vec{\vec{B}}_{rij} \cdot \vec{\vec{B}}_{ijl} \cdot \vec{\vec{B}}_{jl0} \cdot \mathbf{E}_l^{\rm inc} + \sum_{i=1}^{N} \sum_{\substack{j=1 \\ j\neq i}}^{N} \vec{\vec{B}}_{rij} \cdot \vec{\vec{B}}_{iji} \cdot \vec{\vec{B}}_{ji0} \cdot \mathbf{E}_i^{\rm inc}. \tag{64}$$

Higher-order summations in Eq. (63) can be decomposed similarly. Hence, the total field at an observation point $\mathbf{r}$ consists of the incident field and single- and multiple-scattering

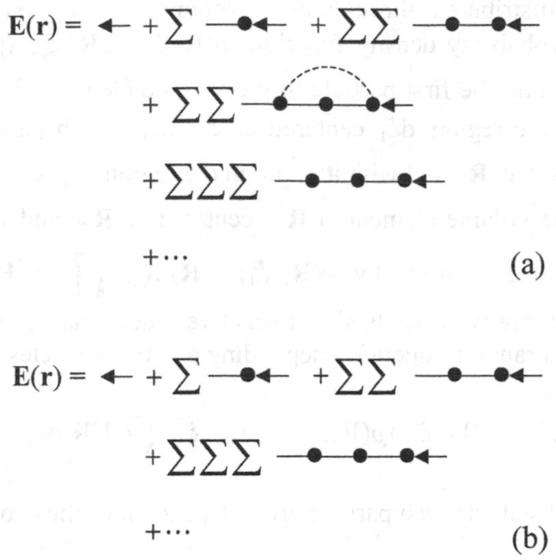

$$\mathbf{E(r)} = \leftarrow + \sum \text{—●←} + \sum\sum \text{—●—●←}$$
$$+ \sum\sum \text{—●—●—●←}$$
$$+ \sum\sum\sum \text{—●—●—●←}$$
$$+ \cdots \qquad \text{(a)}$$

$$\mathbf{E(r)} = \leftarrow + \sum \text{—●←} + \sum\sum \text{—●—●←}$$
$$+ \sum\sum\sum \text{—●—●—●←}$$
$$+ \cdots \qquad \text{(b)}$$

Fig. 5. Diagrammatic representations of (a) Eq. (63) and (b) Eq. (65).

contributions that can be divided into two groups. The first one includes all the terms that correspond to self-avoiding scattering paths, whereas the second group includes all the terms corresponding to the paths that go through a scatterer more than once. The so-called Twersky approximation [27] neglects the terms belonging to the second group and retains only the terms from the first group:

$$\mathbf{E} \approx \mathbf{E}^{\mathrm{inc}} + \sum_{i=1}^{N} \overleftrightarrow{B}_{ri0} \cdot \mathbf{E}_i^{\mathrm{inc}} + \sum_{i=1}^{N} \sum_{\substack{j=1 \\ j \neq i}}^{N} \overleftrightarrow{B}_{rij} \cdot \overleftrightarrow{B}_{ij0} \cdot \mathbf{E}_j^{\mathrm{inc}} + \sum_{i=1}^{N} \sum_{\substack{j=1 \\ j \neq i}}^{N} \sum_{\substack{l=1 \\ l \neq i \\ l \neq j}}^{N} \overleftrightarrow{B}_{rij} \cdot \overleftrightarrow{B}_{ijl} \cdot \overleftrightarrow{B}_{jl0} \cdot \mathbf{E}_l^{\mathrm{inc}}$$

$$+ \sum_{i=1}^{N} \sum_{\substack{j=1 \\ j \neq i}}^{N} \sum_{\substack{l=1 \\ l \neq i \\ l \neq j}}^{N} \sum_{\substack{m=1 \\ m \neq i \\ m \neq j \\ m \neq l}}^{N} \overleftrightarrow{B}_{rij} \cdot \overleftrightarrow{B}_{ijl} \cdot \overleftrightarrow{B}_{jlm} \cdot \overleftrightarrow{B}_{lm0} \cdot \mathbf{E}_m^{\mathrm{inc}} + \cdots . \tag{65}$$

It is straightforward to show that the Twersky approximation includes the majority of multiple-scattering paths and thus can be expected to yield rather accurate results provided that the number of particles is sufficiently large.

Panel (a) of Fig. 5 visualizes the full expansion (63), whereas panel (b) illustrates the Trwersky approximation (65). The symbol ⟵ represents the incident field, the symbol —● denotes multiplying a field by a $\vec{B}$ dyad, and the dashed connector indicates that two scattering events involve the same particle.

3.4. COHERENT FIELD

Let us now consider electromagnetic scattering by a large group of N arbitrarily oriented particles randomly distributed throughout a volume V. The particle ensemble is characterized by a probability density function $p(\mathbf{R}_1,\xi_1;...;\mathbf{R}_i,\xi_i;...;\mathbf{R}_N,\xi_N)$ such that the probability of finding the first particle in the volume element $d^3\mathbf{R}_1$ centered at $\mathbf{R}_1$ and with its state in the region $d\xi_1$ centered at ξ_1, …, the ith particle in the volume element $d^3\mathbf{R}_i$ centered at $\mathbf{R}_i$ and with its state in the region $d\xi_i$ centered at ξ_i, …, and the Nth particle in the volume element $d^3\mathbf{R}_N$ centered at $\mathbf{R}_N$ and with its state in the region $d\xi_N$ centered at ξ_N is given by $p(\mathbf{R}_1,\xi_1;...;\mathbf{R}_N,\xi_N)\prod_{i=1}^{N} d^3\mathbf{R}_i\, d\xi_i$. The state of a particle can collectively indicate its size, refractive index, shape, orientation, etc. The statistical average of a random function f depending on all N particles is given by

$$\langle f \rangle = \int f(\mathbf{R}_1,\xi_1;...;\mathbf{R}_N,\xi_N)\, p(\mathbf{R}_1,\xi_1;...;\mathbf{R}_N,\xi_N) \prod_{i=1}^{N} d^3\mathbf{R}_i d\xi_i \,. \tag{66}$$

If the position and state of each particle are independent of those of all other particles then

$$p(\mathbf{R}_1,\xi_1;...;\mathbf{R}_N,\xi_N) = \prod_{i=1}^{N} p_i(\mathbf{R}_i,\xi_i)\,. \tag{67}$$

This is a good approximation when particles are sparsely distributed so that the finite size of the particles can be neglected. In this case the effect of size appears only in the particle scattering characteristics. If, furthermore, the state of each particle is independent of its position then

$$p_i(\mathbf{R}_i,\xi_i) = p_{\mathbf{R}i}(\mathbf{R}_i)\, p_{\xi i}(\xi_i)\,. \tag{68}$$

Finally, assuming that all particles have the same statistical characteristics, we have

$$p_i(\mathbf{R}_i,\xi_i) \equiv p(\mathbf{R}_i,\xi_i) = p_{\mathbf{R}}(\mathbf{R}_i)\, p_{\xi}(\xi_i)\,. \tag{69}$$

Obviously,

$$p_{\mathbf{R}}(\mathbf{R}) = n_0(\mathbf{R})/N\,. \tag{70}$$

If the spatial distribution of the N particles throughout the volume V is statistically uniform then

$$n_0(\mathbf{R}) \equiv n_0 = N/V\,, \qquad p_{\mathbf{R}}(\mathbf{R}) = 1/V\,. \tag{71}$$

The electric field $\mathbf{E}(\mathbf{r})$ at a point $\mathbf{r}$ in the scattering medium is a random function of $\mathbf{r}$ and of the coordinates and states of the particles and can be decomposed into the average (coherent) field $\mathbf{E}_c(\mathbf{r})$ and the fluctuating field $\mathbf{E}_f(\mathbf{r})$:

$$\mathbf{E}(\mathbf{r}) = \mathbf{E}_c(\mathbf{r}) + \mathbf{E}_f(\mathbf{r})\,, \quad \mathbf{E}_c(\mathbf{r}) = \langle \mathbf{E}(\mathbf{r}) \rangle\,, \quad \langle \mathbf{E}_f(\mathbf{r}) \rangle = 0\,. \tag{72}$$

Assuming that the particles are sparsely distributed and have the same statistical characteristics, we have from Eqs. (65), (67), and (69):

$$\mathbf{E}_c = \mathbf{E}^{\rm inc} + \sum_{i=1}^{N} \int \left\langle \vec{A}(\hat{\mathbf{r}}_i, \hat{\mathbf{s}}) \right\rangle \cdot \mathbf{E}_i^{\rm inc} G(r_i) p_{\mathbf{R}}(\mathbf{R}_i)\, \mathrm{d}^3\mathbf{R}_i$$
$$+ \sum_{i=1}^{N} \sum_{\substack{j=1 \\ j\neq i}}^{N} \int \left\langle \vec{A}(\hat{\mathbf{r}}_i, \hat{\mathbf{R}}_{ij}) \right\rangle \cdot \left\langle \vec{A}(\hat{\mathbf{R}}_{ij}, \hat{\mathbf{s}}) \right\rangle \cdot \mathbf{E}_j^{\rm inc} G(r_i) G(R_{ij})$$
$$\times p_{\mathbf{R}}(\mathbf{R}_i) p_{\mathbf{R}}(\mathbf{R}_j)\, \mathrm{d}^3\mathbf{R}_i \mathrm{d}^3\mathbf{R}_j + \cdots, \tag{73}$$

where $\langle \vec{A}(\hat{\mathbf{m}}, \hat{\mathbf{n}}) \rangle$ is the average of the scattering dyad over the particle states. Finally, recalling Eq. (70), we obtain in the limit $N \to \infty$:

$$\mathbf{E}_c \underset{N\to\infty}{=} \mathbf{E}^{\rm inc} + \int \left\langle \vec{A}(\hat{\mathbf{r}}_i, \hat{\mathbf{s}}) \right\rangle \cdot \mathbf{E}_i^{\rm inc} G(r_i) n_0(\mathbf{R}_i)\, \mathrm{d}^3\mathbf{R}_i$$
$$+ \int \left\langle \vec{A}(\hat{\mathbf{r}}_i, \hat{\mathbf{R}}_{ij}) \right\rangle \cdot \left\langle \vec{A}(\hat{\mathbf{R}}_{ij}, \hat{\mathbf{s}}) \right\rangle \cdot \mathbf{E}_j^{\rm inc} G(r_i) G(R_{ij})\; n_0(\mathbf{R}_i) n_0(\mathbf{R}_j)\, \mathrm{d}^3\mathbf{R}_i\, \mathrm{d}^3\mathbf{R}_j + \cdots, \tag{74}$$

where we have replaced all factors $(N-n)!/N!$ by N^n. This is the general vector form of the expansion derived by Twersky [27] for scalar waves.

Assume now that the particles are distributed uniformly throughout the volume so that $n_0(\mathbf{R}) \equiv n_0$ and that the scattering medium has a concave boundary. The latter assumption ensures that all points of a straight line connecting any two points of the medium lie inside the medium. It is convenient to introduce an s-axis parallel to the incidence direction and going through the observation point (Fig. 6). This axis enters the volume V at the point A such that $s(A) = 0$ and exits it at the point B. One can then use the asymptotic expansion of a plane wave in spherical waves [28],

$$\exp(\mathrm{i}k_1 \hat{\mathbf{s}} \cdot \mathbf{R}'_i) \underset{k_1 R'_i \to \infty}{=} \frac{\mathrm{i}2\pi}{k_1 R'_i} [\delta(\hat{\mathbf{s}} + \hat{\mathbf{R}}'_i) \exp(-\mathrm{i}k_1 R'_i) - \delta(\hat{\mathbf{s}} - \hat{\mathbf{R}}'_i) \exp(\mathrm{i}k_1 R'_i)],$$

and assume that the observation point is in the far-field zone of any particle to derive [25]:

$$\mathbf{E}_c(\mathbf{r}) = \exp[\mathrm{i}2\pi n_0 k_1^{-1} s(\mathbf{r}) \langle \vec{A}(\hat{\mathbf{s}}, \hat{\mathbf{s}}) \rangle] \cdot \mathbf{E}^{\rm inc}(\mathbf{r}). \tag{75}$$

Since $\mathbf{r} = \mathbf{r}_A + s(\mathbf{r})\hat{\mathbf{s}}$, we have

$$\mathbf{E}_c(\mathbf{r}) = \exp[\mathrm{i}\vec{\kappa}(\hat{\mathbf{s}}) s(\mathbf{r})] \cdot \mathbf{E}^{\rm inc}(\mathbf{r}_A) = \vec{\eta}[\hat{\mathbf{s}}, s(\mathbf{r})] \cdot \mathbf{E}^{\rm inc}(\mathbf{r}_A), \tag{76}$$

where

$$\vec{\kappa}(\hat{\mathbf{s}}) = k_1 \vec{I} + \frac{2\pi n_0}{k_1} \left\langle \vec{A}(\hat{\mathbf{s}}, \hat{\mathbf{s}}) \right\rangle \tag{77}$$

is the dyadic propagation constant for the propagation direction $\hat{\mathbf{s}}$ and

$$\vec{\eta}(\hat{\mathbf{s}}, s) = \exp[\mathrm{i}\vec{\kappa}(\hat{\mathbf{s}}) s] \tag{78}$$

is the coherent transmission dyad. This is the general vector form of the Foldy

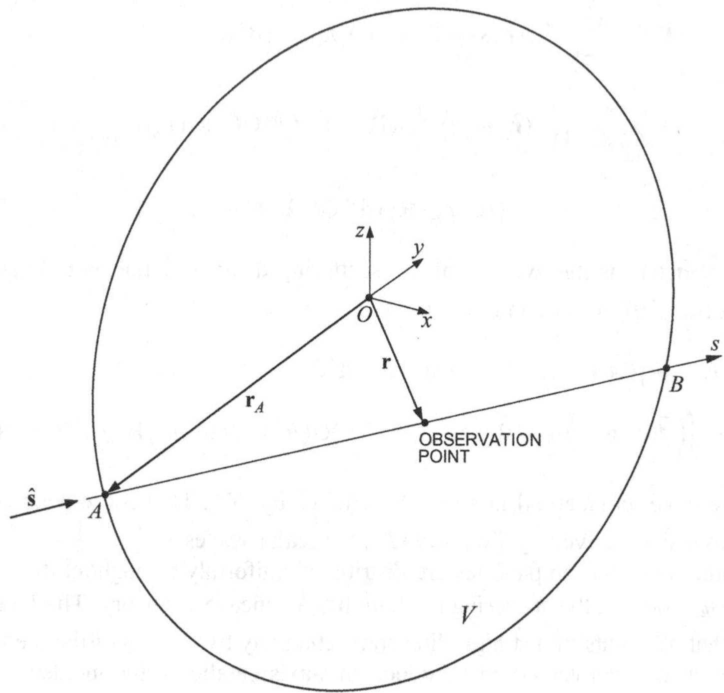

Fig. 6. Scattering volume

approximation for the coherent field. Another form of Eq. (76) is

$$\frac{\mathrm{d}\mathbf{E}_{\mathrm{c}}(\mathbf{r})}{\mathrm{d}s} = \mathrm{i}\vec{\kappa}(\hat{\mathbf{s}}) \cdot \mathbf{E}_{\mathrm{c}}(\mathbf{r}). \tag{79}$$

The coherent field also satisfies the vector Helmholtz equation

$$\nabla^2 \mathbf{E}_{\mathrm{c}}(\mathbf{r}) + k_1^2 \vec{\varepsilon}(\hat{\mathbf{s}}) \cdot \mathbf{E}_{\mathrm{c}}(\mathbf{r}) = 0, \tag{80}$$

where $\vec{\varepsilon}(\hat{\mathbf{s}}) = \vec{I} + 4\pi n_0 k_1^{-2} \left\langle \vec{A}(\hat{\mathbf{s}}, \hat{\mathbf{s}}) \right\rangle$ is the effective dyadic dielectric constant.

These results have several important implications. First, they show that the coherent field is a wave propagating in the direction of the incident field $\hat{\mathbf{s}}$. Second, since the products $\left\langle \vec{A}(\hat{\mathbf{s}}, \hat{\mathbf{s}}) \right\rangle \cdot \mathbf{E}_0^{\mathrm{inc}}$, $\left\langle \vec{A}(\hat{\mathbf{s}}, \hat{\mathbf{s}}) \right\rangle \cdot \left\langle \vec{A}(\hat{\mathbf{s}}, \hat{\mathbf{s}}) \right\rangle \cdot \mathbf{E}_0^{\mathrm{inc}}$, etc. always give electric vectors normal to $\hat{\mathbf{s}}$, the coherent wave is transverse: $\mathbf{E}_{\mathrm{c}}(\mathbf{r}) \cdot \hat{\mathbf{s}} = 0$. Third, Eq. (77) generalizes the optical theorem to the case of many scatterers by expressing the dyadic propagation constant in terms of the forward-scattering amplitude matrix averaged over the particle ensemble.

We can now make use of the transverse character of the coherent wave to rewrite the above equations in a simpler matrix form. As usual, we characterize the propagation direction $\hat{\mathbf{s}}$ at the observation point $\mathbf{r}$ using the corresponding zenith and azimuth angles in the local coordinate system centered at the observation point and having the same spatial orientation as the laboratory coordinate system $\{x, y, z\}$ (Fig. 6). Then the coherent field can be written as $\mathbf{E}_c(\mathbf{r}) = E_{c\theta}(\mathbf{r})\hat{\boldsymbol{\theta}} + E_{c\varphi}(\mathbf{r})\hat{\boldsymbol{\varphi}}$. Denoting, as always, the two-component electric column-vector of the coherent field by $\mathsf{E}_c(\mathbf{r})$, we have

$$\frac{\mathrm{d}\mathsf{E}_c(\mathbf{r})}{\mathrm{d}s} = \mathrm{i}\mathsf{k}(\hat{\mathbf{s}})\mathsf{E}_c(\mathbf{r}), \tag{81}$$

where $\mathsf{k}(\hat{\mathbf{s}})$ is the 2×2 matrix propagation constant with elements

$$\begin{aligned} k_{11}(\hat{\mathbf{s}}) &= \hat{\boldsymbol{\theta}}(\hat{\mathbf{s}})\cdot\vec{\kappa}(\hat{\mathbf{s}})\cdot\hat{\boldsymbol{\theta}}(\hat{\mathbf{s}}), \qquad & k_{12}(\hat{\mathbf{s}}) &= \hat{\boldsymbol{\theta}}(\hat{\mathbf{s}})\cdot\vec{\kappa}(\hat{\mathbf{s}})\cdot\hat{\boldsymbol{\varphi}}(\hat{\mathbf{s}}), \\ k_{21}(\hat{\mathbf{s}}) &= \hat{\boldsymbol{\varphi}}(\hat{\mathbf{s}})\cdot\vec{\kappa}(\hat{\mathbf{s}})\cdot\hat{\boldsymbol{\theta}}(\hat{\mathbf{s}}), \qquad & k_{22}(\hat{\mathbf{s}}) &= \hat{\boldsymbol{\varphi}}(\hat{\mathbf{s}})\cdot\vec{\kappa}(\hat{\mathbf{s}})\cdot\hat{\boldsymbol{\varphi}}(\hat{\mathbf{s}}). \end{aligned} \tag{82}$$

Obviously,

$$\mathsf{k}(\hat{\mathbf{s}}) = k_1 \mathrm{diag}[1,1] + \frac{2\pi n_0}{k_1}\langle \mathsf{S}(\hat{\mathbf{s}}, \hat{\mathbf{s}})\rangle, \tag{83}$$

where $\langle \mathsf{S}(\hat{\mathbf{s}}, \hat{\mathbf{s}})\rangle$ is the forward-scattering amplitude matrix averaged over the particle states.

It is not surprising that the propagation of the coherent field is controlled by the forward-scattering amplitude matrix. Indeed, the fluctuating component of the total field is the sum of the partial fields generated by different particles. Random movements of the particles involve large phase shifts in the partial fields and cause the fluctuating field to vanish when it is averaged over particle positions. The exact forward-scattering direction is different because in any plane parallel to the incident wave-front, the phase of the partial wave forward-scattered by a particle in response to the incident wave does not depend on the particle position. Therefore, the interference of the incident wave and the forward-scattered partial wave is always the same irrespective of the particle position, and the result of the interference does not vanish upon averaging over all particle positions.

3.5. TRANSFER EQUATION FOR THE COHERENT FIELD

We will now switch to quantities that have the dimension of monochromatic energy flux and can thus be measured by an optical device. We first define the coherency column vector of the coherent field according to

$$\mathsf{J}_c = \frac{1}{2}\sqrt{\frac{\varepsilon_1}{\mu_0}} \begin{bmatrix} E_{c\theta}E_{c\theta}^* \\ E_{c\theta}E_{c\varphi}^* \\ E_{c\varphi}E_{c\theta}^* \\ E_{c\varphi}E_{c\varphi}^* \end{bmatrix} \tag{84}$$

and derive from Eqs. (81) and (83) the following transfer equation:

$$\frac{\mathrm{d}\mathsf{J}_c(\mathbf{r})}{\mathrm{d}s} = -n_0\langle \mathsf{K}^J(\hat{\mathbf{s}})\rangle \mathsf{J}_c(\mathbf{r}), \tag{85}$$

where K^J is the coherency extinction matrix given by Eq. (23). The Stokes-vector representation of this equation is obtained using the definition $\mathbf{I}_c = \mathbf{D}\mathbf{J}_c$ and Eq. (25):

$$\frac{\mathrm{d}\mathbf{I}_c(\mathbf{r})}{\mathrm{d}s} = -n_0 \langle \mathbf{K}(\hat{\mathbf{s}}) \rangle \mathbf{I}_c(\mathbf{r}), \tag{86}$$

where $\mathbf{K}$ is the Stokes extinction matrix. Both $\mathbf{J}_c$ and $\mathbf{I}_c$ have the dimension of monochromatic energy flux. The formal solution of Eq. (86) can be written in the form

$$\mathbf{I}_c(\mathbf{r}) = \mathbf{H}[\hat{\mathbf{s}}, s(\mathbf{r})]\, \mathbf{I}_c(\mathbf{r}_A), \tag{87}$$

where

$$\mathbf{H}(\hat{\mathbf{s}}, s) = \exp\{-n_0 \langle \mathbf{K}(\hat{\mathbf{s}}) \rangle s\} \tag{88}$$

is the coherent transmission Stokes matrix.

The interpretation of Eq. (87) is most obvious when the average extinction matrix is given by Eq. (38):

$$\mathbf{I}_c(\mathbf{r}) = \exp[-n_0 \langle C_{\mathrm{ext}} \rangle s(\mathbf{r}))]\, \mathbf{I}_c(\mathbf{r}_A) = \exp[-\alpha_{\mathrm{ext}} s(\mathbf{r})]\, \mathbf{I}_c(\mathbf{r}_A), \tag{89}$$

which means that the Stokes parameters of the coherent wave are exponentially attenuated as the wave travels through the discrete random medium. The attenuation rates for all four Stokes parameters are the same, which means that the polarization state of the wave does not change. Equation (89) is the standard Beer's law, in which α_{ext} is the extinction coefficient. The attenuation is a combined result of scattering of the coherent field by particles in all directions and, possibly, absorption inside the particles and is an inalienable property of all scattering media, even those composed of nonabsorbing particles with $\langle C_{\mathrm{abs}} \rangle = 0$. In general, the extinction matrix is not diagonal and can explicitly depend on the propagation direction. This occurs, for example, when the scattering medium is composed of non-randomly oriented nonspherical particles. Then the coherent transmission matrix $\mathbf{H}$ in Eq. (87) can also have non-zero off-diagonal elements and cause a change in the polarization state of the coherent wave as it propagates through the medium.

3.6. DYADIC CORRELATION FUNCTION

An important statistical characteristic of the multiple-scattering process is the so-called dyadic correlation function defined as the ensemble average of the dyadic product $\mathbf{E}(\mathbf{r}) \otimes \mathbf{E}^*(\mathbf{r}')$. Obviously, the dyadic correlation function has the dimension of monochromatic energy flux. Recalling the Twersky approximation (65) and Fig. 5(b), we conclude that the dyadic correlation function can be represented diagrammatically by Fig. 7. To classify different terms entering the expanded expression inside the angular brackets on the right-hand side of this equation, we will use the notation illustrated in Fig. 8(a). In this particular case, the upper and the lower scattering paths go through different particles. However, the two paths can involve one or more common particles, as shown in panels (b)–(d) by using the dashed connectors. Furthermore, if the number of common particles is two or more, they can enter the upper and lower paths in the same order, as in panel (c), or in reverse order, as in panel (d). Panel (e) is a mixed diagram in which two common particles appear in the same order and two other common particles appear in reverse order.

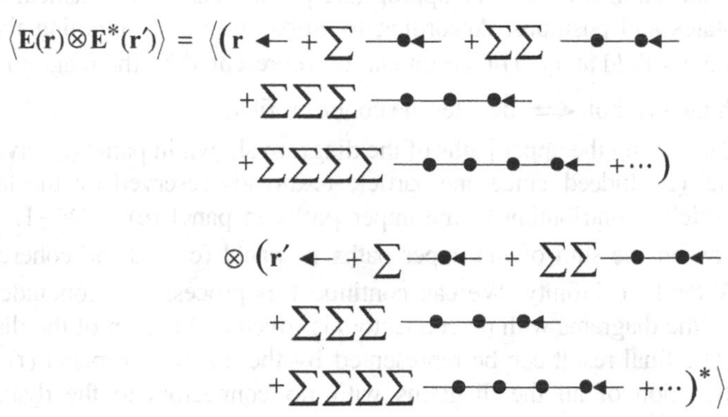

$$\begin{aligned}\langle \mathbf{E}(\mathbf{r})\otimes\mathbf{E}^*(\mathbf{r}')\rangle = \Big\langle \Big(\mathbf{r} \leftarrow &+ \sum -\!\bullet\!\!\leftarrow + \sum\sum -\!\bullet\!-\!\bullet\!\!\leftarrow \\ &+ \sum\sum\sum -\!\bullet\!-\!\bullet\!-\!\bullet\!\!\leftarrow \\ &+ \sum\sum\sum\sum -\!\bullet\!-\!\bullet\!-\!\bullet\!-\!\bullet\!\!\leftarrow + \cdots\Big) \\ \otimes \Big(\mathbf{r}' \leftarrow &+ \sum -\!\bullet\!\!\leftarrow + \sum\sum -\!\bullet\!-\!\bullet\!\!\leftarrow \\ &+ \sum\sum\sum -\!\bullet\!-\!\bullet\!-\!\bullet\!\!\leftarrow \\ &+ \sum\sum\sum\sum -\!\bullet\!-\!\bullet\!-\!\bullet\!-\!\bullet\!\!\leftarrow + \cdots\Big)^*\Big\rangle\end{aligned}$$

Fig. 7. The Twersky representation of the dyadic correlation function.

According to the Twersky approximation, no particle can be the origin of more than one connector.

To simplify the problem, we will neglect all diagrams with crossing connectors and will take into account only the diagrams with vertical or no connectors. This approximation will allow us to sum and average large groups of diagrams independently and eventually derive the radiative transfer equation.

We begin with diagrams that have no connectors. Since these diagrams do not involve common particles, the ensemble averaging of the upper and lower paths can be performed independently. Consider first the sum of the diagrams shown in Fig. 9(a), in which the Σ

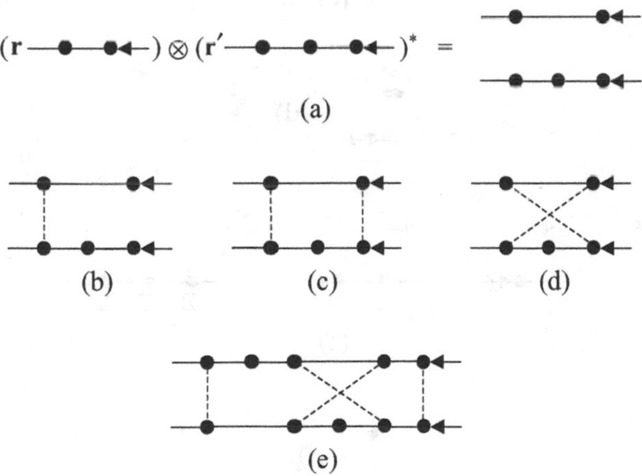

Fig. 8. Classification of terms entering the Twersky expansion of the dyadic correlation function.

indicates both the summation over all appropriate particles and the statistical averaging over the particle states and positions. According to Subsection 3.4, summing the upper paths yields the coherent field at $\mathbf{r}_1$. This result can be represented by the diagram shown in Fig. 9(b), in which the symbol ⇐ denotes the coherent field.

Similarly, summing the upper paths of the diagram shown in panel (c) gives the diagram shown in panel (d). Indeed, since one particle is already reserved for the lower path, the number of particles contributing to the upper paths in panel (c) is $N-1$. However, the difference between the sum of the upper paths in panel (c) and the coherent field at $\mathbf{r}_1$ vanishes as N tends to infinity. We can continue this process and conclude that the total contribution of the diagrams with no connectors is given by the sum of the diagrams shown in panel (e). The final result can be represented by the diagram in panel (f), which means that the contribution of all the diagrams with no connectors to the dyadic correlation

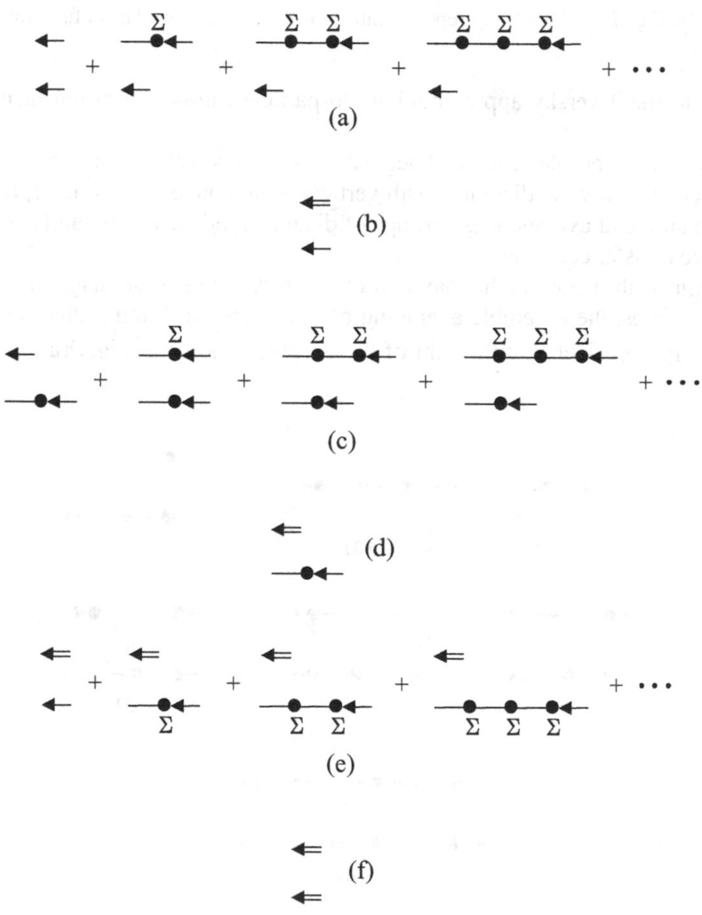

Fig. 9. Calculation of the total contribution of the diagrams with no connectors.

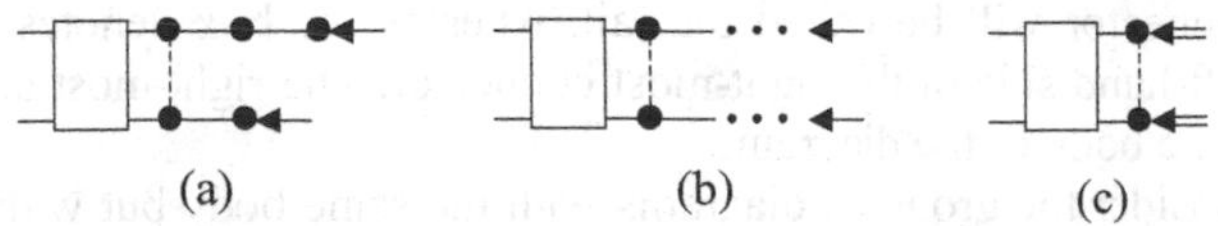

Fig. 10. Diagrams with one or more vertical connectors.

function is simply the dyadic product of the coherent fields at the points $\mathbf{r}$ and $\mathbf{r}'$: $\mathbf{E}_c(\mathbf{r}) \otimes \mathbf{E}_c^*(\mathbf{r}')$.

All other diagrams contributing to the dyadic correlation function have at least one

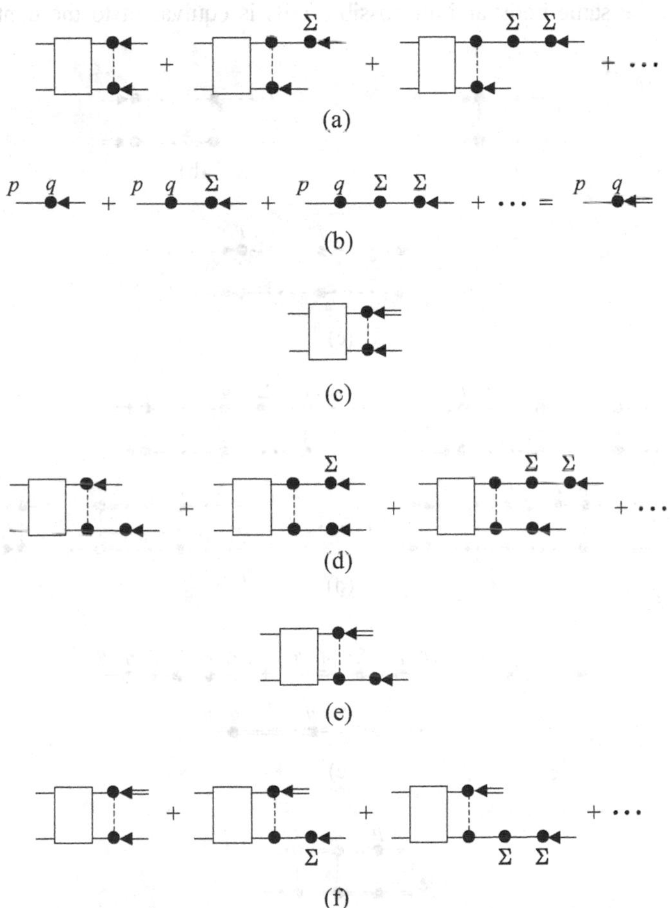

Fig. 11. Summation of the tails.

vertical connector, as shown in Fig. 10(a). The part of the diagram on the right-hand side of the right-most connector will be called the tail, whereas the box denotes the part of the diagram on the left-hand side of the right-most connector. The right-most common particle and the box form the body of the diagram.

Let us first consider the group of diagrams with the same body but with different tails, as shown in Fig. 10(b). We can repeat the derivation of subsection 3.4 and verify that the sum of all diagrams in Fig. 11(a) gives the diagram shown in Fig. 11(c). Indeed, let particle q be the right-most connected particle and particle p be the right-most particle on the left-hand side of particle q in the upper scattering paths of the diagrams shown in panel (a). The electric field created by particle q at the origin of particle p is represented by the sum of the diagrams on the left-hand side of panel (b). This result is summarized by the right-hand side of panel (b). Analogously, the sum of the diagrams in panel (d) is given by the diagram in panel (e), and so on. We can now sum up all diagrams in panel (f) and obtain the diagram shown in Fig. 10(c). Thus the total contribution to the dyadic correlation function of all the diagrams with the same body and all possible tails is equivalent to the contribution of a

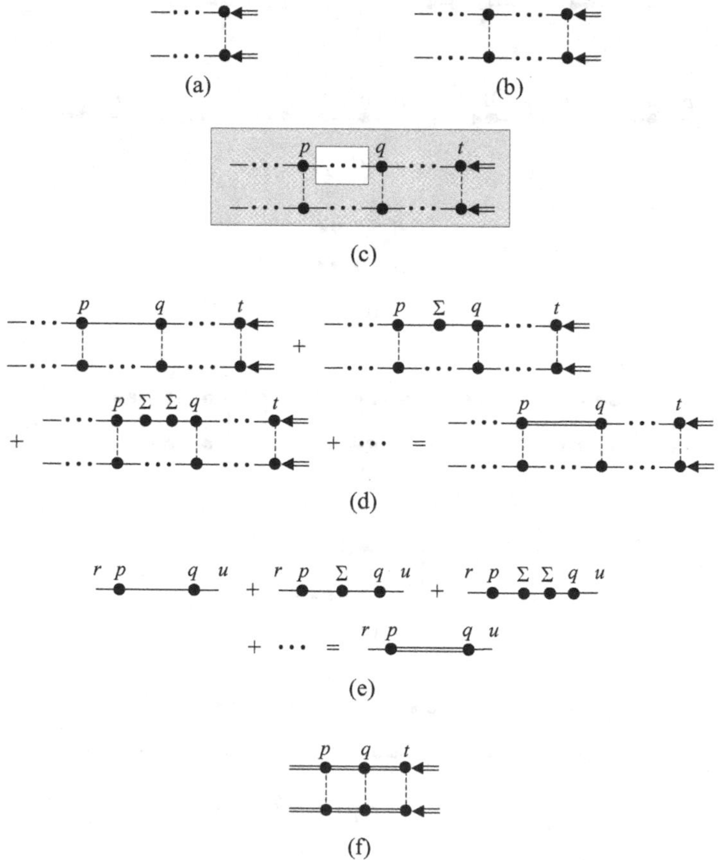

Fig. 12. Derivation of the ladder approximation for the dyadic correlation function.

single diagram formed by the body alone, *provided that the right-most common particle is excited by the coherent field rather than by the external incident field.* Thus we can cut off all tails and consider only truncated diagrams like those shown in Fig. 10(c).

Thus, the dyadic correlation function is equal to $\mathbf{E}_c(\mathbf{r})\otimes\mathbf{E}_c^*(\mathbf{r}')$ plus the statistical average of the sum of all connected diagrams of the type illustrated by panels (a)–(c) of Fig. 12, where the $\cdots$ denotes all possible combinations of unconnected particles. Let us, for example, consider the statistical average of the sum of all diagrams of the kind shown in panel (c) with the same fixed shaded part. We thus must evaluate the left-hand side of the equation shown in panel (d). Let particle r be the right-most particle on the left-hand side of particle p in the upper scattering paths of the diagrams on the left-hand side of panel (d) and u be the left-most particle on the right-hand side of particle q. The electric field created by particle p at the origin of particle r via all the diagrams shown on the left-hand side of panel (d) is given by the left-hand side of the equation shown diagrammatically in panel (e) and can be written in expanded form as

$$\begin{aligned}\mathbf{E}_r &= G(R_{rp})G(R_{pq})\ddot{A}_p(\hat{\mathbf{R}}_{rp},\hat{\mathbf{R}}_{pq})\cdot\ddot{A}_q(\hat{\mathbf{R}}_{pq},\hat{\mathbf{R}}_{qu})\cdot\mathbf{E}_q\\ &+\sum_i G(R_{rp})\left\langle G(R_{pi})G(R_{iq})\ddot{A}_p(\hat{\mathbf{R}}_{rp},\hat{\mathbf{R}}_{pi})\cdot\ddot{A}_i(\hat{\mathbf{R}}_{pi},\hat{\mathbf{R}}_{iq})\cdot\ddot{A}_q(\hat{\mathbf{R}}_{iq},\hat{\mathbf{R}}_{qu})\right\rangle\cdot\mathbf{E}_q\\ &+\sum_{ij} G(R_{rp})\left\langle G(R_{pi})G(R_{ij})G(R_{jq})\ddot{A}_p(\hat{\mathbf{R}}_{rp},\hat{\mathbf{R}}_{pi})\cdot\ddot{A}_i(\hat{\mathbf{R}}_{pi},\hat{\mathbf{R}}_{ij})\cdot\ddot{A}_j(\hat{\mathbf{R}}_{ij},\hat{\mathbf{R}}_{jq})\right.\\ &\left.\cdot\ddot{A}_q(\hat{\mathbf{R}}_{jq},\hat{\mathbf{R}}_{qu})\right\rangle\cdot\mathbf{E}_q+\cdots,\end{aligned}\tag{90}$$

where $\mathbf{E}_q$ is the field at the origin of particle q created by particle u and the summations and integrations are performed over all appropriate unconnected particles. In the limit $N\to\infty$, Eq. (90) takes the form

$$\begin{aligned}\mathbf{E}_r &= G(R_{rp})G(R_{pq})\ddot{A}_p(\hat{\mathbf{R}}_{rp},\hat{\mathbf{R}}_{pq})\cdot\ddot{A}_q(\hat{\mathbf{R}}_{pq},\hat{\mathbf{R}}_{qu})\cdot\mathbf{E}_q\\ &+n_0G(R_{rp})\int_V \mathrm{d}^3\mathbf{R}_i G(R_{pi})G(R_{iq})\ddot{A}_p(\hat{\mathbf{R}}_{rp},\hat{\mathbf{R}}_{pi})\cdot\left\langle\ddot{A}(\hat{\mathbf{R}}_{pi},\hat{\mathbf{R}}_{iq})\right\rangle\cdot\ddot{A}_q(\hat{\mathbf{R}}_{iq},\hat{\mathbf{R}}_{qu})\cdot\mathbf{E}_q\\ &+n_0^2G(R_{rp})\int_V \mathrm{d}^3\mathbf{R}_i\mathrm{d}^3\mathbf{R}_j G(R_{pi})G(R_{ij})G(R_{jq})\ddot{A}_p(\hat{\mathbf{R}}_{rp},\hat{\mathbf{R}}_{pi})\cdot\left\langle\ddot{A}(\hat{\mathbf{R}}_{pi},\hat{\mathbf{R}}_{ij})\right\rangle\\ &\cdot\left\langle\ddot{A}(\hat{\mathbf{R}}_{ij},\hat{\mathbf{R}}_{jq})\right\rangle\cdot\ddot{A}_q(\hat{\mathbf{R}}_{jq},\hat{\mathbf{R}}_{qu})\cdot\mathbf{E}_q\ \cdot\left\langle\ddot{A}(\hat{\mathbf{R}}_{pq},\hat{\mathbf{R}}_{pq})\right\rangle\cdot\ddot{A}_q(\hat{\mathbf{R}}_{pq},\hat{\mathbf{R}}_{qu})\cdot\mathbf{E}_{qu}\,.\end{aligned}\tag{91}$$

The integrals on the right-hand side of Eq. (91) can be evaluated using the method of stationary phase. The final result is [25]

$$\mathbf{E}_r = G(R_{rp})\ddot{A}_p(\hat{\mathbf{R}}_{rp},\hat{\mathbf{R}}_{pq})\cdot\frac{\ddot{\eta}(\hat{\mathbf{R}}_{pq},R_{pq})}{R_{pq}}\cdot\ddot{A}_q(\hat{\mathbf{R}}_{pq},\hat{\mathbf{R}}_{qu})\cdot\mathbf{E}_q,\tag{92}$$

where the coherent transmission dyad $\ddot{\eta}$ is given by Eq. (78). Obviously, this equation describes the coherent propagation of the wave scattered by particle q towards particle p through the scattering medium. The presence of other particles on the line of sight causes attenuation and, potentially, a change in polarization state of the wave.

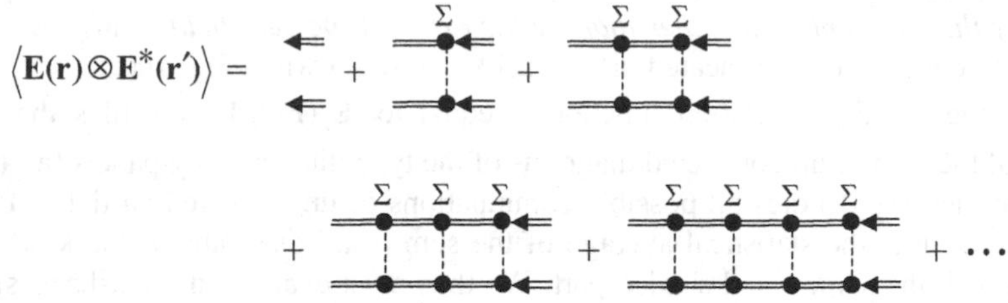

$$\langle \mathbf{E}(\mathbf{r}) \otimes \mathbf{E}^*(\mathbf{r}') \rangle = \cdots$$

Fig. 13. Ladder approximation for the dyadic correlation function.

Equation (92) can be summarized by the diagram on the right-hand side of Fig. 12(e), where the double line indicates that the scalar factor $\exp[ik_1 R_{pq}]/R_{pq}$ has been replaced by the dyadic factor $\exp[\mathrm{i}\overleftrightarrow{\kappa}(\hat{\mathbf{R}}_{pq})R_{pq}]/R_{pq}$. Thus the total contribution of all diagrams with three fixed common particles t, q, and p to the dyadic correlation function can be represented by the diagram in Fig. 12(f).

It is now clear that the final expression for the dyadic correlation function can be represented graphically by Fig. 13. Owing to their appearance, the diagrams on the right-hand side are called ladder diagrams, and this entire formula is called the ladder approximation for the dyadic correlation function.

3.7. INTEGRAL EQUATION FOR THE SPECIFIC COHERENCY DYAD

The coherency dyad is defined as $\overleftrightarrow{C}(\mathbf{r}) = \mathbf{E}(\mathbf{r}) \otimes \mathbf{E}^*(\mathbf{r})$. The expanded form of the ladder approximation for the coherency dyad follows from Figs. 13 and 14:

$$\begin{aligned}\overleftrightarrow{C}(\mathbf{r}) &= \overleftrightarrow{C}_\mathrm{c}(\mathbf{r}) + n_0 \int \mathrm{d}^3\mathbf{R}_1 \mathrm{d}\xi_1 \frac{\overleftrightarrow{\eta}(\hat{\mathbf{r}}_1, r_1)}{r_1} \cdot \overleftrightarrow{A}_1(\hat{\mathbf{r}}_1, \hat{\mathbf{s}}) \cdot \overleftrightarrow{C}_\mathrm{c}(\mathbf{R}_1) \cdot \overleftrightarrow{A}_1^{\mathrm{T}*}(\hat{\mathbf{r}}_1, \hat{\mathbf{s}}) \cdot \frac{\overleftrightarrow{\eta}^{\mathrm{T}*}(\hat{\mathbf{r}}_1, r_1)}{r_1} \\ &+ n_0^2 \int \mathrm{d}^3\mathbf{R}_1 \mathrm{d}\xi_1 \int \mathrm{d}^3\mathbf{R}_2 \mathrm{d}\xi_2 \frac{\overleftrightarrow{\eta}(\hat{\mathbf{r}}_1, r_1)}{r_1} \cdot \overleftrightarrow{A}_1(\hat{\mathbf{r}}_1, \hat{\mathbf{R}}_{12}) \cdot \frac{\overleftrightarrow{\eta}(\hat{\mathbf{R}}_{12}, R_{12})}{R_{12}} \cdot \overleftrightarrow{A}_2(\hat{\mathbf{R}}_{12}, \hat{\mathbf{s}}) \\ &\cdot \overleftrightarrow{C}_\mathrm{c}(\mathbf{R}_2) \cdot \overleftrightarrow{A}_2^{\mathrm{T}*}(\hat{\mathbf{R}}_{12}, \hat{\mathbf{s}}) \cdot \frac{\overleftrightarrow{\eta}^{\mathrm{T}*}(\hat{\mathbf{R}}_{12}, R_{12})}{R_{12}} \cdot \overleftrightarrow{A}_1^{\mathrm{T}*}(\hat{\mathbf{r}}_1, \hat{\mathbf{R}}_{12}) \cdot \frac{\overleftrightarrow{\eta}^{\mathrm{T}*}(\hat{\mathbf{r}}_1, r_1)}{r_1} + \cdots, \end{aligned} \tag{93}$$

where $\overleftrightarrow{C}_\mathrm{c}(\mathbf{r}) = \mathbf{E}_\mathrm{c}(\mathbf{r}) \otimes \mathbf{E}_\mathrm{c}^*(\mathbf{r})$ is the coherent part of the coherency dyad. It is convenient to integrate over all positions of particle 1 using a local coordinate system with origin at the observation point, integrate over all positions of particle 2 using a local coordinate system with origin at the origin of particle 1, etc. Using the notation introduced in Fig. 14 yields

$$\overleftrightarrow{C}(\mathbf{r}) = \int_{4\pi} \mathrm{d}\hat{\mathbf{p}} \overleftrightarrow{\Sigma}(\mathbf{r}, -\hat{\mathbf{p}}), \tag{94}$$

where $\overleftrightarrow{\Sigma}(\mathbf{r}, -\hat{\mathbf{p}})$ is the specific coherency dyad defined by

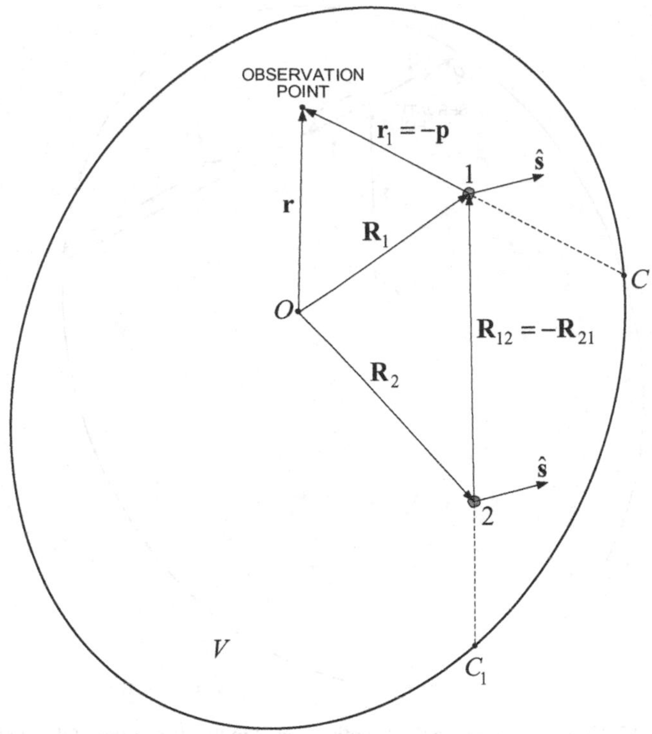

Fig. 14. Geometry showing the quantities used in Eq. (93).

$$\begin{aligned}
\ddot{\Sigma}(\mathbf{r},-\hat{\mathbf{p}}) &= \delta(\hat{\mathbf{p}}+\hat{\mathbf{s}})\ddot{C}_{\mathrm{c}}(\mathbf{r}) \\
&+ n_0 \int \mathrm{d}p\, \mathrm{d}\xi_1\, \ddot{\eta}(-\hat{\mathbf{p}},p)\cdot \ddot{A}_1(-\hat{\mathbf{p}},\hat{\mathbf{s}})\cdot \ddot{C}_{\mathrm{c}}(\mathbf{r}+\mathbf{p})\cdot \ddot{A}_1^{\mathrm{T}*}(-\hat{\mathbf{p}},\hat{\mathbf{s}})\cdot \ddot{\eta}^{\mathrm{T}*}(-\hat{\mathbf{p}},p) \\
&+ n_0^2 \int \mathrm{d}p\, \mathrm{d}\xi_1 \int \mathrm{d}R_{21}\mathrm{d}\hat{\mathbf{R}}_{21}\mathrm{d}\xi_2 \ddot{\eta}(-\hat{\mathbf{p}},p)\cdot \ddot{A}_1(-\hat{\mathbf{p}},-\hat{\mathbf{R}}_{21})\cdot \ddot{\eta}(-\hat{\mathbf{R}}_{21},R_{21}) \\
&\quad \cdot \ddot{A}_2(-\hat{\mathbf{R}}_{21},\hat{\mathbf{s}})\cdot \ddot{C}_{\mathrm{c}}(\mathbf{r}+\mathbf{p}+\mathbf{R}_{21})\cdot \ddot{A}_2^{\mathrm{T}*}(-\hat{\mathbf{R}}_{21},\hat{\mathbf{s}})\cdot \ddot{\eta}^{\mathrm{T}*}(-\hat{\mathbf{R}}_{21},R_{21}) \\
&\quad \cdot \ddot{A}_1^{\mathrm{T}*}(-\hat{\mathbf{p}},-\hat{\mathbf{R}}_{21})\cdot \ddot{\eta}^{\mathrm{T}*}(-\hat{\mathbf{p}},p)+\cdots . \qquad (95)
\end{aligned}$$

Note that p ranges from zero at the observation point to the corresponding value at the point where the straight line in the $\hat{\mathbf{p}}$-direction crosses the boundary of the medium (point C in Fig. 14), R_{21} ranges from zero at the origin of particle 1 to the corresponding value at point C_1, etc. The specific coherency dyad has the dimension of specific intensity ($\mathrm{Wm^{-2}sr^{-1}}$) rather than that of monochromatic energy flux.

It is straightforward to verify that $\ddot{\Sigma}$ satisfies the following integral equation:

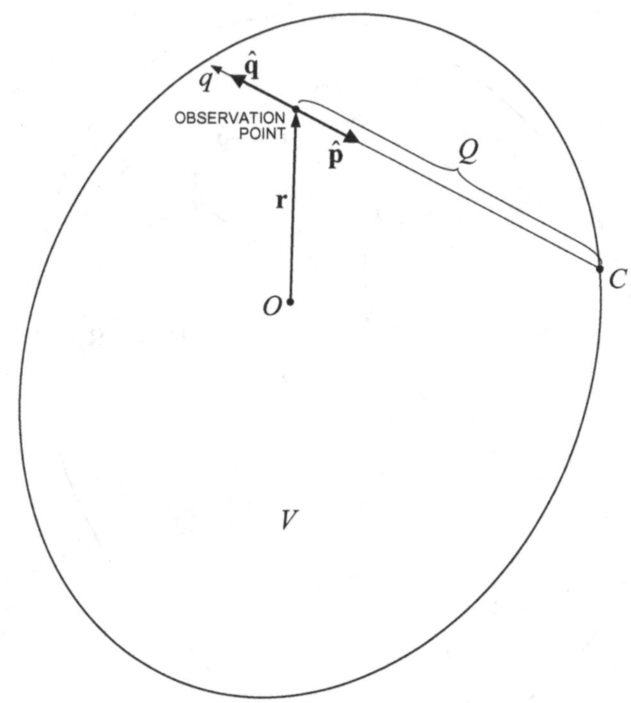

Fig. 15. Geometry showing the quantities used in the derivation of the RTE.

$$\overleftrightarrow{\Sigma}(\mathbf{r},-\hat{\mathbf{p}}) = \delta(\hat{\mathbf{p}}+\hat{\mathbf{s}})\overleftrightarrow{C}_{\mathrm{c}}(\mathbf{r}) + n_0 \int \mathrm{d}p\, \mathrm{d}\xi\, \mathrm{d}\hat{\mathbf{p}}'\, \overleftrightarrow{\eta}(-\hat{\mathbf{p}},p)\cdot \overleftrightarrow{A}(-\hat{\mathbf{p}},-\hat{\mathbf{p}}')\cdot \overleftrightarrow{\Sigma}(\mathbf{r}+\mathbf{p},-\hat{\mathbf{p}}')$$
$$\cdot \overleftrightarrow{A}^{\mathrm{T}*}(-\hat{\mathbf{p}},-\hat{\mathbf{p}}')\cdot \overleftrightarrow{\eta}^{\mathrm{T}*}(-\hat{\mathbf{p}},p)\,. \tag{96}$$

Indeed, Eq. (95) is reproduced by iterating Eq. (96). Equation (95) is simply an order-of-scattering expansion of the specific coherency dyad *with coherent field serving as the source of multiple scattering.*

The interpretation of Eq. (96) is clear: the specific coherency dyad for a direction $-\hat{\mathbf{p}}$ at a point $\mathbf{r}$ consists of a coherent part and an incoherent part. The latter is a cumulative contribution of all particles located along the straight line in the $\hat{\mathbf{p}}$-direction and scattering radiation coming from all directions $-\hat{\mathbf{p}}'$ into the direction $-\hat{\mathbf{p}}$.

3.8 RTE FOR SPECIFIC COHERENCY DYAD

We now introduce a q-axis as shown in Fig. 15 and rewrite Eq. (96) as

$$\overleftrightarrow{\Sigma}(Q,\hat{\mathbf{q}}) = \delta(\hat{\mathbf{q}}-\hat{\mathbf{s}})\overleftrightarrow{C}_{\mathrm{c}}(Q) + n_0 \int_0^Q \mathrm{d}q \int \mathrm{d}\xi \int_{4\pi} \mathrm{d}\hat{\mathbf{q}}'\, \overleftrightarrow{\eta}(\hat{\mathbf{q}},Q-q)\cdot \overleftrightarrow{A}(\hat{\mathbf{q}},\hat{\mathbf{q}}')$$

$$\cdot \ddot{\Sigma}(q,\hat{\mathbf{q}}') \cdot \ddot{A}^{\mathrm{T}*}(\hat{\mathbf{q}},\hat{\mathbf{q}}') \cdot \ddot{\eta}^{\mathrm{T}*}(\hat{\mathbf{q}},Q-q) . \tag{97}$$

Defining the diffuse specific coherency dyad as $\ddot{\Sigma}_{\mathrm{d}}(Q,\hat{\mathbf{q}}) = \ddot{\Sigma}(Q,\hat{\mathbf{q}}) - \delta(\hat{\mathbf{q}}-\hat{\mathbf{s}})\ddot{C}_{\mathrm{c}}(Q)$, we obtain

$$\begin{aligned}\ddot{\Sigma}_{\mathrm{d}}(Q,\hat{\mathbf{q}}) = {} & n_0 \int_0^Q \mathrm{d}q \int \mathrm{d}\xi\, \ddot{\eta}(\hat{\mathbf{q}},Q-q) \cdot \ddot{A}(\hat{\mathbf{q}},\hat{\mathbf{s}}) \ddot{C}_{\mathrm{c}}(q) \cdot \ddot{A}^{\mathrm{T}*}(\hat{\mathbf{q}},\hat{\mathbf{s}}) \cdot \ddot{\eta}^{\mathrm{T}*}(\hat{\mathbf{q}},Q-q) \\ & + n_0 \int_0^Q \mathrm{d}q \int \mathrm{d}\xi \int_{4\pi} \mathrm{d}\hat{\mathbf{q}}'\, \ddot{\eta}(\hat{\mathbf{q}},Q-q) \cdot \ddot{A}(\hat{\mathbf{q}},\hat{\mathbf{q}}') \cdot \ddot{\Sigma}_{\mathrm{d}}(q,\hat{\mathbf{q}}') \\ & \qquad \cdot \ddot{A}^{\mathrm{T}*}(\hat{\mathbf{q}},\hat{\mathbf{q}}') \cdot \ddot{\eta}^{\mathrm{T}*}(\hat{\mathbf{q}},Q-q) .\end{aligned} \tag{98}$$

Differentiating both sides of Eq. (98) yields

$$\begin{aligned}\frac{\mathrm{d}\ddot{\Sigma}_{\mathrm{d}}(Q,\hat{\mathbf{q}})}{\mathrm{d}Q} = {} & \mathrm{i}\ddot{\kappa}(\hat{\mathbf{q}}) \cdot \ddot{\Sigma}_{\mathrm{d}}(Q,\hat{\mathbf{q}}) - \mathrm{i}\ddot{\Sigma}_{\mathrm{d}}(Q,\hat{\mathbf{q}}) \cdot \ddot{\kappa}^{\mathrm{T}*}(\hat{\mathbf{q}}) \\ & + n_0 \int \mathrm{d}\xi \int_{4\pi} \mathrm{d}\hat{\mathbf{q}}' \ddot{A}(\hat{\mathbf{q}},\hat{\mathbf{q}}') \cdot \ddot{\Sigma}_{\mathrm{d}}(Q,\hat{\mathbf{q}}') \cdot \ddot{A}^{\mathrm{T}*}(\hat{\mathbf{q}},\hat{\mathbf{q}}') \\ & + n_0 \int \mathrm{d}\xi \ddot{A}(\hat{\mathbf{q}},\hat{\mathbf{s}}) \cdot \ddot{C}_{\mathrm{c}}(Q) \cdot \ddot{A}^{\mathrm{T}*}(\hat{\mathbf{q}},\hat{\mathbf{s}}) .\end{aligned} \tag{99}$$

For further use, it is more convenient to rewrite Eq. (99) in the following form:

$$\begin{aligned}\frac{\mathrm{d}\ddot{\Sigma}_{\mathrm{d}}(\mathbf{r},\hat{\mathbf{q}})}{\mathrm{d}q} = {} & \mathrm{i}\ddot{\kappa}(\hat{\mathbf{q}}) \cdot \ddot{\Sigma}_{\mathrm{d}}(\mathbf{r},\hat{\mathbf{q}}) - \mathrm{i}\ddot{\Sigma}_{\mathrm{d}}(\mathbf{r},\hat{\mathbf{q}}) \cdot \ddot{\kappa}^{\mathrm{T}*}(\hat{\mathbf{q}}) \\ & + n_0 \int \mathrm{d}\xi \int_{4\pi} \mathrm{d}\hat{\mathbf{q}}' \ddot{A}(\hat{\mathbf{q}},\hat{\mathbf{q}}') \cdot \ddot{\Sigma}_{\mathrm{d}}(\mathbf{r},\hat{\mathbf{q}}') \cdot \ddot{A}^{\mathrm{T}*}(\hat{\mathbf{q}},\hat{\mathbf{q}}') \\ & + n_0 \int \mathrm{d}\xi \ddot{A}(\hat{\mathbf{q}},\hat{\mathbf{s}}) \cdot \ddot{C}_{\mathrm{c}}(\mathbf{r}) \cdot \ddot{A}^{\mathrm{T}*}(\hat{\mathbf{q}},\hat{\mathbf{s}}) ,\end{aligned} \tag{100}$$

where $\mathrm{d}q$ is measured along the unit vector $\hat{\mathbf{q}}$. Equation (100) is the integro-differential RTE for the diffuse specific coherency dyad.

3.9. RTE FOR SPECIFIC INTENSITY VECTOR

It follows from Eq. (98) that $\hat{\mathbf{q}} \cdot \ddot{\Sigma}_{\mathrm{d}}(\mathbf{r},\hat{\mathbf{q}}) = \ddot{\Sigma}_{\mathrm{d}}(\mathbf{r},\hat{\mathbf{q}}) \cdot \hat{\mathbf{q}} = 0$, which allows us to introduce the 2×2 diffuse specific coherency matrix $\tilde{\boldsymbol{\rho}}_{\mathrm{d}}$ using the local coordinate system with origin at the observation point and orientation identical to that of the laboratory coordinate system:

$$\tilde{\boldsymbol{\rho}}_{\mathrm{d}}(\mathbf{r},\hat{\mathbf{q}}) = \frac{1}{2}\sqrt{\frac{\varepsilon_1}{\mu_0}} \begin{bmatrix} \hat{\boldsymbol{\theta}}(\hat{\mathbf{q}}) \cdot \ddot{\Sigma}_{\mathrm{d}}(\mathbf{r},\hat{\mathbf{q}}) \cdot \hat{\boldsymbol{\theta}}(\hat{\mathbf{q}}) & \hat{\boldsymbol{\theta}}(\hat{\mathbf{q}}) \cdot \ddot{\Sigma}_{\mathrm{d}}(\mathbf{r},\hat{\mathbf{q}}) \cdot \hat{\boldsymbol{\varphi}}(\hat{\mathbf{q}}) \\ \hat{\boldsymbol{\varphi}}(\hat{\mathbf{q}}) \cdot \ddot{\Sigma}_{\mathrm{d}}(\mathbf{r},\hat{\mathbf{q}}) \cdot \hat{\boldsymbol{\theta}}(\hat{\mathbf{q}}) & \hat{\boldsymbol{\varphi}}(\hat{\mathbf{q}}) \cdot \ddot{\Sigma}_{\mathrm{d}}(\mathbf{r},\hat{\mathbf{q}}) \cdot \hat{\boldsymbol{\varphi}}(\hat{\mathbf{q}}) \end{bmatrix} . \tag{101}$$

We can now rewrite Eq. (100) in the form of the RTE for the diffuse specific coherency matrix:

$$\frac{\mathrm{d}\tilde{\boldsymbol{\rho}}_{\mathrm{d}}(\mathbf{r},\hat{\mathbf{q}})}{\mathrm{d}q} = \mathrm{i}\mathbf{k}(\hat{\mathbf{q}})\tilde{\boldsymbol{\rho}}_{\mathrm{d}}(\mathbf{r},\hat{\mathbf{q}}) - \mathrm{i}\tilde{\boldsymbol{\rho}}_{\mathrm{d}}(\mathbf{r},\hat{\mathbf{q}})\mathbf{k}^{\mathrm{T}*}(\hat{\mathbf{q}})$$
$$+ n_0 \int \mathrm{d}\xi \int_{4\pi} \mathrm{d}\hat{\mathbf{q}}' \mathbf{S}(\hat{\mathbf{q}},\hat{\mathbf{q}}')\tilde{\boldsymbol{\rho}}_{\mathrm{d}}(\mathbf{r},\hat{\mathbf{q}}')\mathbf{S}^{\mathrm{T}*}(\hat{\mathbf{q}},\hat{\mathbf{q}}')$$
$$+ n_0 \int \mathrm{d}\xi \mathbf{S}(\hat{\mathbf{q}},\hat{\mathbf{s}})\boldsymbol{\rho}_{\mathrm{c}}(\mathbf{r})\mathbf{S}^{\mathrm{T}*}(\hat{\mathbf{q}},\hat{\mathbf{s}}), \tag{102}$$

where $\mathbf{S}$ is the amplitude matrix, $\mathbf{k}$ is the matrix propagation constant given by Eq. (82), and

$$\boldsymbol{\rho}_{\mathrm{c}}(\mathbf{r}) = \frac{1}{2}\sqrt{\frac{\varepsilon_1}{\mu_0}} \begin{bmatrix} \hat{\boldsymbol{\theta}}(\hat{\mathbf{s}})\cdot\vec{C}_{\mathrm{c}}(\mathbf{r})\cdot\hat{\boldsymbol{\theta}}(\hat{\mathbf{s}}) & \hat{\boldsymbol{\theta}}(\hat{\mathbf{s}})\cdot\vec{C}_{\mathrm{c}}(\mathbf{r})\cdot\hat{\boldsymbol{\varphi}}(\hat{\mathbf{s}}) \\ \hat{\boldsymbol{\varphi}}(\hat{\mathbf{s}})\cdot\vec{C}_{\mathrm{c}}(\mathbf{r})\cdot\hat{\boldsymbol{\theta}}(\hat{\mathbf{s}}) & \hat{\boldsymbol{\varphi}}(\hat{\mathbf{s}})\cdot\vec{C}_{\mathrm{c}}(\mathbf{r})\cdot\hat{\boldsymbol{\varphi}}(\hat{\mathbf{s}}) \end{bmatrix}. \tag{103}$$

The next obvious step is to introduce the corresponding coherency column vectors $\tilde{\mathbf{J}}_{\mathrm{d}}$ and $\mathbf{J}_{\mathrm{c}}$:

$$\tilde{\mathbf{J}}_{\mathrm{d}}(\mathbf{r},\hat{\mathbf{q}}) = \begin{bmatrix} \tilde{\rho}_{\mathrm{d}11}(\mathbf{r},\hat{\mathbf{q}}) \\ \tilde{\rho}_{\mathrm{d}12}(\mathbf{r},\hat{\mathbf{q}}) \\ \tilde{\rho}_{\mathrm{d}21}(\mathbf{r},\hat{\mathbf{q}}) \\ \tilde{\rho}_{\mathrm{d}22}(\mathbf{r},\hat{\mathbf{q}}) \end{bmatrix}, \qquad \mathbf{J}_{\mathrm{c}}(\mathbf{r}) = \begin{bmatrix} \rho_{\mathrm{c}11}(\mathbf{r}) \\ \rho_{\mathrm{c}12}(\mathbf{r}) \\ \rho_{\mathrm{c}21}(\mathbf{r}) \\ \rho_{\mathrm{c}22}(\mathbf{r}) \end{bmatrix}. \tag{104}$$

Lengthy, but simple algebraic manipulations yield

$$\frac{\mathrm{d}\tilde{\mathbf{J}}_{\mathrm{d}}(\mathbf{r},\hat{\mathbf{q}})}{\mathrm{d}q} = -n_0\langle \mathbf{K}^J(\hat{\mathbf{q}})\rangle \tilde{\mathbf{J}}_{\mathrm{d}}(\mathbf{r},\hat{\mathbf{q}}) + n_0 \int_{4\pi} \mathrm{d}\hat{\mathbf{q}}' \langle \mathbf{Z}^J(\hat{\mathbf{q}},\hat{\mathbf{q}}')\rangle \tilde{\mathbf{J}}_{\mathrm{d}}(\mathbf{r},\hat{\mathbf{q}}')$$
$$+ n_0\langle \mathbf{Z}^J(\hat{\mathbf{q}},\hat{\mathbf{s}})\rangle \mathbf{J}_{\mathrm{c}}(\mathbf{r}), \tag{105}$$

where $\langle \mathbf{K}^J(\hat{\mathbf{q}})\rangle$ is the coherency extinction matrix averaged over the particle states and $\langle \mathbf{Z}^J(\hat{\mathbf{q}},\hat{\mathbf{q}}')\rangle$ is the ensemble average of the coherency phase matrix. The column vector $\mathbf{J}_{\mathrm{c}}(\mathbf{r})$ satisfies the transfer equation (85).

The final step in the derivation of the RTE is to define the diffuse specific intensity column vector, $\tilde{\mathbf{I}}_{\mathrm{d}}(\mathbf{r},\hat{\mathbf{q}}) = \mathbf{D}\tilde{\mathbf{J}}_{\mathrm{d}}(\mathbf{r},\hat{\mathbf{q}})$, and the coherent Stokes column vector, $\mathbf{I}_{\mathrm{c}}(\mathbf{r}) = \mathbf{D}\mathbf{J}_{\mathrm{c}}(\mathbf{r})$, and rewrite Eq. (105) in the form

$$\frac{\mathrm{d}\tilde{\mathbf{I}}_{\mathrm{d}}(\mathbf{r},\hat{\mathbf{q}})}{\mathrm{d}q} = -n_0\langle \mathbf{K}(\hat{\mathbf{q}})\rangle \tilde{\mathbf{I}}_{\mathrm{d}}(\mathbf{r},\hat{\mathbf{q}}) + n_0 \int_{4\pi} \mathrm{d}\hat{\mathbf{q}}' \langle \mathbf{Z}(\hat{\mathbf{q}},\hat{\mathbf{q}}')\rangle \tilde{\mathbf{I}}_{\mathrm{d}}(\mathbf{r},\hat{\mathbf{q}}') + n_0\langle \mathbf{Z}(\hat{\mathbf{q}},\hat{\mathbf{s}})\rangle \mathbf{I}_{\mathrm{c}}(\mathbf{r}), \tag{106}$$

where $\langle \mathbf{K}(\hat{\mathbf{q}})\rangle$ is the ensemble average of the Stokes extinction matrix and $\langle \mathbf{Z}(\hat{\mathbf{q}},\hat{\mathbf{q}}')\rangle$ is the ensemble average of the Stokes phase matrix. The coherent Stokes column vector $\mathbf{I}_{\mathrm{c}}(\mathbf{r})$ satisfies the transfer equation (86).

3.10. DISCUSSION

Equations (86) and (106) represent the classical form of the RTE applicable to arbitrarily shaped and arbitrarily oriented particles. The microphysical derivation of these equations outlined above is based on fundamental principles of statistical electromagnetics and naturally replaces the original incident field as the source of multiple scattering by the decaying coherent field and leads to the introduction of the diffuse specific intensity vector describing the photometric and polarimetric characteristics of the multiply scattered light. The physical interpretation of $\tilde{\mathbf{I}}_{\mathrm{d}}(\mathbf{r},\hat{\mathbf{q}})$ is rather transparent. Imagine a collimated detector centered at the observation point and aligned along the direction $\hat{\mathbf{q}}$ $(\neq \hat{\mathbf{s}})$ (Fig. 16). Let ΔS be the detector area and $\Delta\Omega$ its acceptance solid angle. Each infinitesimal element of the detector surface responds to the radiant energy coming from the directions confined to a narrow cone with the small solid-angle aperture $\Delta\Omega$ centered around $\hat{\mathbf{q}}$. On the other hand, we can use Eq. (98) to write

$$\Delta\Omega\, \tilde{\mathbf{I}}_{\mathrm{d}}(\mathbf{r},\hat{\mathbf{q}}) \approx n_0 \int_{\Delta V} \mathrm{d}^3\mathbf{p} \frac{1}{p^2} \mathsf{H}(\hat{\mathbf{q}}, p) \times\Big(\langle \boldsymbol{Z}(\hat{\mathbf{q}},\hat{\mathbf{s}}) \rangle \mathbf{I}_{\mathrm{c}}(\mathbf{r}+\mathbf{p}) + \int_{4\pi} \mathrm{d}\hat{\mathbf{q}}' \langle \mathbf{Z}(\hat{\mathbf{q}},\hat{\mathbf{q}}') \rangle \tilde{\mathbf{I}}_{\mathrm{d}}(\mathbf{r}+\mathbf{p},\hat{\mathbf{q}}') \Big), \tag{107}$$

where $\mathbf{p}$ originates at the observation point $\mathbf{r}$ (Fig. 15) and the integration is performed

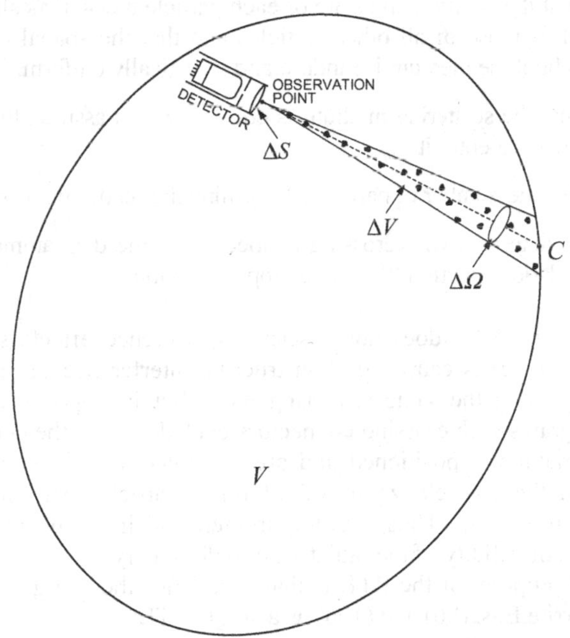

Fig. 16. Physical meaning of the diffuse specific intensity vector.

over the conical volume element ΔV having the solid-angle aperture $\Delta\Omega$ and extending from the observation point to point C (Fig. 16). The right-hand side of Eq. (107) is simply the integral of the scattering signal per unit surface area perpendicular to $\hat{\mathbf{q}}$ per unit time over all particles contained in the conical volume element. It is now clear what quantity describes the total polarized signal measured by the detector per unit time: it is the product $\Delta S \Delta\Omega\, \tilde{\mathbf{I}}_{\mathrm{d}}(\mathbf{r},\hat{\mathbf{q}})$, which has the dimension of power (W). The first element of $\tilde{\mathbf{I}}_{\mathrm{d}}(\mathbf{r},\hat{\mathbf{q}})$ is the standard diffuse specific intensity $\tilde{I}_{\mathrm{d}}(\mathbf{r},\hat{\mathbf{q}})$ defined such that the product $\Delta t \Delta S \Delta\Omega\, \tilde{I}_{\mathrm{d}}(\mathbf{r},\hat{\mathbf{q}})$ gives the amount of radiant energy transported in a time interval Δt through an element of surface area ΔS normal to $\hat{\mathbf{q}}$ in directions confined to a solid angle element $\Delta\Omega$ centered around $\hat{\mathbf{q}}$. The fact that the diffuse specific intensity vector can be measured by an optical device and computed theoretically by solving the RTE explains the practical usefulness of this quantity.

The microphysical derivation of the RTE was based on the following fundamental approximations:

- We assumed that each particle is located in the far-field zones of all other particles and that the observation point is also located in the far-field zones of all the particles forming the scattering medium.
- We neglected all scattering paths going through a particle two and more times (the Twersky approximation).
- We assumed that the position and state of each particle are statistically independent of each other and of those of all other particles and that the spatial distribution of the particles throughout the medium is random and statistically uniform.
- We assumed that the scattering medium is convex, which assured that a wave exiting the medium cannot re-enter it.
- We assumed that the number of particles N forming the scattering medium is large.
- We ignored all diagrams with crossing connectors in the diagrammatic expansion of the dyadic correlation function (the ladder approximation).

As a consequence, the RTE does not describe interference effects such as coherent backscattering. The latter is caused by constructive interference of pairs of conjugate waves propagating along the same scattering paths but in opposite directions and is represented by diagrams with crossing connectors excluded from the derivation [21, 23]. Particles that are randomly positioned and are separated widely enough that each of them is located in the far-field zones of all other particles are traditionally called independent scatterers [26]. Thus the requirement of independent scattering is a necessary condition of validity of the radiative transfer theory.

A fundamental property of the RTE is that it satisfies the energy conservation law. Indeed, we can rewrite Eqs. (86) and (106) as a single RTE:

$$\hat{\mathbf{q}}\cdot\nabla\tilde{\mathbf{I}}(\mathbf{r},\hat{\mathbf{q}}) = \nabla\cdot[\hat{\mathbf{q}}\tilde{\mathbf{I}}(\mathbf{r},\hat{\mathbf{q}})] = -n_0\langle\mathbf{K}(\hat{\mathbf{q}})\rangle\tilde{\mathbf{I}}(\mathbf{r},\hat{\mathbf{q}}) + n_0\int_{4\pi}\mathrm{d}\hat{\mathbf{q}}'\langle\mathbf{Z}(\hat{\mathbf{q}},\hat{\mathbf{q}}')\rangle\tilde{\mathbf{I}}(\mathbf{r},\hat{\mathbf{q}}'), \quad (108)$$

where $\widetilde{\mathbf{I}}(\mathbf{r},\hat{\mathbf{q}})=\delta(\hat{\mathbf{q}}-\hat{\mathbf{s}})\mathbf{I}_c(\mathbf{r})+\widetilde{\mathbf{I}}_d(\mathbf{r},\hat{\mathbf{q}})$ is the full specific intensity vector. The flux density vector is defined as $\mathbf{F}(\mathbf{r})=\int_{4\pi}\mathrm{d}\hat{\mathbf{q}}\,\hat{\mathbf{q}}\widetilde{I}(\mathbf{r},\hat{\mathbf{q}})$. The product $\hat{\mathbf{p}}\cdot\mathbf{F}(\mathbf{r})\mathrm{d}S$ gives the amount and the direction of the net flow of power through a surface element dS normal to $\hat{\mathbf{p}}$. Integrating both sides of Eq. (108) over all directions $\hat{\mathbf{q}}$ and recalling the definitions of the extinction, scattering, and absorption cross sections (Subsection 2.5), we derive

$$-\nabla\cdot\mathbf{F}(\mathbf{r})=n_0\int_{4\pi}\mathrm{d}\hat{\mathbf{q}}\langle C_{\mathrm{abs}}(\hat{\mathbf{q}})\rangle\widetilde{I}(\mathbf{r},\mathbf{q}). \tag{109}$$

This means that the net inflow of electromagnetic power per unit volume is equal to the total power absorbed per unit volume. If the particles forming the scattering medium are nonabsorbing so that $\langle C_{\mathrm{abs}}(\hat{\mathbf{q}})\rangle=0$, then the flux density vector is divergence-free: $\nabla\cdot\mathbf{F}(\mathbf{r})=0$.

For macroscopically isotropic and mirror-symmetric media, Eq. (108) can be significantly simplified (see Subsection 2.7):

$$\frac{\mathrm{d}\widetilde{\mathbf{I}}(\mathbf{r};\theta,\varphi)}{\mathrm{d}\tau}=-\widetilde{\mathbf{I}}(\mathbf{r};\theta,\varphi)+\frac{\varpi}{4\pi}\int_{-1}^{+1}\mathrm{d}(\cos\theta')\int_0^{2\pi}\mathrm{d}\varphi'\,\widetilde{\mathbf{Z}}(\theta,\theta',\varphi-\varphi')\,\widetilde{\mathbf{I}}(\mathbf{r};\theta',\varphi'), \tag{110}$$

where $\mathrm{d}\tau=n_0\langle C_{\mathrm{ext}}\rangle\mathrm{d}q$ is the optical pathlength element. By writing the normalized phase matrix in the form $\widetilde{\mathbf{Z}}(\theta,\theta',\varphi-\varphi')$, we explicitly indicate that it depends on the difference of the azimuth angles of the scattering and incident directions rather than on their specific values. Equation (110) can be made even simpler by neglecting polarization and replacing the specific intensity vector by its first element (i.e., specific intensity), and the normalized phase matrix by its (1, 1) element (i.e., the phase function):

$$\frac{\mathrm{d}\widetilde{I}(\mathbf{r};\theta,\varphi)}{\mathrm{d}\tau(\mathbf{r})}=-\widetilde{I}(\mathbf{r};\theta,\varphi)+\frac{\varpi}{4\pi}\int_{-1}^{+1}\mathrm{d}(\cos\theta')\int_0^{2\pi}\mathrm{d}\varphi'\,a_1(\Theta)\,\widetilde{I}(\mathbf{r};\theta',\varphi'), \tag{111}$$

where Θ is the scattering angle (Fig. 3). Although ignoring the vector nature of light and replacing the exact vector radiative transfer equation by its approximate scalar counterpart has no rigorous physical justification, this simplification is widely used when the medium is illuminated by unpolarized light and only the intensity of multiply scattered light is required. The scalar approximation gives poor accuracy when the size of the scattering particles is much smaller than the wavelength [29], but provides acceptable results for particles comparable to and larger than the wavelength [30].

4. Adding equations

In order to apply the RTT to analyses of laboratory measurements or remote sensing observations, one needs efficient theoretical techniques for solving the RTE. Unfortunately, like many integro-differential equations, the RTE is difficult to study mathematically and numerically. In order to facilitate the analysis, we will need several simplifying assumptions. The most important of them are that the scattering medium (i)

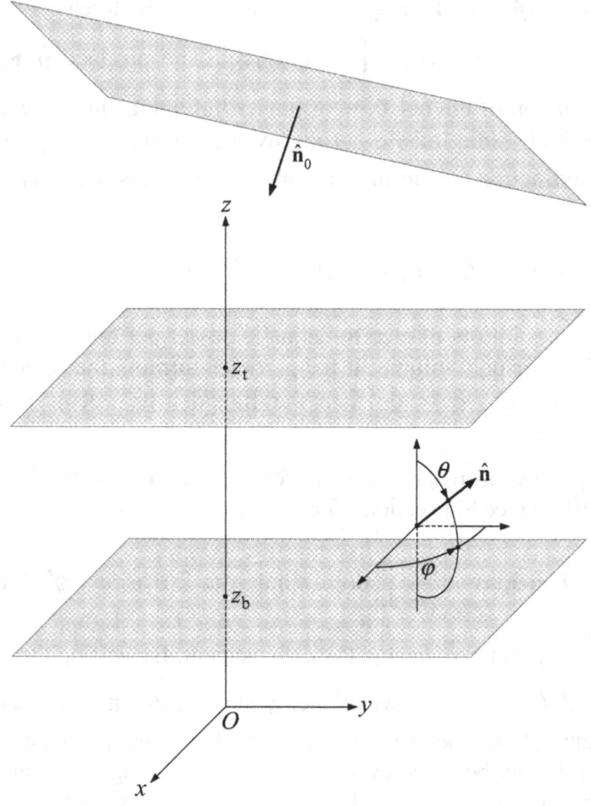

Fig. 17. Plane-parallel scattering medium illuminated by a parallel quasi-monochromatic beam of light.

is plane parallel, (ii) has an infinite horizontal extent, and (iii) is illuminated from above by a parallel quasi-monochromatic beam of light. These assumptions mean that all properties of the medium and of the radiation field may vary only in the vertical direction and are independent of the horizontal coordinates. Taken together, these assumptions specify the so-called *standard problem* of atmospheric optics and provide a model relevant to a great variety of applications in diverse fields of science and technology. In this section we will not make any further assumptions and will derive several important equations describing the internal diffuse radiation field as well as the diffuse radiation exiting the medium.

4.1. THE STANDARD PROBLEM

Let us consider a plane-parallel layer extending in the vertical direction from $z = z_b$ to $z = z_t$, where the z-axis of the laboratory coordinate system is perpendicular to the boundaries of the medium and is directed upwards, and "b" and "t" stand for "bottom" and "top," respectively (Fig. 17). A propagation direction $\hat{\mathbf{n}}$ at a point in space will be specified by a couplet $\{u, \varphi\}$, where $u = -\cos\theta \in [-1, +1]$ is the direction cosine, and θ and φ are the corresponding polar and azimuth angles with respect to the local coordinate system having the same spatial orientation as the laboratory coordinate system. It is also convenient to introduce a non-negative quantity $\mu = |u| \in [0, 1]$. In order to make many formulas of this section more compact, we will denote by $\hat{\mu}$ the pair of arguments (μ, φ) and by $-\hat{\mu}$ the pair of arguments $(-\mu, \varphi)$ (note that $\hat{\mu}$ and $-\hat{\mu}$ are not unit vectors). A $\hat{\mu}$ always corresponds to a downward direction and a $-\hat{\mu}$ always corresponds to an upward direction. We also denote

$$\int \mathrm{d}\hat{\mu} = \int_0^1 \mathrm{d}\mu \int_0^{2\pi} \mathrm{d}\varphi. \tag{112}$$

Let us assume that the scattering layer is illuminated from above by a parallel quasi-monochromatic beam of light propagating in the direction $\hat{\mathbf{n}}_0 = \{\mu_0, \varphi_0\}$. The uniformity and the infinite transverse extent of the beam ensure that all parameters of the internal radiation field and those of the radiation leaving the scattering layer are independent of the coordinates x and y. Therefore, Eq. (108) can be rewritten in the form

$$-u \frac{\mathrm{d}\tilde{\mathbf{I}}(z, \hat{\mathbf{n}})}{\mathrm{d}z} = -n_0(z)\mathbf{K}(z, \hat{\mathbf{n}})\tilde{\mathbf{I}}(z, \hat{\mathbf{n}}) + n_0(z) \int_{4\pi} \mathrm{d}\hat{\mathbf{n}}' \mathbf{Z}(z, \hat{\mathbf{n}}, \hat{\mathbf{n}}')\tilde{\mathbf{I}}(z, \hat{\mathbf{n}}') \tag{113}$$

and must be supplemented by the boundary conditions

$$\tilde{\mathbf{I}}(z_t, \hat{\mu}) = \delta(\mu - \mu_0)\delta(\varphi - \varphi_0)\mathbf{I}_0, \tag{114}$$

$$\tilde{\mathbf{I}}(z_b, -\hat{\mu}) = \mathbf{0}, \tag{115}$$

where $\tilde{\mathbf{I}}(z, \hat{\mathbf{n}}) = \delta(\hat{\mathbf{n}} - \hat{\mathbf{n}}_0)\mathbf{I}_c(z) + \tilde{\mathbf{I}}_d(z, \hat{\mathbf{n}})$ is the full specific intensity vector including both the coherent and the diffuse component, $\mathbf{K}$ and $\mathbf{Z}$ are the ensemble-averaged extinction and phase matrices, respectively (note that we have omitted the angular brackets for the sake of brevity), $\mathbf{I}_0$ is the Stokes vector of the incident beam, and $\mathbf{0}$ is a zero four-element column. The boundary conditions follow directly from the integral form of the RTE and mean that the downwelling radiation at the upper boundary of the layer consists only of the incident parallel beam and that there is no upwelling radiation at the lower boundary. Equations (113)–(115) collectively represent what we have called the *standard problem*.

Since $n_0(z)$ is a common factor in both terms on the right-hand side of Eq. (113), it is convenient to eliminate it by introducing a new vertical "coordinate" $\psi(z)$ according to $\mathrm{d}\psi = -n_0(z)\mathrm{d}z$ or

$$\psi(z) = \int_z^{\infty} n_0(z')\mathrm{d}z'. \tag{116}$$

The $\psi(z)$ has the dimension m^{-2} and is the number of particles in a vertical column having a unit cross section and extending from $z' = z$ to infinity. It is, therefore, natural to call it the "particle depth." Unlike the z-coordinate, which increases in the upward direction, the ψ-coordinate increases in the downward direction. We then have

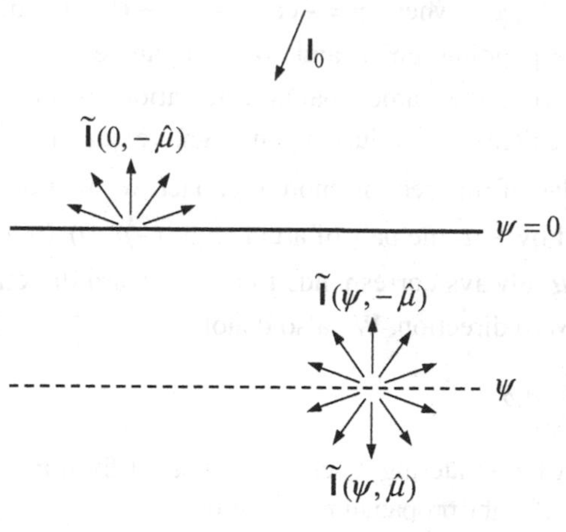

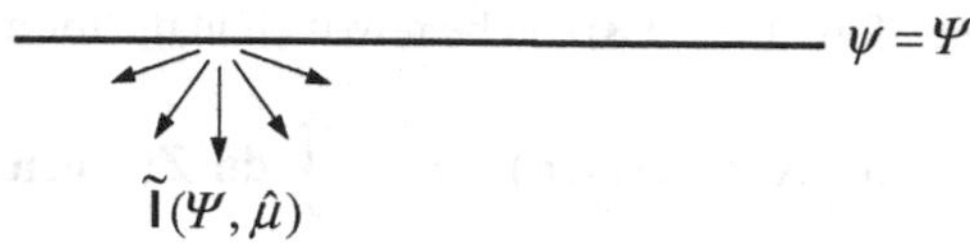

Fig. 18. The standard problem.

$$u\frac{\mathrm{d}\widetilde{\mathbf{I}}(\psi,\hat{\mathbf{n}})}{\mathrm{d}\psi}=-\mathbf{K}(\psi,\hat{\mathbf{n}})\widetilde{\mathbf{I}}(\psi,\hat{\mathbf{n}})+\int_{4\pi}\mathrm{d}\hat{\mathbf{n}}'\mathbf{Z}(\psi,\hat{\mathbf{n}},\hat{\mathbf{n}}')\widetilde{\mathbf{I}}(\psi,\hat{\mathbf{n}}'), \tag{117}$$

$$\widetilde{\mathbf{I}}(0,\hat{\mu})=\delta(\mu-\mu_0)\delta(\varphi-\varphi_0)\mathbf{I}_0, \tag{118}$$

$$\widetilde{\mathbf{I}}(\Psi,-\hat{\mu})=\mathbf{0}, \tag{119}$$

where $\Psi=\psi(z_b)$ is the "particle thickness" of the layer (Fig. 18).

4.2. THE MATRIZANT

Consider first the solution of the differential transfer equation

$$\mu\frac{\mathrm{d}\widetilde{\mathbf{I}}(\psi,\hat{\mu})}{\mathrm{d}\psi}=-\mathbf{K}(\psi,\hat{\mu})\widetilde{\mathbf{I}}(\psi,\hat{\mu}),\qquad \psi\geq\psi_0 \tag{120}$$

supplemented by the initial condition

$$\widetilde{\mathbf{I}}(\psi_0,\hat{\mu})=\widetilde{\mathbf{I}}_0. \tag{121}$$

It is convenient to express $\widetilde{\mathbf{I}}(\psi,\hat{\mu})$ in terms of the solution of the following auxiliary initial-value problem:

$$\mu \frac{\mathrm{d}\mathbf{X}(\psi, \psi_0, \hat{\mu})}{\mathrm{d}\psi} = -\mathbf{K}(\psi, \hat{\mu})\mathbf{X}(\psi, \psi_0, \hat{\mu}), \qquad \psi \geq \psi_0, \tag{122}$$

$$\mathbf{X}(\psi_0, \psi_0, \hat{\mu}) = \mathbf{\Delta}, \tag{123}$$

where $\mathbf{X}(\psi, \psi_0, \hat{\mu})$ is a 4×4 real matrix called the matrizant and $\mathbf{\Delta} = \mathrm{diag}[1,1,1,1]$ is the 4×4 unit matrix. Specifically, if the matrizant is known then the solution of Eqs. (120)–(121) is simply

$$\tilde{\mathbf{I}}(\psi, \hat{\mu}) = \mathbf{X}(\psi, \psi_0, \hat{\mu})\tilde{\mathbf{I}}_0 . \tag{124}$$

The matrizant has the obvious property

$$\mathbf{X}(\psi, \psi_0, \hat{\mu}) = \mathbf{X}(\psi, \psi_1, \hat{\mu})\mathbf{X}(\psi_1, \psi_0, \hat{\mu}), \tag{125}$$

where $\psi_0 \leq \psi_1 \leq \psi$.

If the scattering layer is homogeneous then $\mathbf{K}(\psi, \hat{\mu}) \equiv \mathbf{K}(\hat{\mu})$, and the matrizant can be written in the form of a matrix exponent:

$$\mathbf{X}(\psi, \psi_0, \hat{\mu}) = \exp\left[-(\psi - \psi_0)\mathbf{K}(\hat{\mu})/\mu\right]. \tag{126}$$

If the layer is inhomogeneous, one should exploit the property (125) by subdividing the interval $[\psi_0, \psi]$ into a number N of equal subintervals $[\psi_0, \psi_1]$, …, $[\psi_{n-1}, \psi_n]$, …, $[\psi_{N-1}, \psi]$ and calculating the matrizant in the limit $N \to \infty$:

$$\begin{aligned} \mathbf{X}(\psi, \psi_0, \hat{\mu}) = \lim_{N\to\infty} \{ & [\mathbf{\Delta} - (\Delta\psi/\mu)\mathbf{K}(\psi_{N-1} + \Delta\psi/2, \hat{\mu})] \cdots \\ & \times [\mathbf{\Delta} - (\Delta\psi/\mu)\mathbf{K}(\psi_{n-1} + \Delta\psi/2, \hat{\mu})] \cdots \\ & \times [\mathbf{\Delta} - (\Delta\psi/\mu)\mathbf{K}(\psi_0 + \Delta\psi/2, \hat{\mu})] \}, \end{aligned} \tag{127}$$

where $\Delta\psi = (\psi - \psi_0)/N$.

Similarly, the solution of the equation

$$-\mu \frac{\mathrm{d}\tilde{\mathbf{I}}(\psi, -\hat{\mu})}{\mathrm{d}\psi} = -\mathbf{K}(\psi, -\hat{\mu})\tilde{\mathbf{I}}(\psi, -\hat{\mu}), \qquad \psi \leq \psi_0 \tag{128}$$

supplemented by the initial condition

$$\tilde{\mathbf{I}}(\psi_0, -\hat{\mu}) = \tilde{\mathbf{I}}_0 \tag{129}$$

can be expressed in terms of the solution of the auxiliary initial-value problem

$$-\mu \frac{\mathrm{d}\mathbf{X}(\psi, \psi_0, -\hat{\mu})}{\mathrm{d}\psi} = -\mathbf{K}(\psi, -\hat{\mu})\mathbf{X}(\psi, \psi_0, -\hat{\mu}), \qquad \psi \leq \psi_0, \tag{130}$$

$$\mathbf{X}(\psi_0, \psi_0, -\hat{\mu}) = \mathbf{\Delta} \tag{131}$$

as

$$\tilde{\mathbf{I}}(\psi, -\hat{\mu}) = \mathbf{X}(\psi, \psi_0, -\hat{\mu})\tilde{\mathbf{I}}_0 . \tag{132}$$

The matrizant $\mathbf{X}(\psi, \psi_0, -\hat{\mu})$ has the property

$$\mathbf{X}(\psi, \psi_0, -\hat{\mu}) = \mathbf{X}(\psi, \psi_1, -\hat{\mu})\mathbf{X}(\psi_1, \psi_0, -\hat{\mu}), \qquad \psi \le \psi_1 \le \psi_0 \tag{133}$$

and is given by

$$\mathbf{X}(\psi, \psi_0, -\hat{\mu}) = \exp\left[-(\psi_0 - \psi)\mathbf{K}(-\hat{\mu})/\mu\right] \tag{134}$$

if the layer is homogeneous and by

$$\begin{aligned}\mathbf{X}(\psi, \psi_0, -\hat{\mu}) = \lim_{N\to\infty}\{&\left[\mathbf{\Delta} - (\Delta\psi/\mu)\mathbf{K}(\psi_{N-1} - \Delta\psi/2, -\hat{\mu})\right]\cdots \\ &\times\left[\mathbf{\Delta} - (\Delta\psi/\mu)\mathbf{K}(\psi_{n-1} - \Delta\psi/2, -\hat{\mu})\right]\cdots \\ &\times\left[\mathbf{\Delta} - (\Delta\psi/\mu)\mathbf{K}(\psi_0 - \Delta\psi/2, -\hat{\mu})\right]\}\end{aligned} \tag{135}$$

if the layer is inhomogeneous, where $\Delta\psi = (\psi_0 - \psi)/N$ and $\psi_n = \psi_0 - n\Delta\psi$.

4.3. THE GENERAL PROBLEM

The standard problem (117)–(119) implies that the scattering layer is illuminated only from above and only by a parallel beam of light. It is useful, however, to consider mathematically the following more general boundary values, which include the boundary conditions (118) and (119) as a particular case:

$$\tilde{\mathbf{I}}(0, \hat{\mu}) = \tilde{\mathbf{I}}_{\downarrow}(\hat{\mu}), \tag{136}$$

$$\tilde{\mathbf{I}}(\Psi, -\hat{\mu}) = \tilde{\mathbf{I}}_{\uparrow}(-\hat{\mu}), \tag{137}$$

where $\tilde{\mathbf{I}}_{\downarrow}(\hat{\mu})$ and $\tilde{\mathbf{I}}_{\uparrow}(-\hat{\mu})$ are arbitrary. We will call Eqs. (117), (136), and (137) the *general problem*.

The linearity of the RTE allows us to express the radiation field $\tilde{\mathbf{I}}(\psi, \hat{\mathbf{n}})$ for $\psi \in [0, \Psi]$ in terms of the specific intensity vectors $\tilde{\mathbf{I}}_{\downarrow}(\hat{\mu})$ and $\tilde{\mathbf{I}}_{\uparrow}(-\hat{\mu})$ as follows:

$$\begin{aligned}\tilde{\mathbf{I}}(\psi, \hat{\mu}) = \mathbf{X}(\psi, 0, \hat{\mu})\tilde{\mathbf{I}}_{\downarrow}(\hat{\mu}) &+ \tfrac{1}{\pi}\int d\hat{\mu}'\mu'\,\mathbf{D}(\psi, \hat{\mu}, \hat{\mu}')\tilde{\mathbf{I}}_{\downarrow}(\hat{\mu}') \\ &+ \tfrac{1}{\pi}\int d\hat{\mu}'\mu'\,\mathbf{U}^{\dagger}(\psi, \hat{\mu}, \hat{\mu}')\tilde{\mathbf{I}}_{\uparrow}(-\hat{\mu}'),\end{aligned} \tag{138}$$

$$\begin{aligned}\tilde{\mathbf{I}}(\psi, -\hat{\mu}) = \mathbf{X}(\psi, \Psi, -\hat{\mu})\tilde{\mathbf{I}}_{\uparrow}(-\hat{\mu}) &+ \tfrac{1}{\pi}\int d\hat{\mu}'\mu'\,\mathbf{U}(\psi, \hat{\mu}, \hat{\mu}')\tilde{\mathbf{I}}_{\downarrow}(\hat{\mu}') \\ &+ \tfrac{1}{\pi}\int d\hat{\mu}'\mu'\,\mathbf{D}^{\dagger}(\psi, \hat{\mu}, \hat{\mu}')\tilde{\mathbf{I}}_{\uparrow}(-\hat{\mu}'),\end{aligned} \tag{139}$$

where the 4×4 matrices $\mathbf{D}$ and $\mathbf{U}$ describe the response of the scattering layer to the radiation incident on the upper boundary from above, while the 4×4 matrices $\mathbf{D}^{\dagger}$ and $\mathbf{U}^{\dagger}$

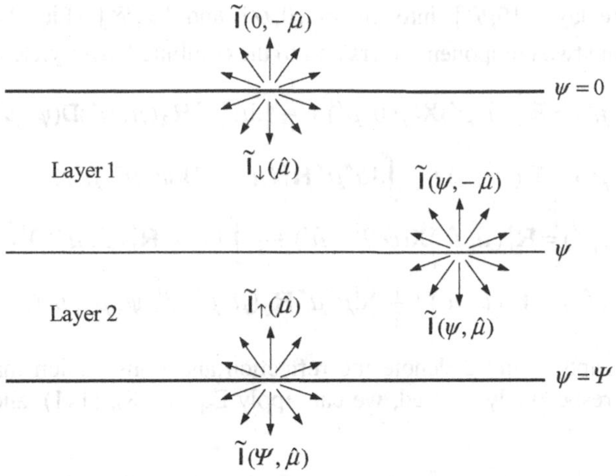

Fig. 19. Illustration of the adding principle.

describe the response to the radiation illuminating the bottom boundary of the layer from below. The first terms on the right-hand side of Eqs. (138) and (139) describe the coherent propagation of the incident light, whereas the remaining terms describe the result of multiple scattering. The corresponding reflection and transmission matrices determine the Stokes parameters of the radiation exiting the layer and are defined as

$$\mathbf{R}(\hat{\mu},\hat{\mu}')=\mathbf{U}(0,\hat{\mu},\hat{\mu}'), \tag{140}$$

$$\mathbf{T}(\hat{\mu},\hat{\mu}')=\mathbf{D}(\Psi,\hat{\mu},\hat{\mu}'), \tag{141}$$

$$\mathbf{R}^{\dagger}(\hat{\mu},\hat{\mu}')=\mathbf{U}^{\dagger}(\Psi,\hat{\mu},\hat{\mu}'), \tag{142}$$

$$\mathbf{T}^{\dagger}(\hat{\mu},\hat{\mu}')=\mathbf{D}^{\dagger}(0,\hat{\mu},\hat{\mu}'). \tag{143}$$

The matrices $\mathbf{R}$ and $\mathbf{T}$ describe the response of the layer to the external radiation falling from above, whereas the matrices $\mathbf{R}^{\dagger}$ and $\mathbf{T}^{\dagger}$ describe the response to the external radiation falling from below.

The reader can easily verify that the solution of the standard problem can now be expressed as

$$\tilde{\mathbf{I}}(\psi,\hat{\mu})=\delta(\mu-\mu_0)\delta(\varphi-\varphi_0)\mathbf{X}(\psi,0,\hat{\mu}_0)\mathbf{I}_0+\tfrac{1}{\pi}\mu_0\mathbf{D}(\psi,\hat{\mu},\hat{\mu}_0)\mathbf{I}_0, \tag{144}$$

$$\tilde{\mathbf{I}}(\psi,-\hat{\mu})=\tfrac{1}{\pi}\mu_0\mathbf{U}(\psi,\hat{\mu},\hat{\mu}_0)\mathbf{I}_0, \tag{145}$$

$$\tilde{\mathbf{I}}(\Psi,\hat{\mu})=\delta(\mu-\mu_0)\delta(\varphi-\varphi_0)\mathbf{X}(\Psi,0,\hat{\mu}_0)\mathbf{I}_0+\tfrac{1}{\pi}\mu_0\mathbf{T}(\hat{\mu},\hat{\mu}_0)\mathbf{I}_0, \tag{146}$$

$$\tilde{\mathbf{I}}(0,-\hat{\mu})=\tfrac{1}{\pi}\mu_0\mathbf{R}(\hat{\mu},\hat{\mu}_0)\mathbf{I}_0. \tag{147}$$

4.4. ADDING EQUATIONS

In this subsection we will describe an elegant mathematical scheme for computing the matrices $\mathbf{D}$, $\mathbf{U}$, $\mathbf{D}^{\dagger}$, $\mathbf{U}^{\dagger}$, $\mathbf{R}$, $\mathbf{T}$, $\mathbf{R}^{\dagger}$, and $\mathbf{T}^{\dagger}$ based on so-called adding equations. Let us divide the entire layer $[0,\Psi]$ into layers $[0,\psi]$ and $[\psi,\Psi]$ (Fig. 19). Applying Eqs. (138)–(143) to the two component layers and to the combined layer yields

$$\mathbf{U}(\psi,\hat{\mu},\hat{\mu}') = \mathbf{R}_2(\hat{\mu},\hat{\mu}')\mathbf{X}(\psi,0,\hat{\mu}') + \tfrac{1}{\pi}\int \mathrm{d}\hat{\mu}''\mu''\,\mathbf{R}_2(\hat{\mu},\hat{\mu}'')\mathbf{D}(\psi,\hat{\mu}'',\hat{\mu}'), \quad (148)$$

$$\mathbf{D}(\psi,\hat{\mu},\hat{\mu}') = \mathbf{T}_1(\hat{\mu},\hat{\mu}') + \tfrac{1}{\pi}\int \mathrm{d}\hat{\mu}''\mu''\,\mathbf{R}_1^{\dagger}(\hat{\mu},\hat{\mu}'')\mathbf{U}(\psi,\hat{\mu}'',\hat{\mu}'), \quad (149)$$

$$\mathbf{U}^{\dagger}(\psi,\hat{\mu},\hat{\mu}') = \mathbf{R}_1^{\dagger}(\hat{\mu},\hat{\mu}')\mathbf{X}(\psi,\Psi,-\hat{\mu}') + \tfrac{1}{\pi}\int \mathrm{d}\hat{\mu}''\mu''\,\mathbf{R}_1^{\dagger}(\hat{\mu},\hat{\mu}'')\mathbf{D}^{\dagger}(\psi,\hat{\mu}'',\hat{\mu}'), \quad (150)$$

$$\mathbf{D}^{\dagger}(\psi,\hat{\mu},\hat{\mu}') = \mathbf{T}_2^{\dagger}(\hat{\mu},\hat{\mu}') + \tfrac{1}{\pi}\int \mathrm{d}\hat{\mu}''\mu''\,\mathbf{R}_2(\hat{\mu},\hat{\mu}'')\mathbf{U}^{\dagger}(\psi,\hat{\mu}'',\hat{\mu}'), \quad (151)$$

where the subscripts 1 and 2 denote the reflection and transmission matrices of isolated layers 1 and 2, respectively. Indeed, we can apply Eqs. (138), (141), and (142) to layer 1 and write

$$\begin{aligned}\tilde{\mathbf{I}}(\psi,\hat{\mu}) &= \mathbf{X}(\psi,0,\hat{\mu})\tilde{\mathbf{I}}_{\downarrow}(\hat{\mu}) + \tfrac{1}{\pi}\int \mathrm{d}\hat{\mu}'\mu'\,\mathbf{T}_1(\hat{\mu},\hat{\mu}')\tilde{\mathbf{I}}_{\downarrow}(\hat{\mu}') \\ &\quad + \tfrac{1}{\pi}\int \mathrm{d}\hat{\mu}'\mu'\,\mathbf{R}_1^{\dagger}(\hat{\mu},\hat{\mu}')\tilde{\mathbf{I}}(\psi,-\hat{\mu}') \\ &= \mathbf{X}(\psi,0,\hat{\mu})\tilde{\mathbf{I}}_{\downarrow}(\hat{\mu}) + \tfrac{1}{\pi}\int \mathrm{d}\hat{\mu}'\mu'\,\mathbf{T}_1(\hat{\mu},\hat{\mu}')\tilde{\mathbf{I}}_{\downarrow}(\hat{\mu}') \\ &\quad + \tfrac{1}{\pi}\int \mathrm{d}\hat{\mu}'\mu'\,\mathbf{R}_1^{\dagger}(\hat{\mu},\hat{\mu}')\Big[\mathbf{X}(\psi,\Psi,-\hat{\mu}')\tilde{\mathbf{I}}_{\uparrow}(-\hat{\mu}') + \tfrac{1}{\pi}\int \mathrm{d}\hat{\mu}''\mu''\,\mathbf{U}(\psi,\hat{\mu}',\hat{\mu}'')\tilde{\mathbf{I}}_{\downarrow}(\hat{\mu}'') \\ &\qquad + \tfrac{1}{\pi}\int \mathrm{d}\hat{\mu}''\mu''\,\mathbf{D}^{\dagger}(\psi,\hat{\mu}',\hat{\mu}'')\tilde{\mathbf{I}}_{\uparrow}(-\hat{\mu}'')\Big], \end{aligned} \quad (152)$$

which, after comparison with Eq. (138), gives Eqs. (149) and (150). Similarly, Eqs. (148) and (151) follow from

$$\begin{aligned}\tilde{\mathbf{I}}(\psi,-\hat{\mu}) &= \mathbf{X}(\psi,\Psi,-\hat{\mu})\tilde{\mathbf{I}}_{\uparrow}(-\hat{\mu}) + \tfrac{1}{\pi}\int \mathrm{d}\hat{\mu}'\mu'\,\mathbf{T}_2^{\dagger}(\hat{\mu},\hat{\mu}')\tilde{\mathbf{I}}_{\uparrow}(-\hat{\mu}') \\ &\quad + \tfrac{1}{\pi}\int \mathrm{d}\hat{\mu}'\mu'\,\mathbf{R}_2(\hat{\mu},\hat{\mu}')\tilde{\mathbf{I}}(\psi,\hat{\mu}') \\ &= \mathbf{X}(\psi,\Psi,-\hat{\mu})\tilde{\mathbf{I}}_{\uparrow}(-\hat{\mu}) + \tfrac{1}{\pi}\int \mathrm{d}\hat{\mu}'\mu'\,\mathbf{T}_2^{\dagger}(\hat{\mu},\hat{\mu}')\tilde{\mathbf{I}}_{\uparrow}(-\hat{\mu}') \\ &\quad + \tfrac{1}{\pi}\int \mathrm{d}\hat{\mu}'\mu'\,\mathbf{R}_2(\hat{\mu},\hat{\mu}')\Big[\mathbf{X}(\psi,0,\hat{\mu}')\tilde{\mathbf{I}}_{\downarrow}(\hat{\mu}') + \tfrac{1}{\pi}\int \mathrm{d}\hat{\mu}''\mu''\,\mathbf{D}(\psi,\hat{\mu}',\hat{\mu}'')\tilde{\mathbf{I}}_{\downarrow}(\hat{\mu}'') \\ &\qquad + \tfrac{1}{\pi}\int \mathrm{d}\hat{\mu}''\mu''\,\mathbf{U}^{\dagger}(\psi,\hat{\mu}',\hat{\mu}'')\tilde{\mathbf{I}}_{\uparrow}(-\hat{\mu}'')\Big] \end{aligned} \quad (153)$$

and Eq. (139). By analogy, one can derive

$$\begin{aligned}\mathbf{R}(\hat{\mu},\hat{\mu}') &= \mathbf{R}_1(\hat{\mu},\hat{\mu}') + \mathbf{X}(0,\psi,-\hat{\mu})\mathbf{U}(\psi,\hat{\mu},\hat{\mu}') \\ &\quad + \tfrac{1}{\pi}\int \mathrm{d}\hat{\mu}''\mu''\,\mathbf{T}_1^{\dagger}(\hat{\mu},\hat{\mu}'')\mathbf{U}(\psi,\hat{\mu}'',\hat{\mu}'), \end{aligned} \quad (154)$$

$$\begin{aligned}\mathbf{T}(\hat{\mu},\hat{\mu}') &= \mathbf{T}_2(\hat{\mu},\hat{\mu}')\mathbf{X}(\psi,0,\hat{\mu}') + \mathbf{X}(\Psi,\psi,\hat{\mu})\mathbf{D}(\psi,\hat{\mu},\hat{\mu}') \\ &\quad + \tfrac{1}{\pi}\int \mathrm{d}\hat{\mu}''\mu''\,\mathbf{T}_2(\hat{\mu},\hat{\mu}'')\mathbf{D}(\psi,\hat{\mu}'',\hat{\mu}'), \end{aligned} \quad (155)$$

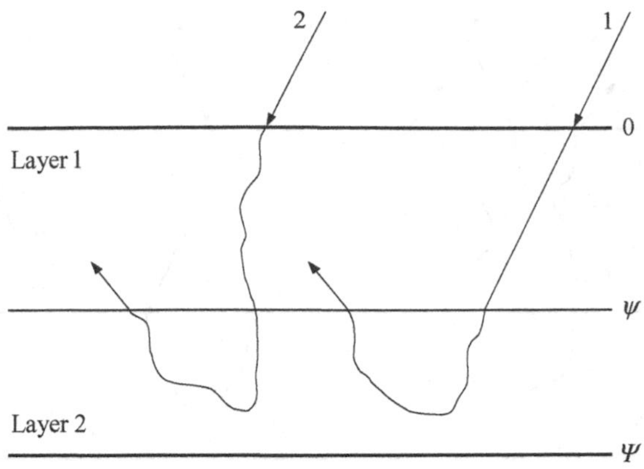

Fig. 20. Physical interpretation of Eq. (148).

$$\mathbf{R}^{\dagger}(\hat{\mu},\hat{\mu}')=\mathbf{R}_2^{\dagger}(\hat{\mu},\hat{\mu}')+\mathbf{X}(\Psi,\psi,\hat{\mu})\mathbf{U}^{\dagger}(\psi,\hat{\mu},\hat{\mu}') + \frac{1}{\pi}\int \mathrm{d}\hat{\mu}''\mu''\,\mathbf{T}_2(\hat{\mu},\hat{\mu}'')\mathbf{U}^{\dagger}(\psi,\hat{\mu}'',\hat{\mu}'), \tag{156}$$

$$\mathbf{T}^{\dagger}(\hat{\mu},\hat{\mu}')=\mathbf{T}_1^{\dagger}(\hat{\mu},\hat{\mu}')\mathbf{X}(\psi,\Psi,-\hat{\mu}')+\mathbf{X}(0,\psi,-\hat{\mu})\mathbf{D}^{\dagger}(\psi,\hat{\mu},\hat{\mu}') + \frac{1}{\pi}\int \mathrm{d}\hat{\mu}''\mu''\,\mathbf{T}_1^{\dagger}(\hat{\mu},\hat{\mu}'')\mathbf{D}^{\dagger}(\psi,\hat{\mu}'',\hat{\mu}'). \tag{157}$$

The interpretation of Eqs. (148)–(151) and (154)–(157) is clear. For example, Eq. (148) indicates that the upwelling radiation at the interface between layers 1 and 2 in response to the beam incident on the combined layer from above is simply the result of the reflection of the corresponding downwelling radiation by layer 2. This downwelling radiation consists of the attenuated direct component represented by the matrizant $\mathbf{X}(\psi,0,\hat{\mu}')$ (photon trajectory 1 in Fig. 20) and the diffuse component represented by the matrix $\mathbf{D}(\psi,\hat{\mu}'',\hat{\mu}')$ (photon trajectory 2 in Fig. 20). Similarly, Eq. (154) shows that the reflected radiation in response to the beam illuminating the combined layer from above consists of three components: (i) the photons that never reached the interface between layers 1 and 2 (the first term on the right-hand side of Eq. (154) and photon trajectory 1 in Fig. 21); (ii) the photons reflected by layer 2 and transmitted by layer 1 without scattering (the second term on the right-hand side of Eq. (154) and photon trajectory 2 in Fig. 21); and (iii) the photons reflected by layer 2 and diffusely transmitted by layer 1 (the third term on the right-hand side of Eq. (154) and photon trajectory 3 in Fig. 21). The reader may find it a useful exercise to give similar graphical interpretations of Eqs. (149)–(151) and (155)–(157).

Equations (148)–(151) and (154)–(157) are called adding equations because they allow one to compute the scattering properties of the combined layer provided that the scattering properties of each component layer are known. Indeed, if the matrices $\mathbf{R}_1$, $\mathbf{T}_1$, $\mathbf{R}_1^{\dagger}$, and $\mathbf{T}_1^{\dagger}$ for layer 1 in isolation from layer 2 and the matrices $\mathbf{R}_2$, $\mathbf{T}_2$, $\mathbf{R}_2^{\dagger}$, and $\mathbf{T}_2^{\dagger}$ for layer 2 in isolation from layer 1 are known then one can solve Eqs. (148)–(151) and find the

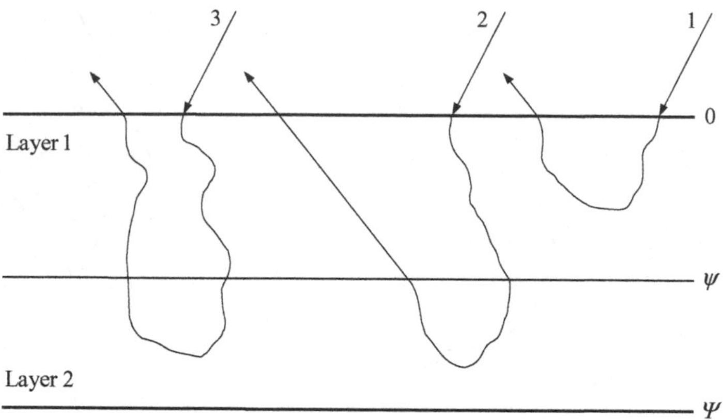

Fig. 21. Physical interpretation of Eq. (154).

matrices $\mathbf{D}$, $\mathbf{U}$, $\mathbf{D}^\dagger$, and $\mathbf{U}^\dagger$ describing the radiation field at the interface between the layers in the combined slab. This procedure involves replacing the angular integrals by appropriate quadrature sums. For example, Eq. (148) becomes

$$\mathbf{U}(\psi; \mu_i, \varphi_j; \mu_k, \varphi_l) = \mathbf{R}_2(\mu_i, \varphi_j; \mu_k, \varphi_l)\mathbf{X}(\psi, 0; \mu_k, \varphi_l) + \frac{1}{\pi}\sum_{m=1}^{N_\mu}\sum_{n=1}^{N_\varphi} w_m u_n \mu_m \mathbf{R}_2(\mu_i, \varphi_j; \mu_m, \varphi_n)\,\mathbf{D}(\psi; \mu_m, \varphi_n; \mu_k, \varphi_l),$$

where μ_i and w_i $(i = 1, ..., N_\mu)$ are quadrature division points and weights on the interval $[0, 1]$ and φ_i and u_i $(i = 1, ..., N_\varphi)$ are quadrature division points and weights on the interval $[0, 2\pi]$. The resulting system of linear algebraic equations for the unknown values of the matrices $\mathbf{D}, \mathbf{U}$, $\mathbf{D}^\dagger$, and $\mathbf{U}^\dagger$ at the quadrature division points can be solved using one of many available numerical techniques. After the matrices $\mathbf{D}$, $\mathbf{U}$, $\mathbf{D}^\dagger$, and $\mathbf{U}^\dagger$ at the quadrature division points are found, the reflection and transmission matrices of the combined layer can be calculated using the discretized version of Eqs. (154)–(157). Adding two identical layers is traditionally called the doubling procedure.

Furthermore, let us assume that the matrices $\mathbf{U}_1$, $\mathbf{D}_1$, $\mathbf{U}_1^\dagger$, and $\mathbf{D}_1^\dagger$ for a vertical level inside layer 1 are known, where the subscript 1 indicates that these matrices pertain to layer 1 taken in isolation from layer 2. Then the matrices $\mathbf{U}$, $\mathbf{D}$, $\mathbf{U}^\dagger$, and $\mathbf{D}^\dagger$ for the same level in the combined layer can also be easily calculated. Indeed, applying Eqs. (138) and (139) to each component layer and to the combined layer, we derive

$$\mathbf{U}(\psi', \hat{\mu}, \hat{\mu}') = \mathbf{U}_1(\psi', \hat{\mu}, \hat{\mu}') + \mathbf{X}(\psi', \psi, -\hat{\mu})\mathbf{U}(\psi, \hat{\mu}, \hat{\mu}') + \frac{1}{\pi}\int \mathrm{d}\hat{\mu}''\mu''\,\mathbf{D}_1^\dagger(\psi', \hat{\mu}, \hat{\mu}'')\mathbf{U}(\psi, \hat{\mu}'', \hat{\mu}'), \tag{158}$$

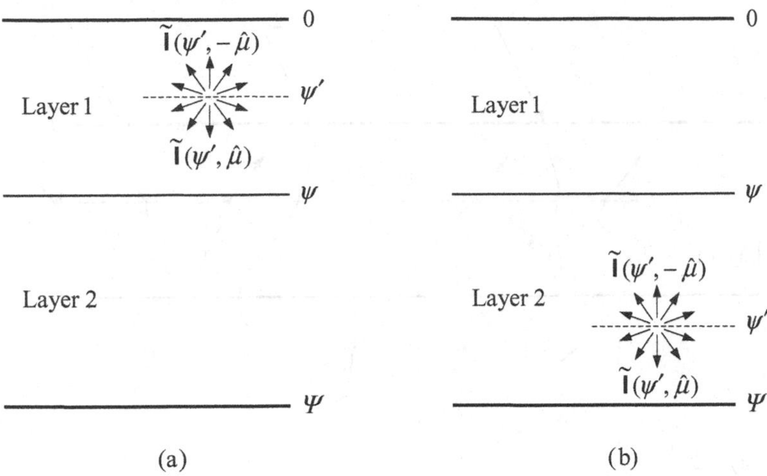

Fig. 22. Internal radiation field.

$$\mathbf{D}(\psi', \hat{\mu}, \hat{\mu}') = \mathbf{D}_1(\psi', \hat{\mu}, \hat{\mu}') + \tfrac{1}{\pi} \int d\hat{\mu}'' \mu'' \mathbf{U}_1^\dagger(\psi', \hat{\mu}, \hat{\mu}'') \mathbf{U}(\psi, \hat{\mu}'', \hat{\mu}'), \tag{159}$$

$$\mathbf{U}^\dagger(\psi', \hat{\mu}, \hat{\mu}') = \mathbf{U}_1^\dagger(\psi', \hat{\mu}, \hat{\mu}') \mathbf{X}(\psi, \Psi, -\hat{\mu}') + \tfrac{1}{\pi} \int d\hat{\mu}'' \mu'' \mathbf{U}_1^\dagger(\psi', \hat{\mu}, \hat{\mu}'') \mathbf{D}^\dagger(\psi, \hat{\mu}'', \hat{\mu}'), \tag{160}$$

$$\mathbf{D}^\dagger(\psi', \hat{\mu}, \hat{\mu}') = \mathbf{D}_1^\dagger(\psi', \hat{\mu}, \hat{\mu}') \mathbf{X}(\psi, \Psi, -\hat{\mu}') + \mathbf{X}(\psi', \psi, -\hat{\mu}) \mathbf{D}^\dagger(\psi, \hat{\mu}, \hat{\mu}') + \tfrac{1}{\pi} \int d\hat{\mu}'' \mu'' \mathbf{D}_1^\dagger(\psi', \hat{\mu}, \hat{\mu}'') \mathbf{D}^\dagger(\psi, \hat{\mu}'', \hat{\mu}') \tag{161}$$

for $\psi' \in [0, \psi]$ (Fig. 22(a)). Similarly, if we know the matrices $\mathbf{U}_2$, $\mathbf{D}_2$, $\mathbf{U}_2^\dagger$, and $\mathbf{D}_2^\dagger$ for a vertical level inside layer 2 taken in isolation from layer 1 then

$$\mathbf{U}(\psi', \hat{\mu}, \hat{\mu}') = \mathbf{U}_2(\psi' - \psi, \hat{\mu}, \hat{\mu}') \mathbf{X}(\psi, 0, \hat{\mu}') + \tfrac{1}{\pi} \int d\hat{\mu}'' \mu'' \mathbf{U}_2(\psi' - \psi, \hat{\mu}, \hat{\mu}'') \mathbf{D}(\psi, \hat{\mu}'', \hat{\mu}'), \tag{162}$$

$$\mathbf{D}(\psi', \hat{\mu}, \hat{\mu}') = \mathbf{D}_2(\psi' - \psi, \hat{\mu}, \hat{\mu}') \mathbf{X}(\psi, 0, \hat{\mu}') + \mathbf{X}(\psi', \psi, \hat{\mu}) \mathbf{D}(\psi, \hat{\mu}, \hat{\mu}') + \tfrac{1}{\pi} \int d\hat{\mu}'' \mu'' \mathbf{D}_2(\psi' - \psi, \hat{\mu}, \hat{\mu}'') \mathbf{D}(\psi, \hat{\mu}'', \hat{\mu}'), \tag{163}$$

$$\mathbf{U}^\dagger(\psi', \hat{\mu}, \hat{\mu}') = \mathbf{U}_2^\dagger(\psi' - \psi, \hat{\mu}, \hat{\mu}') + \mathbf{X}(\psi', \psi, \hat{\mu}) \mathbf{U}^\dagger(\psi, \hat{\mu}, \hat{\mu}') + \tfrac{1}{\pi} \int d\hat{\mu}'' \mu'' \mathbf{D}_2(\psi' - \psi, \hat{\mu}, \hat{\mu}'') \mathbf{U}^\dagger(\psi, \hat{\mu}'', \hat{\mu}'), \tag{164}$$

$$\mathbf{D}^\dagger(\psi', \hat{\mu}, \hat{\mu}') = \mathbf{D}_2^\dagger(\psi' - \psi, \hat{\mu}, \hat{\mu}') + \tfrac{1}{\pi} \int d\hat{\mu}'' \mu'' \mathbf{U}_2(\psi' - \psi, \hat{\mu}, \hat{\mu}'') \mathbf{U}^\dagger(\psi, \hat{\mu}'', \hat{\mu}') \tag{165}$$

for $\psi' \in [\psi, \Psi]$ (Fig. 22(b)). The physical meaning of these formulas is rather transparent.

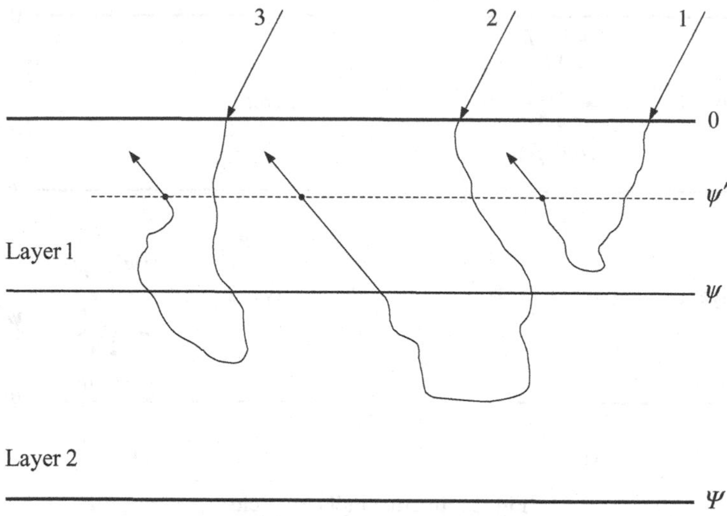

Fig. 23. Physical interpretation of Eq. (158).

For example, the first term on the right-hand side of Eq. (158) represents the contribution of photons that never reached the interface between layers 1 and 2, as shown schematically by photon trajectory 1 in Fig. 23. The second term describes the contribution of the photons that crossed the interface, exited layer 2 in the direction $\hat{\mu}$, and reached the level ψ' without scattering, as illustrated by photon trajectory 2 in Fig. 23. The last term gives the contribution of the photons that crossed the interface and were scattered at least once inside layer 1 before they reached the level ψ' (trajectory 3 in Fig. 23).

A practical implementation of the adding method can involve the following basic steps.

(1) A vertically inhomogeneous layer of particle thickness Ψ is approximated by a stack of N partial homogeneous layers having particle thicknesses $\Psi_1, ..., \Psi_N$ such that $\Psi = \sum_{n=1}^{N} \Psi_n$ (Fig. 24). The number of partial layers and their partial thicknesses can depend on the degree of vertical inhomogeneity of the original layer as well as on the desired numerical accuracy of computations.

(2) The reflection and transmission matrices $\mathbf{R}_n, \mathbf{T}_n, \mathbf{R}_n^{\dagger}$, and $\mathbf{T}_n^{\dagger}$ of partial layer n in isolation from all other layers are computed by using the doubling method (Fig. 25). The doubling process can be started with a layer having a particle thickness $\Delta\Psi_n = \Psi_n/2^{k_n}$ small enough that the reflection and transmission matrices for this layer can be computed by considering only the first order of scattering. Specifically, choosing the number of doubling events k_n sufficiently large that all elements of the matrices $\Delta\Psi_n\mathbf{Z}_n$ and $\Delta\Psi_n\mathbf{K}_n$ are much smaller than unity, using Eqs. (117) and (136)–(143), and neglecting all terms proportional to $(\Delta\Psi_n)^m$ with $m > 1$, we derive

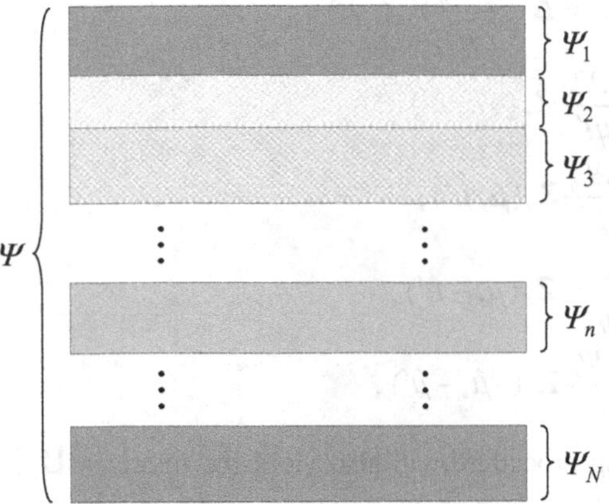

Fig. 24. Representation of a vertically inhomogeneous scattering layer by a stack of N homogeneous sublayers.

$$\mathbf{X}_n(\Delta\Psi_n, 0, \hat{\mu}') = \mathbf{\Delta} - \frac{\Delta\Psi_n}{\mu'}\mathbf{K}_n(\hat{\mu}'), \tag{166}$$

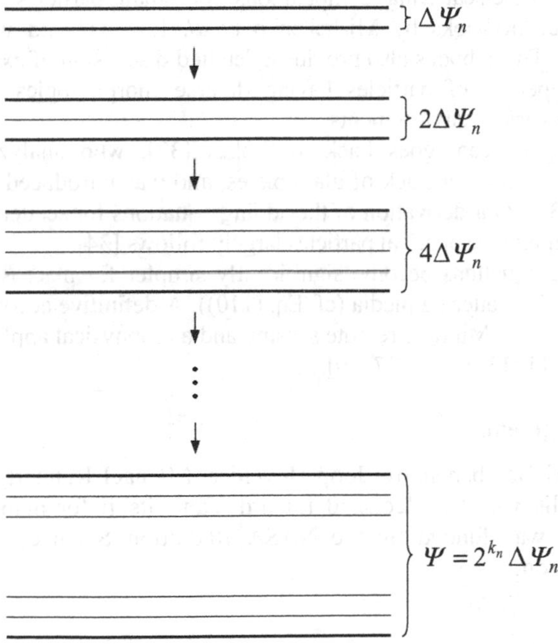

Fig. 25. The doubling procedure.

$$\mathbf{X}_n(0, \Delta\Psi_n, -\hat{\mu}) = \mathbf{\Delta} - \frac{\Delta\Psi_n}{\mu}\mathbf{K}_n(-\hat{\mu}). \tag{167}$$

$$\mathbf{R}_n(\hat{\mu}, \hat{\mu}') = \frac{\pi\Delta\Psi_n}{\mu\mu'}\mathbf{Z}_n(-\hat{\mu}, \hat{\mu}'), \tag{168}$$

$$\mathbf{T}_n(\hat{\mu}, \hat{\mu}') = \frac{\pi\Delta\Psi_n}{\mu\mu'}\mathbf{Z}_n(\hat{\mu}, \hat{\mu}'), \tag{169}$$

$$\mathbf{R}_n^{\dagger}(\hat{\mu}, \hat{\mu}') = \frac{\pi\Delta\Psi_n}{\mu\mu'}\mathbf{Z}_n(\hat{\mu}, -\hat{\mu}'), \tag{170}$$

$$\mathbf{T}_n^{\dagger}(\hat{\mu}, \hat{\mu}') = \frac{\pi\Delta\Psi_n}{\mu\mu'}\mathbf{Z}_n(-\hat{\mu}, -\hat{\mu}'), \tag{171}$$

Obviously, the doubling procedure will also yield the matrices $\mathbf{U}_n$, $\mathbf{D}_n$, $\mathbf{U}_n^{\dagger}$, and $\mathbf{D}_n^{\dagger}$ at $2^{k_n} - 1$ equidistant levels inside the *n*th partial layer (Fig. 25).

(3) The N partial homogeneous layers are recursively added starting from layer 1 and moving down or starting from layer N and moving up. This process gives the reflection and transmission matrices of the combined slab and the matrices $\mathbf{U}$, $\mathbf{D}$, $\mathbf{U}^{\dagger}$, and $\mathbf{D}^{\dagger}$ at the $N-1$ interfaces between the partial layers as well as at the $\sum_{n=1}^{N}(2^{k_n}-1)$ levels inside the partial layers rendered by the doubling procedure.

Numerical solution of the adding equations requires the knowledge of the ensemble-averaged extinction and phase matrices. The exact and approximate theoretical methods applicable to single-scattering computations for small particles have been extensively reviewed in recent books by Mishchenko *et al.* [26, 31] and will not be specifically discussed here. Those books also provide a detailed discussion of extinction, scattering, and absorption properties of particles having diverse morphologies and compositions and encountered in various environments.

The adding concept goes back to Stokes [32], who analyzed the reflection and transmission of light by a stack of glass plates, and was introduced to radiative transfer by van de Hulst [33]. Our derivation of the adding equations for scattering layers consisting of arbitrarily oriented nonspherical particles largely follows [34].

The adding equations become significantly simpler for macroscopically isotropic and mirror-symmetric scattering media (cf. Eq. (110)). A definitive account of this situation can be found in [35, 36]. Multiple remote sensing and astrophysical applications of the RTT can be found in [6–11, 13, 15, 31, 37–40].

5. Acknowledgments

We thank Yuri Barabanenkov, Joop Hovenier, Michael Kahnert, Andrew Lacis, Larry Travis, Cornelis van der Mee, and Edgard Yanovitskii for many fruitful discussions. This research was funded by the NASA Radiation Sciences Program managed by Donald Anderson.

References

1. O. D. Khvolson, Grundzüge einer mathematischen Theorie der inneren Diffusion des Lichtes, *Bull. St. Petersburg Acad. Sci.* **33**, 221–256 (1890).
2. A. Schuster, Radiation through a foggy atmosphere, *Astrophys. J.* **21**, 1–22 (1905).
3. J. E. Hansen and L. D. Travis, Light scattering in planetary atmospheres, *Space. Sci. Rev.* **16**, 527–610 (1974).
4. J. W. Hovenier and C. V. M. van der Mee, Fundamental relationships relevant to the transfer of polarized light in a scattering atmosphere, *Astron. Astrophys.* **128**, 1–16 (1983).
5. J. Lenoble, ed., *Radiative Transfer in Scattering and Absorbing Atmospheres* (A. Deepak, Hampton, Va., 1985).
6. A. K. Fung, *Microwave Scattering and Emission Models and Their Applications* (Artech House, Boston, 1994).
7. A. Z. Dolginov, Yu. N. Gnedin, and N. A. Silant'ev, *Propagation and Polarization of Radiation in Cosmic Media* (Gordon and Breach, Basel,1995).
8. E. G. Yanovitskij, *Light Scattering in Inhomogeneous Atmospheres* (Springer, Berlin, 1997).
9. G. E. Thomas and K. Stamnes, *Radiative Transfer in the Atmosphere and Ocean* (Cambridge Univ. Press, New York, 1999).
10. K. N. Liou, *An Introduction to Atmospheric Radiation* (Academic Press, San Diego, 2002).
11. J. W. Hovenier, C. V. M. van der Mee, and H. Domke, *Transfer of Polarized Light in Planetary Atmospheres* (Kluwer, Dordrecht, 2003).
12. V. Kourganoff, *Basic Methods in Transfer Problems* (Clarendon, Oxford, 1952).
13. S. Chandrasekhar, *Radiative Transfer* (Dover, New York, 1960).
14. V. V. Sobolev, *Light Scattering in Planetary Atmospheres* (Pergamon, Oxford, 1974).
15. H. C. van de Hulst, *Multiple Light Scattering* (Academic Press, New York, 1980).
16. L. Mandel and E. Wolf, *Optical Coherence and Quantum Optics* (Cambridge University Press, Cambridge, 1995).
17. A. G. Borovoy, Method of iterations in multiple scattering: the transfer equation, *Izv. Vuzov. Fizika*, No. 6, 50–54 (1966).
18. Yu. N. Barabanenkov and V. M. Finkel'berg, Radiation transport equation for correlated scatterers, *Soviet Phys. JETP* **26**, 587–591 (1968).
19. A. Z. Dolginov, Yu. N. Gnedin, and N. A. Silant'ev, Photon polarization and frequency change in multiple scattering, *J. Quant. Spectrosc. Radiat. Transfer* **10**, 707–754 (1970).
20. L. A. Apresyan and Yu. A. Kravtsov, *Radiation Transfer* (Gordon and Breach, Basel, 1996).
21. A. Lagendijk and B. A. van Tiggelen, Resonant multiple scattering of light, *Phys. Rep.* **270**, 143–215 (1996).
22. A. Ishimaru, *Wave Propagation and Scattering in Random Media* (IEEE Press, New York, 1997).
23. L. Tsang and J. A. Kong, *Scattering of Electromagnetic Waves: Advanced Topics* (Wiley, New York, 2001).
24. V. P. Tishkovets, Multiple scattering of light by a layer of discrete random medium: backscattering, *J. Quant. Spectrosc. Radiat. Transfer* **72**, 123–137 (2002).
25. M. I. Mishchenko, Vector radiative transfer equation for arbitrarily shaped and arbitrarily oriented particles: a microphysical derivation from statistical electromagnetics, *Appl. Opt.* **41**, (2002).
26. M. I. Mishchenko, L. D. Travis, and A. A. Lacis, *Scattering, Absorption, and Emission of Light by Small Particles* (Cambridge University Press, Cambridge, 2002).
27. V. Twersky, On propagation in random media of discrete scatterers, *Proc. Symp. Appl. Math.* **16**, 84–116 (1964).
28. D. S. Saxon, Lectures on the scattering of light (Science Report No. 9, Department of Meteorology, University of California, Los Angeles, 1955).
29. M. I. Mishchenko, A. A. Lacis, and L. D. Travis, Errors introduced by the neglect of

polarization in radiance calculations for Rayleigh-scattering atmospheres, *J. Quant. Spectrosc. Radiat. Transfer* **51**, 491–510 (1994).

30. J. E. Hansen, Multiple scattering of polarized light in planetary atmospheres. II. Sunlight reflected by terrestrial water clouds, *J. Atmos. Sci.* **28**, 1400–1426 (1971).
31. M. I. Mishchenko, J. W. Hovenier, and L. D. Travis, L. D., eds., *Light Scattering by Nonspherical Particles: Theory, Measurements, and Applications* (Academic Press, San Diego, 2000).
32. G. G. Stokes, On the intensity of the light reflected from or transmitted through a pile of plates, *Proc. R. Soc. London* **11**, 545–556 (1862).
33. H. C. van de Hulst, A new look at multiple scattering (Technical Report, NASA Institute for Space Studies, New York, 1963).
34. M. I. Mishchenko, Multiple scattering of light in anisotropic plane-parallel media, *Transp. Theory Stat. Phys.* **19**, 293–316 (1990).
35. J. F. de Haan, P. B. Bosma, and J. W. Hovenier, The adding method for multiple scattering calculations of polarized light, *Astron. Astrophys.* **183**, 371–391 (1987).
36. P. Stammes, J. F. de Haan, and J. W. Hovenier, The polarized internal radiation field of a planetary atmosphere, *Astron. Astrophys.* **225**, 239–259 (1989).
37. G. Asrar, ed., *Theory and Applications of Optical Remote Sensing* (John Wiley, New York, 1989).
38. G. L. Stephens, *Remote Sensing of the Lower Atmosphere* (Oxford University Press, New York, 1994).
39. C. D. Mobley, *Light and Water: Radiative Transfer in Natural Waters* (Academic Press, San Diego, 1994).
40. M. Mishchenko, J. Penner, and D. Anderson, eds., Special issue on Global Aerosol Climatology Project, *J. Atmos. Sci.* **59**, 249–783.

TRANSFER OF TRAPPED ELECTROMAGNETIC RADIATION IN AN ENSEMBLE OF RESONANT MESOSCOPIC SCATTERERS

YURII N. BARABANENKOV
Institute of Radioengineering and Electronics
of the Russian Academy of Sciences
Mokhovaya 11, 103907 Moscow GSP–3, Russia

Abstract. A mesoscopic analogy to Compton-Milne radiation trapping is discussed. The phenomenon of radiation trapping is known since 1923 and refers to 'trapping' of resonant non-stationary optical radiation in gases, whose atoms posses a resonant absorption line. We discuss this phenomenon in connection with the modern study of the energy transport velocity reduction for light or microwave pulse propagation in a medium consisting of randomly placed small resonant dielectric scatterers. The mesoscopic analogy to the Compton-Milne phenomenon is considered with the aid of the improved Sobolev radiative transfer equation with delay at resonant single-scattering events. Some preliminary general comments are given on the range of validity of the radiative transfer theory from the view point of the statistical multiple scattering theory, and on the known advances in the study of this range in terms of the Van Hove limit. As a starting point in the theory of Compton-Milne phenomenon we take the exact kinetic equation for the coherence function of the electromagnetic pulse propagating in a discrete dielectric random medium. The equation includes a term accounting for energy accumulation inside the scatterers due to their polarization. The appearance of this term is imposed by Ward identity, the key identity in the context of the present study. According to Ward identity, the dynamic extinction of the ensemble averaged field results from the partially-coherent scattering and the energy accumulation/absorption inside the scatterers. Two approaches towards the solution of the exact kinetic equation are considered: the first one assumes diffusion regime of pulse evolution on large spatio-temporal scale, while the second one relies on the radiative transfer equation with delay and is valid on arbitrary spatio-temporal scales. The diffusion asymptotics for the coherence function of the wave field allows us to obtain an exact expression for the diffusion constant of electromagnetic

B.A. van Tiggelen and S.E. Skipetrov (eds.),
Wave Scattering in Complex Media: From Theory to Applications, 415–459.

radiation in a random medium in the form of the Green-Kubo relation written in terms of the energy flux correlation function and the density of states of the wave field. In the diffusion regime, the electromagnetic energy is proportionally distributed inside and outside the scatterers in accordance to the parameter of energy accumulation inside the scatterers. A modified radiative transfer equation, accounting for time delay at resonant single-scattering events and derived from the exact kinetic equation under certain approximations, is applied to study the two effects of radiation trapping on a short pulse reflected from a resonant semi-infinite random medium: (a) the redistribution of the pulse energy from the front to the rear of the pulse, and (b) the broadening of the coherent backscattering cone as compared to the the case of non-resonant random medium.

1. Introduction

1.1. PREFACE

One of the most interesting and challenging problems in the theory of multiple scattering of waves in random media concerns the range of validity ('status') of the phenomenological radiative transfer theory (see, e.g., Chandrasekhar, 1960) that appeared in its initial form in 19th century. Radiative transfer theory has ever increasing applications in optics, radiophysics, acoustics, medical physics, and elaborated mathematical techniques. This problem still remains partially unresolved, though many its essential peculiarities have been studied in the second half of the last century. It was shown, in particular, that neglecting the repeated scattering of a monochromatic wave by just the same scatterer — so-called independent scattering approximation (Lagendijk and Van Tiggelen, 1996) or, more generally, by just the same group of correlated scatterers coupled with the aid of one correlation function (scatterer cluster) — so-called single-group approximation (Finkelberg, 1968), together with the far-field approximation for fields scattered by scatterers/scatterer clusters, lead to radiative transfer theory (Barabanenkov and Finkelberg, 1968; Watson, 1969; Barabanenkov, 1976). The basic significance of the single-group approximation for the radiative transfer theory was demonstrated more definitely in the study of multiple scattering of non-stationary wave radiation in a random medium model consisting of scatterers with a finite 'lifetime'. In this model, the radiative transfer theory was derived (after the Papanicolaou paper, 1972, on a kinetic theory of power transport in stochastic systems, and the Tatarskii paper, 1969, concerning the Markov approximation for light propagation in a random medium) as an exact Van Hove (Van Hove, 1955) limit

(Barabanenkov, 1988), i.e., under conditions that (i) the ratio of lifetime of a scatterer/scatterer cluster (interaction time of radiation with the scatterer/scatterer cluster) to a mean free time of radiation between scattering events tends to zero and (ii) the ratio of propagation time of radiation to the mean free time is fixed. The second condition of the Van Hove limit means that repeated scattering of radiation by just the same scatterer/scatterer cluster may occur on a long temporal or spatial scale of wave propagation, and one should then take into account the multi-group or dependent scattering events. In this case, the usual radiative transfer equation, being linear and local with respect to temporal and spatial variables, is no longer valid and needs to be modified substantially. The best-known effect of the multi-group scattering events is the coherent backscattering enhancement caused by the contribution of so-called cyclical (or maximally crossed) diagrams. This effect leads to a correction to the transfer equation for backward scattering. It was shown (Barabanenkov, 1973; Akkermans *et al.*, 1986) that the relative value of this correction is of the order of unity in a narrow cone of directions of backward scattering, the cone width being of the order of the ratio of the wavelength to the size of the scattering volume or to extinction length.

It is worth noting that the 1973 paper of Barabanenkov was published in order to discuss a paper of Gazaryan (Gazaryan, 1969) where on the basis of the exact solution of the problem of strong (Anderson) wave localization in a one-dimensional random medium (Anderson, 1958; Gertsenshtein and Vasil'ev, 1959), it was concluded that radiative transfer theory is not valid at all for multiple scattering of waves by an ensemble of randomly placed dielectric scatterers in three dimensions.

Certainly, a modification of radiative transfer theory is a difficult problem, though some such modifications have been performed in a study of multiple scattering of non-stationary radiation taking into account effects of time delay or memory caused by repeated scattering of ray intensity by the same scatterer (Barabanenkov and Ozrin, 1985), and by coherent backscattering enhancement (Strinati *et al.*, 1986).

Perhaps, the simplest modified radiative transfer equation has been introduced, on the basis of phenomenological assumptions, by Sobolev (Sobolev, 1956) taking into account the time delay at single-scattering events in order to study, in particular, the Compton-Milne phenomenon of transient resonant optical radiation 'trapping' in gases where atoms posses a resonant absorption line (Compton, 1923; Milne, 1926). It should be noted that the theory of radiation trapping developed rather dramatically between 1920 and 1970. Milne and Holstein (Holstein, 1947) gave the clearest physical description of this phenomenon. The term 'resonant' is applied to the radiation emitted by an atom due to a transition from an excited to the ground state. Since resonant radiation is efficiently absorbed by gas atoms in their ground state, it is clear that when a resonant quantum is emitted by a given gas atom, it is unlikely that it can reach the boundaries of the cell containing the gas and then escape the cell without being absorbed by another atom, exciting the latter. This process of emission and reabsorption of light quanta by different atoms results in the transfer of excitation energy from atom to atom. Resonant quanta spend most of their time being 'trapped' by atoms. To study the transfer of energy in a gas of resonant atoms, Milne suggested a modified diffusion equation including the effect of (weak) delay in the diffusion coefficient. But even the Milne's modified diffusion theory did not agree satisfactory with experiments (Zemansky, 1927; Mitchell and Zemansky, 1934). The reason for this disagreement was in the redistribution of radiated energy in the frequency domain due to the Doppler shift caused by the motion of atoms. Holstein used a Boltzmann-type equation to describe the resonant energy transfer in atomic gases. The latter equation takes into account the redistribution of radiated energy in the frequency domain. Both the Milne's modified diffusion theory and the Holstein's Bolzmann-type equation may be obtained from the Sobolev modified radiative transfer theory with delay mentioned above. In the lecture, we will discuss a mesoscopic analogy to the Compton-Milne phenomenon in connection with the study of reduction of the energy transport velocity for light (Van Albada *et al.*, 1991) and microwave (Kuga *et al.*, 1993) pulses in a resonant dielectric medium consisting of randomly placed dielectric scatterers, under condition that the carrier frequency of the pulse is in the Mie-resonance scattering region of a dielectric scatterer. Microscopic analysis shows that the Sobolev modified radiative transfer equation with delay may be, generally speaking, in conflict with the Van Hove limit mentioned above. Nevertheless, there exists at present an improved version (Barabanenkov and Barabanenkov, 1997) of the Sobolev modified radiative transfer equation in the independent scattering approximation with delay, which takes into account not only the time delay at resonant single-scattering events but also the energy accumulation inside the scatterers due to their polarization. This improved Sobolev equation with delay enables us, as will be shown in the lecture, to apply the technique of the generalized Chandrasekhar H-function to study effects of pulse trapping on the pulse energy redistribution (Barabanenkov *et al.*, 1998) and weak localization (Barabanenkov and Barabanenkov, 1999) for a pulse reflected from a semi-infinite resonant random medium.

1.2. RADIATIVE TRANSFER AS AN ASYMPTOTIC LIMIT OF MULTIPLE SCATTERING THEORY OF WAVES

Let us make some preliminary comments on the general problem concerning the range of validity of the radiative transfer theory starting with the stochastic Helmholtz equation

$$(L - V)\mathcal{G} = I \tag{1.1}$$

for the retarded Green function $\mathcal{G}(\mathbf{r}, \mathbf{r}') \equiv \mathcal{G}(\mathbf{r}, \mathbf{r}'; \Omega + i0)$ of monochromatic wave field in a discrete dielectric random medium with a 'scattering potential' $V(\mathbf{r})$ of the form

$$V(\mathbf{r}) = \sum_i V_0(\mathbf{r} - \mathbf{r}_i) \tag{1.2}$$

The scatterers of the medium are assumed to be identical non-overlapping spheres with given radius r_0, dielectric permittivity, scattering potential $V_0(\mathbf{r} - \mathbf{r}_i)$, and centered at random positions $\mathbf{r}_i$. For brevity, the Helmholtz equation (1.1) is written in the symbolic operator notation where the differential operator L has the form $L = L^0 + \Omega^2/C_0^2$, with L^0 being in the scalar case merely the Laplacian, $L^0 = \nabla^2$, and Ω and C_0 being the frequency and the phase velocity in a background medium, respectively. The symbol I denotes the unit operator with kernel $\delta(\mathbf{r} - \mathbf{r}')$. Multiplying (in operator sense) equation (1.1) by J on the right, with $J(\mathbf{r})$ being a source distribution, gives Helmholtz equation for the wave field $E = \mathcal{G}J$.

We are interested in the ensemble averaged Green function $G = \langle \mathcal{G} \rangle$ and in the coherence function $F(\mathbf{r}_1, \mathbf{r}_2; \mathbf{r}'_1, \mathbf{r}'_2) = \langle \mathcal{G}(\mathbf{r}_1, \mathbf{r}'_1)\mathcal{G}^*(\mathbf{r}_2, \mathbf{r}'_2) \rangle$, which we will write symbolically as $F = \langle \mathcal{G} \otimes \mathcal{G}^* \rangle$. The tensor product $A \otimes B$ of two operators $A(\mathbf{r}, \mathbf{r}')$ and $B(\mathbf{r}, \mathbf{r}')$ is defined by

$$(A \otimes B)(\mathbf{r}_1, \mathbf{r}_2; \mathbf{r}'_1, \mathbf{r}'_2) = A(\mathbf{r}_1, \mathbf{r}'_1)B(\mathbf{r}_2, \mathbf{r}'_2) \tag{1.3}$$

If in this definition the points $\mathbf{r}_1$ and $\mathbf{r}_2$ coincide, we will use the notation $A \hat{\otimes} B$. From the definition (1.3) one can prove a simple algebraic relation $(A \hat{\otimes} B)(C \otimes D) = AC \hat{\otimes} BD$ (see problem 1.1).

The ensemble averaged quantities satisfy the Dyson equation

$$(L - M)G = I \tag{1.4}$$

and the Bethe-Salpeter one

$$F = G \otimes G^* + G \otimes G^* K F \tag{1.5}$$

with the mass and intensity operators M and K, respectively. The Dyson and Bethe-Salpeter equations describe (see Fig. 1) the coherent and incoherent scattering, respectively, of wave radiation by effective inhomogeneities

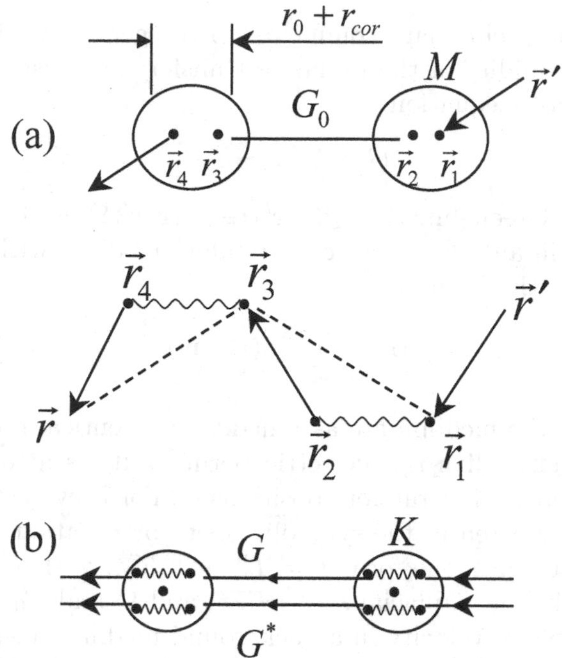

Figure 1. Solutions to Dyson (a) and Bethe-Salpeter (b) equations in the form of iterative series in powers of mass M and intensity K operators, respectively.

of the random medium, the mass and intensity operators defining the spatial scale and optical features of an inhomogeneity. Energy conservation imposes on M and K a constraint in the form of the optical theorem which for the case of non-absorbing scatterers (and non-absorbing background medium) may be written as (Barabanenkov and Finkelberg, 1968)

$$M\hat{\otimes}I - I\hat{\otimes}M^* = (G\hat{\otimes}I - I\hat{\otimes}G^*)K \tag{1.6}$$

and will also appear as a consequence of the Ward identity discussed below. The relation of the optical theorem to energy conservation can be inferred from the Bethe-Salpeter equation written in a 'transport' form:

$$\begin{aligned}(L \otimes I - I \otimes L)F &= \{M \otimes I - I \otimes M^* + (I \otimes G^* - G \otimes I) \cdot K\}F \\ &+ I \otimes G^* - G \otimes I \end{aligned} \tag{1.7}$$

This equation, when transformed to an equation for the coherence function $\langle E(\mathbf{r}_1)E^*(\mathbf{r}_2)\rangle$, leads directly to the energy conservation equation (see problem 1.2), once $\mathbf{r}_1 = \mathbf{r}_2$ is taken and the optical theorem (1.6) is applied.

1.2.1. *Neglecting the repeated scattering by the same scatterer/scatterer cluster*

In the so-called independent scattering approximation (Lagendijk and Van Tiggelen, 1996) or, more generally, in the so-called single-group approximation (Finkelberg, 1968), one neglects the repeated scattering of waves by just the same scatterer or by just the same group of correlated scatterers coupled with the aid of one correlation function (scatterer cluster). In this approximation, only those Feynman diagrams of infinite sums for mass and intensity operators M and K are counted that are *linear* either in number density g_1 of scatterers or in their correlation function $g_n(\mathbf{r}_1, ..., \mathbf{r}_n)$ $(n = 1, 2, \ldots)$ (Frisch, 1965); see diagrams 1, 2, and 4 for M and 1–4 for K in Fig. 2. The single-group diagrams decrease quickly as the different elements separate, much like the scattering potential of a single scatterer or the correlation function of a set of scatterers, in sharp contrast to multiple-group diagrams (diagrams 3 and 5 for M and 5–7 for K in Fig. 2) that decrease slowly (as a power of the Green function). It is interesting to note that Feynman diagrams of Fig. 2 involve contributions of two-particle scattering events into the mass and intensity operators. One may sum over all possible two-particle scattering events and express their contribution into the mass and intensity operators in terms of scattering operators of one and two scatterers. This can be done in both the general case, as we show below in the single-group approximation (see also Barabanenkov and Finkelberg, 1968) and in the case of point-like scatterers (Van Tiggelen and Lagendijk, 1994). It is also demonstrated in the two cited papers that the mass and intensity operators, written in the second order in the number density g_1 of scatterers, satisfy the optical theorem (1.6) in the second order in g_1. We note that Van Tiggelen and Lagendijk (1994) used two-particle corrections to the mass and intensity operators to calculate the contribution of resonantly induced dipole-dipole interactions into the energy accumulation (stored energy) inside resonant dipoles.

In the single-group approximation, the mass and intensity operators M_1 and K_1 may be represented in the following compact form (Finkelberg, 1968):

$$M_1 = \sum_{i=1}^{\infty} \frac{1}{n!} \int d1 \cdots \int dn\, g_n(1, \cdots, n) T^{gr}_{1,\cdots,n} \tag{1.8}$$

and

$$K_1 = \sum_{i=1}^{\infty} \frac{1}{n!} \int d1 \cdots \int dn\, g_n(1, \cdots, n) \left(T_{1,\cdots,n} \otimes T^*_{1,\cdots,n}\right)^{gr} \tag{1.9}$$

The integers $1, \cdots, n$ denote the points $\mathbf{r}_1, \cdots, \mathbf{r}_n$, and $T_{1,\cdots,n}$ is the scattering operator of a set of n scatterers located at these points. The scattering

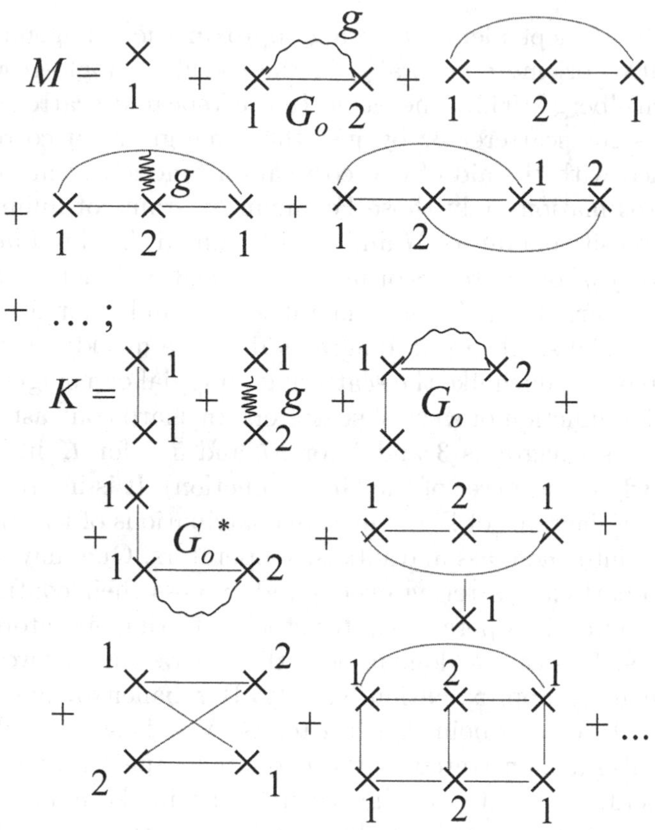

Figure 2. Feynman diagrams for mass M and intensity K operators in the case of correlated scatterers. Vertices 1 and 2 denote single-scattering operators T_1 and T_2 at points 1 and 2, respectively.

operator satisfies the Lippmann-Schwinger equation

$$T_{1,\cdots,n} = V_{1,\cdots,n} + V_{1,\cdots,n} G_0 T_{1,\cdots,n} \tag{1.10}$$

with scattering potential $V_{1,\cdots,n}$ given by Eq. (1.2) in the case of $n = 1, 2, \cdots$ scatterers and $G_0(r) = -\exp(ik_0 r)/(4\pi r)$ being the Green function in background with wave number $k_0 = \Omega/C_0$. The superscript 'gr' denotes the group operation which is defined according to

$$\psi_s^{gr}(1, \cdots, s) = \sum_{q=1}^{s} \sum_{1 \leq j_1 < \cdots < j_q \leq s} (-1)^{s-q} \psi_q(j_1, \cdots, j_q), \quad s \geq 1 \tag{1.11}$$

where integers $1, \cdots, s$ denote, again, a set of points $\mathbf{r}_1, \cdots, \mathbf{r}_s$ and the sum is over the subset of $q = 1, \cdots, s$ scatterers whose indices are $j_1, \ldots, j_q$. The group operation excludes from operators $T_{1,\cdots,n}$ and $T_{1,\cdots,n} \otimes T^*_{1,\cdots,n}$ the coherent and partially coherent scattering events, respectively, by all subsets of the set of n scatterers.

The single-group approximation includes many approximations known in the multiple scattering theory (see problem 1.3). In particular, the first terms on the right hand side (r.h.s.) of Eqs. (1.8) and (1.9) disregard correlations between scatterer centers and coincide with the independent scattering approximation considered by Foldy (Foldy, 1945) and Lax (Lax, 1951). The second terms of these equations, expanded in the multiplicity of scattering, account for the second order correlations of scatterer centers (see, e.g., Hirschfelder *et al.*, 1954). Further expansion of the first and second terms leads to Bourret and ladder approximations for the mass and intensity operators (Bourret, 1962), valid for optically weak scatterers.

The single-group approximation (1.8) and (1.9) may be obtained as a first order approximation for the mass and intensity operators in a small parameter γ (see problem 1.4). For this reason, the mass and intensity operators M_1 and K_1 in the single-group approximation satisfy the optical theorem (1.6) into first order in γ, that is the optical theorem (1.6) with the averaged Green function G on the r.h.s. replaced by the Green function G_0 in the background (see problem 1.5).

It is worth noting that the mass and intensity operators (1.8) and (1.9) in the single-group approximation allow the introduction of the concept of the specific operators M_0 and K_0 by the equations

$$M_1 = \int d\mathbf{r}_1 M_0(\mathbf{r}_1), \; K_1 = \int d\mathbf{r}_1 K_0(\mathbf{r}_1) \tag{1.12}$$

where $\mathbf{r}_1$ is the center of one of the scatterers of a correlation group. In the case of independent scatterers, the specific operators $M_0(\mathbf{r}_1)$ and $K_0(\mathbf{r}_1)$ go over into $g_1 T(\mathbf{r}_1)$ and $g_1 T(\mathbf{r}_1) \otimes T(\mathbf{r}_1)^*$, where $T(\mathbf{r}_1)$ is the scattering operator of a single scatterer centered at $\mathbf{r}_1$. This makes it possible to interpret the operators M_0 and K_0 as scattering operators of independent effective inhomogeneities of the medium.

1.2.2. *Far-field approximation*

In the single-group approximation, an effective inhomogeneity of the random medium has a well defined spatial scale of the order of the sum $r_0 + r_{cor}$ of the scatterer radius r_0 and the range r_{cor} of scatterer correlations. This allows one to consider the far-field approximation of fields scattered by scatterers/scatterer clusters assuming that the latter are placed, on the average, in the Fraunhofer zone of each other. In the far-field approximation, the solution of the Dyson equation (1.4) gives the averaged Green function in the form (see problem 1.6)

$$G(r) = -\exp(ik_1 r)/4\pi r \tag{1.13}$$

which is obtained from the Green function of the background by formally replacing the wave number in the background by an effective complex wave number k_1 in the random medium defined as

$$k_1^2 = k_0^2 - M_1(k_0) \tag{1.14}$$

Here $M_1(\mathbf{p})$ is the spatial Fourier transform of the mass operator $M_1(\mathbf{r})$ in the single-group approximation:

$$M_1(\mathbf{r}) = \int_{\mathbf{p}} \exp(i\mathbf{pr}) M_1(\mathbf{p}) \tag{1.15}$$

with $\int_{\mathbf{p}}$ denoting $(2\pi)^{-3}\int d\mathbf{p}$. The solution of the Bethe-Salpeter equation (1.5) in the far-field approximation can be written (Barabanenkov and Finkelberg, 1968) in terms of a propagator $\tilde{F}(\mathbf{R},\mathrm{s};\mathbf{R}',\mathrm{s}')$ of the radiative transfer equation (see problem 1.7)

$$\begin{aligned} \mathbf{s}\nabla\tilde{F}(\mathbf{R},\mathrm{s};\mathbf{R}',\mathrm{s}') &= -\frac{1}{\ell}\tilde{F}(\mathbf{R},\mathrm{s};\mathbf{R}',\mathrm{s}') + \int_{4\pi} ds'' W(\mathbf{s},\mathbf{s}'')\tilde{F}(\mathbf{R},\mathrm{s}'';\mathbf{R}',\mathrm{s}') \\ &+ \frac{1}{(4\pi)^2}\delta(\mathbf{R}-\mathbf{R}')\delta(\mathbf{s}-\mathbf{s}') \end{aligned} \tag{1.16}$$

The extinction coefficient $1/\ell$, with ℓ being the mean free path, and the phase function $W(\mathbf{s},\mathbf{s}')$ are defined by

$$\frac{1}{\ell} = -\frac{\operatorname{Im} M_1(k_0)}{k_0} \tag{1.17}$$

and

$$W(\mathbf{s},\mathbf{s}') = \frac{1}{(4\pi)^2} K_1(k_0\mathbf{s}, k_0\mathbf{s}'; 0) \tag{1.18}$$

and satisfy the optical theorem

$$\frac{1}{\ell} = \int_{4\pi} d\mathbf{s}\, W(\mathbf{s},\mathbf{s}') \tag{1.19}$$

which is a consequence of the general optical theorem (1.6) evaluated in first order of the small parameter γ (see problem 1.8); $\mathbf{s}$, $\mathbf{s}'$, and $\mathbf{s}''$ denote unit vectors. In the definition (1.18), $K_1(\mathbf{p},\mathbf{p}';\mathbf{q})$ denotes the spatial Fourier transform using the Wigner variables of the intensity operator

$K_1(\mathbf{r}_1, \mathbf{r}_2; \mathbf{r}_1', \mathbf{r}_2')$ in the single-group approximation:

$$K_1(\mathbf{R}+\mathbf{r}/2, \mathbf{R}-\mathbf{r}/2; \mathbf{R}'+\mathbf{r}'/2, \mathbf{R}'-\mathbf{r}'/2) = \int_{\mathbf{p}}\int_{\mathbf{p}'}\int_{\mathbf{q}} \exp\left[i\mathbf{q}\left(\mathbf{R}-\mathbf{R}'\right)+i\left(\mathbf{p}\mathbf{r}-\mathbf{p}'\mathbf{r}'\right)\right] K_1(\mathbf{p}, \mathbf{p}'; \mathbf{q}) \quad (1.20)$$

One may also introduce the spatial Fourier transform of the coherence function F of the wave field using the mixed Wigner variables:

$$F(\mathbf{R}+\mathbf{r}/2, \mathbf{R}-\mathbf{r}/2; \mathbf{R}'+\mathbf{r}'/2, \mathbf{R}'-\mathbf{r}'/2) = \int_{\mathbf{p}}\int_{\mathbf{p}'} \exp(i\mathbf{p}\mathbf{r} - i\mathbf{p}'\mathbf{r}') F(\mathbf{R}, \mathbf{p}; \mathbf{R}', \mathbf{p}') \quad (1.21)$$

where $F(\mathbf{R}, \mathbf{p}; \mathbf{R}', \mathbf{p}')$ has the meaning of the spatial spectral density of the field coherence function. In the far-field approximation, this spectral density becomes (Barabanenkov, 1969; see also problem 1.9) a sharp function of the momentum moduli p and p':

$$F(\mathbf{R}, \mathbf{p}; \mathbf{R}', \mathbf{p}') \simeq \tilde{F}(\mathbf{R}, \mathbf{s}; \mathbf{R}', \mathbf{s}')(2\pi)^6 k_0^{-4} \delta(p-k_0)\delta(p'-k_0) \quad (1.22)$$

where $\mathbf{s} = \mathbf{p}/p$ and $\mathbf{s}' = \mathbf{p}'/p'$ are unit vectors.

Let us note that aside from the described microscopic derivation of the radiative transfer equation (1.16) based on the far-field approximation, there exists a macroscopic derivation based on the assumption of a slow variation of the wave field spectral density (1.21) as a function of $\mathbf{R}-\mathbf{R}'$ within the spatial scale $r_0 + r_{cor}$ of the effective inhomogeneity in the single-group approximation (see problem 1.10).

1.2.3. *Example of situation where the radiative transfer is an exact Van Hove limit of the multiple scattering theory: pulse propagation in a non-stationary random medium*

One can easily recognize the basic role of the single-group and, in particular, the independent scattering approximation for the status of the radiative transfer theory in the theory of multiple scattering of waves. However, this approximation is difficult to be asymptotically proved.

The single-group approximation could be justified with the aid of an exact, so-called Bethe-Salpeter non-uncoupled relation (Barabanenkov and Kalinin, 1992):

$$\begin{aligned}(L \otimes I - I \otimes L)\langle \mathcal{G} \otimes \mathcal{G}^* \rangle &= \langle [\mathcal{M} \otimes I - I \otimes \mathcal{M}^* - (\mathcal{G} \otimes I - I \otimes \mathcal{G}^*)\mathcal{K}]\, \mathcal{G} \otimes \mathcal{G}^* \rangle \\ &\quad + I \otimes \langle \mathcal{G}^* \rangle - \langle \mathcal{G} \rangle \otimes I \end{aligned} \quad (1.23)$$

In this relation, derived by using Furutsu-Donsker-Novikov formalism (see, e.g., Rytov, Kravtsov and Tatarskii, 1989), one encounters a random 'mass' operator $\mathcal{M}$ and a random 'intensity' operator $\mathcal{K}$ defined as

$$\mathcal{M} = \sum_{i=1}^{\infty} \frac{1}{n!} \int d1 \cdots \int dn \; g_n(1, \cdots, n) \; \mathcal{T}^{gr}_{1,\cdots,n} \tag{1.24}$$

and

$$\mathcal{K} = \sum_{i=1}^{\infty} \frac{1}{n!} \int d1 \cdots \int dn \; g_n(1, \cdots, n) \left(\mathcal{T}_{1,\cdots,n} \otimes \mathcal{T}^*_{1,\cdots,n} \right)^{gr} \tag{1.25}$$

where the self-consistent random scattering operator $\mathcal{T}_{1,\cdots,n}$ satisfies a Lippman–Schwinger equation in a random medium:

$$\mathcal{T}_{1,\cdots,n} = V_{1,\cdots,n} + V_{1,\cdots,n} \, \mathcal{G} \, \mathcal{T}_{1,\cdots,n} \tag{1.26}$$

The latter equation is obtained from the Lippmann-Schwinger equation (1.10) by a formal replacement of the Green function G_0 by the Green function $\mathcal{G}$ of the random medium.

Let us assume that the random Green function $\mathcal{G}(\mathbf{r}, \mathbf{r}')$ as well as the averaged Green function $G(\mathbf{r}, \mathbf{r}')$ obey the local property

$$\mathcal{G}(\mathbf{r}, \mathbf{r}') \simeq G(\mathbf{r}, \mathbf{r}') \simeq G_0(\mathbf{r} - \mathbf{r}'), \quad |\mathbf{r} - \mathbf{r}'| \simeq r_0 + r_{cor} \tag{1.27}$$

i.e. are both close to the Green function of the background, provided that the distance between the point of observation $\mathbf{r}$ and source $\mathbf{r}'$ is of the order of the sum of the scatterer radius and the range of scatterer correlations. Under this assumption, one may replace the random Green function $\mathcal{G}$ in the self-consistent Lippmann-Schwinger equation (1.26) by the Green function G_0 of the background, thus transforming immediately the random mass and intensity operators (1.24) and (1.25) into (1.8) and (1.9) evaluated in the single-group approximation. Using the local property (1.27), one can also replace both random Green functions in $\mathcal{G} \otimes I - I \otimes \mathcal{G}^*$ on the r.h.s. of the non-uncoupled Bethe-Salpeter relation (1.23) by the averaged Green functions, provided that the distance between the observation points $\mathbf{r}_1$ and $\mathbf{r}_2$ (coherence radius) of the coherence function $F(\mathbf{r}_1, \mathbf{r}_2; \mathbf{r}_1', \mathbf{r}_2')$ is of the order of $r_0 + r_{cor}$. This leads to the Bethe-Salpeter equation in the transport form (1.7).

At first sight, the above reasoning justifies the single-group approximation at the cost of the local property (1.27). But this local property may actually not be realized because of the contribution of the loop-like scattering events, involving scatterers from the whole volume of the random medium. To sum up, the single-group and, in particular, the independent scattering approximation (and hence the radiative transfer theory) is not asymptotically justified, strictly speaking, at present. The known derivation (Ryzhik *et al.*, 1996) of radiative transfer (transport) equations for waves in random media using a multiscale expansion in powers of small parameters is a formal derivation assuming that radiative transfer theory is justified already.

What is the best known achievement in the study the problem concerning the asymptotic validity of radiative transfer theory? According to our knowledge, this problem has been solved only in the case of pulse propagation in a strictly non-stationary random medium, where a scattering potential $V(\mathbf{r}, t)$ fluctuates both in space and time, the temporal fluctuations being fast enough. In such a medium, one can imagine the scatterers to have an effective finite time τ_0 of their 'life' and then use a technique elaborated for stochastic dynamic systems (Papanicolaou, 1972; Tatarskii, 1969; Barabanenkov, 1988) to derive a dynamic radiative transfer theory in the exact Van Hove (Van Hove, 1955) limit

$$\tau_0/t_M \to 0, \; t/t_M = \text{const} \tag{1.28}$$

This limit means, first, that repeated scattering of the pulse by the same scatterer is practically impossible if the scatterer lifetime τ_0 is small enough as compared to a specific lower estimation t_M of the mean free time between scattering events. Second, such repeated scattering may be possible if the observation time t is large enough as compared to t_M.

To clarify the meaning of the Van Hove limit, let us consider propagation of an acoustic (the electromagnetic case is more complicated because of the effect of longitudinal fields) quasi-monochromatic pulse in a non-stationary medium with phase velocity $C(\mathbf{r}, t)$ fluctuating in space and time according to $1/C^2(\mathbf{r}, t) = (1/C_0^2)[1 + \delta\epsilon(\mathbf{r}, t)]$, where $\delta\epsilon(\mathbf{r}, t)$ is a stationary zero-mean Gaussian random process. We substitute the pressure $p(\mathbf{r}, t) = \mathrm{Re}\left[\exp(-i\Omega t)\psi(\mathbf{r}, t)\right]$ of the acoustic wave field (with Ω and ψ being the carrier frequency and the complex amplitude) into the wave equation (Landau and Lifshitz, 1986) and assume $\delta\epsilon(\mathbf{r}, t)$ to have small magnitude and also, as well as $\psi(\mathbf{r}, t)$, to vary slowly in time. We obtain then the following approximate parabolic equation (Barabanenkov *et al.*, 1985):

$$i\frac{2k_0}{C_0}\frac{\partial\psi}{\partial t} = \left[-\nabla^2 - k_0^2 + V(\mathbf{r}, t)\right]\psi \tag{1.29}$$

where the scattering potential is $V(\mathbf{r}, t) = -k_0^2\delta\epsilon(\mathbf{r}, t)$. The object of interest is the density matrix $\rho(\mathbf{p}_1, \mathbf{p}_2, t) = \psi(\mathbf{p}_1, t)\psi^*(\mathbf{p}_2, t)$ of the wave field which we consider in the spatial Fourier representation. The density matrix

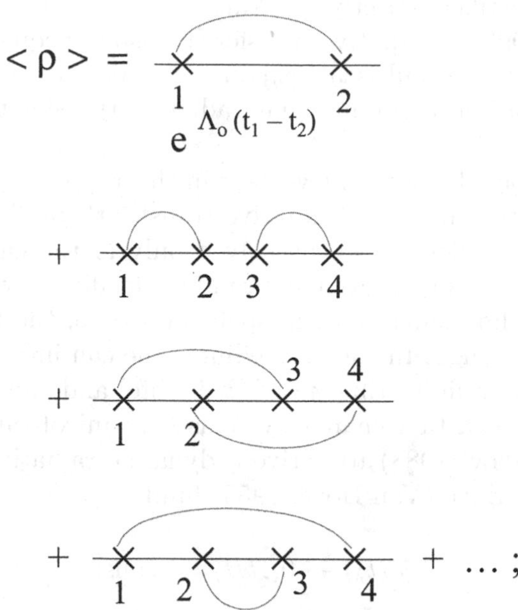

Figure 3. Feynman diagrams for the averaged density matrix $\langle\rho(t)\rangle$. Vertices 1, 2, ... denote spatial Fourier transforms of the scattering potentials $V(\mathbf{k}_1, t_1)$, $V(\mathbf{k}_2, t_2)$, ... Arcs connecting two vertices denote cumulants of scattering potentials.

satisfies the Liouville equation

$$\frac{d}{dt}\rho(t) = [\Lambda_0 + \Lambda(t)]\,\rho(t) \tag{1.30}$$

with a non-perturbed Λ_0 and a perturbed $\Lambda(t)$ parts of the Liouville operator (see problem 1.11).

The ensemble averaged density matrix $\langle\rho(t)\rangle$ can be represented as a sum of Feynman diagrams (see Fig. 3). We consider each cumulant $\langle V(\mathbf{k},t)V(\mathbf{k}',t')\rangle$ of the spatial Fourier transform $V(\mathbf{k},t)$ of the scattering potential as a quantity corresponding to an effective inhomogeneity. The diagrams 1 and 2 of Fig. 3 then describe the single and double scattering of a pulse by one and two inhomogeneities, respectively, without accounting for the repeated scattering. Summing all such diagrams leads to a sort of Dyson equation with the mass operator in Bourret approximation:

$$\frac{d}{dt}\rho_1(t) = \Lambda_0\rho_1(t) + \int_0^t dt'\,\langle\Lambda(t)\exp\left[\Lambda_0(t-t')\right]\Lambda(t')\rangle\,\rho_1(t') \tag{1.31}$$

The diagrams 3 and 4 of Fig. 3 describe the repeated scattering of pulse by inhomogeneities. Diagram 3, for example, describes the scattering of the pulse by the first inhomogeneity, then by the second, then again by the first, and then again by the second. The principal result is that the contribution of all diagrams with repeated scattering can be estimated by the following inequality (Barabanenkov *et al.*, 1985; Barabanenkov, 1988):

$$\parallel \langle \rho(t) \rangle - \rho_1(t) \parallel / \parallel \rho_0 \parallel \leq \frac{\tau_0}{t_M} \frac{t}{t_M} \tag{1.32}$$

This strict inequality in Hilbert-Schmidt norm (Kato, 1996) estimates the difference between the exact ensemble averaged density matrix and solution of the Dyson equation (1.31) relative to the norm of the initial matrix density ρ_0 in terms of the effective scatterer lifetime τ_0 and the lower estimate of the mean free time t_M given by

$$\frac{1}{t_M} = 4 \int_0^\infty dt \int d\mathbf{k} \, |B(\mathbf{k}, t)| \tag{1.33}$$

and

$$\frac{\tau_0}{t_M} = \int_0^\infty dt \, t \int d\mathbf{k} \, |B(\mathbf{k}, t)| \tag{1.34}$$

The function $B(\mathbf{k}, t)$ is related to the cumulant of the spatial Fourier transform of the scattering potential $V(\mathbf{k}, t)$ by

$$\langle V(\mathbf{k}, t) V(\mathbf{k}, t') \rangle = B(\mathbf{k}, t) \delta(\mathbf{k} + \mathbf{k}') \tag{1.35}$$

The Dyson equation (1.31), being detailed in the mixed Wigner variables $\rho_1(\mathbf{R}, \mathbf{p}, t)$ (see problem 1.12), includes the specific effects of time delay and spatial dispersion. Disregarding these effects leads to a kinetic equation

$$\left(\frac{\partial}{\partial t} + 2\mathbf{p} \cdot \nabla \right) \rho_1(\mathbf{R}, \mathbf{p}, t) = -\frac{1}{t_r} \rho_1(\mathbf{R}, \mathbf{p}, t) + \int d\mathbf{p}' W(\mathbf{p}, \mathbf{p}') \rho_1(\mathbf{R}, \mathbf{p}', t) \tag{1.36}$$

where the phase function $W(\mathbf{p}, \mathbf{p}')$ and the mean free time t_r are given by

$$W(\mathbf{p}, \mathbf{p}') = 2 \int_0^\infty dt \, B(\mathbf{p} - \mathbf{p}', t) \cos \left[(p^2 - p'^2) t \right] \tag{1.37}$$

and

$$\frac{1}{t_r} = \int d\mathbf{p}' W(\mathbf{p}, \mathbf{p}') < \frac{1}{t_M} \tag{1.38}$$

Note that in the derived kinetic equation a 'natural' (for parabolic equation) variable $C_0 t / 2k_0$ is used instead of time t.

In the case of a stationary random medium, where the scattering potential $V(\mathbf{r})$ does not depend on time, the kinetic equation (1.36) formally becomes a transport equation derived using a multiscale expansion (Ryzhik *et al.*, 1996). But in the case of a stationary random medium, the principal inequality (1.32) and definitions (1.33) and (1.34) lose their meaning. Although it is still possible that the Van Hove limit (1.28), with the scatterer lifetime τ_0 replaced by a time of pulse interaction with a scatterer, remains valid for a stationary random medium as well, this hypothesis has to be proved.

1.2.4. *Radiative transfer equation with time delay: different speeds in a resonant random medium*

As we have mentioned above, the simplest modified radiative transfer equation was introduced by Sobolev (Sobolev, 1956) in the form

$$\frac{1}{v_g}\frac{\partial}{\partial t}I(\mathbf{s};\mathbf{R},t) + \mathbf{s}\nabla I(\mathbf{s};\mathbf{R},t) = -\frac{1}{l}I(\mathbf{s};\mathbf{R},t) + \int_0^t \frac{dt'}{t_{delay}}\int_{4\pi} d\mathbf{s}' \exp\left(-\frac{t-t'}{t_{delay}}\right) W(\mathbf{s},\mathbf{s}')I(\mathbf{s}';\mathbf{R},t') \quad (1.39)$$

where v_g is the group velocity and $I(\mathbf{s};\mathbf{R},t)$ is a specific intensity. The second term on the r.h.s., commonly referred to as a source term (Jefferies, 1968), takes into account the time delay at single scattering events. Starting from (1.39), one can obtain the Holstein's Boltzmann-type equation (Holstein, 1947) that enables one to consider the time delay t_{delay} as a lifetime of the 'excited state' of a scatterer. On the other hand, in the case of small time delay, the Taylor expansion of the source term with respect to $t-t'$ transforms the Sobolev equation (1.39) into the equation used by Kogan and Kaveh (Kogan and Kaveh, 1992). We intend to apply equation (1.39) to study the mesoscopic analogue to the Compton-Milne phenomenon of radiation trapping for a pulse propagating in a resonant dielectric random medium. Here one should note that this intention, generally speaking, may be difficult to realize. Indeed, according to Compton (Compton, 1923), the pulse trapping implies

$$t_{delay}/t_l \geq 1 \quad (1.40)$$

i.e. that the single-scattering delay time t_{delay} exceeds the mean free time $t_l = l/v_g$ between scattering events and, therefore, the radiation spends most of its propagation time being 'trapped' inside resonant scatterers. As one can see, the trapping condition (1.40) is in conflict with the Van Hove limit (1.28) required for the radiative transfer theory to be valid. To avoid this difficulty, we first make the following remarks concerning the Sobolev equation (1.39) with delay:

(i) Sobolev equation does not take into account the accumulation of energy inside scatterers due to their polarization.
(ii) Sobolev equation is consistent with the Poynting's theorem for non-stationary radiation only on average, i.e. after being integrated over the time t from $t = 0$ to $t = \infty$.
(iii) Sobolev equation does not take into account the time delay in the first term on its r.h.s.

The above remarks force us to improve the Sobolev equation with delay starting, for example, with a two-frequency Bethe-Salpeter equation. The improved Sobolev radiative transfer equation with delay will include several different wave speeds in a resonant random medium, such as phase and energy transport velocity, and therefore several different mean free times.

2. Exact Mesoscopic Kinetic Equation for a Pulse in a Dielectric Random Medium: Accounting for Energy Accumulation inside Scatterers

We start with Maxwell equations (Stratton, 1941) for a non-stationary electromagnetic field in a discrete dielectric random medium consisting of randomly placed dielectric scatterers, and apply the Laplace transform in the time domain. The tensor Green function $\mathcal{G}_{\alpha\beta}(\mathbf{r}, \mathbf{r}'; \Omega)$ of the electric field satisfies the stochastic tensor Helmholtz equation (1.1) with a tensor differential operator $L^0_{\alpha\beta}(i\nabla) = \delta_{\alpha\beta}\nabla^2 - \nabla_\alpha\nabla_\beta$ and the scattering potential $V(\mathbf{r}, t)$ of the form (1.2) which we rewrite as

$$V(\mathbf{r}, \Omega) = g(\Omega)v(\mathbf{r}) \tag{2.1}$$

The factors on the r.h.s. are $g(\Omega) = (\Omega^2/C_0^2)(1 + i\sigma_1/\Omega\epsilon')$ and $v(\mathbf{r}) = -[\epsilon(\mathbf{r}) - \epsilon^{bac}]/\epsilon^{bac}$, where $\epsilon(\mathbf{r})$ and ϵ^{bac} are the dielectric permittivity of the random medium and the background, respectively, $\epsilon' = \epsilon_1 - \epsilon^{bac}$, and ϵ_1 and σ_1 denote the dielectric permittivity and the conductivity of a single scatterer. In the system of units we use following Stratton, the phase velocity in the background is $C_0 = 1/(\epsilon^{bac}\mu)^{1/2}$, where μ is the (uniform) magnetic permeability. It is worth noting that in the representation (2.1), the spatial $\mathbf{r}$ and the frequency Ω variables are separated (separation property). This will be used below in the derivation of the Ward identity.

The two-frequency Bethe-Salpeter equation for the two-frequency coherence function $F_{12} = \langle \mathcal{G}_1 \otimes \mathcal{G}_2^* \rangle$ can be written similarly to the Bethe-Salpeter equation (1.5) for a monochromatic wave field, but now one has to supply quantities by additional subscripts 1 or 2 which relate to frequencies

$$\Omega_{1,2} = \pm\Omega + \omega/2 + i0 \tag{2.2}$$

The two following 'plus' and 'minus' equations are simple algebraic consequences of the Dyson and the two-frequency Bethe-Salpeter equations (see problem 2.1):

$$
\begin{aligned}
&(L_1 \otimes I \pm I \otimes L_2)F_{12} \\
&\quad = \{M_1 \otimes I \pm I \otimes M_2 + (I \otimes G_2 \pm G_1 \otimes I) \cdot K_{12}\} F_{12} \\
&\quad +I \otimes G_2 \pm G_1 \otimes I
\end{aligned} \tag{2.3}
$$

The 'minus' equation is known as the Bethe-Salpeter equation in a 'kinetic' form. We have already met this kinetic equation in the transport (1.7) and non-uncoupled (1.23) forms.

2.1. SPECTRAL DECOMPOSITION OF POYNTING'S THEOREM

The Poynting's theorem in the two-frequency representation becomes (Barabanenkov *et al.*, 1995; see also problem 2.2)

$$-i\omega(W_\epsilon + W_\mu) + \nabla \cdot \mathbf{\Pi} + \Theta = 0 \tag{2.4}$$

where the spectral densities W_ϵ and W_μ of electric and magnetic energy, the Poynting vector $\mathbf{\Pi}$, and the power density Θ of the field energy absorption, including the work of the field source, are given by

$$W_\epsilon = \frac{1}{4}\epsilon \mathbf{E}_1 \cdot \mathbf{E}_2 \tag{2.5}$$

$$W_\mu = \frac{1}{4}\mu \mathbf{H}_1 \cdot \mathbf{H}_2 \tag{2.6}$$

$$\mathbf{\Pi} = \frac{1}{4}(\mathbf{E}_1 \times \mathbf{H}_2 + \mathbf{E}_2 \times \mathbf{H}_1) \tag{2.7}$$

$$\Theta = \frac{1}{4}(\mathbf{E}_1 \cdot \mathbf{j}_2 + \mathbf{E}_2 \cdot \mathbf{j}_1) \tag{2.8}$$

Here the current density $\mathbf{j} = \sigma\mathbf{E} + \mathbf{j}^{src}$ consists of the conductivity part $\sigma\mathbf{E}$, resulting from the scatterer conductivity $\sigma(\mathbf{r})$, and the electromagnetic field source part $\mathbf{j}^{src}$.

Independently of (2.4), one can obtain the following 'reduced' but exact version of the Poynting's theorem (see problem 2.3):

$$-i\omega w + \nabla \cdot \mathbf{S} + Q = 0 \tag{2.9}$$

where

$$w = \frac{1}{2}\epsilon \mathbf{E}_1 \cdot \mathbf{E}_2, \tag{2.10}$$

$$\mathbf{S} = \frac{1}{4i\Omega\mu}\left[-\mathbf{E}_1 \times (\nabla \times \mathbf{E}_2) + \mathbf{E}_2 \times (\nabla \times \mathbf{E}_1)\right] \tag{2.11}$$

$$Q = \frac{1}{4\Omega}(-\Omega_2 \mathbf{E}_1 \cdot \mathbf{j}_2 + \Omega_1 \mathbf{E}_2 \cdot \mathbf{j}_1) \tag{2.12}$$

The reduced Poynting's theorem (2.9) includes $w = 2W_\epsilon$ that can be termed the spectral density of the electromagnetic field energy. Let us introduce a tensor $f_{\alpha\beta}(\mathbf{p}; \mathbf{q}, \omega)$ with the meaning of the spatio-temporal spectral density of the coherence function of the electric field of the wave:

$$\begin{aligned} &\left\langle E_\alpha \left(\mathbf{R} + \frac{\mathbf{r}}{2}, \Omega_1\right) E_\beta \left(\mathbf{R} - \frac{\mathbf{r}}{2}, \Omega_2\right)\right\rangle \\ &\qquad = \int_{\mathbf{p}} \int_{\mathbf{q}} \exp\left[i\left(\mathbf{q}\mathbf{R} + \mathbf{p}\,\mathbf{r}\right)\right] f_{\alpha\beta}(\mathbf{p}; \mathbf{q}, \omega) \end{aligned} \tag{2.13}$$

with Greek subscripts $\alpha, \beta = 1, 2, 3$ referring to Cartesian projections of vectors. One may verify that the 'reduced' Poynting's theorem (2.9) gives a good approximation for the Poynting's theorem (2.4) in a range of momenta $\mathbf{p}$, $\mathbf{q}$ and frequencies Ω, ω obeying the conditions

$$q \ll p \sim k_0, \qquad |\omega|/\Omega \ll 1 \tag{2.14}$$

Usually, a random wave field that satisfies these conditions is referred as weakly inhomogeneous in space and quasi-monochromatic in time (Apresyan and Kravtsov, 1983), with the 'fast' variables $\mathbf{p}$ and Ω relating to carrier momentum and frequency of a pulse propagating in the medium, and the 'slow' variables $\mathbf{q}$ and ω relating to the pulse envelope (Lagendijk and Van Tiggelen, 1996).

2.2. WARD IDENTITY AND POYNTING'S THEOREM

The 'plus' and 'minus' equations (2.3) enable us to derive two physically useful identities called the pre-Ward identities (see problem 2.4):

$$\langle (V_1 \hat{\otimes} I \pm I \hat{\otimes} V_2) \mathcal{G}_1 \otimes \mathcal{G}_2 \rangle = A_{12}^{\pm} F_{12} \tag{2.15}$$

where the 'plus' and 'minus' operators $A_{12}^{\pm}$ are defined in terms of the mass and intensity operators and the averaged Green function by

$$A_{12}^{\pm} = M_1 \hat{\otimes} I \pm I \hat{\otimes} M_2 \pm (G_1 \hat{\otimes} I \pm I \hat{\otimes} G_2) K_{12} \tag{2.16}$$

It is reminded here that a tensor product sign with a hat means that the first points of the tensor product (1.3) coincide.

The pre-Ward identities (2.15) have several worthy consequences. First of all, they lead to a key Ward identity (Barabanenkov and Ozrin, 1995; see also problem 2.5):

$$g_2 M_1 \hat{\otimes} I - I \hat{\otimes} g_1 M_2 = (G_1 \hat{\otimes} I g_1 - g_2 I \hat{\otimes} G_2) K_{12} \tag{2.17}$$

In the case of one frequency, $g_1 = g_2$, and non-absorbing scatterers, this Ward identity coincides with the optical theorem (1.6) since $M_2 = M_1^*(\Omega - \omega/2 + i0)$ and similarly for G_2. The Ward identity (2.17) is directly related to the reduced but exact Poynting's theorem (2.9) (see problem 2.6).

Consider the 'plus' pre-Ward identity (2.15) which by the separation property (2.1) of the scattering potential takes the form

$$\langle v\mathbf{E}_1 \otimes \mathbf{E}_2 \rangle = A^+ \langle \mathbf{E}_1 \otimes \mathbf{E}_2 \rangle \tag{2.18}$$

where the operator A^+ is given by

$$A^+ = \frac{1}{g_1 + g_2} A_{12}^+ \tag{2.19}$$

Remembering the definition of the factor $v(\mathbf{r})$, one can see that the operator $(-A^+)$ defines the difference between the averaged spectral density of electric energy inside the scatterers (the energy accumulated inside the scatterers due to their polarization) and the one of the background. The same operator defines the averaged terms (2.10) and (2.12) of the spectral density of the field energy and the field energy absorption per unit volume in the 'reduced' Poynting's theorem because of the equations

$$\langle w \rangle = \frac{1}{2} \epsilon^{bac} \left[(I \hat{\otimes} I - A^+) \langle \mathbf{E}_1 \otimes \mathbf{E}_2 \rangle \right]_{\alpha\alpha} \tag{2.20}$$

and

$$\langle Q^{con} \rangle = -\frac{1}{2} \frac{\epsilon^{bac} \sigma_1}{\epsilon'} \left(A^+ \langle \mathbf{E}_1 \otimes \mathbf{E}_2 \rangle \right)_{\alpha\alpha} \tag{2.21}$$

the repeated Greek subscripts implying summation.

The energetic balance of the Ward identity (2.17) becomes physically most transparent in the spatial Fourier representation using the Wigner variables as in (1.20):

$$\begin{aligned} \triangle M_{\alpha\beta,\alpha'\beta'}(\mathbf{p};\mathbf{q},\omega) \quad &- \quad \int_{\mathbf{p}'} \triangle G_{\alpha\beta,\alpha''\beta''}(\mathbf{p}';\mathbf{q},\omega) K_{\alpha''\beta'',\alpha'\beta'}(\mathbf{p}',\mathbf{p};\mathbf{q},\omega) \\ &= \quad \frac{1}{2i} (g_1 - g_2) A^+_{\alpha\beta,\alpha'\beta'}(\mathbf{p};\mathbf{q},\omega) \end{aligned} \tag{2.22}$$

The first tensor in the l.h.s. of this identity is defined by

$$\triangle M_{\alpha\beta,\alpha'\beta'}(\mathbf{p};\mathbf{q},\omega) = \frac{1}{2i} \left[M_{\alpha\alpha'}(\mathbf{p}_+,\Omega_1)\delta_{\beta\beta'} - \delta_{\alpha\alpha'} M_{\beta\beta'}(-\mathbf{p}_-,\Omega_2) \right] \tag{2.23}$$

with $\mathbf{p}_\pm = \mathbf{p} \pm \mathbf{q}/2$. The tensor (2.23) may be called the 'dynamic extinction tensor' in contrast to the static extinction coefficient (1.17); the tensor

$\Delta G_{\alpha\beta,\alpha'\beta'}(\mathbf{p};\mathbf{q},\omega)$ in the r.h.s. has the same structure with M being replaced by G. Bearing in mind the definition of $g(\Omega)$, one gets $(g_1 - g_2)/2i$ $= (\Omega/C_0^2)(\sigma_1/\epsilon' - i\omega)$. Now, one may consider the Ward identity (2.22) as an identity describing the dynamic extinction of the averaged field as resulting from the partially-coherent scattering of the field and active and reactive energy absorption and accumulation inside scatterers, respectively.

2.3. MESOSCOPIC KINETIC EQUATION WITH ENERGY ACCUMULATION

Let us return to the two-frequency Bethe-Salpeter equation in the kinetic form (2.3). Performing the spatial Fourier transform in terms of Wigner variables and using the Ward identity (2.22), one gets

$$\begin{aligned}
&\left\{-i\frac{\Omega\omega}{C_0^2}\left[\delta_{\alpha\alpha'}\delta_{\beta\beta'} - A^+_{\alpha\beta,\alpha'\beta'}(\mathbf{p};\mathbf{q},\omega)\right] - i\mathcal{P}_{\alpha\beta,\alpha'\beta'}(\mathbf{p};\mathbf{q})\right\} f_{\alpha'\beta'}(\mathbf{p};\mathbf{q},\omega)\\
&\quad = \frac{\Omega}{C_0}\frac{\sigma_1}{C_0\epsilon'}A^+_{\alpha\beta,\alpha'\beta'}(\mathbf{p};\mathbf{q},\omega)f_{\alpha'\beta'}(\mathbf{p};\mathbf{q},\omega)\\
&\quad + \int_{\mathbf{p}'}\Delta G_{\alpha\beta,\alpha'\beta'}(\mathbf{p}';\mathbf{q},\omega)K_{\alpha'\beta',\alpha''\beta''}(\mathbf{p}',\mathbf{p};\mathbf{q},\omega)f_{\alpha''\beta''}(\mathbf{p};\mathbf{q},\omega)\\
&\quad -\Delta G_{\alpha\beta,\alpha'\beta'}(\mathbf{p};\mathbf{q},\omega)\int_{\mathbf{p}'}K_{\alpha'\beta',\alpha''\beta''}(\mathbf{p},\mathbf{p}';\mathbf{q},\omega)f_{\alpha''\beta''}(\mathbf{p}';\mathbf{q},\omega)\\
&\quad -\Delta G_{\alpha\beta,\alpha'\beta'}(\mathbf{p};\mathbf{q},\omega)J_{\alpha'\beta'}(\mathbf{p};\mathbf{q},\omega) \qquad (2.24)
\end{aligned}$$

The tensor $(-i)\mathcal{P}_{\alpha\beta,\alpha'\beta'}(\mathbf{p};\mathbf{q})$ has the structure of the tensor (2.23) with $M_{\alpha\beta}(\mathbf{p}_\pm,\Omega_{1,2})$ being replaced by $L^0_{\alpha\beta}(\mathbf{p}_\pm)$. The equation (2.24) is an exact kinetic equation for the spatio-temporal spectral density $f_{\alpha\beta}(\mathbf{p};\mathbf{q},\omega)$ defined in (2.13). All energetic quantities entering the averaged reduced Poynting's theorem (2.9) may be written in terms of the spatio-temporal spectral density of the coherence function of the electric field of the wave, as can be seen from the following equations:

$$\langle w(\mathbf{R},\omega)\rangle = \frac{1}{2}\epsilon^{bac}\int_{\mathbf{p}}\int_{\mathbf{q}} e^{i\mathbf{q}\cdot\mathbf{R}}\left[\delta_{\alpha'\beta'} - A^+_{\alpha\alpha,\alpha'\beta'}(\mathbf{p};\mathbf{q},\omega)\right] f_{\alpha'\beta'}(\mathbf{p};\mathbf{q},\omega) \qquad (2.25)$$

$$\langle S_\alpha(\mathbf{R},\omega)\rangle = \frac{1}{2\Omega\mu}\int_{\mathbf{p}}\int_{\mathbf{q}} e^{i\mathbf{q}\cdot\mathbf{R}}C_{\alpha'\beta',\alpha}(\mathbf{p};\mathbf{q})f_{\alpha'\beta'}(\mathbf{p};\mathbf{q},\omega) \qquad (2.26)$$

$$\langle Q^{con}(\mathbf{R},\omega)\rangle = -\frac{1}{2}\frac{\epsilon^{bac}\sigma_1}{\epsilon'}\int_{\mathbf{p}}\int_{\mathbf{q}} e^{i\mathbf{q}\cdot\mathbf{R}}A^+_{\alpha\alpha,\alpha'\beta'}(\mathbf{p};\mathbf{q},\omega)f_{\alpha'\beta'}(\mathbf{p};\mathbf{q},\omega) \qquad (2.27)$$

The tensor $C_{\alpha\beta,\alpha}(\mathbf{p};\mathbf{q})$ in equation (2.26) for the averaged Poynting vector is defined by

$$C_{\alpha\beta,\gamma}(\mathbf{p};\mathbf{q}) = \delta_{\alpha\beta}p_\gamma - \frac{1}{2}(p^-_\alpha\delta_{\beta\gamma} + p^+_\beta\delta_{\alpha\gamma}) \qquad (2.28)$$

Multiplying the kinetic equation (2.24) by $\exp(i\mathbf{q}\cdot\mathbf{R})$, integrating with respect to $\mathbf{p}$ and $\mathbf{q}$, and making use of the sum with respect to $\alpha=\beta$, shows (see problem 2.7) the correspondence between Eq. (2.24) and the averaged reduced Poynting's theorem (2.9). It is worth noting that the ensemble averaged spectral density of the electromagnetic field energy (2.25) consists of two parts (see also Lagendijk and Van Tiggelen, 1996):

$$\langle w(\mathbf{R},\omega)\rangle = \langle w_{outside}(\mathbf{R},\omega)\rangle + \langle w_{inside}(\mathbf{R},\omega)\rangle \tag{2.29}$$

The first part, $\langle w_{outside}(\mathbf{R},\omega)\rangle$, related to the first term with the Kronecker symbol in the brackets on the r.h.s. of (2.25), is the averaged spectral density of the field energy of the background. The second part, $\langle w_{inside}(\mathbf{R},\omega)\rangle$, related to the second term with the operator A^{+}, is the averaged spectral density of the field energy inside the scatterers.

3. Diffusion Approach to Mesoscopic Kinetic Equation

In this section, we intend to apply the exact kinetic equation (2.24) to consider a quasi-monochromatic pulse evolution on a large spatio-temporal scale in an infinite dielectric random medium. We assume that the system is not in the localized regime, and therefore that diffusive behavior of the averaged energy density is expected to occur at long times and large distances from sources. In this case, a spatio-temporal spectral density $F_{\alpha\beta,\alpha'\beta'}(\mathbf{p},\mathbf{p}';\mathbf{q},\omega)$ of the coherence function of the wave field, defined as in (1.20) and (2.13), should be singular when the slow variables $\mathbf{q}$ and ω tend to zero: $\mathbf{q},\omega\to 0$. This singularity was investigated (Barabanenkov and Ozrin, 1995) by solving the kinetic equation (2.24) in terms of the eigenfunctions which satisfy the homogeneous equation (2.24) with eigenvalues $i\Omega\omega/C_0^2=\lambda_n(q,\omega)$.

According to the paper cited above, the diffusion asymptotics of the spectral density of the coherence function has the form

$$F_{\alpha\beta,\alpha'\beta'}(\mathbf{p},\mathbf{p}';\mathbf{q},\omega)\simeq\frac{C_0^2}{\Omega}\frac{f_{\alpha\beta}(\mathbf{p},\mathbf{q})f_{\alpha'\beta'}(\mathbf{p}',\mathbf{q})}{Dq^2-i\omega} \tag{3.1}$$

with

$$\sqrt{\pi N}f_{\alpha\beta}(\mathbf{p},\mathbf{q})=-\operatorname{Im}G_{\alpha\beta}(\mathbf{p})-iq_\gamma f_{\alpha\beta,\gamma}(\mathbf{p}) \tag{3.2}$$

and

$$f_{\alpha\beta,\gamma}(\mathbf{p})=\lim_{\omega\to 0}\int_{\mathbf{p}'}F_{\alpha\beta,\alpha'\beta'}(\mathbf{p},\mathbf{p}';\mathbf{0},\omega)C_{\alpha'\beta',\gamma}(\mathbf{p}')-\frac{\partial}{2\partial p_\gamma}\operatorname{Re}G_{\alpha\beta}(\mathbf{p}) \tag{3.3}$$

The quantity N in the l.h.s. of (3.2) is given by $N=N_0(1+a)$, where

$$\pi N_0=-\int_{\mathbf{p}}\operatorname{Im}G_{\alpha\alpha}(\mathbf{p}) \tag{3.4}$$

and the coefficient a describing the energy accumulation inside the scatterers is

$$a = \frac{1}{\pi N_0} \int_{\mathbf{p}} A^+_{\gamma\gamma,\alpha\beta}(\mathbf{p}; \mathbf{0}, 0) \mathrm{Im}\, G_{\alpha\beta}(\mathbf{p}) \tag{3.5}$$

The diffusion constant D takes the form $D = D_K/N$ with

$$D_K = \frac{C_0^2}{\Omega} \frac{1}{3\pi} \int_{\mathbf{p}} C_{\alpha\beta,\gamma}(\mathbf{p}) f_{\alpha\beta,\gamma}(\mathbf{p}) \tag{3.6}$$

where a tensor $C_{\alpha\beta,\gamma}(\mathbf{p})$ coincides with (2.28) at $\mathbf{q} = 0$.

There are two physically meaningful representations for the coefficient a. The first one follows from definition (2.16) of the operator A^+_{12} using (2.19) and (3.5) and has the form (see problem 3.1):

$$a = \frac{1}{k_0^2} \frac{1}{\pi N_0} \int_{\mathbf{p}} \mathrm{Im}\, [M_{\alpha\beta}(\mathbf{p}) G_{\alpha\beta}(\mathbf{p})] \tag{3.7}$$

that allows to write the equation for N as

$$\pi N = -\frac{1}{k_0^2} \int_{\mathbf{p}} \mathrm{Im} \left\{ [\delta_{\alpha\beta} \frac{\Omega^2}{C_0^2} - M_{\alpha\beta}(\mathbf{p})] G_{\alpha\beta}(\mathbf{p}) \right\} \tag{3.8}$$

The representation (3.7) has been used in the scalar case (Van Tiggelen and Lagendijk, 1994) to study delay effects in the model of induced dipole-dipole interactions. Note that by Dyson equation (1.4), the factor in brackets can be replaced by $\delta_{\alpha\beta} p^2 - p_\alpha p_\beta$. In the scalar case, it then seems natural to use (Barabanenkov and Ozrin, 1992) the delta-approximation (5.19) for the imaginary part of the spatial Fourier transform of the averaged Green function on the r.h.s. of (3.8). However, this approximation appears to be a rather rough one, as noted by Van Tiggelen *et al.* (1993) and Barabanenkov and Ozrin (1993).

Alternatively, in the spirit of the approach developed by Lagendijk and Van Tiggelen (1996), the coefficient a can be treated in terms of inelastic scattering by substituting the Ward identity (2.22) into the r.h.s. of (3.5):

$$\begin{aligned} a &= \frac{1}{k_0^2} \frac{1}{\pi N_0} \frac{\epsilon'}{\mathrm{Im}\, \hat{\epsilon}_1} \lim_{\mathrm{Im}\, \hat{\epsilon}_1 \to 0} \int_{\mathbf{p}} [\mathrm{Im}\, M_{\alpha\beta}(\mathbf{p}) \\ &- \int_{\mathbf{p}'} \mathrm{Im}\, G_{\alpha'\beta'}(\mathbf{p}') K_{\alpha'\beta',\alpha\beta}(\mathbf{p}', \mathbf{p}; \mathbf{0}, 0)] \mathrm{Im}\, G_{\alpha\beta}(\mathbf{p}) \end{aligned} \tag{3.9}$$

where $\hat{\epsilon}_1 = \epsilon_1 + i\sigma_1/\Omega$ is the complex dielectric permittivity of a single conducting scatterer. In the low density limit this formula takes the form (Van Tiggelen *et al.*, 1992; see also problem 3.2):

$$a = \frac{\epsilon'}{k_0} \lim_{\mathrm{Im}\, \hat{\epsilon}_1 \to 0} \frac{n\sigma_a}{\mathrm{Im}\, \hat{\epsilon}_1} \tag{3.10}$$

with n the number density of scatterers and σ_a the absorption cross section (Van de Hulst, 1981) of a single scatterer.

3.1. GREEN-KUBO FORMULA FOR THE DIFFUSION CONSTANT

One can verify that the terms D_K and N in the expression for the diffusion constant D are associated with the energy flux correlation function and with the spectral density of states of the electromagnetic wave field. To see this, one can consider the eigenfunctions $\mathbf{E}^{(j)}(\mathbf{r})$ for the eigenvalues $\Omega = \Omega_j$ which obey the vector Helmholtz equation inside a volume $\mathcal{V}$:

$$L^0 E^{(j)} + \Omega_j^2 \epsilon\mu E^{(j)} = 0 \tag{3.11}$$

with an appropriate boundary condition on the surface of $\mathcal{V}$. Here we use matrix notation. Now the expressions for D_K and N become (see problem 3.3):

$$\begin{aligned} D_K &= \frac{2\pi}{3}\frac{C_0^2}{\Omega}\mu^2 \int d\mathbf{R} \int_0^\infty dt\, e^{-\eta t} \\ &\times \left\langle \sum_{ij} [S_\alpha^{(ij)}(\mathbf{R},t) S_\alpha^{(ji)}(\mathbf{0},0) + S_\alpha^{(ij)}(\mathbf{0},0) S_\alpha^{(ji)}(\mathbf{R},t)]\delta(\Omega-\Omega_i) \right\rangle \end{aligned} \tag{3.12}$$

where $\eta \to 0$ and (see problem 3.4)

$$N = \left\langle \frac{C_0^2}{\mathcal{V}} \sum_i \delta(\Omega^2 - \Omega_i^2) \right\rangle \tag{3.13}$$

In equation (3.12), $\mathbf{S}^{(ij)}(\mathbf{R},t)$ denotes the matrix elements of the Poynting vector (2.11), evaluated with the aid of the above eigenfunctions.

3.2. ENERGY REDISTRIBUTION

As one may readily verify, the diffusion asymptotics (3.1) satisfies the ensemble averaged Poynting's theorem (2.9) and Fick's law is valid: $\langle \mathbf{S}(\mathbf{R},\omega)\rangle = -D\nabla \langle w(\mathbf{R},\omega)\rangle$. The part of the energy density $\langle w_{inside}(\mathbf{R},t)\rangle$ accumulated inside the scatterers due to their polarization is defined by the coefficient a of the energy accumulation inside scatterers, given by equation (3.5). This part of the energy density and a are related by

$$\langle w_{inside}(\mathbf{R},t)\rangle = \frac{a}{1+a}\langle w(\mathbf{R},t)\rangle \tag{3.14}$$

3.3. RADIATIVE TRANSFER LIMIT FOR THE DIFFUSION CONSTANT

It is worth noting that the diffusion asymptotics (3.1) has been derived without using radiative transfer theory. Nevertheless, one can consider the radiative transfer limit for the exact diffusion constant D in the diffusion asymptotics (3.1). In this limit, we disregard the second term on the r.h.s. of (3.3) (including the momentum derivative of the real part of the spatial Fourier transform of the averaged tensor Green function), and apply the transverse field approximation as in the discussion of problem 3.2. However, we now use a more accurate δ-approximation for the imaginary part of the transverse component of the averaged tensor Green function:

$$\operatorname{Im} G_{tr}(p) \simeq -\frac{\pi}{2}\frac{\delta(p - k_{eff})}{k_0 n_{eff}} \tag{3.15}$$

Here the effective real wave number k_{eff} in the random medium is the root of the equation

$$k^2 - k_0^2 + \operatorname{Re} M_{tr}(k) = 0 \tag{3.16}$$

and a quantity $n_{eff} = (k_{eff}/k_0)[1 + \partial \operatorname{Re} M_{tr}(k)/\partial(k^2)]$, with the derivative being taken at $k = k_{eff}$. Under these assumptions, the diffusion constant takes the form

$$D = \frac{V_E l_{tr}}{3} \tag{3.17}$$

In this well-known formula, obtained first by Van Albada *et al.* (1992), the energy transport velocity is defined by $V_E = C_0^2/(1 + a)C_{ph}$, where $C_{ph} = \Omega/k_{eff}$ is the phase velocity in the random medium. The transport mean free path l_{tr} is given by $1/l_{tr} = (1 - \langle\cos\theta\rangle)/l$ with the mean free path l defined, in contrast to the scalar case (1.17), as $1/l = -\operatorname{Im} M_{tr}(k_{eff})/k_{eff}$, and $\langle\cos\theta\rangle$ is the mean cosine of the single scattering angle for scattering on an effective inhomogeneity.

4. Radiative Transfer Approach to Mesoscopic Kinetic Equation

The aim of this section is to apply the exact kinetic equation (2.24) to a quasi-monochromatic pulse evolution on an arbitrary spatio-temporal scale in a semi-infinite dielectric resonant random medium. We intend to reach this aim with the aid of an improved version of the Sobolev radiative transfer equation (1.39), derived from the exact kinetic equation under the following approximations:

(i) Quasi-monochromatic approximation
(ii) Slow spatial variation of the radiance within the scatterer radius
(iii) Allowing strong delay at single scattering events
(iv) Transverse field approximation

(v) Far-field approximation for scattered fields
(vi) Low-density limit
(vii) Resonant expansion of Mie single-scattering multipole coefficients

The approximations (ii) and (v) have already been used for the macroscopic derivation of the stationary radiative transfer equation (1.16) in the discussion of problem 1.10. The approximations (iv) and (v) have been used to get the low-density limit (3.10) for the coefficient a of the energy accumulation inside the scatterers and the radiative transfer limit (3.17) for the diffusion constant. The approximations (iii) and (vii) are specific for the study of a quasi-monochromatic pulse scattering in a dielectric resonant random medium.

4.1. IMPROVED SOBOLEV EQUATION WITH LORENTZIAN DELAY KERNELS

Approximations (i)–(v) lead to a general modified radiative transfer equation with delay (Barabanenkov and Barabanenkov, 1997) which is similar in form to the kinetic equation (2.24). Additional application of the approximations (vi) and (vii), the last approximation meaning in the simplest case to address the resonant point-like scatterer model (Nieuwenhuizen *et al.*, 1992), yields, after the inverse Laplace transform with respect to the slow frequency variable ω, the following improved Sobolev radiative transfer equation with three Lorentzian delay kernels for a quasi-monochromatic pulse with carrier frequency Ω:

$$\frac{c_{ph}}{c_0^2}\frac{\partial}{\partial t}\left[I_{\alpha\beta}(\mathbf{s};\mathbf{R},t)+a\int_0^t\frac{dt'}{t_{delay}}\mathrm{e}^{-(t-t')/t_{delay}}I_{\alpha\beta}(\mathbf{s};\mathbf{R},t')\right]$$
$$+(\mathbf{s}\cdot\nabla_{\mathbf{R}})I_{\alpha\beta}(\mathbf{s};\mathbf{R},t)=-\frac{1}{l}\int_0^t\frac{dt'}{t_{delay}}\mathrm{e}^{-(t-t')/t_{delay}}I_{\alpha\beta}(\mathbf{s};\mathbf{R},t')$$
$$+\int_0^t\frac{dt'}{t_{delay}}\int_{4\pi}d\mathbf{s}'\mathrm{e}^{-(t-t')/t_{delay}}W_{\alpha\beta,\alpha'\beta'}(\mathbf{s},\mathbf{s}')I_{\alpha'\beta'}(\mathbf{s}';\mathbf{R},t') \qquad (4.1)$$

where $I_{\alpha\beta}(\mathbf{s};\mathbf{R},t)$ is the radiance tensor. Only the magnetic dipole resonant multiple scattering effects have been accounted for in this equation. A quantity $W_{\alpha\beta,\alpha'\beta'}(\mathbf{s},\mathbf{s}')$ denotes the static Rayleigh phase tensor

$$W_{\alpha\beta,\alpha'\beta'}(\mathbf{s},\mathbf{s}')=\frac{1}{(4\pi)^2}na_{\alpha\alpha'}(\mathbf{s},\mathbf{s}')a^*_{\beta\beta'}(\mathbf{s},\mathbf{s}') \qquad (4.2)$$

where the single-scattering Rayleigh amplitude tensor $a_{\alpha\beta}(\mathbf{s},\mathbf{s}')=\tilde{t}(\Omega)$ $P^{tr}_{\alpha\mu}(\mathbf{s})P^{tr}_{\mu\beta}(\mathbf{s}')$ is defined using the $\tilde{t}$-matrix of a point-like scatterer with the tensor scattering operator $T_{\alpha\beta}(\mathbf{r},\mathbf{r}';\Omega)=\tilde{t}(\Omega)\delta(\mathbf{r})\delta(\mathbf{r}')\delta_{\alpha\beta}$. The $\tilde{t}$-matrix is expressed by Mie magnetic dipole scattering coefficient b_1 (Van de Hulst, 1981): $\tilde{t}(\Omega) = (6\pi/ik_0)b_1^*(\Omega)$. Near the resonant frequency Ω_r, defined for a non-absorbing scatterer with radius r_0 by the equation $\pi = (\epsilon_1/\epsilon^{bac})^{1/2}(\Omega_r/C_0)r_0$, the $\tilde{t}$-matrix takes the resonant form

$$\tilde{t}(\Omega) = \frac{-4\pi\kappa k_0^2}{1 - \frac{k_0^2}{k_r^2} - \frac{2}{3}i\kappa k_0^3} \tag{4.3}$$

Here $k_r = \Omega_r/C_0$ and $\kappa = (3/\pi^2)r_0^3$. One can rewrite the resonant $\tilde{t}$-matrix (4.3) as follows:

$$\tilde{t}(\Omega) = \frac{\tilde{g}(\Omega)}{1 - \tilde{g}(\Omega)B(\Omega)} \tag{4.4}$$

where $\tilde{g}(\Omega) = -4\pi g(\Omega)\kappa$ and $B(\Omega) = -(\Lambda + ik_0)/6\pi$ with $\Lambda = \epsilon_1/2r_0$. Comparison with resonant point-like scatterer models (Nieuwenhuizen *et al.*, 1992; Barabanenkov and Barabanenkov, 1997) shows that Λ is similar to a cut-off parameter to regularize the resonant point-like scatterer model. The representation (4.4) leads immediately to the relations

$$\tilde{a}_{12}^{\pm} \equiv \tilde{t}_1 \pm \tilde{t}_2 \pm (B_1 \pm B_2)\tilde{t}_1\tilde{t}_2 = \frac{\tilde{g}_1 \pm \tilde{g}_2}{\tilde{g}_1\tilde{g}_2}\tilde{t}_1\tilde{t}_2 \tag{4.5}$$

and then to

$$\tilde{a}_{12}^{-} = \frac{g_1 - g_2}{g_1 + g_2}\tilde{a}_{12}^{+} \tag{4.6}$$

This is the Ward identity (5.30) applied to a resonant point-like scatterer. The quantity

$$\tilde{a}^{+} = \frac{1}{g_1 + g_2}\tilde{a}_{12}^{+} \tag{4.7}$$

defines the energy accumulated inside a point-like scatterer due to its polarization and is similar to the operator (2.19). The coefficient a of the energy accumulation inside the scatterers (3.5) takes in the case of resonant point-like scatterers the form $a = -n\tilde{a}^{+}(\omega = 0)$. The time delay t_{delay} at resonant point-like scattering is defined by the approximate expression for dynamic scattering cross section of the point-like scatterer:

$$\frac{\tilde{t}_1\tilde{t}_2}{|\tilde{t}|^2} \cong \frac{1}{1 - i\omega t_{delay}} \tag{4.8}$$

and equals

$$t_{delay} = \frac{4r_e}{3c_0}\left[4\left(\frac{\Omega - \Omega_r}{\Omega_r}\right)^2 + \left(\frac{2r_e}{3c_0}\Omega_r\right)^2\right]^{-1} \tag{4.9}$$

where $r_e = 3/(2\Lambda)$. As can be readily verified using equation (4.8), the delay time t_{delay} coincides with the Wigner delay time $\partial \arg \tilde{t}/\partial\Omega$ which was first introduced in the quantum scattering theory (Wigner, 1955) and is widely discussed in the context of resonant multiple scattering of light (Lagendijk and Van Tiggelen, 1996; Kogan and Kaveh, 1992). The relation

$$\frac{\tilde{a}^+}{\tilde{a}^+(\omega=0)} = \frac{\Omega^4}{\Omega_1^2\Omega_2^2}\frac{\tilde{t}_1\tilde{t}_2}{|\tilde{t}|^2} \tag{4.10}$$

together with (4.8) and conditions $\Omega \cong \Omega_r$ and $|\omega| \ll \Omega_r$ gives the expression for the dynamic quantity (4.7) of the energy accumulation inside the point-like scatterer in the form

$$\frac{\tilde{a}^+}{\tilde{a}^+(\omega=0)} \cong \frac{1}{1-i\omega t_{delay}} \tag{4.11}$$

The dynamic expressions (4.8) and (4.11) call for the appearance of Lorentzian kernels in the improved Sobolev modified radiative transfer equation (4.1). The averaged energy density and the Pointing vector then take the form

$$\begin{aligned}\langle w(\mathbf{R},t)\rangle &= \frac{1}{2}\epsilon^{bac}\int_{4\pi} d\mathbf{s}\int_0^t dt'\left[\delta(t-t')+\frac{a}{t_{delay}}\mathrm{e}^{-(t-t')/t_{delay}}\right]\\ &\times\; I_{\alpha\alpha}(\mathbf{s};\mathbf{R},t')\end{aligned} \tag{4.12}$$

$$\langle S(\mathbf{R},t)\rangle = \frac{1}{2}\epsilon^{bac}\frac{c_0^2}{c_{ph}}\int_{4\pi} d\mathbf{s}\,\mathbf{s}\,I_{\alpha\alpha}(\mathbf{s};\mathbf{R},t) \tag{4.13}$$

The improved Sobolev radiative transfer equation (4.1) will be applied in the following (Barabanenkov *et al.*, 1998; Barabanenkov and Barabanenkov, 1999) to study, in a simplified scalar version, the effects of pulse trapping on the pulse energy redistribution and on the dynamic weak localization in reflection from a semi-infinite resonant random medium.

4.1.1. *Pulse reflection from a semi-infinite resonant random medium*
Let a plane monochromatic wave $E_0(k\mathbf{s}_0,\mathbf{r}) = \exp(ik\mathbf{s}_0\mathbf{r})u_0(\Omega)$, $k = \Omega/C_0$, propagating along the direction of the unit vector $\mathbf{s}_0$, be incident on a confined random medium. Denote by $E_{1,2}(\mathbf{s}_0,\mathbf{r})$ the solutions to the Helmholtz equation (1.1) with the boundary conditions

$$E_{1,2}(\mathbf{s}_0,\mathbf{r})\,|_{\mathbf{s}_0\cdot\mathbf{r}\to-\infty} \to E_0(k_{1,2}\mathbf{s}_0,\mathbf{r}) \tag{4.14}$$

with wave numbers $k_{1,2} = \Omega_{1,2}/C_0$. The two-frequency albedo $\Sigma(\mathbf{s},\mathbf{s}_0;\Omega,\omega)$ of the random medium, which characterizes the intensity of the fluctuation component of the scattered field propagating along the direction of the unit vector $\mathbf{s}$, is given by the general formula

$$\Sigma(\mathbf{s}, \mathbf{s}_0; \Omega, \omega) = \frac{1}{(4\pi)^2 L_\perp^2} \int d\mathbf{r}_1 \int d\mathbf{r}_2 \int d\mathbf{r}'_1 \int d\mathbf{r}'_2 \langle E_1(-\mathbf{s}, \mathbf{r}_1)\rangle \\ \times \langle E_2(-\mathbf{s}, \mathbf{r}_2)\rangle \Gamma_{12}(\mathbf{r}_1, \mathbf{r}_2; \mathbf{r}'_1, \mathbf{r}'_2) \langle E_1(\mathbf{s}_0, \mathbf{r}'_1)\rangle \langle E_2(\mathbf{s}_0, \mathbf{r}'_2)\rangle \tag{4.15}$$

where $L_\perp^2$ is a geometric cross section of the medium and $\Gamma_{12}(\mathbf{r}_1, \mathbf{r}_2; \mathbf{r}'_1, \mathbf{r}'_2)$ is the two-frequency vertex function for the two-frequency Bethe-Salpeter equation.

Formula (4.15) is exact. We are interested in the single-group approximation for the mass (1.8) and intensity (1.9) operators denoted by $M^{(L)}$ and $K_{12}^{(L)}$ and refer to this as the 'ladder' approximation. The two-frequency intensity operator in the ladder approximation satisfies, similar to the one-frequency case (Barabanenkov, 1973), a specific reciprocity equation in the form

$$K_{12}^{(L)}(\mathbf{r}_1, \mathbf{r}_2; \mathbf{r}'_1, \mathbf{r}'_2) = K_{12}^{(L)}(\mathbf{r}'_1, \mathbf{r}_2; \mathbf{r}_1, \mathbf{r}'_2) = K_{12}^{(L)}(\mathbf{r}_1, \mathbf{r}'_2; \mathbf{r}'_1, \mathbf{r}_2) \tag{4.16}$$

The usual reciprocity relation for the ensemble-averaged Green function $G^{(L)}$ in the ladder approximation and the specific reciprocity relation for the two-frequency intensity operator K_{12}^L enable us to state an equivalence relation (Barabanenkov, 1973) between cyclical (maximally crossed) and ladder diagrams upon inversion of the points on either upper or lower diagram line. This brings us to the conclusion that the cyclical diagrams $K_{12}^{(C)}$ can be expressed in terms of the ladder diagrams $\Gamma_{12}^{(L)}$ in the following way:

$$K_{12}^{(C)}(\mathbf{r}_1, \mathbf{r}_2; \mathbf{r}'_1, \mathbf{r}'_2) = \Gamma_{12}^{(L)}(\mathbf{r}'_1, \mathbf{r}_2; \mathbf{r}_1, \mathbf{r}'_2) = \Gamma_{12}^{(L)}(\mathbf{r}_1, \mathbf{r}'_2; \mathbf{r}'_1, \mathbf{r}_2) \tag{4.17}$$

In turn, the vertex function $\Gamma_{12}^{(L)}$ is expressed in terms of the Green function $F_{12}^{(L)}$ of the two-frequency Bethe-Salpeter equation, taken in the ladder approximation, and the intensity operator $K_{12}^{(L)}$ as

$$\Gamma_{12}^{(L)} = K_{12}^{(L)} F_{12}^{(L)} K_{12}^{(L)} \tag{4.18}$$

where operator notation is used.

To sum up, the full vertex function can be written in the form

$$\Gamma_{12} \simeq K_{12}^{(L)} + \Gamma_{12}^{(L)} + K_{12}^{(C)} \tag{4.19}$$

Substituting this approximate equation to the r.h.s. of (4.15) gives

$$\Sigma \simeq \Sigma^{(1)} + \Sigma^{(L)} + \Sigma^{(C)} \tag{4.20}$$

where the first, second, and third terms on the r.h.s. correspond to the first, second, and third terms on the r.h.s. of (4.19) and account for the contributions of incoherent single scattering, ladder, and cyclical diagrams, respectively, to the two-frequency albedo.

Our next aim is to express these contributions in terms of the solution of a scalar version of the improved Sobolev modified radiative transfer equation (4.1). Returning to the slow frequency variable ω, one can write for the propagator $F(\mathbf{s}, \mathbf{s}'; \mathbf{R}, \mathbf{R}', \omega)$ of this equation the following equation, analogous to the static radiative transfer equation (1.16):

$$\left[1 + \frac{l}{f(\tilde{\omega})}(\mathbf{s} \cdot \nabla_{\mathbf{R}})\right] F(\mathbf{s}, \mathbf{s}'; \mathbf{R}, \mathbf{R}', \omega) \tag{4.21}$$

$$= \lambda(\tilde{\omega}) \int_{4\pi} \frac{d\mathbf{s}''}{4\pi} F(\mathbf{s}'', \mathbf{s}'; \mathbf{R}, \mathbf{R}', \omega) + \frac{1}{(4\pi)^2} \frac{l}{f(\tilde{\omega})} \delta(\mathbf{R} - \mathbf{R}')\delta(\mathbf{s} - \mathbf{s}')$$

where $\tilde{\omega} = \omega t_0$ is a dimensionless slow frequency variable with a mean free time $t_0 = lC_{ph}/C_0^2$ in terms of the phase velocity of the random medium. An effective single scattering albedo $\lambda(\tilde{\omega})$ and a function $f(\tilde{\omega})$ are

$$\lambda(\tilde{\omega}) = \frac{1}{1 - i\tilde{\omega}(1 + a - i\tilde{\omega}a_{kav})} \tag{4.22}$$

and

$$f(\tilde{\omega}) = \frac{1}{(1 - i\tilde{\omega}a_{kav})\lambda(\tilde{\omega})} \tag{4.23}$$

with the dimensionless parameter $a_{kav} = t_{delay}/t_0$ describing single-scattering delay. Starting with the Ward identity (4.6), one may verify (Barabanenkov and Barabanenkov, 1997) the following relation between the parameter a_{kav} and the coefficient a of energy accumulation inside scatterers: $a_{kav} = a + n\partial Re\tilde{t}/\partial E$ with $E = k_0^2$. This relation can be rewritten as $a + 1 = C_0^2/(C_{ph}v_{coh}) + a_{kav}$, where the energy transport velocity of the coherent component v_{coh} satisfies the equation $C_0^2/(C_{ph}v_{coh}) = 1 - n\partial Re\tilde{t}/\partial E$ and practically coincides with the group velocity v_g. The above relation between a_{kav} and a enables one to represent the energy transport velocity V_E in a physically transparent form: $V_E = l/(t_{delay} + t_l)$, where $t_l = l/v_{coh}$ is the mean free time. This expression for the energy transport velocity has been discussed in several papers (Holstein, 1947; Van Tiggelen *et al.*, 1992; Barabanenkov and Barabanenkov, 1997).

As one can see, the modified radiative transfer equation (4.21) contains two dimensionless physical parameters: the coefficient a, describing energy accumulation inside a scatterer due to its polarization, and the parameter a_{kav}, describing the single-scattering delay. These parameters are depicted in Fig. 4 versus the carrier frequency Ω. The r.h.s. of equation (4.15)

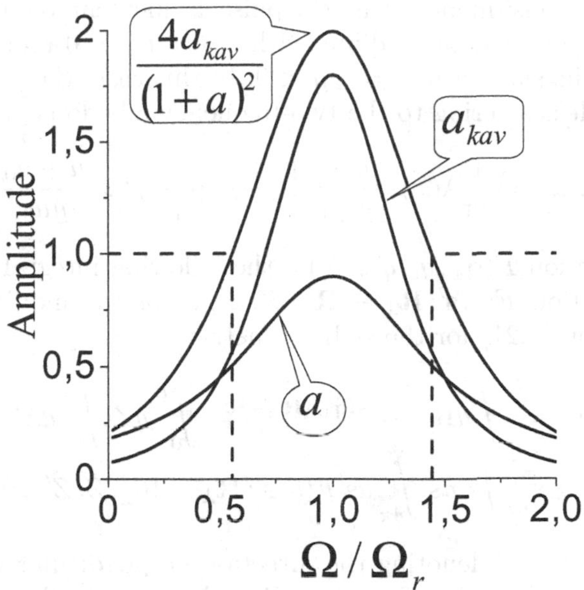

Figure 4. The dependence of the coefficient a (lower curve), a_{kav} (middle curve), and the ratio $4a_{kav}/(1+a)^2$ (upper curve) on the pulse carrier frequency Ω. The resonance frequency Ω_r of a scatterer of radius r_0 and dielectric permittivity $\epsilon_1/\epsilon^{bac} = \pi^2$, is $\Omega_r = C_0/r_0$. The packing fraction of scatterers is $n4\pi r_0^3/3 = 0.3$.

contains a product of ensemble averaged fields for direction $\mathbf{s}_0$ and a similar product for the direction $(-\mathbf{s})$. Both products may be calculated with the aid of the modified radiative transfer equation (4.21) with zero r.h.s. Assuming that the random medium is a plane-parallel slab occupying the region $0 < z < L$, one finds

$$\langle E_1(\mathbf{s},\mathbf{r}_1)\rangle\langle E_2(\mathbf{s},\mathbf{r}_2)\rangle \cong \exp\left[i\,k_{eff}\,\mathbf{s}\cdot(\mathbf{r}_1-\mathbf{r}_2)\right]\exp\left[-\frac{f(\tilde{\omega})}{|s_z|}\frac{z_1+z_2}{2l}\right] \tag{4.24}$$

In order to use equation (4.18), it is also necessary to give an expression for the two-frequency intensity operator. According to expression (4.8) for the dynamic scattering cross section of a point-like scatterer, the intensity operator after spatial Fourier transformation becomes

$$K^{(L)}(\mathbf{p},\mathbf{p}';\mathbf{q},\omega) \cong \frac{4\pi}{l}\,\frac{1}{1-i\omega t_{delay}} \tag{4.25}$$

We are now ready to represent the contributions of the incoherent single scattering, ladder, and cyclical diagrams to the two-frequency albedo in terms of the solutions of the modified radiative transfer equation (4.21). Assume that a quasi-monochromatic pulse is incident on the boundary $z = 0$ of a slab of random medium with $s_{0z} = \mu_0 > 0$ and is diffusively reflected from the slab with $s_z = -\mu < 0$. In this case, the contribution of incoherent single scattering to the two-frequency albedo is given by

$$\Sigma^{(1)}(\mathbf{s}, \mathbf{s}_0; \Omega, \omega) = \frac{1}{4\pi}\lambda(\tilde{\omega})\frac{\mu\mu_0}{\mu + \mu_0}\left\{1 - \exp\left[-f(\tilde{\omega})\frac{\mu + \mu_0}{\mu\mu_0}\frac{L}{l}\right]\right\} \tag{4.26}$$

Let the function $F(\mathbf{q}_\perp, q_z, q'_z, \omega)$ be the following integral transform of the Green function $F(\mathbf{s}, \mathbf{s}'; \mathbf{R}_\perp - \mathbf{R}'_\perp, Z, Z', \omega)$ of the modified radiative transfer equation (4.21) for the slab geometry:

$$\begin{aligned} F(\mathbf{q}_\perp, q_z, q'_z, \omega) &= \int d\mathbf{R}_\perp \cdot \mathrm{e}^{-i\mathbf{q}_\perp \cdot (\mathbf{R}_\perp - \mathbf{R}'_\perp)} \int_0^L dZ \int_0^L dZ' \cdot \mathrm{e}^{i(q_z Z + q'_z Z')} \\ &\times \int_{4\pi} d\mathbf{s} \int_{4\pi} d\mathbf{s}' F(\mathbf{s}, \mathbf{s}'; \mathbf{R}_\perp - \mathbf{R}'_\perp, Z, Z', \omega) \end{aligned} \tag{4.27}$$

with the subscript '$\perp$' denoting the direction perpendicular to the z axis. The contributions of the ladder and cyclical diagrams to the two-frequency albedo then may be written as

$$\Sigma^{(L)}(\mathbf{s}, \mathbf{s}_0; \Omega, \omega) = \frac{1}{l^2}\left(\frac{1}{1 - i\tilde{\omega}a_{kav}}\right)^2 F\left[\mathbf{0}, \frac{if(\tilde{\omega})}{\mu l}, \frac{if(\tilde{\omega})}{\mu_0 l}, \omega\right] \tag{4.28}$$

and

$$\Sigma^{(C)}(\mathbf{s}, \mathbf{s}_0; \Omega, \omega) = \frac{1}{l^2}\left(\frac{1}{1 - i\tilde{\omega}a_{kav}}\right)^2 F\left[-\frac{\boldsymbol{\tau}}{l}, \frac{if(\tilde{\omega})}{lm_+}, \frac{if(\tilde{\omega})}{lm_-}, \omega\right] \tag{4.29}$$

where the vector $\boldsymbol{\tau}$ and the parameters $m_\pm$ are defined by

$$\boldsymbol{\tau} = k_{eff}\, l(\mathbf{s}_\perp + \mathbf{s}_{0\perp}) \tag{4.30}$$

and

$$\frac{1}{m_\pm} = \pm\frac{k_{eff}\, l(-\mu + \mu_0)}{if(\tilde{\omega})} + \frac{\mu + \mu_0}{2\mu\mu_0} \tag{4.31}$$

For the 'exactly backward' scattering direction $\mathbf{s} = -\mathbf{s}_0$, both expressions (4.28) and (4.29) coincide in accordance with the equivalence relation for cyclical and ladder diagrams (Barabanenkov, 1973).

To obtain exact analytic expressions for the contributions of ladder (4.28) and cyclical (4.29) diagrams to the two-frequency albedo, we shall consider from now on a semi-infinite resonant random medium: $L \to \infty$. In this case, the integral transform (4.27) of the propagator of the modified radiative transfer equation (4.21) can be calculated exactly by applying the Sobolev method (Sobolev, 1956) for solution of an integral equation with a kernel depending on variable difference, and may be written in terms of the generalized Chandrasekhar H-function

$$H(m,\lambda|\rho) = \exp\left[-\frac{m}{\pi}\int_0^\infty \frac{dx}{1+m^2x^2}\ln\left(1-\lambda\frac{\operatorname{arctg}\sqrt{\rho^2+x^2}}{\sqrt{\rho^2+x^2}}\right)\right] \quad (4.32)$$

For $\rho = 0$ this function coincides with the 'ordinary' Chandrasekhar H-function (Chandrasekhar, 1960). The mentioned solution of the modified radiative transfer equation (4.21) gives the desired analytic expressions for ladder and cyclical diagrams contributing to the two-frequency albedo:

$$\Sigma^{(incoh)}(\mathbf{s},\mathbf{s}_0;\Omega,\omega) = \frac{1}{4\pi}\frac{\mu\mu_0}{\mu+\mu_0}\lambda(\tilde{\omega})H\left[\mu,\lambda(\tilde{\omega})|0\right]H\left[\mu_0,\lambda(\tilde{\omega})|0\right] \quad (4.33)$$

$$\begin{aligned}\Sigma^{(C)}(\mathbf{s},\mathbf{s}_0;\Omega,\omega) &= \frac{1}{4\pi}\frac{\mu\mu_0}{\mu+\mu_0}\lambda(\tilde{\omega}) \\ &\times \left\{H\left[m_+,\lambda(\tilde{\omega})|\frac{\tau}{f(\tilde{\omega})}\right]H\left[m_-,\lambda(\tilde{\omega})|\frac{\tau}{f(\tilde{\omega})}\right]-1\right\}\end{aligned} \quad (4.34)$$

where $\Sigma^{(incoh)}$ is the sum of incoherent single scattering (4.26) and the ladder diagrams (4.28), and $\tau = |\boldsymbol{\tau}|$.

4.1.2. *Pulse energy redistribution in reflection from a semi-infinite resonant random medium*

Consider the sum (4.33) of incoherent single scattering and ladder diagrams to study the redistribution of pulse energy when it is reflected from a semi-infinite resonant random medium. We assume that the duration Δt_0 of the incident quasi-monochromatic pulse is shorter than the mean free time t_0 and therefore the evolution of the reflected pulse may be described by the dynamic incoherent albedo $\Sigma^{incoh}(\mu,\mu_0;\tilde{t})$ obtained with the aid of the inverse Laplace transform of (4.33) with respect to the slow frequency ω in favor of the dimensionless temporal variable $\tilde{t} = t/t_0$. The exact result of this inverse Laplace transform is depicted in Fig. 5 (curve 1). In the limit of small $\tilde{t}$, one can apply the single-scattering approximation, which leads to (curve 2 in Fig. 5)

$$\Sigma^{incoh}(\mu,\mu_0;\tilde{t}) \to \frac{\mu\mu_0}{\mu+\mu_0}\lambda(\tilde{t}) \quad (4.35)$$

Here the function

$$\lambda(\tilde{t}) = \frac{e^{\nu_1\tilde{t}}-e^{\nu_2\tilde{t}}}{a_{kav}(\nu_1-\nu_2)} \quad (4.36)$$

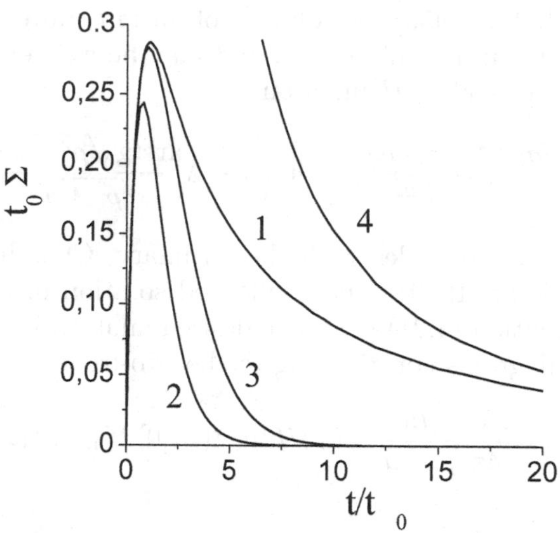

Figure 5. Dependence of the dimensionless dynamic albedo $t_0\Sigma(1,1;\tilde{t})$ on $\tilde{t}$ in different approximations for $a = a^* = 0.51$ and $a_{kav} = a^*_{kav} = 0.57$: exact value (1), single- and double- scattering approximations (2 and 3), diffusion asymptotics (4). The packing fraction is 0.3 as in figure 4.

is the inverse temporal Laplace transform of the complex single-scattering albedo (4.22) with the poles $i\nu_1$ and $i\nu_2$ given by

$$\nu_{1,2} = -\frac{1+a}{2a_{kav}}\left[1 \mp \sqrt{1 - \frac{4a_{kav}}{(1+a)^2}}\right] \tag{4.37}$$

Equation (4.36) explicitly reveals the energy redistribution from the front part towards the rear part of the reflected pulse with increasing the parameters a and a_{kav}. As a matter of fact, $\lambda(\tilde{t}) \to \exp(-\tilde{t})$ and, consequently, $\lambda(\tilde{t}=0) \to 1$ as $a \to 0$ and $a_{kav} \to 0$. But in the general case one gets

$$\lambda(\tilde{t}) \to \frac{\tilde{t}}{a_{kav}} \tag{4.38}$$

as $\tilde{t} \to 0$ and, consequently, $\lambda(\tilde{t}=0) = 0$, while the integral of $\lambda(\tilde{t})$ with respect to $\tilde{t}$ from zero to infinity remains constant and equal to unity. This means that in the single-scattering approximation, an 'energy centroid' $\tilde{t}_\lambda$ of the diffusely reflected pulse

$$\tilde{t}_\lambda = \int_0^\infty d\tilde{t}\,\tilde{t}\,\lambda(\tilde{t}) \Big/ \int_0^\infty d\tilde{t}\,\lambda(\tilde{t}) \tag{4.39}$$

is progressively shifted towards the positive direction of the time axis with increasing a and a_{kav}. In fact, substitution of (4.36) into the r.h.s. of (4.39) gives

$$\tilde{t}_\lambda = 1 + a \tag{4.40}$$

that is the 'energy centroid' of the diffusely reflected pulse is defined by the coefficient a, describing energy accumulation inside a scatterer due to its polarization.

In the limit of large $\tilde{t}$, the dynamic albedo can be evaluated by applying the diffusion approximation that gives (curve 4 in Fig. 5)

$$\Sigma^{incoh}(\mu, \mu_0; \tilde{t}) \to \frac{1}{4\pi}\frac{1}{2}\sqrt{\frac{3}{\pi}} H(\mu, 1|0) H(\mu_0, 1|0) \mu\mu_0 \frac{\sqrt{1+a}}{\tilde{t}^{3/2}} \tag{4.41}$$

as $\tilde{t} \to \infty$. We see from (4.41) that the energy of the diffusely reflected pulse is indeed redistributed towards the rear of the pulse as the coefficient a is increased, i.e. as the trapping of pulse energy inside the scatterers becomes more significant.

4.1.3. *Dynamic weak localization in reflection from a semi-infinite resonant random medium*

Let us now apply the result (4.34) for the cyclical diagrams to study dynamic weak localization of a short pulse in reflection from a semi-infinite resonant random medium. This effect may be described by the dynamic coherent albedo $\Sigma^{(C)}(\mu, \mu_0; \tilde{t})$ obtained from the inverse Laplace transform of (4.34) with respect to ω. We consider the limit of a large value of the temporal variable $\tilde{t}$ that facilitates the following asymptotics of the generalized Chandrasekhar H-function:

$$H(m, \lambda|\rho) \to H(m, 1|0) \left[1 - m\sqrt{3(1-\lambda) + \rho^2}\right] \tag{4.42}$$

for $1 - \lambda \to 0$ and $\rho \to 0$. With the aid of this asymptotics and by summing the incoherent Σ^{incoh} and the coherent $\Sigma^{(C)}$ albedo, we get for the total dynamic albedo in the limit of $\tilde{t} \to \infty$ the following simple result:

$$\begin{aligned} \Sigma(\mathbf{s}, \mathbf{s}_0; \tilde{t}) \quad &\to \quad \frac{1}{4\pi}\frac{1}{2}\sqrt{\frac{3}{\pi}} H^2(\mu_0, 1|0)\ \mu_0^2 \frac{\sqrt{1+a}}{\tilde{t}^{3/2}} \\ &\times \quad \left\{1 + \exp\left[-\frac{\tau^2 \tilde{t}}{3(1+a)}\right]\right\} \end{aligned} \tag{4.43}$$

where $\boldsymbol{\tau}$ is defined by Eq. (4.30). One can introduce the backscattering enhancement factor $\eta(\mathbf{s}, \mathbf{s}_0; \tilde{t})$ as a ratio

$$\eta(\mathbf{s}, \mathbf{s}_0; \tilde{t}) = \Sigma(\mathbf{s}, \mathbf{s}_0; \tilde{t}) / \Sigma^{(incoh)}(\mu, \mu_0; \tilde{t}) \tag{4.44}$$

of total dynamic albedo to its incoherent part, the latter coinciding with the value of the total dynamic albedo on the plateau, i.e. for a large value of τ. In the limit of large $\tilde{t}$, the enhancement factor takes the form

$$\eta(\mathbf{s}, \mathbf{s}_0; \tilde{t}) \to 1 + \exp\left[-\frac{\tau^2 \tilde{t}}{3(1+a)}\right] \tag{4.45}$$

For normal incidence of a pulse on a semi-infinite resonant random medium, when $\tau \simeq k_{eff}\, l\theta$ and $\theta = |\mathbf{s}_\perp|$, the exponential term on the r.h.s. of (4.45) can be rewritten as $\exp(-\theta^2/\theta_t^2)$ with $\theta_t = (1/k_{eff}\, l)\sqrt{3(1+a)/\tilde{t}}$ giving the effective angular width of the coherent backscattering enhancement cone. θ_t can also be rewritten in terms of the diffusion constant D as $\theta_t = (1/k_{eff})(1/\sqrt{Dt})$. As one can see, the angular width of dynamic coherent backscattering cone increases with increase of the coefficient a, describing the energy accumulation inside a scatterer. This result is not surprising. Indeed, let us consider the interference between two scattered waves traveling through the same set of scatterers but in opposite directions. The phase shift $\Delta\Phi$ between these waves may be estimated as (Barabanenkov *et al.*, 1991) $\Delta\Phi = k_{eff}\theta\, |\mathbf{r}_1 - \mathbf{r}_2|$, where $\mathbf{r}_1$ and $\mathbf{r}_2$ are positions of the first and last scattering events. We now estimate the distance between $\mathbf{r}_1$ and $\mathbf{r}_2$ using the Einstein diffusion law: $|\mathbf{r}_1 - \mathbf{r}_2| \simeq (Dt)^{1/2}$, where D is the diffusion constant (3.17) of the resonant random medium. Assuming $\Delta\Phi$ to be of the order of unity: $\Delta\Phi \simeq 1$, we obtain the above formula for θ_t.

5. Discussion and Conclusions

Multiple scattering of an electromagnetic pulse by an ensemble of resonant dielectric particles is studied with the aid of the diffusion and radiative transfer approaches, both of which remain phenomenological to a certain degree. The diffusion approach describes the pulse evolution on a large spatio-temporal scale and yields some useful general expressions, such as those for the density of states, the coefficient of energy accumulation inside scatterers due to their polarization, the correlation function of energy flux, and the diffusion constant written in terms of these expressions. The radiative transfer approach aims to describe the pulse evolution on an arbitrary spatio-temporal scale. In the framework of the resonant point-like scatterer model, radiative transfer allows us to study effects of resonant scattering on pulse delay and dynamic weak localization of a quasi-monochromatic pulse in reflection from a semi-infinite resonant random medium. On temporal scales large compared to the mean free time, these new effects are determined by the coefficient of energy accumulation inside the resonant scatterers and are consistent with the results of the diffusion approach. It is also interesting to study the pulse evolution on small and intermediate spatio-temporal scales, when the resonant single scattering, and not multiple scattering, governs peculiarities of the time evolution. For example, the reflection of a short pulse from a semi-infinite resonant random medium is described by the single scattering time delay, according to subsection 4.1.2. In addition, equation (4.37) shows that in single scattering the delay parameter a_{kav} and the coefficient a of energy accumulation inside a scatterer should satisfy the specific inequality $4a_{kav}/(1+a)^2 \leq 1$ which is difficult to anticipate beforehand.

Acknowledgements

I am grateful to the Scientific Directors Dr. B.A. van Tiggelen and Dr. S.E. Skipetrov and to the Scientific Committee for the invitation to give this lecture at the NATO ASI 'Wave Scattering in Complex Media: From Theory to Applications' (Cargèse, Corsica, France, June 10–12, 2002).

I am also grateful to Dr. M.Yu. Barabanenkov for the help in preparation of this material for publication.

Appendix: Problems

1.1. Generalize the tensor product (1.3) for the case of two tensor operators $A_{\alpha\alpha'}(\mathbf{r},\mathbf{r}')$ and $B_{\beta\beta'}(\mathbf{r},\mathbf{r}')$ with Greek subscripts $\alpha, \beta, \ldots = 1,2,3$ enumerating Cartesian projections of vectors, and bearing in mind the direct tensor product of matrices (see, e.g., Lankaster, 1969). Prove the simple algebraic relation $(A\hat{\otimes}B)(C\otimes D) = AC\hat{\otimes}BD$ introduced in the main text between equations (1.3) and (1.4). Prove the trivial but important identity

$$(v\hat{\otimes}I - I\hat{\otimes}v)A\otimes B = 0 \tag{5.1}$$

where $v = v(\mathbf{r})$ is an arbitrary function and I denotes the tensor unit operator with kernel $\delta_{\alpha\alpha'}\delta(\mathbf{r}-\mathbf{r}')$, $\delta_{\alpha\alpha'}$ being the Kronecker symbol.

1.2. Derive the Bethe-Salpeter equation in transport form (1.7) as an algebraic consequence of the Dyson (1.4) and Bethe-Salpeter (1.5) equations. Starting with (1.7) and using the optical theorem (1.6), derive the following energy conservation equation

$$\nabla\langle\mathbf{S}\rangle + \langle Q\rangle = 0 \tag{5.2}$$

where $\mathbf{S} = -E\nabla E^* + E^*\nabla E$ and $Q = EJ^* - E^*J$ are the vector of energy flux and the power density of the source, respectively.

1.3. By substituting the expansion of the scattering operator $T_{1,2}$ in multiplicity of scattering events:

$$T_{1,2} = T_1 + T_2 + T_1G_0T_2 + T_2G_0T_1 + \cdots \tag{5.3}$$

into the second terms on the r.h.s. of equations (1.8) and (1.9), and by disregarding all terms with higher order correlation functions of scatterer centers, derive the following expressions:

$$M_1 = T_1g_1 + T_1G_0T_2g_2(1,2) + \cdots \tag{5.4}$$

$$K_1 = T_1\otimes T_1^*g_1(1) + (T_1\otimes T_2^* + T_1\otimes T_1^*G_0^*T_2^* + T_1G_0T_2\otimes T_2^*)\,g_2(1,2) + \cdots \tag{5.5}$$

for the mass and intensity operators M_1 and K_1, respectively. Here integration is assumed over $\mathbf{r}_1$ and $\mathbf{r}_2$ denoted by 1 and 2, respectively. Show that the derived expressions for M_1 and K_1

become, if the scattering operators T_1 and T_2 are expanded in the scattering potential,

$$M_1 = (V_1 + V_1 G_0 V_1) g_1 + V_1 G_0 V_2 g_2(1,2) + \cdots \quad (5.6)$$
$$K_1 = V_1 \otimes V_1 g_1(1) + V_1 \otimes V_2 g_2(1,2) + \cdots \quad (5.7)$$

respectively. Equations (5.6) and (5.7) are valid in the limit of weak scattering.

1.4. By bearing in mind the symmetric property of the functions $\psi_s(1,\cdots,s)$ and by using the following group expansion:

$$\psi_n(1,\cdots,n) = \sum_{s=1}^{n} \frac{1}{s!} \sum_{j_1,\cdots,j_s=1}^{n}{}' \psi_s^{gr}(j_1,\cdots,j_s) \quad (5.8)$$

for $n = 1,2,3,\cdots$, derive the group operation (1.11) as an inversion of the group expansion. The primed sum (second sum in Eq. (5.8)) is over mutually different $j_1,\cdots,j_s$. By applying the above group expansion to the Green function $\mathcal{G}$ and by using the inversion of the Dyson and Bethe-Salpeter equations written in the form

$$M = G_0^{-1} - G^{-1}, \qquad K = (G \otimes G^*)^{-1} - F^{-1} \quad (5.9)$$

derive the single-group approximations (1.8) and (1.9) for the mass and intensity operators, respectively. Note that the distribution function of scatterers $f_n(\mathbf{r}_1,\cdots,\mathbf{r}_n)$ is expressed in terms of their correlation functions (Frisch, 1965) as

$$f_n(\mathbf{r}_1,\cdots,\mathbf{r}_n) = \sum_{\nu}\sum_{div} g_{s_1}(\Gamma_1)\cdots g_{s_\nu}(\Gamma_\nu) \quad (5.10)$$

where the sum is over all possible sets of groups $\Gamma_1,\cdots,\Gamma_\nu$ into which the set of points $\mathbf{r}_1,\cdots,\mathbf{r}_n$ can be split. The small parameter γ can be introduced in Eq. (5.10) by replacing $g_s(\Gamma) \to \gamma g_s(\Gamma)$. The single-group approximation for M_1 and K_1 is then given by the first terms of expansions of M and K in power series in γ.

1.5. Use the optical theorem for the scattering operator of a set of n non-absorbing scatterers:

$$T_{1,\cdots,n}\hat{\otimes} I - I\hat{\otimes} T^*_{1,\cdots,n} = (G_0\hat{\otimes} I - I\hat{\otimes} G_0^*) T_{1,\cdots,n} \otimes T^*_{1,\cdots,n} \quad (5.11)$$

to verify that in the single-group approximation, the mass M_1 and intensity K_1 operators, as well as their simplified forms obtained in problem 1.3, satisfy the optical theorem (1.6) in the first order of the small parameter γ.

1.6. Let us consider the Dyson equation (1.4) with mass operator M_1 in the single-group approximation, and let us represent its solution in the form of an iterative series in powers of M_1. Consider, for example, the third term of this series $G_0 M_1 G_0 M_1 G_0$ depicted by the diagram (a) of Fig. 1. We assume integration over the repeated arguments $\mathbf{r}_1, \mathbf{r}_2, \mathbf{r}_3$, and $\mathbf{r}_4$ of operators. First, we may carry out the integration over $\mathbf{r}_4$, assuming all other points to be fixed. By virtue of the compactness of the mass operator $M_1(\mathbf{r}_4 - \mathbf{r}_3)$, $\mathbf{r}_4$ as well as $\mathbf{r}_3$ are localized inside an inhomogeneity with a linear dimension of the order of $r_0 + r_{cor}$. If the point of observation $\mathbf{r}$ lies in the region of Fraunhoffer diffraction of this inhomogeneity, the Green function $G_0(\mathbf{r} - \mathbf{r}_4)$ can be approximated by

$$G_0(\mathbf{r} - \mathbf{r}_4) \cong G_0(\mathbf{r} - \mathbf{r}_4)\exp\left[-ik_0 \mathbf{s}_{\mathbf{r}\mathbf{r}_3}(\mathbf{r}_4 - \mathbf{r}_3)\right] \quad (5.12)$$

where the unit vector $\mathbf{s}_{\mathbf{r}\mathbf{r}_3}$ is directed along $\mathbf{r} - \mathbf{r}_3$. Proceeding analogously, we integrate over $\mathbf{r}_2$ and obtain for the iterative term under consideration the expression $M_1^2(k_0) G_0 G_0 G_0$. By summing all the terms of the iterative series transformed in this way, derive the Green function (1.13).

1.7. Consider the Bethe-Salpeter equation (1.5) with mass M_1 and intensity K_1 operators in the single-group approximation. Represent its solution in the form of an iterative series in powers of the specific intensity operator K_0. Consider, for example, the third term of the series:

$$(G \otimes G^*) K_0 (G \otimes G^*) K_0 (G \otimes G^*) \tag{5.13}$$

This term is depicted by the diagram (b) of Fig. 1. Upper and lower lines of the diagram are associated with the operators G and G^*, respectively. The circles with four points are associated with operators K_0, and the fifth point inside the circles is associated with the center of the inhomogeneity. The diagram describes the successive, partially coherent scattering of waves by two inhomogeneities. Substitute the Green function (1.13) in (5.13) and consider the region of Fraunhoffer diffraction, as in problem 1.6. Derive, after integrating the above expressions over all repeated arguments, the following equation:

$$\begin{aligned} &\int d\mathbf{R}_1 \int d\mathbf{R}_2 G(\mathbf{r}_1 - \mathbf{R}_1) G^*(\mathbf{r}_2 - \mathbf{R}_1) \tilde{K}_0 \left(k_0 \mathbf{s}_{\mathbf{r}_1 \mathbf{R}_1}, k_0 \mathbf{s}_{\mathbf{r}_2 \mathbf{R}_1}; k_0 \mathbf{s}_{\mathbf{R}_1 \mathbf{R}_2}, k_0 \mathbf{s}_{\mathbf{R}_1 \mathbf{R}_2} \right) \\ &\times |G(\mathbf{R}_1 - \mathbf{R}_2)|^2 \tilde{K}_0 \left(k_0 \mathbf{s}_{\mathbf{R}_1 \mathbf{R}_2}, k_0 \mathbf{s}_{\mathbf{R}_1 \mathbf{R}_2}; k_0 \mathbf{s}_{\mathbf{R}_2 \mathbf{r}'_1}, k_0\, \mathbf{s}_{\mathbf{R}_2 \mathbf{r}'_2} \right) G(\mathbf{R}_2 - \mathbf{r}'_1) G^*(\mathbf{R}_2 - \mathbf{r}'_2) \end{aligned} \tag{5.14}$$

where $\tilde{K}_0$ is the spatial Fourier transform $K_0(\mathbf{p}_1, \mathbf{p}_2; \mathbf{p}'_1, \mathbf{p}'_2)$ of the specific intensity operator K_0, with all momenta $\mathbf{p}_1$, $\mathbf{p}_2$, $\mathbf{p}'_1$ and $\mathbf{p}'_2$ on the sphere $p = k_0$. Further transformation of (5.14) can be carried out by introducing operators

$$\tilde{F}_0(\mathbf{R}, \mathbf{s}; \mathbf{R}', \mathbf{s}') = \left| G(\mathbf{R} - \mathbf{R}') \right|^2 \delta(\mathbf{s} - \mathbf{s}_{\mathbf{R}\mathbf{R}'}) \delta(\mathbf{s} - \mathbf{s}') \tag{5.15}$$

and

$$\tau(\mathbf{R}, \mathbf{s}; \mathbf{R}', \mathbf{s}') = \delta(\mathbf{R} - \mathbf{R}') \tilde{K}_0(k_0 \mathbf{s}, k_0 \mathbf{s}; k_0 \mathbf{s}', k_0 \mathbf{s}') \tag{5.16}$$

acting on functions of vector $\mathbf{R}$ and unit vector $\mathbf{s}$, and an operator

$$\mathcal{R}(\mathbf{R}, \mathbf{s}; r'_1, r'_2) = \tilde{K}_0(k_0 \mathbf{s}, k_0 \mathbf{s}; k_0 \mathbf{s}_{\mathbf{r}\mathbf{r}'_1}, k_0 \mathbf{s}_{\mathbf{r}\mathbf{r}'_2}) G(\mathbf{R} - \mathbf{r}'_1) G^*(\mathbf{R} - \mathbf{r}'_2) \tag{5.17}$$

transforming a function of $\mathbf{r}'_1$ and $\mathbf{r}'_2$ into a function of vector $\mathbf{R}$ and unit vector $\mathbf{s}$. Show that (5.14) can be rewritten as $\mathcal{R}^T \tilde{F}_0 \tau \tilde{F}_0 \mathcal{R}$, where $\mathcal{R}^T$ denotes a transpose of $\mathcal{R}$. $\mathcal{R}^T$ transforms a function of vector $\mathbf{R}$ and unit vector $\mathbf{s}$ into a function of $\mathbf{r}_1$ and $\mathbf{r}_2$. Transforming other terms in the iterative series for solution of the Bethe-Salpeter equation (1.5) similar to (5.13), show that the sum of all terms may be written as

$$F = G \otimes G^* + (G \otimes G^*) \tilde{K}_0 (G \otimes G^*) + \mathcal{R}^T \tilde{F} \mathcal{R} \tag{5.18}$$

Here $\tilde{F}(\mathbf{R}, \mathbf{s}; \mathbf{R}', \mathbf{s}')$ is the propagator of the radiative transfer equation (1.16).

1.8. Recall that the imaginary part of the spatial Fourier transform of the averaged Green function (1.13) is a sharp function given approximately by

$$\operatorname{Im} G(p) \simeq -\frac{\pi}{2} \frac{\delta(p - k_0)}{k_0} \tag{5.19}$$

Derive the optical theorem (1.19) for the extinction coefficient and the phase function of the radiative transfer equation from the optical theorem (1.6) simplified by using the single-group approximation for mass and intensity operators.

1.9. Consider the solution of the Bethe-Salpeter equation in the Fraunhoffer approximation, obtained in problem 1.7. Rewrite this solution as $F(\mathbf{R} + \mathbf{r}/2, \mathbf{R} - \mathbf{r}/2; \mathbf{R}' + \mathbf{r}'/2, \mathbf{R}' - \mathbf{r}'/2)$ and apply the Fraunhoffer expansion like in problem 1.6 with respect to the vectors $\mathbf{r}$ and $\mathbf{r}'$. Derive the representation (1.22) for the spatial spectral density of the field coherence function, including the two δ-functions of momenta on the r.h.s.

1.10. Let us denote by $F(\mathbf{p}, \mathbf{p}'; \mathbf{q})$ the spatial Fourier transform of the field coherence function $F(\mathbf{R}, \mathbf{p}; \mathbf{R}', \mathbf{p}')$ with respect to $\mathbf{R} - \mathbf{R}'$, defined as in (1.20). Rewrite the Bethe–Salpeter equation in the transport form (1.7) but now in terms of the function $F(\mathbf{p}, \mathbf{p}'; \mathbf{q})$. This gives a scalar one-frequency version of the kinetic equation (this equation will appear somewhat later; see (2.24)). Simplify this version of the kinetic equation in an approximate way, treating the slow variable $\mathbf{q}$ as a small parameter and using the approximation (5.19) for the imaginary part of the averaged Green function. Derive, by performing the inverse Fourier transformation with respect to $\mathbf{q}$, the representation (1.22) for the field coherence function and the radiative transfer equation (1.16).

1.11. Starting with the parabolic equation (1.29), show that the non-perturbed and perturbed parts of the Liouville operator in equation (1.30) act according to

$$i\Lambda_0 \rho(\mathbf{p}_1, \mathbf{p}_2) = (p_1^2 - p_2^2)\rho(\mathbf{p}_1, \mathbf{p}_2) \tag{5.20}$$

and

$$i\Lambda(\mathbf{k})\rho(\mathbf{p}_1, \mathbf{p}_2) = \rho(\mathbf{p}_1 - \mathbf{k}, \mathbf{p}_2) - \rho(\mathbf{p}_1, \mathbf{p}_2 + \mathbf{k}) \tag{5.21}$$

with $\Lambda(t)$ defined by

$$\Lambda(t) = \int d\mathbf{k} V(\mathbf{k}, t)\Lambda(\mathbf{k}) \tag{5.22}$$

1.12. Show that the Dyson equation (1.31) for the ensemble averaged density matrix, expressed in the mixed Wigner variables similar to definition (1.21), takes the form

$$\begin{aligned}\left(\frac{\partial}{\partial t} + 2\mathbf{p}\cdot\nabla\right)\rho_1(\mathbf{R}, \mathbf{p}, t) = -2\int_0^t \int d\mathbf{p}' B(\mathbf{p} - \mathbf{p}', t - t') \cos\left[(p^2 - p'^2)t\right] \\ \times \left[\rho_1(\mathbf{R} - (\mathbf{p} + \mathbf{p}')(t - t'), \mathbf{p}, t') - \rho_1(\mathbf{R} - (\mathbf{p} + \mathbf{p}')(t - t'), \mathbf{p}', t')\right]\end{aligned} \tag{5.23}$$

where the 'natural' time variable is used (see main text). Specific effects of time delay and space dispersion may be disregarded in this equation under conditions that, respectively, $\tau_0/t_M \ll 1$ and

$$\frac{|\mathbf{p} + \mathbf{p}'|\,(t - t')}{\Delta R} \sim \frac{p\tau_0}{\Delta R} \to \frac{p}{k_0}\frac{C_0\tau_0}{\Delta R} \ll 1 \tag{5.24}$$

where ΔR is a typical scale of variation of $\rho_1(\mathbf{R}, \mathbf{p}, t)$ with $\mathbf{R}$.

2.1. Derive the 'plus' and 'minus' equations (2.3) as an algebraic consequence of the Dyson and the two-frequency Bethe-Salpeter equations, performing manipulations similar to those in the problem 1.2.

2.2. Starting with Maxwell equations in the temporal Laplace transform representation, derive the two-frequency Poynting's theorem (2.4).

2.3. Acting similar to the case of preceding problem, derive the reduced but exact two-frequency Poynting's theorem (2.9).

2.4. Starting with the Helmholtz equation (1.1), derive the two following relations for the tensor Green function:

$$(L_1 \otimes I - V_1 \otimes I)\mathcal{G}_1 \otimes \mathcal{G}_2 = I \times \mathcal{G}_2 \tag{5.25}$$

$$(I \otimes L_2 - I \otimes V_2)\mathcal{G}_1 \otimes \mathcal{G}_2 = \mathcal{G}_1 \times I \tag{5.26}$$

Upon adding and subtracting these equations with further ensemble averaging and using the 'plus' and 'minus' equations (2.3), derive the pre-Ward identities (2.15).

2.5. Apply the separation property of the scattering potential (2.1) to the pre-Ward identities (2.15) and use the trivial identity (5.1) of problem 1.1, to obtain the relation

$$\mathcal{A}F_{1,2} = 0 \tag{5.27}$$

with

$$\mathcal{A} \equiv A^{-}_{1,2} - \frac{g_1 - g_2}{g_1 + g_2} A^{+}_{1,2} \tag{5.28}$$

By employing the reciprocity of the tensor Green function: $\mathcal{G}_{\alpha\beta}(\mathbf{r},\mathbf{r}';\Omega) = \mathcal{G}_{\beta\alpha}(\mathbf{r}',\mathbf{r};\Omega)$, and by rewriting the Dyson and the two-frequency Bethe-Salpeter equations in a transposed form, transform (5.27) to

$$\mathcal{A} + \mathcal{A}F_{12}K_{12} = 0 \tag{5.29}$$

to give

$$\mathcal{A} = 0 \tag{5.30}$$

Transform the last identity to the form (2.17).

2.6. Use the separation property (2.1) of the scattering potential and the trivial identity (5.1) to prove the identity

$$\tilde{\mathcal{P}} \equiv (V_1 \hat{\otimes} I - I \hat{\otimes} V_2)\mathcal{G}_1 \otimes \mathcal{G}_2 - \frac{g_1 - g_2}{g_1 + g_2}(V_1 \hat{\otimes} I - I \hat{\otimes} V_2)\mathcal{G}_1 \otimes \mathcal{G}_2 = 0 \tag{5.31}$$

By transforming the second equality of this identity using the Helmholtz equation (1.1), derive the relation

$$\frac{1}{4i\Omega\mu}(\tilde{\mathcal{P}}\mathbf{J}_1 \otimes \mathbf{J}_2)_{\alpha\alpha} = \tilde{P} \tag{5.32}$$

Here the source vector is $\mathbf{J}(\mathbf{r},\Omega) = -i\mu\Omega\mathbf{j}^{src}(\mathbf{r},\Omega)$, $\tilde{P}$ denotes the expression on the l.h.s. of the reduced Poynting's theorem (2.9), and repeated Greek subscripts imply summation. Using the pre-Ward identities (2.15), derive from (5.31) and (5.32) the desired relation

$$(\mathcal{A}\langle \mathbf{E}_1 \otimes \mathbf{E}_2 \rangle)_{\alpha\alpha} = 4i\Omega\mu \left\langle \tilde{P} \right\rangle \tag{5.33}$$

where the fourth-rank tensor $\mathcal{A}$ is given by (5.28) and the electric field $\mathbf{E}(\mathbf{r},\Omega)$ is related to the tensor Green function $\mathcal{G}_{\alpha\beta}(\mathbf{r},\mathbf{r}';\Omega)$ and to the source vector $\mathbf{J}(\mathbf{r},\Omega)$ by the symbolic equation $\mathbf{E} = \mathcal{G}\mathbf{J}$.

2.7. By performing the operations mentioned after equation (2.28) and by dividing by $2\Omega\mu$, show from (2.25)–(2.27) that the term in square brackets on the l.h.s. of (2.24) gives $(-i)\omega\langle w \rangle$, and that the term containing the tensor $\mathcal{P}_{\alpha\alpha,\alpha'\beta'}$ gives $\langle \nabla \cdot \mathbf{S} \rangle$ because of the relation

$$q_\gamma C_{\alpha\beta,\gamma}(\mathbf{p},\mathbf{q}) = -\mathcal{P}_{\gamma\gamma,\alpha\beta}(\mathbf{p},\mathbf{q}) \tag{5.34}$$

which should also be proved. Also show that the term containing the tensor $A^{+}_{\alpha\beta,\alpha'\beta'}$ on the r.h.s. of (2.24) gives $(-i)\langle Q^{con} \rangle$, that the next two terms involving the intensity operator cancel each other (and therefore may be thought of as describing elastic scattering), and that the last term gives $(-1)\langle Q^{src} \rangle$ with the power of the source given by

$$\langle Q^{src}(\mathbf{R},\omega) \rangle = \frac{1}{2\Omega\mu}\int_{\mathbf{p}}\int_{\mathbf{q}} e^{i\mathbf{q}\mathbf{R}} \triangle G_{\alpha\alpha,\alpha'\beta'}(\mathbf{p};\mathbf{q},\omega) J_{\alpha'\beta'}(\mathbf{p};\mathbf{q},\omega) \tag{5.35}$$

Here $J_{\alpha\beta}(\mathbf{p};\mathbf{q},\omega) = J_\alpha(\mathbf{p}_+,\Omega_1)J_\beta(-\mathbf{p}_-,\Omega_2)$ and $\mathbf{J}(\mathbf{p},\Omega)$ is the spatial Fourier transform of the source vector $\mathbf{J}(\mathbf{r},\Omega)$.

3.1. Derive the representation (3.7) for the coefficient a, describing the energy accumulation inside scatterers, by substituting the definitions (2.16) and (2.19) of operators A^{+}_{12} and A^{+}, respectively, into the r.h.s. of (3.5) and by using the optical theorem (1.6).

3.2. In the low density limit, the mass M_1 and intensity K_1 operators of the single-group approximation (1.8) and (1.9) coincide with those in the independent scattering approximation and are given by the first terms in the expansions (5.4) and (5.5), respectively. Substituting the independent scattering approximation for M_1 and K_1 into the r.h.s. of (3.9) and taking into account only the transverse components of all tensor quantities (so-called 'transverse field' approximation), making use of the δ-approximation (5.19) for $\operatorname{Im} G_{tr}(p)$, derive the low density representation (3.10) for the coefficient a in terms of the absorption cross section σ_a of a single conducting spherical scatterer. To this end, recall that the solution of the tensor Dyson equation (1.4) may be written (after spatial Fourier transform) in the following matrix form:

$$G(\mathbf{k},\Omega) = P_{tr}(\hat{k})G_{tr}(k,\Omega) + P_l(\hat{k})G_l(k,\Omega). \tag{5.36}$$

Here the transverse $P_{tr}(\hat{k})$ and longitudinal $P_l(\hat{k})$ orthogonal projections on the direction $\hat{k}$ of the wave vector $\mathbf{k}$ are defined by $P^{tr}_{\alpha\beta}(\hat{k}) = \delta_{\alpha\beta} - \hat{k}_\alpha\hat{k}_\beta$ and $P^{l}_{\alpha\beta}(\hat{k}) = \hat{k}_\alpha\hat{k}_\beta$. The transverse $G_{tr}(k,\Omega)$ and longitudinal $G_l(k,\Omega)$ components of the averaged tensor Green function are given by

$$G_{tr}(k,\Omega) = \frac{1}{k_0^2 - k^2 - M_{tr}(k,\Omega)} \tag{5.37}$$

and

$$G_l(k,\Omega) = \frac{1}{k_0^2 - M_l(k,\Omega)} \tag{5.38}$$

with the transverse $M_{tr}(k,\Omega)$ and longitudinal $M_l(k,\Omega)$ components of the mass operator defined as in (5.36). In the transverse field approximation, one encounters, in particular, the tensor single scattering amplitude

$$a_{\alpha\beta}(\mathbf{s},\mathbf{s}';\Omega) = P^{tr}_{\alpha\mu}(\mathbf{s})T_{\mu\nu}(k_0\mathbf{s},k_0\mathbf{s}';\Omega)P^{tr}_{\nu\beta}(\mathbf{s}') \tag{5.39}$$

where $T_{\alpha\beta}(\mathbf{p},\mathbf{p}';\Omega)$ is the tensor single scattering operator in the spatial Fourier-transform representation. The absorption cross section of a single conducting spherical scatterer is defined by

$$2\sigma_a = -\frac{1}{k_0}\operatorname{Im} a_{\alpha\alpha}(\mathbf{s}',\mathbf{s}') - \frac{1}{(4\pi)^2}\int d\mathbf{s} a_{\alpha\beta}(\mathbf{s},\mathbf{s}')a^*_{\alpha\beta}(\mathbf{s},\mathbf{s}') \tag{5.40}$$

In the low density limit and transverse field approximation, the part N_0 of the density of states turns out to be $N_0 = k_0/2\pi^2$.

3.3. Starting with the relations (5.25) and (5.26) for the tensor Green function, derive the pre-Ward identity

$$\left(g_2L_1\hat{\otimes}I - I\hat{\otimes}g_1L_2\right)\langle\mathcal{G}_1\otimes\mathcal{G}_2\rangle = g_2I\hat{\otimes}G_2 - G_1\hat{\otimes}g_1I \tag{5.41}$$

The tensor operators $L_{1,2}$ on the l.h.s. may be replaced by $L^0_{1,2}$. Performing the spatial Fourier transform, show that the second term on the r.h.s. of (3.3) takes the form similar to that of the first term with $F_{\alpha\beta,\alpha'\beta'}(\mathbf{p},\mathbf{p}';\mathbf{0},\omega)$ replaced by

$$\frac{1}{2}\left[F_{\alpha\beta,\alpha'\beta'}(\mathbf{p},\mathbf{p}';\mathbf{0},\Omega+i0,\Omega+i0) + F_{\alpha\beta,\alpha'\beta'}(\mathbf{p},\mathbf{p}';\mathbf{0},-\Omega+i0,-\Omega+i0)\right] \tag{5.42}$$

Substituting this expression into (3.3) and then into (3.6), show that

$$D_K = \frac{C_0^2}{\Omega}\frac{2}{3\pi}\int_{\mathbf{p}} C_{\alpha\beta,\gamma}(\mathbf{p})\Delta F_{\alpha\beta,\alpha'\beta'}(\mathbf{p},\mathbf{p}')C_{\alpha'\beta',\gamma}(\mathbf{p}') \tag{5.43}$$

where $\Delta F_{\alpha\beta,\alpha'\beta'}(\mathbf{p},\mathbf{p}')$ is the spatial Fourier transform (1.20) of

$$\langle \operatorname{Im}\mathcal{G}_{\alpha\alpha'}(\mathbf{R}+\frac{\mathbf{r}}{2},\mathbf{R}'+\frac{\mathbf{r}'}{2};\Omega+i0)\operatorname{Im}\mathcal{G}_{\beta\beta'}(\mathbf{R}-\frac{\mathbf{r}}{2},\mathbf{R}'-\frac{\mathbf{r}'}{2};\Omega+i0)\rangle \tag{5.44}$$

evaluated at $\mathbf{q} = 0$. Further, let us assume that the eigenfunctions of the problem (3.11) obey the orthogonality and completeness conditions:

$$\int d\mathbf{r}\mu\epsilon(\mathbf{r})E_{\alpha}^{(j)*}(\mathbf{r})E_{\alpha}^{(j')}(\mathbf{r}) = \delta_{jj'} \tag{5.45}$$

$$\mu\epsilon(\mathbf{r})\sum_j E_{\alpha}^{(j)*}(\mathbf{r})E_{\alpha'}^{(j)}(\mathbf{r}') = \delta_{\alpha\alpha'}\delta(\mathbf{r}-\mathbf{r}') \tag{5.46}$$

Verify on the basis of these conditions that the imaginary part of the tensor Green function can be represented as

$$\mathrm{Im}\,\mathcal{G}_{\alpha\beta}(\mathbf{r},\mathbf{r}';\Omega+i0) = -\pi\sum_j E_{\alpha}^{(j)}(\mathbf{r})E_{\beta}^{(j)*}(\mathbf{r}')\delta(\Omega^2-\Omega_j^2) \tag{5.47}$$

To obtain equation (3.12) for D_K, one substitutes the expression (5.47) into (5.44) and then into (5.43) and performs simple algebraic manipulations.

3.4. The expression (5.47) for the imaginary part of the tensor Green function and the orthogonality condition (5.45) yield

$$\begin{aligned}\langle\mu\epsilon(\mathbf{r})\mathrm{Im}\,\mathcal{G}_{\alpha\alpha}(\mathbf{r},\mathbf{r};\Omega+i0)\rangle &\equiv \left\langle\frac{1}{\mathcal{V}}\int d\mathbf{r}\mu\epsilon(\mathbf{r})\mathrm{Im}\,\mathcal{G}_{\alpha\alpha}(\mathbf{r},\mathbf{r};\Omega+i0)\right\rangle \\ &= -\pi\left\langle\frac{1}{\mathcal{V}}\sum_i\delta(\Omega^2-\Omega_j^2)\right\rangle\end{aligned} \tag{5.48}$$

We recall that the ensemble averaging implies thermodynamic limit for a finite system and restores translational invariance. Taking into account the relations $\langle V\mathcal{G}\rangle = MG$ and $V = \Omega^2/C_0^2 - \mu\epsilon\Omega^2$, show that equations (5.48) and (3.8) lead to the representation (3.13) for the density of states N.

References

1. Akkermans, E., Wolf, P.E., and Maynard, R. (1986) Coherent backscattering of light by disordered media: Analysis of the peak line shape, *Phys. Rev. Lett.* **56,** 1471–1474.
2. Anderson P.W. (1958) Absence of diffusion in certain random lattices, *Phys. Rev.* **109,** 1492–1505.
3. Apresyan, L.A. and Kravtsov, Yu.A. (1983) Theory of Radiative Transfer: Statistical and Wave Aspects, Nauka, Moscow (in Russian).
4. Barabanenkov, Yu.N. and Finkelberg, V.M. (1968) Radiation transport equation for correlated scatterers, *Sov. Phys. JETP* **26,** 587—591.
5. Barabanenkov, Yu.N. and Finkelberg, V.M. (1968) Optical theorem in the theory of multiple wave scattering, *Izv. Vyssch. Uchebn. Zaved. Radiofiz.* **11,** 719–725.
6. Barabanenkov, Yu.N. (1969) On the spectral theory of radiation transport equation, *Sov. Phys. JETP* **29,** 679–684.
7. Barabanenkov, Yu.N. (1973) Wave corrections to the transport equation for backscattering, *Izv. Vyssh. Uchebn. Zaved. Radiofiz.* **16,** 88–96.
8. Barabanenkov, Yu.N. (1976) Multiple scattering of waves by ensemble of particles and the radiative transport, *Sov. Phys. Uspekhi* **18,** 673–689.
9. Barabanenkov, Yu.N. and Ozrin, V.D. (1985) Backscattering effect of the nonstationary radiation in the theory of transfer with memory, *Izv. Vyssh. Uchebn. Zaved. Radiofiz.* **28,** 450–459.

10. Barabanenkov, Yu.N., Ozrin, V.D., and Kalinin, M.I. (1985) Asymptotic Method in the Theory of Stochastic Linear Dynamic Systems, Energoatomizdat, Moscow (in Russian).
11. Barabanenkov, Yu.N. (1988) Asymptotically exact model of the theory of stationary radiation transport in a randomly varying medium, *Sov. Phys. Dokl.* **32,** 556–558.
12. Barabanenkov, Yu.N., Kravtsov, Yu.A., Ozrin, V.D., and Saichev, A.I. (1991) Enhanced Backscattering in Optics (Progress in Optics, vol. 29), ed. Wolf, E., Elsevier, Amsterdam, 65–197.
13. Barabanenkov, Yu.N. and Kalinin, M.I. (1992) Weakly non-uncoupled relations in wave multiple scattering theory for dense discrete random media, *Phys. Lett. A* **163,** 214–218.
14. Barabanenkov, Yu.N. and Ozrin, V.D. (1992) Problem of light diffusion in strongly scattering media, *Phys. Rev. Lett.* **69,** 1364–1366.
15. Barabanenkov, Yu.N. and Ozrin, V.D. (1993) Reply to comment by Van Tiggelen *et al.* (1993), *Phys. Rev. Lett.* **71,** 1285.
16. Barabanenkov, Yu.N. and Ozrin, V.D. (1995) Diffusion asymptotics of the Bethe-Salpeter equation for electromagnetic waves in discrete random media, *Phys. Lett. A* **206,** 116–122.
17. Barabanenkov, Yu.N., Zurk, L.M., and Barabanenkov, M.Yu. (1995) Poynting theorem and electromagnetic wave multiple scattering in dense media near resonance: modified radiative transfer equation, *J. Electr. Waves and Applications* **9,** 1393–1420.
18. Barabanenkov, Yu.N. and Barabanenkov, M.Yu. (1997) Radiative transfer theory with time delay for the effect of pulse entrapping in a resonant random medium: general transfer equation in a point–like scatterer model, *Waves in Random Media* **7,** 607–633.
19. Barabanenkov, Yu.N., Barabanenkov, M.Yu., and Winebrenner, D.P. (1998) Effect of pulse entrapping on diffuse reflection from a resonant random medium: exact solution to the scalar albedo problem, *Waves in Random Media* **8,** 451–463.
20. Barabanenkov, Yu.N. and Barabanenkov, M.Yu. (1999) Effect of a pulse entrapping on weak localization of waves in a resonant random medium, *Waves in Random Media* **9,** 13–26.
21. Bourret, R.C. (1962) Stochastically perturbed fields with applications to wave propagation in random media, *Nuovo Cim.* **26,** 1.
22. Chandrasekhar, S. (1960) Radiative Transfer, Dover, New York.
23. Compton, K.T. (1923) Some properties of resonance radiation and excited atoms, *Phil. Mag.* **45,** 750–760.
24. Foldy, L.L. (1945) The multiple scattering of waves. I. General theory of isotropic scattering by randomly distributed scatterers, *Phys. Rev.* **67,** 107.
25. Finkelberg, V.M. (1968) Propagation of waves in a random medium: the correlated group method, *Sov. Phys. JETP* **26,** 268–277.
26. Frisch, U. (1965) Wave propagation in random media. II. Multiple scattering by N bodies (provisional version), Institut d'Astrophysique, Paris.
27. Gazaryan, Yu.L. (1969) The one-dimensional problems of propagation of waves in a medium with random inhomogeneities, *Sov. Phys. JETP* **56,** 1856-1863.
28. Gertsenshtein, M.E. and Vasil'ev, V.B. (1959) *Theor. Probab. Appl.* **4,** 424.
29. Hirschfelder, J.O., Cartis, Ch.F., and Bird, R.B. (1954) Theory of Gases and Liquids, Wiley, New York.
30. Holstein, T. (1947) Imprisonment of resonance radiation in gases, Phys. Rev. **72,** 1212–1233.
31. Jefferies, J.T. (1968) Spectral Line Formation, MA: Blaisdell, Waltham.
32. Kato, T. (1966) Perturbation Theory for Linear Operators, Springer-Verlag, New York.
33. Kogan, E. and Kaveh, M. (1992) Diffusion constant in a random system near

resonance, *Phys. Rev. B* **46,** 10636–10640.

34. Kuga, Y., Ishimaru, A., and Rice, D. (1993) The velocity of coherent and incoherent electromagnetic waves in a dense strongly scattering media, *Phys. Rev. B* **48,** 13155–13158.
35. Lagendijk, A. and Van Tiggelen, B.A. (1996) Resonant multiple scattering of light, *Physics Reports* **270,** 143–215.
36. Lankaster, P. (1969) Theory of Matrices, Academic Press, New York.
37. Landau, L.D. and Lifshitz, E.M. (1986) Hydrodynamics, Nauka, Moscow (in Russian).
38. Lax, M. (1951) Multiple scattering of waves, *Rev. Mod. Phys.* **23,** 287–310.
39. Milne, E.A. (1926) The diffusion of imprisoned radiation through a gas, *J. London Math. Soc.* **1,** 40–51.
40. Mitchell, A.C. and Zemansky, K.W. (1934) Resonance Radiation and Excited Atoms, Macmillan, New York.
41. Nieuwenhuizen, T.M., Lagendijk, A., and Van Tiggelen, B.A. (1992) Resonant point scatterers in multiple scattering of classical waves, *Phys Lett. A* **169,** 191-194.
42. Papanicolaou, G.C. (1972) A kinetic theory for power transfer in stochastic systems, *J. Math. Phys.* **13,** 1912–1918.
43. Rytov, S.M., Kravtsov, Yu.A., and Tatarskii, V.I. (1989) Principles of Statistical Radiophysics. Vol. 3. Wave Propagation through Random Media, Springer, Berlin.
44. Ryzhik, L., Papanicolaou, G., and Keller, J.K. (1996) Transport equations for elastic and other waves in random media, *Wave Motion* **24,** 327–370.
45. Sobolev, V.V. (1956) Transfer of Radiant Energy in the Atmospheres of Stars and Planets, State Publisher of Technical-Theoretical Literature, Moscow (in Russian); A Treatise on Radiative Transfer (Engl. Transl.), Van Nostrand-Reinhold, Princeton.
46. Stratton, J.A. (1941) Electromagnetic Theory, McGraw-Hill, New York.
47. Strinati, G., Castellam, G., and Di Castro, C. (1986) Kinetic equation for strongly disordered systems: Non-interacting electrons, *Phys. Rev. B* **40,** 12273–12524.
48. Tatarskii, V.I. (1969) Light propagation in a medium with random inhomogeneities of refractive index in approximation of Markov random process, *Zh. Eksp. Theor. Fiz.* **56,** 2106–2117 (in Russian).
49. Van Albada, M.P, Van Tiggelen, B.A., Lagendijk, A., and Tip A. (1992) Speed of propagation of classical waves in strongly scattering media, *Phys. Rev. Lett.* **66,** 3132–3135.
50. Van de Hulst, H.C. (1981) Light Scattering by Small Particles, Dover, New York.
51. Van Hove, L. (1955) Quantum-mechanical perturbations and kinetic equations, *Physica* **21**, 517.
52. Van Tiggelen, B.A., Lagendijk, A., Van Albada, M.P., and Tip, A. (1992) The speed of light in random media, *Phys. Rev. B.* **45,** 12233–12243.
53. Van Tiggelen, B.A., Lagendijk, A., and Tip, A. (1993) Comment on the paper by Barabanenkov and Ozrin (1992), *Phys. Rev. Lett.* **71,** 1284–1285.
54. Van Tiggelen, B.A. and Lagendijk, A. (1994) Resonantly induced dipole-dipole interactions of scalar waves **Phys. Rev. B 50,** 16729–16732.
55. Watson, K.M. (1969) Multiple scattering of electromagnetic waves in an underdense plasma, *J. Math. Phys.* **10,** 688–702.
56. Wigner, E.P. (1955) Lower limit for the energy derivative of the scattering phase shift, *Phys. Rev.* **98,** 145–147.
57. Zemansky, M.W. (1927) The diffusion of imprisoned resonance radiation in mercury vapor, *Phys. Rev.* **29,** 513–523.

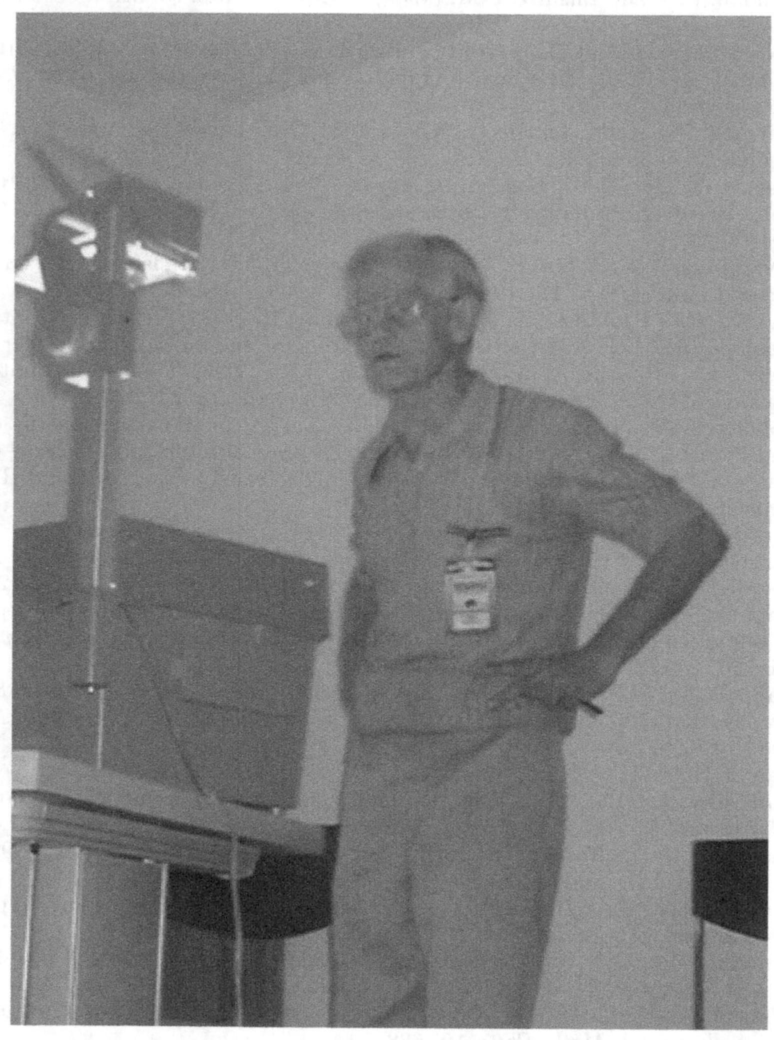

Yurii Barabanenkov. Photo by Gabriel Cwilich

BOUNDARY CONDITIONS IN RADIATIVE TRANSFER

VLADIMIR L. KUZMIN
Institute of Commerce and Economics
St. Petersburg 194021, Russia
and
Department of Physics, St. Petersburg State University
St. Petersburg, 198504 Russia.

Abstract. The averaging procedure of the electrodynamic wave equation is outlined. The results are applied for calculation of permittivity of suspensions; the solution for the boundary Green's function is discussed. The Bethe-Salpeter equation describing the field correlation transfer is derived. The Milne solution is generalized for the multiple scattering correlation and interference phenomena, accounting for the single-scattering anisotropy. Within the P_1 approximation, the initial decay rate of the temporal correlation function appears to be universal. For larger time intervals, stronger single-scattering anisotropy yields higher decay rate. Within the P_2 approximation, which means that the first and second order Legendre polynomials are taken into account in the expansion of the phase function, the obtained solution agrees rather well with the known measurement data. The width of the coherent backscattering cone is found to be scaled by the transport length, as generally accepted.

1. Preface

Studying the correlation phenomena in highly inhomogeneous media, one realizes that the boundedness of the system appears to be of great importance since incident as well as outgoing waves travel inevitably through surface layers. The multiple scattering regime gives rise to the long range diffusion of scattered light. Within the diffusion approximation, the mirror image method is widely used to satisfy the mixed Dirichlet boundary conditions. However the diffusion approximation is known to be at least unjustified for the boundary region. We develop generalizations of the exact Milne

B.A. van Tiggelen and S.E. Skipetrov (eds.),
Wave Scattering in Complex Media: From Theory to Applications, 461–474.

approach for description of the multiple scattering correlation phenomena taking into account the finite size of scatterers. Correlation and interference phenomena occur due to the wave nature of the light. Intensity, as well as binary correlations, are quadratic forms of the field. Thus one needs to bear in mind that in addition to averaging a quadratic expression, the procedure of obtaining the averaged field itself is to be understood properly. Therefore linear field equations are to be discussed first. Averaging them one obtains macroscopic fields used in the treatment of the correlation transfer.

Briefly, the content of the lecture is as follows. In the first part I outline the main features of the averaging procedure for the linear wave equation; as applications I describe the method of calculation of optical constants, the photon mean free path, or scattering length, l and transport mean free path l^*, and derive the Green's function for a system with a boundary. In the second part I derive the Bethe-Salpeter equation describing the correlation transfer. Finally, in the third part I consider generalizations of the exact Milne solution for the coherent backscattering and temporal correlation functions in turbid systems with anisotropic single scattering. For better insight into the problems discussed, I also provide a number of exercises.

2. Macroscopic Electrodynamics: Averaging the Wave Equation

I begin with the wave equation for a medium with random permittivity $\varepsilon(\mathbf{r},t) = \varepsilon_0 + \delta\varepsilon(\mathbf{r},t)$. For multiple scattering, one neglects polarization effects and reduces the Maxwell wave equation to the Helmholtz one:

$$E(\mathbf{r},t) = E_0(\mathbf{r},t) + \frac{1}{4\pi}\int d\mathbf{r}_1 T_0(\mathbf{r}-\mathbf{r}_1)\delta\varepsilon(\mathbf{r}_1,t)E(\mathbf{r}_1,t), \qquad (1)$$

where $E_0(\mathbf{r},t)$ is the mean field of some reference system; for scattering problems such a system is chosen as a non-fluctuating medium with permittivity $\varepsilon_0 = \mathrm{Re}\langle\varepsilon(\mathbf{r},t)\rangle$; for problems linear in the field (like corrections to the Lorenz-Lorentz relation) it may be chosen as a vacuum, then $\delta\varepsilon \to \varepsilon(\mathbf{r},t) - 1$. $T_0(\mathbf{r}) = k_0^2\exp(i\sqrt{\varepsilon_0}k_0 r)/r$ is the Green's function of a scalar wave equation, $k_0 = \omega/c = 2\pi/\lambda$ is the wave number in vacuum. I omit the monochromatic factor $\exp(i\omega t)$ since the time it takes the light to propagate through the medium is sufficiently less than the characteristic scale of the random permittivity variations. The dependence on t is kept due to random permittivity $\varepsilon(\mathbf{r},t)$, which is assumed to be a slow function of time. This equation is a starting point for calculation of optical parameters of random systems, such as photon and transport mean free paths, and also for calculation of the field itself. Immediate averaging of this equation appears to be nonproductive since it will become unclosed, containing the mean field on the l.h.s. and the average of a product of the random field and permittivity on the r.h.s. Thus, one has to iterate the equation and, afterwards, to average the arising iteration series. However, the averaged iteration series is also inappropriate for calculation, since the terms of the series turn out to be spatial integrals that diverge with the size of the system. Therefore, the iteration series has to be resummed, or renormalized. The main features of this derivation are as follows. Let the iteration series be schematically represented as

$$E = \left(I + T_0 \frac{\delta\varepsilon}{4\pi} + \left(T_0 \frac{\delta\varepsilon}{4\pi} \right)^2 + \cdots \right) E_0, \tag{2}$$

where products mean integral convolutions. Averaging over disorder, one obtains a series of terms, which are, generally, spatial integrals containing the field Green's functions and multiple permittivity correlators $\langle \delta\varepsilon_1 \ldots \delta\varepsilon_n \rangle$. These integrals diverge since the multiple permittivity correlators remain constant at large distances and the Green's functions exhibit a long-range behavior. Various resummation procedures have been developed: see, e.g., (Felderhof, 1974; Boots *et al.*, 1975; Kuzmin *et al.*, 1994). They result in a partially summed, or renormalized, series for the macroscopic field that can be represented in the form of the Dyson equation:

$$E\left(\mathbf{r}\right) = E_0\left(\mathbf{r}\right) + \int T_0(\mathbf{r},\mathbf{r}')\Pi(\mathbf{r}',\mathbf{r}'')E(\mathbf{r}'')d\mathbf{r}'d\mathbf{r}''. \tag{3}$$

A similar equation is derived for the macroscopic, or 'dressed', Green's function.

Function $\Pi(\mathbf{r}',\mathbf{r}'')$, known as a polarization operator, can be interpreted as a nonlocal susceptibility. Diagrammatically, $\Pi(\mathbf{r}',\mathbf{r}'')$ is represented by a series of irreducible Feynman diagrams. Note, that similar renormalization procedures have been successfully developed (Stell and Lebowitz, 1968; Hoye and Stell, 1976; Wertheim, 1978; Kuzmin, 1985) in statistical mechanical theories of systems with long range electrostatic interactions. The Fourier transform of the polarization operator is easily identified as

$$\widetilde{\Pi}(q) = \frac{\varepsilon(q) - 1}{4\pi}. \tag{4}$$

A dependence on $\mathbf{q}$ describes the spatial dispersion of permittivity. The first nontrivial contribution, known as Born, or Bourret, approximation, is

$$\Pi_2(q) = \int \frac{d\mathbf{q}'}{(2\pi)^3} T(q')G(\mathbf{q}-\mathbf{q}') = \int \frac{d\mathbf{q}'}{(2\pi)^2} \frac{k_0^2 G(\mathbf{q}-\mathbf{q}')}{q'^2 - k_0^2\left(1 + 4\pi\Pi(q')\right)}. \tag{5}$$

In the case of weak scattering $\lambda \ll l$, the imaginary part of $\Pi(\mathbf{q})$, responsible for the wave extinction, is small, and the integral can be calculated by the residue theorem, yielding

$$l^{-1} = k_0^4 \int d\Omega' \tilde{G}(\mathbf{k} - \mathbf{k}'). \tag{6}$$

This is the famous optical theorem, crucial for multiple scattering problems. The expansion parameter of the multiple- scattering series for the scattered intensity turns out to be exactly unity due to this relationship. For the electromagnetic field, there appears an additional factor $(1 + \overline{\cos^2\theta})/2$ on the r.h.s. of Eq. (6), turning to 2/3 for point-like scatterers. The optical theorem relates the total single-scattering cross-section

$$G(q) = (4\pi)^{-2} \int \langle \Delta\epsilon(0)\Delta\epsilon(\mathbf{r})\rangle \exp(-i\mathbf{qr})d\mathbf{r} \tag{7}$$

to the photon mean free path. In the case of strong scattering, one would need to take into account higher-order terms of the diagram series.

2.1. OPTICAL PARAMETERS OF SUSPENSIONS

The obtained results readily permit to calculate the optical parameters of turbid suspensions. We describe a suspension as a system of spheres with diameter d, randomly distributed in a liquid background medium. The random permittivity of such a system can be represented as

$$\varepsilon(\mathbf{r}) = \varepsilon_0 + \delta\varepsilon \sum_{i=1}^{N} \theta\left(\frac{d}{2} - |\mathbf{r} - \mathbf{R}_i|\right), \tag{8}$$

where ε_0 is the solvent permittivity, $\delta\varepsilon = \varepsilon_s - \varepsilon_0$, ε_s is the permittivity of (solid) particles, and $\theta(r)$ is the Heaviside step function. The binary correlation function of fluctuations $\Delta\varepsilon(\mathbf{r}) = \varepsilon(\mathbf{r}) - \langle\varepsilon(\mathbf{r})\rangle$ then takes the form

$$\begin{aligned} G(\mathbf{r}_1, \mathbf{r}_2) &= \langle \Delta\varepsilon(\mathbf{r}_1)\Delta\varepsilon(\mathbf{r}_2)\rangle \\ = (\delta\varepsilon)^2 &\left[\left\langle \sum_{i,j} \theta\left(\frac{d}{2} - |\mathbf{r}_1 - \mathbf{R}_i|\right)\theta\left(\frac{d}{2} - |\mathbf{r}_2 - \mathbf{R}_j|\right)\right\rangle \right. \\ &\left. - \langle \sum_i \theta\left(\frac{d}{2} - |\mathbf{r}_1 - \mathbf{R}_i|\right)\rangle^2\right]. \end{aligned} \tag{9}$$

For small concentrations, this is equivalent to the Rayleigh-Gans approximation. The angular brackets denote statistical averaging over the distribution of hard spheres. Using the Percus-Yevick solution (Croxton, 1974), well-known in the theory of liquids, for the hard sphere system, we have calculated (Kuzmin *et al.*, 2001) the imaginary part of the permittivity in the Born approximation and found the optical parameters of suspension: scattering length l, transport mean free path l^*, and parameters of cross-section $\overline{\cos^n \theta}$. We compared the results for l and l^* with measurements (Van der Mark *et al.*, 1988) and found that a discrepancy occurs only in the strong scattering region, beyond the range of applicability of the present theory.

2.2. THE GREEN'S FUNCTION FOR A SYSTEM WITH PLANE BOUNDARY

Considering the multiple scattering problem, either for backscattered or transmitted light, one needs to know the field Green's function for a bounded system. For a system with a plane boundary between the phases a and b described by

$$\varepsilon(z) = \varepsilon_a \theta(-z) + \varepsilon_b \theta(z) = \varepsilon_a + \Delta\varepsilon\theta(z),$$

the macroscopic wave equation to be solved is

$$T(\mathbf{r}_1, \mathbf{r}_2) = T_a(\mathbf{r}_1 - \mathbf{r}_2) + \frac{\Delta\varepsilon}{4\pi} \int_{z_3>0} T_a(\mathbf{r}_1 - \mathbf{r}_3) T(\mathbf{r}_3, \mathbf{r}_2) d\mathbf{r}_3. \tag{10}$$

(Solving a similar equation for plane waves, one readily reproduces the Fresnel formulas). The solution is easily found for the 2D Fourier transform of the field Green's function in the scalar case [see (Nieuwenhuizen and Luck, 1993)]:

$$\tilde{T}(\mathbf{q}, z_1, z_2) = \int T(\mathbf{r}_1, \mathbf{r}_2) \exp(-i\mathbf{q}\boldsymbol{\rho}) d\boldsymbol{\rho},$$

where $\boldsymbol{\rho} = (x_1 - x_2, y_1 - y_2)$. Asymptotically, for $\mathbf{r}_1$ and $\mathbf{r}_2$ both inside the phase b and $q \ll 1/\lambda$, the solution can be rewritten as

$$\tilde{T}(\mathbf{q}, z_1, z_2) \longrightarrow \exp(-k_b |z_1 - z_2|) + \left(\frac{n_b - n_a}{n_b + n_a} \right) \exp(-k_b |z_1 + z_2|) \tag{11}$$

and interpreted as a sum of bulk and mirror image terms.

The second term in Eq. (11) is responsible for the index mismatch effects in the multiple scattering phenomena. The Bethe-Salpeter equation has been considered and solved with account for this term (Nieuwenhuizen and Luck, 1993); this solution is used when one takes into account the internal reflection phenomenon caused by the mismatch of refractive indices.

3. Field Correlation Transfer: Bethe-Salpeter Equation

Intensity, as well as binary field correlations, are quadratic forms of the field. Iterating, one represents the field correlation function as

$$\langle \delta E^* \delta E \rangle = \langle E^* \left(\frac{\Delta\varepsilon}{4\pi} T^* + (\frac{\Delta\varepsilon}{4\pi} T^*)^2 + \cdots \right) \left(T \frac{\Delta\varepsilon}{4\pi} + (T \frac{\Delta\varepsilon}{4\pi})^2 + \cdots \right) E \rangle,$$

which is usually illustrated by Feynman diagram series. In the weak scattering limit $\lambda \ll l$, only ladder diagrams are to be considered. The ladder diagrams are summed as a geometric series to produce the Bethe-Salpeter equation. The Bethe-Salpeter equation describes the radiation intensity transfer. However, when the characteristics (temporal, spatial, frequency, angular) of the fields E and E^* are different, it describes also the field correlation transfer. We assume that the permittivity fluctuations obey the Brownian diffusion law with characteristic time $\tau = 1/(Dk^2)$, where D is the self-diffusion coefficient and there is no absorption. Then the Fourier transform of the permittivity correlator can be represented as

$$G(\mathbf{q},t) = \int d\mathbf{r}\,\frac{1}{(4\pi)^2}\langle\delta\varepsilon(\mathbf{r}_1,t_1)\delta\varepsilon(\mathbf{r}_2,t_2)\rangle\exp(-i\mathbf{q}\mathbf{r}) = G_0(\mathbf{q})\exp(-Dq^2t),$$

where $G_0(\mathbf{q})$ is the single-scattering cross-section.

The Bethe-Salpeter equation can be represented in the form

$$\Gamma(\mathbf{R}_2,\mathbf{R}_1,t|\mathbf{k}_s,\mathbf{k}_i) = k_0^4\tilde{G}(\mathbf{k}_s-\mathbf{k}_i,t)\delta(\mathbf{R}_2-\mathbf{R}_1)$$

$$+k_0^4\int d\mathbf{R}_3\tilde{G}(-\mathbf{k}_s+\mathbf{k}_{23},t)\Lambda(R_{23})\Gamma(\mathbf{R}_3,\mathbf{R}_1,t|\mathbf{k}_{23},\mathbf{k}_i), \tag{12}$$

where $\mathbf{k}_{ij} = k\mathbf{R}_{ij}R_{ij}^{-1}$ is the wave vector directed from $\mathbf{R}_j$ to $\mathbf{R}_i$, $\mathbf{R}_{ij} = \mathbf{R}_i - \mathbf{R}_j$, and the propagator $\Lambda(R) = R^{-2}\exp(-R/l)$ stems from the product of two complex-conjugated Green's functions $T(\mathbf{R})$ and $T^*(\mathbf{R})$: $k_0^4\Lambda(R) = T(\mathbf{R})T^*(\mathbf{R})$. The binary field correlation function can be represented as a sum of the ladder term $C^{(L)}(t|\mathbf{k}_f,\mathbf{k}_i)$ and the interference term $C^{(V)}(t|\mathbf{k}_f,\mathbf{k}_i)$:

$$C_E(t|\mathbf{k}_f,\mathbf{k}_i) = \langle\delta E_{i\to f}(\mathbf{r},t)\delta E^*_{i\to f}(\mathbf{r},0)\rangle = C^{(L)}(t|\mathbf{k}_f,\mathbf{k}_i) + C^{(V)}(t|\mathbf{k}_f,\mathbf{k}_i).$$

Let us consider the backscattering from the half-space $z > 0$, where z is the Cartesian coordinate normal to the boundary. Vectors $\mathbf{k}_f$ and $\mathbf{k}_i$ are the scattered and incident wave vectors, respectively, both lying both in the (x,z) plane. The term $C^{(L)}$ describes the temporal field correlations outside the backscattering cone. The interference term $C^{(V)}$ becomes essential within a narrow angular range around the backward direction, describing (at $t=0$) the intensity of the coherent backscattering. These terms can be represented as

$$C^{(L)}(t\,|\,\mathbf{k}_f,\mathbf{k}_i) = \int d\mathbf{R}_1 d\mathbf{R}_2\Gamma(\mathbf{R}_2,\mathbf{R}_1,t\,|\,\mathbf{k}_f,\mathbf{k}_i)\exp\left(-\frac{z_1}{l\cos\theta_i}-\frac{z_2}{l\cos\theta_f}\right), \tag{13}$$

$$C^{(V)}(t\,|\,\mathbf{k}_f,\mathbf{k}_i)=\int d\mathbf{R}_1 d\mathbf{R}_2\left[\Gamma\left(\mathbf{R}_2,\mathbf{R}_1,t\left|\frac{\mathbf{k}_f-\mathbf{k}_i}{2},\frac{\mathbf{k}_i-\mathbf{k}_f}{2}\right.\right)\right.$$

$$\left.-k_0^4\tilde{G}(\mathbf{k}_f-\mathbf{k}_i,t)\delta(\mathbf{R}_2-\mathbf{R}_1)\right]\exp\left[-\frac{z_1+z_2}{2l}\left(\frac{1}{\cos\theta_i}+\frac{1}{\cos\theta_f}\right)\right.$$

$$\left.+in_1k_0(z_1-z_2)(\cos\theta_i-\cos\theta_f)+in_1k_0(x_1-x_2)(\sin\theta_i-\sin\theta_f)\right], \quad (14)$$

where θ_i is the angle of incidence, θ_f is the scattering angle with respect to the backward direction. The anisotropy of single scattering introduces the dependence on direction of incident $\mathbf{k}_i$, as well as scattered $\mathbf{k}_f$ wave vectors. However, the scattered radiation is believed to become nearly isotropic due to the multiple scattering.

4. Generalized Milne Equations

We perform the 2D Fourier transform of Eq. (12) with respect to transversal variables and the Laplace transform with respect to normal variables:

$$\frac{1}{4\pi}\gamma_{\mathbf{q}_\perp}(s,s_i,t|\mathbf{k}_f,\mathbf{k}_i)=\int d\mathbf{R}_{2\perp}\int_0^\infty dz_2\int_0^\infty dz_1\Gamma(\mathbf{R}_2,\mathbf{R}_1,t\,|\,\mathbf{k}_f,\mathbf{k}_i)\times$$

$$\times\exp[-(z_2s+z_1s_i)/l+i\mathbf{q}_\perp(\mathbf{R}_2-\mathbf{R}_1)_\perp],\quad s_i=1/\cos\theta_i. \quad (15)$$

The Bethe-Salpeter equation is diagonal with respect to $\mathbf{k}_i$ (this vector conserves in the r.h.s and l.h.s.), so one approximates the dependence on the orientation of $\mathbf{k}_f$ using an expansion in spherical functions. We consider the dependence on orientation of $\mathbf{k}_f$ within the P_1 approximation, which means the account for terms linear in $\cos\theta_f$ and $\cos\theta_t=(q_\perp k)^{-1}\mathbf{q}_\perp\mathbf{k}_f$:

$$\gamma_{\mathbf{q}_\perp}(s,s_i,t|\mathbf{k}_f,\mathbf{k}_i)=\gamma^{(0)}_{\mathbf{q}_\perp}(s,s_i,t)+\gamma^{(t)}_{\mathbf{q}_\perp}(s,s_i,t)\cos\theta_t-\gamma^{(n)}_{\mathbf{q}_\perp}(s,s_i,t)\cos\theta_f. \quad (16)$$

The term $\gamma^{(t)}_{\mathbf{q}_\perp}(s,s_i,t)$ describes the dependence on the azimuthal angle, while $\gamma^{(n)}_{\mathbf{q}_\perp}(s,s_i,t)$ describes the dependence on the polar one. The theory predicts quite different angular shapes of the coherent backscattering peak if one accounts either only for the polar term or for both the polar and azimuthal terms.

Let us define a linear combination

$$\gamma_{\mathbf{q}_\perp}(s,s_i,t)=\gamma^{(0)}_{\mathbf{q}_\perp}(s,s_i,t)-\gamma^{(n)}_{\mathbf{q}_\perp}(s,s_i,t)/s. \quad (17)$$

This function describes observable quantities for small angles of backscattering. Indeed, substituting (16) into (13) and (14), one obtains the temporal field correlation function as

$$C^{(L)}(t\,|\,\mathbf{k}_f,\mathbf{k}_i) \sim \gamma_0(s_f,s_i,t)$$

and the coherent backscattering intensity as

$$C^{(V)}(0\,|\,\mathbf{k}_f,\mathbf{k}_i) \sim \gamma_{\mathbf{q}_\perp}(s_f,s_i,t) - 4\pi\tilde{G}(k_f-k_i,t)\Big(\int d\Omega_f \tilde{G}(\mathbf{k}_f-\mathbf{k}_i)\Big)^{-1},$$

where $s_f = 1/\cos\theta_f$.

Using the orthogonality of components of expansion (16), one can rearrange the Bethe-Salpeter equation into a set of three integral equations with respect to components $\gamma^{(0)}_{\mathbf{q}_\perp}(s,s_i,t)$, $\gamma^{(t)}_{\mathbf{q}_\perp}(s,s_i,t)$, and $\gamma^{(n)}_{\mathbf{q}_\perp}(s,s_i,t)$. For the function $\gamma_{\mathbf{q}_\perp}(s,s_i,t)$ the equation set reduces to a closed equation:

$$\psi(\mathbf{q}_\perp,s,t)\gamma_{\mathbf{q}_\perp}(s,s_i,t) = \left[1-\frac{\overline{\cos\theta}s^{*2}}{s(1-\overline{\cos\theta})}\right]\frac{1}{s_i+s}$$

$$-\frac{1}{2}\int_1^\infty \frac{ds'}{s'(s'-s)}\left[1+\frac{\overline{\cos\theta}s^{*2}}{s's(1-\overline{\cos\theta})}\right]\gamma_{\mathbf{q}_\perp}(s',s_i,t), \tag{18}$$

where

$$\psi(\mathbf{q}_\perp,s,t) = s^{*2}/3 - s^2 m(s), \tag{19}$$

$$m(s) = \frac{1}{s^2}\left\{\frac{1}{2s}\ln\frac{1+s}{1-s}-1\right\},$$

$$s^* = (1-\overline{\cos\theta})\sqrt{6t\tau^{-1}+(q_\perp l^*)^2}. \tag{20}$$

Equation (18) is valid in leading orders in $t/\tau \ll 1$ and $q_\perp l \ll 1$, where $q_\perp \approx k\cos\theta_i|\Delta\theta|$, $|\Delta\theta| = |\theta_f-\theta_i|$.

The specific features of the considered phenomena originate, mathematically, from the fact that the determinant $\psi(\mathbf{q}_\perp,s,t)$ of the equation set turns to zero at $s^*\to 0$ and $s\to 0$. Therefore, deriving Eq. (18) we neglect the terms containing simultaneously $(q_\perp l)^2$ and s^2 simultaneously, namely, we put $(q_\perp l)^2 m(s) \approx (q_\perp l)^2 m(0)$. For isotropic scattering, one can control the validity of such a substitution by comparing the results with the known exact solution (Gorodnichev *et al.*, 1990; Nieuwenhuizen and Luck, 1993).

We consider first the Milne equation in the isotropic case:

$$\psi_0(s,t)\gamma_0(s,1,t) = a_-(s,t) \tag{21},$$

where $\psi_o(s,t) = 2(t/\tau) + 1 - (2s)^{-1}\ln[(1+s)/(1-s)]$ and

$$a_-(s,t) = \frac{1}{1+s} - \frac{1}{2}\int_1^{\infty} \frac{ds'}{s'(s'-s)}\gamma_0(s',1,t).$$

Function $\psi_0(s,t)$ can be represented as a ratio of functions $(1+s)a_-(s,t)$ and $(1+s)\gamma_0(s,1,t)$, regular in the left (Re$s < -1$) and right (Re$s > 0$) half-planes, respectively. Since $\psi_0(s,t)$ is an even function of s, it can also be represented as a ratio of functions regular in the right and left half-planes, respectively: $\psi_0(s,t) = h_+(s,t)/h_-(s,t)$. The products $(1+s)\gamma_0(s,1,t)h_+(s,t) = (1+s)a_-(s,t)h_-(s,t)$ are therefore regular at any s, finite at $s \to \infty$, and, hence, constant due to the Liouville theorem. These are the main features of the Wiener-Hopf method [see (Chandrasekhar, 1960)] that permits to obtain the solution of Eq. (21) as

$$\gamma_0(s,1,t)_{\text{Milne}} = \frac{\text{Const}}{1+s}\exp\left[-\frac{s}{2\pi}\int_{-\infty}^{+\infty}\frac{ds'}{s'^2+s^2}\ln\psi_0(is',t)\right]. \qquad (22)$$

Equation (18) also permits the application of the Wiener-Hopf method. Now products $s(1+s)\gamma_0(s,1,t)h_+(s,t) = s(1+s)a_-(s,t)h_-(s,t)$ are regular in the whole plane and equal a polynomial of the first order in s, accounting for an extra factor s, due to the generalized Liouville theorem. This allows us to obtain

$$\gamma_{\mathbf{q}}(s,s_i,t) \approx \frac{(C_1 + C_2 s)}{s(s+s^*)(s+s_i)}\exp[-J(s)], \qquad (23)$$

where $J(s) = s/(2\pi)\int\limits_{-\infty}^{\infty} ds'(s'^2+s^2)^{-1}\ln[3m(is')]$ is the generalized Chandrasekhar function.

Formula (23) generalizes the well-known Milne solution with account for the single-scattering anisotropy within the P_1 approximation. Coefficients C_1 and C_2 are independent of s. We find them by substituting the solution (23) into Eq. (18) and expanding both sides of the resulting equation in series in orders of s. As a result, we find (up to linear terms in a small parameter s^*):

$$\gamma_{\mathbf{q}}\left(\frac{1}{\cos\theta_i},\frac{1}{\cos\theta_f},t\right) = \frac{3\cos^2\theta_i\cos^2\theta_f}{(\cos\theta_i+\cos\theta_f)}\exp\left[-J\left(\frac{1}{\cos\theta_i}\right) - J\left(\frac{1}{\cos\theta_f}\right)\right]$$
$$\times\left(1 - \frac{\kappa s^*}{1-\overline{\cos\theta}}\right), \qquad (24)$$

where $\kappa = (\cos\theta_i + \cos\theta_f)$ is the initial slope coefficient.

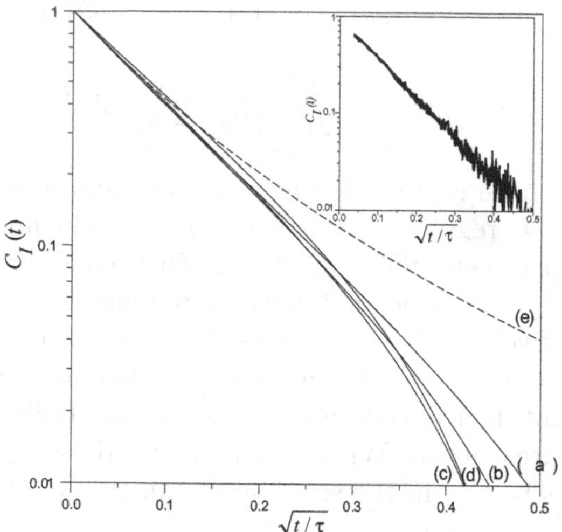

Figure 1. Intensity correlation function vs. $(t/\tau)^{1/2}$ in the P_2 approximation, $P_2 = (1/2)(3\overline{\cos^2\theta} - 1) = \overline{\cos\theta}^2$; (a) $\overline{\cos\theta} = 0$, (b) $\overline{\cos\theta} = 0.2$, (c) $\overline{\cos\theta} = 0.4$, (d) $\overline{\cos\theta} = 0.6$, (e) diffusion approximation. The inset shows the experimental data (Pine, 1988).

Multiple scattering is known to cause specific peculiarities of correlation and interference phenomena. For temporal correlations, this leads to a non-analytic dependence on time of the form $\sqrt{t}$ for short times t, while for the coherent backscattering it involves a triangular shape of the peak, i.e. linear dependence of backscattered intensity on the scattering angle in vicinity of the backward direction. Therefore, the initial slope coefficients, describing the rate of linear decay of temporal correlation or coherent backscattering, appear to be important experimentally accessible parameters.

In case of normal incidence, for $\cos\theta_f \approx \cos\theta_i = 1$, Eq. (24) yields

$$\gamma_0(1,1,t) = \frac{3}{2}\exp\left[-2J(1)\right]\left(1 - \frac{2s^*}{1-\overline{\cos\theta}}\right) \approx 4.23(1 - 2\sqrt{6t/\tau}). \quad (25)$$

Hence, the temporal correlation function decays universally within the P_1 approximation, independently of the anisotropy of single-scattering cross-section, and the slope coefficient $\kappa = 2$ exactly. The temporal correlation function has also been calculated in the P_2 approximation (Kuzmin *et al.*, 2002). In the temporal range where the experimental dispersion of data is still small, theoretical curves appear to practically coincide with the measured ones (Pine *et al.*, 1988) for various $\overline{\cos\theta}$ (see Figure 1).

The diffusion approximation yields for the temporal correlation function

(MacKintosh and John, 1989)

$$C(t) \sim \frac{1}{(1+s^*)^2}\{1 + \frac{1}{s^*}[1 - \exp(-1.4208\sqrt{6t/\tau})]\}. \qquad (26)$$

This equation gives $\kappa \approx 2.4$ in the case of isotropic scattering ($\overline{\cos\theta} = 0$), and $\kappa \approx 0.7$ for $\overline{\cos\theta} \to 1$. Thus, the diffusion approximation fails to predict the universal slope contrary to the Milne-type solution (25).

Now we turn to the coherent backscattering intensity $I(\Delta\theta) = C^{(V)}(0|\mathbf{k}_f, \mathbf{k}_i)$. Setting $t = 0$ in (24) and omitting the term describing the single-scattering contribution, we obtain for small scattering angles and normal incidence:

$$I(\Delta\theta) \sim 1 - 2\frac{kl\Delta\theta}{1 - \overline{\cos\theta}}. \qquad (27)$$

In the case of isotropic scattering, the Milne solution have been generalized for the coherent backscattering (Gorodnichev *et al.*, 1990; Nieuwenhuizen and Luck, 1993) [see also the review (van Rossum and Nieuwenhuizen, 1999)]. The problem of multiple scattering from a half-space was considered in the limit of very strong forward single scattering (Amic *et al.*, 1996). The solution obtained in this limit predicts the same angular dependence as Eq. (27) obtained within the P_1 approximation.

Deriving the diffusion approximation, one considers the Bethe-Salpeter equation (12) far from the boundary, i.e. deep inside the medium. Expanding the Green's function $\Gamma(\mathbf{R}_3, \mathbf{R}_1, t\,|\,\mathbf{k}_{23}, \mathbf{k}_i)$ in powers of $\mathbf{R}_3 - \mathbf{R}_2$, one approximates an integral equation by a second order differential equation; afterwards, the obtained solution is applied to the boundary region. A similar approximation has been made by (Amic *et al.*, 1996) [see Eqs. (2.59) in the cited paper] to consider the coherent backscattering peak. Thus the shape of the coherent backscattering cone obtained by (Amic *et al.*, 1996) appears to be less rigorous than the general consideration, since in this problem it goes along the diffusion approximation lines.

The diffusion approximation yields (Akkermans *et al.*, 1988):

$$I^{(dif)}(\Delta\theta) \sim 1 - 2\frac{(1 + (1 - \overline{\cos\theta})^{-1}z^*)^2}{1 + 2(1 - \overline{\cos\theta})^{-1}z^*}kl\Delta\theta. \qquad (28)$$

Note, that quantitatively Eqs. (27) and (28) predict quite different dependence on $\overline{\cos\theta}$.

Setting the azimuthal term $\gamma^{(t)}_{\mathbf{q}\perp}(s, s_i, t) = 0$ in the P_1 representation (16), we obtained (Kuzmin and Romanov, 2002) $I(\Delta\theta) \sim 1 - 2kl\Delta\theta(1 - \overline{\cos\theta})^{-1/2}$ for the coherent backscattering cone. This result appears to be erroneous, since the integral term in the Bethe-Salpeter equation automatically generates the azimuthal dependence, once the representation (16) is substituted, either with or without an explicit azimuthal term.

5. Conclusions

I outlined the results obtained by solving the multiple scattering boundary problems using the Wiener-Hopf method and taking into account the single-scattering anisotropy. The P_1 approximation is limited, generally, to the case of weak anisotropy. It describes the single-scattering anisotropy by means of a single parameter $\overline{\cos\theta}$. The results (Amic *et al.*, 1996) obtained in the limit of very strong forward single scattering, contain (surprisingly) no other parameters of anisotropy except of $\overline{\cos\theta}$, independently of the shape of the single-scattering cross-section. It was shown (Kuzmin *et al.*, 2002), however, that within the P_2 approximation the temporal correlation function depends on two parameters of anisotropy: $\overline{\cos\theta}$ and $\overline{\cos^2\theta - 1/3}$. Thus, in the case of intermediate anisotropy, the solution of the boundary problem beyond the P_1 approximation is needed.

I considered the multiple scattering phenomena within the scalar field approach. Generally, the vector nature of the electromagnetic field has to bee included into consideration. In particular, the polarized component of the coherent backscattering is known to exhibit the specific triangular shape, while a much weaker depolarized one is shaped as a Lorentzian. The Wiener-Hopf method outlined here have been successfully applied (Amic *et al.*, 1997; Mishchenko, 1996; Mishchenko *et al.*, 2000; Kuzmin, 2002) to the multiple Rayleigh scattering of electromagnetic waves. I have also obtained some results on the temporal correlation function within the P_1 approximation. I have shown that the polarized and depolarized components of the temporal correlation function both exhibit the temporal dependence of the form (25). However, the slope coefficients for these components appear to be different even in the case of small anisotropy. One can expect that a solution of the boundary problem including polarization effects can be of the same importance as the anisotropy effect considered here. The consideration of boundary problems for liquid crystal systems and, generally, for systems with anisotropic permittivity, along similar lines would be also of great interest.

Acknowledgements

I acknowledge thankfully the financial support by the NATO Science Program. I thank scientific directors Bart van Tiggelen and Sergey Skipetrov, for inviting me as a lecturer to the 2002 NATO ASI "Wave Scattering in Complex Media: From Theory to Applications". I acknowledge the collaboration with V. Romanov and E. Aksenova, with whom a lot of work presented here was done. This work was partially supported by the Russian Foundation for Basic Research (Grant 02-02-16577).

Appendix: Exercises

1. The optical theorem for a scalar field reads

$$l^{-1} = k_0^4 \int d\Omega' G(\mathbf{k} - \mathbf{k}').$$

Let the binary permittivity correlation function be of Ornstein-Zernike form:

$$G(q) = G_0/(\xi^2 + q^2),$$

where G_0 is a constant. Find the photon and transport mean free paths, l and l^*. Find $\overline{\cos^n \theta}$ for this phase function.

2. As in Ex. 1, for the Heyney-Greenstein phase function

$$G\left(2k \sin\frac{\theta}{2}\right) \sim \frac{1 - \mu^2}{[1 + \mu^2 - 2\mu\cos\theta]^{3/2}}.$$

3. Let the permittivity of a two-phase system with a plane boundary be

$$\varepsilon(z) = \varepsilon_a \theta(-z) + \varepsilon_b \theta(z) = \varepsilon_a + \Delta\varepsilon\theta(z).$$

Solve the wave equation

$$\mathbf{E}(\mathbf{r}) = \mathbf{E}_a(\mathbf{r}) + \frac{\Delta\varepsilon}{4\pi} \int_{z_3 > 0} \hat{T}_a(\mathbf{r} - \mathbf{r}_3)\mathbf{E}(\mathbf{r}_3) d\mathbf{r}_3,$$

where $\hat{T}_a(r) = (\hat{I}k_0^2 - \nabla\nabla)\exp(-\sqrt{\varepsilon_a}k_0 r)/(\varepsilon_a r)$. Obtain the Ewald-Oseen extinction theorem, Snell's rule, and Fresnel formulas as results.

4. As in Ex. 3, the permittivity of a system with a plane boundary is

$$\varepsilon(z) = \varepsilon_a \theta(-z) + \varepsilon_b \theta(z) = \varepsilon_a + \Delta\varepsilon\theta(z).$$

Find the scalar field Green's function for this system by solving the wave equation (10). (Hint: perform 2D Fourier transform over coordinates tangential to the boundary). The Green's function of an infinite system with permittivity ε_a is $T_a(\mathbf{r}) = k_0^2 r^{-1} \exp(-\sqrt{\varepsilon_a} k_0 r)$.

5. Random electric field of a system of point-like dipoles is

$$E(\mathbf{r}) = E_0(\mathbf{r}) + \alpha \sum_i T_0(\mathbf{r} - \mathbf{r}_i) E_{eff}(\mathbf{r}_i),$$

where $T_0(R)$ is the Green's function. The effective, or local, field acting upon a particle is

$$E_{eff}(\mathbf{r}_i) = E_0(\mathbf{r}_i) + \alpha \sum_{j \neq i} T_0(\mathbf{r}_i - \mathbf{r}_j) E_{eff}(\mathbf{r}_j), \quad \lambda \to \infty.$$

The random polarization vector is $P(\mathbf{r}) = \alpha \sum_i E_{eff}(i)\delta(\mathbf{r} - \mathbf{r}_i)$. The averaged quantities are related by $\langle P \rangle = \frac{\epsilon - 1}{4\pi} \langle E \rangle$. Derive

$$\langle E_{eff} \rangle = \frac{\epsilon - 1}{4\pi\alpha\rho} \langle E \rangle.$$

6. The Milne solution (22) can be written explicitly as

$$\gamma_M(s,1,t) = \frac{3}{(1+s^*)(s+s^*)(1+s)} \exp[-J(s,t) - J(1,t)],$$

where

$$J(s,t) = \frac{s}{2\pi} \int_{-\infty}^{\infty} \frac{ds'}{s'^2+s^2} \ln\left[\frac{s^{*2} + \left(1 - 3s'^{-1}\arctan s'\right)}{s^{*2}+s'^2}\right], \quad s* \approx \sqrt{6t/\tau}.$$

Calculate $\gamma_M(s,1,t)$ numerically as a function of $\sqrt{t/\tau}$ for $s = 1$. Calculate $\gamma_M(s,1,t)$ as a function of the scattering angle θ_f defined by $s = \cos\theta_f$.

7. Calculate the generalized Milne extrapolation parameter

$$z^*(t) = \frac{-1}{2\pi} \int_{-\infty}^{\infty} \frac{ds'}{s'^2+s^2} \ln\left[\frac{s^{*2} + \left(1 - 3\frac{\arctan(s')}{s'}\right)}{s^{*2}+s'^2}\right]$$

as a function of t.

8. Substitute the Milne solution $\gamma_M(s,1,t)$ into the Milne equation (21) and expand the l.h.s and r.h.s. into series in the orders of s. Numerically test the obtained identities.

References

Akkermans, E., Wolf, P.E., Maynard, R., and Maret., G. (1988), *J. Phys. (Paris)* **49,** p. 77.

Amic, E., Luck, J.M., and Nieuwenhuizen, Th. M. (1996), *J. Phys. A* **29,** p. 4915.

Amic, E., Luck, J.M., and Nieuwenhuizen, Th. M. (1997), *J. Phys.* **17,** p. 445.

Boots, H.M.L., Bedeaux, D., and Mazur, P. (1975) *Physica* **79A,** p. 397.

Chandrasekhar, S. (1960)*Radiation Transfer*. Dover, New York.

Croxton, C.A. (1974) *Liquid state physics — a statistical mechanical introduction*. Cambridge University Press, Cambridge.

Felderhof, B.V. (1974) *Physica* **76,** p. 486.

Gorodnichev, E.E., Dudarev, S.L., and Rogozkin, D.B. (1990) Phys. Lett. A., **144,** p. 48.

Hoye, J.S., and Stell, G. (1976) *J. Chem. Phys.* **64,** p. 1952.

Kuzmin, V.L. (1985) *Phys. Rep.* **123,** p. 365.

Kuzmin, V.L., Romanov, V.P., and Zubkov, L.A. (1994) *Phys. Rep.* **248,** pp. 71–365.

Kuzmin, V.L., Romanov, V.P., and Obraztsov, E.P. (2001) *Optics and Spectroscopy* **91,** p. 913.

Kuzmin, V.L., and Romanov, V.P. (2002) *Europhys. Lett.* **59,** p. 206.

Kuzmin, V.L., Romanov V.P., and Aksenova E.V. (2002) *Phys. Rev. E* **65,** p. 016601.

Kuzmin, V.L. (2002) *Optics and Spectroscopy* **93,** p. 439.

MacKintosh, F.C., and John, S. (1989) *Phys. Rev. B* **40,** p. 2383.

Mishchenko, M. I. (1996) *J. Quant. Spectrosc. Radiat. Transf.* **56,** p. 673.

Mishchenko, M. I., Luck, J.-M., and Nieuwenhuizen, Th.M. (2000) *J. Opt. Soc. Am.A* **17,** p. 888.

Nieuwenhuizen, Th.M., and Luck, J.-M. (1993) *Phys. Rev. E* **48,** p. 569.

Pine D.J., Weitz, D.A., Chaikin, P.M., and Herbolzheimer, E. (1988) *Phys. Rev. Lett.* **60,** p. 1134.

Stell, G., and Lebowitz, J.L. (1968) *J. Chem. Phys.* **49,** p. 3706.

Van der Mark, M.B., Van Albada, M., and Lagendijk, A. (1988) *Phys. Rev. B* **37,** p. 3575.

Van Rossum, M.C.W., and Nieuwenhuizen, Th.M. (1999) *Rev. Mod. Phys.* **71,** p. 313.

Wertheim, M.S. (1978) *Mol. Phys.* **36,** p. 1217.

CHAPTER VI

OPTICS OF SOFT CONDENSED MATTER

Stripes of SmA (8CB) liquid crystal on a glass substrate. Size of the imaged area is 170 × 170 μm. Image provided by Alenka Mertelj, J. Stefan Institute, Ljubljana, Slovenia.

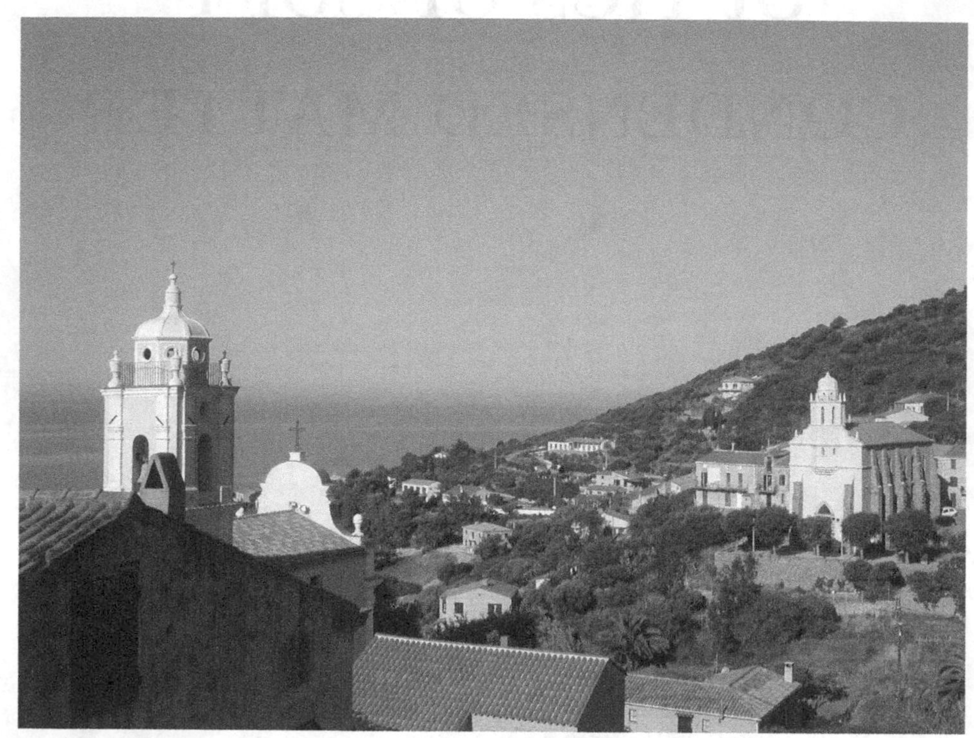

Cargèse village. Photo by Valentin Freilikher

RADIATION TRANSFER IN NEMATIC LIQUID CRYSTALS

V.P. ROMANOV
Department of Physics, St. Petersburg State University
Petrodvorets, St. Petersburg, 198504 Russia

Abstract. Optical properties of nematic liquid crystals are considered. We analyze the light scattering by director fluctuations, the singularities of the single light scattering and extinction coefficients of the ordinary and extraordinary beams. The multiple light scattering is considered in the eikonal approximation and in the framework of radiation transfer equation for anisotropic media. This equation is analyzed in small angle and diffusion approximation. For multi-domain liquid crystals the effect of the coherent backscattering is discussed.

1. Introduction

Liquid crystals are intermediate phases between liquids and crystalline solids. They can flow like an ordinary liquid but at the same time liquid crystals possess such a property as birefringence. As a rule, liquid crystals consist of strongly elongated molecules and this determines their unusual properties. The liquid crystalline phase is characterized by the orientational order although the centers of mass of the molecules are not ordered as they are in liquids. Two classes of liquid crystals can be distinguished: thermotropic and liotropic liquid crystalline phases. Single-component systems that exhibit anisotropic behavior in a certain temperature range are called thermotropic. Liotropic liquid crystalline phases show their anisotropic properties in solutions.

In what follows, we will consider thermotropic liquid crystals. The main classification of thermotropic liquid crystals with elongated molecules is into nematic and smectic mesophases. In the nematic phase the centers of mass of the molecules are distributed at random. In this sense they may be called anisotropic liquids. Smectic phases are characterized, in addition, by a positional order in at least one dimension: the centers of the molecules are

B.A. van Tiggelen and S.E. Skipetrov (eds.),
Wave Scattering in Complex Media: From Theory to Applications, 477–497.

located on average in equidistant planes. There exist many types of smectic phases designated as A, C, C^*, B etc. They differ in the molecular orientation with respect to the layers and also in the ordering of the molecular centers of mass in layers.

2. Order Parameter

From the general statistical point of view, the order in liquid crystals is described by the distribution function of molecules over angles and positions of the centers of mass [1]. Here we are mainly interested in optical properties of liquid crystals and, in particular, in light scattering. Therefore one can describe the ordering by the second rank tensor $R_{\alpha\beta}(\mathbf{r},t)$. It is convenient to consider separately the orientational order parameter

$$S_{\alpha\beta}(\mathbf{r},t) = R_{\alpha\beta}(\mathbf{r},t) - \frac{1}{3}\delta_{\alpha\beta}R_{\gamma\gamma}(\mathbf{r},t), \tag{1}$$

which is a traceless tensor, and the scalar parameter

$$\rho(\mathbf{r},t) = \frac{1}{3}R_{\gamma\gamma}(\mathbf{r},t), \tag{2}$$

characterizing the microscopic density. For optical problems it is convenient to choose $S_{\alpha\beta}(\mathbf{r},t)$ as the traceless part of permittivity tensor $\varepsilon_{\alpha\beta}(\mathbf{r},t)$ at optical frequency, neglecting the temporal and spatial dispersions of this tensor. We will assume that the tensor $S_{\alpha\beta}(\mathbf{r},t)$ is real and symmetric.

The equilibrium tensor $S^0_{\alpha\beta}(\mathbf{r})$ can be diagonalized in the local coordinate frame $\{\mathbf{e}_1(\mathbf{r}), \mathbf{e}_2(\mathbf{r}), \mathbf{e}_3(\mathbf{r})\}$:

$$S^0_{\alpha\beta}(\mathbf{r}) = \sum_{j=1}^{3} S_j(\mathbf{r})e_{j\alpha}(\mathbf{r})e_{j\beta}(\mathbf{r}), \tag{3}$$

where $S_j(\mathbf{r})$ are the eigenvalues of the tensor $S^0_{\alpha\beta}$ satisfying the relation $S_1 + S_2 + S_3 = 0$. If two of the three eigenvalues are equal, the medium is said to be uniaxial, otherwise it is biaxial. If $\mathbf{e}_j$ and S_j do not depend on $\mathbf{r}$, then the system is spatially homogeneous. This system represents a nematic liquid crystal.

In smectic liquid crystals, the density $\rho_0(\mathbf{r})$ is one-dimensionally periodic. Usually, it is taken as

$$\rho_0(\mathbf{r}) = \rho_0 + \Psi_0 \cos(\mathbf{q}_0\mathbf{r} - \phi_0), \tag{4}$$

where $\frac{2\pi}{q_0}$ is the period of smectic layers.

For uniaxial nematics and smectics-A, the equilibrium value of the tensor order parameter $S^0_{\alpha\beta}$ can be conveniently written in the form

$$S^0_{\alpha\beta} = S_0(n_\alpha n_\beta - \frac{1}{3}\delta_{\alpha\beta}). \tag{5}$$

where $\mathbf{n}^0$ is the equilibrium unit vector director, $\mathbf{n}^0 = \mathbf{e}_3$, $S_1 = S_2 = -\frac{1}{3}S_0$, $S_3 = \frac{2}{3}S_0$. The scalar S_0 is a measure of the alignment of the molecules, $S_0 = \frac{1}{2}(3\langle\cos^2\theta\rangle - 1)$, where θ is the angle between the long molecular axis and the director.

Fluctuations in orientation are described by

$$\phi_{\alpha\beta}(\mathbf{r},t) = S_{\alpha\beta}(\mathbf{r},t) - S^0_{\alpha\beta}. \tag{6}$$

The tensor $\phi_{\alpha\beta}(\mathbf{r},t)$ has, in general, five independent fluctuation modes: ξ_1, ξ_2, ..., ξ_5. It contributes to fluctuations of the permittivity tensor $\delta\varepsilon_{\alpha\beta}$ and the pair correlation function

$$G_{\alpha\beta\gamma\delta}(\mathbf{r}_1,\mathbf{r}_2,t_2 - t_1) = \langle\delta\varepsilon_{\alpha\beta}(\mathbf{r}_1,t_1)\delta\varepsilon_{\gamma\delta}(\mathbf{r}_2,t_2)\rangle\,. \tag{7}$$

This function determines, in particular, the intensity of single-scattered light at a point $\mathbf{r}$ in Born approximation (see, e.g., [9]):

$$\begin{aligned} I(\mathbf{e}^{(s)},\mathbf{e}^{(i)},\omega) &= I_0\int_{V_s} d\mathbf{r}_1 d\mathbf{r}_2 e^{(s)}_\alpha e^{(s)}_\beta T_{\alpha\gamma}(\mathbf{r},\mathbf{r}_1,\omega+\omega_0)T^*_{\beta\delta}(\mathbf{r},\mathbf{r}_2,\omega+\omega_0) \\ &\times\ G_{\gamma\mu\delta\nu}(\mathbf{r}_1,\mathbf{r}_2,\omega)E^{(i)}_0(\mathbf{r}_1)E^{(i)}_0(\mathbf{r}_2)e^{(i)}_\mu e^{(i)}_\nu. \end{aligned} \tag{8}$$

Here V_s is the scattering volume, $\mathbf{e}^{(i)}$ and $\mathbf{e}^{(s)}$ are the polarization vectors of the incident and scattered light,

$$\mathbf{E}^{(i)}_0(\mathbf{r},t) = \mathbf{e}^{(i)}E^{(i)}_0(\mathbf{r})e^{-i\omega_0 t} \tag{9}$$

is the incident field, $T_{\alpha\beta}(\mathbf{r},\mathbf{r}_1,\omega)$ is the Green function of Maxwell equations in $(\mathbf{r},\omega)$-representation. For repeated indices, the summation rule is assumed.

3. Fluctuations of Orientation

In uniaxial nematic liquid crystals with equilibrium director $\mathbf{n}^0$, it is convenient to parametrize the fluctuations of the tensor order parameter $\phi_{\alpha\beta}$ in the form [2, 3]

$$\phi_{\alpha\beta} = \phi^{\perp(1)}_{\alpha\beta} + \phi^{\perp(2)}_{\alpha\beta} + \phi^{\parallel}_{\alpha\beta} \tag{10}$$

where

$$\begin{aligned}\phi^{\perp(1)}_{\alpha\beta} &= \xi_1(\mathbf{r})(n^0_\alpha e_{1\beta} + n^0_\beta e_{1\alpha}) + \xi_2(\mathbf{r})(n^0_\alpha e_{2\beta} + n^0_\beta e_{2\alpha}),\\ \phi^{\perp(2)}_{\alpha\beta} &= \xi_3(\mathbf{r})(e_{1\alpha}e_{2\beta} + e_{1\beta}e_{2\alpha}) + \xi_4(\mathbf{r})(e_{1\alpha}e_{1\beta} - e_{2\beta}e_{2\alpha}),\\ \phi^{\|}_{\alpha\beta} &= \sqrt{3}\xi_5(\mathbf{r})(n^0_\alpha n^0_\beta - \frac{1}{3}\delta_{\alpha\beta}). \end{aligned} \quad (11)$$

Here $\mathbf{e}_1, \mathbf{e}_2, \mathbf{n}^0$ are the basis vectors. The modes ξ_1 and ξ_2 are transverse uniaxial, the modes ξ_3 and ξ_4 are transverse biaxial, and the mode ξ_5 is longitudinal. All types of fluctuations are illustrated in Fig. 1 by possible motions of the dielectric ellipsoid. The modes ξ_1 and ξ_2 describe the shift of the ellipsoid axis with respect to $\mathbf{n}^0$. They correspond to the director fluctuations. The modes ξ_3 and ξ_4 are responsible for the occurrence of random biaxiality, and the mode ξ_5 describes elongation and contraction of the ellipsoid.

The main modes are the transverse uniaxial modes, or director fluctuations:

$$\mathbf{n}(\mathbf{r}) = \mathbf{n}^0 + \delta\mathbf{n}. \quad (12)$$

Since the director is a unit vector, the equality $(\mathbf{n}^0 \cdot \delta\mathbf{n}) = 0$ holds in the linear approximation in $\delta\mathbf{n}$. ξ_1/S_0 and ξ_2/S_0 are projections of the director fluctuations onto the axes $\mathbf{e}_1$ and $\mathbf{e}_2$, and

$$\delta\mathbf{n} = \frac{\xi_1\mathbf{e}_1 + \xi_2\mathbf{e}_2}{S_0}. \quad (13)$$

The director fluctuations are calculated using the elastic free energy, or Frank energy [1]:

$$F_e = \frac{1}{2}\int dV[K_{11}(\mathrm{div}\mathbf{n})^2 + K_{22}(\mathbf{n}\ \mathrm{curl}\ \mathbf{n})^2 + K_{33}(\mathbf{n}\times \mathrm{curl}\ \mathbf{n})^2 - \chi_a(\mathbf{nH})^2]. \quad (14)$$

Here K_{jj} are Frank modules, $j = 1, 2, 3$, χ_a is magnetic susceptibility, $\mathbf{H}$ is magnetic field. If we consider the electric field, the following substitutions are in order:

$$\mathbf{H} \to \mathbf{E}, \qquad \chi_a \to \frac{\varepsilon}{4\pi}. \quad (15)$$

Director fluctuations are calculated using the distribution function

$$\rho \sim \exp\left(-\frac{F_e}{k_B T}\right). \quad (16)$$

In the Gaussian approximation and after Fourier transformation we get [1, 4]

$$\left\langle |\delta n_{j\mathbf{q}}|^2 \right\rangle = \frac{V k_B T}{q_\perp^2 K_{jj} + q_\|^2 K_{33} + \chi_a H^2}, \quad (17)$$

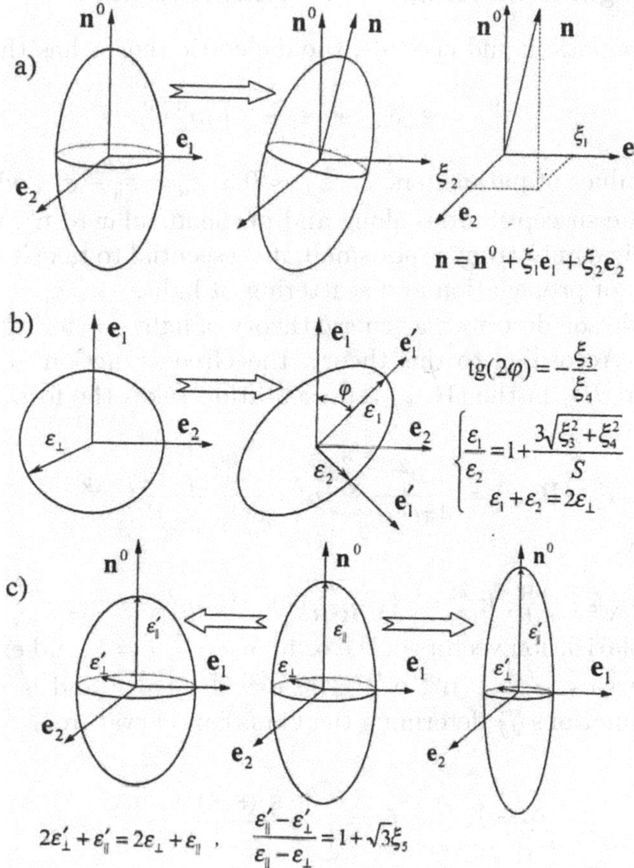

Figure 1. Three types of the dielectric ellipsoid fluctuations for a nematic liquid crystal: a) fluctuations of the director, b) biaxial fluctuations, c) longitudinal fluctuations, or the fluctuations of the ordering degree. Arrows denote the form of deviations from the equilibrium state.

for $j = 1, 2$. It it is assumed here that $\chi_a > 0$ and $\mathbf{H} \parallel \mathbf{n}^0$. This formula is written in the coordinate frame

$$\mathbf{e}_1 = \frac{\mathbf{q}_\perp}{q_\perp}, \ \mathbf{e}_2 = \mathbf{n}_0 \times \frac{\mathbf{q}_\perp}{q_\perp}, \ \mathbf{e}_3 = \mathbf{n}^0. \tag{18}$$

We note that the relation $\langle|\delta n_{j\mathbf{q}}|^2\rangle \sim 1/q^2$ for $H = 0$ resembles the fluctuations in the points of a second-order phase transition. However, external field makes $\langle|\delta n_{j\mathbf{q}}|^2\rangle$ finite for $q = 0$.

4. Single Light Scattering in Anisotropic Media

In uniaxial nematic liquid crystals, the dielectric tensor has the form [1]

$$\varepsilon^0_{\alpha\beta} = \varepsilon_\perp \delta_{\alpha\beta} + (\varepsilon_\parallel - \varepsilon_\perp) n^0_\alpha n^0_\beta. \tag{19}$$

For typical values of parameters, $\varepsilon_a/\varepsilon_\perp \sim 0.3$, $\varepsilon_a = \varepsilon_\parallel - \varepsilon_\perp$, where $\varepsilon_a = \varepsilon_\parallel$ and $\varepsilon_\perp$ are the susceptibilities along and perpendicular to $\mathbf{n}^0$, respectively. Since the optical anisotropy is not small, it is essential to take it into account in the theory of propagation and scattering of light.

Lax and Nelson described a general theory of light scattering in anisotropic media [5, 6]. According to this theory, the Green function of the electromagnetic field $T_{\alpha\beta}$ in the $(\mathbf{R}, \omega)$-representation takes the form

$$T_{\alpha\beta}(\mathbf{R}, \omega) = \frac{\omega^2}{4\pi c^2 R} \sum_{j=1}^{2} n^2_{(j)} \frac{e^{(j)}_\alpha e^{(j)}_\beta}{e^{(j)}_\mu \varepsilon_{\mu\nu} e^{(j)}_\nu} f_j e^{i\mathbf{k}^{(j)}_{st}\mathbf{R}}, \tag{20}$$

where $\mathbf{k}^{(1)}_{st} = \sqrt{\varepsilon_\perp}\frac{\omega}{c}\frac{\mathbf{R}}{R}$, $\mathbf{k}^{(2)}_{st} = \frac{\omega}{c}\sqrt{\frac{\varepsilon_\parallel \varepsilon_\perp}{\mathbf{R}\hat{\varepsilon}\mathbf{R}}}(\hat{\varepsilon})^{-1}\mathbf{R}$, $n^2_{(1)} = \varepsilon_\perp$, $n^2_{(2)} = \frac{\varepsilon_\parallel \varepsilon_\perp}{\mathbf{s}\hat{\varepsilon}\mathbf{s}}$, $\mathbf{s} = \frac{\mathbf{k}}{k}$, $\mathbf{e}^{(j)}$ are polarization vectors of the ordinary (o), $j = 1$, and extraordinary (e), $j = 2$, waves; $\mathbf{e}^{(1)} \perp \mathbf{n}^0$, $\mathbf{e}^{(1)} \perp \mathbf{s}$, $\mathbf{e}^{(2)} \perp (\hat{\varepsilon})^{-1}\mathbf{s}$ and is in the plane $(\mathbf{s}, \mathbf{n}^0)$; the functions f_j determine the Gaussian curvature:

$$f_1 = 1, \qquad f_2 = \left[\frac{(\mathbf{s}\hat{\varepsilon}\mathbf{s})(\mathbf{s}(\hat{\varepsilon})^2\mathbf{s}))}{\varepsilon_\parallel \varepsilon^2_\perp}\right]^{1/2} \tag{21}$$

Equation (8) for single light scattering in uniaxial anisotropic media can be rewritten as

$$I(\mathbf{e}^{(i)}, \mathbf{e}^{(s)}) = I^{(i)}_0 \frac{n_{(s)} f^2_s}{n_{(i)} \cos\delta^{(i)} \cos^3\delta^{(s)}} e^{(s)}_\nu e^{(s)}_\mu G_{\nu\rho\mu\eta} e^{(i)}_\rho e^{(i)}_\eta, \tag{22}$$

where $\cos\delta^{(j)} = (\mathbf{e}^{(j)}, \hat{\varepsilon}^0\mathbf{e}^{(j)})^{1/2}/n_{(j)}$ is the cosine of the angle between the vectors $\mathbf{e}^{(j)}$ and $\hat{\varepsilon}^0\mathbf{e}^{(j)}$ directed along the electric field and induction vectors, respectively. For the ordinary beam, $\delta^{(1)} = 0$.

The correlation function

$$G_{\nu\rho\mu\eta} = \langle \delta\varepsilon_{\nu\rho} \delta\varepsilon_{\mu\eta} \rangle_{\mathbf{q}_{sc} = \mathbf{k}^{(i)} - \mathbf{k}^{(s)}} \tag{23}$$

is determined mainly by the director fluctuations

$$\delta\varepsilon_{\alpha\beta} = \delta(\varepsilon_\perp \delta_{\alpha\beta} + \varepsilon_a n_\alpha n_\beta) = \varepsilon_a(n_\alpha \delta n_\beta + n_\beta \delta n_\alpha) \tag{24}$$

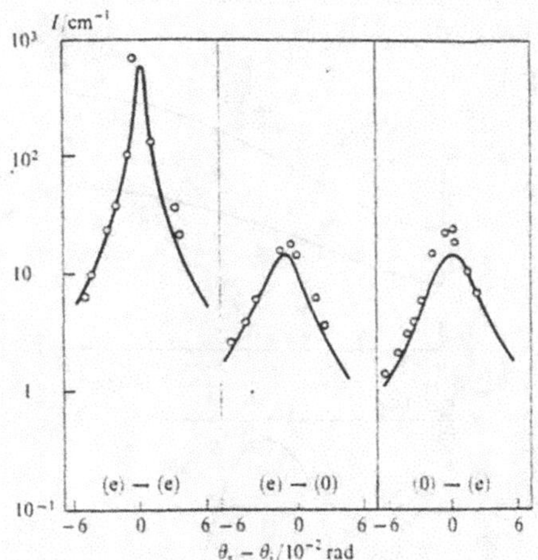

Figure 2. Angular dependence of the normalized intensity I of the scattered light for three types of scattering $(e) \to (e)$, $(e) \to (o)$, and $(o) \to (e)$ in a 1 mm-thick MBBA sample; θ_i (θ_s) is the angle between the wave vectors of the incident (scattered) light and the director. The angle θ_i is fixed at 25°. Circles are experimental data, lines were calculated using Eq. (22) for known values of the Frank moduli [7].

and has the form

$$\langle \delta\varepsilon_{\nu\rho}\delta\varepsilon_{\mu\eta}\rangle = \varepsilon_a^2 \left\langle (n_\nu \delta n_\rho + n_\rho \delta n_\nu)(n_\mu \delta n_\eta + n_\eta \delta n_\mu)\right\rangle \tag{25}$$

Note that if the incident and the scattered beams are ordinary, $\mathbf{e}^{(1)} \perp \mathbf{n}^0$, then according to Eq. (22) the intensity of single-scattered light is zero. Hence, three types of scattering are possible:

$$o \to e, \qquad e \to o, \qquad e \to e.$$

The angular dependence of all types of light scattering is shown in Fig. 2. It is seen that forward $e \to e$ scattering is much stronger than the other types of scattering.

Due to its strong fluctuations, the effective susceptibility $\varepsilon_{\alpha\beta}^{eff}$ is nonlocal [8]. In the lowest order, the nonlocal part of $\varepsilon_{\alpha\beta}^{eff}$ has the form

$$\varepsilon_{\alpha\beta}^{eff}(\mathbf{k}) = \varepsilon_{\alpha\beta}^0 + \int \frac{d\mathbf{q}}{(2\pi)^3} G_{\alpha\mu\beta\nu}(\mathbf{k}-\mathbf{q}) T_{\mu\nu}(\mathbf{q}). \tag{26}$$

The imaginary part of this expression determines the extinction coefficient of light. Figure 3 shows the dependence of the extinction coefficients on the

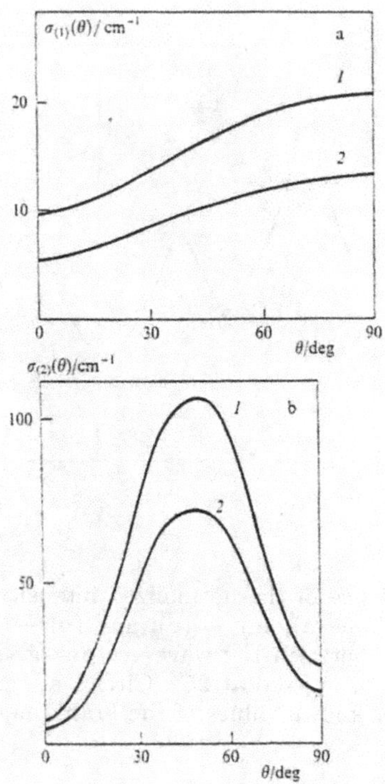

Figure 3. Calculated angular dependence of extinction coefficients of ordinary (a) and extraordinary (b) beams for PAA (curve 1) and MBBA (curve 2) nematic liquid crystals.

angle θ between the wave vector and the director for ordinary and extraordinary beams [9]. As it is seen from the figure, the extinction coefficient $\sigma^{(e)}$ is much greater than $\sigma^{(o)}$ and exhibits a sharp dependence on θ.

The difference behavior of ordinary and extraordinary beams may be explained in the following way. The ordinary beam has losses due to scattering of $o \rightarrow e$ type only. In this case, the scattering vector $q_{sc} = |\mathbf{k}^{(i)} - \mathbf{k}^{(s)}|$ does not become zero even for forward scattering, when $\mathbf{k}^{(i)} \parallel \mathbf{k}^{(s)}$, since the refraction indices $n_0 \neq n_e$. As far as the director fluctuations depend on q as $1/q^2$, the scattered intensity of the ordinary beam is finite for all scattering angles.

For extraordinary beam, the scattering of $e \rightarrow e$ type can take place when $q_{sc} = 0$ in the case of forward scattering. It leads to an infinite light scattering intensity and to the divergence of the extinction coefficient $\sigma^{(e)}$. In practice its value is bounded either by a geometrical factor or by an external field. In particular, in the presence of external field, the extinction coefficient $\sigma^{(e)}$ has the form

$$\sigma^{(e)} = \sigma_0 U(\theta) \frac{t_1(F_1+F_2)}{F_1(t_1F_1+t_2F_2)} \ln \frac{L_H}{2\lambda}, \tag{27}$$

where

$$U(\theta) = (\varepsilon_\| \varepsilon_\perp)^{3/2} \frac{\sin^2 2\theta}{2(\varepsilon_\| \sin^2\theta + \varepsilon_\perp \cos^2\theta)^2},$$

$$F_j(\theta) = (t_j^2 \varepsilon_\|^2 \cos^2\theta + t_j \varepsilon_\perp^2 \sin^2\theta)^{1/2}, \tag{28}$$

and $t_j = \frac{K_{jj}}{K_{33}}$, $j = 1, 2$, $\sigma_0 = \frac{k_B T \omega^2 (\varepsilon_\| - \varepsilon_\perp)^2}{8\pi c^2 K_{33} (\varepsilon_\| \varepsilon_\perp)^{1/2}}$. Here $L_H = 2\pi[K_{33}/(\chi_a H^2)]^{1/2}$ is the magnetic coherence length. For a finite sample in the absence of external field, L_H is replaced by L.

The obtained results are sufficient for the description of extinction in real optical experiments. However the fundamental question of the extinction coefficient $\sigma^{(e)}$ behavior in an unbounded nematic liquid crystal in the absence of external field remains obscure. It was found that the divergence of the extinction coefficients for $H = 0$ and $L \to \infty$ is due to the fact that the true asymptotic behavior of the mean field involves a superexponential attenuation (see, e.g., [10]):

$$\left\langle E^{(e)}(\mathbf{r}) \right\rangle = \mathbf{e}^{(e)} E_0^{(e)} e^{i\mathbf{k}^{(e)}\mathbf{r}} e^{-k''^{(e)} r \ln(k''^{(e)} r)}.$$

This superexponential attenuation has a transparent physical meaning. Two mechanisms exist for the attenuation of incident light. The usual mechanism of the mean field attenuation which leads to the exponential Beer law: $E^{(e)}(\mathbf{r}) \sim e^{-k''^{(e)} r}$ due to scattering. The second mechanism is due to the anomalously strong forward scattering. The decay of the coherent radiation is due to the superposition of waves with randomly displaced phases scattered in the (almost) forward direction. A correct calculation of the mean field in directions close to $\theta_{sc} = 0$ can be carried out in the eikonal approximation [11].

5. Multiple Light Scattering in Nematics

It may appear that relatively small extinction coefficient for o-waves compared to e-waves leads to a much slower decay of the coherent o-beam. The large extinction coefficient $\sigma^{(e)}$ is caused by random phase shifts of extraordinary waves scattered in the forward direction, so that almost all the energy of light is scattered at $\theta_{sc} \approx 0$ with the polarization of extraordinary beam. Thus the incident coherent laser beam with e polarization is transformed into a diffuse beam with a relatively small beam expansion and without significant change neither in total intensity nor in polarization. The reduced 'extinction coefficient' of this diffuse beam is of the same order of magnitude as $\sigma^{(o)}$ due to o-scattering at large angles. This effect has been confirmed in the following experiment [12]. The cell having 2 mm thickness and 3 cm width filled with liquid crystal N-106 was prepared. The orientation of the nematic was planar. Quality control has been carried out with the help of the polarization microscope. If disclinations appeared, the sample was placed in a magnetic field $H \sim 3 \times 10^3$ G, which led to a total disappearance of disclination within 3–5 min. The beam of a He-Ne laser at $\lambda = 633$ nm was transmitted through the cell at various angles of incidence. The beams with o and e polarizations were studied. The transmitted waves are presented in Fig. 4. It is seen (Fig. 4, left panel) that the shape of the ordinary beam does not depend noticeably on the angle θ between the beam and the director. The bright central spot with o polarization is surrounded by a weak background with e polarization. This result is understandable since the extinction of the ordinary beam is not large and there is no $o \to o$ scattering.

Right panel of Fig. 4 illustrates the results for the extraordinary beam. In the center there is a diffuse spot with an angle width increasing up to 1° with θ variation from 90° to 60° (the limit refraction angle in this experiment is equal to 52°). The transmitted beam consists of small bright spots, this structure slowly varies in time. The weak background consists of o and e polarizations.

The theoretical study of multiple scattering in an ordered nematic for $\theta_{sc} \approx 0$ in the absence of external field, $H \to 0$, is of considerable interest. Examination of the analogous scalar problem by the eikonal method [11], by the Glauber method [13], and on the basis of the radiation transfer equation in the small angle approximation [14] showed that the intensity of multiple scattered light $I_{mult}(\theta)$ for small angles differs significantly from the single scattering result $I \sim \theta^{-2}$. According to Refs. [11, 13–15], the singularity of $I_{mult}(\theta)$ at $\theta \to 0$ has the form

$$I_{mult}(\theta) \sim \theta^{-(2-z/z_0)}, \tag{29}$$

where z is the depth of light penetration in the medium and z_0 is the length of the order of $(\sigma^{(o)})^{-1}$. In Ref. [16], this result has been extended to the case of realistic nematics, where it is necessary to take into account the vector nature of the electromagnetic field and the tensor nature of the correlation function. The form of the singularity defined by Eq. (29) in the multiple scattering regime at $z/z_0 < 2$ is retained in this case also. The reason is that the anomalously intense forward scattering is associated primarily with the extraordinary wave and therefore the presence of the ordinary beam may be neglected, i.e. the problem is essentially scalar.

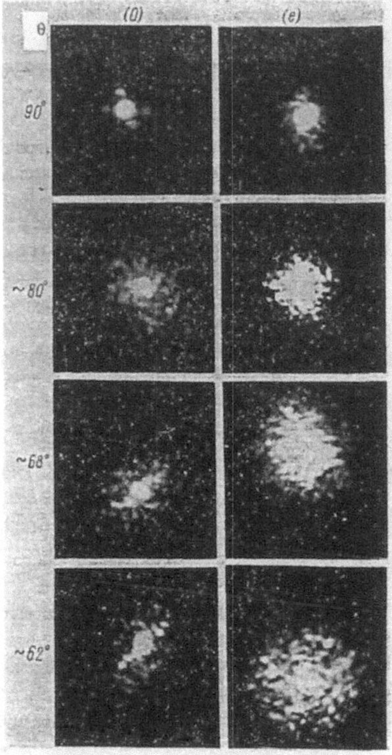

Figure 4. Image of a beam which has passed through a cell with planary oriented 2 mm-thick H-106 liquid crystal for four angles of incidence θ of the ordinary (o) and extraordinary (e) rays.

On the other hand if the problem of multiple scattering in an ordered nematic liquid crystal in the presence of external field H is considered, the correlation length r_c of the director fluctuations becomes finite and the ordinary beam cannot be neglected. The principal problem in this case consists in the extension of the eikonal approximation to the two-mode problem when the so-called linear interactions of the modes arise [17]. In this case, waves with one type of polarization enter the medium and, after multiple scattering by tensor fluctuations, waves with both types of polarization exit from the medium.

6. Radiation Transfer Equation

There exists another approach to description of the angular distribution of the scattered radiation which is closely related to the eikonal approximation in the one-mode problem. However, in contrast to the latter, it permits a direct extension to the two-mode case. This approach is based on the use of the radiation transfer equation for description of propagation of both the extraordinary and ordinary beams. In order to apply the radiation transfer equation, the inequality $\sigma^{(i)} r_c \ll 1$ should hold [8, 18]. In experiments with ordered nematics of 1–5 cm thickness, the magnetic field $H \sim 10^3$ G is generally applied. If we calculate the extinction coefficient using the magnetic coherence length L_H as r_c, the conditions of applicability of the radiation transfer equation are fulfilled. For a scalar field in the stationary case this equation has the form

$$[(\mathbf{m}\nabla) + \sigma]a(\mathbf{r}, \mathbf{m}) = \int d\Omega_{\mathbf{m}'} F(\mathbf{m}, \mathbf{m}')a(\mathbf{r}, \mathbf{m}') + b(\mathbf{r}, \mathbf{m}). \tag{30}$$

where $\mathbf{m}$ is a unit vector, σ is the extinction coefficient, F is the phase function, $a(\mathbf{r}, \mathbf{m})$ is the radiation intensity at the point $\mathbf{r}$ in direction $\mathbf{m}$, b is the radiation source. The coherence function characterizing the radiation [8] has in the case of electromagnetic field four independent components due to the transverse nature of electromagnetic waves. In this case four Stokes parameters are usually employed. One then obtains, instead of Eq. (30), a set of four equations. As far as in anisotropic medium there are two types of waves with their own polarizations and wave vectors, the number of independent components of the coherence function reduces to two [19]:

$$[(\mathbf{s}^{(i)}\nabla) + \sigma_{(i)}]a_{(i)}(\mathbf{r}, \mathbf{m}) = \sum_{j=1}^{2} \int d\Omega_{\mathbf{m}'} F_{(ij)}(\mathbf{m}, \mathbf{m}')a_{(ij)}(\mathbf{r}, \mathbf{m}' + b_{(i)}(\mathbf{r}, \mathbf{m}). \tag{31}$$

Here $i = 1, 2$, $\mathbf{s}^{(i)}$ is the direction of the group velocity of a wave of type (i) with the direction of wave vector $\mathbf{m}$, $F_{(ij)}(\mathbf{m}, \mathbf{m}')$ is the phase function for scattering of a wave of type (j) into a wave of type (i), $b_{(i)}(\mathbf{r}, \mathbf{m})$ is the source of type-(i) waves. Note that in contrast to isotropic medium, the difference between the directions of the group velocity and the wave vector is taken into account.

Equations (31) have a simple physical meaning. The variation of the energy along the beam for a wave of type (i), term $(\mathbf{s}^{(i)}\nabla)a_{(i)}(\mathbf{r}, \mathbf{m})$, takes place as a result of (i) scattering losses in other directions and transformation into other types of waves, the term $\sigma_{(i)}a_{(i)}(\mathbf{r}, \mathbf{m})$; (ii) scattering of waves of type (j) with wave vectors $k_{(j)}\mathbf{m}'$ in waves of type (i) with wave vector $k_{(i)}\mathbf{m}$, the integral term; (iii) emission from the source, the term $b_{(i)}$.

A characteristic feature of this set of equations for a nematic liquid crystal is that the phase functions $F_{(ij)}$ for distinct types of waves, i and j, are significantly different. The phase function $F_{(22)}$ has a sharp peak at $\mathbf{m} = \mathbf{m}'$ since the extraordinary ray is scattered mainly in the forward direction. The phase functions $F_{(12)}$ and $F_{(21)}$ do not have such a maximum

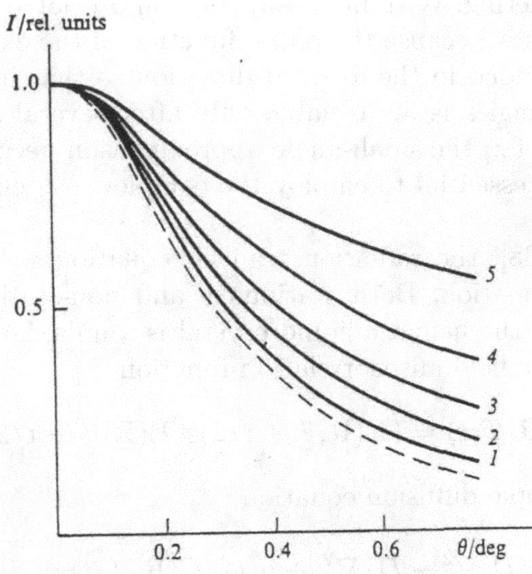

Figure 5. Angular distribution of the intensity of scattered light with polarization of the extraordinary ray at different depths $l = \sigma_{(e)}z$, where the extiction coefficient of the extraordinary beam $\sigma_{(e)}$ is calculated in the presence of external field: $l = 2$ (1), $l = 4$ (2), $l = 6$ (3), $l = 8$ (4), $l = 10$ (5). Dashed line is the intensity of single-scattered light.

for forward scattering because the wave numbers $k_{(1)}$ and $k_{(2)}$ are different. The quantity $F_{(11)}$ is identically zero since the scattering of an ordinary beam to an ordinary beam due to director fluctuations, $(o \to o)$, does not occur. Equations (31) describe the joint propagation of the ordinary and extraordinary waves in nematics taking into account multiple scattering. However, in the general case, integration of Eqs. (31) is possible only numerically since the extinction coefficients depend on the direction $\mathbf{m}$, while the phase functions $F_{(ij)}$ depend on vectors $\mathbf{m}$ and $\mathbf{m}'$ separately.

The problem is significantly simplified when the extraordinary beam is incident in $\mathbf{m}_0$ direction. Then in the region of small scattering angles $F_{(22)} \gg F_{(21)}$ and the term taking into account the scattering of ordinary waves into extraordinary waves in the equation for $a_{(2)}(\mathbf{r}, \mathbf{m})$ may be neglected. A closed equation is obtained then for $a_{(2)}(\mathbf{r}, \mathbf{m})$ which can be considered in terms of the small-angle approximation [8, 18], i.e. with a replacement $(\mathbf{s}, \nabla) \to (\mathbf{m}_0, \nabla)$.

Angular intensity distributions of the extraordinary waves for different penetration depths z, obtained in [19] on the basis of the small-angle approximation, are shown in Fig. 5. Evidently, for z less than several extinction lengths, the angular intensity distribution is approximately the same as for single scattering. With increasing of z, appreciable broadening of the beam starts. This is because the phase function of the extraordinary beam is extremely extended in the forward direction so that its variation in the region of small angles is appreciable only after several scattering events. For large values of z, the small-angle approximation becomes inapplicable. In this case, it is essential to employ the complete system of the radiation transfer equations.

In Refs. [33–38], the radiation transfer equation was obtained in the diffusion approximation. Both stationary and non-stationary cases were considered. Since the nematic liquid crystal is a uniaxial system the time-dependent electric field autocorrelation function

$$G(\mathbf{R}, T, t) = \left\langle \mathbf{E}(\mathbf{R}, T + t/2)\hat{\varepsilon}^0 \mathbf{E}(\mathbf{R}, T - t/2) \right\rangle, \tag{32}$$

obeys an anisotropic diffusion equation

$$\left[\frac{\partial}{\partial T} - D_{\parallel}\nabla_{\parallel}^2 - D_{\perp}\nabla_{\perp}^2 + \mu(t) \right] G(\mathbf{R}, T, t) = J(\mathbf{R}, T), \tag{33}$$

where $J(\mathbf{r}, T)$ is the source term, $D_{\parallel}$ and $D_{\perp}$ are the photon diffusion coefficients for the directions along and perpendicular to $\mathbf{n}_0$, respectively; $\mu(t)$ is the dynamic absorption coefficient measured in the diffusing wave spectroscopy (DWS) experiments.

In an isotropic system the photon diffusion coefficient may be defined as

$$D \equiv \frac{1}{6}\frac{\langle R^2 \rangle}{\langle T \rangle} = \frac{1}{3}\bar{c}l^*, \tag{34}$$

where $\langle T \rangle$ is the average time, $\langle R^2 \rangle$ is the mean square displacement of a photon, $\bar{c}$ is the average velocity of light, l^* is the transport length. Diffusion of multiple-scattered light was studied in static and time resolved experiments. In static experiments one measures the average 'step' of the photon random walk l^*, while dynamic experiments give the time evolution of the diffusion process, i.e. measures the diffusion coefficient.

7. Coherent Backscattering in Multi-Domain Liquid Crystals

Multiple light scattering is of great importance in multi-domain liquid crystals. Such a liquid crystal is thought to consist of small domains of the order of several microns with random local ordering. These domains exhibit no sharp boundaries, changing their orientations continuously. Therefore the light is assumed to be scattered by local fluctuations of the order parameter.

The features of multiple light scattering are very pronounced in the phenomenon of coherent backscattering. This effect was experimentally observed for the first time by Van Albada and Lagendijk [20] and Wolf and

Maret [21]. It manifests itself as a sharp enhancement of scattered intensity in the backward direction, in a narrow angular range $\theta \sim \lambda/l_{tr}$, where l_{tr} is the transport length. The physical mechanism underlying coherent backscattering is quite simple. Coherent plane waves incident upon a turbid sample become generally incoherent due to random inhomogeneities except for wave paths that pass some sequence of scatterers in opposite directions. These waves can interfere constructively. However, their interference is important only in backward direction. Such an interference can exist for any wave phenomena. The theory of this phenomenon is based on the summation of series in order of scattering. This series corresponds to an infinite sequence of ladder and cyclic diagrams. Ladder diagrams are important for all scattering angles, whereas cyclic diagrams are essential for backward direction only and determine the coherent backscattering phenomenon. The backscattering problem was solved analytically for pointlike scatterers in Refs. [22–25] for a scalar wave field and in Refs. [23, 26–28] for the electromagnetic field. Stephen and Cwilich [23] took into consideration the polarization effects. They showed that the backscattering peak occurs in both polarized and depolarized components. The magnitude of the peak of the polarized component is five to seven times higher than that of the depolarized component, and its shape is close to triangular, while the shape of the depolarized component is close to Lorentzian. These results have been confirmed in numerous light scattering experiments in latex suspensions [20, 21, 29–32].

We consider the scattering of light in a multi-domain liquid crystal, assuming that the fluctuations of the permittivity are described by a traceless tensor order parameter. Its correlation function has the form

$$\langle \Delta\hat{\epsilon}(\mathbf{r}_i)\Delta\hat{\epsilon}(\mathbf{r}_j)\rangle = \delta(\mathbf{r}_i - \mathbf{r}_j)\,\langle \Delta\hat{\epsilon}\Delta\hat{\epsilon}\rangle\,, \tag{35}$$

where

$$\langle \Delta\epsilon_{\alpha\beta}\Delta\epsilon_{\gamma\delta}\rangle = \frac{a_0}{2}\left(\delta_{\alpha\gamma}\delta_{\beta,\delta} + \delta_{\alpha\delta}\delta_{\beta\gamma} - \frac{2}{3}\delta_{\alpha\beta}\delta_{\gamma\delta}\right). \tag{36}$$

This function describes the correlation of orientation fluctuations in an isotropic medium.

To reveal the main features of coherent backscattering from purely anisotropic fluctuations, we first consider the scattering in the strictly backward direction within the double scattering approximation. Ladder diagram contributes

$$\begin{aligned} I^{\beta}_{2\alpha}(L) &= \int d\mathbf{r}_1 d\mathbf{r}_2 d\mathbf{r}'_1 d\mathbf{r}'_2 T_{\alpha\alpha_1}(\mathbf{r}_0 - \mathbf{r}_2)T^*_{\alpha\alpha_2}(\mathbf{r}_0 - \mathbf{r}'_2)\delta(\mathbf{r}_2 - \mathbf{r}'_2) \\ &\times \langle \Delta\epsilon_{\alpha_1\nu_1}\Delta\epsilon_{\alpha_2\nu_2}\rangle\, T_{\nu_1\gamma_1}(\mathbf{r}_2 - \mathbf{r}_1)T^*_{\nu_2\gamma_2}(\mathbf{r}'_2 - \mathbf{r}'_1)\delta(\mathbf{r}_1 - \mathbf{r}'_1) \\ &\times \langle \Delta\epsilon_{\gamma_1\beta}\Delta\epsilon_{\gamma_2\beta}\rangle\, E_{\beta}(\mathbf{r}_1)E^*_{\beta}(\mathbf{r}'_1). \end{aligned} \tag{37}$$

where the product

$$T_{\alpha\alpha_1}(\mathbf{r}_0-\mathbf{r}_1)T_{\alpha\alpha_2}(\mathbf{r}_0-\mathbf{r}_1') = \frac{k_0^4}{r_0^2}\left(\delta_{\alpha\alpha_1}-\frac{k_{s\alpha}k_{s\alpha_1}}{k_0^2}\right)\left(\delta_{\alpha\alpha_2}-\frac{k_{s\alpha}k_{s\alpha_2}}{k_0^2}\right) \times \exp[-i\mathbf{k}_s(\mathbf{r}_1-\mathbf{r}_1')] \quad (38)$$

represents the pair of propagators containing the observation point $\mathbf{r}_0$, written in the far-field approximation. Indices α and β in Eq. (37) are fixed, whereas summation over all other indices is assumed .

If we take into account the extinction, the product of the intrinsic propagators can be written in the form

$$T_{\phi\mu}(\mathbf{r})T^*_{\gamma\rho}(\mathbf{r}) \equiv \Lambda_{\phi\gamma,\mu\rho}(\mathbf{r}) = \frac{k_0^4\exp(-\sigma r)}{r^2}\left(\delta_{\phi\mu}-\frac{r_\phi r_\mu}{r^2}\right) \times \left(\delta_{\gamma\rho}-\frac{r_\gamma r_\rho}{r^2}\right), \quad (39)$$

where $\sigma = l_{ext}^{-1}$, l_{ext} is the extinction length. Substituting Eqs. (35) and (38) into Eq. (37) for the double-scattered intensity, we rewrite the latter equation in the form

$$I_{2\alpha}^{\beta}(L) = A\int d\mathbf{r}_1 d\mathbf{r}_2\left(\Lambda_{\nu\nu,\gamma\gamma}+\frac{1}{3}\Lambda_{\alpha\alpha,\gamma\gamma}+\frac{1}{3}\Lambda_{\nu\nu,\beta\beta}+\frac{1}{9}\Lambda_{\alpha\alpha,\beta\beta}\right), \quad (40)$$

where

$$\hat{\Lambda} = \hat{\Lambda}(\mathbf{r}_1-\mathbf{r}_2), \quad A = \frac{k_0^4\,|E|^2}{r_0^2}\frac{a_0^2}{4}. \quad (41)$$

The contribution of cyclic diagram to the double-scattered intensity for backward scattering equals

$$I_{2\alpha}^{\beta}(C) = A\int d\mathbf{r}_1 d\mathbf{r}_2\,\langle\Delta\epsilon_{\alpha\nu_1}\Delta\epsilon_{\gamma_2\beta}\rangle\,\Lambda_{\nu_1\nu_2,\gamma_1\gamma_2}\,\langle\Delta\epsilon_{\gamma_1\beta}\Delta\epsilon_{\alpha\nu_2}\rangle\,|E|^2. \quad (42)$$

After a convolution over indices with account for Eqs.(35) and (38), we get

$$I_{2\alpha}^{\beta}(C) = \frac{k_0^4\,|E|^2}{r_0^2}\frac{a_0^2}{4}\int d\mathbf{r}_1 d\mathbf{r}_2[\delta_{\alpha\beta}(\Lambda_{\beta\gamma_1,\gamma_1\alpha}+\Lambda_{\nu_1\beta,\alpha\nu_1}+\Lambda_{\nu_1\gamma_1,\gamma_1\nu_1} - \frac{2}{3}\Lambda_{\nu_1\alpha,\beta\nu_1}-\frac{2}{3}\Lambda_{\alpha\gamma_1,\gamma_1\beta})-\frac{4}{3}\Lambda_{\beta\alpha,\beta\alpha}+\frac{13}{9}\Lambda_{\beta\beta,\alpha\alpha}]. \quad (43)$$

Hence the contributions to polarized I_{2V}^V and depolarized I_{2V}^H components are

$$I_{2V}^V(L) = A\int d\mathbf{r}_1 d\mathbf{r}_2\left(\Lambda_{\nu\nu,\gamma\gamma}+\frac{1}{3}\Lambda_{\nu\nu,22}+\frac{1}{3}\Lambda_{22,\gamma\gamma}+\frac{1}{9}\Lambda_{22,22}\right),$$

$$
\begin{aligned}
I_{2V}^{V}(C) &= A\int d\mathbf{r}_1 d\mathbf{r}_2\left(\Lambda_{\nu\gamma,\nu\gamma}+\frac{1}{3}\Lambda_{2\nu,2\nu}+\frac{1}{3}\Lambda_{\nu 2,\nu 2}\right),\\
I_{2V}^{H}(L) &= A\int d\mathbf{r}_1 d\mathbf{r}_2\left(\Lambda_{\nu\nu,\gamma\gamma}+\frac{1}{3}\Lambda_{\nu\nu,22}+\frac{1}{3}\Lambda_{11,\gamma\gamma}+\frac{1}{9}\Lambda_{11,22}\right),\\
I_{2V}^{H}(C) &= A\int d\mathbf{r}_1 d\mathbf{r}_2\left(\Lambda_{22,11}-\frac{2}{3}\Lambda_{21,21}-\frac{2}{3}\Lambda_{12,12}+\frac{4}{9}\Lambda_{11,22}\right). \quad (44)
\end{aligned}
$$

Upon calculating these integrals in the case of disordered half-space, we obtain

$$
I_{2V}^{V}(L) = I_{2V}^{V}(C) = \frac{1252}{135}AV\pi l_{ext}, \quad (45)
$$

$$
I_{2V}^{H}(L) = \frac{1228}{135}AV\pi l_{ext}, \qquad I_{2V}^{H}(C) = -\frac{236}{135}AV\pi l_{ext},
$$

where $V \sim S l_{ext}$ is the scattering volume, S is the illuminated area.

As is seen from Eq. (45), the contributions of ladder and cyclic diagrams to the polarized component are equal. It means that the polarized component exhibits the coherent backscattering peak. As for the depolarized component, the contribution of cyclic diagram is small as compared to that of the ladder one and is even negative. It means that, at least within the double-scattering approximation, there is no backscattering enhancement for the depolarized component. Note that the negative sign of cyclic diagram for the depolarized component does not contradict the requirement for the intensity to be positive.

In the general case, one has to sum a full series in scattering orders. This sum can be represented in the operator form as

$$
\hat{S} = \hat{N} + \hat{M}\hat{N} + \hat{M}\hat{M}\hat{N} + \ldots, \quad (46)
$$

where the tensors $\hat{N}$ and $\hat{M}$ are defined as

$$
\begin{aligned}
N_{\mu_1\mu_2,\alpha_1\alpha_2}(\mathbf{r}_1-\mathbf{r}_2) &= \langle\Delta\epsilon_{\mu_1\gamma_1}\Delta\epsilon_{\mu_2\gamma_2}\rangle\,\Lambda_{\gamma_1\gamma_2,\beta_1\beta_2}(\mathbf{r}_1-\mathbf{r}_2)\\
&\times\ \langle\Delta\epsilon_{\beta_1\alpha_1}\Delta\epsilon_{\beta_2\alpha_2}\rangle,\\
M_{\mu_1\mu_2,\alpha_1\alpha_2}(\mathbf{r}_1-\mathbf{r}_2) &= \langle\Delta\epsilon_{\mu_1\gamma_1}\Delta\epsilon_{\mu_2\gamma_2}\rangle\,\Lambda_{\gamma_1\gamma_2,\alpha_1\alpha_2}(\mathbf{r}_1-\mathbf{r}_2). \quad (47)
\end{aligned}
$$

The series (46) can be summed to get a closed equation for the radiation transfer propagator:

$$
\hat{S} = \hat{N} + \hat{M}\hat{S}, \quad (48)
$$

or, in analytical form,

$$
S_{\alpha\beta,\gamma\delta}(\mathbf{r}_1,\mathbf{r}_2) = N_{\alpha\beta\gamma\delta}(\mathbf{r}_1-\mathbf{r}_2) + \int d\mathbf{r}_3 M_{\alpha\beta,\mu\nu}(\mathbf{r}_1-\mathbf{r}_3)S_{\mu\nu,\gamma\delta}(\mathbf{r}_3,\mathbf{r}_2). \quad (49)
$$

Considering the problem of scattering from a medium occupying the half-space $z \geq 0$, where z is Cartesian coordinate, $\mathbf{r} = \{x, y, z\}$, one has to take into account the boundary conditions explicitly. In accordance with Ref. [7], we require the propagator $\hat{S}$ to be zero at the boundary $z = 0$. Using the image method, one finds the solution of Eq. (47) in the form

$$\hat{S}(\mathbf{r}_1, \mathbf{r}_2) = \hat{S}_0(\mathbf{r}_1 - \mathbf{r}_2) - S_0(\mathbf{r}_1 - \mathbf{r}^{(s)}), \tag{50}$$

where $S_0(\mathbf{r})$ is the corresponding solution for an infinite homogeneous system, $\mathbf{r}_2^{(s)}$ is the mirror image with respect to $\mathbf{r}_2$ ($\mathbf{r}_2^{(s)} = (x_2, y_2, -z_2)$). The radiation transfer propagator $\hat{S}(\mathbf{r}_1, \mathbf{r}_2)$ given by Eq. (50) turns out to be zero if at least one of two points $\mathbf{r}_1$ and $\mathbf{r}_2$ lies in the $z = 0$ plane. For a homogeneous system, the propagator $\hat{S}_0$ is easily obtained by means of Fourier transformation of Eq. (49).

Assuming that the wave vector of the scattered light lies in the (x, y) plane, different contributions to the scattered intensity can be represented as

$$\begin{aligned} I_\alpha^\beta(L) &= \int d\mathbf{r}_1 \int d\mathbf{r}_2 T_{\alpha\alpha_1}(\mathbf{r}_0 - \mathbf{r}_2) T^*_{\alpha\alpha_2}(\mathbf{r}_0 - \mathbf{r}_2) S_{\alpha_1\alpha_2,\beta\beta}(\mathbf{r}_1, \mathbf{r}_2) \\ &\times \exp\left[i\mathbf{r}_1(\mathbf{k}_i - \mathbf{k}_i^*) - i\mathbf{r}_2(\mathbf{k}_s - \mathbf{k}_s^*)\right] |E_\beta|^2, \qquad (51) \\ I_\alpha^\beta(C) &= \int d\mathbf{r}_1 \int d\mathbf{r}_2 T_{\alpha\alpha_1}(\mathbf{r}_0 - \mathbf{r}_2) T^*_{\alpha\alpha_2}(\mathbf{r}_0 - \mathbf{r}_1) S_{\alpha_1\beta,\beta\alpha_2}(\mathbf{r}_1, \mathbf{r}_2) \\ &\times \exp\left[i\mathbf{r}_1(\mathbf{k}_i + \mathbf{k}_s^*) - i\mathbf{r}_2(\mathbf{k}_s + \mathbf{k}_i^a st)\right] |E_\beta|^2, \qquad (52) \end{aligned}$$

where the imaginary part of the wave number is $k'' = \sigma/2$, indices α and β are fixed, and summation over indices α_1 and α_2 is assumed . As in the double-scattering approximation, we assume that the incident light with $\mathbf{e}_y$ polarization is normal to the boundary. Taking into account the translational invariance in the (x, y) plane for scattering angles close to 180°, we obtain from Eqs. (51) and (52)

$$\begin{aligned} I_\alpha^\beta(L) &= \frac{S|E_\beta|^2 k_0^4}{r_0^2} \left(\delta_{\alpha\alpha_1} - \frac{r_{0\alpha} r_{0\alpha_1}}{r_0^2}\right) \left(\delta_{\alpha\alpha_2} - \frac{r_{0\alpha} r_{0\alpha_2}}{r_0^2}\right) \\ &\times \int_{z_1 \geq 0} dz_1 \int_{z_2 \geq 0} d\mathbf{r}_2 S_{\alpha_1\alpha_2,\beta\beta}(x_2, y_2, z_1, z_2) \\ &\times \exp\left[-\sigma(z_1 + z_2)\right], \qquad (53) \\ I_\alpha^\beta(C) &= \frac{S|E_\beta|^2 k_0^4}{r_0^2} \left(\delta_{\alpha\alpha_1} - \frac{r_{0\alpha} r_{0\alpha_1}}{r_0^2}\right) \left(\delta_{\alpha\alpha_2} - \frac{r_{0\alpha} r_{0\alpha_2}}{r_0^2}\right) \\ &\times \int_{z_1 \geq 0} dz_1 \int_{z_2 \geq 0} d\mathbf{r}_2 S_{\alpha_1\beta,\beta\alpha_2}(x_2, y_2, z_1, z_2) \\ &\times \exp\left[i(\mathbf{k}_i + \mathbf{k}_s)(\mathbf{r}_1 - \mathbf{r}_2) - \sigma(z_1 + z_2)\right], \qquad (54) \end{aligned}$$

where $\mathbf{r}_1 = (0, 0, z_1)$, and S is the illuminated area.

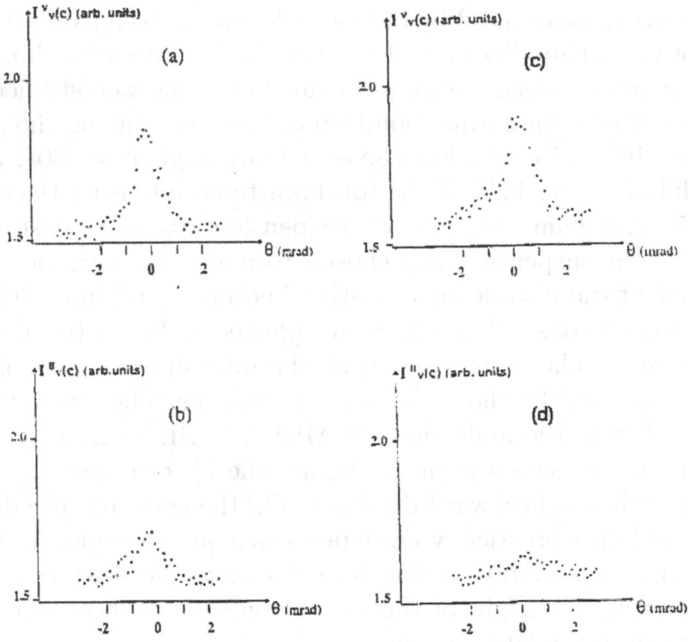

Figure 6. Coherent backscattering: polarized (a) and depolarized (b) components in the 0.55 mkm-diam latex suspension; polarized (c) and depolarized (d) components in the multi-domain liquid crystal.

The calculated shape of the backscattering peak for the polarized component is close to triangular and its height is approximately equal to the intensity of the background. The behavior of the depolarized component is quite different. $I_V^H(C)$ appears to be negative and its absolute value is 75 times smaller than the value of the background $I_V^H(L)$. Thus the model of point-like anisotropic scatterers predicts a triangular shape of the backscattering peak for polarized component and a very small, of the order of 1–2%, decrease in the depolarized component of backscattered intensity. These results have been tested experimentally. A continuous mode He-Ne laser was used as a source of radiation. The principal optical element of the set up was a beam splitter. A part of the incident beam was reflected from the beam splitter and hit the cell, whereas the second part was used to adjust the optical system. The backscattered light passed through the splitter and was detected by a photomultiplier. The resulting signal was then analyzed by a multichannel analyzer. The cell was filled with a disordered liquid crystal, a MBBA+EBBA mixture, at a temperature of 25° C. The polarized I_V^V and depolarized I_V^H components were both measured. Since the expected peak height for the I_V^V component is quite moderate, special attention was paid to remove noises caused by diffuse reflection as well as scattering from elements of the set up. To control the obtained data and to improve their validity, test measurements were performed on some well-studied systems. The cell was filled with a water solution of black ink having the absorption coefficient $\sim 10^4$–10^5 cm^{-1}. For the scattering angle $\theta = 180°$, the diffuse intensity did not exceed 2% of the total scattered intensity. Backscattering from a 0.55 mkm-diam latex particle suspension was also measured. A concentration of the suspension was chosen to make the backscattering peak close to that of the disordered nematic. The measured intensities I_V^V and I_V^H of components of scattered light are plotted in Fig. 6 (a, b) for a 2.3% latex suspension. The points represent the intensity averages of the scattered light measured by the multichannel analyzer. The intensities of light backscattered from the multi-domain MBBA+EBBA mixture are plotted in Fig. 6 (c, d). As is seen from the figure, the I_V^V component exhibits the expected peak in the backward direction. On the contrary, the depolarized component exhibits practically no dependence on the angle. A small peak, of order of 3%, is probably produced by the isotropic scattering. However, if this were true, the height of the peak should have been at least several times less than in the latex suspension.

8. Conclusion

We have considered the multiple scattering of light in nematic liquid crystals. Although it is the simplest anisotropic liquid, there are a lot of unusual optical effects in this system. Among them are the difference in propagation conditions of ordinary and extraordinary beams, significant sensitivity to the external electric or magnetic field, anisotropy of light diffusion, etc. Also of interest is the study of multiple light scattering in liquid crystals of more complex structure, such as cholesterics and biaxial nematics. Moreover, interesting phenomena may be observed in the vicinity of phase transition points, near thresholds, such as the Fredericksz transition, etc. These investigations extend our knowledge on multiple light scattering in the systems with complex optical structure.

References

1. P.G. de Gennes and J. Prost, The Physics of Liquid Crystals (Clarendon Press, Oxford, 1993).
2. V.I. Pokrovskii, E.I. Kats, Zh. Eksp. Teor. Fiz. 73, 774 (1977) [Sov. Phys. JETP 46, 405 (1977)].
3. A.Yu. Valkov, V.P. Romanov, Zh. Eksp. Teor. Fiz. 83, 1777 (1982) [Sov. Phys. JETP 56, 1028 (1982)].
4. L.D. Landau, E.M. Lifshitz, Statistical Physics, Oxford (Pergamon Press, 1980).

5. M. Lax, D.F. Nelson, Phys. Rev. B 4, 3694 (1971).
6. M. Lax, D.F. Nelson, in Proceedings of III Rochester Conference on Coherent and Quantum Optics, Eds. I. Mandel, E. Wolf (New York, Plenum, 1973), p. 415.
7. E. Miraldi, L. Trossi, P. Taverna Valabreda, Nuovo Cimento B 60, 165 (1980).
8. S.M. Rytov, Yu.A. Kravtsov, V.I. Tatarskii, Introduction to Statistical Radiophysics. Part II, (Moscow, Nauka, 1978).
9. A.Yu. Valkov, V.P. Romanov, Zh. Eksp. Teor. Fiz. 90, 1264 (1986) [Sov. Phys. JETP 63, 737 (1986)].
10. A.Yu. Valkov, V.P. Romanov, T.I. Tipyasova, Zh. Eksp. Teor. Fiz. 99, 1283 (1991) [Sov. Phys. JETP 72, 714 (1991)].
11. L.Ts. Adzhemyan, A.N. Vasil'ev, M.M. Perekalin, Kh.Yu. Reittu, Opt. Spektrosk. 69, 640 (1990) [Opt. Spectrosc. (USSR) 69, (1990)].
12. A.Yu. Valkov, L.A. Zubkov, A.P. Kovshik, V.P. Romanov, Pisma Zh. Eksp. Teor. Fiz. 40, 281 (1984) [JETP Lett. 40, 1064 (1984)].
13. H.A. Gersch, W.A. Holm, Phys. Rev. B 14, 1307 (1976).
14. Yu.N. Barabanenkov, E.G. Stainova, Opt. Spektrosk. 63, 362 (1987) [Opt. Spectrosc. (USSR) 63, (1987)].
15. A.Yu. Valkov, V.P. Romanov, T.I. Tupyasova, Zh. Eksp. Teor. Fiz. 99, 1283 (1991) [Sov. Phys. JETP 72, 714 (1991)].
16. L.Ts. Adzhemyan, A.N. Vasil'ev, M.M. Perekalin, Opt. Spektrosk. 74, 1139 (1993) [Opt. Spectrosc. (USSR) 74, (1993)].
17. V.V. Zheleznyakov, V.V. Kocharovskii, Vl.V. Kocharovskii, Usp. Fiz. Nauk 141, 257 (1983) [Sov. Phys. Usp. 26, 877 (1983)].
18. L.A. Apresyan, Yu.A. Kravtsov, Radiation Transfer Theory (Moscow, Nauka, 1983).
19. V.P. Romanov, A.N. Shalaginov, Opt. Spektrosk. 64, 1299 (1988) [Opt. Spectrosc. (USSR) 64, 774(1988)].
20. M.P. van Albada, A. Lagendijk, Phys. Rev. Lett. 55, 2692 (1985).
21. P.E. Wolf, G. Maret, Phys. Rev. Lett. 55, 2696 (1985).
22. E. Akkermans, P.E. Wolf, R. Maynard, Phys. Rev. Lett. 56, 1471 (1986).
23. M.J. Stephen, G. Cwilich, Phys. Rev. B 34, 7564 (1986).
24. Yu.N. Barabanenkov, V.D. Ozrin, Zh. Eksp. Teor. Fiz. 94, 56 (1988) [Sov. Phys. JETP 67, 30 (1988)].
25. A.A. Golubentsev, Zh. Eksp. Teor. Fiz. 86, 47 (1984) [Sov. Phys. JETP 59, 26 (1984)].
26. E. Akkermans, P.E. Wolf, R. Maynard, G. Maret, J. Phys. (Fr) 49, 77 (1988).
27. V.L. Kuzmin, V.P. Romanov, L.V. Kuzmin, Opt. Spektrosk. 72, 227 (1992) [Opt. Spectrosc. (USSR) 72, 125 (1992)].
28. V.L. Kuzmin, V.P. Romanov, L.V. Kuzmin, Opt. Spektrosk. 73, 376 (1992) [Opt. Spectrosc. (USSR) 73, 220 (1992)].
29. S. Etemad, R. Thompson, H.J. Andreiko, Phys. Rev. Lett. 57, 575 (1986).
30. Y. Kuga, A. Ishimaru, J. Opt. Soc. Am. A 8, 831 (1984).
31. D.V. Vlasov, L.A. Zubkov, N.V. Orekhova, V.P. Romanov, Pisma. Zh. Eksp. Teor. Fiz. 48 86 (1988) [JETP Lett. 48, 91 (1988)]
32. A. Almamuri, D.V. Vlasov, L.A. Zubkov, V.P. Romanov, Zh. Eksp. Teor. Fiz. 99, 1431 (1991) [Sov. Phys. JETP, 72, 798 (1991)].
33. H. Stark, M. Kao, K. Jester, T. Lubensky and A. Yodh, J. Opt. Soc. Am. A 14, 156 (1997).
34. H. Stark and T. Lubensky, Phys. Rev. E 55, 514 (1997).
35. B. van Tiggelen, R. Maynard, and A. Heiderich, Phys. Rev. Lett. 77, 639 (1996).
36. B. van Tiggelen, A. Heiderich, and R. Maynard, Mol. Cryst. Liq. Cryst. 293, 205 (1997).
37. B. van Tiggelen and H. Stark, Rev. Mod. Phys. 72, 1017 (2000).
38. D.S. Wiersma, A. Muzzi, M. Colocci, and R. Righini, Phys. Rev. Lett. 83, 4321 (1999).

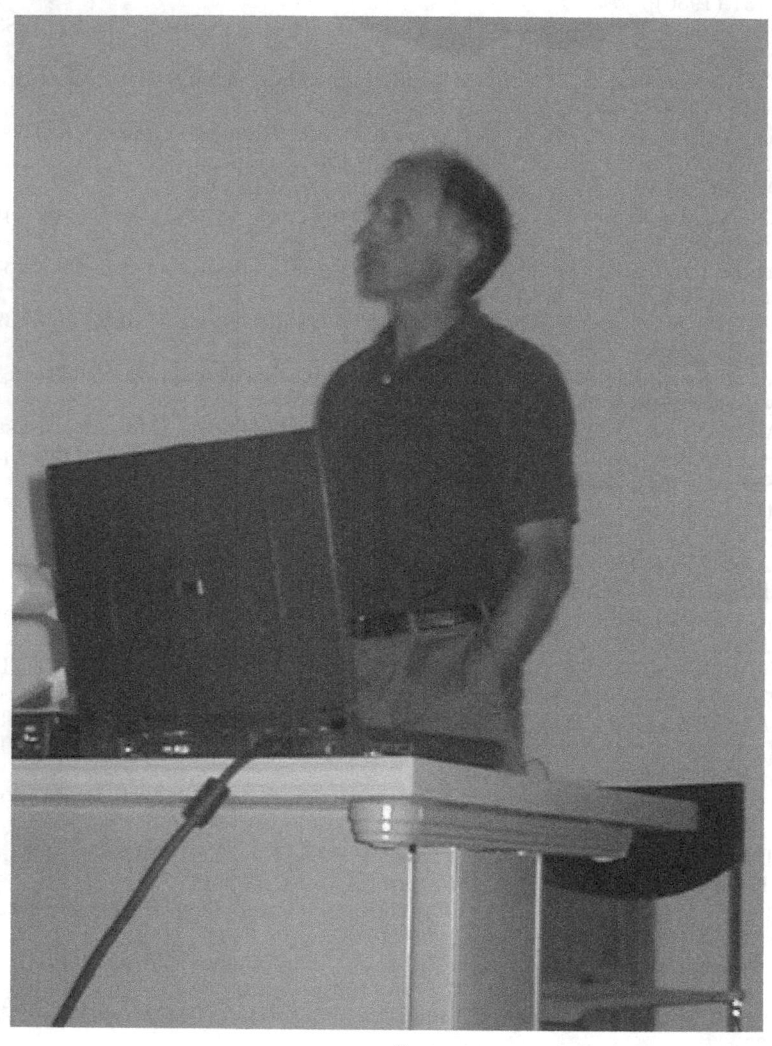

Martin Čopič. Photo by Gabriel Cwilich

DYNAMIC LIGHT SCATTERING IN CONFINED LIQUID CRYSTALS

M. ČOPIČ[1,2] AND A. MERTELJ[1]

[1] *J. Stefan Institute, Jamova 39, 1001 Ljubljana, Slovenia*
[2] *Department of Physics, University of Ljubljana 1000 Ljubljana, Slovenia*

Abstract. After a brief introduction to light scattering and liquid crystal dynamics, that is, bulk orientational fluctuation modes and their coupling to light, several examples of confined liquid crystal systems like thin slabs, cylindrical cavities, droplets dispersed in polymers, and liquid crystals in random matrices are presented. The structure of fluctuation modes in confined geometries and light scattering on localized modes are discussed. In confined liquid crystal systems multiple scattering often occurs and the methods of diffusive wave spectroscopy must be employed. As examples study of interaction of liquid crystals with surfaces via light scattering and rotational diffusion of droplet orientation in polymer dispersed liquid crystals are presented.

1. Introduction

Liquid crystalline phases are more or less fluid states of matter characterized by partial orientational and translational order. The simplest one is the nematic phase usually formed by elongated molecules which are translationally completely disordered like in a normal fluid but locally they orient along a particular direction — the director. The molecules can further order into fluid layers forming the smectic phases. The director can be either perpendicular to the smectic layers (smectic A phase) or tilted with respect to the layer normal (smectic C and other tilted phases). Due to their optical and mechanical properties, partly explained below, liquid crystals are very useful for electrooptic applications. The richness of phases and structures formed by liquid crystals also makes them very interesting as model

B.A. van Tiggelen and S.E. Skipetrov (eds.),
Wave Scattering in Complex Media: From Theory to Applications, 499–517.

systems for a broad spectrum of fundamental phenomena in physics. For a good introduction to the fundamentals of liquid crystals see [1].

In the following we will only deal with nematic liquid crystals. The ordering of the molecules along the director causes the phase to be locally optically uniaxial, with the optic axis along the director. Due to the fluid nature of the phase, the orientation of the director can change as a function of position. There is an orientational elastic energy associated with such deformations, but it is comparatively small so that in a sample with a size of a millimeter or more in the absence of external influences the director will be disordered. External fields, either electric or magnetic, can easily change the orientation of the director. This, together with large optical anisotropy, is the basis for the usefulness of nematic liquid crystals in display devices.

Nematic liquid crystals are always milky in their appearance. This is due to strong scattering of light on the intrinsic thermal fluctuations of the orientation of the director. The energy associated with the local perturbation of the director is small, so the amplitude of the thermal fluctuations is large. As the director is also the optic axis, the fluctuations of the director give rise to large fluctuations of the optical dielectric constant and to strong scattering of light.

The dynamics of orientational fluctuations are governed by elastic and viscous properties of nematics. Local perturbations of the director are driven back to equilibrium by elastic torques which are opposed by a viscous drag on molecular rotation. This viscous drag is relatively large, making inertial effects negligible, and so the orientational fluctuations are always overdamped. Light scattering investigations are a very good way to obtain the elastic and viscous properties of nematic liquid crystals [2, 3]. Both the elastic constants and viscosities are among the most important parameters for the application of liquid crystals in display devices.

In recent years there has been considerable interest in the properties of composite systems of fluids embedded in a solid matrix or confined to small cavities of different morphologies. Such systems containing liquid crystals are particularly interesting [4]. They are excellent model systems to study the effects of confinement and disorder on phase transitions, influence of surfaces and other properties. Some of these liquid crystal systems like nematic droplets in polymer matrices (polymer dispersed liquid crystals — PDLC) have also found important display applications.

In this article we will focus on the properties of fluctuations in nematic liquid crystals confined in different pores and matrices, studied mainly by dynamic light scattering. Confinement changes the behavior of fluctuations and associated light scattering in several important ways. In a finite geometry the spectrum of fluctuations is no longer continuous but becomes discrete. When liquid crystal is embedded in many small cavities, the surface to volume ratio becomes large and interactions of liquid crystal with surfaces become important or even dominant. In some important cases this allowed us to obtain parameters describing surface interactions via light scattering, as will be described below [5–7]. These parameters are of paramount importance in many display applications.

In the following we will first provide the reader with an overview of the continuous theory of nematic liquid crystals and basic description of light scattering, then we will describe some cases of confined liquid crystals that we have studied in recent years, from simple planar geometry to liquid crystals embedded in random silica aerogel structure. In the next section we will show the effects of confinement on the fluctuations of the director and on light scattering. Then we will discuss two specific cases that we have investigated, that is, nematics confined to cylindrical pores and PDLCs.

2. Continuum Description of Nematic Liquid Crystals

The main characteristic of nematic liquid crystals is the appearance of uniaxial anisotropy which can be described by saying that any traceless second rank tensor property is nonzero in the nematic phase. Such a tensor is an order parameter of the nematic phase. Assuming that the molecules are cylindrically symmetric so that their orientation can be described by a single unit vector **s**, the most convenient tensor for the order parameter is

$$Q_{\alpha\beta} = \left\langle s_\alpha s_\beta - \frac{1}{3}\delta_{\alpha\beta} \right\rangle \tag{1}$$

where the average is taken over the distribution of molecular orientations. As the nematic phase is uniaxial, $Q_{\alpha\beta}$ can be written in the form

$$Q_{\alpha\beta} = S\left(n_a n_\beta - \frac{1}{3}\delta_{\alpha\beta}\right) \tag{2}$$

where **n** is a unit vector in the direction of the local average molecular orientation and is also the local optic axis direction. S is called the scalar order parameter describing the degree of orientational ordering.

Any second rank tensor property of the system can be expressed in terms of **Q**. For example the dielectric tensor is

$$\begin{aligned} \varepsilon_{\alpha\beta} &= \overline{\varepsilon}\,\delta_{\alpha\beta} + \varepsilon_{as}\,Q_{\alpha\beta} \\ &= \overline{\varepsilon}\,\delta_{\alpha\beta} + \varepsilon_a\left(n_a n_\beta - \frac{1}{3}\delta_{\alpha\beta}\right) \end{aligned} \tag{3}$$

where $\overline{\varepsilon}$ is the isotropic part of the dielectric tensor, ε_{as} would be the value of the dielectric anisotropy if the molecules were perfectly ordered, and ε_a is the actual anisotropy which is proportional to S: $\varepsilon_a = S\,\varepsilon_{as}$.

In the nematic phase, when the gradient is not too large, it is usually sufficient to take the scalar order parameter S to be constant. Then it is only necessary to consider the elastic energy associated with the spatial changes in the director $\mathbf{n}$. This is the well-known Frank elastic energy

$$F_{el} = \frac{1}{2} \int \left[K_1 (\nabla \cdot \mathbf{n})^2 + K_2 (\mathbf{n} \cdot (\nabla \times \mathbf{n}))^2 + K_3 (\mathbf{n} \times (\nabla \times \mathbf{n}))^2 \right] dV \tag{4}$$

The three terms describe splay, twist and bend deformations, respectively. To expression (4) we must add the contribution of the external fields and surfaces confining the sample. The most important external field term is the dielectric one: $1/2\varepsilon_a (\mathbf{E} \cdot \mathbf{n})^2$, which for $\varepsilon_a > 0$ causes the director to orient along the field. For the surface contribution it is usually assumed that at the surface there is a preferred direction ν of the director minimizing the surface energy. The simplest model for the surface anchoring energy is the Rapini-Papoular form

$$W_s = -\frac{1}{2} W (\mathbf{n} \cdot \nu)^2 \tag{5}$$

The magnitude of the elastic constants in the Frank free energy is of the order of 10^{-11} N, a comparatively small value, so that on a length scale of a fraction of a millimeter a sample of nematic liquid crystal can easily deform. A typical magnitude of the surface anchoring energy coefficient is 10^{-5} N/m. The ratio $\lambda = K/W$ has the unit of length and is called the extrapolation length.

In an incompressible nematic there are five independent viscosity coefficients. Three are needed to describe the dissipative stress due to the fluid velocity gradient, one couples the director and flow and one gives the dissipation due to rotation of the director with respect to the fluid. In the following, we will neglect the fluid flow and so will only use an effective rotational viscosity η.

A viscous dissipation term can also be added to the boundary conditions. If the molecules close to the surface can rotate, there could be additional dissipation different from the bulk orientational viscosity. This can be described by a surface torque proportional to the rotation rate of the director at the surface. The proportionality constant is called the surface viscosity [8]. The equations of motion for the director are given by the requirement that the viscous dissipation torque equals the torques due to elasticity, external fields, and surfaces. The effect of molecular moment of inertia is negligible on time and space scales where continuum theory is applicable, i.e. nearly always except when one considers single molecular motions on time scales below 1 ns. The torque due to elasticity and external fields is given by the functional derivative of free energy, so the equation of motion is

$$\eta \frac{\partial \mathbf{n}(\mathbf{r},t)}{\partial t} = \frac{\delta}{\delta \mathbf{n}} F_{el} \tag{6}$$

In the one elastic constant approximation that we will mostly use, Eq. (6) gives

$$\eta \frac{\partial \mathbf{n}(\mathbf{r},t)}{\partial t} = K\left[\nabla^2 \mathbf{n} - \left(\mathbf{n}\cdot\nabla^2\mathbf{n}\right)\mathbf{n}\right] \tag{7}$$

3. Fluctuations and Light Scattering

One of the most prominent features of nematic liquid crystals is that they strongly scatter light so that a bulk sample always looks milky. This is due to thermally excited fluctuations of the director — fundamental excitation modes of the nematic order that try to restore the full rotational symmetry of the isotropic phase. As **n** is a unit vector, the fluctuations $\delta\mathbf{n}$ are always perpendicular to **n**. The amplitude of these fluctuations is sufficiently small that they can be derived from the linearized equations of motion following from Eq. (6). The resulting equations are essentially diffusion equations. In an infinite sample the eigenmodes have the form of plane waves with two possible polarizations, one going from pure twist for wave-vector **q** perpendicular to **n** to pure bend for **q** parallel to **n**, and the other from pure splay again to pure bend. The relaxation rate of the modes is

$$\frac{1}{\tau} = \frac{K_i(\mathbf{q})}{\eta(\mathbf{q})} q^2 \tag{8}$$

where K_i is a combination of twist and bend ($i = 1$) or splay and bend elastic constants ($i = 2$) that depend on the direction of **q** with respect to **n**, and $\eta(\mathbf{q})$ is an effective rotational viscosity which also depends on the direction of **q** due to the coupling to fluid flow. In the following we will usually make the one elastic constant approximation and neglect the fluid flow effects. The amplitude of the fluctuations is by the equipartition theorem

$$\left\langle \delta n^2(q) \right\rangle = \frac{k_B T}{K q^2} = \frac{k_B T\, \tau}{\eta} \tag{9}$$

It is important to note that in an infinite sample the eigenmodes are characterized by their wave-vector which is an eigenvalue of the spatial part of the diffusion equation — the Helmholtz equation. In an infinite sample the spectrum of these eigenvalues is continuous — any **q** is allowed.

Director **n** is also the direction of the optic axis so that local fluctuations of **n** produce fluctuations of the dielectric tensor. From Eq. (3) we get

$$\delta\varepsilon = \mathbf{n}\otimes\delta\mathbf{n} + \delta\mathbf{n}\otimes\mathbf{n} \tag{10}$$

In the single scattering (Born) approximation, the scattered field can be written as the sum of dipole radiation induced by the incoming light field:

$$E_s(\mathbf{k}_s, t) = \frac{E_0 k^2}{4\pi\varepsilon_0 R} \int_{V_s} e^{-i\mathbf{q}\cdot\mathbf{r}} \left(\mathbf{f} \cdot \varepsilon(\mathbf{r},t) \cdot \mathbf{i}\right) dV \tag{11}$$

Here $\mathbf{q} = \mathbf{k}_s - \mathbf{k}_i$ is the scattering vector, and $\mathbf{i}$ and $\mathbf{f}$ the initial and scattered wave polarization directions.. Usually the scattering volume dimensions are much larger than $1/q$. In that case the integral in Eq. (11) gives simply the Fourier component of the fluctuations of ε at the scattering vector, projected on the initial and scattered polarization directions: $E_s \propto (\mathbf{f} \cdot \varepsilon(\mathbf{q},t) \cdot \mathbf{i})$. Then from Eq. (10) $E_s \propto (\mathbf{i} \cdot \mathbf{n})(\mathbf{f} \cdot \delta\mathbf{n}) + (\mathbf{f} \cdot \mathbf{n})(\mathbf{i} \cdot \delta\mathbf{n})$. Obviously, by a proper choice of the incoming and scattered polarizations it is possible to observe either of the two orientational fluctuation branches.

In a dynamic light scattering experiment we usually measure the autocorrelation function of the detected light intensity. In the cases discussed in this chapter we nearly always get a large amount of statically scattered light which acts as a local oscillator and beats with the component scattered on dynamic fluctuations. In that case the relaxing part of the intensity correlation function is proportional to the autocorrelation function of the dynamically scattered light field $G^{(1)}(\tau) = \mathrm{Re}\left\langle \mathrm{E_s}(0)\, \mathrm{E_s^*}(\tau)\right\rangle$, which is further proportional to $\langle \delta\mathbf{n}(0)\, \delta\mathbf{n}(\tau)\rangle$. This allows us to directly obtain the relaxation rate of the orientational fluctuations at the scattering vector $\mathbf{q}$. Using Eq. (8) one can then determine the ratio of the elastic constants and effective rotational viscosities. In principle, from the amplitude of $G^{(1)}(\tau)$ one can also get the elastic constants themselves, using Eq. (9), but as it is difficult to measure absolute intensities in light scattering, the elastic constants are usually obtained by other methods.

4. Confined Liquid Crystals

In the following we will discuss the spectrum of orientational fluctuations in cases where liquid crystals are confined to finite spaces of different geometries. So we first give an overview of the most important systems of confined liquid crystals.

4.1. PLANAR CELLS

The simplest and by far the most common case is the planar confinement which includes the display devices. A nematic liquid crystal is put between to solid flat surfaces, usually glass, with some surface treatment which induces a prescribed anchoring condition for the nematic director. This can be either planar, where the director is aligned in a certain direction in the

Figure 1. Schematic picture of a bipolar structure of nematic director in a droplet.

plane of the surface, or homeotropic, with the director perpendicular to the surface. If both confining surfaces have parallel alignment we get homogeneous planar or homeotropic orientation of the sample. In most displays, twisted planar structure is used. It is obtained with the two surfaces having mutually perpendicular planar alignment. In hybrid cells one surface induces planar orientation while the other gives homeotropic one. In this case we get a homogeneous bend deformation throughout the sample.

4.2. POLYMER DISPERSED LIQUID CRYSTALS

An important and nowadays common case of confinement are liquid crystal droplets in a polymer matrix — polymer dispersed liquid crystals (PDLCs). The droplets usually have sizes in the micrometer range. Such a system is obtained by mixing a liquid crystal with a suitable monomer which is miscible with the liquid crystal. Then polymerization of the monomer is induced. The liquid crystal is insoluble or much less soluble in the polymer and therefore separates from the polymer and forms droplets. Their size can be controlled by the conditions of polymerization and phase separation process.

The structure of the nematic director field $\mathbf{n}(\mathbf{r})$ inside a droplet can not be uniform. Depending on the boundary conditions on the surface a number of structures can in principle exist [9]. In the seemingly prevailing case of planar anchoring the most common structure is shown in Figure 1. Two singular points must exist and the droplet has an average orientation along the axis joining the two defects. Such a structure is called bipolar. A concentric configuration in which the director is aligned along circles around an axis through the droplet, is also compatible with the boundary conditions, but has higher energy due the axis being a defect line (disclination), and is therefore not observed. For homeotropic anchoring again several structures are possible [10].

Without an applied field the droplets in the bipolar configuration are

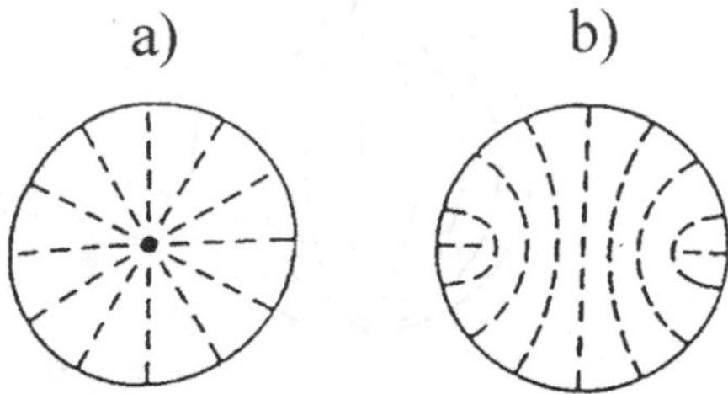

Figure 2. Schematic picture of two director configurations in cylinders in the case of homeotropic anchoring. a) planar radial and b) planar polar.

oriented randomly and strongly scatter light so that a 20 μm thick PDLC film is opaque. If the liquid crystal has positive dielectric anisotropy, an applied electric field orients the droplets so that their axis is along the field, perpendicular to the PDLC film. If the polymer is chosen so that its index of refraction is equal to the ordinary index of the liquid crystal, light propagating in the direction of the normal to the film is not scattered and the film becomes transparent. This is the basis of PDLC optical shutters which can have very large area and are cheap.

4.3. CYLINDRICAL CAVITIES

Liquid crystals can also be embedded in cylindrical pores formed in polymer membranes or in similar structures [11]. Our experiments were done on polycarbonate membranes (Nuclepore) in which cylindrical pores of well-defined diameter are formed by ion bombardment and etching. We used pore diameters from 50 nm to 400 nm.

Deuterium NMR experiments on liquid crystals in Nuclepore membranes have shown that the boundary conditions are homeotropic [12]. In such case the director structure can be radial, or , if the anchoring strength is not too large, it can also be planar with two defect lines running along opposite sides of the cylinder. These two lines can also move out of the pore and become virtual defects [Figure 2: a) radial and b) planar].

4.4. RANDOM STRUCTURES

From the point of view of fundamental physics a very interesting group of confined liquid crystal systems is formed by porous glasses and silica aerogels filled with a liquid crystal. In porous glasses liquid crystal fills an interconnected random system of tubular pores of nanometer size. In aerogels, a dried silica gel, consisting of a random network of silica strands, is filled with a liquid crystal. Compared to porous glasses, in aerogels the volume fraction of liquid crystal is larger and more strongly interconnected, so that we can also consider this system as bulk liquid crystal strongly perturbed by the permeating silica structure. In aerogels the volume fraction of liquid crystal is from 0.7 to more than 0.9. The silica structure is still rigid although quite brittle.

Typical pore size or void lengths in these systems is of the order of a few times 10 nm. This is comparable to the nematic correlation length at the nematic to isotropic transition. Both in porous glasses and in aerogel the director field is strongly deformed so that the system has a substantial elastic energy density. This lowers the N-I phase transition temperature T_{NI} and also, due to randomness, smears it out. The changes in the thermodynamic properties have been studied in [13, 14].

5. Fluctuation Spectrum in Confined Systems

In all confined systems the spectra of orientational fluctuations (and any other wavelike excitations) have many common features. As we are no longer dealing with an infinite homogeneous system, the modes are no longer plane waves characterized by a continuum set of values for the wavevector $\mathbf{q}$. Instead, the eigenmodes are standing waves with shape depending on the geometry of the cavity, i.e., sinusoidal standing waves in the rectangular geometry [15], similar to Bessel and Neumann functions in the cylinders [16], to spherical Bessel functions in the droplets [17] etc. The allowed values of the wave vectors, or more precisely, of the eigenvalues are discrete and depend on the boundary conditions.

The eigenmodes of director orientational fluctuations are exponentially relaxing modes with relaxation rate which is in one elastic constant approximation given by expression (8) [1], using in the place of q eigenvalues k_N of the equation

$$\nabla^2\delta\mathbf{n} - \left(\mathbf{n}_0\cdot\nabla^2\mathbf{n}_0\right)\delta\mathbf{n} - \left(\mathbf{n}_0\cdot\nabla^2\delta\mathbf{n}\right)\mathbf{n}_0 - \left(\delta\mathbf{n}\cdot\nabla^2\mathbf{n}_0\right)\mathbf{n}_0 = -k_N^2\delta\mathbf{n} \quad (12)$$

which is obtained from Eq. (7) by writing $\mathbf{n} = \mathbf{n}_0 + \delta\mathbf{n}$, where $\mathbf{n}_0$ is the static configuration of the director, and linearizing in fluctuations $\delta\mathbf{n}$. The solutions of Eq. (12) depend on the shape of the cavity and boundary conditions [15],

$$\begin{aligned} K(\nu\cdot\nabla)\,\delta\mathbf{n} \;&-\; W\left(\mathbf{n}_p\cdot\delta\mathbf{n}\right)\mathbf{n}_p \\ &+\; 2W\left(\mathbf{n}_p\cdot\mathbf{n}_0\right)\left(\mathbf{n}_p\cdot\delta\mathbf{n}\right)\mathbf{n}_0 + W\left(\mathbf{n}_p\cdot\mathbf{n}_0\right)^2\delta\mathbf{n} = -\zeta\delta\,\dot{\mathbf{n}} \quad (13)\end{aligned}$$

where ν is unit vector normal to the surface, $\mathbf{n}_p$ is the orientation of the director preferred by the surface, W is the surface anchoring strength, and

ζ is the surface orientational viscosity. The origin of the surface viscosity, however, is not very clear. It can be due to dissipation coming from surface processes like adsorption-desorption or molecular slipping on the surface. What are the possible values for it is an open question and it is usually neglected.

The most simple one dimensional confined system, i.e., a nematic layer, has been treated both theoretically [15] and experimentally [18]. It has been shown how anchoring strength affects the relaxation rates of the orientational fluctuations, but the influence of the surface orientational viscosity has not been examined yet. In a uniform nematic layer placed between two equally treated glass plates the shape of the eigenmode of the orientational fluctuations is $\cos(kz)$ and $\sin(kz)$ for even and odd modes, respectively. The z axis is perpendicular to a nematic layer and $z = 0$ lays in the middle of the layer. The layer thickness is d and eigenvalues of the eigenmodes of the orientational fluctuations are denoted by k. In the case of a strong anchoring regime, i.e., the director orientation on the surface does not deviate from the orientation that is preferred by the surface, the amplitude of the orientational fluctuations on the surface is equal to 0, eigenvalues to $k = (N+1)\pi/d$ $(N = 0, 1, \ldots)$, and the relaxation time depends only on the bulk viscoelastic properties of liquid crystal and the thickness of the layer,

$$\tau = \frac{\eta d^2}{K(N+1)^2\pi^2}. \tag{14}$$

Last expression simply reflects the fact, that due to the fluctuation a deformation of the director appears and therefore an elastic torque appears that forces the director back to its equilibrium orientation. A viscous torque opposes this motion.

In weak anchoring regime the amplitude of the director orientational fluctuations at the surface is not zero, so in addition to the bulk elastic torque also surface elastic torque contributes to the restoring torque on the director. This affects also the relaxation rates of the fluctuation eigenmodes through the eigenvalues k, which are in this case smaller. The mode which can be easily used to measure the properties of the surface interaction is the fundamental mode. For $d \lesssim \lambda$, Eq. (13) yields approximate solution for the relaxation time of the fundamental mode

$$\tau \approx \frac{\eta d^2}{12K} + \frac{\eta d}{2W} + \frac{\zeta}{W} \tag{15}$$

The first term is usual bulk term and it prevails upon the second one when the thickness of the layer is much larger than the surface extrapolation length λ, i.e., in the strong anchoring regime where the relaxation time is given by Eq. (14). The second term describes the contribution of the surface elastic torque to the relaxation of the director in the layer, while the last term comes from the relaxation of the director at the surface and it does not depend on the thickness of the layer. Details of this analysis are explained in Ref. [6]. By measuring the thickness dependence of the relaxation rate of the fundamental mode one can, in principle, obtain both the anchoring strength and the surface viscosity.

In other confined systems the situation is more complex, the equilibrium configuration of the director is not uniform and the calculation of the eigenmodes is more involved. The most simple and experimentally accessible confined system in two dimensions is liquid crystal embedded in cylindrical pores of either polycarbonate (Nuclepore) or silica (Anopore) membranes. Also in the cylinder the properties of the fundamental mode are similar to those in the layer. We focus on the liquid crystal embedded in Nuclepore membranes, since those were used in our experiments. The equilibrium configuration of the director field in the pores of Nuclepore membranes is known to be either escaped radial with point defects or, for small anchoring strength, planar-polar where the director is in the plane perpendicular to the cylinder axis and nearly perpendicular to the pore walls except at two opposite points on the perimeter [12, 19]. In the one elastic constant approximation, the analytic form of static director configuration of the latter is known [19]. From this solution we deduced the linearized equation of the fluctuations in the axial direction of the static structure [21]. For $\lambda > R$, we then get for the relaxation time of the fundamental mode of the planar structure with weak anchoring

$$\tau_1 \approx \frac{\eta R^2}{8K} + \frac{R\eta}{W} + \frac{\zeta}{W}. \tag{16}$$

Eq. (16) is very similar to the situation in the layer, i.e., Eq. (15). As in the case of a nematic layer, the second term in Eq. (16) dominates when $\lambda \gtrsim R$ so the orientational fluctuations in this case are governed by surface anchoring and not by bulk orientational elasticity. The contribution of the surface viscosity is the same in both cases.

In a similar way the fluctuations for radial structure of spherical nematic droplets in PDLC can be worked out. This has been done for the case of strong anchoring [17]. In radial spherical droplets there is a singularity in the center of the droplet and therefore the lowest eigenvalue of the fluctuations is of the order of $1/R$ for any anchoring strength so that bulk elasticity always dominates the relaxation rate. For bipolar structure in spherical droplets a new low frequency mode appears, in which the average direction of a droplet reorients. As there is no restoring torque for this kind of motion in a spherical droplet without applied external field, this mode is free diffusion of the droplet orientation. More about this mode will be said later.

In other, more complicated confinement geometries, where one can not obtain analytically the solution to Eq. (8) for the static structure, the fluctuation spectrum can only be found numerically [20].

6. Experiments

6.1. CYLINDERS

As shown in the previous sections well defined geometric static properties of the nematic liquid crystal embedded in cylindrical pores of the Nuclepore membranes allow us to fully analyze the light scattering data. In the following we will present results of the DLS experiments performed in this system.

Our samples were prepared similarly as described in Ref. [12]. A piece of membrane was wetted with the liquid crystal 4-pentyl-4'-cyanobiphenyl (5CB) that filled the cylindrical pores of the membrane due to the capillary action. The remaining liquid crystal on the surface of the membrane was removed by pressing the membrane between two Whatman filtration papers and then the filled membrane was placed between two glass plates. The refractive index of the polycarbonate Nuclepore membranes approximately matches the average refractive index of the 5CB, so multiple scattering in the samples was negligible.

The light source was a He-Ne laser with the wavelength of 632.8 nm. The intensity correlation function was measured using an ALV5000 correlator that enables measurements over a time range of 10^{-8}–10^{3} seconds. We have measured the normalized intensity correlation function $g^{(2)}(\tau) = \langle I(t) I(t+\tau)\rangle / \langle I(t)\rangle \langle I(t+\tau)\rangle$ of light exiting the sample as a function of scattering angle, the radius of cavities and the temperature. The incoming and scattered directions were chosen so that the scattering vector was perpendicular to the axis of the pores. Following the selection rules, by which the orientational fluctuations are coupled to the off-diagonal elements of the dielectric tensor, we chose orthogonal polarizations of incident and scattered light.

The measured intensity correlation functions showed a well defined relaxation component which is due to the orientational director fluctuations in the pores. The amount of statically scattered light off the sample and pore surfaces was such that the measurements were in the heterodyne regime and the relaxation rate of the intensity correlation function is equal to the relaxation rate of the fluctuations.

In Fig. 3 measured relaxation time τ vs. R with its fit to Eq. (16) is shown for several temperatures. The coefficient of the linear term, together with the bulk values of K/η, gives the extrapolation length λ, plotted vs. T in Fig. 4. The constant term can be used to get h, and it is shown in

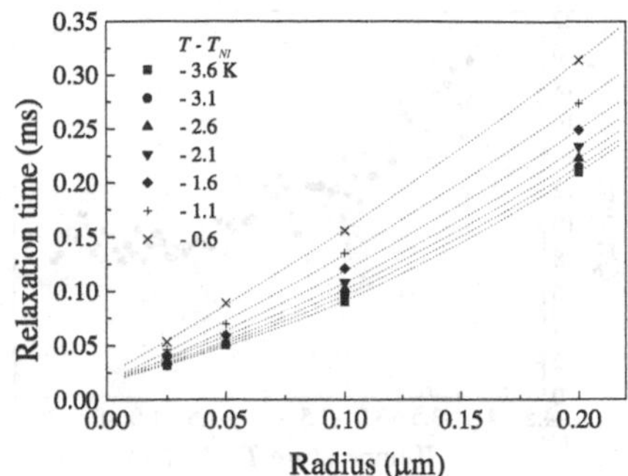

Figure 3. Dependence of the relaxation rate of the fundamental mode on the cylinder radius at different temperatures. Lines are the second order polynomial fits.

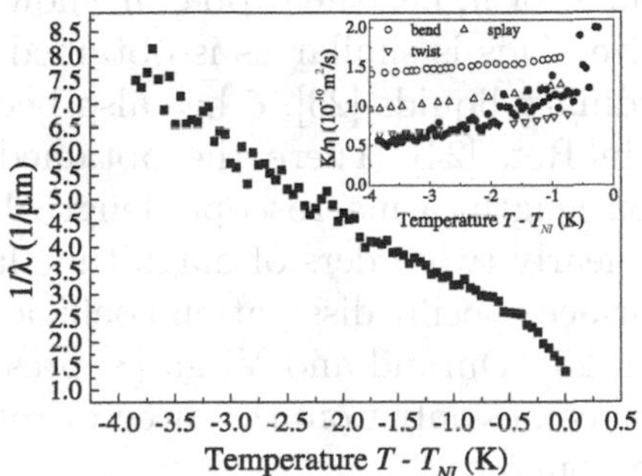

Figure 4. Temperature dependence of the inverse penetration depth. Inset: comparison of measured diffusivity (full circles) with the diffusivities of pure modes where the backflow is not considered.

Fig. 5. The quadratic coefficient, giving the bulk orientational diffusivity K/η, can only be obtained for $T_{NI} - T > 0.5$ K and then is quite scattered, but the average value is of the order of $10^{-10}\text{m}^2/\text{s}$, which is close to the known bulk values for 5CB if we assume that backflow does not contribute to the effective viscosity (inset of Fig. 4). The correct value for the quadratic coefficient gives a strong support to our analysis.

The dependence of λ^{-1} on T in Fig. 4 is nearly linear, so that $\lambda^{-1} \propto S^2$, where S is the scalar order parameter. As $K \propto S^2$ [1], and $\lambda = K/W$, $W \propto S^4$. That λ increases as T approaches T_{NI} has also been observed by other authors [22].

Using the known values for K, we get that the surface anchoring strength is $3 \times 10^{-6}\text{J/m}^2$ close to T_{NI} and $5 \times 10^{-5}\text{J/m}^2$ 4 K below T_{NI}. At these anchoring strengths the deuterium NMR experiments have shown that the configuration of the director is planar-polar [19].

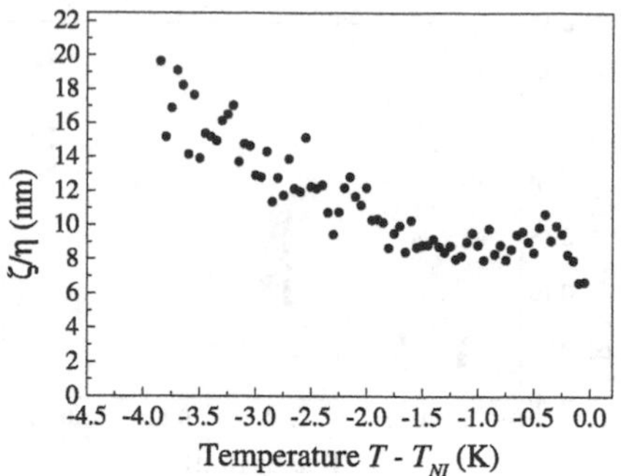

Figure 5. Temperature dependence of h, i.e., the ratio ζ/η.

Our measured values for h, i.e., the ratio ζ/η, shown in Fig. 5, are in the range of molecular size. This is similar as is obtained for the translational surface friction in ordinary liquids [23]. ζ has also been obtained in rather different conditions in Ref. [24]. There the obtained ratio ζ/η is of the order of extrapolation length, a macroscopic length having nothing to do with dissipation and nearly two orders of magnitude larger than our value. The existence of a surface specific dissipation coefficient has been a matter of some debate [8, 24, 25]. Durand and Virga [8] presented a model where the surface dissipation comes only from the director rotation and associated backflow close to the boundary.

6.2. PDLC

We performed similar experiments also in polymer dispersed nematic liquid crystals, where liquid crystal is embedded in a polymer matrix in the form of more or less spherical droplets that form as a result of phase separation of the monomer-liquid crystal solution upon polymerization. The size of the droplets depends on the polymerization process and ranges from $\sim 0.3\mu$m to 10 μm [26]. The static properties and internal structure of PDLCs are well known and this makes PDLCs very suitable to study the effects of surface on orientational dynamics.

Dynamic light scattering was performed on a 20 μm thick PDLC foil sandwiched between two glass plates coated with indium-tin oxide electrodes. The liquid crystal was E7 mixture, embedded in the polymer matrix in droplets with average diameter around 0.5 μm, as measured by scanning electron microscopy and atomic force microscopy. The electron microscope picture also shows that droplets, while having relatively narrow size distribution, are not perfectly spherical, but have a somewhat irregular shape. The system showed the transition from the nematic, scattering state, to the

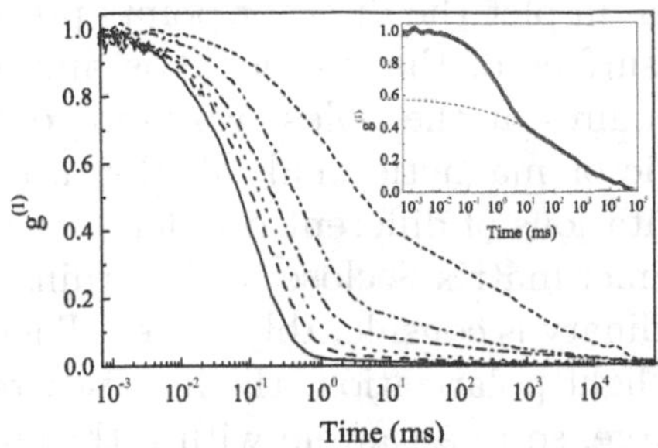

Figure 6. The dependence of the normalized correlation function on the applied voltage, 0 V (short dash), 64 V (dash-double-dotted), 80 V (dash-dotted), 100 V (dots), 120 V (dash), and 160 V (line). The temperature and outer scattering angle are 315K and 30 degrees, respectively. Inset: The measured zero field correlation function (circles), its fit (line) [7], and the stretched exponential part of the fit (dash) that describes the restricted rotational diffusion. The stretching exponent is 0.26.

isotropic transparent state at 328 K.

Measurements were performed at different scattering angles and as a function of applied field. To avoid screening by impurity charges AC field with frequency of 44 kHz was used. The directions of the incoming and scattered light were chosen so that the scattering vector was in the plane of the sample, that is, perpendicular to the applied field. The incoming polarization was perpendicular to the scattering plane and the scattered polarization was in the scattering plane.

Fig. 6 shows $g^{(1)}(\tau)$ at different strengths of the external field in the nematic phase. Two components are observed, a fast and a slow one. The fast component is due to the ordinary orientational fluctuations within the droplets. At zero field, when the sample is multiply scattering, it consists of distribution of relaxation rates due to the contributions at different scattering vectors. With the application of the electric field, this mode becomes monodisperse with increasing relaxation rate as expected for the orientational fluctuations of the director in external field. The behavior of this mode is thoroughly analyzed in Ref. [27]. Here we want to focus on the slow mode which is highly nonexponential. Its decay extends over four or more decades, as seen in Fig. 6. Over that range it can be described as a stretched exponential function with a small stretching exponent $s \sim 0.2$, which means that the spectrum of this process has a $\sim 1/\omega^{1+s}$ form in the interval from ~ 0.1 Hz to 10^3Hz. The amplitude of the process decreases with the applied field but its general form remains unchanged.

To analyze the slow relaxation process we first note that the director configuration in the PDLC droplets in our sample is bipolar as was determined by optical observations and deuterium NMR measurements [28, 29].

So in the interior of a droplet the director points predominantly along one direction **n**. At the surface of the droplet it is approximately parallel to the surface which requires at the poles two point defects. In the absence of an external electric or magnetic field, all the directions are equivalent and the average orientations of different droplets are random. The index of refraction of the polymer matrix is close to the ordinary index of the liquid crystal while extraordinary is considerably larger. Therefore, for a generally oriented droplet and light polarization, the light scattering cross section of a droplet is rather large, so in a sample with a thickness of 20 μm light is multiply scattered.

In strongly scattering systems the usual single scattering approximation used in analysis of photon correlation spectroscopy experiments is not applicable and the approach of diffusing wave spectroscopy (DWS) [30, 31] has to be used. In the regime of DWS, the observed decay of $g^{(1)}(\tau)$ is shifted to shorter times. For a process with an autocorrelation function of the stretched exponential form $\exp\left(-\left(t/\tau_0\right)^s\right)$, the characteristic relaxation time is $\tau_0\left(l^*/L\right)^{2/s}$ [32]. In our sample the transport length $l^* \sim 5\mu$m, so for the sample thickness $L/l^* \sim 4$.

The slow part of the zero field data in Fig. 6 can be well fit with a stretched exponential with $s \sim 0.26$, while the corresponding time can not be determined with any precision due to masking by the fast fluctuations within the droplets. From the data in Fig. 6 we get a very rough estimate for $\tau_0 \sim 4000$ s.

To explain the highly stretched correlation function of the slow part of the dynamics of PDLC, we proposed the following model. In the absence of external field, the average direction of the nematic director in perfectly spherical droplet can freely diffuse. The mechanism for this diffusion is motion of the point defects; the torque needed to turn the director is approximately γV, where γ is the rotational viscosity and V the volume of the droplet. The diffusion time is then approximately $1/\tau_{diff} = k_B T/\gamma V$, that is of the order of 1 s^{-1}. Thermally induced reorientation in samples with droplet size a few micrometers has been actually observed visually under the microscope by Amundson [33]. As those droplets were larger than in our sample, the reorientation was only observed close to the isotropic phase, where the effective viscosity is smaller and the characteristic time a few seconds.

Free rotational diffusion gives a simply exponentially decaying correlation function. The droplets, however, are not of perfectly spherical shapes and also the anchoring of the director on the surface is probably not perfectly homogeneous. In our model, based on the theories of $1/f$ noise and stretched exponential relaxation in condensed matter [34], in particular on the random walk in random potential [35], these irregularities were modeled

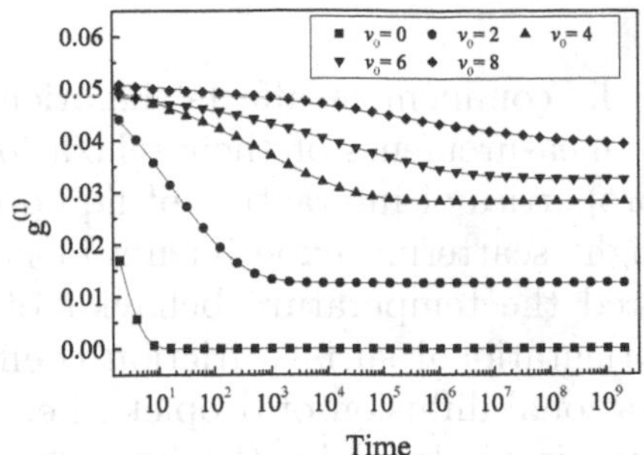

Figure 7. Autocorrelation function as described in text computed from the model of droplet reorientation in random potential for different variances of the potential v_0 (in units of k_BT). At $v_0 = 6$, $s = 0.2$. Lines are the fits to the stretched exponential function.

by a random potential in which the average director of each droplet moves.

To see what is the effect of the random potential, we performed a model simulation for the diffusion of a point on a lattice on the sphere. The lattice was taken to be vertices of the dodecahedron. In each vertex a value V_i for the random potential was assigned. They were taken to have Gaussian distribution with zero mean and variance v_0. Then the single step transition probabilities were calculated according to an adaptation of the Metropolis scheme. The director $\mathbf{n}$ is determined by the coordinates (x, y, z) of the currently occupied vertex, so as the component of the dielectric tensor which gives the main contribution to the scattering is ε_{xz}, we compute $\langle n_x(0)\, n_z(0)\, n_x(\tau)\, n_z(\tau)\rangle$, with τ taking values 2^i. This quantity should be proportional to $g^{(1)}(\tau)$ [21].

The correlation function must be averaged over the distribution of the random potential, which we do by repeating the calculation for a 100–1000 samples of the random potential. The necessity to average both over time and over the ensemble of droplets of course implies that the system is nonergodic. Experimentally that means that a sufficient number of droplets must contribute to scattering to ensure a true ensemble average.

The resulting correlation functions for different choices of the variance of the potential distribution v_0 (in units of k_BT) are shown in Fig. 7. The time unit is of the order of the free diffusion time τ_{diff}, as can be seen from the $v_0 = 0$ curve. The decay can be very well described with a stretched exponential, with a stretching exponent s from $s = 0.45$ for $v_0 = 2$ to $s = 0.12$ for $v_0 = 8$. The best agreement with the measurements is obtained for $v_0 = 5$, where $s = 0.25$. The corresponding parameter τ_0 is $\sim 3000\tau_{diff}$. The experimental value from the zero field data in Fig. 6 is roughly $4000\tau_{diff}$, so the agreement is quite good.

7. Conclusions

We have shown how the confinement affects the orientational fluctuations eigenmodes and how measurements of their relaxation rates can be then used to obtain the properties of interaction of liquid crystal with surface. From the dynamic light scattering experiments in nematic in cylindrical pores we have deduced the temperature behavior of the surface extrapolation length and orientational surface friction coefficient. In PDLC we observed the free rotational diffusion of droplets, i.e., the 0th mode of orientational fluctuations, is hindered by the irregularities of the droplets' shape and this can be explained by the model of thermally driven jumps of director field configuration in a random potential.

References

1. De Gennes, P.G. and Prost, J. (1993) *The Physics of Liquid Crystals*, (Clarendon, Oxford).
2. Orsay Liquid Crystal Group (1969) *Phys. Rev. Lett.* **22**, 1361.
3. Orsay Liquid Crystal Group (1969) *J. Chem. Phys.* **51**, 816.
4. *Liquid Crystals in Complex Geometries Formed by Polymer and Porous Networks*, edited by Crawford, G.P. and Žumer, S. (1996) (Taylor and Francis, London).
5. Mertelj, A. and Čopič, M. (1998) *Phys. Rev. Lett.* **81**, 5844.
6. Mertelj, A. and Čopič, M. (2000) *Phys. Rev. E* **61**, 1622.
7. Vilfan, M.,. Mertelj, A., and Čopič, M. (2002) *Phys. Rev. E* **65**, 041712.
8. Durand, G. E. and Virga, E. G. (1999) *Phys. Rev. E* **59**, 4137.
9. Dubois-Violette, E. and Parodi, O. (1969) *J. Phys. C* **4**, 57.
10. Doane, J. W. (1990), in *Liquid Crystals — Applications and Uses*, edited by Bahadur, B. (World Scientific, Singapore), Ch. 14.
11. Iannacchione, G. S. and Finnotello, D. (1992) *Phys. Rev. Lett.* **69**, 2094.
12. Crawford, G. P., Allender, D. W., Doane, J. W., Vilfan, M., and Vilfan, I. *Phys. Rev. A* **44**, 2571 (1991).
13. Kutnjak, Z. and Garland, C. W. (1997) *Phys. Rev. E* **55**, 488.
14. Bellini, T., Clark, N. A., Muzny, C. D., Wu, L., Garland, C.W., Schaefer, D.W., and Oliver, B. J. (1992) *Phys. Rev. Lett.* **69**, 788.
15. Stallinga, S. , Wittebrood, M. M., Luijendijk, D. H., and Rasing Th. (1996) *Phys. Rev. E* **53**, 6085.
16. Ziherl, P. and Žumer, S. (1996) *Phys. Rev. E* **54**, 1592 .
17. Kelly, J. R. and Palffy-Muhoray P. (1997) *Phys. Rev. E* **55**, 4378.
18. Wittebrood, M. M., Rasing, Th., Stallinga, S., and Muševič, I. (1998) *Phys. Rev. Lett.* **80**, 1232.
19. Allender, D. W., Crawford, G. P., and Doane, J. W. (1991) *Phys. Rev. Lett.* **67**, 1442.
20. Čopič, M., Maclennan, J. E., and Clark, N. A. (2002) *Phys. Rev. E* **65**, 021708.
21. Čopič, M. and Mertelj, A. (1998) *Phys. Rev. Lett.* **80**, 1449.
22. Dai-Shik Seo, Yasufumi Limura, and Shunsuke Kobayashi (1992) *Appl. Phys. Lett.* **61**, 234.
23. Bocquet, L. and Barrat, J.-L. (1993) *Phys. Rev. Lett.* **70**, 2726.
24. Petrov, A. G., Ionescu, A. Th., Versace, C., and Scaramuzza, N. (1995), *Liq. Cryst.* **19**, 169.
25. Nobili, M., Barberi, R., and Durand, G. (1995) *J. Phys. II France* **5**, 531.
26. Crawford, G. P. and Doane, J. W. (1993) *Mod. Phys. Lett.* **7**,1785.

27. Mertelj, A., Spindler, L., and Čopič, M. (1997), *Phys. Rev. E* **56**, 549.
28. Ondris-Crawford, R., Boyko, E. B., Wagner, B. G., Erdmann, J. H., Žumer, S., and Doane, J. W. (1991) *J. Appl. Phys.* **69**, 6380.
29. Golemme, A., Žumer, S., Allender, D. W., and Doane, J. W. (1988) *Phys. Rev. Lett.* **61**, 2937.
30. Maret, G. and Wolf, P. E. (1987) *Z. Phys. B* **65**, 409.
31. Pine, D. J., Weitz, D. A., Chaikin, P. M., and Herbolzheimer, E. (1988) *Phys. Rev. Lett.* **60**, 1134.
32. Weitz, D. A. and Pine, D. J. (1992) in *Dynamic Light Scattering*, edited by Brown, W. (Oxford University Press, Oxford), 652–720.
33. Amundson, K. (1996) *Phys. Rev. E* **53**, 2412.
34. Weissman, M. B. (1988) *Rev. Mod. Phys.* **60**, 537.
35. Marinari, E., Parisi, G., Ruelle, D., and Windey, P. (1983) *Phys. Rev. Lett.* **50**, 1223.

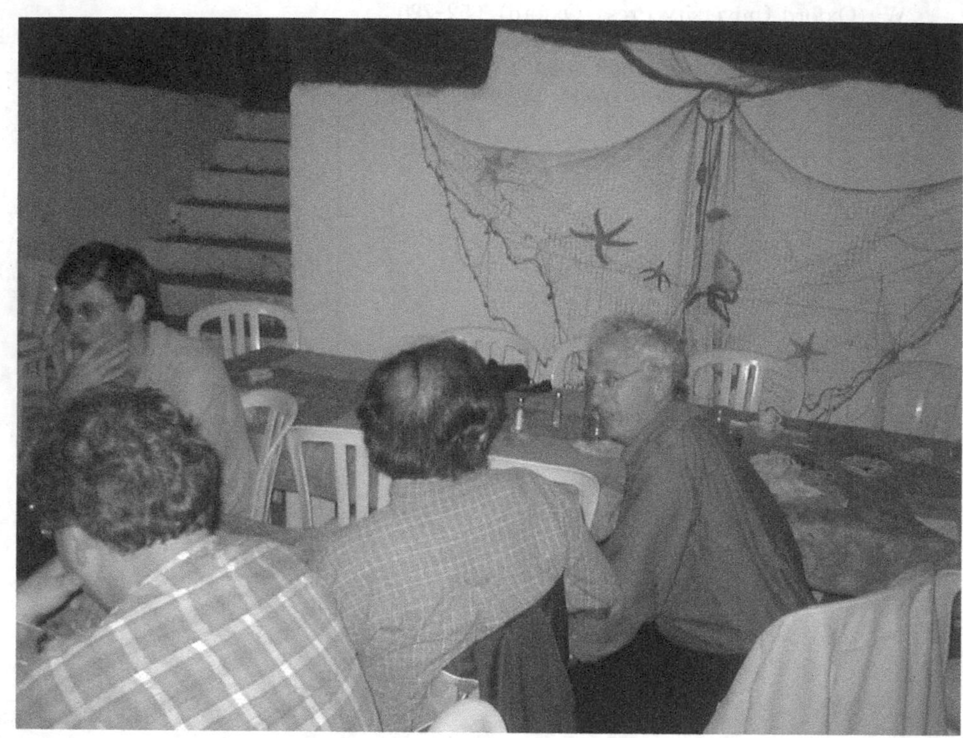

Richard Weaver, Rudolf Sprik, Costas Soukoulis and Ad Lagendijk (from left to right). Photo by Gabriel Cwilich

OPTICAL PROPERTIES AND FLUCTUATIONS IN LIQUID CRYSTALS WITH ONE-DIMENSIONAL LARGE-SCALE PERIODICITY

E.V. AKSENOVA
Institute for High Performance Computing and Data Bases
198005, Fontanka, 120, Saint-Petersburg, Russia
AND
V.P. ROMANOV AND A.YU. VAL'KOV
Department of Physics, Saint-Petersburg State University
198504, Ul'yanovskaya, 1, Saint-Petersburg, Russia

Abstract. The Green's function of electromagnetic field and the correlation function of director fluctuations in cholesteric liquid crystals with a pitch being large compared to the wavelength are considered. The wave propagation and the fluctuations spectrum are analyzed in the framework of unique approach based on the modified WKB method. The periodic system is distinguished from an anisotropic medium by a discontinuity of the wave vector surface and a break of the beam vector surface. Trajectories of beams in such a medium are not plane. The forbidden zone corresponds to a capture of beams and formation of a wave channel. The director fluctuations near turn points where the WKB method is not valid are considered.

1. Introduction

Liquid crystals (LC) with one-dimensional periodicity are systems with unusual optical and statistical properties [1]. Among these LC are cholesteric liquid crystals (CLC) and twisted nematic liquid crystals. The problems of electromagnetic wave propagation and fluctuation spectrum in such media are important both for the theory of optical and statistical properties of complex systems and for applications of helical LC for information mapping.

B.A. van Tiggelen and S.E. Skipetrov (eds.),
Wave Scattering in Complex Media: From Theory to Applications, 519–534.

Considerable study has been given to CLC with a pitch d comparable with the wavelength of light λ or with the wavelength of fluctuation. In this case, diffraction phenomena, selective reflection of circularly polarized light and the anomalously large optical activity are investigated in detail [2, 3]. In Ref. [4], the case of long-wavelength fluctuations, when the pitch is much less than the wavelength of fluctuation q^{-1}, was considered in $(\mathbf{q}\,, z)$-representation. At the same time, cholesterics with a pitch exceeding the wavelength significantly and weakly twisted nematics attract particular interest. Although these systems appear to be simpler, unusual effects are observed here too.

In order to study the optical properties of these systems, it is convenient to investigate the Green's function of the electromagnetic field. The point is that the Green's function is actually a field of a point source. By now, a formal algorithm exists for construction of the Green's function in the CLC [5]. It provides a numerical procedure only and does not permit derivation of an explicit analytic solution. For solving the problems of scattering, the correlation function of director fluctuations has to be known. In a homogeneous unbounded medium, the correlation radius of director fluctuations is infinite. When passing to helical structures, essential difficulties are associated with the necessity of taking into account two characteristic length scales: the correlation radius and the characteristic size of the structure. Up to now, an exact solution of the problem for such a medium has not been found.

The present work is devoted to a study of the Green's function and the spatial correlation function in CLC with a large helix pitch. We use a unique approach based on the modified WKB method. We investigate the propagation of waves in such media and fluctuation spectrum for the case of short-wave fluctuating modes. We analyze in detail the behavior of fluctuations in the vicinity of turn points where the WKB approximation is not applicable.

2. The Green's Function

In cholesterics, the optical axis is parallel to the equilibrium unit vector of the director $\mathbf{n}_0$ that rotates along the axis of cholesteric $\mathbf{e}_z$:

$$\mathbf{n}_0(\mathbf{r}) = \mathbf{n}_0(z) = (\cos\phi;\ \sin\phi;\ 0), \tag{1}$$

where $\phi = p_0 z + \psi_0$, $p_0 = \pi/d$, ψ_0 defines the direction of $\mathbf{n}_0$ in $z = 0$ plane. The optical properties of CLC are determined by the locally uniaxial tensor of permittivity [1, 2]

$$\varepsilon_{\alpha\beta}(\mathbf{r}) = \varepsilon_\perp \delta_{\alpha\beta} + \varepsilon_a n_{0\alpha}(\mathbf{r}) n_{0\beta}(\mathbf{r})\,. \tag{2}$$

Here $\varepsilon_a = \varepsilon_\parallel - \varepsilon_\perp$; $\varepsilon_\parallel$, $\varepsilon_\perp$ are permittivities in the directions parallel and perpendicular to $\mathbf{n}_0$, respectively.

The Maxwell equations in such a medium have the form

$$\nabla \times \mathbf{E}(\mathbf{r}) = ik_0\mu\mathbf{H}(\mathbf{r})\,, \quad \nabla \times \mathbf{H}(\mathbf{r}) = -ik_0\hat{\varepsilon}(z)\mathbf{E}(\mathbf{r})\,, \tag{3}$$

where $\mathbf{E}(\mathbf{r})$ and $\mathbf{H}(\mathbf{r})$ are the electric and magnetic vectors, $k_0 = \omega/c$, ω is a circular frequency, c is the velocity of light in vacuum. Hereinafter we suppose the magnetic permeability $\mu = 1$. The Green's function (field of a point source) obeys the equation

$$\left(\nabla \times \nabla \times -k_0^2\,\hat{\varepsilon}(z)\right)\widehat{T}(\boldsymbol{\rho} - \boldsymbol{\rho}_1, z, z_1) = \delta(\mathbf{r} - \mathbf{r}_1)\widehat{I}\,, \tag{4}$$

where $\boldsymbol{\rho} = (x, y)$, $\delta(\mathbf{r} - \mathbf{r}_1)$ is the delta-function, $\widehat{I}$ is the unit matrix.

As $\hat{\varepsilon}(z)$ depends on the z variable only, we perform a two-dimensional Fourier transform in Eq. (4):

$$\widehat{T}(\mathbf{q}; z, z_1) = \int d\boldsymbol{\rho}\exp(-i\mathbf{q}\boldsymbol{\rho})\widehat{T}(\boldsymbol{\rho}, z, z_1)\,. \tag{5}$$

Let us choose the directions of x and y axes as follows: $\mathbf{x} \parallel \mathbf{q}$, $\mathbf{y} \perp \mathbf{q}$. Excluding elements of the third line and the third column of the $\widehat{T}(\mathbf{q}; z, z_1)$ matrix in Eq. (4), we have

$$-\frac{\partial^2}{\partial z^2}\hat{t}(z, z_1) + \hat{b}(z)\hat{t}(z, z_1) = \delta(z - z_1)\widehat{I}, \tag{6}$$

where

$$\hat{b}(z) = k_0^2\begin{pmatrix} -\varepsilon_{xx}(1-\mathcal{H}) & -\varepsilon_{xy}(1-\mathcal{H}) \\ -\varepsilon_{xy} & -\varepsilon_{yy} + \varepsilon_\perp\mathcal{H} \end{pmatrix}, \tag{7}$$

$\mathcal{H} = q^2/k_0^2\varepsilon_\perp$. If the function $\hat{t}(z, z_1)$ is known it is easy to obtain the Green's function $\widehat{T}(\mathbf{q}; z, z_1)$. Thus, the Green's function calculation is reduced to the solution of the system of ordinary linear differential equations (6), with periodically varying coefficients.

The boundary condition for the Green's function $\widehat{T}$ in coordinate representation is a condition of radiation. It leads to the corresponding boundary condition for the function $\hat{t}(z, z_1)$ in the system (6). Then we can write for the function $\hat{t}(z, z_1)$

$$\hat{t}(z, z_1) = \begin{cases} \widehat{V}_1(z)\widehat{V}_1^{-1}(z_1)\widehat{W}^{-1}(z_1)\,, & z \geq z_1 \\ \widehat{V}_2(z)\widehat{V}_2^{-1}(z_1)\widehat{W}^{-1}(z_1)\,, & z < z_1 \end{cases} \tag{8}$$

where $\widehat{W}(z) = \widehat{V}_2'(z)\widehat{V}_2^{-1}(z) - \widehat{V}_1'(z)\widehat{V}_1^{-1}(z)$, $\widehat{V}_1(z) \to 0$ for $z \to +\infty$ and $\widehat{V}_2(z) \to 0$ for $z \to -\infty$. The columns of $\widehat{V}_{1,2}(z)$ matrices are the linearly independent solutions of the system (6) with a zero right hand side.

In order to find these linearly independent solutions we perform the Fourier transform (5) in the Maxwell equations (3). Excluding the components E_z, H_z from the equations and introducing a dimensionless variable $\xi = p_0 z$, we obtain a system of first-order linear differential equations

$$\frac{\partial}{\partial\xi}\begin{pmatrix} E_x \\ E_y \\ H_x \\ -H_y \end{pmatrix} = -i\Omega \begin{pmatrix} 0 & 0 & 0 & 1-\mathcal{H} \\ 0 & 0 & 1 & 0 \\ \varepsilon_{xy} & \varepsilon_{yy}-\mathcal{H}\varepsilon_{\perp} & 0 & 0 \\ \varepsilon_{xx} & \varepsilon_{xy} & 0 & 0 \end{pmatrix} \begin{pmatrix} E_x \\ E_y \\ H_x \\ -H_y \end{pmatrix}. \tag{9}$$

Here $\Omega = k_0/p_0 = 2d/\lambda$ is a large dimensionless parameter. In matrix notation,the system (9) can be written as

$$\mathbf{\Phi}'(\xi) = i\Omega\widehat{A}(\xi)\mathbf{\Phi}(\xi)\,, \tag{10}$$

where $\widehat{A}(\xi + p_0 d) = \widehat{A}(\xi)$ is a periodic function. The solution of the system (10) with an initial condition $\mathbf{\Phi}_0 = \mathbf{\Phi}(\xi_0)$, $\xi_0 = p_0 z_0$ is

$$\mathbf{\Phi}(\xi) = \widehat{M}(\xi,\xi_0)\mathbf{\Phi}(\xi_0)\,, \tag{11}$$

where $\widehat{M}(\xi,\xi_0)$ is a matrix of evolution. In the first approximation in $1/\Omega$, the matrix of evolution $\widehat{M}(\xi,\xi_0)$ is [6]:

$$\widehat{M}(\xi,\xi_0)=\widehat{U}(\xi)\widehat{\mathrm{diag}}\left\{\exp\left[\int_{\xi_0}^{\xi}\left(i\Omega\lambda_l(x) - \left(\widehat{U}^{-1}(x)\widehat{U}'(x)\right)_{ll}\right)dx\right]\right\}\widehat{U}^{-1}(\xi_0). \tag{12}$$

The columns of the $\widehat{U}(\xi)$ matrix are eigenvectors of the $\widehat{A}(\xi)$ matrix, $\lambda_l(\xi)$ are corresponding eigenvalues. Eq. (12) is many-dimensional analog of the WKB approximation. The structure of the solution (11), (12) agrees with the Floquet theorem [7].

Using the special block structure of the $\widehat{A}(\xi)$ matrix, it is possible to obtain explicit expressions for the eigenvalues λ_l and the $\widehat{U}$ matrix [6]. As a result we have

$$\widehat{V}_1(z) = \hat{u}(z)\exp(\hat{\phi}_+)\,, \quad \widehat{V}_2(z) = \hat{u}(z)\exp(\hat{\phi}_-)\,, \tag{13}$$

where

$$\exp(\hat{\phi}_\pm) = \sqrt{\frac{1-\mathcal{H}\cos^2\tilde{\xi}_0}{1-\mathcal{H}\cos^2\tilde{\xi}}} \times \begin{pmatrix} \exp\left[\pm i\Omega\lambda_1(\xi-\xi_0)\right] & 0 \\ 0 & \sqrt{\frac{\lambda_2(\tilde{\xi}_0)}{\lambda_2(\tilde{\xi})}}\exp\left[\pm i\Omega\int_{\xi_0}^{\xi}\lambda_2(x)\,dx\right] \end{pmatrix}, \tag{14}$$

$$\hat{u}(\xi) = \begin{pmatrix} \sin\tilde{\xi} & (1-\mathcal{H})\cos\tilde{\xi} \\ -\cos\tilde{\xi} & \sin\tilde{\xi} \end{pmatrix}, \; \hat{u}_0 = \hat{u}(\tilde{\xi}_0), \\ \lambda_1 = \sqrt{\varepsilon_\perp(1-\mathcal{H})}, \quad \lambda_2(\xi) = \sqrt{\varepsilon_\| - \varepsilon_{xx}\mathcal{H}}\,. \tag{15}$$

Here $\tilde{\xi} = \xi + \psi_0$ $\tilde{\xi}_0 = \xi_0 + \psi_0$. From Eqs. (8) and (13) we obtain the Green's function $\widehat{T}(\mathbf{q}; z, z_1)$

$$\widehat{T}(\mathbf{q}; z, z_1) = \exp\left(ik_0\lambda_1 \left|z - z_1\right|\right) \widehat{F}^{(1)}(\mathbf{q}; z, z_1) \\ + \exp\left(ik_0 \left|\int_{z_1}^{z} \lambda_2(x)\, dx\right|\right) \widehat{F}^{(2)}(\mathbf{q}; z, z_1), \tag{16}$$

where $\xi_1 = p_0 z_1$, $\tilde{\xi}_1 = p_0 z_1 + \psi_0$. Here $\widehat{F}^{(1)}$ and $\widehat{F}^{(2)}$ matrices can be expressed through tensor products of unit polarization vectors of ordinary and extraordinary waves $\mathbf{e}^{(1)}$ and $\mathbf{e}^{(2)}$:

$$F_{kj}^{(1)} = \frac{i}{2k_0\sqrt{1-\mathcal{H}}\sqrt{\varepsilon_\perp}}\, e_k^{(1)}(\tilde{\xi}) e_j^{(1)}(\tilde{\xi}_1), \\ F_{kj}^{(2)} = \frac{i\sqrt{(\varepsilon_\perp + \varepsilon_a \mathcal{H}\cos^2\tilde{\xi})(\varepsilon_\perp + \varepsilon_a\mathcal{H}\cos^2\tilde{\xi}_1)}}{2k_0\varepsilon_\perp\sqrt{\lambda_2(\tilde{\xi})\lambda_2(\tilde{\xi}_1)}}\, e_k^{(2)}(\tilde{\xi}) e_j^{(2)}(\tilde{\xi}_1), \tag{17}$$

$k, j = 1, 2, 3$. The first and the second terms in Eq. (16) correspond to the ordinary and extraordinary waves, respectively.

The Green's function in the coordinate representation has the form

$$\widehat{T}(\boldsymbol{\rho} - \boldsymbol{\rho}_1, z, z_1) = \int \frac{d\mathbf{q}}{(2\pi)^2} \exp\left[i\mathbf{q}(\boldsymbol{\rho} - \boldsymbol{\rho}_1)\right] \widehat{T}(\mathbf{q}; z, z_1). \tag{18}$$

Let us analyze this function in the far field, i.e., for $|\mathbf{r} - \mathbf{r}_1| \gg \lambda$. The integral (18) can be calculated using the stationary phase method. In the first term of Eq. (16) for the stationary point we get

$$\mathbf{q}_{st}^{(1)} = \sqrt{\varepsilon_\perp} k_0 \frac{\boldsymbol{\rho} - \boldsymbol{\rho}_1}{|\mathbf{r} - \mathbf{r}_1|}. \tag{19}$$

The phase of the wave is equal to $\Psi_1 = \sqrt{\varepsilon_\perp} k_0 |\mathbf{r} - \mathbf{r}_1|$. It means that a surface of wave vectors and a surface of a constant phase are spheres just as in an isotropic medium with permittivity $\varepsilon_\perp$.

The second term of Eq. (16) describes the extraordinary wave and the equation for the stationary point has the form

$$\boldsymbol{\rho} - \boldsymbol{\rho}_1 = \frac{1}{k_0\varepsilon_\perp} \left| \int_z^{z_1} \frac{\hat{\varepsilon}^\perp(z')\mathbf{q}_{st}^{(2)}\, dz'}{\lambda_2(z', \mathbf{q}_{st}^{(2)})} \right|, \tag{20}$$

where $\varepsilon^{\perp}_{\alpha\beta}(z) = \varepsilon_{\alpha\beta}(z)$, $(\alpha, \beta = 1, 2)$ is the transversal part of the $\hat{\varepsilon}(z)$ tensor, $\lambda_2(z, \mathbf{q}) = \sqrt{\varepsilon_{\|} - (\mathbf{q}\hat{\varepsilon}^{\perp}(z)\mathbf{q})/k_0^2\varepsilon_{\perp}}$. The phase of the wave is equal to

$$\Psi_2 = (\boldsymbol{\rho} - \boldsymbol{\rho}_1)\mathbf{q}^{(2)}_{st} + k_0 \left| \int_z^{z_1} dz' \lambda_2(z', \mathbf{q}^{(2)}_{st}) \right| . \tag{21}$$

The Green's function in the far field then becomes

$$\widehat{T}(\mathbf{r}, \mathbf{r}_1) = \frac{-\sqrt{\varepsilon_{\perp}} \exp(i\sqrt{\varepsilon_{\perp}} k_0 \left|\mathbf{r} - \mathbf{r}_1\right|) \left|z - z_1\right| \widehat{F}^{(1)}(\mathbf{q}^{(1)}_{st}; z, z_1)}{4\pi\sqrt{(\left|\mathbf{r} - \mathbf{r}_1\right|^2 - \left|\boldsymbol{\rho} - \boldsymbol{\rho}_1\right|^2 \cos^2\tilde{\xi})(\left|\mathbf{r} - \mathbf{r}_1\right|^2 - \left|\boldsymbol{\rho} - \boldsymbol{\rho}_1\right|^2 \cos^2\tilde{\xi}_1)}}$$
$$- \frac{\varepsilon_{\perp} k_0 \exp(i\Psi_2)(\det \widehat{a}(\mathbf{q}^{(2)}_{st}; z, z_1))^{-1/2} \widehat{F}^{(2)}(\mathbf{q}^{(2)}_{st}; z, z_1)}{4\pi\sqrt{(k_0^2\varepsilon_{\perp} - (\mathbf{q}^{(2)}_{st}\mathbf{n}_0(z))^2)(k_0^2\varepsilon_{\perp} - (\mathbf{q}^{(2)}_{st}\mathbf{n}_0(z_1))^2)}} , \tag{22}$$

where

$$a_{\alpha\beta}(\mathbf{q}; z, z_1) = \frac{-1}{\sqrt{\varepsilon_{\perp}}} \left| \int_z^{z_1} \frac{\varepsilon^{\perp}_{\alpha\beta}(\varepsilon_{\|}\varepsilon_{\perp}k_0^2 - (\mathbf{q}\hat{\varepsilon}^{\perp}\mathbf{q})) + (\hat{\varepsilon}^{\perp}\mathbf{q})_{\alpha}(\hat{\varepsilon}^{\perp}\mathbf{q})_{\beta}}{(\varepsilon_{\|}\varepsilon_{\perp}k_0^2 - (\mathbf{q}\hat{\varepsilon}^{\perp}\mathbf{q}))^{3/2}} dz' \right| .$$

3. Waveguide Propagation

Let us find the stationary point $\mathbf{q}^{(2)}_{st}$ for distances $|z - z_1| \gg d$. Note, the integrand in the right hand side of Eq. (20) is a periodic function. The primitive of a periodic function can be presented as a sum of a linear function and a periodic one. The inclination of the linear part is equal to the average value on the period of the integrand. In the case of $|z - z_1| \gg d$, the periodic contribution can be omitted. Then we can write

$$\Psi_2 = |\boldsymbol{\rho} - \boldsymbol{\rho}_1| q^{(2)}_{st} + (z - z_1) q^{(2)}_z . \tag{23}$$

The wave vector $\mathbf{k}^{(2)}_{st} = (\mathbf{q}^{(2)}_{st}, q^{(2)}_z)$ is real if the following condition is fulfilled:

$$q^{(2)2}_{st} < k_0^2 \min\left(\varepsilon_{\perp}, \varepsilon_{\|}\right) . \tag{24}$$

For other values of $q^{(2)}_{st}$, the phase Ψ_2 becomes complex and the wave does not propagate in the region of large $|z - z_1|$ due to damping. For these values of $q^{(2)}_{st}$ a forbidden zone is formed, i.e., there is a restriction on possible directions of the wave vector $\mathbf{k}^{(2)}_{st}$.

Fig. 1 shows a cross section of the surface of wave vectors by a plane containing the z axis for $\varepsilon_a > 0$. The discontinuities correspond to the forbidden zones. Note, the directions of the wave and the beam vectors in such a medium do not coincide, similarly to a uniaxial anisotropic medium.

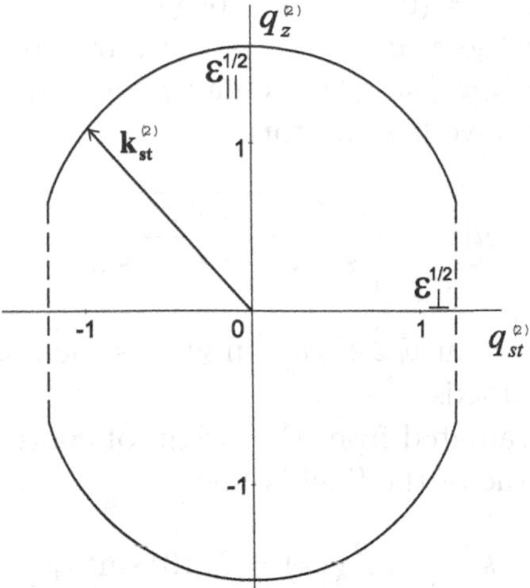

Figure 1. Cross section of the surface of wave vectors by a plane containing the z axis for $\varepsilon_a = 1.0$, $\varepsilon_{\|} = 2.5$. The dotted lines show the forbidden zones. The value of $k_{st}^{(2)}$ depends on the wave vector direction. All wave numbers are expressed in terms of k_0.

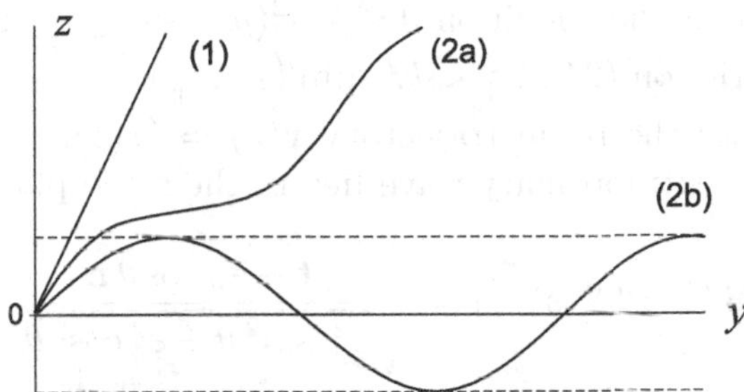

Figure 2. Capture of a beam and formation of a plane wave channel. Curve (1) is the ordinary beam, curve (2a) is the extraordinary beam, curve (2b) is the extraordinary beam captured in the wave channel.

If the condition (24) is violated, a ray begins to turn and the component $q_z^{(2)}$ becomes equal to zero at some point $z = z_t$, then it changes the sign. It means that a ray is reflected from the plane $z = z_t$. As the refractive index is a periodic function, such a ray will be reflected alternately by two planes (see Fig. 2). Thus, a planar wave channel is formed [8, 9]. Inside this channel, the waves can propagate at large distances for the ρ variable, remaining within the limits of one period of z.

In order to explain the origin of the wave channel, we consider an extraordinary wave with the wave vector $\mathbf{k}(z) = k_0 n^{(e)} \mathbf{t}$, lying in the plane yz:

$\mathbf{k}(z) = (0, k_\perp, k_z(z))$, $\mathbf{t} = (0, \sin\chi(z), \cos\chi(z))$, $\chi(z)$ is the angle between the wave vector and the z axis, $n^{(e)}$ is the refractive index. Note, due to the Snell's law $k_\perp = k_0 n^{(e)} \sin\chi(z)$ remains constant. The refractive index of the extraordinary wave has the form

$$n^{(e)} = \sqrt{\frac{\varepsilon_\parallel \varepsilon_\perp}{\varepsilon_\parallel \cos^2\theta + \varepsilon_\perp \sin^2\theta}}, \tag{25}$$

where $\cos\theta = (\mathbf{n}_0, \mathbf{t}) = \sin(p_0 z + \psi_0)\sin\chi(z)$ is the angle between the wave vector and the optical axis.

Let the wave be radiated from the origin of coordinates at an angle χ_0 to the z axis. By virtue of the Snell's law

$$k^2(z)\sin^2\chi(z) = k^2(0)\sin^2\chi_0\,. \tag{26}$$

At the turn point $k_z(z_t) = 0$, the angle $\chi(z_t) = \pi/2$. Then Eq. (26) yields

$$\sin^2(p_0 z_t + \psi_0) = \frac{\varepsilon_\perp}{\varepsilon_a}\cot^2\chi_0 + \sin^2\psi_0\,. \tag{27}$$

Taking into account the condition $0 \le \sin^2(p_0 z_t + \psi_0) \le 1$ in Eq. (27), we come to the restriction (24), $k_\perp^2 < k_0^2 \min\left(\varepsilon_\perp, \varepsilon_\parallel\right)$.

Let us consider the beam trajectory $\mathbf{r}(z) = (x(z), y(z), z)$. The beam vector $\mathbf{S}^{(e)}$ of the extraordinary wave lies in the same plane as the vectors $\mathbf{k}$ and $\mathbf{n}_0$:

$$\mathbf{S}^{(e)} = S^{(e)}\mathbf{s}\,, \quad \mathbf{s} = \frac{\varepsilon_\perp \mathbf{t} + \varepsilon_a \cos\theta\, \mathbf{n}_0}{\sqrt{\varepsilon_\perp^2 \sin^2\theta + \varepsilon_\parallel^2 \cos^2\theta}}\,. \tag{28}$$

Vector $\mathbf{S}^{(e)}$ is parallel to a tangent to the beam trajectory in every point. This leads to relations

$$x'(z) = s_x(z)\ell(z),\ y'(z) = s_y(z)\ell(z),\ 1 = s_z(z)\ell(z)\,, \tag{29}$$

where $\ell(z) = \sqrt{x'^2(z) + y'^2(z) + 1}$. We then obtain from Eqs. (28), (29)

$$\begin{aligned} x'(z) &= \frac{\varepsilon_a}{\varepsilon_\perp}\sin(p_0 z + \psi_0)\cos(p_0 z + \psi_0)\tan\chi(z)\,, \\ y'(z) &= \tan\chi(z) + \frac{\varepsilon_a}{\varepsilon_\perp}\sin^2(p_0 z + \psi_0)\tan\chi(z)\,, \end{aligned} \tag{30}$$

where $\tan\chi(z)$ can be obtained from Eq. (26).

The trajectory of the beam can be found by integrating Eqs. (30). Outside the forbidden zone, the trajectory is calculated by integration of

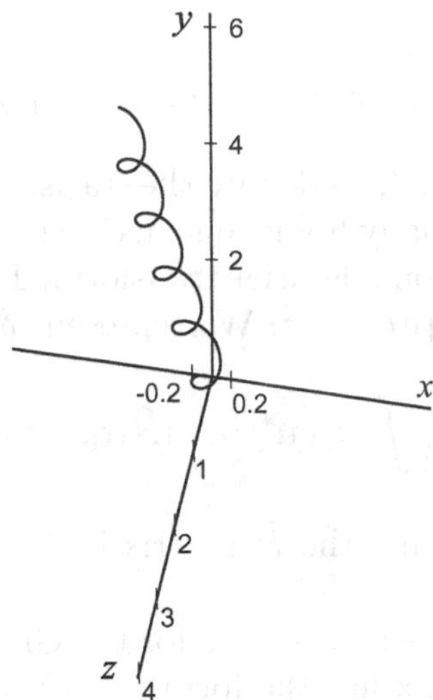

Figure 3. Trajectory of a beam outside the wave channel. The trajectory was calculated for $\varepsilon_a = 2.0$, $\varepsilon_{\|} = 2.5$, $\psi_0 = -\pi/4$, $\chi_0 = \pi/6$. All distances are expressed in units of d.

Eqs. (30) (see Fig. 3). Inside the wave channel, it is necessary to sew correctly the regions with positive and negative k_z. The trajectory of the extraordinary beam is not plane neither inside nor outside the wave channel.

4. Correlation Function of Director Fluctuations

The free energy of CLC has the form [1]

$$F = F_0 + \frac{1}{2}\int d\mathbf{r}\left\{K_{11}(\nabla\mathbf{n})^2 + K_{22}[\mathbf{n}(\nabla\times\mathbf{n}) + p_0]^2 + K_{33}(\mathbf{n}\times(\nabla\times\mathbf{n}))^2\right\}, \tag{31}$$

where F_0 is the free energy of a uniform system, K_{ll} $(l = 1,2,3)$ are the Frank elastic constants, $\mathbf{n}$ is the director.

We are interested in the director fluctuations $\delta\mathbf{n}(\mathbf{r}) = \mathbf{n}(\mathbf{r}) - \mathbf{n}_0(\mathbf{r})$. In quadratic approximation in $\delta\mathbf{n}$, the contribution to the free energy has the form

$$\begin{gathered}\delta F = \tfrac{1}{2}\int d\mathbf{r}\,\{K_{11}(\nabla\delta\mathbf{n})^2 \\ + K_{22}(\mathbf{n}_0(\nabla\times\delta\mathbf{n}))^2 + K_{33}[(\delta\mathbf{n}\nabla)\mathbf{n}_0 + (\mathbf{n}_0\nabla)\delta\mathbf{n})]^2\}\,.\end{gathered} \tag{32}$$

Vector $\delta\mathbf{n}$ can be parameterized by the two variables u_1 and u_2

$$\delta n_x = -u_1\sin\phi, \quad \delta n_y = u_1\cos\phi, \quad \delta n_z = u_2. \tag{33}$$

The u_1 mode determines the director fluctuations in the xy plane, and the u_2 mode is the fluctuations along the z axis. We will study the behavior of the correlation function

$$G_{\beta\gamma}(\mathbf{r},\mathbf{r}_1) = \langle u_\beta(\mathbf{r})u_\gamma(\mathbf{r}_1)\rangle, \tag{34}$$

where $\beta,\gamma = 1,2$, brackets $\langle\ldots\rangle$ denote the statistical averaging. Taking into account that CLC is spatially homogeneous in the xy plane, it is convenient to perform a two-dimensional Fourier transform. Then the distortion energy has the form $\delta F = \int d^2\mathbf{q}\,\delta F_q/4\pi^2$. We represent δF_q as a quadratic form

$$\delta F_q = \frac{1}{2}\int dz \left(\mathbf{u}^*(\mathbf{q},z)\widehat{H}(\mathbf{q},z)\mathbf{u}(\mathbf{q},z)\right). \tag{35}$$

Here vector $\mathbf{u} = (u_1, u_2)$ and the $\widehat{H}$ matrix is a differential operator of the second order.

We choose the x axis as it was done for the Green's function, i.e. $q_x = q$, $q_y = 0$. Then the $\widehat{H}$ matrix has the form

$$\begin{aligned}\widehat{H} = K_{11}\begin{pmatrix} q^2\sin^2\phi & iq\sin\phi\,\partial_z \\ iq\partial_z\sin\phi & -\partial_z^2 \end{pmatrix} + K_{22}\begin{pmatrix} -\partial_z^2 & -iq\partial_z\sin\phi \\ -iq\sin\phi\,\partial_z & q^2\sin^2\phi \end{pmatrix} \\ +K_{33}\begin{pmatrix} q^2\cos^2\phi & -ip_0q\cos\phi \\ ip_0q\cos\phi & q^2\cos^2\phi + p_0^2 \end{pmatrix},\end{aligned} \tag{36}$$

where $\partial_z \equiv \partial/\partial z$, $\partial_z^2 \equiv \partial^2/\partial z^2$. The fluctuation probability is $w \sim \exp[-\delta F_q/k_BT]$, where k_B is the Boltzmann constant and T is the temperature. Therefore the calculation of the correlation function $\widehat{G}$ reduces to inversion of the $\widehat{H}$ matrix. This is equivalent to the solution of the following equation

$$\widehat{H}(\mathbf{q},z)\widehat{G}(\mathbf{q},z,z_1) = k_BT\delta(z-z_1)\widehat{I}. \tag{37}$$

As boundary conditions we take the principle of correlation damping, i.e. $\widehat{G}(\mathbf{q},z,z_1) \to 0$ with $z \to \pm\infty$.

The problem (37) represents a set of two non-homogeneous differential equations of the second order with periodic coefficients. We will solve this set by a WKB-like method with a large parameter $\tilde{\Omega} = q/p_0 \gg 1$. First, we solve homogeneous equations at $z > z_1$ and $z < z_1$. Then, using the conditions of continuity and the derivative jump, we calculate the correlation function.

For the application of the WKB method, it is convenient to reduce Eq. (37) to a set of four equations of the first order. For this purpose we introduce a vector $\mathbf{v}$: $i\tilde{\Omega}\mathbf{v} = d\mathbf{u}/d\xi$. Then the set (37) with the zero right hand side takes the form

$$\frac{d}{d\xi}\Psi = \left(i\tilde{\Omega}\widehat{B} + \widehat{C}\right)\Psi. \tag{38}$$

Here we use the following notations: $\Psi = (\mathbf{u}, \mathbf{v})$,

$$\widehat{B} = \begin{pmatrix} 0 & 0 & 1 & 0 \\ 0 & 0 & 0 & 1 \\ b_1 & 0 & 0 & b_3 \\ 0 & b_2 & b_4 & 0 \end{pmatrix}, \quad \begin{array}{l} b_1 = -(K_{11}\sin^2\phi + K_{33}\cos^2\phi)/K_{22} \\ b_2 = -(K_{22}\sin^2\phi + K_{33}\cos^2\phi)/K_{11} \\ b_3 = (K_{11}/K_{22} - 1)\sin\phi \\ b_4 = -(K_{22}/K_{11} - 1)\sin\phi\,, \end{array} \tag{39}$$

$$\widehat{C} = \begin{pmatrix} 0 & 0 & 0 & 0 \\ 0 & 0 & 0 & 0 \\ 0 & -\dfrac{K_{22}+K_{33}}{K_{22}}\cos\phi & 0 & 0 \\ \dfrac{K_{11}+K_{33}}{K_{11}}\cos\phi & \dfrac{-iK_{33}}{K_{11}\tilde{\Omega}} & 0 & 0 \end{pmatrix}.$$

Then, as it was done for the Green's function, we can obtain for the matrix of evolution of Eq. (38) in the first approximation

$$\begin{array}{c} \widehat{M}(\xi,\xi_0) \approx \widehat{U}(\xi)\widehat{\mathrm{diag}}\left\{\exp\left[\int_{\xi_0}^{\xi}\left(i\tilde{\Omega}\tilde{\lambda}_l(\xi')\right.\right.\right. \\ \left.\left.\left. -\left(\widehat{U}^{-1}(\xi')\widehat{U}'(\xi') - \widehat{U}^{-1}(\xi')\widehat{C}(\xi')\widehat{U}(\xi')\right)_{ll}\right)d\xi'\right]\right\}\widehat{U}^{-1}(\xi_0), \end{array} \tag{40}$$

where the columns of the $\widehat{U}$ matrix are eigenvectors of the $\widehat{B}$ matrix and $\tilde{\lambda}_l$ are corresponding eigenvalues. Considering terms of the next order in $1/\tilde{\Omega}$, we get the condition of applicability of the WKB method as

$$\left|\frac{1}{\tilde{\lambda}_m - \tilde{\lambda}_l}\left(\widehat{U}^{-1}\widehat{U}' - \widehat{U}^{-1}\widehat{C}\widehat{U}\right)_{lm}\right| \ll \tilde{\Omega}, \quad l \neq m. \tag{41}$$

To obtain the correlation function, it is necessary to find the solutions $\mathbf{u}$ which represent the first and the second components of the $\Psi(\xi)$ vector. By virtue of the boundary conditions for $\widehat{G}$ we construct two matrices $\widehat{u}_1(\xi)$ and $\widehat{u}_2(\xi)$ so that $\widehat{u}_1(\xi) \to \widehat{0}$ at $\xi \to +\infty$ and $\widehat{u}_2(\xi) \to \widehat{0}$ at $\xi \to -\infty$. With these matrices the correlation function can be written as

$$\widehat{G}(\xi,\xi_1) = \left\{ \begin{array}{ll} \widehat{u}_1(\xi)\widehat{u}_1^{-1}(\xi_1)(\widehat{u}_2'\widehat{u}_2^{-1} - \widehat{u}_1'\widehat{u}_1^{-1})^{-1}(\xi_1)k_BTp_0^{-1}\widehat{K}^{-1} & \xi \geq \xi_1 \\ \widehat{u}_2(\xi)\widehat{u}_2^{-1}(\xi_1)(\widehat{u}_2'\widehat{u}_2^{-1} - \widehat{u}_1'\widehat{u}_1^{-1})^{-1}(\xi_1)k_BTp_0^{-1}\widehat{K}^{-1} & \xi < \xi_1 \end{array} \right., \tag{42}$$

where $\widehat{K} = \widehat{\mathrm{diag}}(K_{22}, K_{11})$.

Following the procedure similar to that for the Green's function we get

$$\widehat{G}(\xi,\xi_1)=\frac{ik_BT}{2qK_{33}\cos\phi(\xi_1)\cos\phi(\xi)}\begin{pmatrix} s(\xi-\xi_1)\sin\phi(\xi) & s(\xi-\xi_1)\tilde{\lambda}_2(\xi) \\ -\tilde{\lambda}_1(\xi) & \sin\phi(\xi)\end{pmatrix}$$
$$\times\begin{pmatrix} \frac{\exp(-\theta_1(\xi,\xi_1))}{\sqrt{\tilde{\lambda}_1(\xi)\tilde{\lambda}_1(\xi_1)}} & 0 \\ 0 & \frac{\exp(-\theta_2(\xi,\xi_1))}{\sqrt{\tilde{\lambda}_2(\xi)\tilde{\lambda}_2(\xi_1)}}\end{pmatrix}\begin{pmatrix} s(\xi_1-\xi)\sin\phi(\xi_1) & \tilde{\lambda}_1(\xi_1) \\ s(\xi_1-\xi)\tilde{\lambda}_2(\xi_1) & -\sin\phi(\xi_1)\end{pmatrix}, \tag{43}$$

where

$$\tilde{\lambda}_l = i\sqrt{\sin^2\phi + \frac{K_{33}}{K_{ll}}\cos^2\phi}, \quad l=1,2, \quad \tilde{\lambda}_3=-\tilde{\lambda}_1, \quad \tilde{\lambda}_4=-\tilde{\lambda}_2, \tag{44}$$

$\theta_l(\xi,\xi_1) = \tilde{\Omega}\left|\int_{\xi}^{\xi_1}\tilde{\lambda}_l d\xi'\right|$, $l=1,2$, $s(\xi)\equiv \text{sign}(\xi)$. The correlation function consists of a diagonal matrix and two matrices that rotate the reference frame around the points ξ and ξ_1.

Equation (43) loses its meaning if $\cos\phi = 0$ in the points ξ or ξ_1, or if the point ξ_* with $\cos\phi(\xi_*) = 0$ is situated in between ξ and ξ_1. The reason is that the eigenvalues $\tilde{\lambda}_1$ and $\tilde{\lambda}_2$ coincide if $\cos\phi = 0$. As it follows from Eq. (41), if $\tilde{\lambda}_l$ and $\tilde{\lambda}_m$ approach each other the applicability condition of the WKB method can be violated. In our case the WKB method is not applicable in the vicinity of the point $\phi = \pi/2$ where according to Eq. (44) $\tilde{\lambda}_1 = \tilde{\lambda}_2 = i$.

Fig. 4 shows G_{11} and G_{12} components of the correlation function expressed in relative units. The figure is the result of numerical calculations using Eq. (43). Both components of the correlation function decrease exponentially as the distance increases. The G_{11} component is real whereas the G_{12} component is imaginary. Approaching to the point $\phi = \pi/2$, the surface G_{11} begins to bend whereas the G_{12} component increases sharply. It indicates that we approach the region where the WKB method is not valid and Eq. (43) is not applicable.

We now estimate the region of applicability of the WKB method. For this purpose we introduce a new variable $\zeta = \phi - \pi/2$ and expand the eigenvalues $\tilde{\lambda}_l$ in the vicinity of the point $\zeta = 0$ as

$$\tilde{\lambda}_l \approx i\left(1-\frac{1}{2}C_l\zeta^2\right), \quad l=1,2, \quad \tilde{\lambda}_3=-\tilde{\lambda}_1, \quad \tilde{\lambda}_4=-\tilde{\lambda}_2, \tag{45}$$

where $C_l = 1 - K_{33}/K_{ll}$. Then the condition of applicability of the WKB method becomes

$$\tilde{\Omega}|\zeta|^3 \gg 1. \tag{46}$$

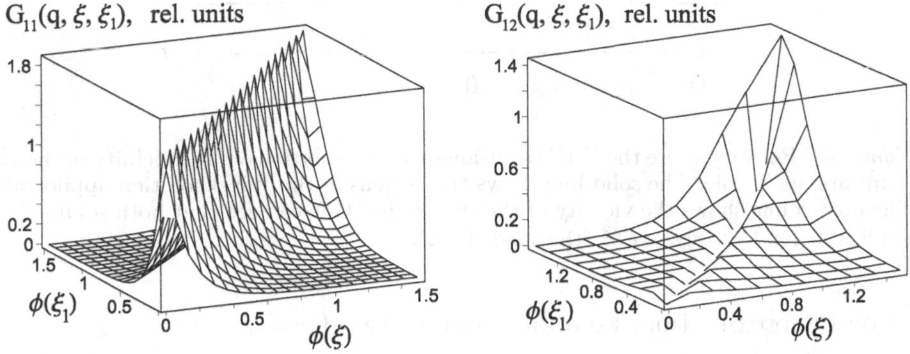

Figure 4. $G_{11}(\mathbf{q},\xi,\xi_1)$ and $G_{12}(\mathbf{q},\xi,\xi_1)$ components of the correlation function expressed in relative units as functions of ξ and ξ_1.

It means that Eq. (43) is valid if $\tilde{\Omega}|\zeta|^3 \gg 1$, $\tilde{\Omega}|\zeta_1|^3 \gg 1$ and there are no points between ζ and ζ_1 where eigenvalues $\tilde{\lambda}_l$ coincide.

5. Solution in the Vicinity of the Turn Point

We construct the solution in the vicinity of the turn point using the approach developed in Ref. [10]. Let us expand Eq. (38) in Taylor series in the vicinity of the point $\zeta = 0$. It is convenient to introduce a new variable $\tau = \tilde{\Omega}^{1/3}\zeta$. So we have

$$\left[i\widehat{B}(0) + \frac{i}{2}\tilde{\Omega}^{-2/3}\tau^2\widehat{B}''(0) + \tilde{\Omega}^{-4/3}\left(\frac{i}{24}\tau^4\widehat{B}^{IV}(0) + \tau\widehat{C}'(0)\right) + \ldots\right]\Psi(\tau) = \tilde{\Omega}^{-2/3}\Psi'(\tau). \tag{47}$$

The terms of this equation represent a series in $\tilde{\Omega}^{-2/3}$ but τ is no longer a small parameter. We will seek the solution of this equation corresponding to eigenvalues $\tilde{\lambda}_1$ and $\tilde{\lambda}_2$ in the form

$$\Psi_1(\tau) = \exp\left(i\tilde{\Omega}^{2/3}\tilde{\lambda}_1(0)\tau\right)\left(\Psi_1^{(0)}(\tau) + \tilde{\Omega}^{-2/3}\Psi_1^{(1)}(\tau) + \tilde{\Omega}^{-4/3}\Psi_1^{(2)}(\tau) + \ldots\right), \tag{48}$$

where $\tilde{\lambda}_1(0) = \tilde{\lambda}_2(0) = i$. Substituting (48) in Eq. (47) we get in the zero approximation an equation for the eigenvector: $\widehat{B}(0)\Psi_1^{(0)} = \tilde{\lambda}_1(0)\Psi_1^{(0)}$. Solution of this equation has the form $\Psi_1^{(0)} = \boldsymbol{\chi}_1\beta(\tau)$, where $\boldsymbol{\chi}_1$ is the eigenvector, $\boldsymbol{\chi}_1 = (1,-i,i,1)$, and $\beta(\tau)$ is an arbitrary function of τ. Using the first approximation of Eq. (48) and the solvability condition of the

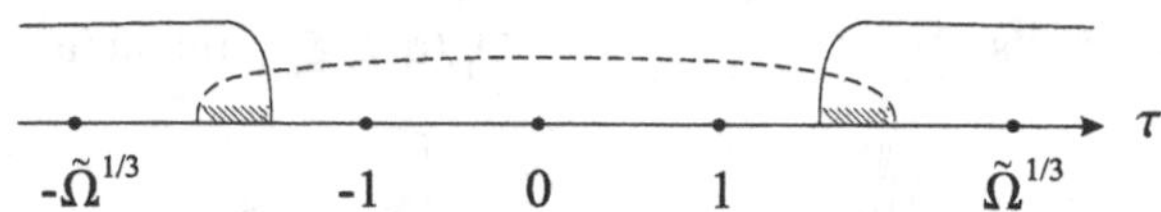

Figure 5. Regions where the WKB solutions and the solutions in the vicinity of the turn point are applicable. The solid line shows the regions of the WKB solution applicability. The dotted line shows the vicinity of the turn point. One can see that both solutions are applicable for $1 \ll |\tau| \ll \tilde{\Omega}^{1/3}$ (the shaded regions).

second approximation, we obtain an equation for $\beta(\tau)$

$$\beta''(\tau) - \beta'(\tau)\tau^2\frac{1}{2}(C_1 + C_2) - \beta(\tau)\left[\frac{1}{2}(C_1 + C_2)\tau - \frac{1}{4}C_1C_2\tau^4\right] = 0. \quad (49)$$

Usually $K_{22} < K_{11} < K_{33}$. Therefore in what follows we will consider $C_1 < 0$, $C_2 < 0$ and $C_1 > C_2$. These conditions do not restrict the generality of the obtained results. Then $\beta(\tau)$ has the form

$$\beta(\tau) = \left[F_1K_{1/6}\left(\frac{1}{12}(C_1 - C_2)\tau^3e^{-3\pi i}\right) + F_2K_{1/6}\left(\frac{1}{12}(C_1 - C_2)\tau^3\right)\right] \times\sqrt{\tau}\exp\left[\frac{1}{12}(C_1 + C_2)\tau^3\right], \quad (50)$$

where $F_{1,2}$ are arbitrary constants and $K_{1/6}(a)$ is the modified Bessel function of the second kind.

The restriction imposed on τ by the expansion of Eq. (38) in Taylor series in ζ has the form $|\tau| \ll \tilde{\Omega}^{1/3}$. The applicability condition of the WKB solution then becomes $|\tau| \gg 1$. So there are regions $1 \ll |\tau| \ll \tilde{\Omega}^{1/3}$ where both the WKB solution and the solution in the vicinity of the turn point are valid (Fig. 5). In these regions both solutions should coincide. Therefore we can write both solutions in these regions and equate them. This produces the binding conditions.

The WKB solution corresponding to the eigenvalues $\tilde{\lambda}_1$ and $\tilde{\lambda}_2$ in the regions $1 \ll |\tau| \ll \tilde{\Omega}^{1/3}$ has the form

$$\Psi_1(\tau) = B_1\Phi_1(\tau) + B_2\Phi_2(\tau). \quad (51)$$

Note, that the coefficients B_1 and B_2 can be different for $\tau > 0$ and $\tau < 0$ since the regions where the WKB solution is valid, $|\tau| \gg 1$, do not intersect. We denote the coefficients B_1 and B_2 at $\tau > 0$ and $\tau < 0$ as $B_l^{(+)}$ and $B_l^{(-)}$ respectively, $l = 1, 2$. If we find the solution $\Psi_1^{(0)}$ at $\tau \to +\infty$ and $\tau \to -\infty$ we match the WKB solution and the solution in the vicinity of the turn point [11]. It gives the relations between the coefficients $B_l^{(+)}$ and $B_l^{(-)}$,

$l = 1, 2,$

$$B_1^{(+)} = 2B_1^{(-)}, \quad B_2^{(+)} = \frac{1}{2}B_2^{(-)}. \tag{52}$$

Thus, there is no exchange between modes at the passage through the turn point. But the amplitudes exhibit jumps. The relation between the amplitude jumps is consistent with the expression for the first integral of the set of differential equations (38), i.e. "the conservation law" is fulfilled.

Let both points ξ and ξ_1, $\xi > \xi_1$, be in the regions where the WKB method is applicable, but let them be separated by the turn point $\xi_* = \pi/2 - \phi_0$. Equation (42) contains the $\widehat{u}_1$ matrix in both points ξ and ξ_1. For this matrix we should write an expression which is valid on the both sides of the turn point. Using Eqs. (52) we get the correlation function (42) in the form

$$\widehat{G}(\xi, \xi_1) = \widehat{u}_1(\xi)\widehat{s}\,\widehat{u}_1^{-1}(\xi_1)(\widehat{u}_2'\widehat{u}_2^{-1} - \widehat{u}_1'\widehat{u}_1^{-1})^{-1}(\xi_1)k_B T p_0^{-1}\widehat{K}^{-1}, \tag{53}$$

where $\widehat{s}$ is a diagonal matrix with elements 2 and 1/2. It describes the change of the solution amplitudes at the passage through the turn point.

If ξ is situated in the vicinity of the turn point and ξ_1 is far from the turn point, $\xi > \xi_1$, we should extend the expression for the $\widehat{u}_1(\xi)$ matrix into vicinity of the turn point [11]. If both points ξ and ξ_1 are in the vicinity of the turn point both matrices $\widehat{u}_1$ and $\widehat{u}_2$ should be extended into this region. Note that in this case we should take into account the corrections of the next order of smallness for constructing the solution in the vicinity of the turn point.

6. Conclusions

In the present paper we study the Green's function and the spatial correlation function of the director fluctuations in CLC with a large helix pitch. For this purpose we use a unique approach based on the modified WKB method. The Green's function contains two contributions due to existence of ordinary and extraordinary waves. For the ordinary wave, the anisotropy and periodicity of cholesterics are not exhibited. For the extraordinary wave, a forbidden zone exists at large distances, i.e., there is a restriction on possible directions of the wave vectors. The phenomenon results from a turn of beams caused by the variation of the refractive index. The beams with large angles of incidence are captured, and a planar wave guide channel is formed. In such a medium, the trajectory of the extraordinary beam does not lie in a plane.

We investigate the fluctuation spectrum for the case of short-wave fluctuating modes. Using the fluctuation spectrum, we construct the spatial correlation function of the director fluctuations. We take into account that the medium is not uniform along the pitch axis. We analyze in detail the behavior of fluctuations in the vicinity of turn points where the WKB approximation is not applicable. We conclude that there is no exchange between modes at the passage through the turn point. But the amplitudes exhibit jumps. So, when studying the correlation function, it is necessary to take into account the vicinity of turn points.

Acknowledgements

This work was partly supported by a joint grant of the Ministry of Education of Russian Federation and Administration of Saint-Petersburg, Russia, Grant No. PD02-1.2-297 and by the Russian Fund for Basic Research through Grant No. 02-02-16577.

References

1. De Gennes, P.G. and Prost, J. (1993) *The Physics of Liquid Crystals.* Clarendon Press, Oxford
2. Belyakov, V.A. (1992) *Diffraction Optics of Periodic Media with a Layered Structure.* Springer–Verlag, Berlin
3. Chandrasekhar, S. (1977) *Liquid Crystals.* Cambridge University Press, Cambridge
4. Lubensky, T.C. (1972) *Phys. Rev. A*, **Vol. no. 6**, pp. 452–470
5. Peterson, M.A. (1983) *Phys. Rev. A*, **Vol. no. 27**, pp. 520–529
6. Val'kov, A.Yu., Romanov, V.P. and Grinin, R.V. (1997) *Optics and Spectroscopy*, **Vol. no. 83**, pp. 221–232
7. Brillouin, L. and Parodi, M. (1956) *Propagation des Ondes dans les Milieux Périodiques* Masson et cie éditeurs, Dunod éditeurs, Paris
8. Felsen, L. and Marcuvitz, N. (1973) *Radiation and Scattering of Waves* Prentice-Hall, Inc., Englewood Cliffs, New Jersey
9. Aksenova, E.V., Romanov, V.P. and Val'kov, A.Yu. (1999) *Phys. Rev. E*, **Vol. no. 59**, pp. 1184–1192
10. Perel, M.V. (1990) *Radiophizika*, **Vol. no. 33**, pp. 1208–1216 (in Russian)
11. Aksenova, E.V., Romanov, V.P. and Val'kov, A.Yu. (2001) *Proceedings of the International Seminar "Day on Diffraction-2001"*, pp. 7–17

LIGHT SCATTERING IN A DISPERSE LAYER WITH PARTIALLY ORDERED SOFT PARTICLES

V.A. LOIKO AND V.V. BERDNIK
B.I. Stepanov Institute of Physics,
National Academy of Sciences of Belarus
F. Scaryna ave. 68, Minsk, 220072, Belarus

1. Introduction

There are a lot of natural and artificial disperse media with a high concentration of particles [1–5]. In such media there is a partial ordering of particles. It grows up with increasing the particle concentration and results in a change of scattering properties. The latter differ qualitatively from those in a dilute medium, when particles are far from each other and ordering is absent. For a rigorous description of the radiative transfer in a partially ordered medium it is necessary to use the theory of multiple scattering of waves [3–6]. Because of the complexity and awkwardness of the mathematical apparatus of this theory, complete solutions can be obtained in an extremely limited number of cases. However, it makes it possible to determine the validity range of another, simpler and widely used, phenomenological approach to the description of the light propagation. It is based on the radiative transfer equation (RTE) and uses the phenomenological scattering characteristics of the medium, namely, scattering and extinction coefficients and phase function [7–10].

The papers devoted to the analysis of the radiative transfer in partially ordered media are restricted, as a rule, to the consideration of either the extinction of a parallel beam of light [11, 12] or the angular structure of single scattered light [13]. In the majority of cases, they are based on equations of wave transfer [14] and numerical simulations by the Monte-Carlo method [15].

In the present paper, based on the radiative transfer equation, we investigate the transmittance and angular structure of scattered light in an optically thick layer with high concentration of particles. Optical interaction of particles and contribution of the multiple scattering are taken into

B.A. van Tiggelen and S.E. Skipetrov (eds.),
Wave Scattering in Complex Media: From Theory to Applications, 535–551.

account. We consider a disperse layer in which the bleaching effect takes place, i.e. the coherent (ballistic) transmittance of the layer increases with the volume concentration of particles in the layer at a constant surface concentration of particles. Such an effect occurs in a system of optically soft particles, for example, in polymer dispersed liquid crystal films.

2. Model of Light Transfer in a Layer

Layers with a high concentration of particles have a wide range of optical thicknesses. The particles can be embedded into a binding medium (matrix) whose refractive index differs from the refractive index of the environment. In analyzing the characteristics of transmitted and reflected light, it is necessary to take into account the effect of multiple scattering and the reflection at the boundaries.

Let an azimuthally symmetric wide light beam of intensity I_0 be incident on a layer (film) of scattering medium (bounded by the planes $z = 0$ and $z = z_0$ from above and below, respectively) at an angle θ_{0e} with respect to the normal (see Fig. 1). The scattering medium contains particles embedded in the matrix with a refractive index n_m. At the boundary, the beam is refracted and enters the medium at angle θ_0. Inside the layer the light is scattered, partially absorbed and leaves the medium through the bounding surfaces. The boundary refraction is described by the Fresnel laws.

To describe the light propagation in the film, we use the radiative transfer equation (RTE). For the azimuth-averaged scattered light intensity under illumination of the layer by a parallel light beam, this equation takes the form

$$\mu\frac{\partial I(z,\mu)}{\partial z} + \varepsilon I(z,\mu) = \sigma\int_{-1}^{1} p(\mu,\mu')\, I(z,\mu')\, d\mu' + I_{1n}^{+}\sigma p(\mu,\mu_0)e^{-\frac{\varepsilon z}{\mu_0}} + I_{2n}^{-}\sigma p(-\mu,\mu_0)e^{-\frac{\varepsilon(z_0-z)}{\mu_0}}, \qquad (1)$$

where $I(z,\mu)$ is the azimuth-averaged intensity of scattered light propagating at depth z at a polar angle $\theta = \arccos\mu$; σ and ε are the scattering and extinction coefficients; $\mu_0 = \cos\theta_0$; $p(\mu,\mu')$ is the azimuth-averaged phase function; $p(\cos\gamma)$ is the elementary volume phase function normalized by the condition $\int_{-1}^{1} p(\cos\gamma)d\cos\gamma = 1$, $\cos\gamma = \mu\mu' + \sqrt{1-\mu^2}\sqrt{1-\mu'^2}\cos\varphi$, γ is the scattering angle, μ and μ' are cosines of the axial angles of the scattered and incident light; φ is the azimuth angle of scattering; I_{1n}^{+} and I_{2n}^{+} are the intensities of the ballistic light at the upper and lower boundaries of the layer.

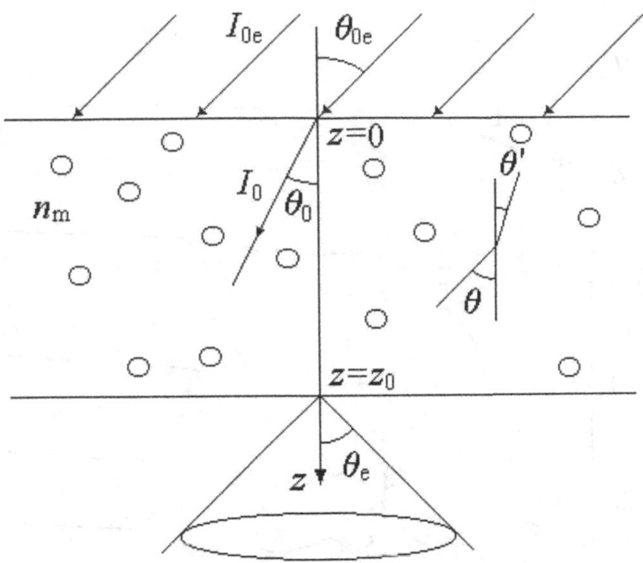

Figure 1. Schematic representation of the illumination and scattering geometry in the layer. Notations are explained in the text.

The RTE is solved with the boundary conditions

$$\begin{array}{l} I(z=0,\ \mu>0)=\kappa(\mu)I(z=0,\mu<0), \\ I(z=z_0,\ \mu<0)=\kappa(\mu)I(z=z_0,\ \mu>0). \end{array} \tag{2}$$

The function $\kappa(\mu)$ stands for the reflectance at the air-matrix interface. It is determined by the relations [16]

$$\kappa(\mu)=\left\{\begin{array}{c} 0.5\left(\frac{tg^2(\theta-\theta_e)}{tg^2(\theta+\theta_e)}+\frac{\sin^2(\theta-\theta_e)}{\sin^2(\theta+\theta_e)}\right),\theta<\theta_r \\ 1,\theta\geq\theta_r. \end{array}\right. \tag{3}$$

Here $\theta=\arccos\mu$ is the angle of incidence on the medium boundary from inside the layer, θ_e is the angle of refraction, and θ_r is the angle of total internal reflection.

3. Calculation of the Elementary Volume Characteristics

The starting point for modeling of the scattering characteristics of concentrated layers is the calculation of the scattering characteristics in the limit of small concentrations, when we deal with a random medium. In calculating the elementary volume characteristics in this limit, we used the Mie theory for spherical particles. The technique for such calculations has been

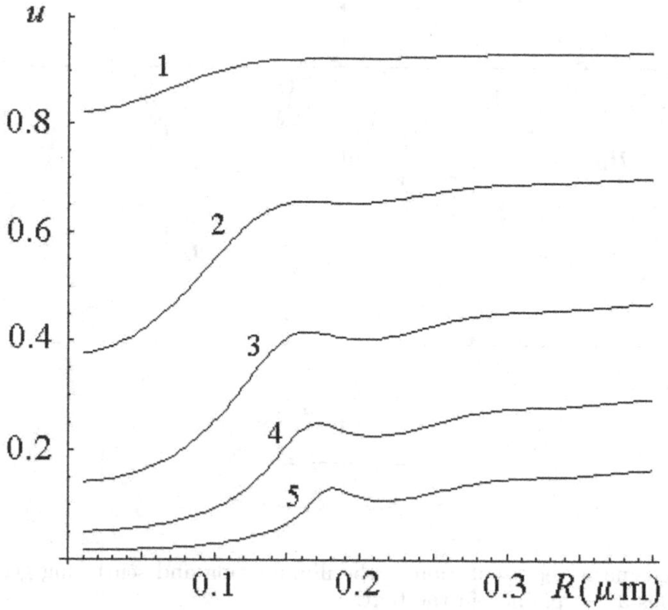

Figure 2. Dependence of u on particle radius for non-absorbing particles with $n = 1.15$ at $\lambda = 0.63$ μm, $w = 0.025$ (curve 1), 0.125 (2), 0.25 (3), 0.375 (4), 0.5 (5).

well developed and reliable methods have been described [17, 18]. We used the algorithm described in [18]. The number N_r of summed terms of Mie series was determined according to the relation [17,19]

$$N_r = x + 4x^{\frac{1}{3}} + 2, \tag{4}$$

where $x = \frac{2\pi R}{\lambda}$ is the size parameter, R is the droplet radius; λ is the wavelength.

To calculate the extinction coefficient and the phase function for large concentrations of particles, it is necessary to solve the problem of light diffraction by a system of many bodies. The rigorous solution of this problem has not been found. Therefore, in calculating the extinction coefficient and the phase function, various approximate methods are being used. The simplest of them is the interference approximation, which is applicable to the case of optically soft particles [12]. In this approximation, for a medium consisting of identical spherical particles, we can write the equations for differential scattering, scattering, and extinction coefficients, respectively, as

$$\sigma_h(\gamma) = w\sigma_{0l}p_l(\gamma)S_3(\gamma, w), \tag{5}$$

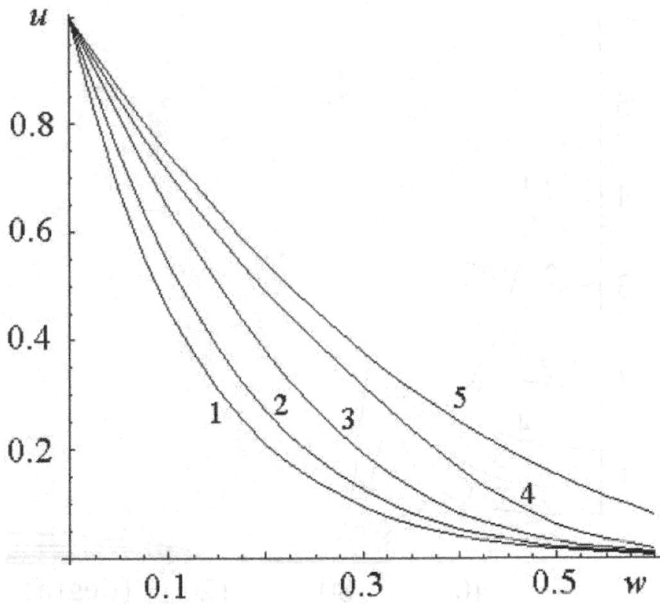

Figure 3. Dependence of u on particle concentration for non-absorbing particles with $n = 1.15$ at $\lambda = 0.63$ μm, $R(\mu\text{m}) = 0.01$ (curve 1), 0.0685 (2), 0.1075 (3), 0.1465 (4), 0.322 (5).

$$\sigma_h = w\sigma_{0l}u, \tag{6}$$

$$\varepsilon_h = w\left(\varepsilon_{0l} - \sigma_{0l} + \sigma_{0l}u\right), \tag{7}$$

where $w = Nv/V$ is the volume concentration of particles; N is the number of particles in volume V; v is the volume of a single particle; $\sigma_h(\gamma)$ is the differential scattering coefficient of the medium with a volume concentration of particles w; $\sigma_{0l} = \Sigma_s/v$; $\varepsilon_{0l} = \alpha_{0l} + \sigma_{0l} = \Sigma_e/v$; $\alpha_0 = \Sigma_a/v$; Σ_a, Σ_s and Σ_e are, respectively, the absorption, scattering and extinction cross sections of an individual particle; $p_l(\gamma)$ is the phase function of an isolated particle normalized by the condition: $\int\limits_0^\pi p_l(\gamma)\sin\gamma d\gamma = 1$; the parameter u is determined by the equation:

$$u = \int\limits_0^\pi p_l(\gamma)S_3(\gamma, w)\sin\gamma d\gamma. \tag{8}$$

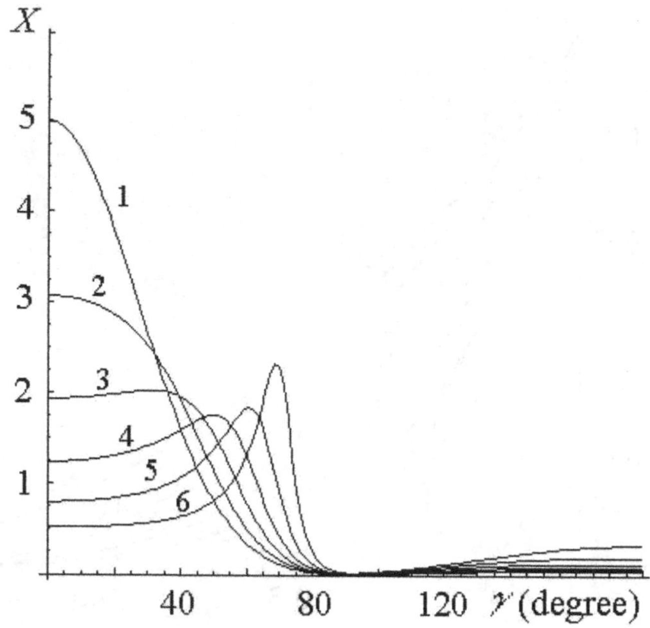

Figure 4. Phase function at $w = 0$ (curve 1), 0.1 (2), 0.2 (3), 0.3 (4), 0.4 (5), 0.5 (6). $R = 0.2\ \mu$m, $n = 1.1$, $\kappa = 0$, $\lambda = 0.42\ \mu$m.

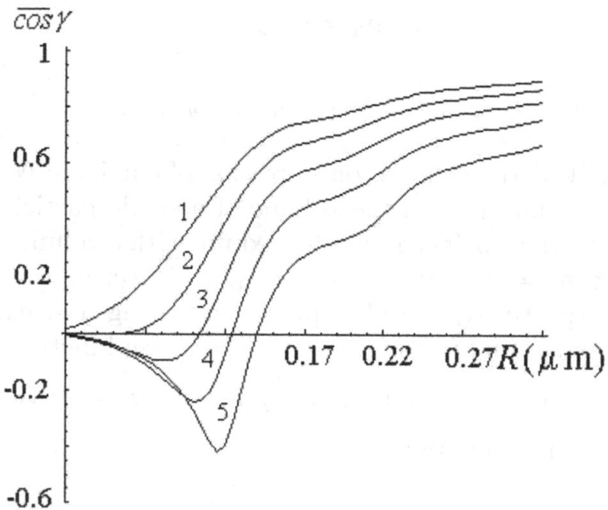

Figure 5. Asymmetry parameter versus the particle radius at $\lambda = 0.43\ \mu$m, $n = 1.1$, $\kappa = 0$, $w = 0.025$ (curve 1), 0.125 (2), 0.25 (3), 0.375 (4), 0.5 (5).

The structure factor $S_3(\gamma, w)$ in Eq. (8) takes into account the influence of light interference in the partially ordered medium:

$$S_3(\gamma, w) = 1 + 4\pi n \int_0^\infty [g(r,w) - 1] \frac{\sin zr}{zr} r^2 dr, \tag{9}$$

where n is the number of particles in a unit volume; $g(r, w)$ is the radial distribution function characterizing the spatial arrangement of particles; $z = 4x \sin \gamma/2$. Figs. 2 and 3 show the value of u as a function of particle size and concentration.

For a system of hard spheres the structure factor can be calculated in the Percus-Yevick approximation [12–15,20–22]:

$$S_3(\gamma, w) = \left(1 - 24n \int_0^1 g(r,w) \frac{\sin zr}{zr} r^2 dr\right)^{-1}, \tag{10}$$

where

$$g(r, w) = \begin{cases} -a - b\frac{r}{2R} - c\left(\frac{r}{2R}\right)^3, r \le 2R \\ 0, r > 2R, \end{cases} \tag{11}$$

$$a = \frac{(1 + 2w)^2}{(1 - w)^4}, \tag{12}$$

$$b = -6w \frac{(1 + 0.5w)^2}{(1 - w)^4}, \tag{13}$$

$$c = 0.5w \frac{(1 + 2w)^2}{(1 - w)^4}. \tag{14}$$

Comparison of our calculations with the experimental results for suspensions of latex particles in water [23] has shown that the interference approximation well describes the dependence of extinction in the layer of fine particles on their volume concentration. The data in Fig. 4 illustrate the dependence of the phase function on the volume concentration of particles w. The intensity of forward-scattered light decreases, and at fairly large concentrations the phase function acquires a characteristic maximum at nonzero scattering angle. Its position is shifted towards large angles with increasing w and towards small angles with increasing the particle size. It is noteworthy that at certain sizes and concentrations of particles the asymmetry parameter of the phase function can be zero and even negative. See Fig. 5 and [13].

In the interference approximation the absorption coefficient $\alpha_h = \varepsilon_h - \sigma_h = w\alpha_{0l}$ is proportional to the concentration. The optical thickness of the layer τ_0, the single scattering albedo Λ, and the phase function $p(\gamma)$ are determined, respectively, by the equations:

$$\tau_0 = \tau_{0l}(1 - \Lambda_l(1-u)), \tag{15}$$

$$\Lambda = \frac{\Lambda_l u}{1 - \Lambda_l(1-u)}, \tag{16}$$

$$p(\gamma) = \frac{p_l(\gamma)S_3(\gamma)}{u}, \tag{17}$$

where τ_{0l} and Λ_l are the optical thickness of the layer and single scattering albedo in the independent scattering approximation.

To solve the RTE, we find the azimuth-averaged phase function:

$$p(\mu, \mu') = \frac{1}{2\pi} \int_0^{2\pi} p(\mu\mu' + \sqrt{1-\mu^2}\sqrt{1-\mu'^2}\cos\varphi)d\varphi. \tag{18}$$

The latter is usually calculated with the use of the Legendre polynomial expansion [8, 10]. However, for extended phase functions, the number of expansion terms to be taken into account reaches a few hundreds and the problem of calculating the expansion coefficients p_l becomes very complicated. If we restrict ourselves by rather a small number of the expansion terms, then some "ripples" will appear on the function $p(\mu, \mu')$ reconstructed by the expansion coefficients. It can assume negative values.

We take use of the spline-approximation method [24, 25] to calculate $p(\mu, \mu')$. We introduce on the $[-1, 1]$ interval a nonuniform grid of nodes u_i, $i = 1, 2, ..., N$, and approximate the phase function by a linear combination of basic n-order splines of the defect $\Delta = 1$:

$$p(x) = \sum_{\alpha=1}^{N+n-1} B_\alpha(x)S_\alpha, \tag{19}$$

where $x = \cos\gamma$, $B_\alpha(x)$ are the basic splines, and S_α is the spline vector.

Thickening the nodes in the region of rapid change in the phase function, we can obtain a good approximation without unduly increasing the number of nodes. Substituting Eq. (19) into Eq. (18), we obtain the following expression for the function $p(\mu, \mu')$:

$$p(\mu, \mu') = \sum_{\alpha=1}^{N+n-1} B_\alpha(\mu, \mu')S_\alpha, \tag{20}$$

where

$$B_\alpha(\mu, \mu') = \frac{1}{\pi} \int_0^{\pi} B_\alpha(\mu\mu' + \sqrt{1-\mu^2}\sqrt{1-\mu'^2}\cos\varphi)d\varphi. \tag{21}$$

In calculating $B_\alpha(\mu, \mu')$, we changed the variables:

$$x = \mu\mu' + \sqrt{1-\mu^2}\sqrt{1-\mu'^2}\cos\varphi. \tag{22}$$

Then

$$B_\alpha(\mu, \mu') = \frac{1}{\pi}\int\limits_{N(\mu,\mu')}^{M(\mu,\mu')} \frac{B_\alpha(x)dx}{\sqrt{(M(\mu,\mu')-x)\,(x-N(\mu,\mu')}}, \tag{23}$$

where

$$M(\mu, \mu') = \mu\mu' + \sqrt{1-\mu^2}\sqrt{1-\mu'^2}, \tag{24}$$

$$N(\mu, \mu') = \mu\mu' - \sqrt{1-\mu^2}\sqrt{1-\mu'^2}. \tag{25}$$

To calculate Eq. (23), we used a Gaussian quadrature, introducing a Gaussian grid x_k on the $[M, N]$ interval:

$$x_k = \mu\nu + \sqrt{1-\mu^2}\sqrt{1-\mu'^2}\cos\frac{k-0.5}{N_x}\pi, k = 1, 2, ..., N_x \tag{26}$$

and replace the integral by the sum:

$$B_\alpha(\mu, \mu') = \frac{1}{N_x}\sum_{k=1}^{N_x} B_\alpha\left(x_k(\mu, \mu')\right). \tag{27}$$

We verified the exactness of this method of approximation by an example of a medium with a Heyney-Greenstein phase function, for which the azimuth-averaged phase function is expressed in terms of an elliptical integral of the second kind. The error of calculation of the $B_\alpha(\mu, \mu')$ functions depends on the number of nodes N_k in the grid x_k and shows up as errors in the approximation of the phase function. The approximation error is maximal when $\mu = \mu'$ and rapidly decreases with going away from this point. The error increases with decreasing μ'. With the asymmetry parameter $\overline{\cos\gamma} < 0.96$, the maximum value of the relative error does not exceed 2% [24]. The greatest stability to variations in the medium parameters is achieved by using first-order splines, i.e. by approximating the phase function by a broken line.

4. Angular Structure of Scattered Light

A homogeneous layer of scattering and absorbing medium with non-reflecting boundaries is characterized by luminance (radiance) [8] factors of light scattered in the backward $[\rho(\mu, \mu')]$ and forward $[\sigma(\mu, \mu')]$ directions. They are

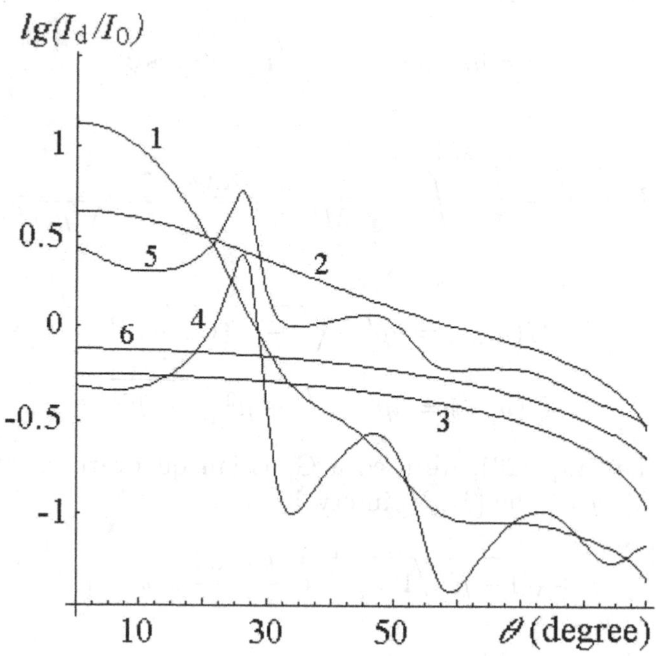

Figure 6. Angular dependence of the forward-scattered light intensity for a layer at $\lambda = 0.42$ μm, $n_m = 1$, $w = 0$ (curves 1, 2, 3), 0.5 (4, 5, 6); $\eta = 1$ (curves 1 and 4), 7 (2 and 5), 100 (3 and 6). $R = 0.5$ μm, $n = 1.1$, $\kappa = 10^{-5}$.

determined by the relations:

$$I^-(z=0,\mu) = \int_0^1 2\rho(\mu,\mu')\mu' I_0(\mu')\,d\mu', \tag{28}$$

$$I^+(z=z_0,\mu) = e^{-\tau_0 e(\mu)/\mu} I_0(\mu) + \int_0^1 2\sigma(\mu,\mu')\,\mu' I_0(\mu')\,d\mu'. \tag{29}$$

Here $I_0(\mu)$ is the intensity of the light incident on the layer, $I^-(z=0,\mu) = I(z=0,\mu<0)$ is the back-scattered intensity at the upper boundary, $I^+(z=z_0,\mu) = I(z=z_0,\mu>0)$ is the forward-scattered intensity at the lower boundary. The luminance factor (radiance coefficient) is determined as the ratio of the luminosity at scattering angle $\arccos\mu$ (for illumination of the film boundary at angle $\arccos\mu'$) to the luminosity of a perfectly reflecting white surface [8]. For example, the luminance factor is unity for a white Lambert surface.

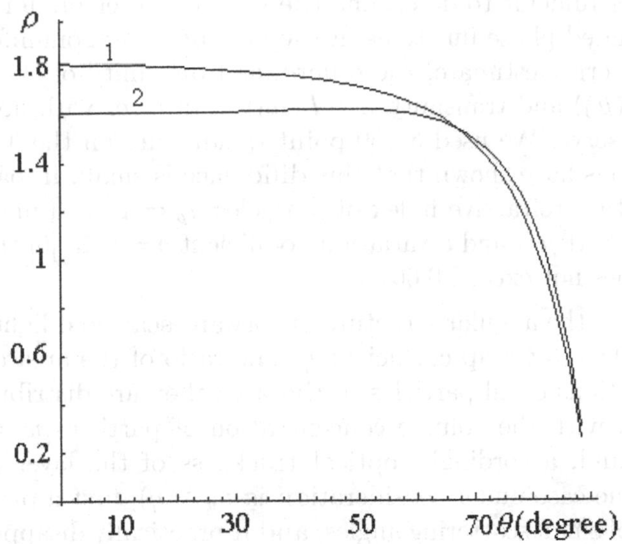

Figure 7. Angular dependence of the luminance factor for back-scattered light at $\eta = 100$, $R = 0.15$ μm, $n = 1.15$, $\kappa = 0$, $n_m = 1.5$, $\lambda = 0.5$ μm, $w = 0$ (curve 1), 0.5 (2).

To determine the luminance factors $\rho(\mu, \mu')$ and $\sigma(\mu, \mu')$, we used the calculation technique based on the layer doubling method [8,10,26–28]. In this method, one begins calculations with choosing a layer of a small enough thickness τ_s, such that $\tau_0 = \tau_s 2^K$, where K is an integer. For a layer with optical thickness τ_s the luminance factors are found approximately. The methods of giving approximate values of $\rho(\mu, \mu')$ and $\sigma(\mu, \mu')$ (initialization) for isotropic media are described in [27]. In this paper, we used initialization of single scattering. The luminance factors of a doubled-thickness layer were found using the known relations of layer doubling obtained from the balance equations at the layer interfaces [8,10,24,25]. The initial optical thickness was chosen to be equal to $\tau_s \approx 10^{-6}$, which permits a fairly high accuracy of calculations.

Using Eqs. (28) and (29), we can obtain the expressions for the reflectance R and transmittance T depending on the cosine of the incident angle μ_0:

$$R(\mu_0) = 2 \int_0^1 \rho(\mu, \mu_0)\mu d\mu, \tag{30}$$

$$T(\mu_0) = e^{-\frac{\tau_0}{\mu_0}} + 2 \int_0^1 \sigma(\mu, \mu_0)\mu d\mu. \tag{31}$$

It is rather difficult to determine the calculation errors for media with strongly extended phase functions in the case of using nonuniform grids of nodes. As an error estimate, the difference from unity of the sum of the reflectance $R(\mu_0)$ and transmittance $T(\mu_0)$ for a layer with non-absorbing particles can serve. We used a 150-point quadrature on the $[0, 1]$ interval. The calculations have shown that this difference is small. In particular, for a medium with a refractive index of particles $n_p = 1.01$, a mean radius of particles $\bar{R} = 3.41$ μm and a variation coefficient $\delta = 0.25$ [6] the difference from unity does not exceed 0.001.

Fig. 6 shows the angular structure of forward-scattered light for various values of the layer overlap coefficient η (the ratio of the area of the cross-section projections of all particles to the area they are distributed within) for a medium with the volume concentration of particles $w = 0.5$. With increasing η and, accordingly, optical thickness of the layer (the optical thickness in the case under consideration is $\tau_{0l} \approx \eta$), first a peak is formed in the region of small scattering angles, and it practically disappears at large η on going to the asymptotic regime. Such a behavior is a consequence of the characteristic maximum in the phase function of the partially ordered medium. Part of the light incident on the layer is scattered at an angle of $\theta = \gamma_m$ (γ_m is the angle at which a phase function maximum is formed) and part of this light, scattered repeatedly at angle γ_m, propagates in the direction of the incident light. Apparently, the formation of a peak of the forward-scattered light takes place in the case where the contribution of the intensity of the double-scattered light to the total scattered light intensity is fairly large. This intensity maximum is observed experimentally (see Fig. 3.8b [6]).

The proposed method allows one to consider Fresnel reflection at the layer boundaries. The intensities of the scattered light can be found from the system of integral equations on the layer boundaries [29]. Calculations are in good agreement with the known experimental data. Note that selecting the particle sizes and the volume concentration so that the characteristic maximum of the phase function is in the vicinity of the angle of total internal reflection, it is possible to increase the homogeneity in the scattering pattern of the reflected light (see Fig. 7).

5. Light Extinction by a Thick Film

Formulas for optical thickness τ_0 of a layer with high concentration of particles can be written in the form:

$$\tau_0 = \tau_{0l} k_h, \tag{32}$$

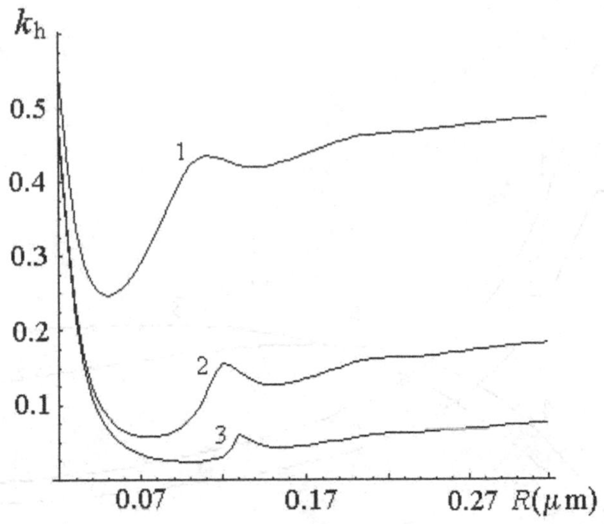

Figure 8. Dependence of k_h on the particle size at $\lambda = 0.43$ μm, $w = 0.25$ (curve 1), 0.5 (2), 0.65 (3). $n = 1.1$, $\kappa = 10^{-5}$.

where

$$k_h = \frac{u}{u + \Lambda(1-u)}. \tag{33}$$

Thus, the change in the extinction of ballistic light with a change in the volume concentration of particles is determined by the value of k_h. The characteristic behavior of k_h versus the particle size and concentration is shown in Fig. 8.

Light attenuation by thick films is determined by the equation:

$$\gamma_d \tau_0 = \tau_{0l} \gamma_d k_h, \tag{34}$$

where γ_d is the dimensionless asymptotic attenuation coefficient depending on the volume concentration of particles.

Light attenuation in the asymptotic regime is characterized by the quantity $k_{hd} = \gamma_d k_h$. To determine the dependence of γ_d on the particle concentration, we used the characteristic transfer equation:

$$(1 - \gamma_d \mu)\varphi(\mu) = \frac{\Lambda}{2} \int_{-1}^{1} p(\mu, \mu')\varphi(\mu')d\mu', \tag{35}$$

where γ_d and $\varphi(\mu)$ are the asymptotic attenuation coefficient and the relative angular luminance distribution for the asymptotic regime, respectively.

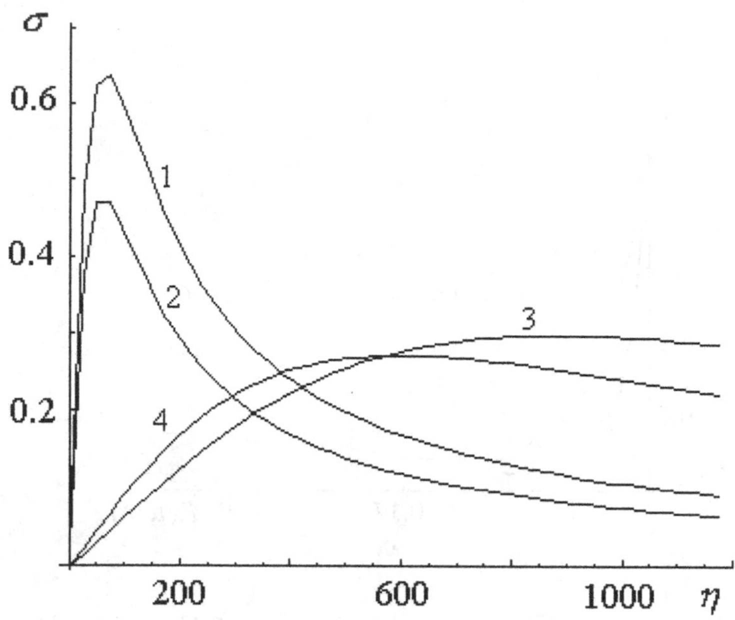

Figure 9. Luminance factors of forward-scattered light versus the overlap coefficient for a layer containing particles with $R = 0.11$ μm, $n = 1.1$ at normal incidence of light. Scattering angle $\theta = 0°$ (curves 1 and 3), 58.2° (2 and 4). $w = 0$ (curves 1 and 2), 0.5 (3 and 4). $n_m = 1$, $\lambda = 0.43$ μm.

In solving this equation, in order to reduce the dimensionality of equations obtained after the discretization, we introduced the functions $\varphi^+(\mu) = \varphi(\mu)$ and $\varphi^-(\mu) = \varphi(-\mu)$ defined on the $[0, 1]$ interval. Calculation shows that the attenuation coefficient decreases with increasing the particle concentration. The influence of concentration effects on the asymptotic attenuation coefficient decreases with increasing the particle size. Such a behavior agrees with experimental data [6].

As mentioned above, in a layer of soft scatterers a bleaching effect takes place, i.e. the ballistic transmittance grows up with the volume concentration of particles. However, if we deal with the scattered light the opposite concentration dependence can be implemented. It can be seen from Fig. 9, where the luminance factors of forward-scattered light are shown. Fig. 10 illustrates the dependence of the luminance factors of the back-scattered light on the overlap coefficient at different scattering angles and volume concentrations.

Up to now the luminance factors were considered. In experiments, we

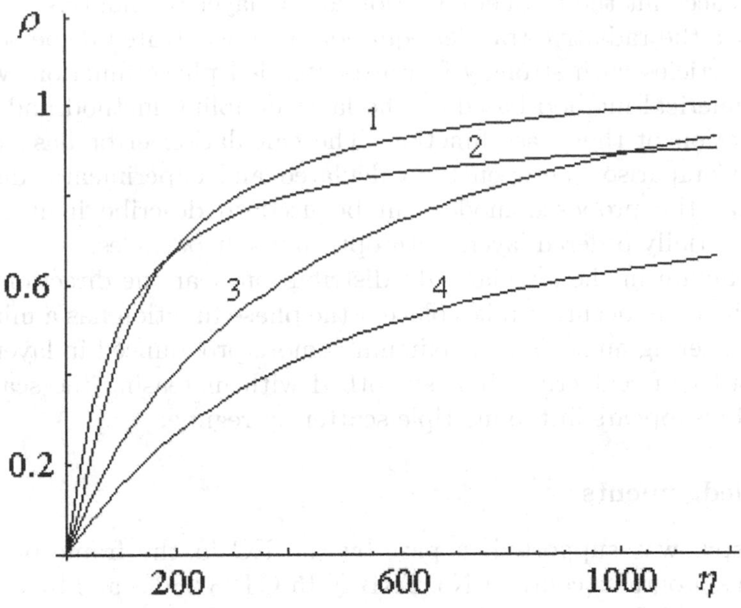

Figure 10. Luminance factors of back-scattered light versus the overlap coefficient for a layer containing particles with $R = 0.11$ μm, $n = 1.1$ at normal incidence of light. Scattering angle $\theta = 2°$ (curves 1 and 3), 58.2° (2 and 4). $w = 0$ (curves 1 and 2), 0.5 (3 and 4). $n_m = 1$, $\lambda = 0.43$ μm.

always deal with a finite aperture of the optical detector. The equation for transmittance with regard to refraction at the layer boundaries can be written in the form:

$$T(\mu_0, \gamma_r) = \frac{(1-\kappa(\mu_0))e^{-\frac{\tau_0}{\mu_0}}}{1-\kappa^2(\mu_0)e^{-\frac{2\tau_0}{\mu_0}}} + \frac{1}{\mu_0 n_m^2}\int\limits_0^{\cos\gamma_r}(1-\kappa(\mu_e))\mu_e I_{2d}^{+}(\mu_e)d\mu_e, \quad (36)$$

where μ_0 is the cosine of the angle of incidence; n_m is the refractive index of the matrix ; γ_r is half the angle of light collection, $\mu_e = \cos\theta_e$. Eq. (36) is in good agreement with the experimental data [6] for a layer of soft particles.

6. Conclusion

We have developed a method for modeling the light transfer in disperse concentrated layers of optically soft particles. The proposed model is based on the radiative transfer theory for describing multiply scattered light, the interference approximation for treating the collective effects of scattering, and the Mie theory for the single-scattering characteristics. The theory takes into account the Fresnel reflection at the layer boundaries.

To solve the radiative transfer equation in concentrated disperse layers of large particles with strongly forward-extended phase function, we propose a numerical method based on the layer doubling method and spline-approximation of the phase function. The calculation error has been estimated. Comparison between the calculated and experimental data has shown that the proposed model can be used to describe light transfer through partially ordered layers with optically soft particles.

A maximum in the angular light distribution near the direction of the incident light can occur. It takes place if the phase function has a minimum at zero scattering angle. This maximum is more pronounced in layers with a small order of scattering. It is smoothed with increasing the scattering order and disappears in the multiple scattering regime.

Acknowledgements

The research was supported in part by the EC in the frame of INCO-Copernicus program (contract No. ERB IC15-CT98-0806) and by the Fond of Fundamental Research of Belarus (Project F01-042).

References

1. Drzaic, P.S. (1995) *Liquid Crystal Dispersions*, World Scientific, Singapore.
2. Ivanov, A.P., Loiko, A.V. (1983) *Optics of Photographic Film*, Nauka i Technika, Minsk.
3. Tsang, L., Kong, J.A., Shin, R.T. (1985) *Theory of Microwave Remote Sensing*, Wiley, New York.
4. Ishimaru, A. (1981) *Propagation and Scattering of Light in a Random Media*, Academic press, New York San Francisco London.
5. Ivanov, A.P., Loiko, V.A., Dick V.P. (1988) *Propagation of Light in Close-Packed Disperse Media*, Nauka i Technika, Minsk.
6. Twersky, V. (1962) Multiple Scattering of Waves and Optical Phenomena, *Journal of the Optical Society of America*, **Vol. no. 52**, pp. 145–171.
7. Chandrasekhar, S. (1960) *Radiative Transfer*, Dover, New York.
8. van de Hulst, H.C. (1980) *Multiple Light Scattering. Tables, Formulas and Applications*, New York.
9. Sobolev, V.V. (1975) *Light Scattering in Planetary Atmospheres*, Pergamon Press, Oxford.
10. Lenoble, J. (1989) *Radiative Transfer in Scattering and Absorbing Atmospheres: Standard Computational Procedures*, A.Deepak Publishing.
11. Dick, V.P., Loiko, V.A. (2001) Model for Coherent Transmittance Calculation for Polymer Dispersed Liquid Crystal Films, *Liquid Crystals*, **Vol. no. 28**, pp. 1193–1198.
12. Dick, V.P., Ivanov, A.P. (1999) Extinction of Light in Dispersion media with a High particle Concentration, *Journal of the Optical Society of America A*, **Vol. no. 16**, pp. 1040–1048.
13. Mischenko, M.I. (1994) Asymmetry Parameters of the Phase Function for Densely packed Scattering Grains, *Journal of Quantitative Spectroscopy and Radiative*

Transfer, **Vol. no. 52**, pp. 95–110.
14. Tsang, L., Ishimaru, A. (1985) Theory of Backscattering Enhancement of Random Discrete Isotropic Scatterers based on the Summation of All Ladder and Cyclical Terms, *Journal of the Optical Society of America A* , **Vol. no. 2**, pp. 1131–1138.
15. Tsang, L., Ding, K.H., Shih, S.E., Kong, J.A. (1998) Scattering of Electromagnetic Waves from Dense Distributions of Spheroidal Particles Based on Monte Carlo Simulation, *Journal of the Optical Society of America A* , **Vol. no. 15**, pp. 2660–2669.
16. Born, M., Wolf, E. (1965) *Principles of Optics*, Pergamon Press, London Edinburgh New York Paris Francfurt.
17. Bohren, C.F., Huffman, D.R. (1983) *Absorption and Scattering of Light by Small Particles*, New York Brisbane Toronto Singapore.
18. Deirmendjian, D. (1969) *Electromagnetic Scattering of Spherical Polydispersions*, American Elsevier Publishing Company Inc, New York.
19. Wiscombe, W.J. (1980) Improved Mie Scattering Algorithms, *Applied Optics* , **Vol. no. 19**, pp. 1505–1509.
20. Percus, J.K., Yevick, G.Y. (1958) Analysis of Classical Statistical Mechanics by Means of Collective Coordinates, *Physical Review*, **Vol. no.110**, pp. 1–13.
21. Ziman, J.M. (1979) *Models of Disorder* University press, Cambridge.
22. Kuzmin, V.L., Romanov, V.P., Obraztsov, E.P. (2001) Fluctuations of Dielectric Constant in a System of hard Shperes, *Optics and Spectroscopy* , **Vol. no. 91**, pp. 972–979.
23. Ishimaru, A., Kuga, Y. (1982) Attenuation constants of a Coherent Field in a Dense Distribution of Particles, *Journal of the Optical Society of America*, **Vol. no. 72**, pp. 1317–1320.
24. Berdnik, V.V., Loiko, V.A. (1999) Modeling of Radiative Transfer in Disperse Layers of a Medium with a Highly Stretched Phase Function, *Journal of Quantitative Spectroscopy and Radiative Transfer* , **Vol. no. 61**, pp. 49–57.
25. Berdnik, V.V., Loiko, V.A. (1996) Light Reflection and Transmission by Layers with Oriented Spheroidal Particles, *Particle and Particle Systems Characterization*, **Vol. no. 13**, pp. 171–176.
26. Plass, G.N., Kattawar, G.W., Catchings, F.F. (1973) Matrix Operator Theory of Radiative Transfer. 1: Rayleigh Scattering, *Applied Optics*, **Vol. no. 12**, pp. 314–329.
27. Wiscombe, W.J. (1976) On Initialization Error and Flux Conversation in the Doubling Method, *Journal of Quantitative Spectroscopy and Radiative Transfer* , **Vol. no. 16**, pp. 636–658.
28. Hunt, G.E. (1971) The Effect of Coarse Angular Discretization on Calculations of the Radiation Energy from a Model Cloudy Atmosphere, *Journal of Quantitative Spectroscopy and Radiative Transfer* , **Vol. no. 11**, pp. 309–321.
29. Berdnik, V.V., Loiko, V.A. (1999) Radiative Transfer in a Layer with Oriented Spheroidal Particles, *Journal of Quantitative Spectroscopy and Radiative Transfer* , **Vol. no. 63**, pp. 369–382.

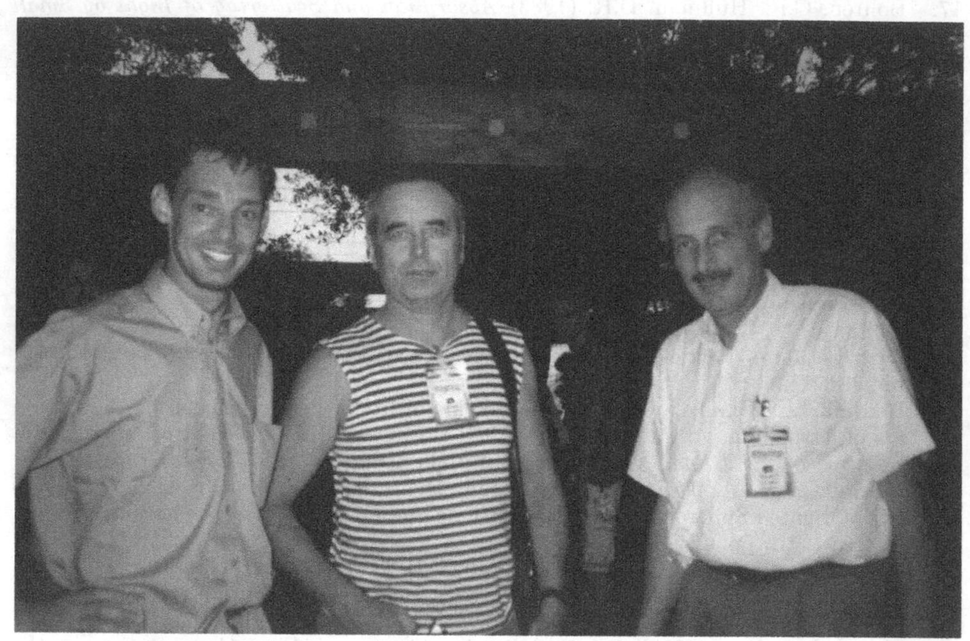

Bart van Tiggelen, Valery Loiko, and Georg Maret (from left to right). Photo by Valery Loiko

OPTICAL MICRORHEOLOGY OF SOFT COMPLEX MATERIALS

FRANK SCHEFFOLD[1], FRÉDÉRIC CARDINAUX,
SARA ROMER AND PETER SCHURTENBERGER
Department of Physics, University of Fribourg
CH-1700 Fribourg, Switzerland

SERGEY E. SKIPETROV
Laboratoire de Physique et Modélisation des Milieux Condensés
CNRS, F-38042 Grenoble, France

AND

LUCA CIPELLETTI
GDPC, Université Montpellier II
F-34095 Montpellier Cedex 05, France

Abstract. Dynamic multiple light scattering (diffusing wave spectroscopy, DWS) has been used to study the viscoelastic properties of soft materials. Several new multiple scattering approaches were implemented to extend the range of application of this optical microrheology technique. Taking advantage of the recently developed "two-cell technique" we show how DWS can be used to investigate the properties of fluid and solid-like complex media. Furthermore, we have significantly extended the range of accessible correlation times to at least 10^{-8}–10^{4} sec using a CCD-based multi-speckle analysis scheme. Excellent agreement is found when comparing the results obtained from DWS to classical oscillatory shear measurements. However, compared to classical rheology, we were able to significantly increase the range of accessible frequencies, thereby opening up a wealth of new possibilities for the study of these fascinating materials.

[1]E-mail: Frank.Scheffold@unifr.ch

B.A. van Tiggelen and S.E. Skipetrov (eds.),
Wave Scattering in Complex Media: From Theory to Applications, 553–563.

1. Introduction

DWS-based optical microrheology uses dynamic light scattering in the multiple scattering regime to obtain information about the viscoelastic properties of soft complex materials (see Refs. [1] and [2] for an overview). This can be done either by direct investigation or by addition of tracer particles to otherwise transparent systems. The underlying idea of optical microrheology is to study the thermal response of small (colloidal) particles embedded in the system under study. Based on the local dynamics, the macroscopic mechanic (viscoelastic) properties are predicted. In recent years, significant progress has been made in development and further understanding about the validity of this approach and its application to fluid and solid soft materials [1]–[12], [16].

Initially, optical microrheology has been mostly restricted to fundamental researchers while it is now becoming increasingly available to both industrial and applied researchers [17]–[20]. One of the most popular techniques to study the thermal motion of the particles is the diffusing wave spectroscopy (DWS), an extension of standard photon correlation spectroscopy (PCS) to turbid media. Here the analysis of (multiply) scattered laser light is used to determine the time evolution of the probe particles mean square displacement [21, 22]. DWS allows access to a broad range of time scales which translates into a large frequency range covered by DWS-based optical microrheology. In this article, we show how modern optical techniques can extend and improve classical rheology both in the sense of frequency range and applicability. Our experiments cover such different materials as polystyrene latex dispersions and gels, giant micelle solutions, as well as ceramic green bodies, casein micellar gels, and biopolymer solutions (for the latter see Refs. [17, 18, 20]) .

2. Diffusing Wave Spectroscopy

Dynamic light scattering (DLS), or photon correlation spectroscopy (PCS), analyzes the fluctuations of the light intensity scattered from a system under study. The light fluctuates due to the local motion of the scatterers. While in "conventional" light scattering experiments the sample has to be almost transparent (and hence often highly diluted), diffusing wave spectroscopy extends conventional dynamic light scattering to media with strong multiple scattering, treating the transport of light as a diffusion process [21, 22]. Provided that the structure factor $S(q) \equiv 1$ (no spatial correlations of the particles) or $k_0 a \gg 1$, it is possible to express the measured intensity autocorrelation function $g_2(\tau) = \langle I(t)I(t+\tau)\rangle / \langle I\rangle^2$ in terms of the mean square displacement $\langle \Delta\mathbf{r}^2(\tau)\rangle$ of the scattering particle (here $k_0 = 2\pi n/\lambda$ is the wave number of light in a medium with refractive index

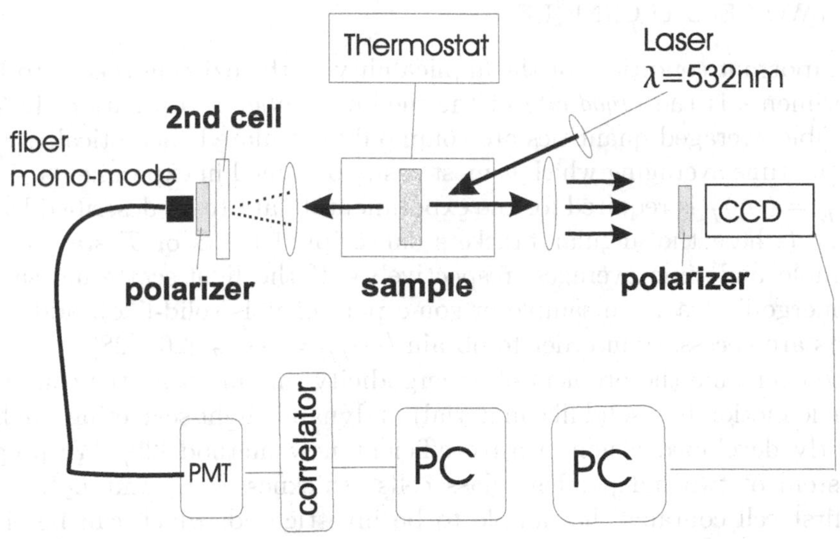

Figure 1. DWS setup: An intense laser beam (Verdi from Coherent) is scattered from a turbid sample contained in a temperature controlled water bath. Two-cell DWS (TCDWS): Light transmitted diffusively from the sample cell is imaged via a lens onto a second cell, containing a highly viscous colloidal suspension of moderate optical density. Subsequently the light is detected with a single mode fiber and analyzed digitally (correlator and PC). Multi-speckle DWS (MSDWS): In backscattering, the fluctuations of the scattered light are analyzed with a CCD camera. MSDWS can also be applied in transmission geometry, in parallel to TCDWS, by placing a beamsplitter between the sample and the lens.

n and a is the particle radius) [21, 22]:

$$g_2(\tau) - 1 = \left| \int_0^\infty ds\, P(s) \exp\left\{ -\frac{1}{3} k_0^2 \left\langle \Delta \mathbf{r}^2(\tau) \right\rangle \frac{s}{l^*} \right\} \right|^2, \tag{1}$$

where $P(s)$ is the distribution of photon trajectories of length s in the sample and it can be calculated within the diffusion model taking into account the experimental geometry (typically, transmission or backscattering). The transport mean free path l^* characterizes the typical step length of the photon random walk, given by the individual particles scattering properties and particle concentration. l^* can be determined independently and enters the analysis as a constant parameter. Equation (1) allows one to calculate the particle mean square displacement $\langle \Delta \mathbf{r}^2(\tau) \rangle$ from the measured autocorrelation function $g_2(\tau)$. Figure 1 sketches a typical experimental setup to measure $g_2(\tau)$ both in transmission and backscattering [23]–[25].

2.1. TWO-CELL TECHNIQUE

An important condition for the applicability of the existing theory to PCS experiments is the *ergodicity* of the medium under investigation. Indeed, ensemble-averaged quantities are commonly calculated theoretically, while it is the time averaging which is most easily obtained in experiments. Thus, $\langle\cdots\rangle_E = \langle\cdots\rangle_T$ is required for the experimental data to be described by the theory (where the angular brackets subscripted by E or T stay for the ensemble and time averages, respectively). If the light-scattering sample is nonergodic (say, the sample or some part of it is solid-like), additional efforts are necessary in order to obtain $\langle\cdots\rangle_E = \langle\cdots\rangle_T$ [26]–[28].

To overcome the problem of nonergodicity (arising from the constraint particle motion in a solid like material) in dynamic light scattering, we have recently developed a non-invasive efficient new method [29]. We prepare a system of two independent glass cells (thicknesses L_1 and L_2), where the first cell contains the sample to be investigated, which can be either a stable ergodic or an arrested non-ergodic sample [29]. The second cell, which serves to properly average the signal of the first cell, contains an ergodic, well-characterized system with very slow internal dynamics and moderate turbidity. A nonergodic speckle pattern at the back surface of the first cell (L_1) is imaged via a lens onto the second cell (L_2), as we show schematically in Fig. 1 (see also Refs. [25] and [30]). Transmission through the second cell diffusely broadens the individual speckles to a size $\sim L_2$. Since the second cell is ergodic, a given speckle spot at the back surface of the second cell (and hence the output signal measured by the photomultiplier) is obtained as a result of a two-dimensional average over many speckle spots at the back surface of the first cell. This is equivalent to the ensemble averaging and thus allows us to achieve the required condition of ergodicity $\langle\cdots\rangle_E = \langle\cdots\rangle_T$ for the two-cell system.

The correlation function $g_2(\tau)-1$ of the two-cell setup can be expressed through the joint distribution function $P_2(s_1, s_2)$ of path segments s_1, s_2 in the cells:

$$\begin{aligned} g_2(\tau)-1 &= \left| \int_0^\infty ds_1 \int_0^\infty ds_2\, P_2(s_1,s_2) \exp\left\{-\frac{1}{3}k_1^2 \left\langle \Delta\mathbf{r}_1^2(\tau)\right\rangle \frac{s_1}{l_1^*}\right\} \right. \\ &\times \left. \exp\left\{-\frac{1}{3}k_2^2 \left\langle \Delta\mathbf{r}_2^2(\tau)\right\rangle \frac{s_2}{l_2^*}\right\}\right|^2 . \end{aligned} \tag{2}$$

In the case that the cells decouple (i.e. no loops of the scattering paths between the two cells, see Ref. [29] for a more detailed discussion), $P_2(s_1, s_2)$ reduces to a product of two one-cell terms: $P_2(s_1, s_2) = P(s_1)P(s_2)$, and therefore we can write the correlation function of the two-cell setup, $g_2(\tau)-1$, as a product of the correlation functions of the two independent cells:

$g_2(\tau) - 1 = [g_2(\tau, L_1) - 1] \times [g_2(\tau, L_2) - 1]$ which we call the *"multiplication rule"* (the general theoretical treatment of the two-cell technique is given in Refs. [29] and [31]). We can now directly determine the contribution of the first cell by dividing the signal of the two-cell setup $g_2(\tau) - 1$ by the separately measured autocorrelation function $g_2(\tau, L_2) - 1$ of the ergodic system in the second cell:

$$g_2(\tau, L_1) - 1 = \frac{g_2(\tau) - 1}{g_2(\tau, L_2) - 1}. \tag{3}$$

2.2. MULTI-SPECKLE DWS

Another very useful extension of the standard DWS is the use of a CCD camera to follow temporal fluctuations of the scattered light (see Fig. 1). Instead of analyzing the fluctuations of the intensity at a single spatial position (one speckle spot) we now analyze a large area of the intensity pattern of the scattered light (hence "multi-speckle") using a CCD camera [6, 12, 14, 15]. The main advantage of this setup for the DWS-based microrheology is the significantly improved data acquisition time, since a large number of DWS-scattering experiments is actually performed simultaneously. In the standard DWS (or DLS) measurements, the data acquisition time has to be several orders of magnitude larger than the typical relaxation time of the autocorrelation function $g_2(t) - 1$, a restriction that does not apply to the multi-speckle DWS. Furthermore, since different configurations of the sample are probed simultaneously, the measured autocorrelation function never suffers the problems of nonergodicity described above. The main drawback of the camera-based DWS is the currently still much limited time resolution of CCD cameras. Typically, correlation times τ down to a few ms can be accessed (as compared to 10 ns with a standard photo-multiplier-digital-correlator setup), which is not sufficient for the fast relaxation processes usually encountered in DWS. However, a combination of the novel two-cell technique and multi-speckle DWS, as shown in Fig. 1, turns out to be a perfect one to overcome most of the commonly encountered experimental limitations. Using both techniques allows to cover a temporal range from ~ 10 ns to at least 10^4 sec (the latter being only limited by the "waiting" time), hence more than twelve orders of magnitude in correlation time [12, 13, 25, 30].

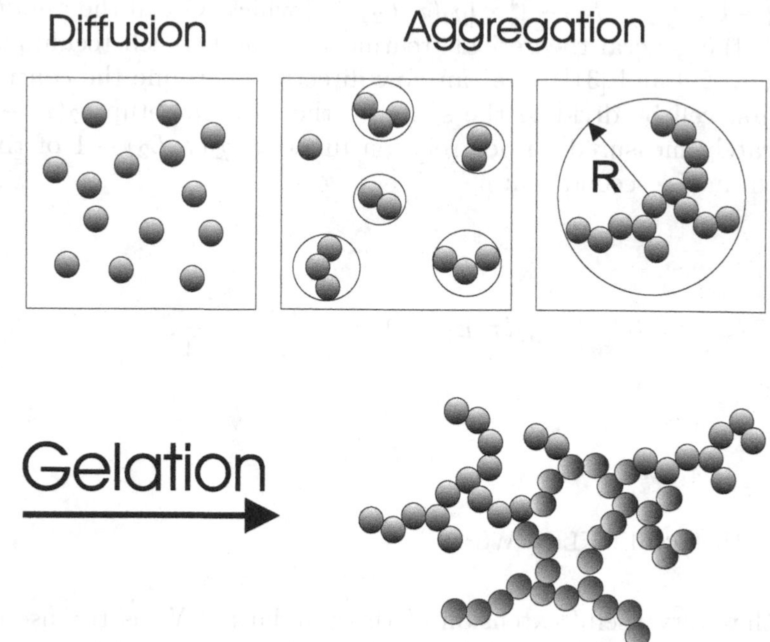

Figure 2. Aggregation and gelation: Attractive inter-particle interactions lead to the growth of individual (fractal) clusters until they fill up the whole accessible volume at $R_c \approx a\,\Phi^{-1/(3-d_f)}$ (Φ is the particle volume fraction).

3. Microrheology

3.1. SOFT SOLIDS: COLLOIDAL AGGREGATES AND GELS

Aggregation and gelation in complex fluids has been for a long time a field of intense research, where both fundamental as well as applied questions are equally important. Applications of gels and sol-gel processing include such different areas as ceramics processing, cosmetics and consumer products, food technology, to name only a few. Gels are formed by chemical or physical reactions of small sub-units (molecules, polymers, or colloids) which can be either reversible or irreversible. The macroscopic features that bring together such different materials are based on the microstructural properties of all gels, which can be described as random networks built up by aggregation of the individual sub-units (see Fig. 2). Starting from a solution of the sub-units, the systems is destabilized, which leads to aggregation, cluster formation, and gelation. At the gel point, a liquid-solid transition is observed which can be characterized by the appearance of a storage modulus

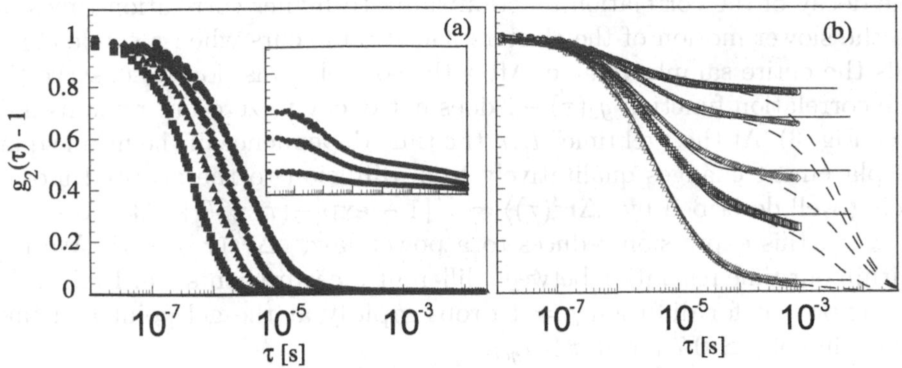

Figure 3. Sol-gel transition in a dense colloidal suspension. Intensity autocorrelation function $g_2(\tau) - 1$ of a two-cell setup during aggregation and gelation at $T = 20°$C. (a) Stable 298 nm suspension (20% vol. fract.) at $t = 0$ min (left curve) and time evolution after destabilization for $t = 11$, 16, 82.5 min (from left to right). Inset: Repeated (time averaged) measurements of $g_2(\tau) - 1$ in the gel state (single cell) show the typical (non-reproducible) nonergodic light scattering signal of a solid-like system. (b) Two-cell correlation function $g_2(\tau, L_1) - 1$ in the gel state after $t = 108$, 256, 734, 4683, 14400 min (from left to right). The solid lines have been recalculated from the fit to the data (see Refs. [7] and [29] for details). The raw-data $g_2(\tau) - 1$ is also displayed (dashed lines).

in rheological measurements. In dilute colloidal gel networks, the individual clusters show a fractal structure leading to a power-law scaling of the structure factor with q: $S(q) \propto q^{-d_f}$ for $a < 1/q < R_c$. The fractal dimension d_f is a measure of the "compactness" of an individual cluster, the higher d_f the more compact the clusters are. For the diffusion-limited cluster aggregation (DLCA) $d_f \simeq 1.8$, while for reaction-limited cluster aggregation (RLCA) $d_f \simeq 2.1$ is expected [32]–[34]. In the most simple picture, the individual clusters grow at a constant rate until they fill up the whole accessible volume, with a critical cluster radius given by

$$R_c \approx a\, \Phi^{-1/(3-d_f)}, \tag{4}$$

where Φ is the particle volume fraction.

Under highly dilute conditions, the structure and dynamics of colloidal gels can be analyzed using single light scattering [35]–[38]. For concentrated colloidal dispersions, we have recently reported the first study of the sol-gel transition based on DWS [7]. In our experiments, a suspension of monodisperse polystyrene latex spheres is destabilized by increasing the solvent ionic strength with a catalytic reaction. Thereby the electrostatic repulsion of the double layer is reduced and the particles aggregate due to Van-der-Waals attraction. Figure 3 shows the measured autocorrelation function as a function of time. At early stages, clusters form due to particle aggregation and the decay of the correlation function shifts to higher correlation times due to the slower motion of the clusters. Gelation occurs when a single cluster fills the entire sample volume. After the sol-gel transition we observe that the correlation function $g_2(\tau) - 1$ does not decay to zero but remains finite (see Fig. 3). At the "gel time" t_{gel}, the time dependence of the mean square displacement changes qualitatively from diffusion to the arrested motion, being well described by $\langle \Delta \mathbf{r}^2(\tau) \rangle = \delta^2\{1 - \exp[-(\tau/\tau_c)^p]\}$. At short times $\tau \ll \tau_c$, this expression reduces to a power law: $\langle \Delta \mathbf{r}^2(\tau) \rangle \propto \tau^p$. We find, within our time resolution between different measurements (ca. 1 min), that the exponent for diffusion $p = 1$ drops rapidly at the gel point and takes the value of $p \simeq 0.7$ for all $t > t_{gel}$.

It is furthermore possible to directly link the results of our DWS measurements to the macroscopic storage modulus, taking advantage of a recent model developed by Krall and Weitz [35]. Once the gel spans over the whole sample, the DWS signal is dominated by a broad distribution of elastic gel modes. In the case of fractal (dilute) gels, the storage modulus is given by $G'_0 = G'(\omega \simeq 0) = 6\pi\eta/\tau_c$ [35]. Recently, we have demonstrated that indeed the macroscopic storage modulus G'_0 deduced from the DWS measurements is in a good agreement with the classical rheological measurements over a large range of particle concentrations up to $\approx 10\%$ volume fraction [39].

3.2. VISCOELASTIC FLUIDS: CONCENTRATED SURFACTANT SOLUTIONS

The previous examples addressed systems where the scattering particles are part of the system under study. Now we want to discuss recent measurements where we have introduced tracer particles (polystyrene spheres, diameter 720 nm and 1500 nm) in an otherwise transparent matrix consisting of a concentrated surfactant solution. Under the chosen conditions these surfactants form giant, polymer-like micelles, which results in a surprisingly strong viscoelastic liquid [12, 40]. In this case, we can take advantage of the formalisms derived for the tracer microrheology, which allows to link directly the particle mean square displacement, as obtained from DWS, to the macroscopic viscoelastic moduli $G'(\omega)$ and $G''(\omega)$ [1]–[4]. Figure 4 shows the good agreement between the classical rheology and the DWS-based microrheology, with a dramatically increased frequency range for the latter technique. For a quantitative comparison, a scaling factor of 3/2 has been introduced. The origin of this factor is not well understood yet but, as we think, likely reflects the coupling of the tracer sphere to the medium. Unfortunately, the latter coupling is not understood sufficiently well by current theoretical approaches. In this context, it is interesting to note that the classical Stokes-Einstein relation differs by a very similar factor of 3/2 depending on the choice of the boundary conditions on the particle surface

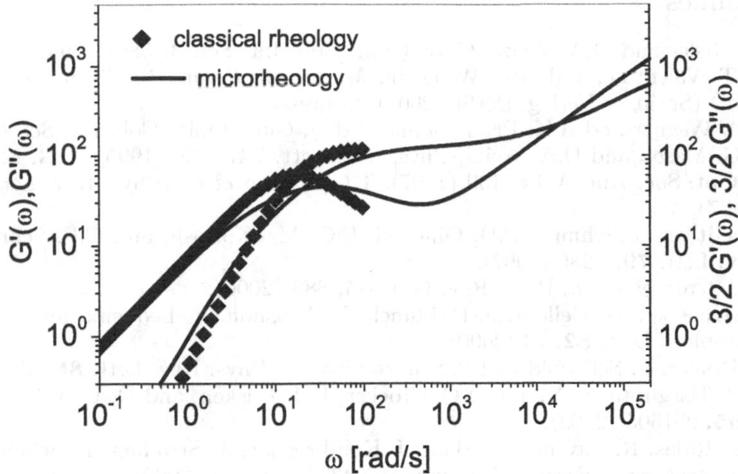

Figure 4. Frequency-dependent elastic moduli of giant micellar solutions (c = 100 mg/ml) at $T = 28°$C obtained from classical rheometry (symbols) and DWS-based microrheology (solid lines, right axis). The microrheology results have been multiplied by a factor 3/2 to allow a quantitative comparison of both experiments. The microrheology results are based on $\langle \Delta r^2(\tau) \rangle$ data, averaged over two tracer sizes and different particle concentrations (< 4%), thereby reducing statistical errors.

(stick or slip) [41]. Qualitatively similar features have been reported by Starrs and Bartlett in a tracer microrheology study of polymer solutions [42]. However, Bellour *et al.* have found a perfect agreement between the DWS-based microrheology and the classical rheometry for a similar system of worm like micelles [16].

4. Conclusions

In conclusion, we have shown two examples of how dynamic multiple light scattering (diffusing wave spectroscopy, DWS) can be applied to study the mechanical properties of soft complex materials. With the advent of new light scattering techniques, such as the two-cell method, and the use of CCD cameras for slow relaxation processes, it is now possible to apply DWS to an increasingly large number of liquid and solid-like media. This is expected to have an equally strong impact on both applied as well as fundamental research. DWS allows one to obtain the information about the medium rapidly and over a large range of frequencies, which is also of primary importance for many industrial applications, e.g., for rapid system characterization and process monitoring.

References

1. T. Gisler and D.A. Weitz, Curr. Opin. Coll. Int. Sci. **3,** 586 (1998); M.L. Gardel, M.T. Valentine, and D.A. Weitz, in: *Microscale Diagnostic Techniques*, K. Breuer (Ed.) (Springer Verlag, Berlin, 2002), in press.
2. N.J. Wagner and R.K. Prud'homme (Eds.), Curr. Opin. Coll. Int. Sci. **6** (2001).
3. T.G. Mason and D.A. Weitz, Phys. Rev. Lett. **74,** 1250 (1995); T.G. Mason *et al.,* J. Opt. Soc. Am. A **14,** 139 (1997); T.G. Mason *et al.,* Phys. Rev. Lett. **79,** 3282 (1997).
4. F. Gittes, B. Schnurr, P.D. Olmsted, F.C. MacKintosh, and C.F. Schmidt, Phys. Rev. Lett. **79,** 3286 (1997).
5. J.C. Crocker *et al.,* Phys. Rev. Lett. **85,** 888 (2000).
6. A. Knaebel, M. Bellour, J.-P. Munch, V. Viasnoff, F. Lequeux, and J.L. Harden, Europhys. Lett. **52,** 73 (2000).
7. S. Romer, F. Scheffold and P. Schurtenberger, Phys. Rev. Lett. **85,** 4980 (2000).
8. B.R. Dasgupta, S.-Y. Tee, J.C. Crocker, B.J. Frisken, and D.A. Weitz, Phys. Rev. E **65,** 051505 (2002).
9. L.F. Rojas, R. Vavrin, C. Urban, J. Kohlbrecher, A. Stradner, F. Scheffold, and P. Schurtenberger, Faraday Discussions **123,** to appear (2002).
10. K.M. Addas, J.X. Tang, A.J. Levine, C.F. Schmidt, Biophys. J. **82,** 2432 (2002).
11. A.J. Levine and T.C. Lubensky, Phys. Rev. Lett. **85,** 1774 (2000); A.J. Levine and T.C. Lubensky, Phys. Rev. E **63,** 041510 (2000).
12. F. Cardinaux, L. Cipelletti, F. Scheffold, and P. Schurtenberger, Europhys. Lett. **57,** 738 (2002).
13. F.A. Erbacher, R. Lenke, and G. Maret, Europhys. Lett. **21,** 551 (1993).
14. S. Kirsch, V. Frenz, W. Schartl, E. Bartsch, and H. Sillescu, J. Chem. Phys. **104,** 1758 (1996).
15. L. Cipelletti, S. Manley, R.C. Ball, and D.A. Weitz, Phys. Rev. Lett. **84,** 2275 (2000).
16. M. Bellour, M. Skouri, J.P. Munch, and P Hebraud, Eur. J. Phys. **8,** 431 (2002).
17. P. Schurtenberger, A. Stradner, S. Romer, C. Urban, and F. Scheffold, CHIMIA **55,** 155 (2001).
18. H. Wyss, S. Romer, F. Scheffold, P. Schurtenberger, and L.J. Gauckler, J. Coll. Int. Sci. **241,** 89 (2001).
19. A.J. Vasbinder, P.J.J.M. van Mil, A. Bot, and K.G. de Kruif, Coll. and Surfaces B — Biointerfaces **21,** 245 (2001).
20. C. Heinemann, F. Cardinaux, F. Scheffold, P. Schurtenberger, F. Escher, and B. Conde-Petit, Tracer microrheology of γ-dodecalactone induced gelation of aqueous starch systems, in preparation.
21. G. Maret and P.E. Wolf, Z. Phys. B **65,** 409 (1987).
22. D.A. Weitz and D.J. Pine, in: *Dynamic Light Scattering,* W. Brown (Ed.) (Oxford Univ. Press, New York, 1993), Chap. 16.
23. Figure 1 schematically dispays the most recent and advanced DWS setup installed in Fribourg. The different setups used in the actual experiments are similar but not identical. For details we refer to the original papers (references given in this article).
24. F. Scheffold, J. of Disp. Sci. and Tech. **23,** 591 (2002).
25. H. Bissig, S. Romer, L. Cipelleti, V. Trappe, and P. Schurtenberger, Intermittent dynamics and hyper-aging in dense colloidal gels, submitted to PhysChemComm (e-journal).
26. J.Z. Xue, D.J. Pine, S.T. Milner, X.L. Wu, and P.M. Chaikin, Phys. Rev. A **46,** 6550 (1992); P.N. Pusey and W. van Megen, Physica A **157,** 705 (1989).
27. G. Nisato, P. Hébraud, J-P. Munch, and S.J. Candau, Phys. Rev. E **61,** 2879 (2000).
28. M. Heckmeier and G. Maret, Progr. Colloid Polym. Sci. **104,** 12 (1997); Opt. Commun. **148,** 1 (1998).
29. F. Scheffold, S.E. Skipetrov, S. Romer, and P. Schurtenberger, Phys. Rev. E **63,**

061404 (2001).
30. V. Viasnoff, F. Lequeux, and D.J. Pine, Rev. Sci. Instrum. **73,** 2336 (2002).
31. S.E. Skipetrov and R. Maynard, Phys. Lett. A **217,** 181 (1996).
32. D.A. Weitz and M. Oliveria, Phys. Rev. Lett. **52,** 1433 (1984); D.A. Weitz, J.S. Huang, M.Y. Lin, and J. Sung, Phys. Rev. Lett. **53,** 1657 (1984); P. Dimon, S.K. Sinha, D.A. Weitz, C.R. Safinya, G.S. Smith, W.A. Varady, and H.M. Lindsay, Phys. Rev. Lett. **57,** 595 (1986); R. Klein, D.A. Weitz, M.Y. Lin, H.M. Lindsay, R.C. Ball, and P. Meakin, Prog. Coll. Int. Sci. **81,** 161 (1990).
33. E. Dickinson, J. Coll. Int. Sci. **225,** 2 (2000) and references therein.
34. C.M. Sorensen, Aerosol Sci. Tech. **35,** 648 (2001).
35. A.H. Krall and D.A. Weitz, Phys. Rev. Lett. **80,** 778 (1998).
36. D.W. Schaefer, J.E. Martin, P. Wiltzius, and D.S. Cannell, Phys. Rev. Lett. **52,** 2371 (1984); Dietler and D.S. Cannell, Phys. Rev. Lett. **60,** 1852 (1988).
37. M. Carpineti and M. Giglio, Phys. Rev. Lett. **68,** 3327 (1992).
38. T. Nicolai, D. Durand, and J. Gimel, Phys. Rev. B **50,** 16357 (1994).
39. H. Bissig, S. Romer, V. Trappe, F. Scheffold, and P. Schurtenberger, unpublished.
40. J.D. Ferry, *Viscoelastic Properties of Polymers*, 3rd ed. (John Wiley and Sons, New York, 1980).
41. J.-P. Hansen and I.R. McDonald, *Theory of Simple Liquids* (Academic Press, London, 1996).
42. L. Starrs and P. Bartlett, Faraday Discussions **123,** to appear (2002).

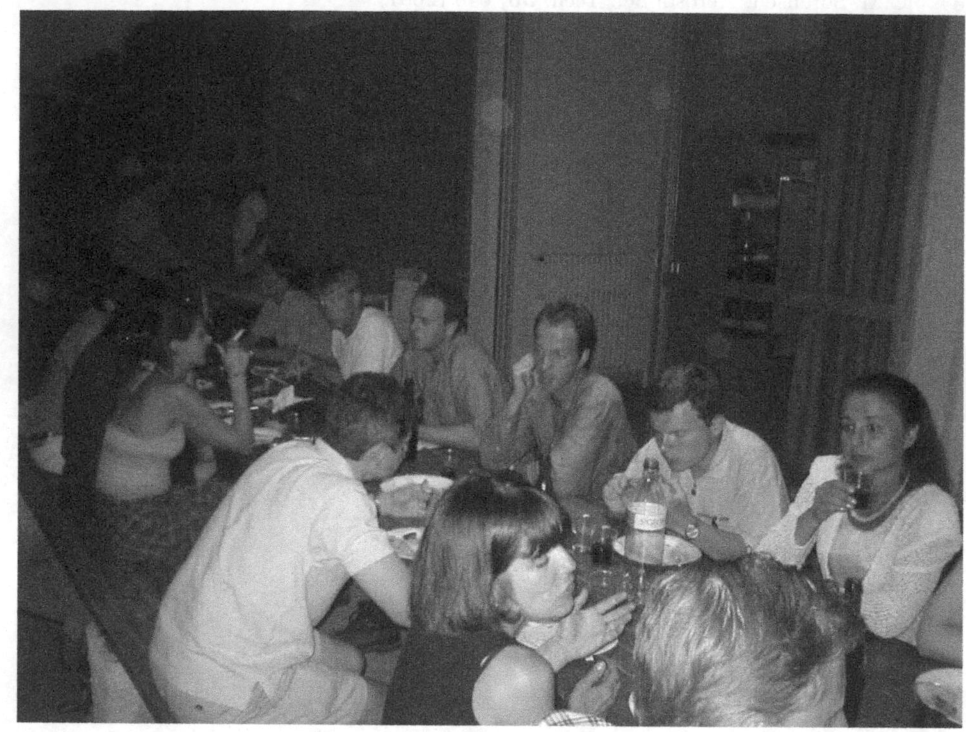

During the barbecue. Photo by Gabriel Cwilich

CHAPTER VII

WAVE SCATTERING IN NATURAL MEDIA

Left: Apollo 17 astronaut Eugene Cernan's photograph of his own shadow cast on the coal black lunar surface December 1972. His shadow is surrounded by a bright aureole. It is an example of the "Opposition Effect". Photo is taken from the book "Full Moon" by Michael Light, 1999/2003, reproduced with a kind permission of the author. Right: Yurii Kravtsov and Konstantin Bliokh trying to observe the opposition effect at the beach in Cargèse. Photo by Konstantin Bliokh

Pavel Litvinov. Photo by Gabriel Cwilich

COHERENT OPPOSITION EFFECT FOR DISCRETE RANDOM MEDIA

P.V. LITVINOV[1]
Institute of Radio Astronomy of NAS of Ukraine
4 Chervonopraporna St., Kharkov, 61002, Ukraine

V.P. TISHKOVETS[2]
Institute of Astronomy of Kharkov National University
35 Sumskaya St., Kharkov, 61022, Ukraine

KARRI MUINONEN[3]
University of Helsinki, Observatory
P.O. Box 14, FIN-00014, Helsinki, Finland

AND

GORDEN VIDEEN[4]
Army Research Laboratory AMSRL-CI-EE
2800 Powder Mill Road, Adelphi, Maryland 20783-1197 USA

Abstract. We examine the coherent opposition effect from a medium containing arbitrary scatterers. A rigorous analysis of all interactions is numerically unwieldy, so we examine first and second-order effects. We proceed to a ray-tracing analysis. While not as elegant, the results are more easily accessible, and the effect of multiple interactions can be examined with relative ease. While scattering from small particles leads to a negative polarization opposition effect, scattering from larger spheres, comparable to the wavelength can lead to a positive or negative polarization opposition effect.

[1]E-mail: Litvinov@ira.kharkov.ua
[2]E-mail: tishkovets@astron.kharkov.ua
[3]E-mail: Karri.Muinonen@helsinki.fi
[4]E-mail: GVideen@arl.army.mil

B.A. van Tiggelen and S.E. Skipetrov (eds.),
Wave Scattering in Complex Media: From Theory to Applications, 567–581.

1. Introduction

The inverse scattering problem of determining the structure and properties of objects by analyzing their scattered radiation arises in different fields of science, especially when the object is inaccessible for direct investigations. Very often measurements of angular dependencies of scattered radiation characteristics are the primary sources of information. The backscattering region is often of great importance. In this region the coherent photometric (weak localization effect) and polarimetric opposition effects are observed (see for example, [1–6] and the literature cited therein). The first one manifests itself as a sharp peak of intensity centered at the exact backscattering direction [1, 2]. The second one manifests itself as a region of negative linear polarization (i.e., the TM polarization state is dominant) at scattering angles just off the exact backscatter, where due to symmetry, the polarization state is zero. The width of the photometric effect is comparable to the minimum position of the polarimetric effect [4–6]. Both effects are the result of constructive interference of multiply scattered waves propagating in a medium along certain direct and reciprocal trajectories [3–6]. Their features (the width of the intensity peak, the position of the minimum of linear polarization degree etc.) depend strongly on the physical, chemical and morphological properties of the scattering system.

The theoretical description of coherent backscattering of light can be an extremely complicated problem. The complete analytical solution of the problem has been obtained in [7, 8] for the case of normal incidence of radiation on a semi-infinite medium composed of nonabsorbing randomly positioned Rayleigh scatterers. Numerical calculations of the polarization opposition effect for such media give good agreement with laboratory data [9]. For a layer of discrete random medium consisting of randomly oriented and positioned arbitrary scatterers, such a solution has been obtained in [10]. The equations derived in [10] allow a detailed study of the influence of microphysical properties of scatterers (size parameter, real and imaginary parts of the refractive index, shape, etc.) on the coherent opposition effect characteristics. Unfortunately, these equations are quite complicated, and a full numerical implementation and analysis has yet to be performed. To obtain some numerical results, the double-scattering approximation and a model of semi-infinite medium was used in [11–13]. Together with an exact theoretical description, various numerical techniques have been developed to analyze backscattering by random media (see for example, [14] and the literature cited therein).

In this chapter, further analysis of the equations describing the coherent part of the radiation backscattered by semi-infinite medium, consisting of randomly oriented and positioned arbitrary scatterers, are carried out in the double-scattering approximation. The equations are reduced to a form that gives a simple physical interpretation of the dependencies of the angular profiles of the scattering matrix elements of the coherent component of the scattered radiation on the microphysical properties of the scatterers. We then examine the phenomena using ray-tracing techniques. We start with second-order analysis and conclude with a general analysis.

2. Coherent Opposition Effect for Semi-Infinite Discrete Random Media in the Double-Scattering Approximation

Radiation scattered by a medium can be presented as the sum of the coherent and incoherent (diffuse) components. For a semi-infinite sparse, homogeneous, isotropic, discrete random media, consisting of randomly oriented particles, the elements of the scattering matrix for the diffuse component of the scattered radiation in the double-scattering approximation can be written as [11]:

$$S_{pn\mu\nu}^{(nc)} = -\frac{n_0 \cos\vartheta}{2k\mathrm{Im}(\varepsilon)(1-\cos\vartheta)} \sum_L d_{M_0N_0}^L(\vartheta) A_L^{(pn)(\mu\nu)}, \tag{1}$$

where

$$\begin{aligned} A_L^{(pn)(\mu\nu)} &= \chi_L^{(pn)(\mu\nu)} + \frac{\pi n_0}{k^3\mathrm{Im}(\varepsilon)} \sum_{qq_1} \chi_L^{(pq)(\mu q_1)} \sum_l \chi_l^{(qn)(q_1\nu)} \\ &\times \left(\int_0^{\pi/2} d_{M_0N}^L(\omega) d_{M_0N}^l(\omega) \frac{\cos\vartheta \sin\omega d\omega}{\cos\vartheta - \cos\omega} \right. \\ &+ \left. \int_0^{\pi/2} d_{M_0N}^L(\pi-\omega) d_{M_0N}^l(\pi-\omega) \frac{\sin\omega d\omega}{1+\cos\omega} \right). \end{aligned} \tag{2}$$

Here the basis of the circular polarization, CP-representation [15], of the scattering matrix elements is used. Matrix $S_{pn\mu\nu}^{(nc)}$ is normalized to a unit of surface area; n_0 is the concentration of particles in the medium; $k = 2\pi/\lambda$; λ is the wavelength of the incident radiation; $d_{M_0N_0}^L(\vartheta)$ is the Wigner function [16]; ϑ is the scattering angle, the angle between the incident beam and the scattering direction; $M_0 = \nu - n; N_0 = \mu - p$ $(p, n, \nu, \mu = \pm 1$ — polarization indexes); and ε is the complex effective refractive index of the medium. Coefficients $\chi_L^{(pn)(\mu\nu)}$ are the expansion coefficients of the scattering matrix elements $S_{pn\mu\nu}^0(\vartheta)$ in the CP-representation for the isolated particles, expanded using the Wigner D-function [11],

$$S_{pn\mu\nu}^0(\vartheta) = \sum_L \chi_L^{(pn)(\mu\nu)} d_{M_0N_0}^L(\vartheta), \tag{3}$$

The first term in Eq. (2) corresponds to the single scattering; the second one is related to the second-order scattering.

The expressions for the elements of the coherent scattering matrix in the double-scattering approximation have the form [11]:

$$\begin{aligned} S^{(co)}_{pn\nu\mu} &= -\frac{\pi n_0^2 \cos\vartheta}{k^4 \mathrm{Im}(\varepsilon)(1-\cos\vartheta)} \\ &\times \sum_{LMlmqq_1} (-1)^L \zeta^{*(q_1\mu)(qp)}_{LM} \zeta^{(qn)(q_1\nu)}_{lm} i^{-|m-M|} B^{(qq_1)}_{LMlm}(\vartheta), \end{aligned} \quad (4)$$

$$\begin{aligned} B^{(qq_1)}_{LMlm}(\vartheta) &= \int_0^\pi d^L_{MN}(\omega) d^l_{mN}(\omega) \int_0^\infty J_{|m-M|}(cx)\exp(-fx)dx \\ &= \int_0^\pi d^L_{MN}(\omega) d^l_{mN}(\omega) \frac{c^{|m-M|}\sin\omega d\omega}{\sqrt{c^2+f^2}(f+\sqrt{c^2+f^2})^{|m-M|}}, \quad (5) \\ c &= \sin\vartheta\sin\omega, N = q_1 - q \\ f &= 2\mathrm{Im}(\varepsilon) + |\cos\omega|\mathrm{Im}(\varepsilon)(1-\frac{1}{\cos\vartheta}) \\ &+ i\cos\omega\Big([\mathrm{Re}(\varepsilon)-1](1+\frac{1}{\cos\vartheta}) + (1+\cos\vartheta)\Big). \end{aligned}$$

Here $J_{|m-M|}(cx)$ are the Bessel functions, and $\zeta^{(pn)(\mu\nu)}_{LM}$ are expansion coefficients. In the exact backscattering direction (at $\vartheta = \pi$) [11]:

$$\zeta^{(pn)(\mu\nu)}_{LM} = -\chi^{(pn)(\mu\nu)}_{L} \delta_{M,\nu-n}. \quad (6)$$

The equations describing the coherent part of the scattered radiation [Eqs. (4) and (5)] can be reduced to a form showing the influence of the microphysical properties of the scatterers (size parameter, real and imaginary parts of the refractive index, etc.) on opposition effects.

Since these effects manifest themselves in a narrow range of scattering angle in the backward direction, it is possible (at least for particles whose size is less than or of the order of the wavelength) to neglect the angular dependence of the expansion coefficients $\zeta^{(pn)(\mu\nu)}_{L}$. Thus, replacing coefficients $\zeta^{(pn)(\mu\nu)}_{LM}$ with $\chi^{(pn)(\mu\nu)}_{L}$ [according to Eq. (6)], using the expansion (3) and symmetry properties of the Wigner d-function [16], Eqs. (4) and (5) are reduced to the following form:

$$\begin{aligned} S^{(co)}_{pn\nu\mu} &= -\frac{\pi n_0^2 \cos\vartheta}{k^4 \mathrm{Im}(\varepsilon)(1-\cos\vartheta)} i^{-|M_0+N_0|} \\ &\times \int_0^\pi \sum_{qq_1} S^{(0)}_{qnq_1\nu}(\omega) S^{(0)}_{qpq_1\mu}(\pi-\omega) I_{|M_0+N_0|} \sin\omega d\omega, \end{aligned} \quad (7)$$

$$I_{|M_0+N_0|} = \mathrm{Re}\left(\frac{c^{|M_0+N_0|}}{\sqrt{c^2+f^2}(f+\sqrt{c^2+f^2})^{|M_0+N_0|}}\right). \tag{8}$$

The linear polarization (LP-representation) state is more commonly measured in experimental systems, so we transform the scattering matrix elements from the CP to LP-representation [15]. For elements R_{11} and R_{21} of the reflection matrix we have [11, 12]:

$$R_{11} = w\sum_{pn} S_{pnpn}, \qquad R_{21} = -w\sum_{pn} S_{pn-pn}, \tag{9}$$

where $w = -1/2k^2\cos\vartheta$. In the case of unpolarized incident radiation, the element R_{11} defines intensity (first Stokes parameter); whereas, R_{21} defines the linear polarization state (second Stokes parameter) of the scattered by medium radiation.

According to (9) and the symmetry properties of the scattering matrix in the CP-representation, $S_{pn\mu\nu} = S_{-p-n-\mu-\nu}$, elements $R_{11}^{(co)}$ and $R_{21}^{(co)}$ can be expressed as

$$R_{11}^{(co)} = \frac{\pi n_0^2}{k^6\mathrm{Im}(\varepsilon)(1-\cos\vartheta)}\int_0^\pi Q_{11}(\omega)I_0\sin\omega d\omega, \tag{10}$$

$$\begin{aligned} Q_{11}(\omega) &= F_{11}^0(\omega)F_{11}^0(\pi-\omega) + F_{21}^0(\omega)F_{21}^0(\pi-\omega) \\ &= F_{11}^0(\omega)F_{11}^0(\pi-\omega)\Big(1+P_0(\omega)P_0(\pi-\omega)\Big), \end{aligned} \tag{11}$$

$$R_{21}^{(co)} = \frac{\pi n_0^2}{k^6\mathrm{Im}(\varepsilon)(1-\cos\vartheta)}\int_0^\pi Q_{21}(\omega)I_2\sin\omega d\omega, \tag{12}$$

$$Q_{21}(\omega) = -\Big(F_{11}^0(\omega)F_{12}^0(\pi-\omega) + F_{21}^0(\omega)F_{22}^0(\pi-\omega)\Big). \tag{13}$$

Here F_{11}^0, F_{12}^0, F_{21}^0, F_{22}^0 are the elements of the scattering matrix in the LP-representation for isolated particles; $P_0 = -F_{21}^0/F_{11}^0$ is the degree of linear polarization of radiation scattered by the isolated particles; I_0 and I_2 are defined by (8).

For spherical particles ($F_{11}^0 = F_{22}^0$):

$$Q_{21}(\omega) = F_{11}^0(\omega)F_{11}^0(\pi-\omega)\Big(P_0(\omega)+P_0(\pi-\omega)\Big). \tag{14}$$

In expressions (7), (10) and (12), the interference of waves scattered by each pair of scatterers in a medium is described by I_m [Eq. (8)], which depends on effective complex refractive index ε. Influence of the microphysical

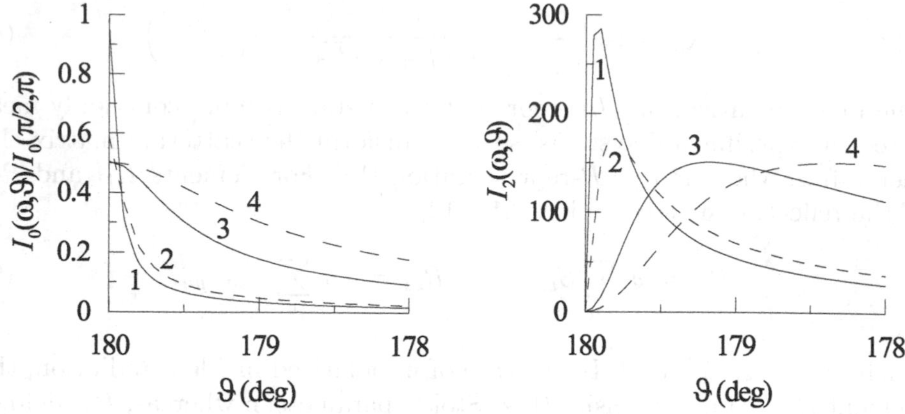

Figure 1. Dependence of I_0 and I_2 on scattering angle. $\mathrm{Im}(\varepsilon) \simeq 0.0002864$. 1: $\omega = 90°$; 2: $\omega = 45°$; 3: $\omega = 10°$; 4: $\omega = 5°$.

properties of particles (at fixed value of ε) on the magnitude of the interference at different scattering direction in medium is defined by means of functions Q_{11} and Q_{21}.

Typical forms of the interference terms I_0 and I_2 at fixed values of ε are presented in Fig. 1. The following value of effective refractive index of the medium was used $\varepsilon = 1 + i\mathrm{Im}(\varepsilon)$, $\mathrm{Im}(\varepsilon) = n_0 C_{ext}/2k$, where C_{ext} is the optical extinction cross section, $n_0 = 3\xi/4\pi\tilde{a}^3$. ξ is a filling factor of the medium, and $\tilde{a}$ is the radius of particles of the medium. A medium consisting of particles with more pronounced lateral scattering leads to a narrower intensity peak than for a medium consisting of particles with predominant forward and backward scattering (see also [12]). This is illustrated in Fig. 2, which shows plots of the normalized total intensity

$$I = \frac{R_{11}^{(co)}(\vartheta) + R_{11}^{(nc)}(\vartheta)}{R_{11}^{(co)}(\pi) + R_{11}^{(nc)}(\pi)} \tag{15}$$

and the degree of linear polarization

$$P = -\frac{R_{21}^{(co)}(\vartheta) + R_{21}^{(nc)}(\vartheta)}{R_{11}^{(co)}(\vartheta) + R_{11}^{(nc)}(\vartheta)}. \tag{16}$$

For example, spherical particles with $\tilde{x} = 4.5$ and $\tilde{m} = 1.35 + 0i$ scatter radiation more effectively in the forward and backward directions than particles with $\tilde{x} = 3$ and $\tilde{m} = 1.5 + 0i$ [curves 1 and 4 on Fig. 2(b)]. As a result, the intensity peak width for a medium composed of these particles is larger [curves 1 and 4 on Fig. 2(a)]. Particles with $\tilde{x} = 3$, $\tilde{m} = 1.35 + 0i$ and $\tilde{x} = 2$, $\tilde{m} = 1.35 + 0i$ display a more symmetric indicatrix, and scatter

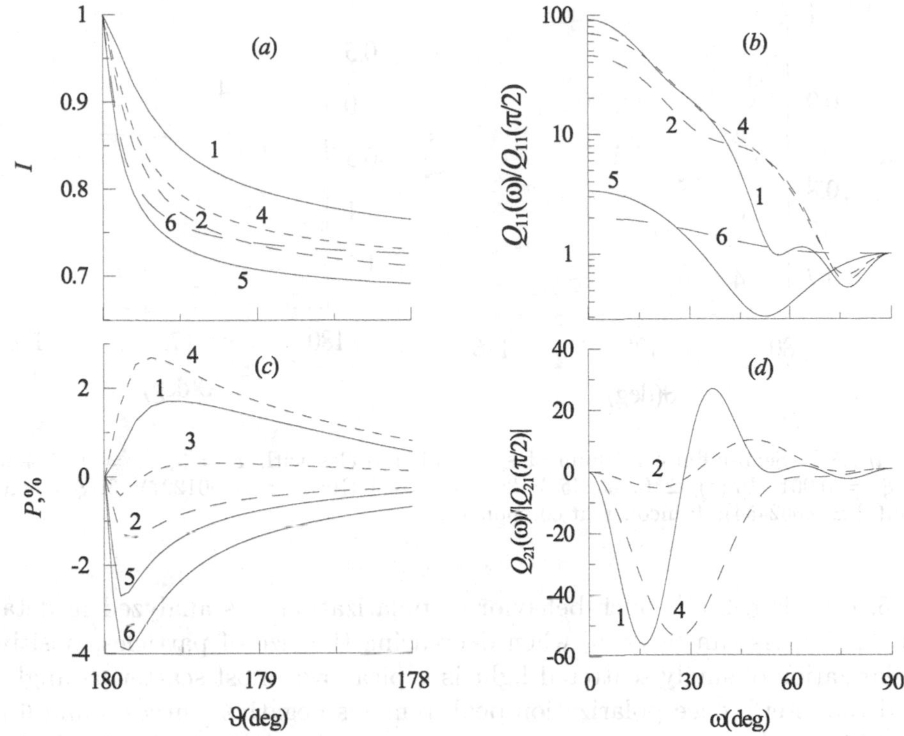

Figure 2. (a) Normalized intensity and (c) degree of linear polarization of radiation scattered by the semi-infinite medium. Normalized single particle scattering properties (b) $Q_{11}(\omega)/Q_{11}(\pi/2)$ and (d) $Q_{21}(\omega)/Q_{21}(\pi/2)$ are also shown. $\text{Im}(\varepsilon) \simeq 0.000286$. 1: $\tilde{x} = 4.5$, $\tilde{m} = 1.35 + 0i$, $\xi = 0.001$; 2: $\tilde{x} = 3$, $\tilde{m} = 1.35 + 0i$, $\xi = 0.0012$; 3: $\tilde{x} = 3.1$, $\tilde{m} = 1.35 + 0.0i$, $\xi = 0.0011$; 4: $\tilde{x} = 3$, $\tilde{m} = 1.5 + 0.0i$, $\xi = 0.0007$; 5: $\tilde{x} = 2$, $\tilde{m} = 1.35 + 0i$, $\xi = 0.0019$; 6: $\tilde{x} = 0.5$, $\tilde{m} = 1.35 + 0i$, $\xi = 0.0505$.

more effectively in lateral directions than particle with $\tilde{x} = 3$, $\tilde{m} = 1.5 + 0i$, and the peak for these particles is sharper [curves 2, 5 and 4 on Fig. 2(b)]. For a medium composed of Rayleigh scatterers, the interference peak is the narrowest on Fig. 2(a) (curve 6).

The results of Fig. 2 demonstrate that the state of polarization of the backscattered radiation is controlled strongly by the state of polarization of singly scattered radiation and the intensity of scattering by the particles of the medium (i.e. by angular profiles of F_{11}^0 and P_0). It is well known that the positive polarization of scattered light by isolated particles leads to a negative polarization peak in the backscattering direction [4, 5]. The angular dependence of the polarization state for spherical particles whose size is of the order of the wavelength is much more complicated and oscillatory. The resulting interference of doubly scattered rays can result in positive polarization [curves 1 and 4 on Fig. 2(c)] as well as negative [curves

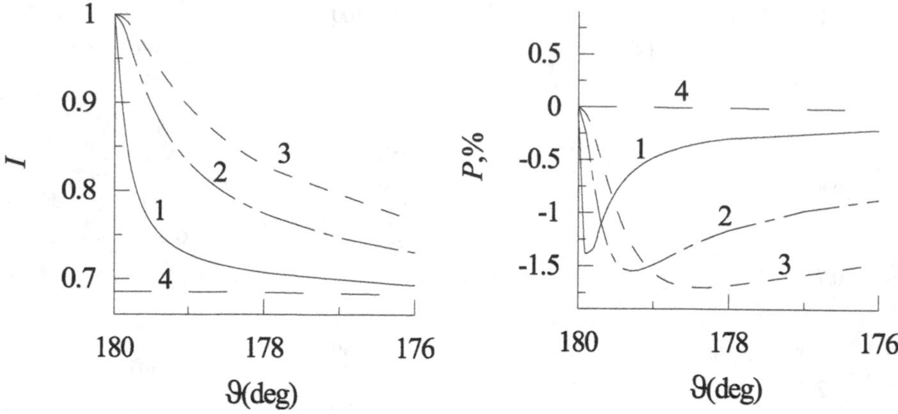

Figure 3. Semi-infinite medium of spherical particles with $\tilde{x} = 3$, $\tilde{m} = 1.35 + 0i$. 1: $\xi = 0.001$ ($\mathrm{Im}(\varepsilon) \simeq 0.0002454$); 2: $\xi = 0.005$ ($\mathrm{Im}(\varepsilon) \simeq 0.001227$); 3: $\xi = 0.01$ ($\mathrm{Im}(\varepsilon) \simeq 0.002454$); 4: incoherent component.

2, 5, 6 on Fig. 2(c)]. Such behavior of polarization was analyzed in detail in [11–13]. As can be seen, when decreasing the size of particles, positive polarization of singly scattered light is typical over most scattering angles, and the interference polarization peak remains negative [curves 5 and 6 in Fig. 2(c)]. Interference may lead to a more complex dependence of polarization on the scattering angle, when positive and negative polarization peaks appear simultaneously [curve 3, Fig. 2(c)].

Figure 3 demonstrates the dependence of the interference effects on $\mathrm{Im}(\varepsilon)$. It can be seen from Fig. 3 that the interference peak width and the position of the linear polarization minimum is almost a linear function of $\mathrm{Im}(\varepsilon)$. This linear dependence has been mentioned in a number of works (see for example, [1, 2, 9]). But it is necessary to note that in some cases, at least in the double-scattering approximation, the linearity may be violated because the element $R_{21}^{(nc)}$ is superimposed on the angular dependence of the element $R_{21}^{(co)}$ [12].

3. Second-Order Ray Tracing

While the method of the previous section is based on a rigorous description of the multiple scattering from complex particle systems, there are several reasons to consider approximate techniques for such analyses. First, a great deal of time and effort must be invested to acquire numerical output from such a method. Second, many times we are interested only in discovering general behaviour of an effect when particular conditions are changed. And third, inherent approximations in a rigorous technique, like assuming scatterers of a particular shape, may be much greater than those involved in the approximate technique. One simple technique that can be used to examine the coherent opposition effect is that of ray-tracing. In this case, the physical processes are transparent and numerical output can be achieved without a great investment of time and resources ([17, 18], cf. [5, 19]). We consider a scattering system composed of particles whose individual phase functions are specified as in the previous section. In this section we consider only the second-order ray-tracing contribution; i.e., only the rays which undergo two scattering events before reaching the detector; and incorporate the interference of reciprocal rays, which serves to enhance the parallel polarization component.

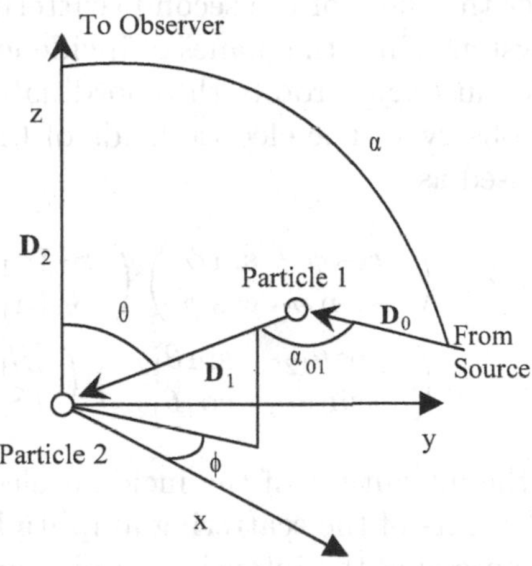

Figure 4. Geometry showing two scatterers.

A schematic showing the orientation of the two particles of the scattering system is shown in Fig. 4. A light ray traveling in a plane parallel to the xz-plane, oriented at angle α from the z axis, strikes Particle 1 located at the point (r, θ, ϕ). This ray travels in the direction of vector $\boldsymbol{D_0}$. After scattering from Particle 1, the ray, traveling in the direction of vector $\boldsymbol{D_1}$, continues to Particle 2, located at the origin, where it is again scattered and travels in a direction parallel to the positive z axis, given by $\boldsymbol{D_2}$, to the observer. The relevant unit ray vectors are given by

$$\begin{aligned} \boldsymbol{D_0} &= (\sin\alpha, 0, -\cos\alpha), \\ \boldsymbol{D_1} &= (-\sin\theta\cos\phi, -\sin\theta\sin\phi, -\cos\theta), \\ \boldsymbol{D_2} &= (0, 0, 1). \end{aligned} \tag{17}$$

The relevant scattering angle from Particle 1 is shown in Fig. 4 and given by

$$\alpha_{01} = \pi - \arccos\left(-\sin\theta\cos\phi\sin\alpha + \cos\theta\cos\alpha\right). \tag{18}$$

The orientations of the planes of incidence are different for each particle. The orientation of the plane of incidence of Particle 1 with respect to Particle 2 is designated θ_{12} and is given by

$$\cos\theta_{12} = \frac{\sin\theta\cos\alpha + \cos\theta\cos\phi\sin\alpha}{\sqrt{\sin^2\theta\sin^2\phi + (\sin\theta\cos\phi\cos\alpha + \cos\theta\sin\alpha)^2}}. \tag{19}$$

The last orientation of the plane of incidence from Particle 2 to the system is designated θ_{20} and is equal to ϕ. We must consider the polarization states of the detected light separately. We assume that the light from the source is unpolarized. This light is scattered by Particle 1 before striking Particle 2. To consider the effect of the second scattering event, we must rotate the coordinate system, since the planes of incidence of the two particles are not the same. We must again rotate the coordinate system to consider the light going to the observer. The electric fields of light scattered to the observer can be expressed as

$$\begin{pmatrix} E_s \\ E_p \end{pmatrix} \sim \begin{pmatrix} \cos\phi & \sin\phi \\ -\sin\phi & \cos\phi \end{pmatrix} \begin{pmatrix} S_1(\alpha_{12}) & S_4(\alpha_{12}) \\ S_3(\alpha_{12}) & S_2(\alpha_{12}) \end{pmatrix} \tag{20}$$

$$\begin{pmatrix} \cos\theta_{12} & \sin\theta_{12} \\ -\sin\theta_{12} & \cos\theta_{12} \end{pmatrix} \begin{pmatrix} [S_1(\alpha_{01}) + S_4(\alpha_{12})]\, E_o \\ [S_2(\alpha_{01}) + S_3(\alpha_{12})]\, E_o \end{pmatrix}, \tag{21}$$

where E_o is the magnitude of the incident electric field on the system and S_n are the elements of the scattering amplitude matrix (Jones vectors) We proceed by expressing the intensities and then incorporating the coherent backscattering factors in these expressions. The last thing that needs to be specified is the distance a ray can be expected to travel between the two scattering sites, which we express as P_x. The intensity is proportional to the sum of all the different relative positions and orientations of the particles. If we consider rotationally symmetric particles, the intensities can be written as

$$\begin{pmatrix} I_s \\ I_p \end{pmatrix} \sim \int\int\int \begin{pmatrix} E_s^* E_s \\ E_p^* E_p \end{pmatrix} P_x \sin\theta d\theta dx d\phi. \tag{22}$$

To obtain results we only need to specify the scattering amplitudes of the isolated particles and the interaction distance, $x = |\boldsymbol{D_1}|$. For our illustrative purposes, we assume an exponential distribution:

$$I_2 = I_1 \exp(-\beta x); \tag{23}$$

i.e., the intensity at Particle 2, I_2, equals the intensity at Particle 1, I_1, multiplied by a decaying exponential, the argument of which is proportional to the distance between the two particles and inversely proportional to a characteristic length of the system $d = 1/\beta$. This particular distribution is characteristic of randomly oriented particles in a volume. At this point, there is no interference mechanism that can include backscattering enhancement. We incorporate this effect explicitly by considering the interference between each set of reciprocal rays. The electric fields used in Eq. (22) contain two reciprocal components:

$$E_s = E_s^{(1)} + E_s^{(2)}; E_p = E_p^{(1)} + E_p^{(2)}; \qquad (24)$$

where $E_s^{(1)}$ is the s-component complex amplitude of the ray from the source that is scattered by Particle 1 to Particle 2 before going to the observer, and $E_s^{(2)}$ is the s-component complex amplitude of the ray from the source that is scattered by Particle 2 to Particle 1 before going to the observer, and likewise for the p-component.

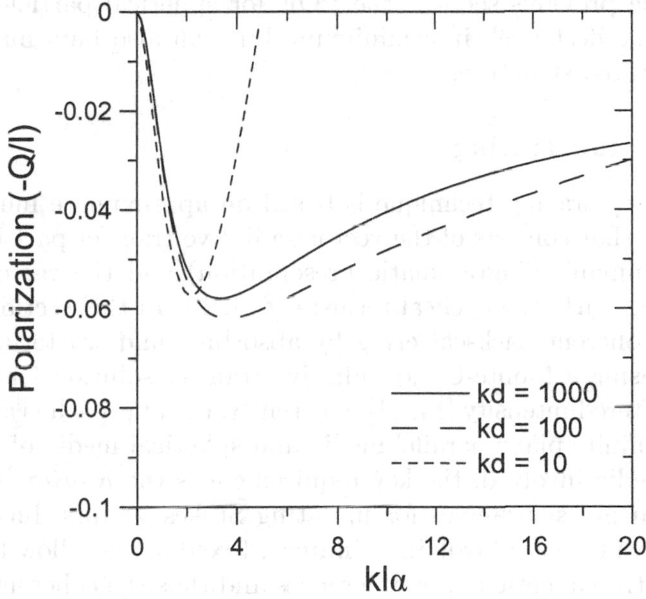

Figure 5. Polarization response for different populations of spherical particles. In this case ($r = \lambda/10$, $m = 1.55$), minima are present. I and Q refer to the first two elements of the Stokes vector; i.e., $I = \langle E_p^* E_p + E_s^* E_s \rangle$ and $Q = \langle E_p^* E_p - E_s^* E_s \rangle$.

The electric field components now vary from each other by a phase difference $\exp(i\Phi)$ and Eq. (24) can be written as

$$E_s^* E_s = 2E_s^{(1)*} E_s^{(1)} (1 + \cos \Phi); \; E_p^* E_p = 2E_p^{(1)*} E_p^{(1)} (1 + \cos \Phi); \qquad (25)$$

where

$$\Phi = k\,|\mathbf{D_1}|\,(-\cos\theta + \sin\alpha\sin\theta\cos\phi + \cos\alpha\cos\theta) = \gamma\,|\mathbf{D_1}|\,. \tag{26}$$

Integrating Eq. (22) over the separation distance between the particles yields

$$\begin{pmatrix} I_s \\ I_p \end{pmatrix} \sim 2\int\int \begin{pmatrix} E_s^{(1)*}E_s^{(1)} \\ E_p^{(1)*}E_p^{(1)} \end{pmatrix} \left(\frac{2\beta^2+\gamma^2}{\beta^2+\gamma^2}\right)\sin\theta d\theta d\phi. \tag{27}$$

Equation (24) provides second-order ray-tracing results of the s and p polarization intensities scattered by a cloud of particles. Sample calculations of the polarization state of a cloud of spherical particles are given in Fig. 5. As the separation distances between particles increase, the minimum position decreases. For finite-size spheres, unlike dipoles, an exact inverse relationship does not exist between these parameters. As demonstrated in the previous section, the POE for spherical particles does not necessarily manifest itself in a minimum, but can also be a maximum, or more complicated structure.

4. General Ray Tracing

The general ray-tracing technique is based on approximate multiple scattering theory that consists of the vector radiative transfer part (the ladder terms of Feynman's diagrammatic presentation) and the vector coherent backscattering part (the cyclical terms; e.g., [22]). In this section the computation of coherent backscattering by absorbing and scattering media is carried out using a Monte-Carlo radiative transfer solution for the Stokes vector of scattered intensity [14, 21]. Currently, the approach is available for finite/semi-infinite plane-parallel media and spherical media of scatterers.

For all media involved, the key requirement is the *a priori* selection of reflection/transmission angles for updating Stokes vectors during Monte-Carlo radiative transfer (avoiding binning). Fixed angles allow the computation of electromagnetic phase differences and thus the coherent backscattering effect. The numerical technique relies on the reciprocity principle for electromagnetic scattering, allowing an *a posteriori* renormalization of the coherent-backscattering contributions at each multiple-scattering event. Renormalization is carried out with the help of amplitude scattering matrices.

The vector radiative transfer part of the technique has been carefully verified using reference results for a plane-parallel Rayleigh-scattering planetary atmosphere [20]. Of general interest in Monte-Carlo radiative transfer computations is the generation of scattering angles in each scattering

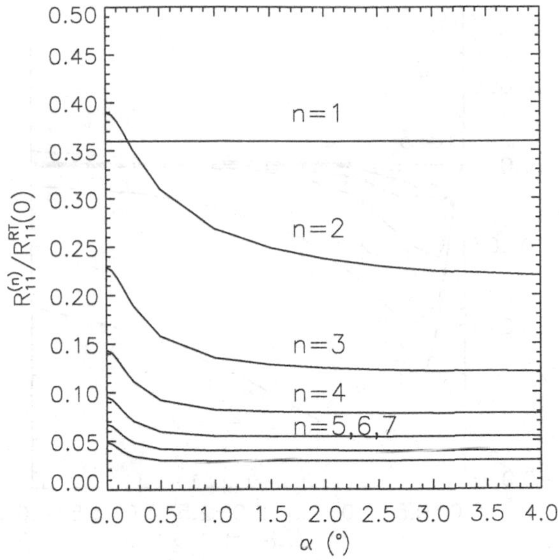

Figure 6. Contributions of various orders of scattering to the photometric opposition effect for a semi-infinite plane-parallel medium of Rayleigh scatterers. The single-scattering albedo is $\tilde{\omega} = 0.9$ and the transport mean free path is $k\ell = 300$.

process. We make use of an efficient technique for Lorenz-Mie type block-diagonal scattering phase matrices based on solving a Kepler-type equation for the azimuthal scattering angle [14].

The coherent backscattering part of the technique has been verified using reference results for semi-infinite plane-parallel media of conservative Rayleigh scatterers [7–9]; whereas the numerical technique currently cannot be used to compute coherent backscattering by conservative media, convergence toward the reference results has been established for increasing single-scattering albedo. So far, single-scattering albedos as high as 0.9999 and scattering orders up to 10000 are within the reach of the numerical technique.

We illustrate the ray-tracing technique by studying the contributions of different scattering orders to the photometric and polarimetric signatures of coherent backscattering for a semi-infinite plane-parallel medium of Rayleigh scatterers with single-scattering albedo $\tilde{\omega} = 0.9$ and transport mean free path $k\ell = 300$. We traced 100,000 rays to obtain the coherent backscattering surges in Figs. 6 and 7. As suggested in [4, 23], coherent backscattering produces both enhanced backscattering and enhanced negative polarization from scattering orders higher than the second.

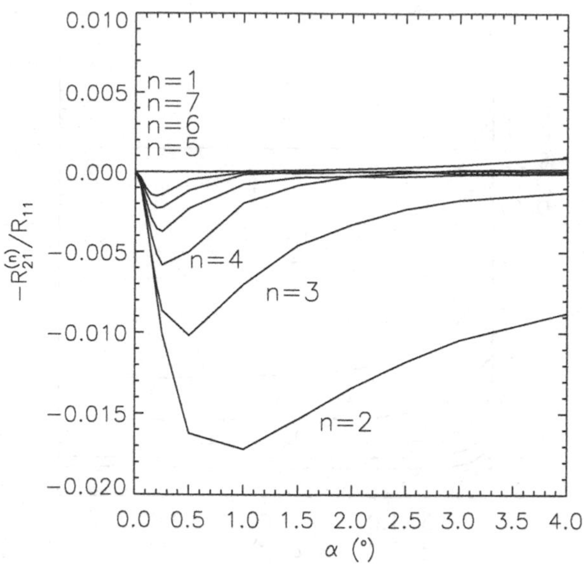

Figure 7. Contributions of various orders of scattering to the polarimetric opposition effect. Parameters are the same as in Fig. 6.

5. Acknowledgements

P. Litvinov and V. Tishkovets thank INTAS (grant N1999-00652) for the support of this work. K. Muinonen acknowledges the support of the Academy of Finland. G. Videen acknowledges support from the TechBase Program on Chemical and Biological Defense.

References

1. Van Albada, M. P., Lagendijk, A. (1985) Observation of weak localization of light in a random medium, *Phys. Rev. Lett.*, **Vol. 55 no. 24**, pp. 2692–2695.
2. Wolf, P.E., Maret., G. (1985) Weak localization and coherent backscattering of photons in disordered media, *Phys. Rev. Lett.*, **Vol. 55 no. 24**, pp. 2696–2699.
3. Barabanenkov, Yu.N., Kravtsov, Yu.A., Ozrin, V.D., Saichev, A.I. (1991) Enhanced backscattering in optics. In: Progress in Optics. Vol. 29, ed. Wolf, E., Elsevier, New York, pp. 65–197.
4. Muinonen, K. (1990) Light scattering by inhomogeneous media: Backward enhancement and reversal of linear polarization (Ph.D. thesis, Univ. Helsinki).
5. Shkuratov, Yu.G. (1989) New mechanism of the negative polarization of light scattered by atmosphereless cosmic bodies, *Astron. Vestnik*, **Vol. 23**, pp. 176–180.
6. Mishchenko, M.I. (1993) On the nature of the polarization opposition effect exhibit by Saturn's rings, *Astrophys. J.*, **Vol. 411**, pp. 351–361.
7. Ozrin, V.D. (1992) Exact solution for backscattering of polarized light from a random medium of Rayleigh scatterers, *Waves Random Med.*, **Vol. 2**, pp. 141–164.
8. Amic, E., Luck, J.M., and Nieuwenhuizen, Th.M. (1997) Multiple Rayleigh scatter-

ing of electromagnetic waves, *J. Phys. I*, **Vol. 7**, pp. 445–483.
9. Mishchenko, M.I., Luck, J.M., and Nieuwenhuizen, Th.M. (2000) Full angular profile of the coherent polarization affect, *J. Opt. Soc. Am. A.*, **Vol. 17**, pp. 888–891.
10. Tishkovets, V.P. (2001) Multiple scattering of light by a layer of discrete random medium. Backscattering, *J. Quant. Spectrosc. Radiat. Transfer*, **Vol. 72**, pp. 123–137.
11. Tishkovets, V.P., Litvinov, P.V., and Lyubchenko, M.V. (2002) Coherent opposition effects for semi-infinite descrete random medium in the double-scattering aproximation, *J. Quant. Spectrosc. Radiat. Transfer*, **Vol. 72**, pp. 803–811.
12. Tishkovets, V.P., Litvinov, P.V., and Tishkovets, S.V. (2002) Intereference effects at backscattering of light by layer of descrete random media, *Optics and Spectroscopy*, **Vol. 93 no. 6**, pp. 976–985 (in Russian).
13. Mishchenko, M.I., Tishkovets, V.P., and Litvinov, P.V. (2002) In: Optics of Cosmic Dust, ed. Videen, G., and Kocifaj, M., *Nato Science Series, II. Mathemetics, Physics and Chemistry*, Kluwer Academic Publishers, **Vol. 79**, pp. 239–260.
14. Muinonen, K., Videen, G., Zubko, E., and Shkuratov, Yu. (2002) In: Optics of Cosmic Dust, ed. Videen, G., and Kocifaj, M., *Nato Science Series, II. Mathemetics, Physics and Chemistry*, Kluwer Academic Publishers, **Vol. 79**, pp. 261–282.
15. Newton, R.G. (1966) Scattering Theory of Waves and Particles, McGraw-Hill, New York.
16. Varshalovich D.A., Moskalev A.N., Khersonskii V.K. (1988) Quantum Theory of Angular Momentum, World Scientific, Singapore.
17. Videen, G. (2002) In: The 6th International Conference on Electromagnetic and Light Scattering by Non-spherical Particles, ed. Gustafson, B., Kolokolova, L., and Videen, G., Army Research Laboratory, Adelphi, Maryland, p. 223.
18. Videen, G. (2002) Polarization opposition effect and second-order ray tracing, *Appl. Opt.*, **Vol. 41**, pp. 5115-5121.
19. Muinonen, K., and Lumme, K. (1991) In: IAU Colloquium 126, Origin and Evolution of Dust in the Solar System, ed. Levasseur-Regourd, A.-C., and Hasekawa, H., Kluwer Academic Publishers, Japan, p. 159.
20. Coulson, K.L., Dave, J.V., and Sekera, Z. (1960) Tables Related to Radiation Emerging from a Planetary Atmosphere with Rayleigh Scattering, University of California Press, Berkeley & Los Angeles.
21. Muinonen, K. (2002) In: The 6th International Conference on Electromagnetic and Light Scattering by Non-spherical Particles, ed. Gustafson, B., Kolokolova, L., and Videen, G., Army Research Laboratory, Adelphi, Maryland, p. 223.
22. Tsang, L., Kong, J.A., and Shin, R.T. (1985) Theory of Microwave Remote Sensing, Wiley, New York.
23. Muinonen, K. (1994) Coherent backscattering by solar system dust particles, In: Asteroids, Comets and Meteors 1993, ed. Milani, A., Di Martino, M., and Cellino, A., Kluwer Academic Publishers, Dordrecht, pp. 271–296.

List of experiments to be done

Exp. 1. Influence of EBS on measurements of dust and aerosol content in turbul. atmosp.

Exp 2. Effect of far correlations under light backscattering in a turbulent atmosphere

Exp. 3. Effect of the partial reverse of the wave front in a turbulent atmosphere

Exp. 4. Remote measurements of the oceanic turbulence

Exp. 5. Shift of the coherent summing point in a turbulent atmosphere due to wind

Exp. 6. Measurements of EBS factor reducing due to time-delay

Exp. 7. Cumulative effect of turbulence and reflecting surface form on EBS factor

Exp. 8. Measurements of air bubbles density in the sea water

Exp. 9. EBS from bodies in the underwater waveguides

Exp. 10. Measurements of $\bar{\bar{I}}/\bar{I}$ ratio (coupled / single)

Exp. 11. Enhanced FORWARD scattering (EFS) and Sun disk profile

Exp. 12. Revealing of the internal structure of half transparent bodies

Exp. 13. Backscattering from the surface with high and steep roughnesses.

Exp. 14. EBS from steep sea waves of mesoscale spectrum

Exp. 15. EBS from aircraft, ship, ...

Exp. 16. Enhanced time-delay and angle of arrival fluctuations

Contact address:

Kravtsov Yury

<kravtsov@asp.iki.rssi.ru>

<kravtsov@wsm.szczecin.pl>

List of "experiments to be done" compiled by Yurii Kravtsov in Cargèse

SCATTERING OF *SH* WAVES IN MEDIA WITH CRACKS AND ELASTIC INCLUSIONS

URSULA ITURRARÁN-VIVEROS
Instituto Mexicano del Petróleo, Eje Central Lázaro Cárdenas #152, C.P. 07730, México D.F., MÉXICO

AND

FRANCISCO J. SÁNCHEZ-SESMA
Instituto de Ingeniería U.N.A.M. Ciudad Universitaria, Apdo. 70-472, C.P. 04510, México D.F., MÉXICO

Abstract. We study scattering effects caused by 2-D cracks and elastic inclusions. To this aim we use two different numerical approaches. On the one hand we apply the Indirect Boundary Element Method (IBEM) to model the diffraction of *SH* waves by arbitrary shaped, zero-thickness cracks. We validate the IBEM with an analytic solution for a simple case. On the other hand we use the Direct Solution Method (DSM) to study scattering of *SH* waves by elastic inclusions. The DSM is based on solving the weak form (Galerkin formulation) of the elastic equation of motion. Although both methods work in the frequency domain, we show results in the time domain obtained by means of the Fast Fourier Transform.

1. Introduction

Fracture systems might provide significant hydrocarbon storage and directional permeability, so in many reservoirs their identification and characterization can be exploited to improve production and increase economic potential. For example, such information can be used to optimize production from horizontal wells by guiding their drilling perpendicular to aligned fracturing which will provide higher yields than wells drilled parallel to fractures. Additional fracture characterization could help to guide waterflood-assisted production and identify compartments with good flow properties. Crack's orientation, size and location may be inferred from their effects

B.A. van Tiggelen and S.E. Skipetrov (eds.),
Wave Scattering in Complex Media: From Theory to Applications, 583–593.

on the seismic waves recorded in controlled experiments. In order to study scattering effects caused by 2-D cracks we use the Indirect Boundary Element Method (IBEM). We validate the method and then we apply it to study scattering by arbitrary-shaped cracks where no analytic solutions are known. This method can be extended to in-plane compression, vertical polarized shear waves (P-SV) and 3-D cases. On the other hand we use the Direct Solution Method (DSM) (Geller & Ominhato 1994) to study diffraction of horizontally polarized shear waves (SH-waves) by 2-D elastic inclusions. The DSM is based on solving the weak form (Galerkin formulation) of the elastic equation of motion. We employ a set of cosines as trial functions enabling us to manipulate the Discrete Cosine/Sine Fourier Transforms (DCT/DST) to speed up computations.

2. The IBEM to Study Scattering by 2-D Cracks

In this section we proceed with the formulation of the problem of scattering by cracks. We use a multi-regional approach that consists of splitting the boundary into two identical boundaries so the crack lies in the interface. Let us consider wave propagation in an elastic homogeneous isotropic medium. If the reference system is taken such that the x_2-axis is perpendicular to the x_1x_3-plane, the antiplane displacement $\mathbf{v}$ on this plane is a solution of the scalar wave equation. The total displacement field can be expressed as superposition of an incident or reference field, plus a diffracted field, i.e.

$$\mathbf{v} = \mathbf{v}^{(0)} + \mathbf{v}^{(d)} ; \tag{1}$$

where the superscripts (0) and (d) represent the reference and diffracted fields, respectively. The incident field is analytically known and for a harmonic plane wave is given by $\mathbf{v}^{(0)} = \mathbf{v}_0(\omega)\exp(ikx\sin\gamma - ikz\cos\gamma)$, where $\mathbf{v}_0(\omega)$ is the wave amplitude, ω is the circular frequency, k is the wavenumber ($k = \omega/\beta$), β is the SH-wave velocity and γ is the incident angle. The time dependence of $\mathbf{v}$, *i.e.* $\exp(i\omega t)$, is omitted here and hereafter. The goal is to find an adequate algorithm to compute the unknown diffracted field or to find $\mathbf{v}$ directly. To this end, the plane in which the crack was embedded is divided into two complementary 2-D subdomains, being S^+ and S^- their boundaries. The illuminated surface S^+ is the boundary which is first struck by the incident wave. The shaded surface is denoted by S^-. The physics of the problem is enforced by appropriate boundary conditions along S^+ and S^-. This coupling of subdomains approach is designed to avoid the appearance of hyper-singular integrals that require a more involved treatment. Therefore, we need to impose suitable boundary conditions for each case. We impose zero tractions at the points $\mathbf{x} \in S$ such that $\mathbf{x}$ is on the crack. On the other hand we impose continuity of tractions

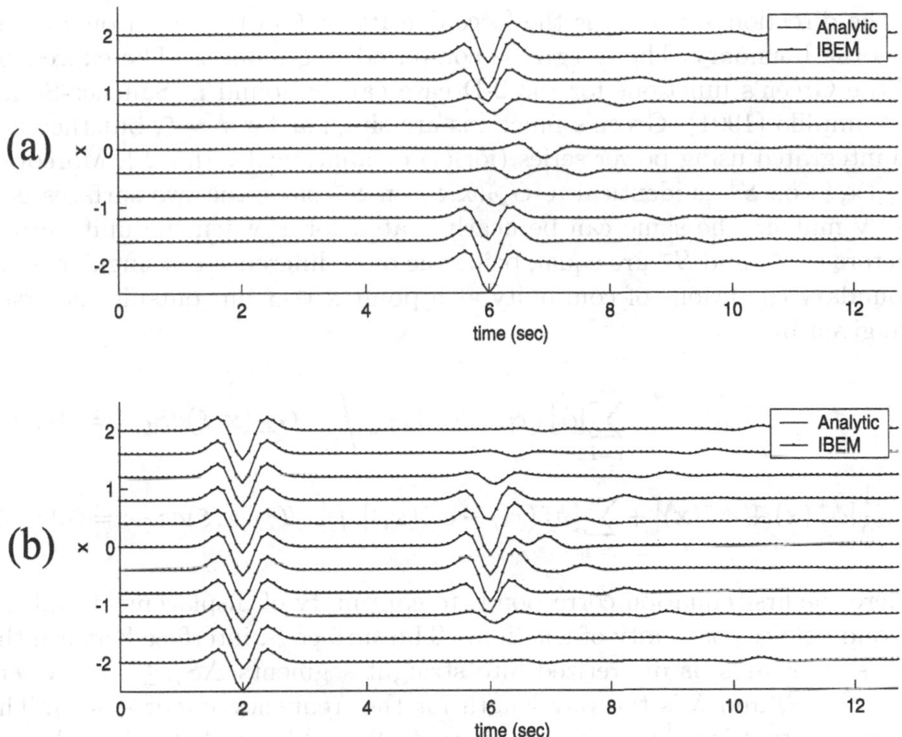

Figure 1. Seismograms for an incident wave-front with incidence angle $\gamma = 0°$. The incident time signal is a Ricker wavelet with characteristic period $t_p = 1.0$ s. The agreement between results obtained with the analytic solution and with the IBEM is excellent. The finite crack is located within the interval $[-1, 1]$. (a) Seismograms for a line of equally spaced receivers located at $z = -2$. Since these stations are on the shaded side of the crack the incident wave-front is perturbed. (b) A line of equally spaced receivers is located at $z = 2$. The incident wave-front reaches stations on the illuminated side before the crack. Therefore the wave-front does not have perturbations. After 4 s the reflected and diffracted waves arrive.

and displacements for those points $\mathbf{x} \in S$ but outside the crack. Following Sánchez-Sesma & Campillo (1991) for each one of the domains, associated to boundaries S^+ and S^-, the IBEM equations are

$$v_2^{(d)}(\mathbf{x}) = \int_S \phi_2(\xi) G_{22}(\mathbf{x}, \xi)\, dS_\xi\,, \tag{2}$$

$$t_2^{(d)}(\mathbf{x}) = \int_S \phi_2(\xi) T_{22}(\mathbf{x}, \xi)\, dS_\xi; \tag{3}$$

where $v_2^{(d)}$ $(t_2^{(d)})$ is the diffracted displacement (traction); $G_{22}(\mathbf{x}, \xi)$ $(T_{22}(\mathbf{x}, \xi))$ is the displacement (traction) Green's function, *i.e.* the displacement (traction) at a point $\mathbf{x}$ along the direction x_2 caused by a unit force at a point ξ in the direction x_2; $\phi_2(\xi)$ is the force density at ξ in the direction x_2; and S is the boundary. The integral is computed in ξ along S. The expression of the Green's functions for the 2-D case can be found in Sánchez-Sesma & Campillo (1991). Green's functions are singular for $x = \xi$, but they can be integrated using power series (for G) or analytically (for T). Moreover, $G_{22}(\mathbf{x}, \xi)$ on S^+ is identical to $G_{22}(\mathbf{x}, \xi)$ on S^- since the two surfaces perfectly match. The same can be nearly stated for T when the unit normal vectors to S^+ and S^- are equal, being the only difference the singular term. Boundary conditions of continuity at a point $\mathbf{x} \in S$ but outside the crack are given by

$$\sum_{i=1}^{N}[\phi_2^+(x_i) - \phi_2^-(x_i)] \int_{\Delta S_i} G_{22}(\mathbf{x}, \xi) dS_\xi = 0 \,, \quad (4)$$

$$\frac{1}{2}[\phi_2^+(\mathbf{x}) + \phi_2^-(\mathbf{x})] + \sum_{i=1}^{N}[\phi_2^+(x_i) - \phi_2^-(x_i)] \int_{\Delta S_i} T_{22}(\mathbf{x}, \xi) dS_\xi = 0 \,; \quad (5)$$

where the first equation corresponds to continuity of displacement and the second defines continuity of traction. The surface S, interface between the two subdomains, is discretized into straight segments $\Delta S_i \le \lambda/6$ (where $S = \sum \Delta S_i$ and λ is the wavelength for the frequency under study). The force density ϕ is assumed to be constant along ΔS_i so that it is evaluated just at its middle point x_i. The integral equation (4) is calculated expanding G in power series when $\mathbf{x} \in \Delta S_i$. The integral in equation (5) is null when $\mathbf{x} \in \Delta S_i$, since the singularity is explicitly solved and expressed by the first term of the equation. Numerical realization of boundary conditions at a point $\mathbf{x}$ on the crack is given by

$$\frac{1}{2}\phi_2^+(\mathbf{x}) + \sum_{i=1}^{N} \phi_2^+(x_i) \int_{\Delta S_i} T_{22}(\mathbf{x}, \xi) dS = -t_2^{(0)+}(\mathbf{x}) \,, \quad (6)$$

$$-\frac{1}{2}\phi_2^-(\mathbf{x}) + \sum_{i=1}^{N} \phi_2^-(x_i) \int_{\Delta S_i} T_{22}(\mathbf{x}, \xi) dS = -t_2^{(0)-}(\mathbf{x}) \,. \quad (7)$$

Equations (4)-(7) allow us to form a non-singular system of linear equations for which there is a unique solution. Once the unknown force densities are found one can substitute them into equation (2), properly discretized, to obtain the displacement at any point $\mathbf{x}$ along the crack. The crack opening displacement (COD) and the Somigliana's representation theorem are used to extend the solution to any point within the medium. Next we consider an analytic solution for a zero-thickness crack (Sánchez-Sesma & Iturrarán-Viveros 2001). This solution is based on the solution for a semi-infinite crack

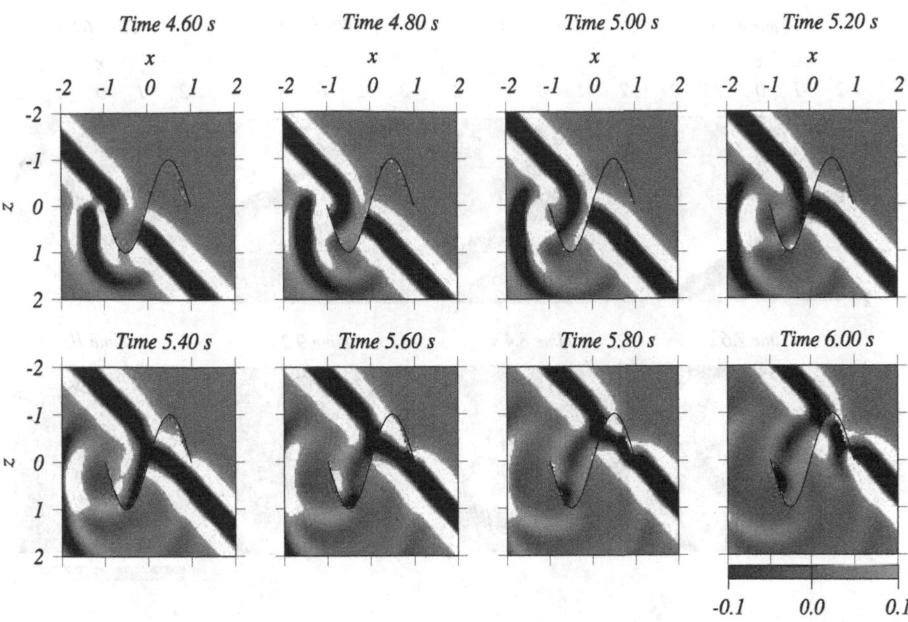

Figure 2. Snapshots for an SH plane wave with incidence angle $\gamma = 45°$ that strikes a sine-shaped crack. A square grid of 101×101 equally spaced receivers within a square of length $4a$. The crack is located at the center of the square in the interval $(-a, a)$, where $a = 1$. The incident time signal is a Ricker wavelet with characteristic period $t_p = 1.0$ s and $t_s = 5.0$ s. The effect of diffraction on the *shaded* side creates a gap or a shadow. On the other hand, diffraction can be observed on the *illuminated* side. As time increases the wavefront recovers and the diffraction effect caused by the crack on the plane wave disappears. The gray-scale is out of range to enhance diffraction effects.

due to Sommerfeld (1949) and enables us to test the IBEM. In Fig. 1 we show comparisons with excellent agreement between the results obtained with the analytic solution and the IBEM. From these comparisons we are confident that the method works fine and we show some simulations for 2-D, arbitrary-shaped cracks. In Figures 2 and 3 we show snapshots for a sine-shaped and semi-circular crack, respectively. Note that reflected and diffracted waves have particular marks depending on the shape of the crack and the incident angle. Since the contrast of medium properties within a cracked medium is very sharp we are not able to use DSM to this purpose. However, to study scattering effects caused by elastic inclusions it is more suitable to use DSM than IBEM. In the next section we briefly explain the DSM and show some results.

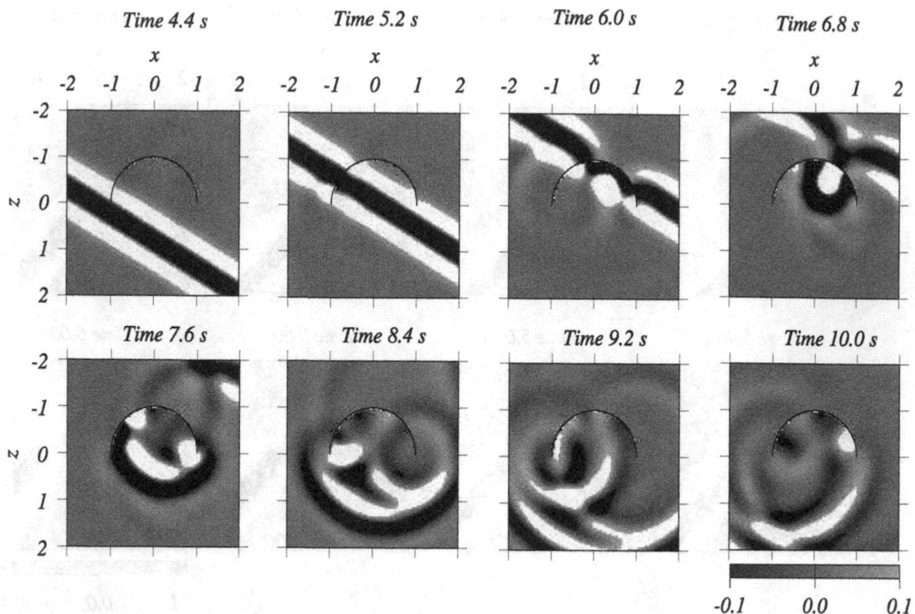

Figure 3. Snapshots for an SH plane wave with incidence angle $\gamma = 30°$ that strikes a semi-circular crack with radius $r = 1$ and zero-thickness. The incident time signal is a Ricker wavelet with characteristic period $t_p = 1.0$ s and $t_s = 5.0$ s. Note that the cylindrical waves created at the edges bounce back and forth along the crack until their energy is lost.

3. The Direct Solution Method

In this section we first consider a 1-D layered medium and show how the DSM can be used to find displacements in the frequency domain under an incident plane wave. We represent the displacement v as

$$v(z) = v^{(0)}(z) + V(z) \tag{8}$$

where $v^{(0)}$ is a known driving field given by

$$v^{(0)} = v_0(\omega)e^{+ik_z z} \tag{9}$$

where the vertical wave-number is given by $k_z = \frac{\omega}{\beta}\cos\gamma$. The dependence on x and t is given by the factor $\mathrm{e}^{+i\omega t - ik_x x}$ which for simplicity, is omitted hereafter. The function $V(z)$ is a function of depth represented as a linear combination of trial functions

$$V(z) = \sum_{n=0}^{N-1} c_n \psi^n(z) \tag{10}$$

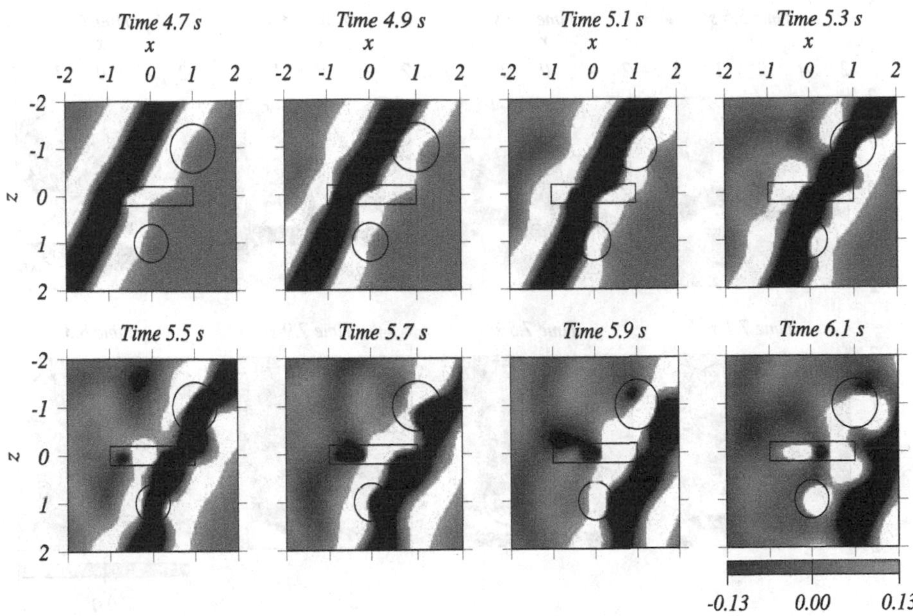

Figure 4. Snapshots for a mesh of 101 × 101 equally spaced receivers. Two cylinders and a rectangular elastic inclusions within a square medium of size $H = L = 4$ m. The elastic parameters are $\beta = 1$ m/s, $\rho = 1$ gm/cm^3 (the surrounding medium), for the cylinders $\beta_1 = \beta_2 = 0.8$ m/s, $\rho_1 = \rho_2 = \rho$, ratios $r_1 = 0.4$ m, $r_2 = 0.52$ m and for the rectangular inclusion we have $\beta_3 = 0.7$ m/s, $\rho_3 = \rho$ for (x, z) such that $-1 \leq x \leq 1$ and $-0.25 \leq z \leq 0.25$. We used 45 × 45 cosine trial functions and 128 × 128 integration points.

where N is the number of trial functions along the z direction. The expansion coefficients c_n are the unknowns and the set of orthogonal trial functions is defined by:

$$\psi^n(z) = \cos\left(\frac{n\pi z}{H}\right) \tag{11}$$

where H is the size of the model along z. Observe that we consider the displacement v in (8) as a series of trial functions plus a driving field (9) which represents a plane wave. Apart from this slight difference we follow the derivations given in Geller & Ohminato (1994) and Geller & Takeuchi (1995). If an antiplane stress field is applied on a direction parallel to the y-axis, the 1-D weak form of the Method of Weighted Residuals (MWR) for the elastic equation of motion, in Cartesian coordinates, in the spectral

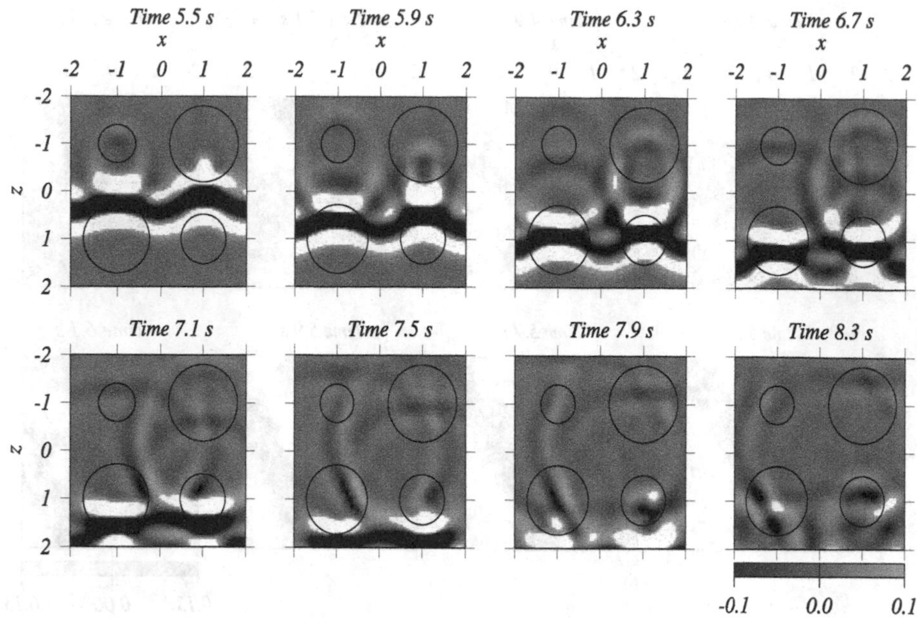

Figure 5. Snapshots for a mesh of 101 × 101 equally spaced receivers. Four cylinders within a square medium of size $H = L = 4$ m. The elastic parameters are $\beta = 1$ m/s, $\rho = 1$ gm/cm^3 (the surrounding medium), for the cylinders $\beta_1 = \beta_2 = \beta_3 = \beta_4 = 0.8$ m/s, $\rho_1 = \rho_2 = \rho_3 = \rho_4 = \rho$, ratios $r_1 = 0.4$ m, $r_2 = 0.52$ m, $r_3 = 0.8$ m and $r_4 = 0.72$ m. We used 45 × 45 cosine trial functions and 128 × 128 integration points. The wave front attenuates while traversing the cylinders as do the interactions between the scattered waves.

domain (ω, k_x, z) is given by

$$\int_0^H [\psi^m(\rho\omega^2 - k_x^2\mu)v - \psi^m_{,z}\,\mu\,v_{,z}]dz + R(m) = -\int_0^H \psi^m f dz \quad (m = 0, \ldots, M-1). \tag{12}$$

where the subscript '$_{,z}$' denotes differentiation with respect to z, v is the particle displacement in the y-direction, f represents the body forces, ρ is the density, μ is a Lamé constant that corresponds to the shear modulus (these four parameters are depth-dependent) and the auxiliary function $R(m)$ is given by

$$R(m) = \psi^m(z)\mu v_{,z}(z)\Big|_{z=H} - \psi^m(z)\mu v_{,z}(z)\Big|_{z=0} \tag{13}$$

Eq. (12) can be written in matrix form, substituting (8) and (10) to obtain

$$(\omega^2\mathbf{T} - \mathbf{H} + \mathbf{R})\mathbf{c} = -\mathbf{g} - (\omega^2\mathbf{T}^0 - \mathbf{H}^0 + \mathbf{R}^0) \tag{14}$$

where $\mathbf{T}$ is the kinetic energy matrix, $\mathbf{H}$ is the potential energy matrix, $\mathbf{R}$ is the matrix operator that includes the natural boundary conditions and $\mathbf{g}$ is the vector containing body forces, which for simplicity are neglected in this paper. The superscript '0' represents the matrices where the incident or reference field $v^{(0)}$ (or its derivative) is involved. The explicit matrix and vector elements for the SH case are given by:

$$T_{mn} = \int_0^H \psi^m \rho \psi^n dz, \qquad T_m^0 = \int_0^H \psi^m \rho v^{(0)} dz \tag{15}$$

$$H_{mn} = k_x^2 H_{mn}^{(1)} + H_{mn}^{(2)}, \qquad H_m^0 = k_x^2 H_m^{0(1)} + H_m^{0(2)} \tag{16}$$

$$g_m = \int_0^H \psi^m f dz \tag{17}$$

where

$$H_{mn}^{(1)} = \int_0^H \psi^m \mu \psi^n dz, \qquad H_m^{0(1)} = \int_0^H \psi^m \mu v^{(0)} dz \tag{18}$$

$$H_{mn}^{(2)} = \int_0^H \psi_{,z}^m \mu \psi_{,z}^n dz, \qquad H_m^{0(2)} = \int_0^H \psi_{,z}^m \mu v_{,z}^{(0)} dz \tag{19}$$

$$R_{mn} = -ik_z \psi^m(H) \mu \psi^n(H) \tag{20}$$

$$R_m^0 = -ik_z \psi^m(H) \mu v^{(0)}(H) + ik_z \psi^m(0) \mu v^{(0)}(0) \tag{21}$$

where Eqs. (20) and (21) serve to change the natural boundary conditions from a free-surface to an absorbing boundary condition. The next step is to compute the integrals and solve a linear system of equations to obtain the unknown coefficients in (10). Since the trial functions used in this scheme are cosine basis functions we are able to apply the Discrete Cosine Transform (DCT) and the Discrete Sine Transform (DST) to compute integrals (15), (16), (17), (18) and (19) in a faster way. Let N_I be the number of integration points along x or z. Using the Discrete Fourier Transform (DFT), the number of operations to compute the integrals is of the order $O(N_I \log N_I)$ whereas with other strategy is of the order $O(N_I^2)$. Similarly, this approach is extended for the 2-D case. In the 2-D formulation four natural absorbing boundary conditions result. The elastic parameters depend now on both x and z. In Fig. 4 we show two cylinders and a rectangular inclusion within a square medium of size $H = L = 4$ m. We observe that the incident wave-front is perturbed and scattering is produced by the presence of elastic inclusions. A similar experiment is presented in Fig. 5 where the cylinders at the top act as a barrier that attenuates the wave that reaches the second row of heterogeneities. Part of the energy trapped inside each cylinder comes out and produces diffraction, the rest produces diffraction that bounces back and forth until the energy is lost. As time increases the wavefront recovers and the scattering effect caused by the inclusions on the plane wave diminishes.

4. Conclusions

We have tested the IBEM to solve scattering of SH waves by a zero-thickness crack. The method was tested against an analytical solution, for a canonical case, obtaining an excellent agreement. The IBEM allows us to study media with arbitrary shaped cracks. Regarding the study of elastic wave propagation by elastic inclusions, we have shown the feasibility of simulate wave propagation in heterogeneous media using the DSM. A current limitation of the DSM is the fact that the minimum wavelength used has to be at least twice the size of the model divided by the number of trial functions. This limits the resolution because the size and number of heterogeneities together with its elastic properties are constrained by the number of trial functions used. If more trial functions are needed the resulting systems of linear equations become very large. The use of cosine basis functions allows us to use the Fast Fourier Transform to compute more efficiently the integrals needed to fill the matrix corresponding to the linear system to be solved. This helps alleviate the computational burden. Parallel computing for methods in the frequency domain is much simpler than time domain computation based on domain decomposition. This is an advantage of the two methods presented here and it will help to study more realistic problems in the future.

Acknowledgments

We thank R. Vai for providing us with an IBEM Fortran program and also for her comments that helped to improve the manuscript. We thank T. Furumura for his valuable comments. This work has been partially supported by Instituto Mexicano del Petróleo under project D.00117, by DGAPA-UNAM, México under grant IN121202 and by Conacyt México under project NC-204.

References

Geller, R. & T. Ohminato: 1994, Computation of synthetic seismograms and their partial derivatives for heterogeneous media with arbitrary natural boundary conditions using the Direct Solution Method, *Geophys. J. Int.* **116**, pp. 421-446.

Geller, R. & N. Takeuchi: 1995, A new method for computing highly accurate DSM synthetic seismograms, *Geophys. J. Int.* **132**, pp. 449-470.

Sánchez-Sesma, F. J., & M. Campillo: 1991, Diffraction of *P*, *SV* and Rayleigh waves by topographic features: a boundary integral formulation, *Bull. Seis. Soc. Am.* **81**, pp. 2234-2253.

Sánchez-Sesma, F. J.,& U. Iturrarán-Viveros: 2001, Scattering and diffraction of *SH* waves by a finite crack: an analytical solution, *Geophys. J. Int.* **145**, pp. 749-758.

Sommerfeld, A.: Optics Vol. IV, Lectures on Theoretical Physics. Academic Press, New York, USA, 1949.

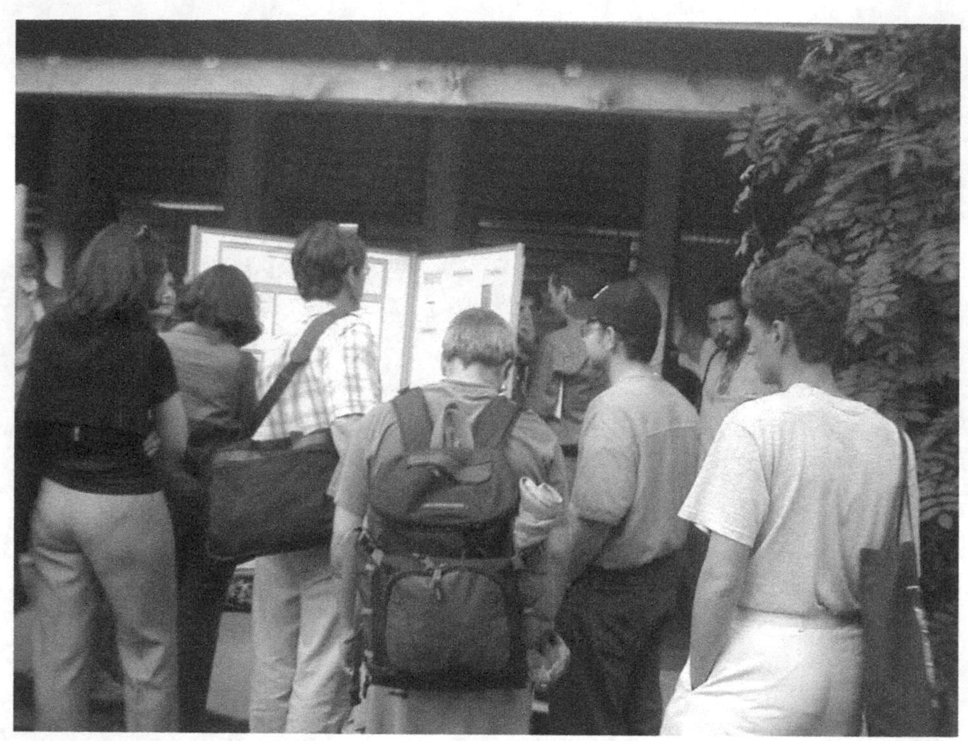

Outdoor poster session. Photo by Gabriel Cwilich

CHAPTER VIII

COMMUNICATION IN A DISORDERED WORLD

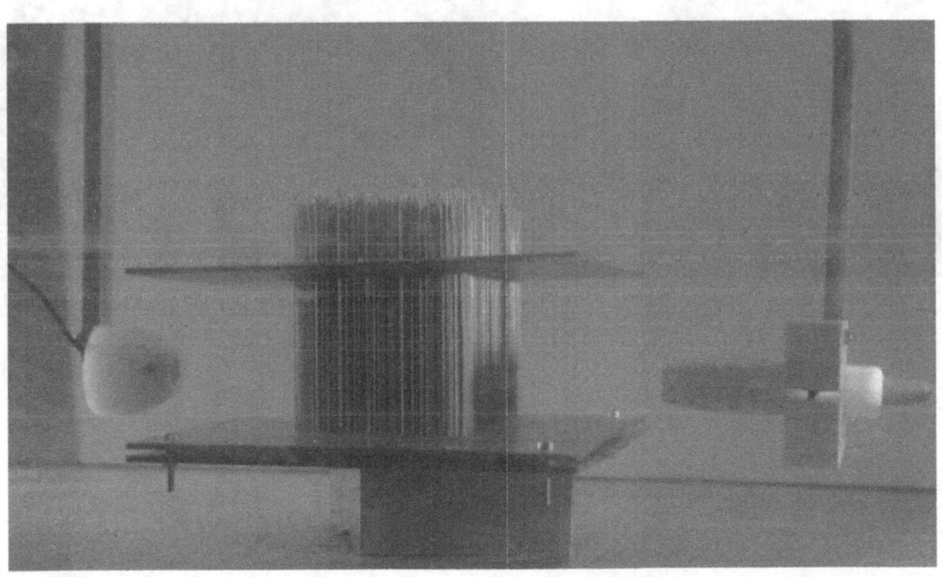

Experimental setup for time-reversal communication through a forest of rods. Image provided by Arnaud Tourin, ESPCI, Paris, France.

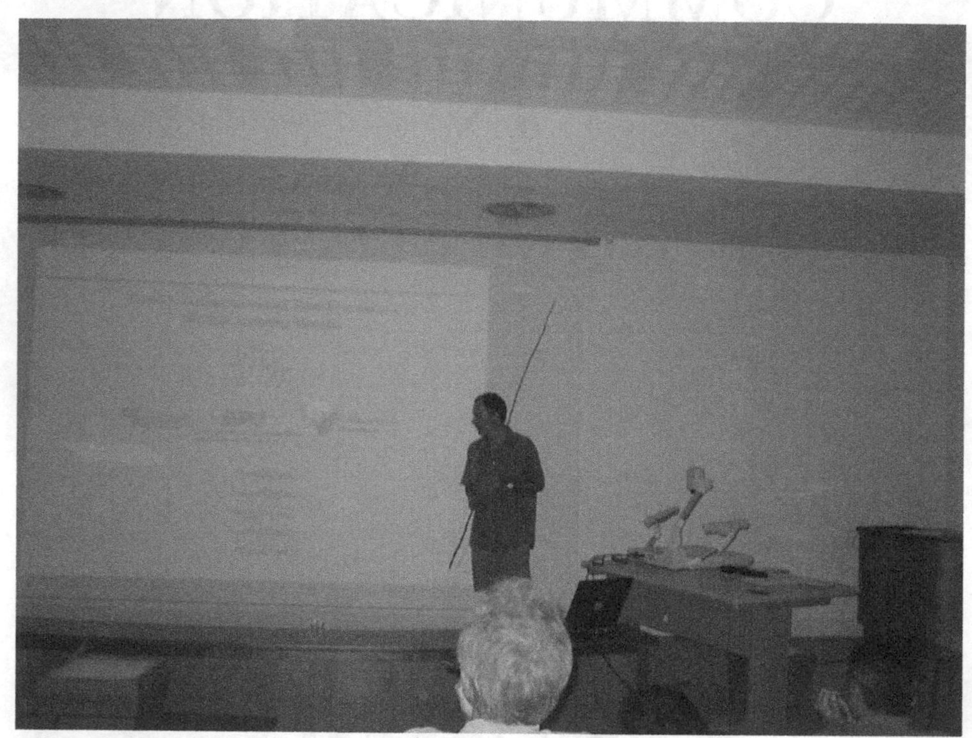

Arnaud Tourin. Photo by Gabriel Cwilich

DIGITAL COMMUNICATION WITH TIME-REVERSAL IN A MULTIPLE SCATTERING MEDIUM

A. TOURIN, A. DERODE, J. DE ROSNY,
M. TANTER AND M. FINK
Laboratoire Ondes et Acoustique, ESPCI, Universite Paris VII UMR 7587, 10 rue Vauquelin 75005 Paris, France

Abstract. We present an experimental demonstration showing that contrary to first intuition the more scattering a mesoscopic medium is, the more information can be conveyed through it. We used a MIMO (Multiple Input Multiple Output) configuration: a multi-channel ultrasonic time-reversal antenna is used to transmit random series of bits simultaneously to different receivers which were only a few wavelengths apart. Whereas the transmission is free of error when multiple scattering occurs in the propagation medium, the error rate is huge in a homogeneous medium. This is discussed in relation with the number of eigenvalues of the time-reversal operator in a homogeneous or multiple scattering medium. This number is strongly related to the number of focal spots that a transmitting antenna can create in a given region of space. For communication purposes, it gives the number of independent receivers an antenna can talk to; more generally in mesoscopic physics, it is related to the conductance of the propagation medium.

1. Introduction

The ever-growing need for faster wireless communication in urban areas has brought to light interesting connections between issues of communication engineering (how can one optimize the data transfer rate between N antennas and M receivers?) and mesoscopic wave physics. Indeed both subjects have to do with the correlations (in space, time, and frequency) of a diffuse wave field propagating in a random medium. When waves propagate through a disordered medium, the key element is the ability of a communication system to exploit independent channels of propagation [6]; very

B.A. van Tiggelen and S.E. Skipetrov (eds.),
Wave Scattering in Complex Media: From Theory to Applications, 597–605.

roughly, the more heterogeneous and scattering the medium is, the more degrees of freedom there are to communicate through it. In the case of radio signals for instance, multiple paths arise because of scattering and multiple reverberation on the buildings or indoors; similar multipath phenomena are also frequently encountered in underwater acoustics. For radio signals, scattering permits to exploit different polarizations of the electro-magnetic waves to convey more information than in free space: Andrews [2] showed that the data transfer rate in a scattering environment could be increased 6 times, as long as the six components of the electromagnetic field are uncorrelated with each other, which therefore yields 6 independent channels. This approach is based on the polarization decorrelation. Another approach is to exploit spatial decorrelation: with N antennas transmitting to M receivers, with $N > M$, Shannon's capacity (i.e. the maximum amount of information conveyed without any error) can be N times higher than with a single-transmitter and single-receiver scheme [6, 8, 10]. We demonstrate here the efficiency of another approach based not only on spatial decorrelation but also on frequency decorrelation due to the randomness of the medium. Indeed, unlike what happens in a homogeneous medium (free space), the response of a highly scattering medium can change dramatically even if the frequency is changed by a small amount $\delta\omega$. This opens up another way to exploit the variety of communication channels in a scattering medium. To that end, the bandwidth $\Delta\omega$ has to be as large as possible compared to the correlation frequency $\delta\omega$. Moreover, the waveforms that are sent or received must be completely controllable both in amplitude and phase all along the bandwidth, in order to perform a coherent treatment. For the time being, in the gigahertz domain, electromagnetic antennas cannot achieve this. But our experimental demonstration uses ultrasonic waves, for which both conditions are met.

2. Experiment

The propagation medium we considered in the experiments is deliberately disordered and highly scattering: it is a forest of parallel steel rods with density $18.75/\text{cm}^2$ and diameter 0.8 mm ($\sim$ 1.7 times the wavelength). The typical distance between two scattering events is measured by the mean free path l. The sample thickness is $L = 40$ mm, much larger than l which was found to be 4 mm [12]. As a consequence, when a short ultrasonic pulse (typically 1 μs) is sent, the transmitted waveforms received on the other side of the slab last several hundreds of times the initial pulse duration. The experiment takes place in a water tank, and we try to communicate to five different receivers with a 23-element array (Fig. 1) at a 3.2-MHz central frequency. One simple way to address simultaneously different receivers is

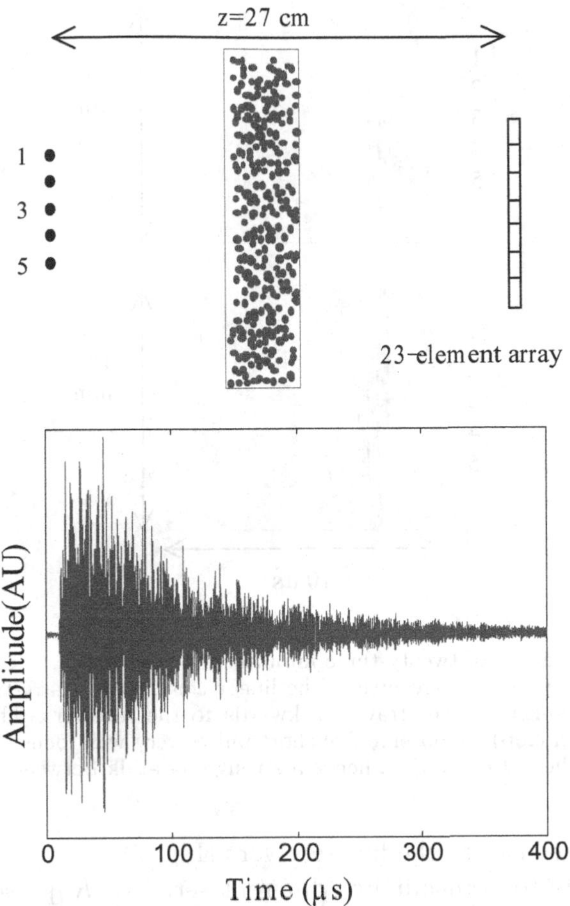

Figure 1. Top: Experimental set-up. An ultrasonic array (central frequency 3.2 MHz, pitch 0.4 mm, total aperture 9.2 mm) communicates to five points simultaneously through a multiple scattering medium. The distance between neighboring receivers is 2 mm. Bottom: typical waveform scattered by the medium when a 1-μs pulse is sent through it by the array to the receivers.

time-reversal (TR) focusing [5]. In a first step, the five receivers fire a 1-μs pulse one by one; five sets of $N = 23$ signals $h_{ij}(t)$ are recorded on the array and digitized. If, for example, the array sends back the time-reversed signals $h_{3j}(-t)$, $1 < j < N$, then the wave will focus back to the third receiver and recover its initial duration, as if it were traveling backwards in time. Since the system is linear, it is also possible to send back any combination of the $h_{ij}(-t)$; for instance, in order to transmit a series of positive and negative impulses like {+1 -1 -1 +1 -1} one has to send back $h_{1j}(-t) - h_{2j}(-t) - h_{3j}(-t) + h_{4j}(-t) - h_{5j}(-t)$. Then five short pulses will

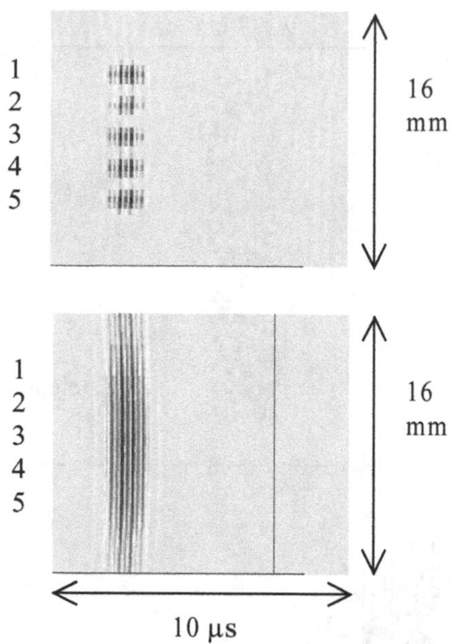

Figure 2. The five sets of twenty-three impulse responses $h_{ij}(t)$, $1 < i < 5$, $1 < j < 23$, have been recorded and time-reversed. The linear combination is re-transmitted by the array, and the resulting waves travel backwards to the receivers. Through a multiple scattering medium (top) we observe five short pulses very well focused on each receiver. Through water, the pulses overlap, hence a strong cross-talk between the receivers.

arrive simultaneously to the five receivers (Fig. 2).

If one wants to transmit five different series of K pulses with various signs, the signal to be sent back by the j-th element is $\sum_{k=1}^{K}\sum_{i=1}^{5} a_{ik}h_{ij}(-t+kT)$, with a_{ik} the amplitude of the k-th pulse to be transmitted to the i-th receiver and T the delay between two pulses. We used this approach to transmit simultaneously 5 random and uncorrelated series of $K = 2000$ bits each to the 5 receivers, with $N = 23$ transmitters. The result was that all bits were correctly transmitted through the scattering medium, whereas the error rate was 28% through water. The reason for these errors is cross-talk between the receivers. Indeed, in a homogeneous medium such as water, the resolution of the array is diffraction-limited: the array size is $D = 9.2$ mm, the wavelength $\lambda = 0.47$ mm, and the distance $z = 27$ cm, so $\lambda z/D = 13.8$ mm.

Since the receivers are only 2 mm apart, the messages overlap. On the contrary, it was shown [3, 4] that due to multiple scattering, a finite-size TR array manages to focus a pulse back to the source with a spatial resolution that beats by far the diffraction limit in the homogeneous medium: it is only limited by the correlation length of the scattered field, and no longer by the

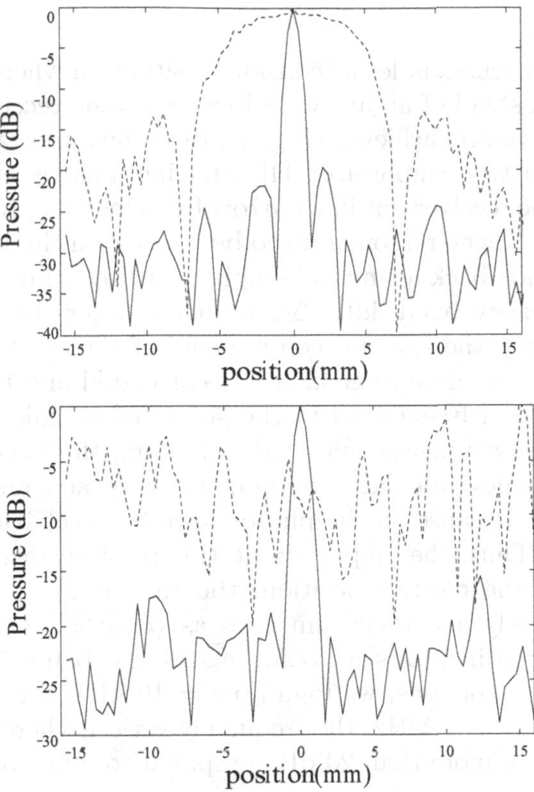

Figure 3. Directivity of the time-reversed waves around the desired focal point. Top: comparison between the multiple scattering medium (full line) and water (dotted line), for a 23-element array; the -6 dB widths are 1.1 mm and 10.3 mm. Bottom: comparison between quasi-monochromatic (3.2 MHz, dotted line) and wideband (full line) focusing through the multiple scattering medium, for a single element.

array aperture; as the wave is completely scrambled by multiple scattering and loses its coherence, the correlation length of the scattered waves is of the order of the wavelength, and so is the spatial resolution of the focal spot. As to the peak-to-noise ratio of the focal spots, it depends obviously on the number of receivers since each of them receives its bitstream, plus noise due to the bitstreams going to all the other and on the number of transmitting elements. It also highly depends on the bandwidth as we now argue.

3. Discussion

To simplify the discussion, let us consider a situation where only one transmitter is used, instead of an array, to focus on some receiver. In this case, TR focusing can be still achieved (Fig. 3). Here the importance of frequency decorrelation must be emphasized [4]. Imagine a single element trying to focus on the same receiver but in a narrow frequency bandwidth; the phase-conjugated wave has no reason at all to be focused on this receiver since the element only sends back a sinusoidal spherical wave through the medium. But if the frequency bandwidth $\Delta\omega$ is much larger than the correlation frequency $\delta\omega$, then the spectral components of the scattered field at two frequencies apart by more than $\delta\omega$ are decorrelated and there are roughly $\Delta\omega/\delta\omega$ decorrelated frequencies in the scattered signals. When we time-reverse (i.e. phase-conjugate coherently all along the bandwidth, and not just at one frequency) all these components, they add up in phase at the receiver position, because all the phases have been set back to 0 all along the bandwidth. Thus, the amplitude at this position increases as $\Delta\omega/\delta\omega$ whereas outside the receiver position, the various frequency components add up incoherently and their sum rises as $\sqrt{\Delta\omega/\delta\omega}$. On the whole, the peak-to-noise ratio increases as $\sqrt{\Delta\omega/\delta\omega}$ as the bandwidth is enlarged. Through the forest of rods, we found $\delta\omega = 10$ kHz; the total bandwidth at half-maximum is 1.5 MHz, the frequency ratio is therefore ~ 150, thus an improvement of more than 20 dB compared to a monochromatic phase conjugation technique.

Therefore, the data transfer rate of a TR antenna highly depends both on the ratio $\Delta\omega/\delta\omega$, giving the number of uncorrelated frequencies, and on the number of independent focal spots it can generate in the receivers plane at one frequency. In a homogeneous medium [11] this number equals the rank of the propagation matrix that appears in Shannon's capacity, which we now discuss.

In the case of a one-to-one communication scheme, Shannon [9] proved the following theorem in 1948: if a receiver receives a complex-valued signal with a zero-mean Gaussian distribution (variance S) plus a complex noise with a zero-mean Gaussian distribution (variance N), then the maximum number of error-free bits that can be decoded is $\log_2(1+S/N)$. This number is expressed in bits per second and per Hertz, indicating that it grows linearly with $\Delta\omega T_0$, with $\Delta\omega$ the frequency bandwidth and T_0 the duration of the transmission. Shannon's formula was recently generalized [6] to the case of N transmitters and M receivers: if the signals transmitted by each element are uncorrelated, Shannon's capacity is given by $C = \log_2(\det[I + \rho^t H H^*])$ bits/s/Hz, with ρ the signal to noise ratio, I the identity matrix and H the $N \times M$ propagation matrix. H is the Fourier-transform of the point-to-point impulse responses $h_{ij}(t)$. Interestingly, ${}^tHH^*$ is the Time-Reversal Operator [11] which would describe a TR sequence between N transmitters and M receivers. Indeed, imagine that the N transmitters send N waveforms $e_i(t)(i = 1,..,N)$, which can be described in the Fourier domain by a set of vectors E with N components for each frequency. The M-component received vectors can be written as a matrix product HE. When the signals are time-reversed (i.e. phase conjugated in the Fourier domain) and re-transmitted, the resulting vector is ${}^tHH^*E^*$.

At this step, it is worth performing a singular value decomposition of H to get a more physical insight into the capacity: H can be written as $H = UD\ {}^tV^*$ where U and V are unitary matrices and D is a diagonal matrix whose elements are the singular values λ_i. Thus ${}^tHH^* = VD^2\ {}^tV^*$, which means that the eigenvalues of the time reversal operator are the squared singular values of the propagation matrix. Then C can be rewritten as $C = \log_2(\det[I + \rho VD^2\ {}^tV^*]) = \sum_{i=1}^{\min(M,N)} \log_2(1 + \rho\lambda_i^2)$. This formula implies that each significant eigenvalue of the time-reversal operator adds an independent channel of propagation which contributes to increase the capacity.

With ultrasonic devices, the propagation matrix H can be easily measured. To that end, we have used two 40-element 0.4 mm-pitch arrays at a distance $z = 27$ cm, and measured the 1600 inter-element impulse responses $h_{ij}(t)$ through water and through the multiple scattering sample described above. After a Fourier transform of $h_{ij}(t)$, H is known for a whole set of frequencies. For each frequency, we applied a singular value decomposition. The singular values of H are represented on figure 4.

Through the scattering medium, the number of singular values is much higher than through water and H has a higher rank. Physically, this increase is related to the possibility of talking to different receivers by focusing on them with an array. The number of significant singular values (or degrees of freedom) is roughly the number of independent receivers we can talk to through the medium in a given region of space and at a given frequency. However Shannon's formula is essentially monochromatic. In a wide-band coherent technique such as TR, the whole bandwidth must be taken into account. Then the total number of degrees of freedom is roughly the number of significant singular values at the center frequency multiplied by frequency ratio $\Delta\omega/\delta\omega$.

4. Conclusion

Our experimental results demonstrate that high-order scattering in a disordered medium can help increasing the information transfer rate, especially if the time-reversal technique is used to naturally focus the different bit-

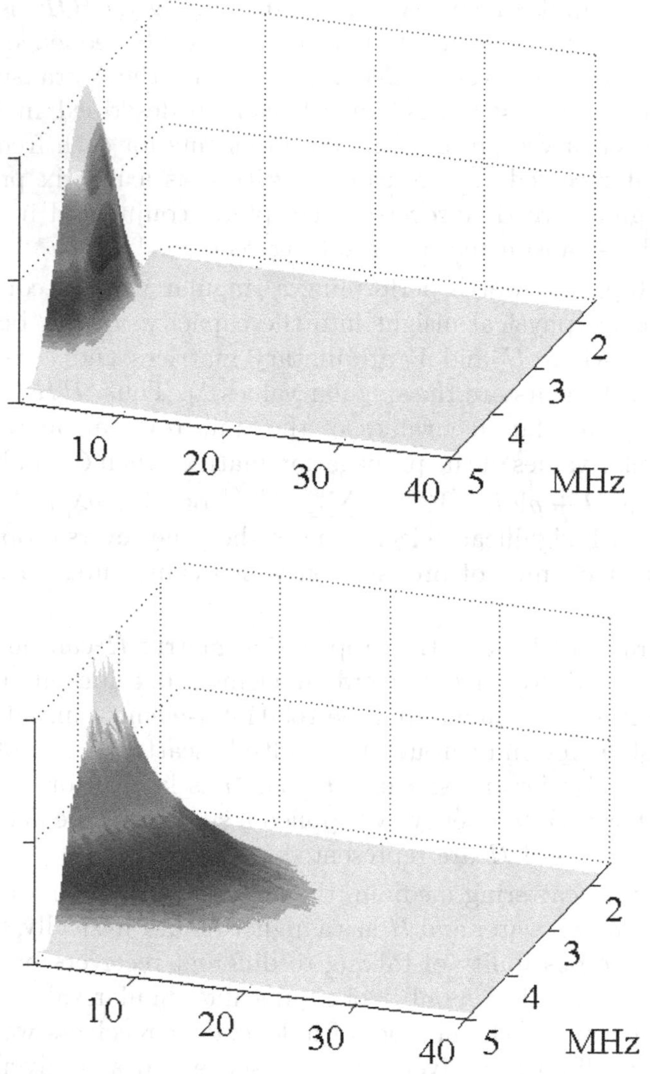

Figure 4. Singular values of the propagation operator H through water (top) and through the multiple scattering medium (bottom). At the central frequency (3.2 MHz), there are 34 singular values for the scattering medium and only 6 through water (with a -32 dB threshold relatively to the first singular value).

streams onto the receivers. The first key parameter in that experiment is the number of independent focal spots that can be created by the transmitting array in the receiving plane, which is also the number of different receivers one can address simultaneously. At a given frequency, this number is directly related to the number of significant singular values of the propagation matrix or the number of eigenvalues of the time-reversal operator which are basically the same. In the ultrasonic range, the time-reversal operator which is studied in mesoscopic physics and whose trace gives the conductance [7] can be easily measured. The second key parameter is the number of uncorrelated frequencies within the bandwidth, which governs the peak-to-noise ratio on each receiver; in the medium we studied, we have $\Delta\omega/\delta\omega \sim 150$. Could these ideas be applied to radio signals? Having a complete control of the field amplitude and phase over a very large frequency band is possible in everyday laboratory life for ultrasonic waves as well as in ocean acoustics [1], but for the time being, applying wide band coherent techniques such as time-reversal to electromagnetic waves remains a technical challenge.

References

1. Akal, T., Edelmann, G., Kim, S., Hodgkiss, W., Kuperman, W., and Song, H.C. (2000) Low and high frequency ocean acoustic phase conjugation experiments, European Conference on Underwater Acoustics.
2. Andrews, M., Mitra, P., and De Carvalho, R. (2001) *Nature* **409,** 316.
3. Blomgren, P., Papanicolaou, G., and Zhao, H.K. (2002) *J. Acoust. Soc. Am.* **111,** 230.
4. Derode, A., Tourin, A., and Fink, M. (2001) *Phys. Rev. E.* **64,** 036606.
5. Fink, M. (1999) *Scientific American* **281,** 67.
6. Foschini, G. and Gans, M. (1998) *Wireless Personal Communications* **6,** 311.
7. Garcia-Martin, A. and Saenz, J. (1999) *Phys. Rev. Lett.* **87,** 116603.
8. Moustakas, A., Baranger, H., Balents, L., Sengupta, A., and Simon, S. (2000) *Science* **287,** 287 (2000).
9. Shannon, C. (1948) *Bell Syst. Tech. J.* **27,** 379.
10. Simon, S., Moustakas, A., Stoychev, M., and Safar, H. (2001) *Physics Today* **54,** 38.
11. Tanter, M., Aubry, J.-F., Gerber, J., Thomas, J.-L., and Fink, M. (2001) *J. Acoust. Soc. Am.* **101,** 37.
12. Tourin, A., Derode, A., Peyre, A., and Fink, M. (2000) *J. Acoust. Soc. Am.* **108,** 503.

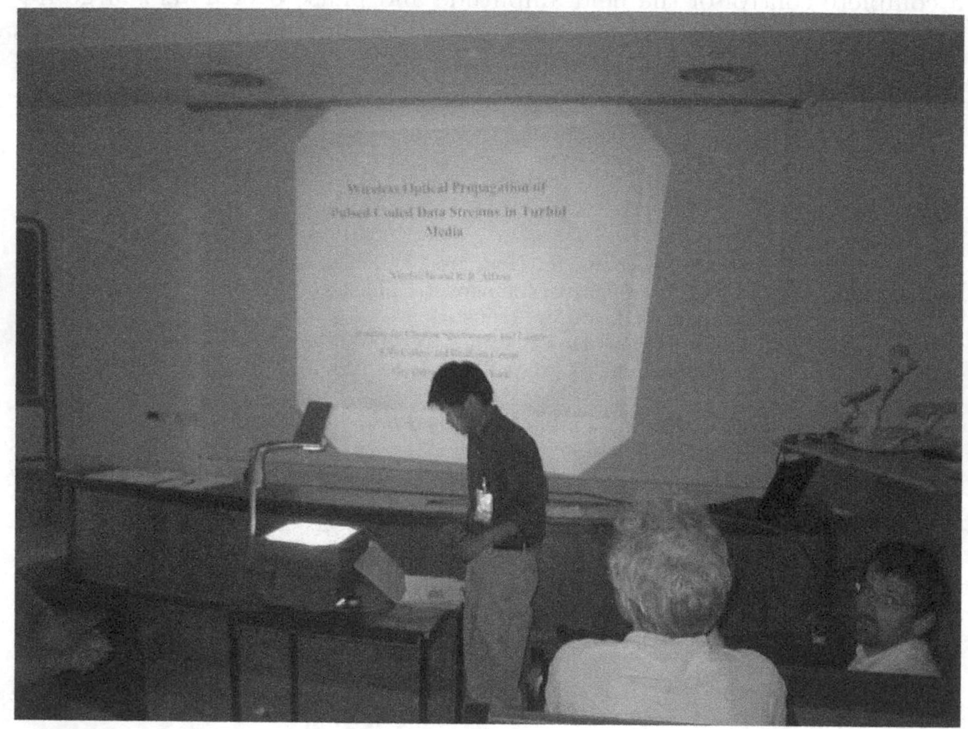

Xiaohui Ni. Photo by Gabriel Cwilich

WIRELESS PROPAGATION OF OPTICAL CODED PULSE STREAMS THROUGH TURBID MEDIA

X.H. NI, Q.R. XING, W. CAI, AND R.R. ALFANO
Institute for Ultrafast Spectroscopy and Lasers
New York State Center of Advanced Technology for Ultrafast Photonic Materials and Applications
Department of Physics
The City College and Graduate Center of the City University of New York
New York, NY 10031

Abstract. Experiments on time-resolved photon transmission through polystyrene scattering solutions were performed. Time-resolved polarization measurements were used to extract coded information buried within the multiple-scattered pulse profile from the early ballistic and snake components passing through a turbid medium. This method has the potential to improve optical wireless communication in a cloudy environment.

1. Introduction

Wireless communication systems of today rely mainly on microwave and radio frequency pulses. There will be a need for more speed [1] in the future for the Internet users requiring larger bandwidth, which can be handled using free space optics. Scattering through smog and cloudy media is the main problem resulting in a mix up of information encoded in overlapping adjacent pulses and leading to a degradation of information. To discriminate a sequence of data signals buried in noise, one needs to improve the way of extracting the coded data. Polarization analysis can provide a novel approach to extract the ballistic and snake components of light. This approach is based on the depolarization property of the diffusive component of the signals and the well-defined polarization property of the early ballistic and snake components [2, 3]. Depolarization of an incident polarized pulse occurs because of multiple scattering processes [4, 5]. The multiple-scattered

B.A. van Tiggelen and S.E. Skipetrov (eds.),
Wave Scattering in Complex Media: From Theory to Applications, 607–612.

component is delayed in time as compared to the ballistic pulse. The directions of light in the scattering plane are randomly changed when photons interact with the randomly distributed scatterers. The time-dependent polarization difference function $\Delta I(t)$ between the intensities of parallel and perpendicular polarized pulses after passage through a turbid medium is given by

$$\Delta I(t) = I_{||}(t) - I_{\perp}(t), \tag{1}$$

where $I_{||}(t)$ and $I_{\perp}(t)$ are the intensities of components polarized parallel and perpendicular to the incident light, respectively. For light completely depolarized by multiple scattering, the polarization difference $\Delta I(t)$ would be

$$\Delta I(t) \approx 0. \tag{2}$$

Depolarized light arises due to the diffusive component of the signal and the unpolarized natural light background. The intensity of the component polarized parallel to the incident light is given by

$$I_{||}(t) \approx I_{B,S}(t) + I_D(t)/2, \tag{3}$$

and the intensity of the perpendicular-polarized component is given by

$$I_{\perp}(t) \approx I_D(t)/2, \tag{4}$$

where $I_{B,S}(t)$ are the intensities of ballistic and snake components, respectively, and $I_D(t)$ is the diffusive component. The difference is

$$\Delta I(t) = I_{B,S}(t), \tag{5}$$

where the non-zero value of $\Delta I(t)$ occurs due to the early ballistic and snake components of the signal. In the past experiments [6–8], the parallel- and perpendicular-polarized signals were separately recorded during two different measurements. The results were then subtracted to obtain ΔI. To extract the coded information from the early component of light signals, one needs to perform a time-resolved polarization analysis of both $I_{||}(t)$ and $I_{\perp}(t)$ simultaneously. Such an analysis can significantly improve the signal to noise ratio degraded due to multiple scattering of light in the turbid medium.

In this preliminary report, time-resolved polarization measurements of multiple pulses passing through a turbid medium and their simultaneous analysis are reported. This work demonstrates the improvement in extracting information from the early ballistic and snake components which can be clearly distinguished and extracted from multiple coded pulses passed through a turbid medium.

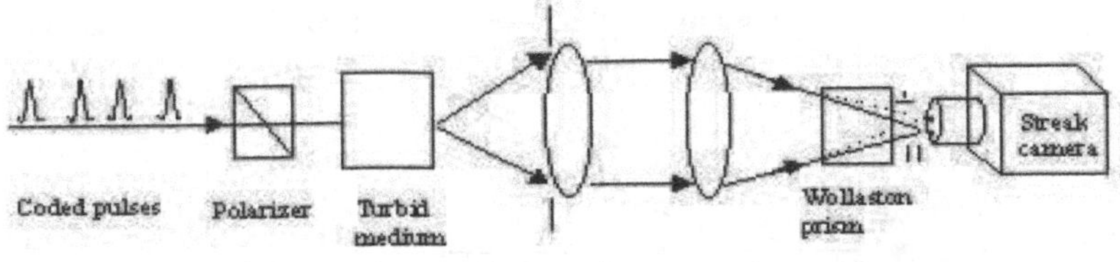

Figure 1. Schematic diagram of the experiment setup.

2. Experimental Setup

The setup is shown in Fig 1. The laser power is 5 mW, the central wavelength is 620 nm, and the pulse width is 100 fs. The laser beam, polarized by a Glan prism, is collimated and is then incident onto a sample of turbid medium contained in a 6 cm(L) $\times$ 6 cm(W) $\times$ 10 cm(H) glass cell. The beam scattered by the cell is measured by a polarization analyzer, which consists of a Wollaston prism combined with a synchroscan streak camera. The Wollaston prism splits the incident light into two polarized beams. One beam is polarized parallel and the another beam is polarized perpendicular to the incident pulse. The time profiles of the two polarized beams of light are recorded by a streak camera and are analyzed using a computer. The intensities $I_{||}(t)$ and $I_{\perp}(t)$ are thus measured simultaneously. A mirror array system is used to produce a coded series of 4 coded pulses. The time interval between two adjacent pulses is set to 200 ps.

The turbid medium consists of a dilute polystyrene water solution with refractive index contrast $m = 1.20$. Two different sizes of polystyrene particles are used in our experiments. The small particles (diameter $d = 0.213$ μm) have the anisotropy factor $g = 0.39$ and the scattering cross-section $\sigma_s = 0.376 \times 10^2$ μm^2 computed according to Mie scattering formulas [9]. The large particles (diameter $d = 0.855$ μm) have the anisotropy factor $g = 0.906$ and the scattering cross-section $\sigma_s = 1.2265$ μm^2. The scattering length is obtained as $l_s = 1/(\rho\sigma_s)$, where ρ is the number density of particles. The transport mean free length is $l_{tr} = l_s/(1-g)$. For both small and large particles, the optical thickness of the sample $L/l_s \gg 1$, where $L = 6$ cm is the length of the turbid medium.

3. Experimental Results

The experimental results are shown in Fig. 2. For 0.213 μm (small) particles, Fig. 2a shows the time-resolved profiles of parallel (solid curve) and perpendicular (dotted curve) polarized light transmitted through the tur-

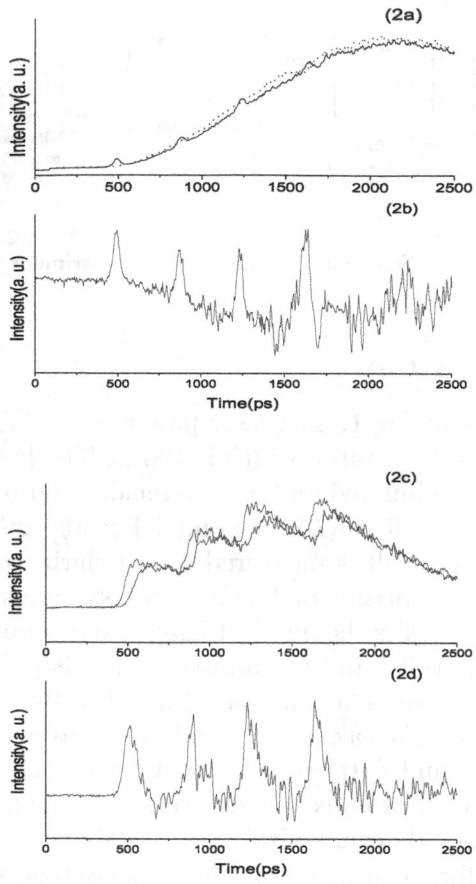

Figure 2. Time-resolved profiles of a pulse train of 4 pulses through a turbid medium. Figs. 2a and 2c show the time-resolved profiles of $I_{||}(t)$ (solid curve) and $I_{\perp}(t)$ (dotted curve). Figs. 2b and 2d show the corresponding $\Delta I(t)$. In the experiment with the particles of diameter $d = 0.213$ μm (Fig. 2a), the scattering length is $l_s = 0.434$ cm and the transport mean free length is $l_{tr} = 0.711$ cm. For the particles of diameter $d = 0.855$ μm (Fig. 2c), $l_s = 0.086$ cm and $l_{tr} = 0.91$ cm.

bid sample. The scattering length l_s is 0.434 cm and the transport mean free length l_{tr} is 0.711 cm. Fig. 2b shows the corresponding $\Delta I(t)$. For the experiment with large particles (diameter $d = 0.855$ μm), Fig. 2c shows the time-resolved profiles of parallel (solid curve) and perpendicular (dotted curve) polarized transmitted light. The scattering length l_s is 0.086 cm and l_{tr} is 0.91 cm. Fig. 2d shows the corresponding $\Delta I(t)$.

4. Conclusion

In our experiments, the early-arriving signals are reduced as compared to the large diffusive component, due to the multiple scattering of light in the turbid solution. When $L/l_s = 13$ for small particles and $L/l_s = 70$ for large particles, the transmitted light is dominated by the diffusive component and the information encoded in the incident pulse train is hardly seen in the transmitted light. As shown in Fig. 2, using the time-resolved polarization analysis, the unpolarized diffusive component and natural light background can be greatly reduced, and the weak early signals can be clearly extracted, thus allowing us to distinguish individual pulses of the pulse train. The experimental results show an evidence that the time-resolved polarization analysis provides a powerful technique for improving the extraction of information from light beams passing through turbid media.

The time-resolved profiles of polarization differences shown in Figs. 2b and 2d are dominated by ballistic signals for small scatterers, while for large scatterers, snake signals give the main contribution to the early-arriving component of transmitted light. Coexistence of ballistic and scattered components is seen when the scatterers are smaller than the wavelength of light, but is hardly seen when the size of the scatterers is comparable to or larger than the wavelength of light. The reason is that the scattering on a small particle is almost isotropic, and is characterized by a small anisotropy factor g, while a large particle scatters mainly in the forward direction and is characterized by $g \sim 1$. The transport mean free path l_{tr} is not much larger than the scattering length l_s in the case of small scatterers, while $l_{tr} \gg l_s$ in the case of large scatterers. Even in the absence of absorption, the ballistic component decays exponentially as $\exp(-L/l_s)$ with the distance L traveled in a turbid medium. The scattered component is characterized by l_{tr} and the diffusion constant $D = cl_{tr}/3$, the parameters arising in the diffusion approximation or in the recently developed cumulant solution of the radiative transfer equation [10]. If one considers two cases with similar l_{tr}, the ballistic component will be stronger (and measurable) in the case of small scatterers, while it will be too weak to be measured in the case of large scatterers. In Fig. 2b, the scatterer size is small, and the ballistic component is seen. When the scatterer size becomes larger (Fig. 2d), the ballistic component becomes too weak to be detectable and only the snake-like component can be measured.

Time-resolved profiles of light transmitted through a random medium can be renormalized using the transport mean free path l_{tr} as a unit of length and l_{tr}/c as a unit of time. One can therefore rescale our results, obtained for a dilute polystyrene solution, to describe the light propagating in the atmosphere, where l_{tr} is larger. This work will be reported elsewhere in more detail.

Acknowledgements

This research is supported in part by NASA Institutional Research Award program, New York State Science Technology and Academic Research.

References

1. Killinger, D. (2002) Free space optics for laser communication through the air, *OPN* **Vol. 13, no. 10**, pp. 36–42.
2. Yoo, K.M., and Alfano, R.R. (1990) Time-resolved coherent and incoherent components of forward light scattering in random media, *Opt. Lett.* **Vol. 15, no. 6**, pp. 320–322.
3. Wang, L., Ho, P.P., Liu, C., Zhang, G., and Alfano, R.R. (1991) Ballistic 2-D imaging through scattering walls using an ultrafast optical Kerr gate, *Science* **Vol. 253**, pp. 769–771.
4. Kim, A.D., and Moscoso, M. (2001) Influence of the relative index on the depolarization of multiple scattered waves, *Phys. Rev. E* **Vol. 64**, 026612.
5. Gorodnichev, E.E., Kuzovlev, A.I., and Rogozkin, D.B. (1998) Diffusion of circularly polarized light in a disordered medium with large-scale inhomogeneities, *JETP Lett.* **Vol. 68, no. 1**, pp. 22–28.
6. Demos, S.G., and Alfano, R.R.(1996) Temporal gating in highly scattering media by the degree of optical polarization, *Opt. Lett.* **Vol. 21, no. 2**, pp. 161–163.
7. Morgan, S.P., Khong, M.P., and Somekh, M.G. (1997) Effects of polarization state and scatterer concentration on optical imaging through scattering media, *Appl. Opt.* **Vol. 36, no. 7**, pp. 1560–1565.
8. Schmitt, J.M., Gandjbakhche, A.H., and Bonner, R.F. (1992) Use of polarized light to discriminate short-path photons in a multiply scattering medium, *Appl. Opt.* **Vol. 31, no. 30**, pp. 6535–6546.
9. Van de Hulst, H.C. (1957) Light Scattering by Small Particles, Wiley, New York.
10. Cai, W., Lax, M., and Alfano, R.R. (2001) Analytical solution of the polarized photon transport equation in an infinite uniform medium using cumulant expansion, *Phys. Rev. E* **Vol. 63**, 016606

ELASTIC LIGHT SCATTERING BY DIELECTRIC MICROSPHERES FOR C-BAND APPLICATIONS

T. BILICI, S. ISCI, A. KURT AND A. SERPENGÜZEL
Microphotonics Research Laboratory,
Koç University, Department of Physics
Rumeli Feneri Yolu, Sariyer, Istanbul 34450 Turkey

1. Introduction

Optical microcavity resonators show great promise for optical communication applications such as filtering, multiplexing, and switching [1, 2]. For optical communication, wavelength division multiplexing (WDM) is important for increasing the bandwidth of existing fiber optic networks. There is a need for an all-optical packet-switching layer at the end of the optical to electronic conversion domain, which consists of all-optical gates, semiconductor optical amplifiers, channel dropping filters, interferometers, resonant cavity enhanced photodetectors, and optical random access memory elements. In these planar lightwave circuits, dielectric microspheres (μ-spheres) with their morphology dependent resonances, can be used as compact optical filtering elements.

In this paper, based on the experiment by BK7 glass μ-spheres, elastic light scattering in dielectric glass μ-spheres are analyzed in the C-band wavelength range that is suitable for optical network communication.

2. Morphology Dependent Resonances

It is known that the elastic scattered light can be used as a sensitive means of detecting dielectric surface-wave resonances of spherical particles [3]. A physical interpretation is that the light is confined by almost total internal reflection (TIR) as it propagates around the inside surface of the μ-sphere. After circumnavigating the sphere, the light wave returns to the starting point in phase to interfere constructively with itself. This constructive interference can occur only at certain discrete resonance wavelengths [4] and they are called as morphology dependent resonances (MDR's) or whispering gallery modes (WGM's). In ray-optics picture, the MDR's correspond to

B.A. van Tiggelen and S.E. Skipetrov (eds.),
Wave Scattering in Complex Media: From Theory to Applications, 613–618.

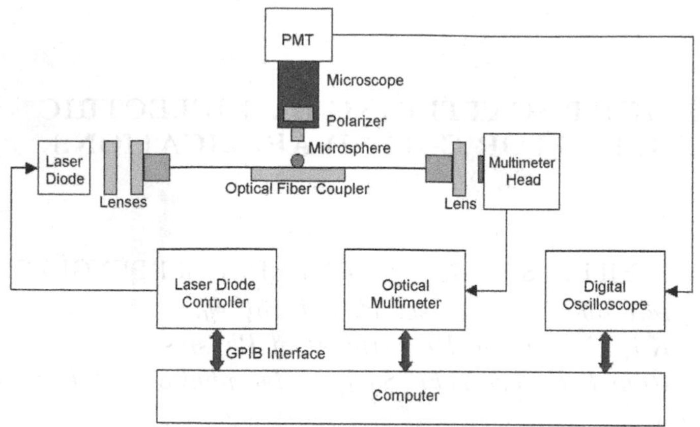

Figure 1. The BK7 glass microsphere in the experimental setup.

light traveling around the great circles of the μ-sphere, constantly reflected back inwards by TIR [5] .

Each MDR is characterized by a mode number (n) and a mode order (l) [6]. For each set of mode numbers, there is a transverse electric (TE) and transverse magnetic (TM) polarization MDR [3]. For a given μ-sphere, the MDR occurs at specific value of the size parameter, $x_{n,l}$, which is given by $2\pi a/\lambda_{n,l}$, where $\lambda_{n,l}$ is the light wavelength in vacuum and a is the radius of the sphere.

Physically, (n) indicates the number of nodes in the internal intensity distribution as the polar angle is varied from 0° to 180°. The mode order (l) indicates the number of poles encountered as x increases from zero for a particular n. On the other hand, physically, l indicates the number of nodes in the internal intensity distribution in the radial direction. For a fixed radius a, the MDR's have n values that are bound by $x < n < mx$, where the upper limit is the maximum number of wavelengths that would fit inside the circumference of the μ-sphere and m is the refractive index of the sphere [7]. The radial electric field distribution of the lowest order modes (i.e. $l = 1$) shows a peak just inside the surface. The higher the mode order (l) becomes, the more the radial mode distribution shifts towards the inside of the μ-sphere [6]. These MDR's have been experimentally verified at optical wavelengths with micrometer-sized spheres.

3. Experimental Results in the M-band

Elastic scattering calculations are based on the optical setup shown in Fig. 1. A BK7 glass μ-sphere with a radius of 500 μm and a refractive index of

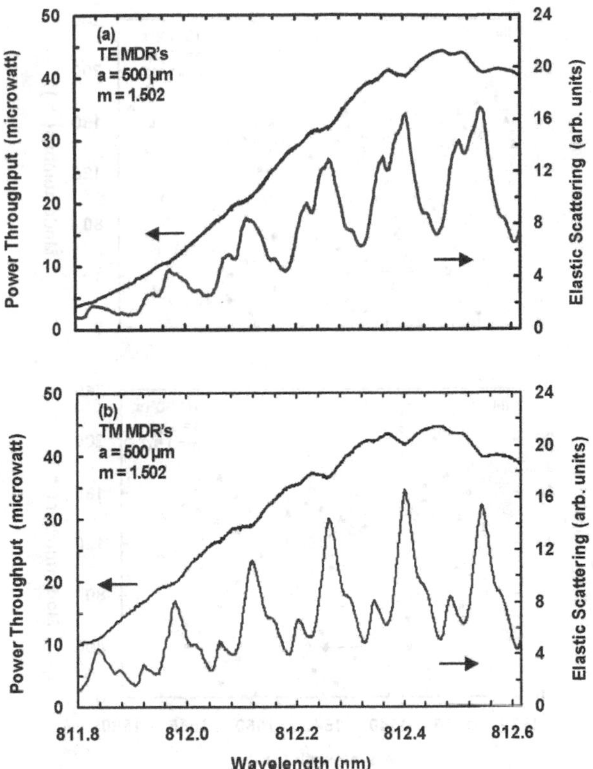

Figure 2. Experimental spectra of elastic scattered and transmitted light by BK7 glass microsphere for (a) TE MDR's (b) TM MDR's.

1.502 is placed on an optical fiber coupler (OFC), and the scattered signal is detected through a polarizing beam splitter [8].

The optical fiber half coupler is fabricated from 800 μm single mode fiber, which is placed into a glass substrate. Fiber is side polished beneath the μ-sphere until the necessary cladding thickness is achieved. Excitation of resonances of the sphere is achieved by using a tunable distributed feedback laser with a laser diode controller. The scattering light is detected by using photomultiplier tube with a microscope. The TE and TM MDR's in the elastic scattering spectra are obtained by a rotating the polarizer inside the microscope. The control of the devices and data acquisition are computerized by using IEEE-488 standard GPIB interface.

Figure 2 shows both the elastically scattered and the transmitted light spectra from the BK7 glass μ-sphere with radius of 500 μm and refractive index of 1.502 for TE and TM MDR's. MDR's in the elastic scattering

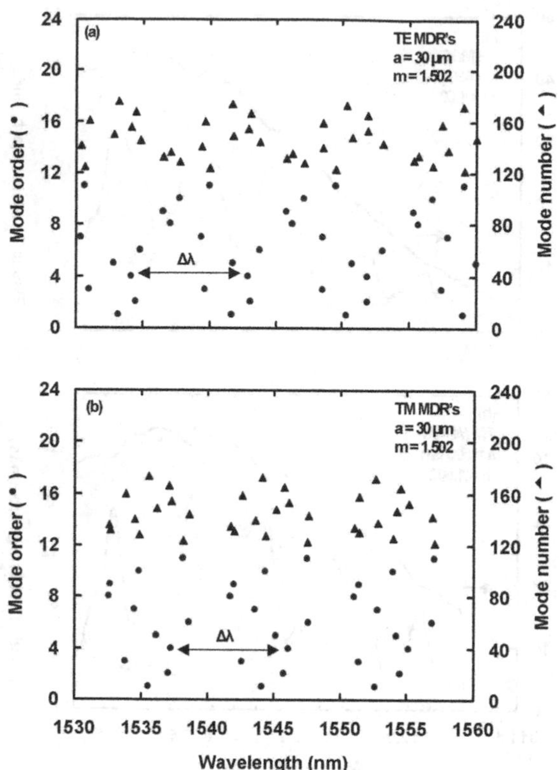

Figure 3. Calculated mode number and mode orders in elastic scattering from the BK7 glass microsphere ($a = 30\ \mu$m and $m = 1.502$) both in (a) TE and (b) TM modes for the C-band wavelengths.

spectra and the associated dips in the power spectrum of the transmitted light are clearly observed. The separation between the same mode order (l) MDR's with consecutive mode numbers (n), $\Delta\lambda$, is measured to be 0.14 nm and the measured Q-factor of the resonances is on the order of 10^4.

4. Calculations in the C-band

For the calculations of both TE and TM elastic scattering mode numbers and mode orders, Gaussian beam (with an infinite skirt length and a beam waist with half-width of 2 mm) propagating just at the edge of the microsphere is used for scattering angle of 90°. Calculations are performed with a step size of 0.035 nm.

Figure 3 shows the calculation of the mode numbers and the mode

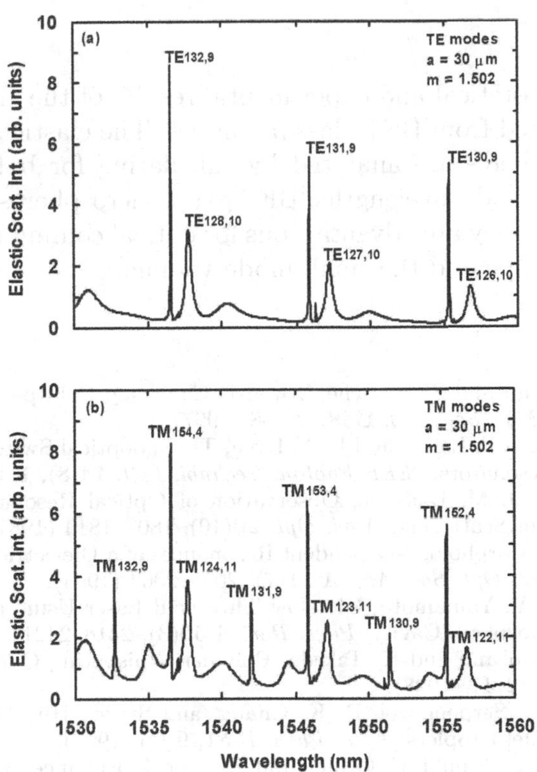

Figure 4. Elastic scattering spectra from from the BK7 glass microsphere ($a = 30$ μm and $m = 1.502$) both in (a) TE and (b) TM modes for the C-band wavelengths.

orders for elastic scattering from BK7 glass microsphere with a radius of 30 μm and refractive index of 1.502 for both in TE and TM MDR's for the C-band. It is clearly seen that mode numbers (n) lie between the size parameter (x) and (mx) and there is a mode order (l) associated with for each mode number (n).

TE and TM elastic scattering spectrum from the BK7 glass microsphere can be seen in Fig. 4. In Fig. 4(a), TE elastic scattering spectrum gives observable MDR's with respect to their linewidths. The MDR at the wavelength of 1536.54 nm is $TE_{132,9}$ mode, which has the linewidth of 0.0208 nm and Q of 1.3×10^4, and the mode at 1537.80 nm is $TE_{128,10}$ MDR, which has the linewidth of 0.2352 nm and Q of 6×10^3. There are some observable modes in TM elastic scattering spectrum in Fig. 4(b). For example, the MDR at the wavelength of 1532.69 nm is $TM_{132,9}$ MDR, which has the linewidth of 0.04 nm and Q of 4×10^4. The separation between the adjacent peak wavelengths of the MDR's, $\Delta\lambda$, is calculated as 9.57 nm.

5. Conclusion

In this paper, theoretical and experimental results of the elastic light scattering are presented from BK7 glass μ-spheres. The elastic scattering spectra by glass μ-spheres are analyzed by calculating for both TE and TM MDR's at the C-band wavelengths. BK7 glass microspheres show very narrow resonances so they are advantageous for optical communication because of their high Q-value and the small mode-volume.

References

1. V. Lefevre-Seguin and S. Haroche, Towards Cavity-QED Experiments with Silica Microspheres, *Mat. Sci. Eng.* **B48,** 53–58 (1997).
2. H. C. Tapalian, J. P. Laine, and P. A. Lane, Thermooptical Switches Using Coated Microsphere Resonators, *IEEE Photon. Technol. Lett.* **14**(8), 1118–1120 (2002).
3. A. Ashkin and J. M. Dziedzic, Observation of Optical Resonances of Dielectric Spheres by Light Scattering, *Appl. Opt.* **20**(10), 1803–1814 (1981).
4. B. R. Johnson, Morphology-Dependent Resonances of a Dielectric Sphere on a Conducting Plane, *J. Opt. Soc. Am. A* **11**(7), 2055–2064 (1994).
5. M. Pelton and Y. Yamamoto, Ultralow Threshold Laser Using a Single Quantum Dot and a Microsphere Cavity, *Phys. Rev. A* **59**(3), 2418–2421 (1999).
6. M. Kuwata-Gonokami and K. Takeda, Polymer Whispering Gallery Mode Lasers, *Opt. Mater.* **9,** 12–17 (1998).
7. W. P. Acker, A. Serpengüzel, R. K. Chang, and Steven Hill, Stimulated Raman Scattering of Fuel Droplets, *Appl. Phys. B* **51,** 9–16 (1990).
8. A. Serpengüzel, S. Arnold, G. Griffel, and J. A. Lock, Enhanced Coupling to Microsphere Resonances with Optical Fibers, *J. Opt. Soc. Am. B* **14**(4), 790–795 (1997).

Subject Index